HANDBOOK OF PETROLEUM REFINING PROCESSES

Other McGraw-Hill Chemical Engineering Books

Chopey • FLUID MOVERS
Chopey • HANDBOOK OF CHEMICAL ENGINEERING CALCULATIONS, SECOND EDITON
Chopey • INSTRUMENTATION AND PROCESS CONTROL
Connell • PROCESS INSTRUMENTATION PROCESS MANUAL
Considine • PROCESS/INDUSTRIAL INSTRUMENTS AND CONTROLS HANDBOOK, FOURTH EDITION
Croom • FILTER DUST COLLECTORS
Datta-Barua • NATURAL GAS MEASUREMENT AND CONTROL
Dean • LANGE'S HANDBOOK OF CHEMISTRY, FOURTEENTH EDITION
Deshotels, Zimmerman • COST EFFECTIVE RISK ASSESSMENT FOR PROCESS DESIGN
Dillon • MATERIALS SELECTION FOR THE PROCESS INDUSTRIES
Fitzgerald • CONTROL VALVES FOR THE CHEMICAL PROCESS INDUSTRIES
Harper • HANDBOOK OF PLASTICS, ELASTOMERS, AND COMPOSITES, THIRD EDITION
Kister • DISTILLATION DESIGN
Kister • DISTILLATION PROCESS APPLICATIONS AND OPERATIONS
Mansfield • ENGINEERING DESIGN FOR PROCESS FACILITIES
Miller • FLOW MEASUREMENT ENGINEERING HANDBOOK, THIRD EDITION
Power • STEAM JET EJECTORS FOR THE PROCESS INDUSTRIES
Reid, Prausnitz, Poling • THE PROPERTIES OF GASES AND LIQUIDS, FIFTH EDITION
Reist • AEROSOL SCIENCE AND TECHNOLOGY, SECOND EDITION
Rhine, Tucker • MODELLING OF GAS-FIRED FURNACES AND BOILERS AND OTHER INDUSTRIAL PROCESSES
Rossiter • WASTE MINIMIZATION THROUGH PROCESS DESIGN
Samdani • SAFETY AND RISK MANAGEMENT TOOLS AND TECHNIQUES IN THE CPI
Samdani • HEAT TRANSFER TECHNOLOGIES AND PRACTICES FOR EFFECTIVE ENERGY MANAGEMENT
Schweitzer • HANDBOOK OF SEPARATION TECHNIQUES FOR CHEMICAL ENGINEERS, THIRD EDITION
Shinskey • FEEDBACK CONTROLLERS FOR THE PROCESS INDUSTRIES
Shinskey • PROCESS CONTROL SYSTEMS, FOURTH EDITION
Shugar, Ballinger • CHEMICAL TECHNICIAN'S READY REFERENCE HANDBOOK, FOURTH EDITION
Shugar, Dean • THE CHEMIST'S READY REFERENCE HANDBOOK
Sikich • EMERGENCY MANAGEMENT PLANNING HANDBOOK
Smallwood • SOLVENT RECOVERY HANDBOOK
Smith • CHEMICAL PROCESS DESIGN
Tatterson • MIXING AND GAS DISPERSION IN AGITATED TANKS
Tatterson • SCALE-UP AND DESIGN OF INDUSTRIAL MIXING PROCESSES
Yokell • A WORKING GUIDE TO SHELL AND TUBE HEAT EXCHANGERS

HANDBOOK OF PETROLEUM REFINING PROCESSES

Robert A. Meyers Editor in Chief

Second Edition

McGraw-Hill

New York San Francisco Washington, D.C. Auckland Bogotá
Caracas Lisbon London Madrid Mexico City Milan
Montreal New Delhi San Juan Singapore
Sydney Tokyo Toronto

Library of Congress Cataloging-in-Publication Data

Handbook of petroleum refining proecesses / Robert A. Meyers, editor in
    chief.—2nd ed.
        p.   cm.
    Includes index.
    ISBN 0-07-041796-2
        1. Petroleum—Refining—Handbooks, manuals, etc.   2. Petroleum
    chemicals—Handbooks, manuals, etc.   I. Meyers, Robert A. (Robert
    Allen), date.
    TP690.H34   1996
    665.5'3—dc20   96-26991
                                                                                              CIP

McGraw-Hill

A Division of The **McGraw·Hill** Companies

Copyright © 1997 by The McGraw-Hill Companies, Inc. All rights reserved. Printed in
the United States of America. Except as permitted under the United States Copyright Act
of 1976, no part of this publication may be reproduced or distributed in any form or by
any means, or stored in a data base or retrieval system, without the prior written permission of the publisher.

1 2 3 4 5 6 7 8 9 0   BKP/BKP   9 0 1 0 9 8 7 6

ISBN 0-07-041796-2

*The sponsoring editor for this book was Zoe G. Foundotos, the editing supervisor was
David E. Fogarty, and the production supervisor was Pamela A. Pelton. It was set in
Times Roman by Donald A. Feldman of McGraw-Hill's Professional Book Group
composition unit.*

Printed and bound by Quebecor/Book Press.

This book is printed on acid-free paper.

Information contained in this work has been obtained by The McGraw-Hill
Companies, Inc. ("McGraw-Hill") from sources believed to be reliable. However,
neither McGraw-Hill nor its authors guarantee the accuracy or completeness of
any information published herein, and neither McGraw-Hill nor its authors shall
be responsible for any errors, omissions, or damages arising out of use of this
information. This work is published with the understanding that McGraw-Hill and
its authors are supplying information but are not attempting to render engineering
or other professional services. If such services are required, the assistance of an
appropriate professional should be sought.

CONTENTS

Contributors xix
Preface xxi
Acknowledgments xxv
Licensing Contacts xxvii

Part 1 Alkylation and Polymerization

Chapter 1.1. Exxon Sulfuric Acid Alkylation Technology
Howard Lerner 1.3

Introduction *1.3*
Alkylation Is a Key Processing Unit *1.4*
Chemistry Overview *1.4*
Process Description *1.5*
Balancing Process Variables Is Critical to Efficient Design and Operation *1.7*
Reactor Cooling via Autorefrigeration Is More Efficient Than Effluent Refrigeration *1.8*
Reaction Staging Results in High Average Isobutane Concentration in Reactor *1.10*
Advantages of the ER&E Reactor Are Numerous *1.11*
Modern ER&E Reactor Is a Vast Improvement over Older Systems *1.11*
Economics of Exxon Sulfuric Acid Alkylation Technology *1.13*
Extensive Commercial Experience Enhances Technology Package *1.13*

Chapter 1.2. The Dow-Kellogg Cumene Process *J. W. Wallace and H. E. Gimpel* 1.15

Introduction *1.15*
History of Cumene Technology *1.15*
History of the Dow-Kellogg Cumene Process *1.16*
Process Features *1.17*
Process Description *1.17*
Yields and Balance *1.19*
Process Economics *1.19*
Wastes and Emissions *1.20*

Chapter 1.3. UOP Catalytic Condensation Process for Transportation Fuels *Diana York, John C. Sheckler, and Daniel G. Tajbl* 1.21

Introduction *1.21*
History *1.21*

Process Chemistry   *1.22*
Process Thermodynamics   *1.23*
Process Description   *1.23*
Commercial Experience   *1.29*

Chapter 1.4. UOP HF Alkylation Technology   John C. Sheckler and B. R. Shah   1.31

Introduction   *1.31*
Process Chemistry   *1.32*
Process Description   *1.34*
Engineering Design   *1.39*
Commercial Information   *1.41*
Environmental Considerations   *1.41*
Mitigating HF Releases—The Texaco-UOP Alkad Process   *1.48*
References   *1.51*

Chapter 1.5. Linear Alkylbenzene (LAB) Manufacture   Peter R. Pujadó   1.53

Introduction   *1.53*
Technology Background   *1.54*
Commercial Experience   *1.55*
Product Quality   *1.60*
Economics   *1.61*
Markets   *1.62*
Environmental Safety   *1.63*
Conclusions   *1.65*
References   *1.66*

Chapter 1.6. UOP Q-Max Process for Cumene Production   M. F. Bentham   1.67

Discussion   *1.67*
Yield Structure   *1.68*
Process Economics   *1.68*
Commercial Experience   *1.69*

Part 2   Base Aromatics Production Processes

Chapter 2.1. Aromatics Complexes   John J. Jeanneret   2.3

Introduction   *2.3*
Configurations   *2.4*
Description of the Process Flow   *2.6*
Feedstock Considerations   *2.8*
Case Study   *2.9*
Commercial Experience   *2.10*
Bibliography   *2.11*

Chapter 2.2. UOP Sulfolane Process   *John J. Jeanneret*   2.13

Introduction   *2.13*
Solvent Selection   *2.15*
Process Concept   *2.15*
Description of the Process Flow   *2.17*
Feedstock Considerations   *2.19*
Process Performance   *2.20*
Equipment Considerations   *2.20*
Case Study   *2.21*
Commercial Experience   *2.21*
Bibliography   *2.22*

Chapter 2.3. UOP Thermal Hydrodealkylation (THDA) Process
W. L. Liggin   2.23

Introduction   *2.23*
Process Description   *2.24*
Process Economics   *2.25*

Chapter 2.4. BP-UOP Cyclar Process   *John J. Jeanneret*   2.27

Introduction   *2.27*
Process Chemistry   *2.28*
Description of Process Flow   *2.29*
Feedstock Considerations   *2.31*
Process Performance   *2.32*
Equipment Considerations   *2.32*
Case Study   *2.33*
Commercial Experience   *2.33*
Bibliography   *2.34*

Chapter 2.5. UOP Isomar Process   *John J. Jeanneret*   2.37

Introduction   *2.37*
Process Chemistry   *2.37*
Description of Process Flow   *2.40*
Feedstock Considerations   *2.41*
Process Performance   *2.41*
Equipment Considerations   *2.42*
Case Study   *2.43*
Commercial Experience   *2.43*
Bibliography   *2.44*

Chapter 2.6. UOP Parex Process   *John J. Jeanneret*   2.45

Introduction   *2.45*
Parex versus Crystallization   *2.46*
Process Performance   *2.48*
Feedstock Considerations   *2.48*
Description of Process Flow   *2.49*
Equipment Considerations   *2.51*

CONTENTS

Case Study   2.52
Commercial Experience   2.52
Bibliography   2.53

Chapter 2.7. UOP Tatoray Process   *John J. Jeanneret*   2.55

Introduction   2.55
Process Chemistry   2.56
Description of the Process Flow   2.57
Feedstock Considerations   2.59
Process Performance   2.60
Equipment Considerations   2.61
Case Study   2.62
Commercial Experience   2.62
Bibliography   2.62

Part 3   Catalytic Cracking

Chapter 3.1. Exxon Flexicracking IIIR Fluid Catalytic Cracking Technology   *Paul K. Ladwig*   3.3

Foreword: Why Select Flexicracking IIIR?   3.3
Introduction: Evolution of Flexicracking IIIR?   3.3
The Flexicracking IIIR Process   3.5
Major Process Features   3.7
Reliability   3.17
Resid Considerations   3.19
Upgrading with Flexicracking IIIR Technology   3.21
Economics of Exxon Flexicracking IIIR Technology   3.22
ER&E Designed Commercial FCC Units   3.22
Summary   3.23

Chapter 3.2. The M.W. Kellogg Company Fluid Catalytic Cracking Process   *T. E. Johnson and P. K. Niccum*   3.29

Introduction   3.29
Feedstocks   3.29
Products   3.30
Process Description   3.31
Process Variables   3.44
Advanced Process Control   3.49
Catalyst and Chemical Consumption   3.51
Investment Utilities and Costs   3.53
Bibliography   3.54

Chapter 3.3. UOP Fluid Catalytic Cracking Process   *Charles L. Hemler*   3.55

Introduction   3.55
Development History   3.56
Process Chemistry   3.60
Thermodynamics of Catalytic Cracking   3.63
Catalyst History   3.64

Process Description   3.64
Modern UOP FCC Unit   3.70
Feedstock Variability   3.72
Process Costs   3.76
Market Situation   3.78
References   3.78

Chapter 3.4. Stone & Webster—Institut Francais du Pétrole RFCC Process   *David A. Hunt*   3.79

History   3.79
Process Description   3.81
RFCC Feedstocks   3.87
Operating Conditions   3.88
RFCC Catalyst   3.88
Two-Stage Regeneration   3.90
S&W-IFP Technology Features   3.93
Mechanical Design Features   3.98
FCC Revamp to RFCC (Second-Stage Regenerator Addition)   3.99
References   3.99

Chapter 3.5. Deep Catalytic Cracking, The New Light Olefin Generator   *David A. Hunt*   3.101

Basis   3.101
Process Description   3.102
Catalyst   3.107
Feedstocks   3.107
Operating Conditions   3.107
DCC Product Yields   3.108
DCC Integration   3.109
References   3.112

Part 4   Catalytic Reforming

Chapter 4.1. UOP Platforming Process   *Natasha Dachos, Aaron Kelly, Don Felch, and Emmanuel Reis*   4.3

Process Evolution   4.3
Process Chemistry   4.7
Process Variables   4.15
Continuous Platforming Process   4.18
Case Studies   4.22
UOP Commercial Experience   4.24

Part 5   Dehydrogenation

Chapter 5.1. UOP Oleflex Process for Light Olefin Production   *Joseph Gregor*   5.3

Introduction   5.3
Process Description   5.3

Dehydrogenation Complexes   5.5
Propylene Production Economics   5.7

Chapter 5.2. UOP Pacol Dehydrogenation Process
Peter R. Pujadó   5.11

Introduction   5.11
Process Description   5.12
Pacol Process Improvements   5.15
Yield Structure   5.17
Commercial Experience   5.18
Process Economics   5.18
Bibliography   5.19

Part 6   Gasification and Hydrogen Production

Chapter 6.1. KRW Fluidized-Bed Gasification Process
W. M. Campbell   6.3

Introduction   6.3
History   6.3
KRW Single-Stage Gasification Process   6.4
Description of the KRW Process Development Unit   6.8
Test Results Obtained in the PDU   6.12
Commercial-Scale Design   6.16
Application   6.17
Conclusions   6.18
References   6.19

Chapter 6.2. FW Hydrogen Production   *James D. Fleshman*   6.21

Introduction   6.21
Uses of Hydrogen   6.21
Hydrogen Production   6.22
Integration into the Modern Refinery   6.41
Heat Recovery   6.46
Economics   6.48
Utility Requirements   6.51
References   6.52

Part 7   Hydrocracking

Chapter 7.1. MAK Moderate-Pressure Hydrocracking
M. G. Hunter, D. A. Pappal, and C. L. Pesek   7.3

Introduction   7.3
Hydrocracking Costs   7.4
Technology Development   7.5
Commercial Results   7.15
MPHC Grassroots Applications   7.17
Revamp of Existing Hydrotreaters   7.19
References   7.20

Chapter 7.2. Chevron Isocracking—Hydrocracking for Superior Fuels and Lubes Production *Alan G. Bridge*   7.21

Isocracking Chemistry   *7.21*
The Importance of Hydrogen   *7.22*
Isocracking Configurations   *7.24*
Isocracking Catalysts   *7.24*
Product Yields and Qualities   *7.28*
Investment and Operating Expenses   *7.36*
Summary   *7.37*
References   *7.38*

Chapter 7.3. UOP Unicracking Process for Hydrocracking *Mark Reno*   7.41

Introduction   *7.41*
Process Applications   *7.42*
Process Description   *7.42*
Yield Patterns   *7.48*
Investment and Operating Expenses   *7.48*

Part 8 Hydrotreating

Chapter 8.1. Chevron RDS/VRDS Hydrotreating—Transportation Fuels from the Bottom of the Barrel *David N. Brossard*   8.3

Introduction   *8.3*
History   *8.4*
Process Description   *8.6*
Process Chemistry   *8.9*
Catalysts   *8.14*
VRDS Hydrotreating   *8.15*
Feed Processing Capability   *8.17*
Commercial Application   *8.18*
The Future   *8.21*
References   *8.26*

Chapter 8.2. Hüls Selective Hydrogenation Process *Scott Davis*   8.27

Process Description   *8.27*
Process Flow   *8.28*
Commercial Experience   *8.28*
Investment and Operating Requirements   *8.28*

Chapter 8.3. UOP Unionfining Technology *James E. Kennedy*   8.29

Introduction   *8.29*
Process Chemistry   *8.29*
Catalyst   *8.34*
Process Flow   *8.35*
Unionfining Applications   *8.35*
Investment   *8.37*
UOP Hydroprocessing Experience   *8.37*
Bibliography   *8.37*

Chapter 8.4. UOP RCD Unionfining Process   *Gregory J. Thompson*   8.39

Introduction   8.39
Market Drivers for RCD Unionfining   8.39
Catalyst   8.40
Process Chemistry   8.42
Process Description   8.42
Operating Data   8.47
Commercial Installations   8.48

Chapter 8.5. UOP Catalytic Dewaxing Process   *Orhan Genis*   8.49

Introduction   8.49
Process Chemistry   8.50
Catalyst   8.50
Process Flow   8.51
Yield Patterns   8.52
Investment and Operating Expenses   8.52
Commercial Experience   8.53

Chapter 8.6. UOP Unisar Process for Saturation of Aromatics
H. W. Gowdy   8.55

Introduction   8.55
Application to Diesel Fuels   8.56
Process Description   8.57
Process Applications   8.60

Chapter 8.7. Exxon Diesel Oil Deep Desulfurization (DODD)
Sam Zaczepinski   8.63

Technical Background of DODD   8.63
Hydrofining Characteristics   8.63
DODD Technology   8.65
DODD Technology Database   8.66
Summary   8.68

Chevron's On-Stream Catalyst Replacement Technology for Processing High-Metal Feeds (see Part 10)

Part 9   Isomerization

Chapter 9.1. UOP BenSat Process   *Dana K. Sullivan*   9.3

Process Discussion   9.4
Process Flow   9.5
Catalyst and Chemistry   9.5
Feedstock Requirements   9.6
Commercial Experience   9.6

CONTENTS   xiii

Chapter 9.2. UOP Butamer Process   *Nelson A. Cusher*   9.7

Introduction   *9.7*
Process Description   *9.8*
Process Chemistry   *9.8*
Process Variables   *9.9*
Process Contaminants   *9.10*
Isomerization Reactors   *9.10*
Process Flow Scheme   *9.11*
Commercial Experience   *9.12*

Chapter 9.3. UOP Penex Process   *Nelson A. Cusher*   9.15

Introduction   *9.15*
Process Discussion   *9.16*
Process Flow   *9.16*
Process Applications   *9.19*
Thermodynamic Equilibrium Considerations, Catalysts, and Chemistry   *9.20*
Feedstock Requirements   *9.23*
Commercial Experience   *9.24*

Chapter 9.4. The UOP TIP and Once-Through Zeolitic Isomerization Processes   *Nelson A. Cusher*   9.29

Introduction   *9.29*
O-T Zeolitic Isomerization Process   *9.31*
TIP Process   *9.35*

Part 10   Separation Processes

Chapter 10.1. Chevron's On-Stream Catalyst Replacement Technology for Processing High-Metal Feeds   *David E. Earls*   10.3

Introduction   *10.3*
Development History   *10.3*
Process Description   *10.4*
Commercial Operation   *10.7*
OCR Applications   *10.9*
Economic Benefits of OCR   *10.11*

Chapter 10.2. FW Solvent Deasphalting   *F. M. Van Tine and Howard M. Feintuch*   10.15

Process Description   *10.16*
Typical Feedstocks   *10.19*
Extraction Systems   *10.20*
Solvent-Recovery Systems   *10.23*
DAO Yields and Properties   *10.28*
Operating Variables   *10.34*
Asphalt Properties and Uses   *10.37*
Integration of SDA in Modern Refineries   *10.38*

Typical Utility Requirements   *10.41*
Estimated Investment Cost   *10.42*
References   *10.43*

Chapter 10.3. UOP Sorbex Family of Technologies
John J. Jeanneret and John R. Mowry   10.45

Introduction   *10.45*
Principles of Adsorptive Separation   *10.46*
The Sorbex Concept   *10.47*
Description of the Process Flow   *10.48*
Comparison with Fixed-Bed Adsorption   *10.50*
Commercial Experience   *10.51*
Bibliography   *10.51*

Chapter 10.4. UOP Demex Process   E. J. Houde   10.53

Introduction   *10.53*
Process Description   *10.53*
Product Yields and Quality   *10.55*
Process Variables   *10.56*
DMO Processing   *10.58*
Process Economics   *10.60*
Demex Process Status   *10.60*

Chapter 10.5. UOP IsoSiv Process   Nelson A. Cusher   10.61

Introduction   *10.61*
General Process Description   *10.63*
Process Perspective   *10.64*
Detailed Process Description   *10.64*
Product and By-Product Specifications   *10.65*
Waste and Emissions   *10.65*
Process Economics   *10.65*

Chapter 10.6. Kerosene IsoSiv Process for Production of Normal Paraffins   Stephen W. Sohn   10.67

General Process Description   *10.68*
Process Perspective   *10.68*
Detailed Process Description   *10.69*
Waste and Emissions   *10.71*
Economics   *10.72*
Bibliography   *10.73*

Chapter 10.7. UOP Molex Process for Production of Normal Paraffins   Stephen W. Sohn   10.75

Discussion   *10.75*
Yield Structure   *10.76*
Economics   *10.76*
Commercial Experience   *10.77*

Chapter 10.8. UOP Olex Process for Olefin Recovery
Stephen W. Sohn 10.79

Discussion *10.79*
Commercial Experience *10.79*
Economics *10.79*

Part 11   Sulfur Compound Extraction and Sweetening

Chapter 11.1. The M.W. Kellogg Company Refinery Sulfur Management   **W. W. Kensell and M. P. Quinlan** 11.3

Introduction *11.3*
Amine *11.4*
Sour Water Stripping *11.7*
Sulfur Recovery *11.9*
Tail Gas Cleanup *11.12*

Chapter 11.2. Exxon Wet Gas Scrubbing Technology: Best Demonstrated Technology for FCCU Emissions Control   **John D. Cunic** 11.15

Introduction *11.15*
Operation *11.16*
Flue Gas and Scrubber Liquid *11.17*
Particulate and SO_2 Revmoval *11.18*
Separation of the Scrubber Liquid from the Clean Flue Gas *11.19*
Clean Gas Emission *11.19*
Purge Liquid Receives Treatment *11.19*
PTU Designs *11.19*
The Aboveground PTU: The Latest Generation *11.20*
Meeting Environmental Goals *11.22*
EPA Testing *11.22*
WGS Background *11.24*
Advantages *11.25*
Summary *11.27*

Chapter 11.3. UOP Merox Process   **D. L. Holbrook** 11.29

Introduction *11.29*
Process Description *11.30*
Process Chemistry *11.35*
Product Specifications *11.37*
Process Economics *11.37*
Process Status and Outlook *11.37*

Part 12   Visbreaking and Coking

Chapter 12.1. Exxon Flexicoking Including Fluid Coking
Eugene M. Roundtree 12.3

Introduction *12.3*
Process Description *12.4*

CONTENTS

Typical Yields and Product Dispositions   *12.5*
Two Specific Process Estimates (Yields, Qualities, Utilities, and Investments)   *12.5*
Low-Btu Gas Utilization   *12.8*
Purge Coke Utilization   *12.15*
Flexicoking Unit Service Factor   *12.15*
Commercial Flexicoking Experience   *12.15*
Flexicoking Options   *12.16*
Fluid Coking Options   *12.19*

Chapter 12.2. FW Delayed-Coking Process   *Howard M. Feintuch and Kenneth M. Negin*   **12.25**

Process Description   *12.27*
Feedstocks   *12.36*
Yields and Product Properties   *12.39*
Operating Variables   *12.45*
Coker Heaters   *12.48*
Hydraulic Decoking   *12.51*
Coke-Handling and -Dewatering Systems   *12.55*
Uses of Petroleum Coke   *12.61*
Integration of Delayed Coking in Modern Refiners   *12.69*
Typical Utility Requirements   *12.78*
Estimated Investment Cost   *12.80*
References   *12.81*
Bibliography   *12.82*

Chapter 12.3. FW/UOP Visbreaking   *Vincent E. Dominici and Gary M. Sieli*   **12.83**

Introduction   *12.83*
Coil versus Soaker Design   *12.84*
Feedstocks   *12.84*
Yields and Product Properties   *12.87*
Operating Variables   *12.88*
Process Flow Schemes   *12.89*
Reaction Product Quenching   *12.94*
Heater Design Considerations   *12.95*
Typical Utility Requirements   *12.97*
Estimated Investment Cost   *12.97*
Bibliography   *12.97*

Part 13   Oxygenates Production Technologies

Chapter 13.1. Hüls Ethers Processes   *Scott Davis*   **13.3**

Introduction   *13.3*
Hüls Ethers Process for MTBE, ETBE, and TAME   *13.4*
Process Flow   *13.5*
Yields   *13.5*
Economics and Operating Costs   *13.8*
Commercial Experience   *13.8*

Chapter 13.2. UOP Ethermax Process for MTBE, ETBE, and TAME Production   *Scott Davis*   13.9

Process Description   *13.9*
Process Flow   *13.10*
Yields   *13.11*
Operating Cost and Economics   *13.11*
Ethermax Commercial Experience   *13.11*

Chapter 13.3. UOP Olefin Isomerization   *Scott Davis*   13.13

Introduction   *13.13*
Description of the Pentesom Process   *13.13*
Description of the Butesom Process   *13.15*
Economics   *13.17*
Commercial Experience   *13.17*

Chapter 13.4. Oxypro Process   *Scott Davis*   13.19

Process Description   *13.19*
Process Flow Scheme   *13.19*
Yields   *13.21*
Operating Costs and Economics   *13.21*
Commercial Experience   *13.22*

Part 14   Hydrogen Processing

Chapter 14.1. Hydrogen Processing   *Alan G. Bridge*   14.3

Introduction   *14.3*
Process Fundamentals   *14.14*
Process Design   *14.35*
Process Capabilities   *14.47*
References   *14.60*

Glossary   G.1

Abbreviations and Acronyms   A.1

Index   I.1

CONTRIBUTORS

Martin F. Bentham *UOP, Des Plaines, Illinois (CHAP. 1.6)*
Alan G. Bridge *Chevron Research and Technology Company, Richmond, California (CHAPS. 7.2, 14.1)*
David N. Brossard *Chevron Research and Technology Company, Richmond, California (CHAP. 8.1)*
W. M. Campbell *The M.W. Kellogg Company, Houston, Texas (CHAP. 6.1)*
John D. Cunic *Exxon Research and Engineering Company, Florham Park, New Jersey (CHAP. 11.2)*
Nelson A. Cusher *UOP, Des Plaines, Illinois (CHAPS. 9.2 TO 9.4, 10.5)*
Natasha Dachos *UOP, Des Plaines, Illinois (CHAP. 4.1)*
Scott Davis *UOP, Des Plaines, Illinois (CHAPS. 8.2, 13.1 TO 13.4)*
Vincent E. Dominici *Foster Wheeler USA Corporation, Clinton, New Jersey (CHAP. 12.3)*
David E. Earls *Chevron Research and Technology Company, Richmond, California (CHAP. 10.1)*
Don Felch *UOP, Des Plaines, Illinois (CHAP. 4.1)*
Howard M. Feintuch *Foster Wheeler USA Corporation, Clinton, New Jersey (CHAPS. 10.2, 12.2)*
James D. Fleshman *Foster Wheeler USA Corporation, Clinton, New Jersey (CHAP. 6.2)*
Orhan Genis *UOP, Guildford, Surrey, England (CHAP. 8.5)*
H. E. Gimpel *The M.W. Kellogg Company, Houston, Texas (CHAP. 1.2)*
H. W. Gowdy *UOP, Des Plaines, Illinois (CHAP. 8.6)*
Joseph Gregor *UOP, Des Plaines, Illinois (CHAP. 5.1)*
Charles L. Hemler *UOP, Des Plaines, Illinois (CHAP. 3.3)*
D. L. Holbrook *UOP, Des Plaines, Illinois (CHAP. 11.4)*
E. J. Houde *UOP, Des Plaines, Illinois (CHAP. 10.4)*
David A. Hunt *Refining Section, Stone & Webster Engineering Corporation, Houston, Texas (CHAPS. 3.4, 3.5)*
M. G. Hunter *The M.W. Kellogg Company, Houston, Texas (CHAP. 7.1)*
John J. Jeanneret *UOP, Des Plaines, Illinois (CHAPS. 2.1, 2.2, 2.4 TO 2.7, 10.3)*
Tiffin E. Johnson *The M.W. Kellogg Company, Houston, Texas (CHAP. 3.2)*
Aaron Kelly *UOP, Des Plaines, Illinois (CHAP. 4.1)*
James E. Kennedy *UOP, Des Plaines, Illinois (CHAP. 8.3)*
W. W. Kensell *The M.W. Kellogg Company, Houston, Texas (CHAP. 11.1)*
Paul K. Ladwig *Exxon Research and Engineering Company, Florham Park, New Jersey (CHAP. 3.1)*
Howard Lerner *Exxon Research and Engineering Company, Florham Park, New Jersey (CHAP. 1.1)*

W. L. Liggin   *UOP, Des Plaines, Illinois (CHAP. 2.3)*

John R. Mowry   *UOP, Des Plaines, Illinois (CHAP. 10.3)*

Kenneth Negin   *Foster Wheeler USA Corporation, Clinton, New Jersey (CHAP. 12.2)*

Phillip K. Niccum   *The M.W. Kellogg Company, Houston, Texas (CHAP. 3.2)*

D. A. Pappal   *Mobil Technology Company, Paulsboro, New Jersey (CHAP. 7.1)*

C. L. Pesek   *Alzo Nobel Catalysts, Houston, Texas (CHAP. 7.1)*

Peter R. Pujadó   *UOP, Des Plaines, Illinois (CHAPS. 1.5, 5.2)*

M. P. Quinlan   *The M.W. Kellogg Company, Houston, Texas (CHAP. 11.1)*

Emanuel Reis   *UOP, Des Plaines, Illinois (CHAP. 4.1)*

Mark Reno   *UOP, Des Plaines, Illinois (CHAP. 7.3)*

Eugene M. Roundtree   *Exxon Research and Engineering Company, Florham Park, New Jersey (CHAP. 12.1)*

B. R. Shah   *UOP, Des Plaines, Illinois (CHAP. 1.4)*

John C. Sheckler   *UOP, Des Plaines, Illinois (CHAPS. 1.3, 1.4)*

Gary M. Sieli   *Foster Wheeler USA Corporation, Clinton, New Jersey (CHAP. 12.3)*

Stephen W. Sohn   *UOP, Des Plaines, Illinois (CHAPS. 10.6 TO 10.8)*

Dana K. Sullivan   *UOP, Des Plaines, Illinois (CHAP. 9.1)*

Daniel G. Tajbl   *UOP, Des Plaines, Illinois (CHAPS. 1.3, 3.3)*

Gregory J. Thompson   *UOP, Des Plaines, Illinois (CHAP. 8.4)*

F. M. Van Tine   *Foster Wheeler USA Corporation, Clinton, New Jersey (CHAP. 10.2)*

J. W. Wallace   *The M.W. Kellogg Company, Houston, Texas (CHAP. 1.2)*

Diana York   *UOP, Des Plaines, Illinois (CHAP. 1.3)*

Sam Zaczepinski   *Exxon Research and Engineering Company, Florham Park, New Jersey (CHAP. 11.3)*

PREFACE

The *Handbook of Petroleum Refining Processes* is a compendium of licensable technologies for the refining of petroleum and production of environmentally acceptable fuels and petrochemical intermediates.

This Second Edition of the *Handbook of Petroleum Refining Processes* is being published just as the petroleum industry is beginning a forecast period of sustained refinery growth. Much of this new capacity will come from the expansion of existing refineries, but many new refineries are also planned, especially in Asia, South America, and the Middle East. It is forecast that the fastest growing classes of products over the next 15 years will be kerosene, jet fuel, diesel, marine diesel, and #2 heating oil, as well as naphtha, LPG, and gasoline.

Environmental regulations, as implemented in the United States, Western Europe, Korea, Taiwan, and Japan, and spreading to other parts of Asia, South America, and other markets, are having a profound effect on the technology and unit operations of modern petroleum refineries. Specifically, since the First Edition of this Handbook, a growing need has originated for oxygenate production and benzene management within the refinery.

Although the Handbook has been prepared by just six of the major licensors of petroleum refining technology, it is actually a compendium of global refining technologies. These firms are UOP, Chevron Research and Technology, Exxon Research and Engineering Company, Foster Wheeler USA Corporation, M.W. Kellogg, and Stone and Webster Engineering Corporation. However, many of the technologies presented here were co-developed with other major petroleum refining firms, including Mobil Oil Company; Dow Chemical; Institut Francais du Petrole; Unocal; Amoco; the Research Institute of Petroleum Processing (RIPP) and Sinopec International—both of the Peoples Republic of China; Hüls, AG and Koch Engineering Company, Inc.—both of Germany; Union Carbide; Instituto Mexicano del Petroleo (IMP), Catalyst Chemicals Industries Company of Japan; Alzo Nobel Catalysts; and British Petroleum.

The Handbook is divided into 14 parts, containing a total of 56 chapters:

Part 1, "Alkylation and Polymerization," presents technologies for combining olefins, or olefins with paraffins, to form clean-burning, high-octane, aromatic-free alkylate, for gasoline or distillate transportation fuel blending, and olefins with benzene to produce cumene and linear alkylbenzene.

Part 2, "Base Aromatics Production Processes," presents technologies to convert petroleum naphtha, LPG, and pyrolysis gasoline into the basic petrochemical intermediates: benzene, toluene, and xylene (BTX). Technologies to recover aromatics from hydrocarbon and aromatic mixtures and processes to convert alkylbenzenes to high-purity benzene, xylene, or naphthalene are also covered.

Part 3, "Catalytic Cracking," covers fluid catalytic cracking technologies for coverting vacuum gas oils, coker gas oils, and some residual oils, as well as aromatic lube extracts to gasoline, C_3 to C_5 olefins and light cycle oil.

Part 4, "Catalytic Reforming," contains information on producing high-octane liquids rich in aromatics from naphtha. By-products include hydrogen, light gas, and LPG.

Part 5, "Dehydrogenation," presents technology for the dehydrogenation of light and heavy paraffins to the corresponding monoolefins.

Part 6, "Gasification and Hydrogen Production," covers production of hydrogen from natural gas by steam reforming, or partial oxidation followed by shift conversion, and also hydrogen production via an initial gasification of bituminous, sub-bituminous, and lignitic coals.

Part 7, "Hydrocracking," presents technologies to convert any petroleum fraction from naphtha to cycle gas oil and coker distillates into LPG, gasoline, diesel, jet fuel, and lubricating oils while removing sulfur, nitrogen, oxygen, and saturating olefins.

Part 8, "Hydrotreating," covers technologies for improving the quality of various oil fractions by removing sulfur, nitrogen, carbon residue, metals, and wax, while increasing the hydrogen content by saturating olefins, dienes, acetylenes, and aromatics. Simultaneous cracking of heavy residua is also covered in this part. See also the Part 10 chapter on the Chevron On-Stream Catalyst Replacement (OCR) Process.

Part 9, "Isomerization," presents technologies for converting light, straight-chain naphthas (C_4 to C_6 and benzene containing reformates to branched and higher octane products while saturating benzene.

Part 10, "Separation Processes," presents technologies for recovery of catalysts from high-metal residua; olefins from mixtures of olefins and paraffins; normal paraffins from isoparaffins, naphthenes, and aromatics; and separation of vacuum residua into an uncontaminated, demetallized oil and highly viscous pitch.

Part 11, "Sulfur Compound Extraction and Sweetening," contains technologies for the removal of sulfur from refinery streams and the production of sulfur products.

Part 12, "Visbreaking and Coking," contains technologies for rejection of metals and coke, gasification of coke, and the recovery of lighter hydrocarbons from distillation unit bottoms.

Part 13, "Oxygenates Production Technologies," presents methods for production of refinery ethers for the oxygenate portion of the gasoline pool.

Part 14, "Hydrogen Processing," is a comprehensive treatment on hydrogen use within the refinery covering fundamentals, design, and process capabilities.

The authors of the various chapters were asked to follow the technology-presentation specification given below insofar as possible (some of the requested information was not deemed to be disclosable by the licensors for certain of the chapters):

1. *General process description.* Including charge and product yield and purity, and a simplified flow diagram
2. *Process chemistry and thermodynamics.* For each major process unit.
3. *Process perspective.* Developers, location, and specification of all test and commercial plants, and near term and long-term plans.
4. *Detailed process description.* Process-flow diagram with mass and energy balances for major process variations, and feeds and details on unique or key equipment.
5. *Product and by-product specifications.* Details analyses of all process products and by-products as a function of processing variations and feeds.
6. *Wastes and emissions.* Process solid, liquid, and gas wastes and emissions as a function of processing variations and feeds.

7. *Process economics.* Installed capital cost by major section, total capital investment operating costs, annualized capital costs with the basis, and a price range for each product.

The back matter contains a list of abbreviations and acronyms, and a glossary of terms, which are meant to be useful for the nonspecialist in understanding the content of the chapters.

Robert A. Meyers
RAMTECH Limited
Tarzana, California

ACKNOWLEDGMENTS

A distinguished group of 51 engineers prepared the 56 chapters of this Handbook. These contributors are listed on pages xix and xx, and we wish to acknowledge the support of their firms.

Our special thanks to Tamara Cochrane, who provided executive secretarial and indexing support and to Robert Meyers, Jr., who provided computer and indexing support. We also wish to thank Adrienne Colvin and Georgia Kramer, who provided typing services.

I thank my wife, Eileen, for her constant encouragement and advice during the years of organizing, editing, and assembling the Handbook.

LICENSING CONTACTS

The following are licensing contacts for technologies presented in this Handbook.

Chevron Research and Technology Company

Bruce Reynolds, Residuum
  Hydroprocessing Technology Manager
Chevron Research and Technology
  Company
100 Chevron Way
Richmond, California 94802-0627
  Telephone: (510) 242-4910
  Fax: (510) 242-3776

Dave Lammel, Isocracking Technology
  Manager
Chevron Research and Technology
  Company
100 Chevron Way
Richmond, California 94802-0627
  Telephone: (510) 242-4276
  Fax: (510)   242-3776

Exxon Research and Engineering Company

Europe, Africa, Middle East
Thomas A. Cavanaugh
Licensing Manager

Far East, Southeast Asia, Canada
Arnold S. Feinberg
Licensing Manager

Latin America
Vincente A. Citarella
Licensing Manager

United States and All Other Regions
Sam Zaczepinski
Manager of Technology Assessment and
  U.S. Licensing

Address for all Exxon Contacts
Exxon Research and Engineering
  Company
Technology Licensing Division
Post Office Box 390
Florham Park, New Jersey 07932

Foster Wheeler USA Corporation

Dr. Howard M. Feintuch, Manager of Technology Licensing
Foster Wheeler USA Corporation
Perryville Corporate Park
Clinton, New Jersey 08809-4000
  Telephone: (908) 730-5388
  Fax: (908) 730-4707

The M.W. Kellogg Technology Company

Joseph Collins, Vice President, Business Development
The M.W. Kellogg Technology Company
601 Jefferson Avenue
Houston, Texas 77002
   Telephone: (713) 753-5655
   Fax: (713) 753-5353

Stone & Webster Technology Corporation

Mr. Warren S. Letzsch, Vice President
Stone & Webster Technology Corporation
1430 Enclave Parkway
Houston, Texas 77077-2023
   Telephone: (713) 368-4353
   Fax: (713) 368-3555

UOP

UOP
25 East Algonquin Road
Des Plaines, Illinois 60017-5017
   Telephone: (847) 391-2000
   Fax: (847) 391-2253

P · A · R · T · 1

ALKYLATION AND POLYMERIZATION

CHAPTER 1.1
EXXON SULFURIC ACID ALKYLATION TECHNOLOGY

Howard Lerner
Exxon Research and Engineering Company
Florham Park, New Jersey

INTRODUCTION

Alkylation is an important process unit used in refineries to convert light olefins (e.g., propylene, butylene), produced in catalytic crackers and cokers, into a more highly valued gasoline component. Alkylate is one of the best gasoline blending components produced in the refinery because of its high octane and low vapor pressure. This chapter describes Exxon Research and Engineering Company's sulfuric acid catalyzed alkylation technology and the reasons why it is able to dependably produce high-quality alkylate at low cost.

Alkylation technology has been around for a long time in the refining industry. The old-time tank reactor system was first installed in Exxon's Baytown refinery in the late 1930s. This sulfuric acid alkylation technology has branched out from those beginnings in a number of directions. Exxon Research and Engineering Company (ER&E) developed the stirred, autorefrigerated reactor design, while others developed the cascade and effluent refrigeration systems.

This chapter concentrates on a number of general sulfuric acid alkylation topics and highlights important differences between the modern ER&E sulfuric acid alkylation reactor design and other designs. First, it discusses how alkylation fits into the refinery and then it describes the chemistry of alkylation. It shows an alkylation process flow sheet, followed by a discussion of the effects of the key process variables and how their interactions are critical to efficient design and operation. It compares the autorefrigeration method of removing the heat of reaction with the indirect effluent refrigeration system and explains why the autorefrigeration method is very efficient. Also, it discusses key advantages and differentiating features of the ER&E reactor system along with commercial experience with the ER&E technology.

ALKYLATION IS A KEY PROCESSING UNIT

The alkylation process is an important unit that is used in refineries to upgrade light olefins and isobutane into a much more highly valued gasoline component. The light olefins are produced mainly from catalytic crackers and also from cokers and visbreakers. Alkylate is one of the best gasoline blending components produced in the refinery because of its high octane, which is typically about 96 research octane number (RON), and low vapor pressure, which allows more lower-value butane to be put into gasoline. The high profitability of upgrading the light olefins and isobutane in the refinery from liquefied petroleum gas (LPG) value to gasoline value explains why it is a very popular process alternative that is used at most locations that have catalytic crackers.

CHEMISTRY OVERVIEW

The primary alkylation reaction involves the reaction of isobutane with a light olefin, such as butylene, in the presence of a strong acid catalyst to form the high octane, trimethyl pentane isomer.

Primary Alkylation Reaction

$$C_4H_{10} + C_4H_8 \xrightarrow{H_2SO_4} 2,2,4\ \text{trimethylpentane} + \text{Heat}$$

Isobutane   Butylene   Isooctane

Although the reaction goes to completion, it does not approach equilibrium. If the C8 isomers formed were to approach equilibrium distribution, the product octane would be about 20 octane numbers lower. This is because the lower-octane, less highly branched isomers, such as dimethyl hexane, would predominate. Although not shown here, similar reactions can be written for the reaction of isobutane with propylene or pentylene, which are other viable alkylation plant feeds that also form high-octane isomers, though not as high as for butylene feeds. If this primary reaction were the only reaction taking place, we would be able to produce alkylate with an octane approaching 100.

Secondary Reactions Produce Wide Spectrum of Compounds

Unfortunately, there are a number of secondary reactions that produce a wide spectrum of compounds that tend to reduce the octane to about the 96 octane level. The examples below show how polymerization, hydrogen transfer, disproportionation, and cracking reactions tend to produce lower octane components.

- Polymerization (e.g., $2C_3H_6 \rightarrow C_6H_{12}$)
- Hydrogen transfer (e.g., $2C_4H_{10} + C_6H_{12} \rightarrow C_8H_{18} + C_6H_{14}$)
- Disproportionation (e.g., $2C_8H_{18} \rightarrow CH_7H_{16} + C_9H_{20}$)
- Cracking (e.g., $C_{12}H_{26} \rightarrow C_7H_{14} + C_5H_{12}$)

Feed Impurities Form Acid-Soluble Compounds

One other important aspect of alkylation chemistry is the effect of feed impurities on the alkylation process. This is because makeup requirements for the acid which is used as reaction catalyst will depend to some extent on the amount of impurities that are present in the alkylation plant feed streams. Many of these impurities form acid-soluble compounds that increase acid makeup requirements, since they must be purged from the plant with spent acid. For example, mercaptan sulfur, which is a common feed impurity, will react with sulfuric acid to form a sulfonic acid and water, which are acid-soluble compounds (that is, $RSH + 3H_2SO_4 \rightarrow RSO_3H + 3H_2O + 3SO_2$). Also formed is some sulfur dioxide, which must be removed in the caustic wash section as discussed later. Roughly 40 pounds of additional makeup acid are needed for each pound of mercaptan sulfur that enters the plant with the feed. Thus caustic treating facilities are always provided to remove sulfur from the olefin feed.

Butadiene is another common impurity in alkylation plant feeds. The butadiene, which is typically at the 1000 to 2000 ppm level, tends to polymerize and form acid-soluble oils which will also increase acid makeup requirements. For every pound of butadiene in the feed, 10 pounds of additional makeup acid will be required. If the feed butadiene level is very high, as might result from a severe catalytic cracker operation or from introducing coker olefins, installing diene saturation facilities to selectively hydrogenate the diolefins without hydrogenating the valuable monoolefins could be considered.

PROCESS DESCRIPTION

A simplified flow diagram for the Exxon Research and Engineering stirred autorefrigerated alkylation process is shown in Fig. 1.1.1. Olefin feed is first mixed with recycled isobutane from the deisobutanizer overhead and cooled before entering the reactor. Insoluble water that is condensed at the lower temperature is removed in the coalescer. The isobutane-olefin mixture, along with recycle acid and refrigerant, is introduced to the reactor. Mixers provide intimate contact between the reactants and the acid catalyst. In the presence of sulfuric acid, the olefin and isobutane react very quickly to form alkylate and release reaction heat. In the autorefrigeration system, the reaction heat is removed by vaporizing some isobutane from the reaction mixture. The vapors leaving the reactor are routed to the refrigeration section where they are compressed, condensed, and sent to the economizer, which is an intermediate-pressure flash drum, before being returned back to the reactor. The intermediate-pressure flash reduces the power requirements of the refrigeration compressor by about 10 percent. Thus the reactor is held at an optimal temperature of about 40°F.

Any propane introduced with the feed concentrates in the refrigeration section and must be removed from the alkylation plant. Therefore a small slipstream of refrigerant is depropanized after being caustic- and water-washed to remove any SO_2. The propane overhead is sent to storage while the isobutane-rich bottoms are returned to the process. Thus the depropanizer operation avoids building up propane in the alkylation plant.

Back at the reactor, the reactor product is routed to the settler, where the acid is settled from the hydrocarbon and is recycled back to the reactor. The hydrocarbon portion of the reactor product which contains alkylate, excess isobutane, and normal butane, is then caustic- and water-washed to remove any acidic components before being fed to the deisobutanizer. Makeup isobutane that is consumed by the alkylation

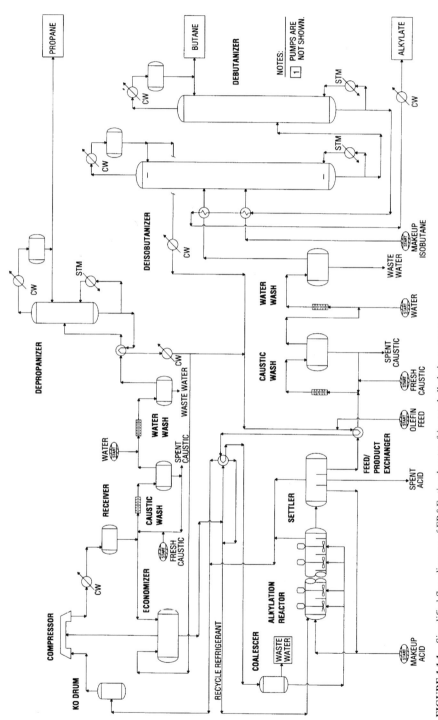

FIGURE 1.1.1 Simplified flow diagram of ER&E stirred, autorefrigerated alkylation process.

reaction is also typically added to the deisobutanizer. The overhead from the tower is an isobutane-rich stream that is recycled back to the reactor, while the bottoms from the tower, which contain normal butane and alkylate, are routed to the debutanizer tower. In the debutanizer, butane is removed overhead while the alkylate bottoms, which meet vapor pressure specifications, are cooled and routed to alkylate storage. These towers can be designed to operate at a low enough pressure to allow the use of low-pressure steam.

There is flexibility in how the flowsheet can be put together to meet the needs of the refiner. For example, a butane sidestream can be taken from the deisobutanizer instead of from a separate tower, thus reducing capital cost. The main disadvantage that usually restricts the use of this option is that the C_5+ content of the butane will usually be higher than the sales specifications allow. It is, however, quite acceptable to use it for gasoline blending.

In the reactor, a small amount of the olefin feed polymerizes to form acid-soluble oils which tend to accumulate in the equilibrium recycle acid. As discussed earlier, feed impurities that also form acid-soluble compounds tend to magnify this effect. Therefore a spent acid purge must be taken from the unit to remove these oils from the system while high-purity, fresh makeup acid is added to replace it in order to keep the spent acid strength high enough to maintain catalytic activity. Providing for the regeneration of the spent acid is a special challenge to a sulfuric acid plant. In the United States, the spent acid is usually regenerated by the supplier of the makeup acid. However, other countries, such as Japan, have relatively small sulfuric acid production capacities. Therefore, some alkylation plant operators in these areas have installed their own acid regeneration facilities. These plants are relatively simple to operate but they add investment cost. The technology is available from a number of licensers.

BALANCING PROCESS VARIABLES IS CRITICAL TO EFFICIENT DESIGN AND OPERATION

Proper specification of process variables is critical to optimize the design and operation of an alkylation plant. Commercial and pilot plant experience provides an understanding of the balances of the process variables and how they affect the economics of the process. This is important since application of this knowledge results in units that are capable of efficiently meeting refinery goals. The primary process variables discussed below are reactor temperature, isobutane recycle rate, reactor space velocity, and spent acid strength.

Low temperature increases octane and reduces acid requirements, which is the goal. Unfortunately, the low temperature also requires higher refrigeration investment and operating costs. Thus a balance must be struck between the benefits of low reactor temperature and the cost of supplying the low temperature.

An increase in isobutane recycle rate increases the octane and lowers the acid requirements. However, the higher isobutane rate also increases the investment required for the deisobutanizer and increases the operating cost of the tower because of the higher steam requirement. The critical choice between the benefits of high isobutane recycle rate and investment must be balanced.

Reactor space velocity is another very important process variable that has to be set during the design. Lowering space velocity increases octane and lowers acid requirements. On the other hand, severe alkylation conditions such as high space velocity increase the formation of acid sulfates which are corrosive to downstream fractionation facilities. The neutral butyl sulfate esters that are formed at high space velocity are not removed by caustic and tend to decompose at the higher temperatures encountered in the fractionation facilities. These decomposition products are sulfur dioxide, which is corrosive to the overhead system, and sulfuric acid, which goes down the tower and chars hydrocarbon when it hits the hot reboiler tubes, thus creating a fouling situation. The low space velocity in ER&E designs requires no special considerations to protect the downstream fractionation facilities other than the caustic and water washes which are provided in all cases. For cases where high esters are a known problem, it would be necessary to install an expensive acid wash step. Thus, the troublesome esters from the reactor product would be extracted before they reach the fractionation facilities.

Finally, spent acid strength impacts not so much on the design as on the operation of the alkylation unit. Increasing spent acid strength tends to increase octane, but also increases makeup acid requirements. While lowering spent acid strength tends to decrease makeup acid requirements, it increases the risk of acid runaway. This acid runaway condition occurs when the strength of the spent acid drops below the critical concentration where the acid is an effective alkylation catalyst. Thus target spent acid strength will be a balance between acid makeup requirements and octane, with an adequate margin of safety above the acid runaway point.

REACTOR COOLING VIA AUTOREFRIGERATION IS MORE EFFICIENT THAN EFFLUENT REFRIGERATION

As shown in Fig. 1.1.2, there are two systems that are used to remove the heat of reaction while maintaining the low reaction temperature needed for alkylation. As mentioned before, ER&E uses the autorefrigeration concept of removing the heat of reaction from reactors. As in Fig. 1.1.2, olefin feed is introduced into a number of mixing compartments where just enough mixing energy is provided to obtain good contacting of acid and hydrocarbon, which is needed to promote good reaction selectivity. The reaction is held at low pressure, about 10 lb/in^2 gage, to keep the reaction temperature at about the 40°F level. Isobutane-rich hydrocarbon vapors boil from the reaction mass and are removed from the top of the reactor and sent to the refrigeration compressor. Thus, there is a 0° temperature difference between the reaction mass and the refrigerant in the direct autorefrigeration system.

The indirect effluent refrigeration system is used by others for removing the heat of reaction. The system pressure is kept high to prevent vaporization of light hydrocarbons in the reactor and settler. In this system, hydrocarbons from the settler are flashed across a control valve into a large number of heat transfer tubes contained within the reactor to provide cooling. The reaction emulsion is pumped across these cooled tubes to keep the reaction temperature down. The reaction emulsion then goes to the settler where acid is recycled and hydrocarbons are flashed through the heat-transfer tubes. The vapors are routed to the refrigeration system by way of a knockout drum. This approach requires that there be a finite temperature difference between the reaction emulsion and the refrigerant in order to transfer heat across the tubes.

Because a temperature difference is required for the indirect effluent refrigeration system, it requires a lower refrigeration temperature than the autorefrigeration system which, as noted, has a zero temperature difference between the reaction mass and the refrigerant. The lower refrigerant temperature in the indirect effluent refrigeration system requires a lower refrigeration compressor suction pressure which in turn results in a higher refrigeration compressor energy requirement to supply the same refrigeration duty. Another factor that increases the energy requirement still further for the indirect effluent refrigeration system is that more mixing power is needed in order to over-

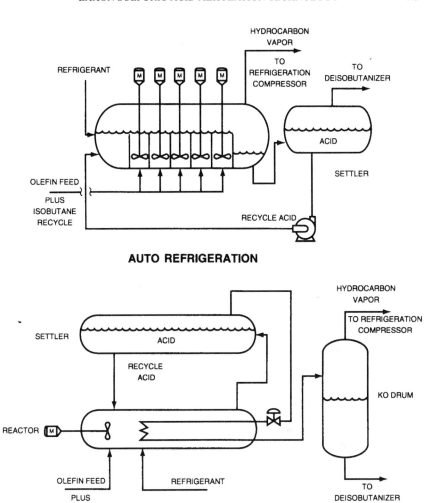

FIGURE 1.1.2 Reactor cooling alternatives.

come the pressure drop across the heat-transfer tubes. In the autorefrigeration system, just enough energy is input to provide good contacting between the acid catalyst and the hydrocarbon, whereas in the effluent refrigeration system, additional mixing power has to be provided to overcome the pressure drop created by the high flow rate of reaction emulsion across the heat-transfer tubes. The high flow rate is needed to produce a reasonable heat-transfer coefficient. Thus, the total mixing energy is higher for this system. Since the heat of mixing has to be removed from the reaction emulsion, a larger refrigeration duty is also required. These two factors make the autorefrigeration system more efficient than the indirect effluent refrigeration system. Thus, the autorefrigeration system is less expensive and has lower operating costs.

REACTION STAGING RESULTS IN HIGH AVERAGE ISOBUTANE CONCENTRATION IN REACTOR

As discussed earlier, vaporization of isobutane-rich refrigerant is suppressed in the effluent refrigeration scheme, while vaporization of refrigerant is allowed to proceed in the autorefrigerated reactor. One might conclude that a higher isobutane recycle rate is required for the autorefrigerated reactor system in order to attain the same isobutane concentration in the reactor. However, a higher isobutane recycle rate is not required in the stirred, autorefrigerated reactor because of the efficient multiple-stage configuration. Multiple stages are provided at low cost due to the nature of the reactor design. Figure 1.1.3 illustrates this point. This is a plot showing how the isobutane concentration and the octane of new alkylate produced change as the reaction emulsion moves through the reactor. As mentioned before, all the refrigerant is returned to the first reaction stage, where only a small fraction of the olefin feed is injected. Since only a small corresponding fraction of the refrigerant is vaporized in that stage, the isobutane concentration in that stage is much higher than the single-stage effluent refrigeration reactor scheme. Therefore, the octane of the alkylate product in that stage is very high. As we move from stage to stage, more refrigerant vaporizes as more olefin is injected and the octane of the alkylate produced in that stage decreases. The average isobutane con-

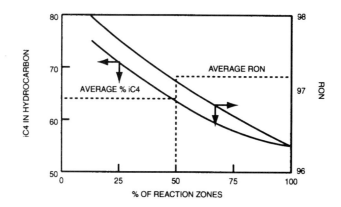

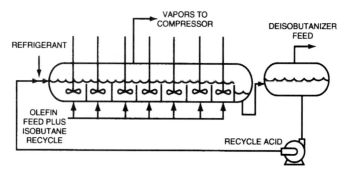

FIGURE 1.1.3 Staged reactor enhances performance.

centration is essentially the same for both reactor types at the same isobutane recycle rate. Thus, the product octane will be essentially the same for the reactors as long as other reactor conditions such as space velocity and temperature are the same. However, as discussed earlier, the space velocity of the stirred, autorefrigerated reactor is lower than the effluent refrigeration reactor, resulting in a higher octane.

ADVANTAGES OF THE ER&E REACTOR ARE NUMEROUS

The ER&E stirred, autorefrigerated reactor design has many advantages. As discussed above, autorefrigeration is efficient. Also, the internals of the ER&E stirred, autorefrigeration reactor are very simple. This results in low reactor cost and equipment that can be constructed in many shops under competitive bidding conditions. In addition, the ER&E reactor can be built very large to accommodate alkylate production of up to 9000 barrels per day (BPD) in a single reactor. This gives ER&E designs a strong advantage in economy of scale over other technologies.

The staged reactor design results in a high-octane product. The octane is high because the reactor design is conducive to maintaining a high isobutane concentration in the reactor, a low space velocity, and low temperature. The low temperature and low space velocity as well as the special feed injection system differentiates the ER&E reactor from reactors offered by others. The net result of applying these reactor design concepts in the stirred, autorefrigeration reactor eliminates corrosion and fouling problems experienced in the fractionation system while producing high-quality process performance.

One further advantage of the stirred, autorefrigerated reactor design is its high reliability. For example, the low-pressure operation means low-maintenance mechanical seals can be used for the low-speed mixers. This is in contrast with the other systems that require a higher reactor pressure, which makes the mechanical seals much more critical and a source of concern for reliability. Also, the seals in the ER&E reactor are in the vapor space instead of immersed in the acidic liquid phase as in the effluent refrigeration reactor, making the consequences of a leaky seal much less severe. Finally, ER&E's autorefrigeration design contains multiple mixers, which translates to higher reliability.

Figure 1.1.4 shows a 1991 alkylation plant using ER&E's stirred, autorefrigerated technology. The ER&E-designed reactor is the horizontal vessel in the foreground.

MODERN ER&E REACTOR IS A VAST IMPROVEMENT OVER OLDER SYSTEMS

The operation of older autorefrigeration reactor designs offered by others in the past has been reported to be troublesome because of producing low-octane product and having the symptoms of high esters discussed previously. Some people mistakenly associate these problems with the autorefrigeration method rather than the specific design features used. In fact, there are very significant improvements encompassed in the modern ER&E autorefrigeration reactor design that contribute to its excellent performance. Some of the more significant differences are summarized in Table 1.1.1.

The combination of better acid-hydrocarbon mixing (to provide the needed intimate contact of the hydrocarbon with the acid catalyst), premixing olefin and isobu-

1.12 ALKYLATION AND POLYMERIZATION

FIGURE 1.1.4 Alkylation plant using ER&E alkylation technology.

TABLE 1.1.1 Comparison of Modern ER&E-Designed Reactor to Older Systems

	Older industry design	Modern ER&E design
Acid/hydrocarbon mixing	Submerged pumps	Special mixers
Isobutane/olefin premixed	No	Yes
Olefin feed injectors	Open pipe	Special nozzles
Acid/hydrocarbon ratio	<0.5	1.5
Space velocity	0.3	0.1
Pressure controllers per reactor	Up to 10	2

tane (to make the isobutane readily available for reaction), using special feed injectors (that distribute the feed uniformly in the reactor), using a much higher acid/hydrocarbon ratio (to assure that the reaction emulsion is in the favorable acid continuous mode), and the low space velocity (to assure a low level of esters) results in high-octane product without any of the problems associated with high ester content.

ECONOMICS OF EXXON SULFURIC ACID ALKYLATION TECHNOLOGY

The estimated costs of a 7.5 thousand barrel per stream day (kBPSD) Exxon stirred, autorefrigerated alkylation plant are listed below.

Direct costs*	
Material	12.0
Labor	1.7
Total	13.7

*In millions of dollars, on U.S. Gulf Coast, second quarter 1993.

Estimated utility requirements and chemical consumption for the same plant are shown in Table 1.1.2.

EXTENSIVE COMMERCIAL EXPERIENCE ENHANCES TECHNOLOGY PACKAGE

Exxon has considerable commercial experience with the stirred, autorefrigerated technology. There are roughly 80,000 barrels per day (BPD) of installed capacity at nine locations using the stirred, autorefrigerated reactor technology and the sizes of the plants vary from 2000 to 30,000 BPD. Details on 10 ER&E-designed sulfuric acid

TABLE 1.1.2 Estimated Utilities and Chemical Consumption, Exxon Stirred, Autorefrigeration Alkylation Process, 7.5 kBPSD

Utilities:	
Power, kW:	
Compressor	2375
Pumps	570
Mixers	400
Total	3345
Cooling water,* gal/min (m^3/h)	10,900 (2476)
Industrial water, MT/D	70
Steam,† klb/h (MT/h)	63 (28.6)
Chemicals:	
Fresh acid,‡ MT/D	65
NaOH (100%), MT/D	0.34

*20°F temperature rise.
†50 lb/in^2 gage.
‡98.0 wt % minimum sulfuric acid.

Note: MT/D = metric tons per day; MT/h = metric tons per hour.

TABLE 1.1.3 ER&E-Designed Sulfuric Acid Alkylation Units

Company	Location	Nominal capacity of alkylate, kBPSD
Exxon unit	United States	25
Exxon unit	United States	25
Licensed unit	United States	10
Exxon unit	Asia	1
Exxon unit	Asia	8
Licensed unit	Asia	4
Exxon unit	Europe	6
Exxon unit	Europe	6
Licensed unit	United States	7
Licensed unit	Asia	8

alkylation units are listed in Table 1.1.3. The ER&E alkylation process utilizes efficient autorefrigeration while incorporating design details that result in low maintenance, low operating cost, and high service factor proven by many years of successful operation. Continuing feedback from Exxon's operating plants is built into the design of new units, thus ensuring low-cost, trouble-free operation.

CHAPTER 1.2
THE DOW-KELLOGG CUMENE PROCESS

J. W. Wallace and H. E. Gimpel
The M.W. Kellogg Company, USA

INTRODUCTION

In the Dow-Kellogg Cumene Process, cumene is produced via alkylation of benzene and propylene in the presence of Dow's unique, shape-selective 3DDM zeolite catalyst. The process is characterized by its low capital cost, superior product yield, high-purity product, corrosion-free environment, low operating cost, and ease of operation.

The process has been demonstrated by an extensive program at Dow that led to the development of the 3DDM zeolite catalyst. This program has provided the pilot and experimental work necessary for the development of the kinetic model by Kellogg for alkylation and transalkylation. A commercial-size transalkylator using the catalyst has been in very successful operation for approximately 2 years at Dow's cumene plant in Terneuzen in the Netherlands. The facility utilizes feed from the existing solid phosphoric acid (SPA) catalyst-based operation. Conversion of the balance of the plant to the new process is planned.

Kellogg has developed an efficient, heat-integrated, low-capital cost process flowsheet for new grassroots plants and for conversions of existing SPA-based plants. These plants, based on the Dow-Kellogg process, provide superior economics to the competing processes. Kellogg is well-known for both cumene and the related phenol technologies, having engineered eight cumene plants worldwide for a total capacity of 4.5 billion pounds (2 million metric tons) a year.

HISTORY OF CUMENE TECHNOLOGY

Cumene (isopropyl benzene) is used now almost exclusively in the production of phenol by the oxidation with air to cumene hydroperoxide (CHP), followed by cleavage to phenol and acetone. This process was developed in the 1940s and is the principal commercial process for phenol production. An earlier use for cumene was as aviation gasoline because of its high octane value.

Cumene always has been commercially produced by the acid-catalyzed alkylation of benzene with propylene. Most of the world's capacity is being produced by processes utilizing SPA catalysts. The relatively low yields and various operating difficulties related to the SPA (acid on clay) catalyst led producers to seek improved types of catalysts.

One alternative approach has been the aluminum chloride–based process applied over the past 10 years to both new plants and revamped SPA units. Prior to this, aluminum chloride catalyst had been used commercially for a number of years to produce ethylbenzene for styrene manufacturing. Although this process offers improved yield—due to the transalkylation ability of the catalyst, high reactor productivity, and constant catalyst activity from continuous catalyst makeup—this process has not gained universal acceptance because of the special materials required to prevent corrosion and concerns related to spent aluminum chloride disposal.

Another more recent approach to improved catalysis for cumene production has been the development of zeolite catalysts, such as Dow's 3DDM special modified mordenite catalyst. This catalyst offers numerous advantages compared to the SPA and aluminum chloride catalyst.

HISTORY OF THE DOW-KELLOGG CUMENE PROCESS

Before combining their efforts, both Kellogg and Dow had a long history of involvement in the design and operation of cumene plants.

Dow has operated the 340,000–metric ton per annum (MTA) cumene plant at Terneuzen for 20 years. This plant originally employed the traditional SPA cumene process. In 1992, Dow added a transalkylator system that utilizes the 3DDM zeolite catalyst. Dow also has been operating several laboratory rigs and pilot plants for this new cumene technology at Terneuzen since 1991. Because of advantages offered by the Dow-Kellogg process, Dow plans to convert the remainder of the plant to the new process in the near future.

Kellogg has engineered eight cumene plants, including six based on the SPA process and one based on the Al_3Cl process. Kellogg has developed energy-saving designs for cumene plants, including direct heat interchange systems between cumene and phenol plants that have been used in three cumene/phenol plant designs. A list of cumene plants engineered by Kellogg is in Table 1.2.1.

TABLE 1.2.1 Cumene Plants Engineered by M. W. Kellogg

Client	Location	Date	Capacity, MTA
Taiwan Prosperity Chemical Co.	Taiwan	1994–95	135,000*
Neofen	Brazil	Delayed	145,000*
Aristech Chemical	Louisiana	Delayed	450,000
Georgia Gulf	Texas	1992	500,000
Shell Chemical No. 2	Texas	1977 (1985)	550,000†
Monsanto No. 2	Texas	1971	180,000
Monsanto No. 1	Texas	1960	55,000
Shell Chemical No. 1	Texas	1959	35,000

*Grassroots $AlCl_3$ process.
†Originally designed for SPA process and revamped to $AlCl_3$.

In 1993, Dow and Kellogg joined together to complete the development of the technology and offer it for license.

PROCESS FEATURES

Key features of the Dow-Kellogg Cumene Process are:

- *High yield.* The very high selectivity of the catalyst and reaction system results in yield of greater than 99 percent to cumene.
- *High-quality cumene.* The same high selectivity together with a very efficient recovery system results in high-quality cumene product, i.e., greater than 99.90 percent purity.
- *All-carbon-steel construction.* The zeolite catalyst is very stable, and no acid enters the plant; therefore, all equipment is fabricated from carbon steel (CS).
- *No environmentally difficult effluents.* The zeolite catalyst is very sturdy and regenerable; thus, there is no catalyst disposal concern as in other commercial processes.
- *Operability and maintainability.* The fixed-bed reactors and distillation train are simple and safe to operate; with no corrosion and with simple equipment, the maintenance costs are very low.
- *Low capital cost.* A simple yet efficient design, with low recycle rates and all-CS equipment, results in a low-capital-cost unit; on the U.S. Gulf Coast, a 550,000,000-lb/yr (250,000-MTA) unit is estimated to cost approximately $17,000,000.

PROCESS DESCRIPTION

As shown in Fig. 1.2.1, benzene and propylene are first reacted in the alkylator (1) in the presence of the 3DDM zeolite catalyst to produce cumene. Complete reaction of propylene occurs. Some di-isopropylbenzene (DIPB) and very little tri-isopropylbenzene (TIPB) and by-products also are formed.

The amount of DIPB depends on the quantity of excess benzene and the reactor operating conditions. The quantity of TIPB formed by further alkylation of DIPB is very small because of the shape selectivity of the catalyst. Alkylation conditions are chosen using the Kellogg Kinetic Model to maximize production of cumene and to avoid formation of impurities, such as n-propylbenzene.

The reactor-section effluent, consisting of unreacted benzene, cumene, and DIPB, is fractionated in a series of four columns (3–6). Light components, including propane present in the propylene feed, are removed; unreacted benzene is recovered and recycled; pure cumene is separated as a product; and, finally, the DIPB is recovered in the distillation section and recycled to the transalkylator (2) to form additional cumene with excess benzene. A small impurity purge is taken to fuel.

The alkylation reactions are rapid and exothermic. In contrast, transalkylation reactions are slower, equilibrium-limited, and thermally neutral. The transalkylator conditions are chosen to favor formation of cumene.

Cumene produced in the fractionation system is very high in purity (see Table 1.2.2) with a bromine index less than 5.

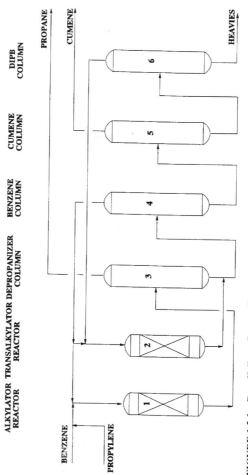

FIGURE 1.2.1 Dow-Kellogg Cumene Process.

THE DOW-KELLOGG CUMENE PROCESS

TABLE 1.2.2 Product Specifications for Cumene Purity

Cumene	99.94 wt % min.
Bromine index	5 max.
Ethylbenzene	100 ppm max.
n-propylbenzene	200 ppm max.
Butylbenzene	100 ppm max.

The Dow-Kellogg process is capable of processing a wide range of feed purities. All grades (refinery, chemical, and polymer) of propylene feed are suitable. Propane content may vary from less than 1 percent to 40 percent or more. The propane is normally recovered as liquefied petroleum gas (LPG), but may be used as fuel if present in small quantities. Normally, no feed pretreatment is necessary.

YIELDS AND MATERIAL BALANCE

Material	Pounds per 1000 lb cumene	Metric tons per MT cumene
Feed:		
Benzene (as 100%)	653	0.653
Propylene (as 100%)	353	0.353
Product:		
Cumene	1000	1.000
By-product:		
Heavies	6	0.006
LPG	*	*
Drag benzene	†	†

*Function of propane contained in propylene feed.
†Function of nonaromatics in benzene feed.

PROCESS ECONOMICS

Utilities	Per 1000 lb cumene	Per metric ton cumene
Fuel (absorbed)*	1.1×10^6 Btu	0.60×10^6 kcal
Steam (export)†	525 lb	525 kg
Cooling water‡	130 gal	1.08 m^3
Electricity‡	7.7 kWh	17.0 kWh

*Hot-oil heating (steam heating and fired reboiler options available).
†Denotes export low-pressure steam, 50 lb/in^2 (3.5 kg/cm^2) gage.
‡Assumes maximum air cooling (cooling water options available).

WASTES AND EMISSIONS

The process produces no liquid or vapor emissions with the exception of normal stack emissions from heaters (hot-oil system), boiler feedwater (BFW) blowdown from steam generators, and vacuum jet or vacuum pump vents. Spent catalyst is benign and requires no special disposal considerations other than normal landfill.

CHAPTER 1.3
UOP CATALYTIC CONDENSATION PROCESS FOR TRANSPORTATION FUELS

Diana York, John C. Scheckler, and Daniel G. Tajbl

UOP
Des Plaines, Illinois

INTRODUCTION

Olefin-producing units such as thermal crackers, visbreakers, fluid catalytic crackers (FCCs), and cokers exist in many refineries. These units produce large quantities of olefinic liquefied petroleum gas (LPG). This olefinic LPG, which typically includes propylene, propane, butylene, and butane, is too volatile to blend directly into transportation fuels.

The olefins in the LPG can be converted to either gasoline or distillate transportation fuels via the UOP* Catalytic Condensation process, also referred to in the industry as the Cat-Poly process. In this process, propylene or butylene or both are oligomerized to yield higher-molecular-weight compounds. The process is versatile and can be used by itself or in combination with downstream processes to produce high-quality transportation-fuel blending stocks. The typical application of the UOP Catalytic Condensation unit for the production of either polymer gasoline or distillate product is discussed in this chapter.

HISTORY

The UOP Catalytic Condensation process appeared relatively early in the development of the petroleum-refining industry. The commercial application of this process is closely tied to the early growth in gasoline demand.

In early refinery operations, simple distillation was used to produce gasoline, kerosene, and lamp fuels from crude oil. As automobiles became more abundant, the demand for gasoline increased. To produce more gasoline, thermal cracking using the

---
*Trademark and/or service mark of UOP.

Dubbs process was introduced. A by-product of thermal cracking is a light-gas stream, which contains methane, ethane, ethylene, propane, propylene, butane, and butylene. Initially, this light gas had little value and was either burned as fuel or flared.

As gasoline demand began to grow, refiners searched for technology to maximize gasoline production. One possible alternative was the oligomerization of C_3 and C_4 olefins in the LPG from thermal cracking units.

The chemistry of olefin reaction was familiar to refiners. A well-known wet chemical analysis for determining the olefin content of a gas sample involved the oligomerization of the gaseous olefins. The oligomerization occurred in a glass Orsat apparatus using liquid acid catalyst. The next logical step was to use the oligomerization mechanism to produce a high-octane gasoline blendstock. In designing a large-scale commercial polymerization process based on this principle, the problem was how to use the acid catalyst in a carbon-steel vessel and avoid excessive acid corrosion. This problem was solved by Dr. Vladimir N. Ipatieff of UOP.

The solution involved fixing phosphoric acid in a silica matrix. Specifically, Ipatieff discovered that a mixture of phosphoric acid and kieselghur (a natural silica) solidified when heated. The resultant solid phosphoric acid catalyst could then be ground up, sized, and loaded into a reactor. Under proper conditions, a gasoline-range product was produced from the oligomerization or condensation reactions of the olefins without the corrosive problems of an aqueous acid. This important discovery led to the development of the UOP Catalytic Condensation process, which was applied commercially in a number of areas, particularly in processing gaseous olefins from thermal reforming units during the early 1930s. The emergence of the FCC process in the 1940s greatly expanded the application of the UOP Catalytic Condensation process.

PROCESS CHEMISTRY

Many refineries have cracking operations that produce a substantial amount of propylene and butylene. These olefins can be oligomerized in a UOP Catalytic Condensation unit to produce gasoline- or distillate-range transportation fuels. The oligomerization reactions that occur depend on the relative concentrations of propylene and butylene in the feed. The primary reactions are

$$2C_3H_6 \rightarrow C_6H_{12} \qquad \Delta H_{av} \approx 1535 \text{ kJ/kg } C_6$$
$$C_3H_6 + C_4H_8 \rightarrow C_7H_{14} \qquad \Delta H_{av} \approx 1303 \text{ kJ/kg } C_7$$
$$2C_4H_8 \rightarrow C_8H_{16} \qquad \Delta H_{av} \approx 1151 \text{ kJ/kg } C_8$$

where ΔH_{av} is the average heat of reaction.

These reactor products are themselves olefinic and can undergo further reaction. Some of the C_6, C_7, and C_8 olefins react with propylene and some with butylene to produce C_9 and C_{10} products. These reactions are

$$C_3H_6 + C_6H_{12} \rightarrow C_9H_{18}$$
$$C_3H_6 + C_7H_{14} \rightarrow C_{10}H_{20}$$
$$C_4H_8 + C_6H_{12} \rightarrow C_{10}H_{20}$$

The resultant C_9 and C_{10} polymers are still olefinic and can react further to yield even higher-molecular-weight products.

In the UOP Catalytic Condensation process, the degree of reaction is controlled to yield products of the desired boiling range. For gasoline production, about 95 percent of the fresh C_3 and C_4 olefins is converted to C_6+ product. Unconverted C_3 or C_4 olefins are generally recovered, and some of them are recycled to the reactor to control reaction temperature and increase overall product yield.

For distillate production, some of the fresh C_3 and C_4 olefins are converted to distillate boiling range in the first pass through the reactor. Distillate production is increased by recycling light-polymer gasoline to the reactor. With the recycle of polymer gasoline, 75 percent or more of the feed C_3-C_4 olefins can be converted to a distillate product.

The UOP Catalytic Condensation process can also be used for the synthesis of specific motor fuels. For example, normal butylene and isobutylene can be selectively reacted to yield a trimethylpentene-rich isooctene. With mild hydrotreating of the product, trimethylpentanes can be produced. This product has a clear research octane number of about 100 and is therefore considered an excellent aviation gasoline.

PROCESS THERMODYNAMICS

Oligomerization reactions are exothermic, as shown by the heats of reaction given previously. These heats of reaction represent the average heat of reaction for all possible isomer combinations. The actual heat of reaction depends on the specific oligomer isomer distribution and plant operating conditions.

The heat released by the reaction must be removed to maintain reactor control. A lack of proper reactor temperature control can result in a loss of either catalyst activity or product yield.

Units for olefin oligomerization have used either chamber-type or tubular reactor design. The standard chamber-type reactor has a series of separate catalyst beds. Temperature control is accomplished by recycling some of the spent propane-butane mix as feed diluent and using some as a quench that is introduced between the catalyst beds. The chamber-type reactor system is less expensive to install than a tubular-type reactor system but has somewhat higher catalyst and utility requirements. One advantage of the chamber-type system is that it can easily be adapted to petrochemical applications, such as alkylation of aromatics.

The UOP tubular-type reactor contains the catalyst in tubes that are surrounded by a jacket of boiling water. This boiling water removes the heat generated by the reaction, and the steam produced can be used to preheat the feed. The tubular-type reactor also provides a somewhat more efficient temperature-control system and allows higher reactor olefin concentrations. The result is lower catalyst and utility requirements.

PROCESS DESCRIPTION

Gasoline Production

The process flow diagram for the production of gasoline using the UOP Catalytic Condensation process is shown in Fig. 1.3.1. The feed stream is typically a mixture of propane plus propylene or butane plus butylene or both. This stream must be free of basic nitrogen compounds, which adversely affect catalyst activity. The nitrogen compounds are typically removed upstream of the unit by a simple water wash. In addition, sulfur compounds usually are removed upstream of the unit. Although the catalyst can tolerate sulfur compounds, these compounds are typically removed upstream to meet product specifications. Also, careful control of the combined-feed moisture level is required to maintain catalyst activity and selectivity.

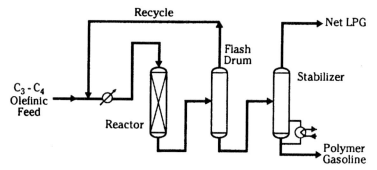

FIGURE 1.3.1 UOP Catalytic Condensation process for motor fuel production.

Before entering the feed drum, the fresh feed is combined with recycled LPG. The LPG stream is recycled to not only increase overall product yields but also control olefin concentration in the reactor feed.

From the feed drum, the combined feed is passed through heat exchange before being charged to the reactor. The heated combined feed enters a chamber-type reactor that contains a series of catalyst beds, as discussed earlier.

The reactor effluent passes through feed-effluent heat exchange and is then charged to a flash drum. The overhead from the flash drum is cooled and then recycled to the feed drum. The bottoms from the flash drum are sent to the product stabilizer.

The product stabilizer separates the polymer gasoline from the LPG, which contains the inert feed propane and butane. The LPG stream also contains a small amount of olefins. This LPG stream is either yielded as a product, recycled to the feed drum, or used as quench in the reactor section. The polymer gasoline is cooled and sent to product blending or storage.

Yields and Product Properties. Two typical operations of the UOP Catalytic Condensation process for gasoline production are summarized in Table 1.3.1. In the first case, the charge to the unit is a C_3-C_4 stream from an FCC unit with a 59 liquid volume percent (LV %) olefin content. The products from the unit include 41.2 LV % LPG and 45.1 LV % C_5+ polymer gasoline. Included in the product LPG are the inert feed components, propane and butane, which pass directly through the unit, and a small amount of unreacted olefins. Overall yield from the UOP Catalytic Condensation unit is 86.3 LV % C_3+ product. This volume loss is due solely to the

TABLE 1.3.1 Product Yields for Gasoline Production

	Feedstock			
	FCC C_3-C_4 LPG		FCC C_4 LPG	
Products	LV % FF*	wt % FF*	LV % FF*	wt % FF*
C_3-C_4 LPG	41.2	41.3	37.4	35.9
C_5 + polymer gasoline	45.1	58.7	51.9	64.1
Total	86.3	100.0	89.3	100.0

*FF = fresh feed.

volume reduction associated with the oligomerization of olefins. The second case describes a C_4 olefin feedstock to the UOP Catalytic Condensation process.

The properties of the C_5+ polymer gasoline are given in Table 1.3.2. Polymer gasoline is a high-quality gasoline blending component. The clear road octane $[(R + M)/2]$ is 89 for the C_3-C_4 olefin case and 91 for the C_4 olefin case. In addition, when blended with aromatic stocks such as those produced in the UOP Platforming* process or with FCC gasolines, the blending octane of polymer gasoline can be even higher.

Polymer gasoline is olefinic and, if necessary, can be hydrotreated to reduce the olefin content. In general, hydrotreating is not done because it reduces the gasoline octane rating. An exception to this rule is when isooctane is the desired product. The UOP Catalytic Condensation process can be used to react normal butylene and isobutylene selectively to produce a product that, when hydrotreated, is isooctane rich and has a research octane number, clear (RONC) of about 100. As a historical note, part of the high-octane aviation gasoline was produced during World War II using the UOP Catalytic Condensation process.

Process Economics. The UOP Catalytic Condensation process is relatively simple to operate and requires a minimum of labor. Its simplicity is reflected in the operating requirements summarized in Table 1.3.3. Utility requirements include only electric power, steam, and cooling water (air cooling can often be substituted for water cooling). Also reflecting the simplicity of the process is the fact that only one operator is generally required. Overall, the cost for operating a UOP Catalytic Condensation unit ranges from $2.30 to $3.30 per barrel of C_5+ polymer gasoline. This cost includes utilities, labor, catalyst, chemicals, and an allowance for process royalty but does not include any direct or indirect capital-related costs.

The total capital required for a plant erected in 1995 on the U.S. Gulf Coast varies from $60 to $85 per metric ton per year ($2500 to $3500 per barrel per day) of C_5+ polymer gasoline. This investment includes material and labor plus an allowance for design engineering and construction engineering costs. These costs cover the range of capacities and feed olefin concentrations typically encountered in most refineries.

---

*Trademark and/or service mark of UOP.

TABLE 1.3.2 Product Qualities for Gasoline Production

Product quality	Olefin feedstock	
	C_3–C_4	C_4
Density:		
°API	60.0	60.2
Specific gravity	0.739	0.738
RONC	95.5	99.0
MONC	82	83
(R + M)/2	88.8	91
RVP, lb/in²	2	2
ASTM distillation, °C (°F):		
10 LV %	100 (212)	100 (212)
50 LV %	125 (257)	120 (248)
90 LV %	190 (374)	180 (356)

Note: °API = degrees on American Petroleum Institute Scale; RONC or R = research octane number, clear; MONC or M = motor octane number, clear; RVP = Reid vapor pressure; ASTM = American Society for Testing and Materials.

TABLE 1.3.3 Operating Costs for Gasoline Production

Utilities	
Electric power:	
kW/bbl C_5 + product	2–3
kW/MT C_5 + product	20–28
Steam:	
lb/bbl C_5 + product	200–300
MT/MT C_5 + product	0.7–1.1
Cooling waters:	
gal/bbl C_5 + product	120–180
m^3/MT C_5 + product	4.4–6.0
Catalyst and chemicals cost	
$/bbl C_5 + product	0.60–1.00
$/MT C_5 + product	5.00–8.20
Labor and operating cost	
Workforce	1 operator-helper
Typical operating cost:	
$/bbl C_5 + product	2.30–3.30
$/MT C_5 + product	19.50–28.10
Investment*	
$/BPD C_5 + product	2500–3500
$/MTA C_5 + product	60–85

*Basis, battery limits plant built on U.S. Gulf Coast in 1995, U.S. dollars.

Note: MT = metric tons; MTA = metric tons per annum; BPD = barrels per day.

Production of Distillate-Type Fuels

One of the attractive features of the UOP Catalytic Condensation process is its ability to produce high-quality commercial jet fuel blending stock. Typically, after being hydrotreated to saturate the olefins, this fuel is characterized by a high smoke point of the order of 40 mm and a low freeze point, generally less than $-70°C$ ($-94°F$).

Although a diesel fuel blending stock can be produced by catalytic condensation, its cetane properties are generally poor. The resulting product can have a cetane number of about 28, which is low compared to most diesel fuel specifications but can be sufficient for blending if enough higher-cetane blending stock is available.

The process flow diagram for the production of distillate fuel using the UOP Catalytic Condensation process is shown in Fig. 1.3.2. The process flow in this case allows for the recycle of gasoline-range polymer as well as LPG to the reactor. The olefinic polymer gasoline is further reacted to produce a distillate-range product.

As in gasoline production, minor feed pretreatment is required to remove feed contaminants. To meet product specifications, feed sulfur components, particularly H_2S, must be removed to prevent mercaptan formation through reaction with feed olefins.

About 95 percent of the feed olefins are reacted through the reactor. Only about 25 percent of the feed olefins are converted to a distillate-range product in the first pass.

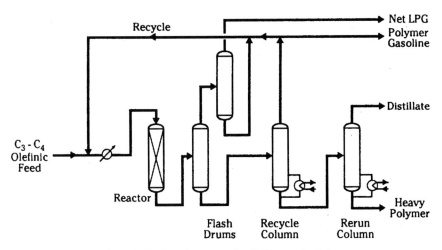

FIGURE 1.3.2 UOP Catalytic Condensation process for distillate production.

Overall conversion to distillate can be increased to 75 percent or higher by recycling light polymer.

The recycle column separates gasoline-range product from distillate-range polymer. The gasoline-range product comes off the recycle column overhead and is partly recycled to the reactor feed and partly yielded as gasoline. The bottoms from the recycle column is sent to a rerun column.

The rerun column separates the final distillate from the heavy-polymer stream. The heavy polymer represents a relatively high-molecular-weight and high-boiling-point by-product. The heavy polymer is separated from the distillate product so that the hydrotreating catalyst generally employed on this product is not fouled. Typically, the recovered heavy polymer is blended into fuel oil.

Yields and Product Properties.   A typical operation of the UOP Catalytic Condensation process for distillate production is summarized in Table 1.3.4. The charge to the unit is a C_3-C_4 LPG stream from an FCC unit with a 59 LV % olefin content. The four products from the unit are LPG, polymer gasoline, distillate, and

TABLE 1.3.4 Product Yields for Distillate Operation

	LV % FF*	wt % FF*
Feedstock:		
FCC C_3-C_4 LPG	100.0	100.0
Products:		
C_3-C_4 LPG	41.2	41.3
Polymer gasoline	8.3	10.5
Distillate	34.0	47.0
Heavy polymer	0.8	1.2
Total	84.3	100.0

*FF = fresh feed.

heavy polymer. The yield of LPG is 41.2 LV % and includes the inert propane and butane from the feed. The LPG also includes a small amount of unreacted propylene and butylene. The yield of polymer gasoline is 8.3 LV %. For maximum distillate production, the polymer gasoline is cut for as low an endpoint as possible. When compared with polymer gasoline from the gasoline mode of operation as discussed earlier, this polymer gasoline is somewhat lighter but is otherwise similar. For the distillate mode of operation, the distillate yield is 34 LV %. This yield can be varied somewhat depending on the amount of gasoline recycled to the reactor. The yield of heavy polymer is about 0.8 LV %.

Overall liquid yield from the unit is 84.3 LV %. As in the gasoline mode of operation, a net volume loss occurs as a result of the oligomerization of olefins. Also, the higher the degree of reactions, or the higher the molecular weight of the products, the greater the volume loss.

The properties of the products are given in Table 1.3.5. Unlike polymer gasoline, which is typically blended directly into the gasoline pool, the distillate product must be hydrotreated. The as-produced distillate is highly olefinic and has poor distillate properties. For example, the cetane index of the as-produced distillate is only 18. In contrast, olefin saturation by hydrotreating greatly improves the polymer distillate properties. As shown in Table 1.3.6, the cetane number improves by approximately 10 with hydrotreating. Even though a cetane number of 28 is low compared to most diesel fuel specifications, it is high enough to allow blending with other high-cetane-number stocks.

Economics. For the production of distillate, the UOP Catalytic Condensation process has moderate operating costs and requires minimum labor, as shown in Table 1.3.7. The utilities are based on maximum air cooling and include electric power, steam, and fired fuel. Table 1.3.7 also gives approximate catalyst and chemical costs. The requirement of one operator for the plant reflects the simplicity of the process.

The total operating costs for the unit range from about $6 to $7 per barrel of C_5+ product. This cost includes utilities, labor, catalyst, chemicals, hydrogen for final product preparation, and an allowance for process royalty. However, it does not include any direct or indirect capital-related charges.

The total erected investment cost, also given in Table 1.3.7, varies from about $97 to $183 per metric ton per year ($4000 to $7500 per barrel per day) of total C_5+ product. This investment includes material and labor costs plus an allowance for design

TABLE 1.3.5 Product Properties for Distillate Operation

Product quality	Gasoline	Distillate
Density:		
°API	65.0	49.9
Specific gravity	0.720	0.780
RONC	94	—
MONC	82	—
RVP, lb/in²	3.0	—
Cetane no.	—	18
ASTM distillation, °C (°F):		
10 LV %	99 (210)	175 (347)
50 LV %	135 (275)	182 (360)
90 LV %	150 (302)	206 (403)

TABLE 1.3.6 Comparison of Hydrotreated and Unhydrotreated Distillate Product Properties

	As-produced distillate	Hydrotreated distillate
Density:		
°API	49.9	52.3
Specific gravity	0.78	0.77
Smoke point, mm	—	40
Freeze point, °C (°F)	—	70 (158)
Viscosity, cSt at 50°C	—	1.2
Cetane no.	18	28
Bromine no.	108	1
ASTM distillation, °C (°F):		
10 LV %	175 (347)	175 (347)
50 LV %	182 (360)	182 (360)
90 LV %	206 (403)	206 (403)

TABLE 1.3.7 Operating Costs for Distillate Production

Utilities, per bbl C_5 + product	
Electric power, kW	7–9
Steam, lb	300–400
Fuel-fired, 10^6 Btu	0.2–0.3
Catalyst and chemicals	
Catalyst: solid phosphoric acid cost, $/bbl C_5+ product	0.80–0.90
Labor and operating cost	
Workforce	1 operator-helper
Typical operating cost, $/bbl C_5+ product	6–7
Investment*	
$/MTA	$97–$183
$/BPSD C_5 + product	$4000–$7500

*Basis: battery limits plant built on U.S. Gulf Coast, in 1995 U.S. dollars.
Note: BPSD = barrels per stream day.

and construction engineering. It also reflects a range of capacities and feed olefin concentration that is typical in the refining industry.

COMMERCIAL EXPERIENCE

The UOP Catalytic Condensation process was commercialized in 1935. Since that time, UOP has licensed and designed more than 200 units worldwide for the production of transportation fuels. Most of the units were designed for the production of polymer gasoline.

CHAPTER 1.4
UOP HF ALKYLATION TECHNOLOGY

John C. Sheckler and B. R. Shah
UOP
Des Plaines, Illinois

INTRODUCTION

The UOP* HF Alkylation process for motor fuel production catalytically combines light olefins, which are usually mixtures of propylene and butylenes, with isobutane to produce a branched-chain paraffinic fuel. The alkylation reaction takes place in the presence of hydrofluoric (HF) acid under conditions selected to maximize alkylate yield and quality. The alkylate product possesses excellent antiknock properties because of its considerable content of highly branched paraffins. Alkylate is a clean-burning, low-sulfur gasoline blending component that does not contain olefinic or aromatic compounds. Alkylate also has excellent lead response, which is important in locations where leaded gasoline is still produced.

The HF Alkylation process was developed in the UOP laboratories during the late 1930s and early 1940s. The process was initially used for the production of high-octane aviation fuels from butylenes and isobutane. In the mid-1950s, the development and consumer acceptance of more-sophisticated high-performance automotive engines placed a burden on the petroleum refiner both to increase gasoline production and to improve motor fuel quality. The advent of catalytic reforming techniques, such as the UOP Platforming* process, made an important tool for the production of high-quality gasolines available to refiners. However, the motor fuel produced in such operations is primarily aromatic-based and is characterized by high sensitivity (that is, the spread between research and motor octane numbers). Because automobile performance is more closely related to road octane rating (approximately the average of research and motor octanes), the production of gasoline components with low sensitivity was required. A natural consequence of these requirements was the expansion of alkylation operations. Refiners began to broaden the range of olefin feeds to both existing and new alkylation units to include propylene and occasionally amylenes as well as butylenes. By the early 1960s, the HF Alkylation process had virtually dis-

---

*Trademark and/or service mark of UOP.

placed motor fuel polymerization units for new installations, and refiners had begun to gradually phase out the operation of existing polymerization plants.

The importance of the HF Alkylation process in the refining situation of the 1990s has not diminished but has actually increased. The contribution of the alkylation process is critical in the production of quality motor fuels. The process provides refiners with a tool of unmatched economy and efficiency, one that will assist refiners in maintaining or strengthening their position in the production and marketing of gasolines.

PROCESS CHEMISTRY

General

In the HF Alkylation process, HF acid is the catalyst that promotes the isoparaffin-olefin reaction. In this process, only isoparaffins with tertiary carbon atoms, such as isobutane or isopentane, react with the olefins. In practice, only isobutane is used because isopentane has a high octane number and a vapor pressure that allows it to be blended directly into finished gasolines. However, where environmental regulations have reduced the allowable vapor pressure of gasoline, isopentane is being removed from gasoline, and refiner interest in alkylating this material with light olefins, particularly propylene, is growing.

The actual reactions taking place in the alkylation reactor are many and are relatively complex. The equations in Fig. 1.4.1 illustrate the primary reaction products that may be expected for several pure olefins.

In practice, the primary product from a single olefin constitutes only a percentage of the alkylate because of the variety of concurrent reactions that are possible in the alkylation environment. Compositions of pilot-plant products produced at conditions to maximize octane from pure-olefin feedstocks are shown in Table 1.4.1.

Reaction Mechanism

Alkylation is one of the classic examples of a reaction or reactions proceeding via the carbenium ion mechanism. These reactions include an initiation step and a propagation step and may include an isomerization step. In addition, polymerization and cracking steps may also be involved. However, these side reactions are generally undesirable. Examples of these reactions are given in Fig. 1.4.2.

Initiation. The initiation step (Fig. 1.4.2a) generates the tertiary butyl cations that will subsequently carry on the alkylation reaction.

Propagation. Propagation reactions (Fig. 1.4.2b) involve the tertiary butyl cation reacting with an olefin to form a larger carbenium ion, which then abstracts a hydride from an isobutane molecule. The hydride abstraction generates the isoparaffin plus a new tertiary butyl cation to carry on the reaction chain.

Isomerization. Isomerization [Eq. (1.4.12), shown in Fig. 1.4.2c] is very important in producing good octane quality from a feed that is high in 1-butene. The isomerization of 1-butene is favored by thermodynamic equilibrium. Allowing 1-butene to isomerize to 2-butene reduces the production of dimethylhexanes (research octane number of 55 to 76) and increases the production of trimethylpentanes.

Equation (1.4.13) is an example of the many possible steps involved in the isomerization of the larger carbenium ions.

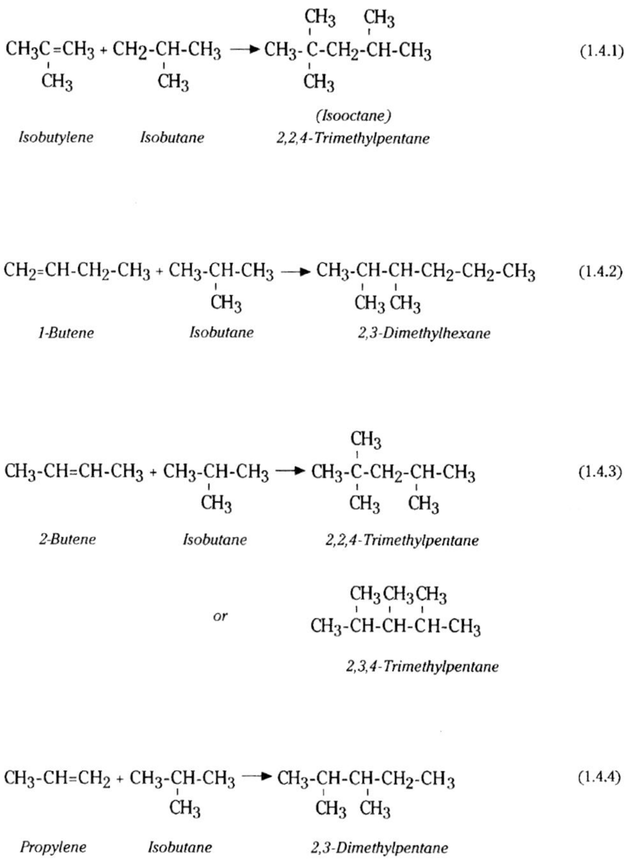

UOP HF ALKYLATION TECHNOLOGY

$$CH_3C=CH_2 + CH_2\text{-}CH\text{-}CH_3 \longrightarrow CH_3\text{-}\underset{\underset{CH_3}{|}}{\overset{\overset{CH_3}{|}}{C}}\text{-}CH_2\text{-}\underset{\underset{CH_3}{|}}{CH}\text{-}CH_3 \quad (1.4.1)$$

Isobutylene   *Isobutane*   (*Isooctane*) 2,2,4-Trimethylpentane

$$CH_2=CH\text{-}CH_2\text{-}CH_3 + CH_3\text{-}\underset{\underset{CH_3}{|}}{CH}\text{-}CH_3 \longrightarrow CH_3\text{-}\underset{\underset{CH_3}{|}}{CH}\text{-}\underset{\underset{CH_3}{|}}{CH}\text{-}CH_2\text{-}CH_2\text{-}CH_3 \quad (1.4.2)$$

1-Butene   *Isobutane*   2,3-Dimethylhexane

$$CH_3\text{-}CH=CH\text{-}CH_3 + CH_3\text{-}\underset{\underset{CH_3}{|}}{CH}\text{-}CH_3 \longrightarrow CH_3\text{-}\underset{\underset{CH_3}{|}}{\overset{\overset{CH_3}{|}}{C}}\text{-}CH_2\text{-}\underset{\underset{CH_3}{|}}{CH}\text{-}CH_3 \quad (1.4.3)$$

2-Butene   *Isobutane*   2,2,4-Trimethylpentane

or

$$CH_3\text{-}\underset{\underset{CH_3}{|}}{CH}\text{-}\underset{\underset{CH_3}{|}}{CH}\text{-}\underset{\underset{CH_3}{|}}{CH}\text{-}CH_3$$

2,3,4-Trimethylpentane

$$CH_3\text{-}CH=CH_2 + CH_3\text{-}\underset{\underset{CH_3}{|}}{CH}\text{-}CH_3 \longrightarrow CH_3\text{-}\underset{\underset{CH_3}{|}}{CH}\text{-}\underset{\underset{CH_3}{|}}{CH}\text{-}CH_2\text{-}CH_3 \quad (1.4.4)$$

Propylene   *Isobutane*   2,3-Dimethylpentane

FIGURE 1.4.1 HF alkylation primary reactions for monoolefins.

Other Reactions. The polymerization reaction [Eq. (1.4.14), shown in Fig. 1.4.2d] results in the production of heavier paraffins, which are undesirable because they reduce alkylate octane and increase alkylate endpoint. Minimization of this reaction is achieved by proper choice of reaction conditions.

The larger polymer cations are susceptible to cracking disproportionation reactions [Eq. (1.4.15)], which form fragments of various molecular weights. These fragments can then undergo further alkylation.

Hydrogen Transfer. The hydrogen transfer reaction is most pronounced with propylene feed. The reaction also proceeds via the carbenium ion mechanism. In the first reaction [Eq. (1.4.16)], propylene is alkylated with isobutane to produce butylene and propane. The butylene is then alkylated with isobutylene [Eq. (1.4.17)] to form trimethylpentane. The overall reaction is given in Eq. (1.4.18). From the viewpoint of octane, this reaction can be desirable because trimethylpentane has substantially higher octane than the dimethylpentane normally formed from propylene. However, two molecules of isobutane are required for each molecule of alkylate, and so this reaction may be undesirable from an economic viewpoint.

TABLE 1.4.1 Compositions of Alkylate from Pure-Olefin Feedstocks

	Olefin			
Component, wt %	C_3H_6	$i\text{-}C_4H_8$	$C_4H_8\text{-}2$	$C_4H_8\text{-}1$
C_5 isopentane	1.0	0.5	0.3	1.0
C_6's:				
Dimethylpentanes	0.3	0.8	0.7	0.8
Methylpentanes	—	0.2	0.2	0.3
C_7's:				
2,3-dimethylpentane	29.5	2.0	1.5	1.2
2,4-dimethylpentane	14.3	—	—	—
Methylhexanes	—	—	—	—
C_8's:				
2,2,4-trimethylpentane	36.3	66.2	48.6	38.5
2,2,3-trimethylpentane	—	—	1.9	0.9
2,3,4-trimethylpentane	7.5	12.8	22.2	19.1
2,3,3-trimethylpentane	4	7.1	12.9	9.7
Dimethylhexanes	3.2	3.4	6.9	22.1
C_9+ products	3.7	5.3	4.1	5.7

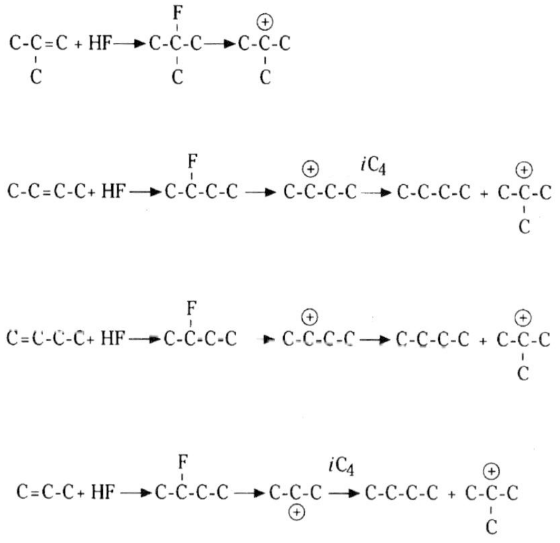

$$\text{C-C=C} + \text{HF} \longrightarrow \text{C-}\underset{\text{C}}{\overset{\text{F}}{\text{C}}}\text{-C} \longrightarrow \overset{+}{\text{C}}\text{-}\underset{\text{C}}{\text{C}}\text{-C} \tag{1.4.5}$$

$$\text{C-C=C-C} + \text{HF} \longrightarrow \text{C-}\overset{\text{F}}{\text{C}}\text{-C-C} \longrightarrow \overset{+}{\text{C}}\text{-C-C-C} \xrightarrow{iC_4} \text{C-C-C-C} + \overset{+}{\text{C}}\text{-}\underset{\text{C}}{\text{C}}\text{-C} \tag{1.4.6}$$

$$\text{C=C-C-C} + \text{HF} \longrightarrow \text{C-}\overset{\text{F}}{\text{C}}\text{-C-C} \longrightarrow \overset{+}{\text{C}}\text{-C-C-C} \longrightarrow \text{C-C-C-C} + \overset{+}{\text{C}}\text{-}\underset{\text{C}}{\text{C}}\text{-C} \tag{1.4.7}$$

$$\text{C=C-C} + \text{HF} \longrightarrow \text{C-}\overset{\text{F}}{\text{C}}\text{-C-C} \longrightarrow \text{C-}\overset{+}{\underset{\oplus}{\text{C}}}\text{-C} \xrightarrow{iC_4} \text{C-C-C-C} + \overset{+}{\text{C}}\text{-}\underset{\text{C}}{\text{C}}\text{-C} \tag{1.4.8}$$

FIGURE 1.4.2a HF alkylation reaction mechanism—initiation reactions.

PROCESS DESCRIPTION

The alkylation of olefins with isobutane is complex because it is characterized by simple addition as well as by numerous side reactions. Primary reaction products are the isomeric paraffins containing carbon atoms that are the sum of isobutane and the corresponding olefin. However, secondary reactions such as hydrogen transfer, polymerization, isomerization, and destructive alkylation also occur, resulting in the formation of secondary products both lighter and heavier than the primary products.

The factors that promote the primary and secondary reaction mechanisms differ, as does the response of each to changes in operating conditions or design options. Not all

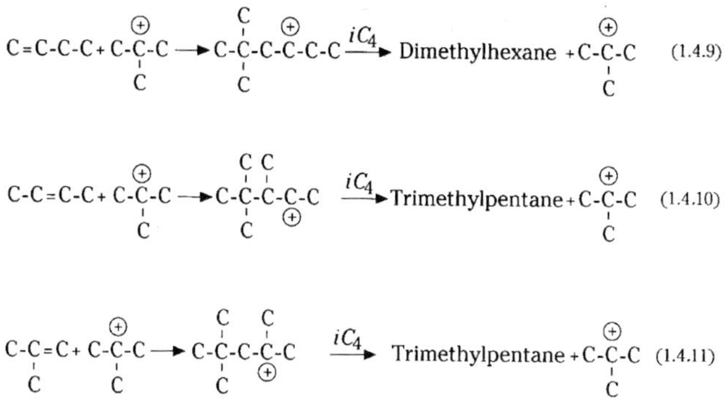

UOP HF ALKYLATION TECHNOLOGY

$$C=C-C-C + C-\overset{+}{C}-C \longrightarrow C-\overset{C}{\underset{C}{C}}-\overset{+}{C}-C-C-C \xrightarrow{iC_4} \text{Dimethylhexane} + C-\overset{+}{\underset{C}{C}}-C \quad (1.4.9)$$

$$C-C=C-C + C-\overset{+}{\underset{C}{C}}-C \longrightarrow C-\overset{C}{\underset{C}{C}}-\overset{C}{\underset{\oplus}{C}}-C-C \xrightarrow{iC_4} \text{Trimethylpentane} + C-\overset{+}{\underset{C}{C}}-C \quad (1.4.10)$$

$$C-C=C + C-\overset{+}{\underset{C}{C}}-C \longrightarrow C-\overset{C}{\underset{C}{C}}-\overset{C}{\underset{\oplus}{C}}-C \xrightarrow{iC_4} \text{Trimethylpentane} + C-\overset{+}{\underset{C}{C}}-C \quad (1.4.11)$$

FIGURE 1.4.2b HF alkylation reaction mechanism—propagation reactions.

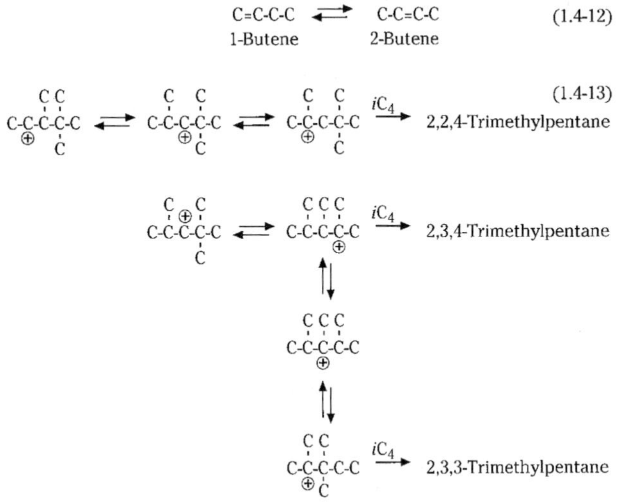

$$C=C-C-C \rightleftharpoons C-C=C-C \quad (1.4-12)$$
$$\text{1-Butene} \qquad \text{2-Butene}$$

$$\underset{\oplus}{C}-\overset{C\;C}{\underset{C}{C-C-C}} \rightleftharpoons \overset{C\;C}{\underset{\oplus\;C}{C-C-C-C}} \rightleftharpoons \overset{C\;C}{\underset{\oplus\;C}{C-C-C-C}} \xrightarrow{iC_4} \text{2,2,4-Trimethylpentane} \quad (1.4\text{-}13)$$

$$\overset{C\;\oplus\;C}{\underset{C}{C-C-C-C}} \rightleftharpoons \overset{C\;C\;C}{\underset{\oplus}{C-C-C-C}} \xrightarrow{iC_4} \text{2,3,4-Trimethylpentane}$$

$$\updownarrow$$

$$\overset{C\;C\;C}{\underset{\oplus}{C-C-C-C}}$$

$$\updownarrow$$

$$\overset{C\;C}{\underset{\oplus\;C}{C-C-C-C}} \xrightarrow{iC_4} \text{2,3,3-Trimethylpentane}$$

FIGURE 1.4.2c HF alkylation reaction mechanism—isomerization.

secondary reactions are undesirable; for example, they make possible the formation of isooctane from propylene or amylenes. In an ideally designed and operated system, primary reactions should predominate, but not to the complete exclusion of secondary ones. For the HF Alkylation process, the optimum combinations of plant economy, product yield, and quality are achieved with the reaction system operating at cooling-water temperature and an excess of isoparaffin and with contaminant-free feedstocks and vigorous, intimate acid-hydrocarbon contact.

To minimize acid consumption and ensure good alkylate quality, the feeds to the alkylation unit should be dry and of low sulfur content. Normally, a simple desiccant-drying system is included in the unit design package. Feed treating in a UOP Merox* unit for mercaptan sulfur removal can be an economic adjunct to the alkylation unit

---
*Trademark and/or service mark of UOP.

Polymerization

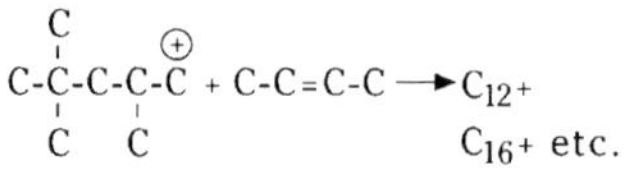

$$\underset{\underset{C}{|}}{\overset{\overset{C}{|}}{C-C-C-C}} + C-C=C-C \xrightarrow{\oplus} \underset{C_{16}+ \text{ etc.}}{C_{12}+} \qquad (1.4.14)$$

Cracking Disproportionation

$$C_{12}^+ \longrightarrow C_5^\bullet + C_7^+ \qquad (1.4.15)$$

Hydrogen Transfer

$$C_3H_6 + iC_4H_{10} \longrightarrow C_4H_8 + C_3H_8 \qquad (1.4.16)$$

$$C_4H_8 + iC_4H_{10} \longrightarrow C_8H_{18} \qquad (1.4.17)$$

Overall Reaction:

$$C_3H_6 + 2\, iC_4H_{10} \longrightarrow C_3H_8 + \text{Trimethylpentane} \qquad (1.4.18)$$

FIGURE 1.4.2d HF alkylation reaction mechanism—other.

for those applications in which the olefinic feed is derived from catalytic cracking or for other operations in which feedstocks of significant sulfur content are processed. Simplified flow schemes for a typical C_4 HF Alkylation unit and a C_3-C_4 HF Alkylation unit are shown in Figs. 1.4.3 and 1.4.4.

Treated and dried olefinic feed is charged along with recycle and makeup isobutane (when applicable) to the reactor section of the plant. The combined feed enters the shell of a reactor–heat exchanger through several nozzles positioned to maintain an even temperature throughout the reactor. The heat of reaction is removed by heat exchange with a large volume of coolant flowing through the tubes having a low temperature rise. If cooling water is used, it is then available for further use elsewhere in the unit. The effluent from the reactor enters the settler, and the settled acid is returned back to the reactor.

The hydrocarbon phase, which contains dissolved HF acid, flows from the settler and is preheated and charged to the isostripper. Saturate field butane feed (when applicable) is also charged to the isostripper. Product alkylate is recovered from the bottom of the column. Any normal butane that may have entered the unit is withdrawn as a sidecut. Unreacted isobutane is also recovered as a sidecut and recycled to the reactor.

The isostripper overhead consists mainly of isobutane, propane, and HF acid. A drag stream of overhead material is charged to the HF stripper to strip the acid. The overhead from the HF stripper is returned to the isostripper overhead system to recover acid and isobutane. A portion of the HF stripper bottoms is used as flushing material. A net bottom stream is withdrawn, defluorinated, and charged to the gas-concentration section (C_3-C_4 splitter) to prevent a buildup of propane in the HF Alkylation unit.

An internal depropanizer is required in an HF Alkylation unit processing C_3-C_4 olefins and may be required with C_4 olefin feedstocks if the quantity of propane enter-

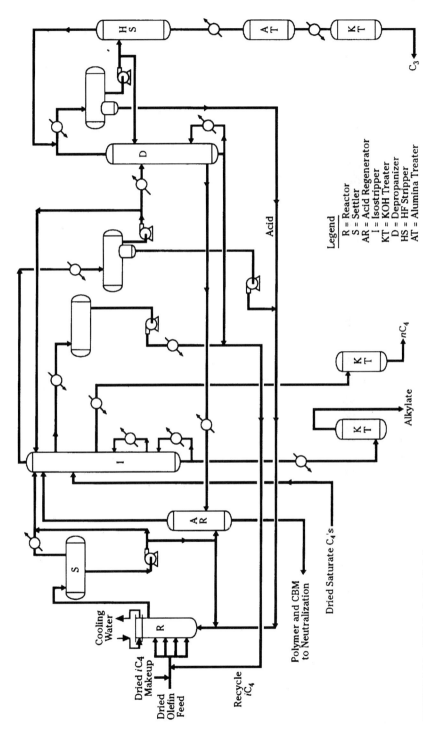

FIGURE 1.4.3 UOP C_3–C_4 HF alkylation process.

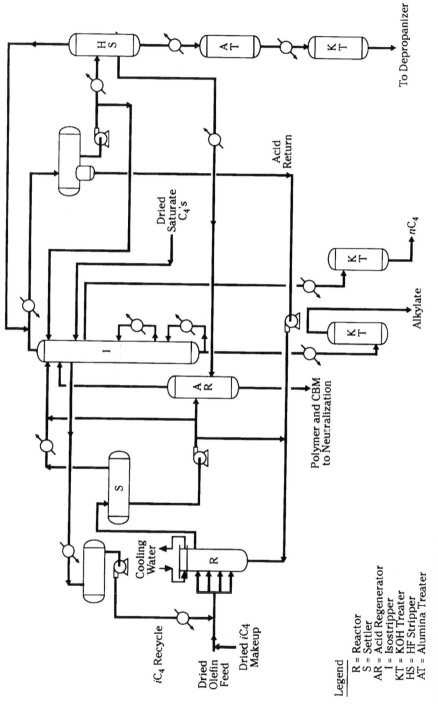

FIGURE 1.4.4 UOP C_4 HF alkylation process.

ing the unit is too high to be rejected economically as previously described. The isostripper overhead drag stream is charged to the internal depropanizer. Overhead from the internal depropanizer is directed to the HF stripper to strip HF acid from the high-purity propane. A portion of the internal depropanizer bottoms is used as flushing material, and the remainder is returned to the alkylation reactor. The HF stripper overhead vapors are returned to the internal depropanizer overhead system. High-purity propane is drawn off the bottom of the HF stripper, passes through a defluorination step, and is then sent to storage.

A small slipstream of circulating HF acid is regenerated internally to maintain acid purity at the desired level. This technique significantly reduces overall chemical consumption. An acid-regenerator column is also provided for start-ups after turnarounds or in the event of a unit upset or feed contamination.

When the propane or normal butane from the HF unit is to be used as liquefied petroleum gas (LPG), defluorination is recommended because of the possible breakdown of combined fluorides during combustion and the resultant potential corrosion of burners. Defluorination is also required when the butane is to be directed to an isomerization unit. After defluorination, the propane and butane products are treated with potassium hydroxide (KOH) to remove any free HF acid that might break through in the event of unit misoperation.

The alkylation unit is built almost entirely of carbon steel although some Monel is used in several minor locations. Auxiliary neutralizing and scrubbing equipment is included in the plant design to ensure that all materials leaving the unit during both normal and emergency operation are acid-free.

ENGINEERING DESIGN

The reactor and distillation systems that UOP uses have evolved through many years of pilot-plant evaluation, engineering development, and commercial operation. The overall plant design has progressed through a number of variations, resulting in the present concepts in alkylation technology.

Reactor Section

In the design of the reactor, the following factors require particular attention:

- Removal of heat of reaction
- Generation of acid surface: mixing and acid-to-hydrocarbon ratio
- Acid composition
- Introduction of olefin feed

The proper control of these factors enhances the quality and yield of the alkylate product.

Selecting a particular reaction-system configuration requires careful consideration of the refiner's production objectives and economics. The UOP reaction system optimizes processing conditions by the introduction of olefin feed through special distributors to provide the desired contact with the continuous-acid phase. Undesirable reactions are minimized by the continuous removal of the heat of reaction in the reaction zone itself. The removal of heat in the reaction zone is advantageous because peak reaction temperatures are reduced and effective use is made of the available cooling-water supply.

Acid-Regeneration Section

The internal acid-regeneration technique has virtually eliminated the need for an acid regenerator and, as a result, acid consumption has been greatly reduced. The acid regenerator has been retained in the UOP design only for start-ups or during periods when the feed has abnormally high levels of contaminants, such as sulfur and water. During normal operation, the acid regenerator is not in service.

When the acid regenerator is in service, a drag stream off the acid-circulation line at the settler is charged to the acid regenerator, which is refluxed on the top tray with isobutane. The source of heat to the bottom of the regenerator for a C_3-C_4 HF Alkylation unit is superheated isobutane from the depropanizer sidecut vapors. For a C_4 HF Alkylation unit, the stripping medium to the acid regenerator is sidecut vapors from the HF stripper bottoms. The regenerated HF acid is combined with the overhead vapor from the isostripper and sent to the cooler.

Neutralization Section

UOP has designed the neutralization section to minimize the amount of waste, offensive materials, and undesirable by-products. Releasing acid-containing vapors to the regular relief-gas system is impractical because of corrosion and odor problems. The system is composed of the relief-gas scrubber, KOH mix tank, circulating pumps, and a KOH regeneration tank.

All acid vents and relief valves are piped to this relief section. Gases pass up through the scrubber and are contacted by a circulating KOH solution to neutralize the HF acid. After the neutralization of the acid, the gases can be safely released into the refinery flare system.

The KOH is regenerated on a periodic basis in the KOH-regeneration tank by using lime to form calcium fluoride (CaF_2) and KOH. The CaF_2 settles to the bottom of the tank and is directed to the neutralizing basin, where acidic water from acid sewers and small amounts of acid from the process drains are treated. Lime is used to convert any fluorides into calcium fluoride before any waste effluent is released into the refinery sewer system.

Distillation System

The distillation and recovery sections of HF Alkylation units have also seen considerable evolution. The modern isostripper recovers high-purity isobutane as a sidecut that is recycled to the reactor. This recycle is virtually acid free, thereby minimizing undesirable side reactions with the olefin feed prior to entry into the reactor. A small rectification section on top of the modern isostripper provides for more efficient propane rejection.

Although a single high-pressure tower can perform the combined functions of isostripper and depropanizer, UOP's current design incorporates two towers (isostripper and depropanizer) for the following reasons:

- Each tower may be operated at its optimum pressure. Specifically, in the isostripper this two-tower design increases the relative volatilities between products and reduces the number of trays required for a given operation, in addition to improving separation between cuts.

- This system has considerably greater flexibility. It is easily convertible to a butylene-only operation because the depropanizer may be used as a feed splitter to separate C_3's and C_4's. The two-tower design permits the use of side feeds to the isostripper column should it be necessary to charge makeup isobutane of low purity. This design also permits the production of lower-vapor-pressure alkylate and a high-purity sidecut nC_4 for isomerizing or blending and the ability to make a clean split of side products.

- The two-tower design permits considerable expanded capacity at low incremental cost by the addition of feed preheat and side reboiling.
- Alkylate octane increases with decreasing reaction temperature. During cooler weather, the unit may be operated at lower recycle ratios for a given product octane, because the isobutane-recycle ratio is fixed by the product requirement and not by the fractionation requirements. The commensurate reduction in utilities lowers operating costs.
- Because of the low isostripper pressure in a two-tower system, this arrangement permits the use of steam for reboiling the isostripper column instead of a direct-fired heater, which is necessary in a single-tower system. In most cases, a stab-in reboiler system is suitable even for withdrawing a sidecut. Using a steam reboiler can be a considerable advantage when refinery utility balances so indicate, and it also represents considerable investment-cost savings.
- The two-tower system has proved its performance in a large number of operating units, and its flexibility has been proved through numerous revamps for increased capacity on existing units.
- The two-tower system also requires less overhead condenser surface, which lowers the investment required for heat exchange.
- Clean isobutane is available for flush, whereas only alkylate flush is available in the single-column operation. This clean-isobutane stream is also available to be taken to storage and is a time-saver during start-ups and shutdowns.
- Although fewer pieces of equipment are required with the single tower, the large number of trays and the high-pressure design necessitate the use of more tons of material and result in a somewhat higher overall cost than does the two-tower system.
- The regenerator column contains no expensive overhead system, and the internal HF-regeneration technique results in improved acid consumption.
- Because a high temperature differential can be taken on most cooling water, cooling-water requirements for the two-tower system are only about two-thirds those of the single-tower system.

COMMERCIAL INFORMATION

Typical commercial yields and product properties for charging various olefin feedstocks to an HF Alkylation unit are shown in Tables 1.4.2 and 1.4.3. Table 1.4.4 contains the detailed breakdown of the investment and production costs for a pumped, settled acid-alkylation unit based on a typical C_4 olefin feedstock.

ENVIRONMENTAL CONSIDERATIONS

The purpose of operating an HF Alkylation unit is to obtain a high-octane motor fuel blending component by reacting isobutane with olefins in the presence of HF acid. In

TABLE 1.4.2 HF Alkylation Yields

Olefin feedstocks	Required vol. iC_4/vol. olefin	Vol. alkylate produced/vol. olefin
C_3-C_4	1.28	1.78
Mixed C_4	1.15	1.77

TABLE 1.4.3 HF Alkylation Product Properties

Property	Propylene-butylene feed	Butylene feed
Specific gravity	0.693	0.697
Distillation temperature, °C (°F):		
IBP	41 (105)	41 (105)
10%	71 (160)	76 (169)
30%	93 (200)	100 (212)
50%	99 (210)	104 (220)
70%	104 (219)	107 (225)
90%	122 (250)	125 (255)
EP	192 (378)	196 (385)
Octanes:		
RONC	93.3	95.5
MONC	91.7	93.5

Note: IBP = initial boiling point; EP = endpoint; RONC = research octane number, clear; MONC = motor octane number, clear.

TABLE 1.4.4 Investment and Production Cost Summary*

Operating cost	$/stream day	$/MT alkylate	$/bbl alkylate
Labor	1,587	0.016	0.176
Utilities	6,609	0.066	0.734
Chemical consumption, laboratory allowance, maintenance, taxes and insurance	5,639	0.056	0.627
Total direct operating costs	13,835	0.138	1.537
Investment, estimated erected cost (EEC), third quarter 1995			$25,600,000

*Basis: 348,120 MTA (9000 BPSD) C_5 + alkylate
Note: MT = metric tons; MTA = metric tons per annum; BPSD = barrels per stream day.

the UOP HF Alkylation process, engineering and design standards have been developed and improved over many years to obtain a process that operates efficiently and economically. This continual process development constitutes the major reason for the excellent product qualities, low acid-catalyst consumption, and minimal waste production obtained by the UOP HF Alkylation process.

As in every process, certain minor process inefficiencies, times of misoperation, and periods of unit upsets occur. During these times, certain undesirable materials can be discharged from the unit. These materials can be pollutants if steps are not taken in the process-waste-management and product-treating areas to render these offensive materials harmless.

In a properly operated HF Alkylation unit, the amount of wastes, offensive materials, or undesirable by-products is minimal, and with proper care, these small quantities can be managed safely and adequately. The potentially offensive nature of the wastes produced in this process as well as the inherent hazards of HF acid has resulted in the development of waste-management and safety procedures that are unique to the UOP HF Alkylation process. The following sections briefly describe these procedures and how the wastes are safely handled to prevent environmental contamination. The refiner must evaluate and comply with any pertinent waste-management regulations. An overall view of the waste-management concept is depicted in Fig. 1.4.5.

Effluent Neutralization

In the alkylation unit's effluent-treating systems, any neutralized HF acid must eventually leave the system as an alkali metal fluoride. Because of its extremely low solubility in water, CaF_2 is the desired end product. The effluent containing HF acid can be treated with a lime [$CaO-Ca(OH)_2$] solution or slurry, or it can be neutralized indirectly in a KOH system to produce the desired CaF_2 product.

The KOH neutralization system currently used in a UOP-designed unit involves a two-stage process. As HF acid is neutralized by aqueous KOH, soluble potassium fluoride (KF) is produced, and the KOH is gradually depleted. Periodically, some of the KF-containing neutralizing solution is withdrawn to the KOH regenerator. In this vessel, KF reacts with a lime slurry to produce insoluble CaF_2 and thereby regenerates KF to KOH. The regenerated KOH is then returned to the system, and the solid CaF_2 is routed to the neutralizing basin.

Waste Disposal

Effluent Gases. The HF Alkylation unit uses two separate gas vent lines to maintain the separation of acidic gases from nonacidic gases until the acidic gases can be scrubbed free of acid.

Acidic Hydrocarbon Gases. Acidic hydrocarbon gases originate from sections of the unit where HF acid is present. These gases may evolve during a unit upset, during a shutdown, or during a maintenance period in which these acidic gases are partially or totally removed from the process vessels or equipment. The gases from the acid vents and from the acid pressure-relief valves are piped to a separate closed relief system for the neutralization of the acid contained in the gas. The acid-free gases are then routed from this acid-scrubbing section to the refinery nonacid flare system, where they are disposed of properly by burning.

The acidic gases are scrubbed in the acid neutralization and caustic regeneration system as shown Fig. 1.4.6. This system consists of the relief-gas scrubber, KOH-mix tank, liquid-knockout drum, neutralization drum, circulating pumps, and a KOH-regeneration tank.

Acidic gases, which were either vented or released, first flow to a liquid-knockout drum to remove any entrained liquid. The liquid from this drum is pumped to the neutralization drum. The acidic gases from the liquid-knockout drum then pass from the drum to the scrubbing section of the relief-gas scrubber, where countercurrent contact with a KOH solution removes the HF acid. After neutralization of the HF acid, the nonacidic gases are released into the refinery flare system.

The KOH used for the acidic-gas neutralization is recirculated by the circulation pumps. The KOH solution is pumped to the top of the scrubber and flows downward

1.44 ALKYLATION AND POLYMERIZATION

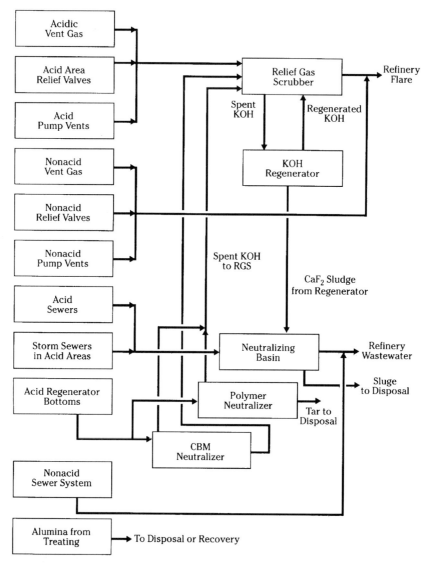

FIGURE 1.4.5 UOP HF alkylation process waste management.

to contact the rising acidic-gas stream and then overflows a liquid-seal pan to the reservoir section of the scrubber. In addition, a slipstream of the circulating KOH contacts the acidic gas just prior to its entry to the scrubber. The circulating KOH removes HF through the following reaction:

$$HF + KOH \rightarrow KF + H_2O \qquad (1.4.19)$$

Maintaining the circulating caustic pH and the correct percent of KOH and KF requires a system to regenerate the caustic. This regeneration of the KOH solution is performed on a batch basis in a vessel separate from the relief-gas scrubber. In this

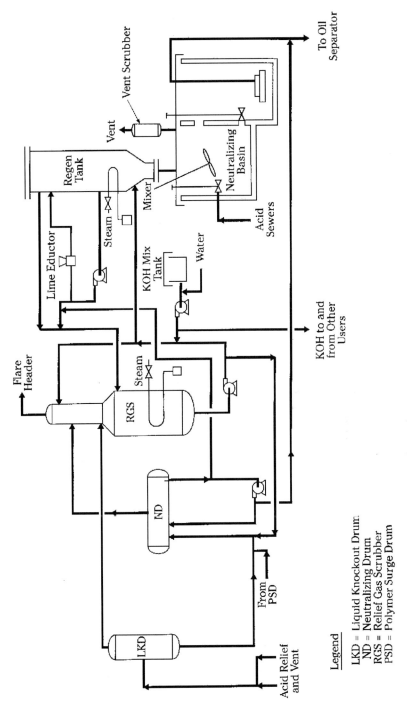

FIGURE 1.4.6 Acid neutralization and caustic regeneration section.

regeneration tank, lime and the spent KOH solution are thoroughly mixed. The regenerated caustic solution is pumped back to the scrubber. The CaF_2 and any unreacted lime are permitted to settle out and are then directed to the neutralization pit. The regeneration of the spent KOH solution follows the Berthollet rule, by which the insolubility of CaF_2 in water permits the complete regeneration of the potassium hydroxide according to the following equation:

$$2\ KF + Ca(OH)_2 \rightarrow 2\ KOH + CaF_2 \qquad (1.4.20)$$

Nonacidic Hydrocarbon Gases. Nonacidic gases originate from sections of the unit in which HF acid is not present. These nonacidic gases from process vents and relief valves are discharged into the refinery nonacid flare system, where they are disposed of by burning. The material that is vented or released to the flare is mainly hydrocarbon in nature. Possibly, small quantities of inert gases are also included.

Obnoxious Fumes and Odors. The only area from which these potentially objectionable fumes could originate is the unit's neutralizing basins. To prevent the discharge of these odorous gases to the surroundings, the neutralizing basins are tightly covered and equipped with a gas scrubber to remove any offensive odors. The gas scrubber uses either water or activated charcoal as the scrubbing agent. However, in the aforementioned neutralizing system, odors from the basin are essentially nonexistent because the main source of these odors (acid-regenerator bottoms) is handled in separate closed vessels.

Liquid Wastes. The HF Alkylation unit is equipped with two separate sewer systems to ensure the segregation of the nonacid from the possibly acid-containing water streams.

Acidic Waters. Any potential HF-containing water streams (rainwater runoff in the acid area and wash water), heavy hydrocarbons, and possibly spent neutralizing media are directed through the acid sewer system to the neutralizing basins for the neutralization of any acidic material. In the basins, lime is used to convert the incoming soluble fluorides to CaF_2.

The neutralizing basins consist of two separate chambers (Fig. 1.4.7). One chamber is filled while the other drains. In this parallel neutralizing-basin design, one basin has the inlet line open and the outlet line closed. As only a few surface drains are directed to the neutralizing basins, inlet flow normally is small, or nonexistent, except when acid equipment is being drained. The operator regularly checks the pH and, if necessary, mixes the lime slurry in the bottom of the basin.

After the first basin is full, the inlet line is closed, and the inlet to the second basin is opened; then lime is added to the second basin. The first basin is then mixed and checked with pH paper after a period of agitation; if it is acidic, more lime is added from lime storage until the basin is again basic. After settling, the effluent from the first basin is drained.

Nonacidic Waters. The nonacid sewers are directed to the refinery wastewater disposal system or to the API separators.

Liquid Process Wastes (Hydrocarbon and Acid). Hydrocarbon and acid wastes originate from some minor undesirable process side reactions and from any feed contaminants that are introduced to the unit. Undesirable by-products formed in this manner are ultimately rejected from the alkylation unit in the acid-regeneration column as a bottoms stream.

The regeneration-column bottoms stream consists mainly of two types of mixtures. One is an acid-water phase that is produced when water enters the unit with the feed streams. The other mixture is a small amount of polymeric material that is formed during certain undesirable process side reactions. Figure 1.4.8 represents the HF acid-regeneration circuit.

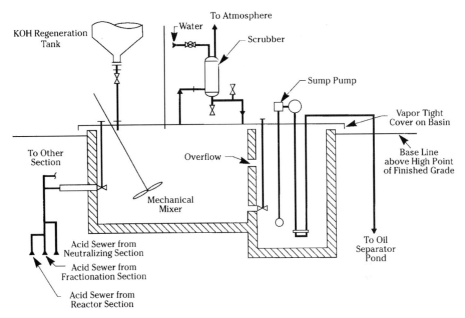

FIGURE 1.4.7 Neutralizing basin.

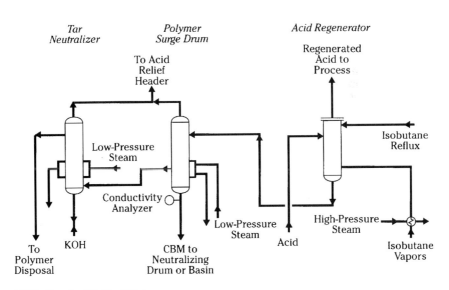

CBM = Constant Boiling Mixture
KOH = Potassium Hydroxide

FIGURE 1.4.8 HF acid regeneration circuit.

The first step in the disposal of these materials is to direct the regenerator bottoms to the polymer surge drum, where the two mixtures separate. The acid and water mixture form an azeotrope, or constant boiling mixture (CBM), which is directed to the neutralizing drum (Fig. 1.4.8) for neutralization of the HF acid. The acid in this CBM ultimately ends up as insoluble CaF_2 (as described previously). The polymer that remains in the polymer surge drum is then transferred to the tar neutralizer, where the free HF acid is removed. The polymer can then be disposed of by burning. However, by the mid-1980s, technology and special operating techniques such as internal acid regeneration had virtually eliminated this liquid-waste stream.

Solid Wastes

Neutralization-Basin Solids. The neutralization-basin solids consist largely of CaF_2 and unreacted lime. As indicated previously, all HF-containing liquids that are directed to the neutralizing basins ultimately have any contained soluble fluorides converted to insoluble CaF_2. The disposal of this solid material is done on a batch basis. A vacuum truck is normally used to remove the fluoride-lime sludge from the pit. This sludge has traditionally been disposed of in a landfill.

Another potential route for sludge disposal is to direct it to a steel-manufacturing company, where the CaF_2 can be used as a neutral flux to lower the slag-melting temperature and to improve slag fluidity. The CaF_2 may possibly be routed back to an HF-acid manufacturer, as the basic step in the HF-manufacturing process is the reaction of sulfuric acid with fluorspar (CaF_2) to produce hydrogen fluoride and calcium sulfate.

Product-Treating Waste Solids. The product-treating waste solids originate when LPG products are defluorinated over activated alumina. Over time, the alumina loses the ability to defluorinate the LPG product streams. At this time, the alumina is considered spent, and it is then replaced with fresh alumina. Spent alumina must be disposed of in accordance with applicable regulations or sent to the alumina vendor for recovery.

Miscellaneous Waste Solids. Porous material such as wiping cloths, wood, pipe coverings, and packings that are suspected of coming into contact with HF acid are placed in specially provided disposal cans for removal and are periodically burned. These wastes may originate during normal unit operation or during a maintenance period. Wood staging and other use of wood in the area are kept to a minimum. Metal staging must be neutralized before being removed from the acid area.

MITIGATING HF RELEASES—THE TEXACO-UOP ALKAD PROCESS

Growing environmental and public safety concerns since the mid-1980s have heightened awareness of hazards associated with many industrial chemicals, including HF acid. Refiners responded to these concerns with the installation of mitigation systems designed to minimize the consequences of accidental releases. Texaco and UOP developed the Alkad* technology[1] to assist in reducing the potential hazards of HF acid and to work in conjunction with other mitigation technology.

HF Acid Concerns and Mitigation

Although HF alkylation was clearly the market leader in motor fuel alkylation by the mid-1980s, growing concerns about public safety and the environment caused HF pro-

---

*Trademark and/or service mark of UOP.

ducers and users to reassess how HF acid was handled and how to respond to accidental releases. In 1986, Amoco and the Lawrence Livermore National Laboratory conducted atmospheric HF release tests at the Department of Energy Liquefied Gaseous Fuels Facility in Nevada. These tests revealed that HF acid could form a cold, dense aerosol cloud that did not rapidly dissipate and remained denser than air. In 1988, another set of tests, the Hawk tests, was conducted to determine the effect of water sprays on an HF aerosol cloud. These tests indicated that a water-to-HF ratio of 40/1 by volume would reduce the airborne HF acid by about 90 percent.[2] As a result of these investigations, many refiners have installed, or are planning to install, water-spray systems in their HF alkylation units to respond to accidental releases.

Other mitigation technology installed by refiners includes acid-inventory reduction, HF-detection systems, isolation valves, and rapid acid transfer systems. These mitigation systems can be described as external, defensive response systems because they depend on an external reaction (for example, spraying water) to a detected leak.

Texaco and UOP chose to develop a system that would respond prior to leak detection. Such a system could be described as an internal, passive response system because it is immediately effective should a leak occur. In 1991, Texaco and UOP began to work together to develop an additive system to reduce the risk associated with the HF alkylation process. The objective was to develop an additive that would immediately suppress the HF aerosol in the event of a leak but would not otherwise interfere with the normal performance of the HF unit.

Aerosol Reduction

Texaco screened a large number of additive materials for aerosol-reduction capability in its R&D facilities in Port Arthur, Texas. The most-promising materials that significantly reduced aerosol and maintained adequate alkylation activity were tested in a large-scale release chamber in Oklahoma.[3]

Release tests with additive demonstrated the potential reduction of airborne HF acid at various additive concentrations. This reduction was determined on the basis of the weight of material collected relative to the weight of material released. The aerosol reduction achieved is described in Fig. 1.4.9. As shown, reductions of airborne HF acid of up to 80 percent may be possible, depending on the additive concentration level at which a refiner is able to operate. Employing the Alkad technology in con-

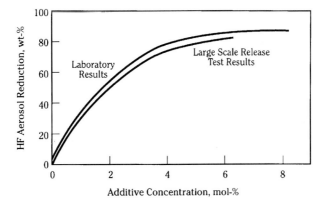

FIGURE 1.4.9 Aerosol reduction results.

junction with water sprays may result in more than 95 percent reduction of the airborne HF acid.

Process Development

Texaco and UOP conducted a trial with the most-interesting additive material in the older of two alkylation units at the Texaco refinery in El Dorado, Kansas, in 1992. During the trial, the alkylation unit operated well, with no changes as a result of the presence of additive in the acid. Following this successful trial, UOP designed facilities to recover the acid-additive complex from the acid regenerator bottoms stream and recycle this material back to the reactor section. The recovery system combines a stripping tower and gravity separators as shown in Fig. 1.4.10.

After cooling, the regenerator bottoms material flows to a separator, where a crude separation between polymer and the acid-additive complex may occur. Separated polymer is sent to neutralization, and the acid-additive complex enters a stripping column, where acid and water are removed with superheated light hydrocarbon. Column overhead material is completely condensed and recycled to the reactor section or used for reflux. Bottoms material is cooled and sent to another separator, where the polymer and acid-additive complex separate into distinct layers. The polymer is sent to neutralization, and the acid-additive complex is recycled to the reactor section.

Commercial Experience

After construction of the modular additive-recovery section was completed, Texaco began operating the Alkad technology in September 1994. The immediate observation when the additive was introduced was an increase in product octane and a reduction in alkylate endpoint. Research octane has been 1.5 or more numbers higher than the baseline operation (Fig. 1.4.11). A comparison of operations with and without additive is shown in Table 1.4.5, which breaks down two alkylate samples from equivalent

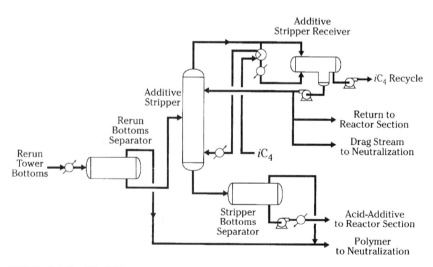

FIGURE 1.4.10 HF additive recovery process.

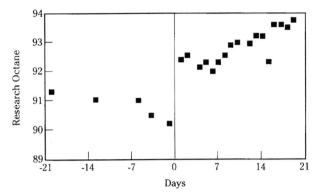

FIGURE 1.4.11 Alkylate octane.

TABLE 1.4.5 Alkylate Composition Comparison

	No additive	With additive
Alkylate RONC (measured)	90.8	92.2
Composition, LV %:		
C_6	2.84	3.58
C_7	14.15	19.06
C_8	45.24	44.35
C_9+	17.49	16.28
Calculated C_9+ RONC	81.6	89.5
Dimethylbutane/methylpentane	1.7	2.5
Dimethylpentane/methylhexane	51.0	77.1

Note: LV % = liquid volume percent.

operating conditions. An analysis of the alkylate components has shown that the increased octane is partially due to a significantly higher octane in the C_9+ material. Increased paraffin branching in the C_7 and lighter fraction is also a contributor to the octane boost. As shown in Fig. 1.4.12, initial data indicated that the alkylate 90 percent distillation point had decreased 14 to 19°C (25 to 35°F) and the endpoint had dropped 17 to 22°C (30 to 40°F). As gasoline regulations change, this distillation improvement may allow refiners to blend in more material from other sources and still meet regulatory requirements in their areas and effectively increase gasoline pool volume. Texaco installed this additive-recovery system for approximately $7 million U.S.

As of 1995, the Alkad technology is the only HF-acid modifier that has been used as a passive mitigation system. The Alkad process significantly reduces the hazards associated with an accidental release of HF-acid and minimizes the refiner's further investment in motor fuel alkylation technology.

REFERENCES

1. Sheckler, J. C., and H. U. Hammershaimb, "UOP Alkylation Technology into the 21st Century," presented at the 1995 UOP Refining Technology Conferences.

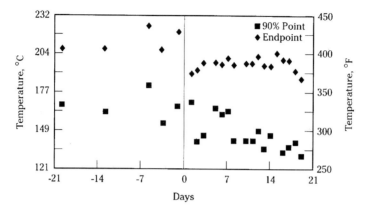

FIGURE 1.4.12 Alkylate distillation.

2. Schatz, K. W., and R. P. Koopman, "Effectiveness of Water Spray Mitigation Systems for Accidental Releases of Hydrogen Fluoride," summary report and volumes I–X, NTIS, Springfield, Va., 1989.
3. Comey, K. R., III, L. K. Gilmer, G. P. Partridge, and D. W. Johnson, "Aerosol Reduction from Episodic Releases of Anhydrous HF Acid by Modifying the Acid Catalyst with Liquid Onium Poly (Hydrogen Fluorides)," AIChE 1993 Summer National Meeting, Aug. 16, 1993.

CHAPTER 1.5
LINEAR ALKYLBENZENE (LAB) MANUFACTURE

Peter R. Pujadó
UOP
Des Plaines, Illinois

INTRODUCTION

The detergent industry originated in the late 1940s with the advent of sodium alkylbenzene sulfonates, which had detergency characteristics far superior to those of natural soaps. Natural soaps are sodium salts of fatty acids obtained by the alkaline saponification of naturally occurring triglycerides from either vegetal or animal sources. The early alkylbenzene sulfonates (ABSs) were essentially sodium dodecylbenzene sulfonates (DDBSs), also known as branched alkylbenzene sulfonates (BASs) obtained by the Friedel-Crafts alkylation of benzene with propylene tetramer, a mixture of branched C_{12} olefins. Dodecylbenzenes (DDBs) are then sulfonated with oleum or sulfur trioxide (SO_3) and neutralized with sodium hydroxide or soda ash.

Because of their lower cost and high effectiveness in a wide range of detergent formulations, DDBS rapidly displaced natural soaps in household laundry and dishwashing applications. However, although excellent from a performance viewpoint, BAS exhibited slow rates of biodegradation in the environment and, in the early 1960s, started to be replaced by linear alkylbenzene sulfonate (LAS or LABS). The linear alkyl chains found in LAS biodegrade at rates that are comparable to those observed in the biodegradation of natural soaps and other natural and semisynthetic detergent products.

The use of DDBS has never been formally banned in the United States, but by the late 1960s, its use had been largely phased out in the United States, Japan, and in several European countries. By the late 1970s, the use of LAS had become more generalized, and new facilities were added in developing countries around the world. In the mid-1990s, LAS accounts for virtually the entire worldwide production of alkylbenzene sulfonates. The demand for linear alkylbenzene (LAB) increased from about 1.0 million metric tons per year (MTA) in 1980 to about 1.7 million in 1990. The LAB production capacity, which exceeded 2.0 million MTA in 1992, is expected to grow to about 2.7 million by the year 2000.

TECHNOLOGY BACKGROUND

Various routes were developed and used in the production of LAB. The first hurdle to be overcome was the recovery, typically from kerosene or gas oil fractions, of linear paraffins (*n*-paraffins) in the C_{10} to C_{14} range. Initial recovery attempts were based on the use of urea adducts, which were soon replaced by adsorptive separation and recovery techniques, either in the vapor or the liquid phase. These techniques used a variety of adsorbents and desorbents. Adsorptive separation techniques based on the molecular sieve action of 5-Å zeolites have dominated this industry since the mid-1960s. Typical commercial process technologies for this separation include the UOP* Molex* process in the liquid phase with a hydrocarbon desorbent that makes use of UOP's Sorbex* simulated moving-bed technology; the UOP IsoSiv* process (formerly Union Carbide's), which operates in the vapor phase also with a hydrocarbon desorbent; Exxon's Ensorb process, which is also in the vapor phase but has an ammonia desorbent; or a similar technology developed in the former German Democratic Republic (East Germany) and known as the GDR Parex process, which also operates in the vapor phase with ammonia desorbent. The GDR Parex process is not to be confused with UOP's Parex* process for the selective recovery of high-purity *p*-xylene from aromatic streams using the Sorbex simulated moving-bed technology.

Once the linear paraffins have been recovered at sufficient purity, typically in excess of about 98 percent, they have to be alkylated with benzene to produce LAB. To date, attempts to alkylate *n*-paraffins with benzene directly have failed, thus necessitating the activation of the *n*-paraffins to a more-reactive intermediate before the alkylation with benzene can take place.

The following routes for the production of LAB emerged during the 1960s:

- Chlorination of *n*-paraffins to form primarily monochloroparaffins. Benzene is then alkylated with monochloroparaffins using an aluminum chloride ($AlCl_3$) catalyst. An example of this route was developed and commercialized by ARCO Technology Inc.[1]

- Chlorination of *n*-paraffins followed by dehydrochlorination and alkylation of the resulting olefins with benzene typically using hydrofluoric (HF) acid as catalyst. Shell's CDC process (for chlorination/dehydrochlorination) is an example of such a process. This type of technology was still used commercially until the mid-1980s by, among others, Hüls AG in Germany.

- Alkylation of linear olefins with benzene also using an HF catalyst. The olefins are usually either linear *alpha*-olefins (LAOs) from wax cracking (now discontinued), *alpha*-olefins from ethylene oligomerization, or linear internal olefins (LIOs) from olefin disproportionation. Various companies, such as Albemarle (formerly Ethyl), Chevron (formerly Gulf), and Shell, offer technologies for the oligomerization of ethylene to LAO; Shell also produces linear internal olefins by disproportionation in their Shell Higher Olefins process (SHOP).

- Dehydrogenation of linear paraffins to a fairly dilute mixture of LIO in unconverted *n*-paraffins, followed by the alkylation of the olefins with benzene also using HF acid catalyst but without the separation and concentration of the LIO. UOP's Pacol* process for the catalytic dehydrogenation of *n*-paraffins and UOP's HF Detergent Alkylate* process for the alkylation of the LIO with benzene are prime examples of this approach. A similar approach is also practiced by Huntsman Corp. (formerly Monsanto's).[2,3]

---

*Trademark and/or service mark of UOP.

During the early days of LAB production, paraffin chlorination followed by alkylation over $AlCl_3$ gained some prominence. However, since the late 1960s, the dehydrogenation and HF alkylation route has been the most prominent because of its economic advantages and higher-quality product. Although LAO and LIO obtained from sources other than dehydrogenation can equally be used, n-paraffin dehydrogenation routes have usually prevailed because of the lower cost of the starting kerosene fractions. Table 1.5.1 shows an approximate 1992 distribution of world LAB production employing these technologies. The dehydrogenation followed by alkylation route accounts for 74 percent of world LAB production. The Detal* process, which replaces HF with a solid heterogenous acid catalyst, was introduced in 1995. The various routes for the production of LAB are illustrated schematically on Fig. 1.5.1.

COMMERCIAL EXPERIENCE

The first commercial operations of UOP's dehydrogenation and alkylation technologies were in Japan and Spain at the end of 1968. Almost all the units built since then throughout the world employ UOP technology. Over the years, UOP has continued

*Trademark and/or service mark of UOP.

TABLE 1.5.1 1992 World LAB Production by Technology Route

Technology route	Production	
	Thousand MTA	%
Chlorination and alkylation	240	12
Dehydrogenation and alkylation	1480	74
High-purity olefins to alkylation	280	14
Total	2000	100

Note: MTA = metric tons per annum.

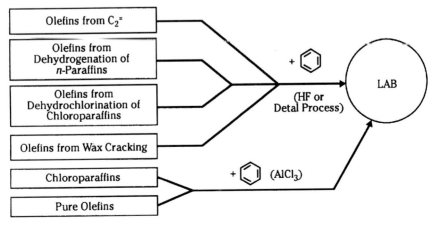

FIGURE 1.5.1 Routes to LAB.

research and development and has introduced numerous improvements that resulted in improved economics of LAB manufacture as well as consistently improved product quality. More than 30 LAB units now operate around the world with this process technology.

The new Detal process was developed jointly by UOP and PETRESA, a wholly owned subsidiary of CEPSA in Spain. The process uses a fixed bed of acidic, noncorrosive catalyst to replace the liquid HF acid used in the present UOP HF Detergent Alkylate process.

The catalyst of choice for LAB production has been HF acid since the first Pacol unit came on-stream in 1968. Its high efficiency, superior product, and ease of use relative to the older $AlCl_3$ catalyst are the reasons for this success. However, in both the HF- and the $AlCl_3$-catalyzed processes, the handling of corrosive catalysts has had implications in terms of the increased capital cost of the plant as well as in the disposal of the small quantities of neutralization products generated in the process. Hence, the advantages of a heterogeneous catalyst in this application have long been recognized.

Aromatic alkylation has been demonstrated over many acidic solids, such as clay minerals, zeolites, metal oxides, and sulfides. Although many of these catalysts are highly active, they are usually lacking in selectivity or stability. The key to a successful solid-bed alkylation process is the development of a catalyst that is active, selective, and stable over prolonged periods of operation. Research at PETRESA and UOP resulted in the development of a solid catalyst for the alkylation of benzene with linear olefins to produce LAB. The resulting Detal process was proven at UOP's pilot plants and at PETRESA's semiworks facility in Spain and is now in commercial operation at a new LAB plant near Montréal in Canada. The process produces a consistent-quality product that meets all detergent-grade LAB specifications.

The simplified flow diagrams in Figs. 1.5.2 and 1.5.3 illustrate the main differences between the HF Detergent Alkylate and Detal processes. Figure 1.5.4 shows an integrated LAB complex that incorporates Pacol, DeFine,* and detergent alkylation units. The flow scheme for the Pacol and DeFine units remains unchanged for either an HF- or a solid-catalyzed, fixed-bed alkylation unit.

In the HF Detergent Alkylate process, olefin feed from the Pacol-DeFine units is combined with makeup and recycle benzene and is cooled prior to mixing with HF acid. The reaction section consists of a mixer reactor and an acid settler. A portion of the HF acid phase from the settler is sent to the HF acid regenerator, where heavy byproducts are removed to maintain acid purity. The hydrocarbon phase from the acid settler proceeds to the fractionation section, where the remaining HF acid, excess benzene, unreacted n-paraffins, heavy alkylate, and LAB product are separated by means of sequential fractionation columns. The HF acid and benzene are recycled to the alkylation reactor. The unreacted n-paraffins are passed through an alumina treater to remove combined fluorides and are then recycled back to the dehydrogenation unit. The flow diagram in Fig. 1.5.2 shows the HF-acid handling and neutralization section, which is required for the safe operation of the plant and is always included within battery limits. This section represents a significant portion of the investment cost of HF alkylation plants.

In the Detal scheme (Fig. 1.5.3), olefin feed combined with makeup and recycle benzene flows through a fixed-bed reactor, which contains the solid catalyst. The reaction occurs at mild conditions in the liquid phase. Reactor effluent flows directly to the fractionation section, which remains the same as for the HF-acid system except that the HF-acid stripper column and the alumina treater are eliminated. Also eliminated is the entire HF reactor section, including the mixer reactor, acid settler, HF-acid regenerator, and associated piping. In addition, all the equipment and special metallur-

---

*Trademark and/or service mark of UOP.

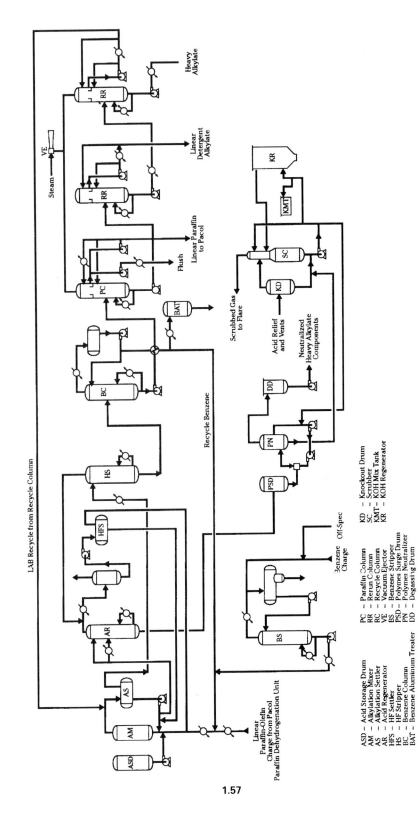

FIGURE 1.5.2 HF Detergent Alkylate process.

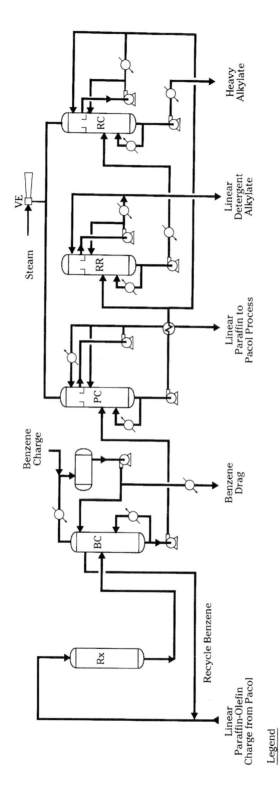

FIGURE 1.5.3 Detal process.

Legend

Rx = Reactor RR = Rerun Column
BC = Benzene Column RC = Recycle Column
PC = Paraffin Column VE = Vacuum Ejector

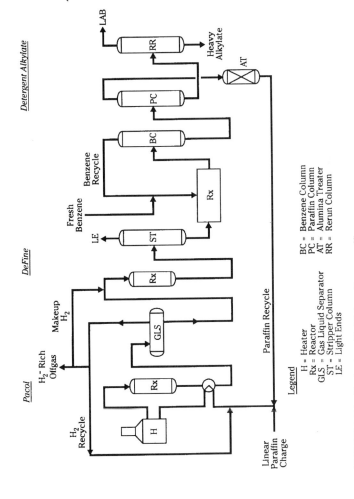

FIGURE 1.5.4 Production of LAB from linear paraffins.

gy required for the safe handling of HF acid, neutralization of waste steams, and disposal of the neutralization products are not required.

Because hydrocarbons such as paraffins, olefins, benzene, and alkylbenzenes are handled in the Detal process, only carbon steel construction is used. Thus, the Monel parts and special pump seals used in HF service are eliminated.

Research on the Detal catalyst showed that diolefins and some other impurities, mostly aromatics, coming from the Pacol dehydrogenation unit have a substantial impact on the stability of the Detal catalyst as well as on LAB quality. Thus, a DeFine process unit must be included to convert all diolefins to monoolefins. Additionally, UOP developed technology to remove aromatics from the alkylation feed. Normally, these aromatics alkylate with olefins and produce a heavy alkylate by-product in the alkylation unit. Thus, aromatics removal has two benefits: increased LAB yield per unit of olefins and improved stability of the Detal catalyst. Because the stability of the Detal catalyst is improved, the alkylate reaction temperature can be lowered. The result is a LAB product with higher linearity.

As shown in Table 1.5.2, the benefits of the aromatics removal unit (ARU) are in lower raw material consumption and improved product quality. The consumption of n-paraffin feed per unit of LAB in weight ratio is reduced from about 0.80 in a HF-acid unit to about 0.78, and product LAB linearity increases from 93 to 95 percent. In a Detal alkylation unit, the specific n-paraffin consumption decreases from about 0.83 to around 0.81, also with a similar increase in product linearity.

PRODUCT QUALITY

Table 1.5.3 compares LAB product properties for the two catalyst systems: HF and Detal. The quality of the two products is similar, but aromatics removal results in an

TABLE 1.5.2 Benefits of Aromatics Removal in the Feed to a Detal Alkylation Unit

	Without ARU	With ARU
Yield, wt LAB per wt paraffin	1.20	1.23
Linearity, %	92–93	94–95

Note: ARU = aromatics removal unit

TABLE 1.5.3 Comparison of HF and Detal LAB

	Typical HF LAB	Typical Detal LAB
Specific gravity	0.86	0.86
Bromine index	<15	<15
Saybolt color	+30	+30
Water, ppm	<100	<100
Tetralins, wt %	<1.0	<0.5
2-phenyl-alkanes, wt %	15–18	>25
n-alkylbenzene, wt %	93	95
Klett color of 5% active LAS solution	20–40	10–30

LAB product with higher linearity. Both processes achieve low levels of tetralins in the LAB. However, the Detal process achieves a lower level (less than 0.5) of tetralins compared to the HF process.

The Detal LAB product also produces a lighter colored sulfonate. As shown in Table 1.5.3, the Klett color of a 5 percent active solution of Detal-derived LAS is typically lower than that of LAS obtained by using HF.

The most significant difference between HF and Detal LAB is in the higher 2-phenyl-alkane content of the LAB obtained in the Detal process. This higher content of 2-phenyl-alkane improves the solubility of the sulfonated LAB. The difference is particularly important in liquid formulations, as illustrated in Fig. 1.5.5, which shows the cloud point of the LAS derived from both systems. Over the range of 13 to 25 percent active solution of sodium LAS, the Detal-derived product exhibits a lower cloud point and is much less sensitive to concentration as compared with the HF-derived product.

ECONOMICS

A comparative economic analysis was prepared for the production of 80,000 MTA of LAB using either the HF Detergent Alkylate or the Detal process. The complex was assumed to consist of Pacol, DeFine, and HF Detergent Alkylate or Detal units (with aromatics removal in the latter) as well as a common hot-oil belt. The equipment was sized on the assumption of 8000 hours on-stream per year, which corresponds to an effective production capacity of 240 metric tons (MT) per stream day.

The erected cost for the complex based on the HF Detergent Alkylate process is estimated at $72 million. The same complex using the Detal process has an estimated erected cost of $67 million. All design, construction, and labor costs were estimated on an open-shop basis for a U.S. Gulf Coast location, and 30 percent off-site allowances were assumed.

The economic analysis is summarized in Table 1.5.4. The yields represent the production of LAB with an average molecular weight of 240. These data were obtained in a pilot plant and, in the case of the complex involving the HF route, in commercial LAB units.

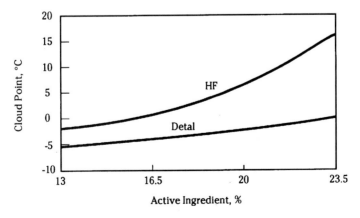

FIGURE 1.5.5 Solubility comparison of HF and Detal LAS.

TABLE 1.5.4 Economic Comparison of HF Detergent Alkylate and Detal Processes in a LAB Complex*

	Unit cost, $	HF Alkylation, per MT LAB		Detal, per MT LAB	
		Quantity	$	Quantity	$
Raw materials:					
n-paraffins, MT	400	0.80	319	0.81	322
Benzene, MT	350	0.34	118	0.34	118
By-product credits, MT	—	—	(39)	—	(41)
Catalysts and chemicals	—	—	18	—	35
Utilities:					
Power, kWh	0.05	270	14	290	14
Steam, MT	7.1	0.02	—	0.02	—
Cooling water, m^3	0.01	87	1	7	—
Fuel fired, GJ	2.32	15.6	36	18.6	43
Labor, maintenance, direct overhead, and supervision	—	—	33	—	28
Overhead, insurance, property taxes	—	—	48	—	44
Cash cost of production	—	—	548	—	563
Cash flow, million $ (LAB at $800/ton)		20.2		19.0	
Estimated erected cost, million $		72		67	
Simple payback, years (on fixed investment)		3.6		3.6	

*Basis: Production cost for 80,000-MTA LAB.

Note: MT = metric tons; MTA = metric tons per annum.

By-product credits include hydrogen at about 95 mol % purity, light ends, heavy alkylate, and HF regenerator bottoms. Utility requirements correspond to a typical modern design of the UOP LAB complex. The cost of effluent treatment and disposal has not been included in this analysis.

The combined investment for the Pacol, DeFine, and the hot-oil units for the two cases is essentially the same. The fixed plant investment for the alkylation section has been reduced by some 15 percent. The absence of HF acid, and hence the absence of the corresponding neutralization facilities for the acidic wastes, is reflected in a lower operating cost.

MARKETS

The evolution in the demand for LAB differs in the various geographic areas. Since the early 1990s, these different growth rates have reflected not only the maturity of the most economically developed markets but also the trend toward a healthier economic future. Table 1.5.5 summarizes the consumption of LAB in various geographic areas for the years 1980, 1985, and 1990. The per capita consumption, in kilograms per year, was used to forecast the potential expected LAB demand worldwide. Figure 1.5.6 reflects the situation in 1991 in these same geographic areas in terms of kilograms per capita per year. The data in the table and the figure highlight the consumption trends in various markets of the world. From these data, scenarios can be established for various parts of the world.

TABLE 1.5.5 Historical Demand for LAB by Geographic Areas

Area	LAB consumption, 10^3 MTA		
	1980	1985	1990
Western Europe	310	345	380
Eastern Europe (including Russia)	105	125	180
Africa	35	50	110
Middle East	30	50	100
Far East (including China)	190	220	305
Southeast Asia (including India)	90	105	180
North America	205	250	295
Latin America	85	105	140
Total	1050	1250	1690

Note: MTA = metric tons per annum.

ENVIRONMENTAL SAFETY

Surfactants are present in the environment because they are commonly used in a variety of household and industrial applications. The presence of a surfactant in the environment is the result of a combination of factors, such as consumption pattern; biodegradability potential; availability of treatment facilities; prevailing physicochemical conditions, for example, water hardness, temperature, pH; and chemical reactions or interactions.

Biodegradation is the primary process to reduce the concentration of a given product in the environment. In fact, the use of products like surfactants is regulated in various countries at least in part based on biodegradation information. Such regulation has been the case in Europe since the mid-1960s. This fact indicates that the detergent industry took steps to protect the environment well before such topics were a global concern. Before any real environmental assessment can be made, the first step is to conduct the appropriate evaluation of biodegradability and aquatic toxicity using laboratory-scale models. However, obtaining universal agreement on a definition and design of a biodegradation test that will adequately predict all possible real-world environments has been difficult. Laboratory testing of biodegradability is a complex activity because of the difficulty of sampling and isolating environmental situations that are the result of many interactions, some of which may still be unknown.

After more than 25 years of methodology development, technology improvement, and progress in analytical techniques, the amount of data available on every aspect of the behavior and disappearance of LAS in the environment is so conclusive that as of mid-1992, no other surfactant, regardless of its origin, could be considered as safe as LAS. This conclusion is based on the results of different laboratory biodegradation tests, summarized in Table 1.5.6, that are used for regulatory purposes. These laboratory tests are fully substantiated with real environmental information as a result of the development and validation of a specific analytical method to detect LAS in trace quantities.[4] This method can be used to monitor LAS in the environment.

Real-world monitoring, when available, should always prevail over laboratory studies, model predictions, or any other theoretical method of assessment. Real environmental monitoring reflects all possible interactions, many of which cannot be reproduced in the laboratory.

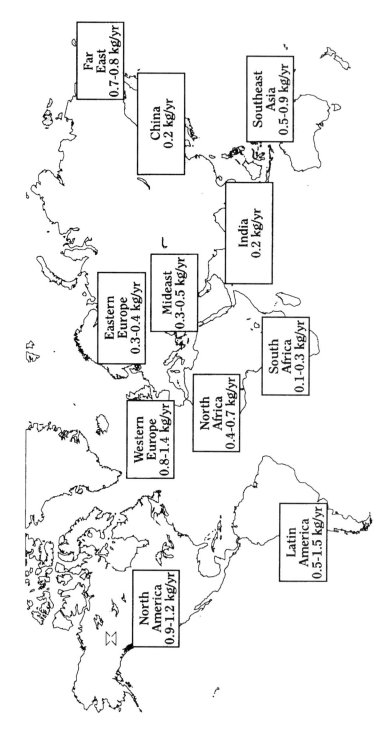

FIGURE 1.5.6 Estimated per capita LAB consumption in 1991.

TABLE 1.5.6 LAS Biodegradability Test Results in the Laboratory

	OECD method	Removal of DOC, %
Inherent biodegradation	302 A	90 to 97
	302 B	>95
Confirmatory test	303 A	95 to 99 (MBAS)
Sewage treatment simulation	303 A	96 to 99 (DOC)

Note: OECD = Organization of Economic Commercial Development; MBAS = methylene blue active substance; DOC = dissolved organic carbon.
Source: PETRESA.

Frequently, environmental safety is determined on the basis of product origin even though no analytical technique to monitor the products in different environmental situations exists. However, a number of techniques do exist for monitoring LAS, and as a result, LAS has a proven history of environmental safety. Table 1.5.7 summarizes the most relevant LAS environmental data based on many monitorings conducted in several countries over the last six years.[5,6]

CONCLUSIONS

LAB continues to be the most cost-effective detergent intermediate, regardless of raw material source. The continuing growth in LAB is spurred by increasing consumption in countries outside the Organization of Economic Commercial Development (OECD). Worldwide LAB consumption is expected to increase by some 650,000 MTA over the next 10 years. Increasing trade between various LAB-producing regions has led to more-uniform, high-quality requirements for the product in different parts of the world.

Developments in LAB technology have addressed the important issues confronting the industry in the 1990s: improved yields and economics, product quality, and environmental and safety considerations.

The use of large volumes of LAS derived from LAB over the last 30 years has resulted in extensive environmental studies of this surfactant by industry and consumer groups. No other surfactant type has undergone such intense scrutiny. This

TABLE 1.5.7 LAS Environmental Safety Summary (Full-Scale Sewage Treatment Water Monitoring)

Types of plants	Activated sludge Lagoons Oxidation ditches
Number of plants monitored	>80 in 10 countries
Removal	99 ± 1%
Specific biodegradation	>80% in 3-h residence time
Half-life	1 to 2 h
Receiving water concentration (worst case)	0.03 ± 0.01 mg/L
Contribution of LAS to total organic carbon	<1%
NOEC (field stream tests)	0.36 to 1.5 mg/L

Note: NOEC = No observed effect concentration.

scrutiny has resulted in the development of improved methods for LAS detection outside of laboratory situations and model predictions. The use of these techniques in real-world monitoring in various countries during the last decade has only confirmed the long-term viability of LAS from the standpoint of environmental safety.

ACKNOWLEDGMENTS

This chapter was adapted from a paper entitled "Growth and Developments in LAB Technologies: Thirty Years of Innovation and More to Come," by J. L. Berna and A. Moreno of PETRESA, Spain, and A. Banerji, T. R. Fritsch, and B. V. Vora of UOP, Des Plaines, Illinois, U.S.A. The paper was presented at the 1993 World Surfactant Congress held in Montreux, Switzerland, on September 23, 1993.

REFERENCES

1. ARCO Technology Inc., Hydrocarbon Processing, **64** (11), 127, 1985.
2. J. F. Roth and A. R. Schaefer, U.S. Patent 3,356,757 (to Monsanto).
3. R. E. Berg and B. V. Vora, *Encyclopedia of Chemical Processing and Design,* vol. 15, Marcel Dekker, New York, 1982, pp. 266–284.
4. E. Matthijs and H. de Henau, "Determination of LAS," *Tenside Surfactant Detergents,* **24**, 193–199, 1987.
5. J. L. Berna et al., "The Fate of LAS in the Environment," *Tenside Surfactant Detergents,* **26**, (2), 101–107, 1989.
6. H. A. Painter et al., "The Behaviour of LAS in Sewage Treatment Plants," *Tenside Surfactant Detergents,* **26**, (2), 108–115, 1989.

CHAPTER 1.6
UOP Q-MAX PROCESS FOR CUMENE PRODUCTION

Martin F. Bentham
UOP
Des Plaines, Illinois

DISCUSSION

The manufacture of cumene (isopropyl benzene) was started in the 1930s as a high-octane component for aviation gasoline. Over the years, the use of cumene in aviation gasoline has decreased progressively to the point that it has disappeared altogether. By the mid-1990s, virtually all cumene produced worldwide is consumed in the manufacture of phenol and acetone. A small amount is consumed in the production of *alpha*-methylstyrene (AMS) by catalytic dehydrogenation. However, most AMS is recovered as a minor by-product of the production of phenol and acetone from cumene.

Cumene is a minor constituent of aromatic fractions obtained from naphtha reforming or pyrolysis operations. However, cumene for industrial uses is never recovered from such aromatic fractions. Rather, it is made synthetically by the direct alkylation of benzene with propylene in the presence of an acidic catalyst.

Many different processes and catalyst systems have been proposed for the alkylation of benzene with propylene. Possible catalyst systems include boron fluoride, aluminum chloride, phosphoric acid, and various silicoaluminates (zeolites). Most current manufacturing operations use the UOP* Catalytic Condensation process for cumene production. This process uses a solid phosphoric acid catalyst that is made by impregnating kieselguhr with phosphoric acid.

Recent research has focused on developing a new process that produces a higher-quality cumene product at lower investment cost. This research effort has resulted in the development by UOP of the Q-Max* process, which is based on a new zeolite catalyst also developed by UOP.

The main features of the Q-Max process are shown in Fig. 1.6.1. The propylene feed can be an almost-pure polymer-grade material or can contain significant amounts of propane, as typically found in refinery-grade propylene. However, producers prefer that the propylene feed be essentially free of ethylene and butylenes, which if present give rise to their own respective alkylation products with benzene and may negatively

---
*Trademark and/or service mark of UOP.

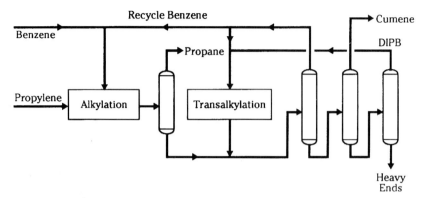

FIGURE 1.6.1 Q-Max process flow.

affect plant investment or product quality. Benzene feed is typically from an extraction unit, where nonaromatics and toluene have been separated to a low level.

The Q-Max process has two reaction steps. The primary reaction, alkylation of benzene with propylene to form cumene, takes place in the first step. This alkylation step occurs in a benzene-rich environment, and propylene conversion is essentially 100 percent. Any propane coming in with the propylene feed is unreacted and is removed from the alkylation reactor effluent by fractionation.

In the alkylation step, some of the cumene produced is alkylated with propylene to diisopropylbenzene. The Q-Max process incorporates a second reaction step (transalkylation) to react the diisopropylbenzene with benzene to form additional cumene. The transalkylation reactor effluent is combined with the depropanized alkylation reactor product and sent to the fractionation section, where recycle benzene, diisopropylbenzene for transalkylation, and a small amount of heavy aromatics are separated from the cumene product.

YIELD STRUCTURE

The Q-Max unit has high raw material utilization and an overall cumene yield of at least 99.0 wt % based on typical propylene and benzene feeds. The remaining 1.0 wt % or less of the overall yield is in the form of a heavy aromatic by-product. At an overall cumene yield of 99.0 wt %, the propylene and benzene feed consumptions are respectively about 0.35 and 0.66 kg of feed per kilogram of cumene product. The cumene product may have a purity in excess of 99.95 percent and a bromine index of less than 5.

Propane entering the unit with the propylene feed is unreactive in the process and is separated in the fractionation section as a propane product.

PROCESS ECONOMICS

The estimated erected cost of a new Q-Max unit with a capacity of 200,000 metric tons per year (MTA) of cumene product is about $12 million (U.S.). This estimate covers the fully erected cost within battery limits of a unit constructed in 1995 at a U.S. Gulf Coast location.

The utility requirements for each unit depend on the project environment for feed, product specifications, and utility availability. Cumene units are often integrated with phenol plants when energy use can be optimized by generating low-pressure steam in the cumene unit for utilization in the phenol plant. An example of utility consumption for such a new unit project is shown below. (Utility consumption is in units per metric ton of cumene produced.)

Electric power, kWh	20
High-pressure steam consumption, MT	1.2
Low-pressure steam production, MT	1.4
Cooling water, m^3	25

COMMERCIAL EXPERIENCE

As previously mentioned, the UOP Catalytic Condensation process unit has been used for most cumene units built around the world. More than 40 units have been designed. Capacities of operating units range from 20,000 to more than 600,000 MTA of cumene produced. Total plant capacity of operating units is approximately 4.8 million MTA of cumene produced. The Q-Max process was introduced in 1992. The first commercial Q-Max unit will start up in 1996.

P · A · R · T · 2

BASE AROMATICS PRODUCTION PROCESSES

CHAPTER 2.1
AROMATICS COMPLEXES

John J. Jeanneret
UOP
Des Plaines, Illinois

INTRODUCTION

An aromatics complex is a combination of process units that can be used to convert petroleum naphtha and pyrolysis gasoline (pygas) into the basic petrochemical intermediates: benzene, toluene, and xylenes (BTX). Benzene is a versatile petrochemical building block used in the production of more than 250 different products. The most important benzene derivatives are ethylbenzene, cumene, and cyclohexane (Fig. 2.1.1). The xylenes product, also known as *mixed xylenes,* contains four different C_8 aromatic isomers: *para*-xylene, *ortho*-xylene, *meta*-xylene, and ethylbenzene. Small

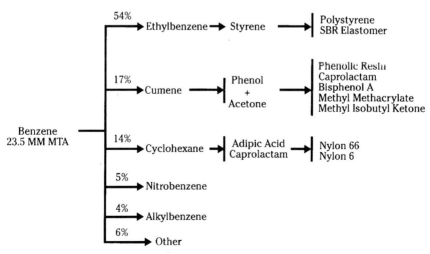

FIGURE 2.1.1 World benzene consumption, 1994.

amounts of mixed xylenes are used for solvent applications, but most xylenes are processed further within the complex to produce one or more of the individual isomers. The most important C_8 aromatic isomer is *para*-xylene, which is used almost exclusively for the production of polyester fibers, resins, and films (Fig. 2.1.2). A small amount of toluene is recovered for use in solvent applications and derivatives, but most toluene is used to produce benzene through hydrodealkylation. Toluene is becoming increasingly important for the production of xylenes through toluene disproportionation and transalkylation with C_9 aromatics.

CONFIGURATIONS

Aromatics complexes can have many different configurations. The simplest complex produces only benzene, toluene, and mixed xylenes (Fig. 2.1.3) and consists of the following major process units:

- *Naphtha hydrotreating* for the removal of sulfur and nitrogen contaminants
- *Catalytic reforming* for the production of aromatics from naphtha
- *Aromatics extraction* for the extraction of BTX

Most new aromatics complexes are designed to maximize the yield of benzene and *para*-xylene and sometimes *ortho*-xylene. The configuration of a modern, integrated

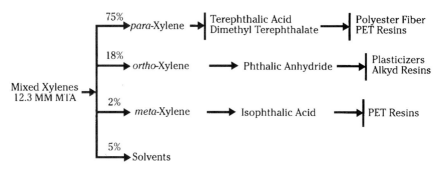

FIGURE 2.1.2 World xylenes consumption, 1994.

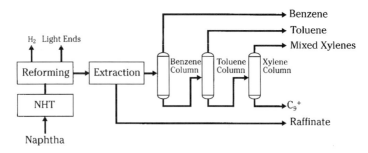

FIGURE 2.1.3 Simple aromatics complex.

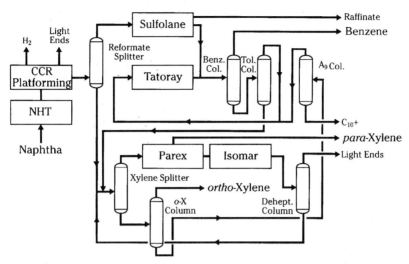

FIGURE 2.1.4 Integrated UOP aromatics complex.

UOP* aromatics complex is shown in Fig. 2.1.4. This complex has been configured for maximum yield of benzene and *para*-xylene and includes the following UOP process technologies:

- *CCR* Platforming** for the production of aromatics from naphtha at high severity
- *Sulfolane** for the extraction of benzene and toluene
- *Parex** for the recovery of *para*-xylene by continuous adsorptive separation
- *Isomar** for the isomerization of xylenes and the conversion of ethylbenzene
- *Tatoray* for the conversion of toluene and heavy aromatics to xylenes and benzene

The Tatoray process is used to produce additional xylenes and benzene by toluene disproportionation and transalkylation of toluene plus C_9 aromatics. The incorporation of a Tatoray unit into an aromatics complex can more than double the yield of *para*-xylene from a given amount of naphtha feedstock. Thus, the Tatoray process is used when *para*-xylene is the principal product. If benzene is the principal product, a thermal hydrodealkylation (THDA) unit may be substituted for the Tatoray unit in the flow scheme. The THDA process is used to dealkylate toluene and heavier aromatics to benzene only. A few aromatics complexes incorporate both the Tatoray and THDA processes to provide maximum flexibility for shifting the product slate between benzene and *para*-xylene. Detailed descriptions of each of these processes are in Chaps. 2.7 and 2.3.

About half of the existing UOP aromatics complexes are configured for the production of both *para*-xylene and *ortho*-xylene. Figure 2.1.4 shows an *ortho*-Xylene (*o*-X) column for recovery of *ortho*-xylene by fractionation. If *ortho*-xylene production is not required, the *o*-X column is deleted from the configuration, and all the C_8 aromatic isomers are recycled through the Isomar unit until they are recovered as *para*-xylene. In those complexes that do produce *ortho*-xylene, the ratio of *ortho*-xylene to *para*-xylene production is usually in the range of 0.2 to 0.6.

---

*Trademark and/or service mark of UOP.

The *meta*-xylene market is currently small but is growing rapidly. In 1995, UOP licensed the first MX Sorbex* unit for the production of *meta*-xylene by continuous adsorptive separation. Although similar in concept and operation to the Parex process, the MX Sorbex process selectively recovers the *meta* rather than the *para* isomer from a stream of mixed xylenes. An MX Sorbex unit can be used alone, or it can be incorporated into an aromatics complex that also produces *para*-xylene and *ortho*-xylene.

An aromatics complex may be configured in many different ways, depending on the available feedstocks, the desired products, and the amount of investment capital available. This range of design configurations is illustrated in Fig. 2.1.5. Each set of bars in Fig. 2.1.5 represents a different configuration of an aromatics complex processing the same full-range blend of straight-run and hydrocracked naphtha. The configuration options include whether a Tatoray or THDA unit is included in the complex, whether C_9 aromatics are recycled for conversion to benzene or xylenes, and what type of Isomar catalyst is used. The xylene-to-benzene ratio can also be manipulated by prefractionating the naphtha to remove benzene or C_9+ aromatic precursors (see the section of this chapter on feedstock considerations). Because of this wide flexibility in the design of an aromatics complex, the product slate can be varied to match downstream processing requirements. By the proper choice of configuration, the xylene-to-benzene product ratio from an aromatics complex can be varied from about 0.6 to 3.8.

DESCRIPTION OF THE PROCESS FLOW

The principal products from the aromatics complex illustrated in Fig. 2.1.4 are benzene, *para*-xylene, and *ortho*-xylene. If desired, a fraction of the toluene and C_9 aromatics may be taken as products, or some of the reformate may be used as a high-octane gasoline blending component. The naphtha is first hydrotreated to remove sulfur and nitrogen compounds and then sent to a CCR Platforming unit, where paraffins and naphthenes are converted to aromatics. This unit is the only one in the

*Trademark and/or service mark of UOP.

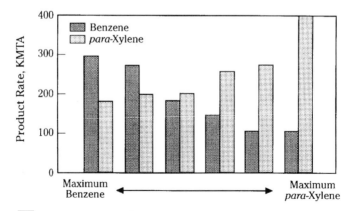

Basis: 790 KMTA (25,000 BPD) Blend of Straight-Run and Hydrocracked Naphtha

FIGURE 2.1.5 Product slate flexibility.

complex that actually creates aromatic rings. The other units in the complex separate the various aromatic components into individual products and convert undesired aromatics into additional high-value products. The CCR Platforming unit is designed to run at high severity, 104 to 106 research octane number, clear (RONC), to maximize the production of aromatics. This high-severity operation also extinguishes virtually all nonaromatic impurities in the C_8+ fraction of the reformate, thus eliminating the need for extraction of the C_8 and C_9 aromatics. The reformate product from the CCR Platforming unit is sent to a debutanizer column within the Platforming unit to strip off the light ends.

The reformate from the CCR Platforming unit is sent to a reformate splitter column. The C_7- fraction from the overhead is sent to the Sulfolane unit for extraction of benzene and toluene. The C_8+ fraction from the bottom of the reformate splitter is clay-treated and then sent directly to the xylene recovery section of the complex.

The Sulfolane unit extracts the aromatics from the reformate splitter overhead and rejects a paraffinic raffinate stream. The aromatic extract is clay-treated to remove trace olefins. Then individual high-purity benzene and toluene products are recovered in the benzene-toluene (BT) fractionation section of the complex. The C_8+ material from the bottom of the toluene column is sent to the xylene recovery section of the complex. The raffinate from the Sulfolane unit may be further refined into paraffinic solvents, blended into gasoline, used as feedstock for an ethylene plant, or converted into additional benzene by an RZ-100* Platforming unit.

Toluene is usually blended with C_9 aromatics (A_9) from the overhead of the A_9 column and charged to a Tatoray unit for the production of additional xylenes and benzene. The effluent from the Tatoray unit is sent to a stripper column within the Tatoray unit to remove light ends. After the effluent is clay-treated, it is sent to the BT fractionation section, where the benzene product is recovered and the xylenes are fractionated out and sent to the xylene recovery section. Toluene and heavy aromatics may also be charged to a THDA unit for the production of benzene. The effluent from the THDA unit is stripped to remove light ends, clay-treated, and sent to the BT fractionation section. The overhead material from the Tatoray stripper or THDA stripper column is separated into gas and liquid products. The overhead gas is exported to the fuel gas system, and the overhead liquid is normally recycled back to the CCR Platforming debutanizer for recovery of residual benzene.

The C_8+ fraction from the bottom of the reformate splitter is clay-treated and then charged to a xylene splitter column. The xylene splitter is designed to rerun the mixed xylenes feed to the Parex unit down to very low levels of A_9 concentration. The A_9 builds up in the desorbent circulation loop within the Parex unit, and removing this material upstream in the xylene splitter is more efficient. The overhead from the xylene splitter is charged directly to the Parex unit. The bottoms is sent to the A_9 column, where the A_9 fraction is rerun and then recycled to the Tatoray or THDA unit. If the complex has no Tatoray or THDA unit, the A_9+ material is usually blended into gasoline or fuel oil.

If *ortho*-xylene is to be produced in the complex, the xylene splitter is designed to make a split between *meta*- and *ortho*-xylene and drop a targeted amount of *ortho*-xylene to the bottoms. The xylene splitter bottoms are then sent to an o-X column where high-purity *ortho*-xylene product is recovered overhead. The bottoms from the o-X column are then sent to the A_9 column.

The xylene splitter overhead is sent directly to the Parex unit, where 99.9 wt % pure *para*-xylene is recovered by adsorptive separation at 97 wt % recovery per pass. Any residual toluene in the Parex feed is extracted along with the *para*-xylene, fractionated out in the finishing column within the Parex unit, and then recycled to the Tatoray or THDA unit. The raffinate from the Parex unit is almost entirely depleted of

---

*Trademark and/or service mark of UOP.

para-xylene, to a level of less than 1 wt %. The raffinate is sent to the Isomar unit, where additional *para*-xylene is produced by reestablishing an equilibrium distribution of xylene isomers. Any ethylbenzene in the Parex raffinate is either converted to additional xylenes or dealkylated to benzene, depending on the type of Isomar catalyst used. The effluent from the Isomar unit is sent to a deheptanizer column. The bottoms from the deheptanizer are clay-treated and recycled back to the xylene splitter. In this way, all the C_8 aromatics are continually recycled within the xylene recovery section of the complex until they exit the aromatics complex as *para*-xylene, *ortho*-xylene, or benzene. The overhead from the deheptanizer is split into gas and liquid products. The overhead gas is exported to the fuel gas system, and the overhead liquid is normally recycled back to the CCR Platforming debutanizer for recovery of residual benzene.

Within the aromatics complex, numerous opportunities exist to reduce overall utility consumption through heat integration. Because distillation is the major source of energy consumption in the complex, the use of cross-reboiling is especially effective. This technique involves raising the operating pressure of one distillation column until the condensing distillate is hot enough to serve as the heat source for the reboiler of another column. In most aromatics complexes, the overhead vapors from the xylene splitter are used to reboil the desorbent recovery columns in the Parex unit. The xylene splitter bottoms are often used as a hot-oil belt to reboil either the Isomar deheptanizer or the Tatoray stripper column. If desired, the convection section of many fired heaters can be used to generate steam.

FEEDSTOCK CONSIDERATIONS

Any of the following streams may be used as feedstock to an aromatics complex:

- Straight-run naphtha
- Hydrocracked naphtha
- Mixed xylenes
- Pyrolysis gasoline (pygas)
- Coke-oven light oil
- Condensate
- Liquid petroleum gas (LPG)

Petroleum naphtha is by far the most popular feedstock for aromatics production. Reformed naphtha, or reformate, accounts for 70 percent of total world BTX supply. The pygas by-product from ethylene plants is the next largest source at 23 percent. Coal liquids from coke ovens account for the remaining 7 percent. Pygas and coal liquids are important sources of benzene that may be used only for benzene production or may be combined with reformate and fed to an integrated aromatics complex. Mixed xylenes are also actively traded and can be used to feed a stand-alone Parex-Isomar loop or to provide supplemental feedstock for an integrated complex.

Condensate is a large source of potential feedstock for aromatics production. Although most condensate is currently used as cracker feedstock to produce ethylene, condensate will likely play an increasingly important role in aromatics production in the future.

Many regions of the world have a surplus of low-priced LPG that could be transformed into aromatics by using the new UOP-BP Cyclar* process. In 1994, UOP

---

*Trademark and/or service mark of UOP.

licensed the first Cyclar unit in Saudi Arabia. This Cyclar unit will be integrated with a downstream aromatics complex to produce *para*-xylene, *ortho*-xylene, *meta*-xylene, and benzene.

Pygas composition varies widely with the type of feedstock being cracked in an ethylene plant. Light cracker feeds such as liquefied natural gas (LNG) produce a pygas that is rich in benzene but contains almost no C_8 aromatics. Substantial amounts of C_8 aromatics are found only in pygas from ethylene plants cracking naphtha and heavier feedstocks. All pygas contains significant amounts of sulfur, nitrogen, and dienes that must be removed by two-stage hydrotreating before being processed in an aromatics complex.

Because reformate is much richer in xylenes than pygas, most *para*-xylene capacity is based on reforming petroleum naphtha. Straight-run naphtha is the material that is recovered directly from crude oil by simple distillation. Hydrocracked naphtha, which is produced in the refinery by cracking heavier streams in the presence of hydrogen, is rich in naphthenes and makes an excellent reforming feedstock but is seldom sold on the merchant market. Straight-run naphthas are widely available in the market, but the composition varies with the source of the crude oil. Straight-run naphthas must be thoroughly hydrotreated before being sent to the aromatics complex, but this pretreatment is not as severe as that required for pygas. The CCR Platforming units used in BTX service are run at a high octane severity, typically 104 to 106 RONC, to maximize the yield of aromatics and eliminate the nonaromatic impurities in the C_8+ fraction of the reformate.

Naphtha is characterized by its distillation curve. The "cut" of the naphtha describes which components are included in the material and is defined by the initial boiling point (IBP) and endpoint (EP) of the distillation curve. A typical BTX cut has an IBP of 75°C (165°F) and an EP of 150°C (300°F). However, many aromatics complexes tailor the cut of the naphtha to fit their particular processing requirements.

An IBP of 75 to 80°C (165 to 175°F) maximizes benzene production by including all of the precursors that form benzene in the reforming unit. Prefractionating the naphtha to an IBP of 100 to 105°C (210 to 220°F) minimizes the production of benzene by removing the benzene precursors from the naphtha.

If a UOP Tatoray unit is incorporated into the aromatics complex, C_9 aromatics become a valuable source of additional xylenes. A heavier naphtha with an EP of 165 to 170°C (330 to 340°F) maximizes the C_9 aromatic precursors in the feed to the reforming unit and results in a substantially higher yield of xylenes or *para*-xylene from the complex. Without a UOP Tatoray unit, C_9 aromatics are a low-value byproduct from the aromatics complex that must be blended into gasoline or fuel oil. In this case, a naphtha EP of 150 to 155°C (300 to 310°F) is optimum because it minimizes the C_9 aromatic precursors in the reforming unit feed. If mixed xylenes are purchased as feedstock for the aromatics complex, they must be stripped, clay-treated, and rerun prior to being processed in the Parex-Isomar loop.

CASE STUDY

An overall material balance for a typical aromatics complex is shown in Table 2.1.1 along with the properties of the naphtha feedstock used to prepare the case. The feedstock is a common straight-run naphtha derived from Arabian Light crude. The configuration of the aromatics complex for this case is the same as that shown in Fig. 2.1.4 except that the o-X column has been omitted from the complex to maximize the production of *para*-xylene. The naphtha has been cut at an endpoint of 165°C (330°F) to include all the C_9 aromatic precursors in the feed to the Platforming unit.

TABLE 2.1.1 Overall Material Balance

Naphtha feedstock properties	
Specific gravity	0.7347
Initial boiling point, °C (°F)	83 (181)
Endpoint, °C (°F)	166 (331)
Paraffins/naphthenes/aromatics, vol %	66/23/11

Overall material balance, kMTA*	
Naphtha	940
Products:	
Benzene	164
para-xylene	400
C_{10}+ aromatics	50
Sulfolane raffinate	140
Hydrogen-rich gas	82
LPG	68
Light ends	36

*MTA = metric tons per annum.

TABLE 2.1.2 Investment Cost and Utility Consumption

Estimated erected cost, million $ U.S.	235
Utility consumption:	
Electric power, kW	12,000
High-pressure steam, MT/h* (klb/h)	63 (139)
Medium pressure steam, MT/h (klb/h)	76 (167)
Cooling water, m³/h (gal/min)	1630 (7180)
Fuel fired, million kcal/h (million Btu/h)	207 (821)

*MT/h = metric tons per hour.

A summary of the investment cost and utility consumption for this complex is shown in Table 2.1.2. The estimated erected cost for the complex assumes construction on a U.S. Gulf Coast site in 1995. The scope of the estimate is limited to equipment inside the battery limits of each process unit and includes engineering, procurement, erection of equipment on the site, and the cost of initial catalyst and chemical inventories. The light-ends by-product from the aromatics complex has been shown in the overall material balance. The fuel value of these light ends has not been credited against the fuel requirement for the complex.

COMMERCIAL EXPERIENCE

UOP is the world's leading licenser of aromatics technology. By 1995, UOP had licensed more than 375 separate process units for aromatics production, including 200 extraction units (Udex,* Sulfolane, Tetra,* and Carom*), 59 Parex units, 1 MX Sorbex unit, 41 Isomar units, 37 Tatoray units, 38 THDA units, and 1 Cyclar unit.

---

*Trademark and/or service mark of UOP.

UOP has designed 59 integrated aromatics complexes, which produce both benzene and *para*-xylene. These complexes range in size from 21,000 to 460,000 metric tons per annum (46 to 1014 million pounds) of *para*-xylene.

BIBLIOGRAPHY

Jeanneret, J. J., "Developments in *p*-Xylene Technology," DeWitt Petrochemical Review, Houston, March 1993.

Jeanneret, J. J., "*para*-Xylene Production in the 1990's," UOP Technology Conferences, various locations, May 1995.

Jeanneret, J. J., C. D. Low, and V. Zukauskas, " New Strategies Maximize *para*-Xylene Production," *Hydrocarbon Processing,* June 1994.

CHAPTER 2.2
UOP SULFOLANE PROCESS

John J. Jeanneret
Marketing Services
UOP
Des Plaines, Illinois

INTRODUCTION

The UOP* Sulfolane* process combines liquid-liquid extraction with extractive distillation to recover high-purity aromatics from hydrocarbon mixtures, such as reformed petroleum naphtha (reformate), pyrolysis gasoline (pygas), or coke-oven light oil. Contaminants that are the most difficult to eliminate in the extraction section are easiest to eliminate in the extractive distillation section and vice versa. This hybrid combination of techniques allows Sulfolane units to process feedstocks of much broader boiling range than would be possible by either technique alone. A single Sulfolane unit can be used for the simultaneous recovery of high-purity C_6-C_9 aromatics; individual aromatic components are recovered downstream by simple fractionation.

The Sulfolane process takes its name from the solvent used: tetrahydrothiophene 1,1-dioxide, or sulfolane. Sulfolane was developed as a solvent by Shell in the early 1960s and is still the most efficient solvent available for the recovery of aromatics. Since 1965, UOP has been the exclusive licensing agent for the Sulfolane process. Many of the process improvements incorporated in a modern Sulfolane unit are based on design features and operating techniques developed by UOP.

The Sulfolane process is usually incorporated in an aromatics complex to recover high-purity benzene and toluene products from reformate. In a modern, fully integrated UOP aromatics complex (Fig. 2.2.1), the Sulfolane unit is located downstream of the reformate splitter column. The C_6-C_7 fraction from the overhead of the reformate splitter is fed to the Sulfolane unit. The aromatic extract from the Sulfolane unit is clay-treated to remove trace olefins, and individual benzene and toluene products are recovered by simple fractionation. The paraffinic raffinate from the Sulfolane unit is usually blended into the gasoline pool or used in aliphatic solvents. A complete description of the entire aromatics complex may be found in Chap. 2.1.

---

*Trademark and/or service mark of UOP.

2.14 BASE AROMATICS PRODUCTION PROCESSES

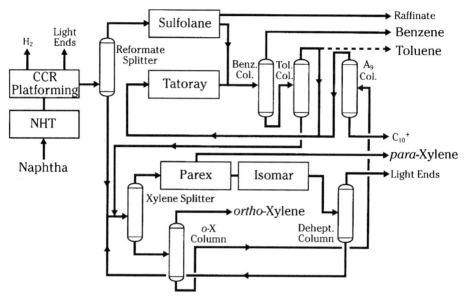

FIGURE 2.2.1 Integrated UOP aromatics complex.

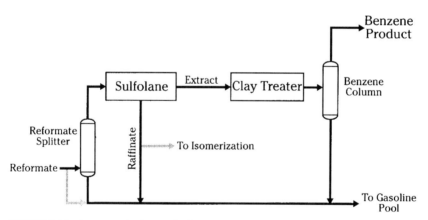

FIGURE 2.2.2 Benzene-reduction application.

The Sulfolane process can also be an attractive way to reduce the benzene concentration in a refinery's gasoline pool so that it meets new reformulated gasoline requirements. In a typical benzene-reduction application (Fig. 2.2.2), a portion of the debutanized reformate is sent to a reformate splitter column. The amount of reformate sent to the splitter is determined by the degree of benzene reduction required. Bypassing some reformate around the splitter and recombining it with splitter bottoms provides control of the final benzene concentration. The benzene-rich splitter overhead is sent to the Sulfolane unit, which produces a high-purity benzene product that can be sold to the petrochemical market. The raffinate from the Sulfolane unit can be blended back into the gasoline pool or upgraded in an isomerization unit.

SOLVENT SELECTION

The suitability of a solvent for aromatics extraction involves the relationship between the capacity of the solvent to absorb aromatics (solubility) and the ability of the solvent to differentiate between aromatics and nonaromatics (selectivity). A study of the common polar solvents used for aromatic extraction reveals the following qualitative similarities:

- When hydrocarbons containing the same number of carbon atoms are compared, solubilities decrease in this order: *aromatics>naphthenes>paraffins.*
- When hydrocarbons in the same homologous series are compared, solubility decreases as molecular weight increases.
- The selectivity of a solvent decreases as the hydrocarbon content, or loading, of the solvent phase increases.

In spite of these general similarities, various commercial solvents used for aromatics recovery have significant quantitative differences. Sulfolane demonstrates better aromatic solubilities at a given selectivity than any other commercial solvent. The practical consequence of these differences is that an extraction unit designed to use sulfolane solvent requires a lower solvent circulation rate and thus consumes less energy.

In addition to superior solubility and selectivity, sulfolane solvent has three particularly advantageous physical properties that have a significant impact on plant investment and operating cost:

- *High specific gravity* (1.26).   High specific gravity allows the aromatic capacity of sulfolane to be fully exploited while maintaining a large density difference between the hydrocarbon and solvent phases in the extractor. This large difference in densities minimizes the required extractor diameter. The high density of the liquid phase in the extractive distillation section also minimizes the size of the equipment required there.
- *Low specific heat* [0.4 cal/g · °C (0.4 Btu/lb · °F)].   The low specific heat of sulfolane solvent reduces heat loads in the fractionators and minimizes the duty on solvent heat exchangers.
- *High boiling point* [287°C (549°F)].   The boiling point of sulfolane is significantly higher than the heaviest aromatic hydrocarbon to be recovered, facilitating the separation of solvent from the aromatic extract.

PROCESS CONCEPT

The Sulfolane process is distinct from both conventional liquid-liquid extraction systems and commercial extractive distillation processes in that it combines both techniques in the same process unit. This mode of operation has particular advantages for aromatic recovery:

- In liquid-liquid extraction systems, light nonaromatic components are more soluble in the solvent than heavy nonaromatics. Thus, liquid-liquid extraction is more effective in separating aromatics from the heavy contaminants than from the light ones.
- In extractive distillation, light nonaromatic components are more readily stripped from the solvent than heavy nonaromatics. Thus, extractive distillation is more

effective in separating aromatics from the light contaminants than from the heavy ones.

Therefore, liquid-liquid extraction and extractive distillation provide complementary features. Contaminants that are the most difficult to eliminate in one section are the easiest to remove in the other. This combination of techniques permits effective treatment of feedstocks with much broader boiling range than would be possible by either technique alone.

The basic process concept is illustrated in Fig. 2.2.3. Lean solvent is introduced at the top of the main extractor and flows downward. The hydrocarbon feed is introduced at the bottom and flows upward, countercurrent to the solvent phase. As the solvent phase flows downward, it is broken up into fine droplets and redispersed into the hydrocarbon phase by each successive tray. The solvent selectively absorbs the aromatic components from the feed. However, because the separation is not ideal, some of the nonaromatic impurities are also absorbed. The bulk of the nonaromatic hydrocarbons remain in the hydrocarbon phase and are rejected from the main extractor as raffinate.

The solvent phase, which is rich in aromatics, flows downward from the main extractor into the backwash extractor. There the solvent phase is contacted with a stream of light nonaromatic hydrocarbons from the top of the extractive stripper. The

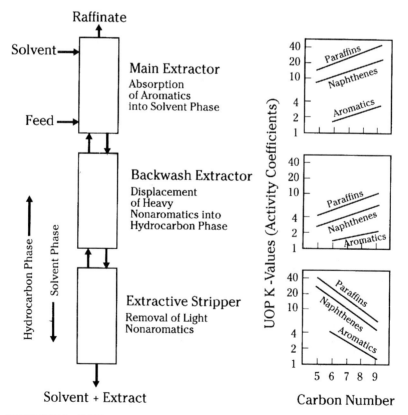

FIGURE 2.2.3 Sulfolane process concept.

light nonaromatics displace the heavy nonaromatic impurities from the solvent phase. The heavy nonaromatics then reenter the hydrocarbon phase and leave the extractor with the raffinate.

The rich solvent from the bottom of the backwash extractor, containing only light nonaromatic impurities, is then sent to the extractive stripper for final purification of the aromatic product. The light nonaromatic impurities are removed overhead in the extractive stripper and recycled to the backwash extractor. A purified stream of aromatics, or extract, is withdrawn in the solvent phase from the bottom of the extractive stripper. The solvent phase is then sent on to the solvent recovery column, where the extract product is separated from the solvent by distillation.

Also shown in Fig. 2.2.3 are the activity coefficients, or K values, for each section of the separation. The K value in extraction is analogous to relative volatility in distillation. K_i is a measure of the solvent's ability to repel component i and is defined as the mole fraction of component i in the hydrocarbon phase, X_i, divided by the mole fraction of component i in the solvent phase, Z_i. The lower the value of K_i, the higher the solubility of component i in the solvent phase.

DESCRIPTION OF THE PROCESS FLOW

Fresh feed enters the extractor and flows upward, countercurrent to a stream of lean solvent, as shown in Fig. 2.2.4. As the feed flows through the extractor, aromatics are selectively dissolved in the solvent. A raffinate stream, very low in aromatics content, is withdrawn from the top of the extractor.

The rich solvent, loaded with aromatics, exits the bottom of the extractor and enters the stripper. The nonaromatic components having volatilities higher than that of benzene are completely separated from the solvent by extractive distillation and removed overhead along with a small quantity of aromatics. This overhead stream is recycled to the extractor, where the light nonaromatics displace the heavy nonaromatics from the solvent phase leaving the bottom of the extractor.

The stripper bottoms stream, which is substantially free of nonaromatic impurities, is sent to the recovery column, where the aromatic product is separated from the solvent. Because of the large difference in boiling point between the sulfolane solvent and the heaviest aromatic component, this separation is accomplished, with minimal energy input. To minimize solvent temperatures, the recovery column is operated under vacuum. Lean solvent from the bottom of the recovery column is returned to the extractor. The extract is recovered overhead and sent on to distillation columns downstream for recovery of the individual benzene and toluene products.

The raffinate stream exits the top of the extractor and is directed to the raffinate wash column. In the wash column, the raffinate is contacted with water to remove dissolved solvent. The solvent-rich water is vaporized in the water stripper by exchange with hot circulating solvent and then used as stripping steam in the recovery column. Accumulated solvent from the bottom of the water stripper is pumped back to the recovery column.

The raffinate product exits the top of the raffinate wash column. The amount of sulfolane solvent retained in the raffinate is negligible. The raffinate product is commonly used for gasoline blending or aliphatic solvent applications.

Under normal operating conditions, sulfolane solvent undergoes only minor oxidative degradation. A small solvent regenerator is included in the design of the unit as a safeguard against the possibility of air leaking into the unit. During normal operation, a small slipstream of circulating solvent is directed to the solvent regenerator for removal of oxidized solvent.

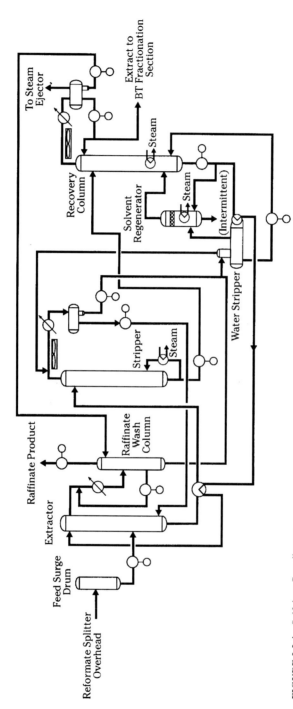

FIGURE 2.2.4 Sulfolane flow diagram.

The extract product from a Sulfolane unit may contain trace amounts of olefins and other impurities that would adversely affect the acid-wash color tests of the final benzene and toluene products. To eliminate these trace impurities, the extract is clay-treated prior to fractionation. Because clay treating is done at mild conditions, clay consumption is minimal.

The treated extract is directed to the aromatics fractionation section, where high-purity benzene, toluene, and sometimes mixed xylenes are recovered. The design of the aromatics fractionation section varies depending on the particular processing requirements of the refiner. The toluene product is often recycled back to a UOP Tatoray* unit for conversion into benzene and xylenes. Mixed xylenes may be routed directly to the xylene recovery section of the plant for separation into *para*-xylene, *ortho*-xylene, and *meta*-xylene products.

Any heavy aromatics in the feed are yielded as a bottoms product from the fractionation section. In most cases, the C_9 aromatics are recovered and recycled to a UOP Tatoray unit for the production of additional xylenes. The heavy aromatics may also be blended back into the refinery gasoline pool or sold as a high-octane blending component.

FEEDSTOCK CONSIDERATIONS

The feed to a Sulfolane unit is usually a benzene-toluene (BT) cut from a naphtha reforming unit. The xylene fraction of the reformate is often already pure enough to sell as mixed xylenes or is sent directly to the *para*-xylene recovery section of the aromatics complex. In many facilities, the pygas by-product from a nearby ethylene plant is also directed to a Sulfolane unit. A few plants also use sulfolane to recover aromatics from coke-oven light oil. Before being sent to a Sulfolane unit, the reformate must first be stripped in a debutanizer column to remove light ends. Pygas and coke-oven light oils must first be hydrotreated to remove dienes, olefins, sulfur, and nitrogen. In general, the feed to a Sulfolane unit should meet the specifications outlined in Table 2.2.1.

---

*Trademark and/or service mark of UOP.

TABLE 2.2.1 Sulfolane Feedstock Specifications

Contaminant	Effect	Limit
Total sulfur	Contaminates product	0.2 ppm max.
Thiophene	Contaminates product	0.2 ppm max.
Total chloride	Contaminates product, causes corrosion	0.2 ppm max.
Bromine index	Causes higher solvent circulation, increased utility consumption	2 max.
Diene index	Causes higher solvent circulation, increased utility consumption	1 max.
Dissolved oxygen	Causes degradation of solvent, irreversible	1.0 ppm max.

PROCESS PERFORMANCE

The performance of the UOP Sulfolane process has been well demonstrated in more than 100 operating units. The recovery of benzene exceeds 99.9 wt %, and recovery of toluene is typically 99.8 wt %. The Sulfolane process is also efficient at recovering heavier aromatics if necessary. Typical recovery of xylenes exceeds 98 wt %, and a recovery of 99 wt % has been demonstrated commercially with rich feedstocks.

UOP Sulfolane units routinely produce a benzene product with a solidification point of 5.5°C or better, and many commercial units produce benzene containing less than 100 ppm nonaromatic impurities. The toluene and C_8 aromatics products from a Sulfolane unit are also of extremely high purity and easily exceed nitration-grade specifications. In fact, the ultimate purities of all of the aromatic products are usually more dependent on the design and proper operation of the downstream fractionation section than on the extraction efficiency of the Sulfolane unit itself.

The purity and recovery performance of an aromatics extraction unit is largely a function of energy consumption. In general, higher solvent circulation rates result in better performance, but at the expense of higher energy consumption. The UOP Sulfolane process demonstrates the lowest energy consumption of any commercial aromatics extraction technology. A typical UOP Sulfolane unit consumes 275 to 300 kcal of energy per kilogram of extract produced, even when operating at 99.99 wt % benzene purity and 99.95 wt % recovery. UOP Sulfolane units are also designed to efficiently recover solvent for recycle within the unit. Expected solution losses of sulfolane solvent are less than 5 ppm of the fresh feed rate to the unit.

EQUIPMENT CONSIDERATIONS

The extractor uses rain-deck trays to contact the upward-flowing feed with the downward-flowing solvent. The rain-deck trays act as distributors to maintain an evenly dispersed "rain" of solvent droplets moving down through the extractor to facilitate dissolution of the aromatic components into the solvent phase. A typical Sulfolane extractor column contains 94 rain-deck trays.

The raffinate wash column is used to recover residual solvent carried over in the raffinate from the extractor. The wash column uses jet-deck trays to provide counter-current flow between the wash water and raffinate. A typical wash column contains eight jet-deck trays.

The stripper column is used to remove any light nonaromatic hydrocarbons in the rich solvent by extractive distillation. The sulfolane solvent increases the relative volatilities between the aromatic and nonaromatic components, thus facilitating the removal of light nonaromatics in the column overhead. A typical stripper column contains 34 sieve trays. The recovery column separates the aromatic extract from the Sulfolane solvent by vacuum distillation. A typical recovery column contains 34 valve trays.

The solvent regenerator is a short, vertical drum that is used to remove the polymers and salts formed as a result of the degradation of solvent by oxygen. The regenerator is operated under vacuum and runs continuously.

The Sulfolane process is highly heat integrated. Approximately 11 heat exchangers are designed into a typical unit.

All of the equipment for the Sulfolane unit, with the exception of the solvent regenerator reboiler, is specified as carbon steel. The solvent regenerator reboiler is constructed of stainless steel.

TABLE 2.2.2 Investment Cost and Utility Consumption*

Estimated erected cost, million $U.S.	12.5
Solvent makeup requirement, $U.S. per day	850
Utility consumption:	
Electric power, kW	390
High-pressure steam, MT/h (klb/h)	27.5 (60.6)
Cooling water, m³/h (gal/min)	274 (1207)

*Basis: 25.0 MT/h of toluene product, 11.8 MT/h of benzene product, 54.5 MT/h (10,400 BPD) of BT reformate feedstock.

Note: MT/h = metric tons per hour; BPD = barrels per day.

CASE STUDY

A summary of the investment cost and utility consumption for a typical Sulfolane unit is shown in Table 2.2.2. The basis for this case is a Sulfolane unit processing 54.5 metric tons per hour (MT/h) [10,400 barrels per day (BPD)] of a BT reformate cut. This case corresponds to the case study for an integrated UOP aromatics complex in Chap. 2.1 of this handbook. The investment cost is limited to the Sulfolane unit itself and does not include downstream fractionation. The estimated erected cost for the Sulfolane unit assumes construction on a U.S. Gulf Coast site in 1995. The scope of the estimate includes engineering, procurement, erection of equipment on the site, and the initial inventory of sulfolane solvent.

COMMERCIAL EXPERIENCE

Since the early 1950s, UOP has licensed four different aromatics extraction technologies, including the Udex,* Sulfolane, Tetra,* and Carom* processes. UOP's experience in aromatics extraction encompasses more than 200 units, which range in size from 2 to 260 MT/h (400 to 50,000 BPD) of feedstock.

In 1952, UOP introduced the first large-scale aromatics extraction technology, the Udex process, which was jointly developed by UOP and Dow Chemical. Although the Udex process uses either diethylene glycol or triethlyene glycol as a solvent, it is similar to the Sulfolane process in that it combines liquid-liquid extraction with extractive distillation. Between 1950 and 1965, UOP licensed a total of 82 Udex units.

In the years following the commercialization of the Udex process, considerable research was done with other solvent systems. In 1962, Shell commercialized the first Sulfolane units at their refineries in England and Italy. The success of these units led to an agreement in 1965 whereby UOP became the exclusive licenser of the Shell Sulfolane process. Many of the process improvements incorporated in modern Sulfolane units are based on design features and operating techniques developed by UOP. By 1995, UOP had licensed a total of 120 Sulfolane units throughout the world.

Meanwhile, in 1968, researchers at Union Carbide discovered that tetraethylene glycol had a higher capacity for aromatics than the solvents being used in existing Udex units. Union Carbide soon began offering this improved solvent as the Tetra process. Union Carbide licensed a total of 17 Tetra units for aromatics extraction; 15

---
*Trademark and/or service mark of UOP.

of these units were originally UOP Udex units that were revamped to take advantage of the improvements offered by the Tetra process.

Union Carbide then commercialized the Carom process in 1986. The Carom flow scheme is similar to that used in the Udex and Tetra processes, but the Carom process takes advantage of a unique two-component solvent system that nearly equals the performance of the sulfolane solvent. In 1988, UOP merged with the CAPS division of Union Carbide. As a result of this merger, UOP now offers both the Sulfolane and Carom processes for aromatics extraction and continues to support the older Udex and Tetra technologies.

The Carom process is ideal for revamping older Udex and Tetra units for higher capacity, lower energy consumption, or better product purity. The Carom process can also be competitive with the Sulfolane process for new-unit applications. By 1995, UOP had licensed a total of five Carom units. Four of these units are conversions of Udex or Tetra units, and one is a new unit.

BIBLIOGRAPHY

Jeanneret, J. J., P. Fortes, T. L. LaCosse, V. Sreekantham, and T. J. Stoodt, "Sulfolane and Carom Processes: Options for Aromatics Extraction," UOP Technology Conferences, various locations, September 1992.

CHAPTER 2.3
UOP THERMAL HYDRODEALKYLATION (THDA) PROCESS

W. L. Liggin
UOP
Des Plaines, Illinois

INTRODUCTION

The importance of benzene as an intermediate in the production of organic-based materials is exceeded only by that of ethylene. Benzene represents the basic building block for direct or indirect manufacture of well over 250 separate products or product classifications.

Historically, the major consumption of benzene has been in the production of ethylbenzene, cumene, and cyclohexane. Significant quantities of benzene are also consumed in the manufacture of aniline, detergent alkylate, and maleic anhydride.

At the present time, approximately 92 percent of the benzene produced worldwide comes directly from petroleum sources. Catalytic reforming supplies most of the petroleum-derived petrochemical benzene. However, toluene is produced in greater quantities than benzene in the reforming operation, and in many areas, low market demand for toluene can make its conversion to benzene via dealkylation economically attractive. Approximately 13 percent of the petrochemical benzene produced in the world is derived from toluene dealkylation.

The thermal hydrodealkylation (THDA) process provides an efficient method for the conversion of alkylbenzenes to high-purity benzene. In addition to producing benzene, the THDA process can be economically applied to the production of quality naphthalene from suitable feedstocks.

PROCESS DESCRIPTION

The UOP* THDA process converts alkylbenzenes and alkylnaphthalenes to their corresponding aromatic rings, benzene and naphthalene. The relation between product distribution and operating severity is such that, for both benzene and naphthalene operations, the conversion per pass of fresh feed is maintained at somewhat less than 100 percent. A simplified process flow diagram for benzene manufacture is presented in Fig. 2.3.1.

The alkyl-group side chains of the alkyl-aromatic feed as well as nonaromatics that may be present in the unit feed are converted to a light paraffinic coproduct gas consisting mainly of methane. The basic hydrodealkylation reaction enables the process to produce a high-purity benzene or naphthalene product without applying extraction or superfractionation techniques even when charging a mixture of alkyl aromatics and nonaromatic hydrocarbons. Excessive nonaromatics in the charge significantly add to hydrogen consumption.

Product yields approach stoichiometric with benzene yield from toluene approximating 99 percent on a molal basis. A small amount of heavy-aromatic material consisting of biphenyl-type compounds is coproduced.

In a benzene unit, fresh toluene feedstock is mixed with recycle toluene and recycle and fresh hydrogen gases, heated by exchange in a fired heater, and then charged to the reactor. Alkyl aromatics are hydrodealkylated to benzene and nonaromatics, and paraffins and naphthalenes are hydrocracked. The effluent from the reactor is cooled and directed to the product separator, where it separates into a liquid phase and gas phase. The hydrogen-rich gas phase is recycled to the reactor, and the separator liquid

---

*Trademark and/or service mark of UOP.

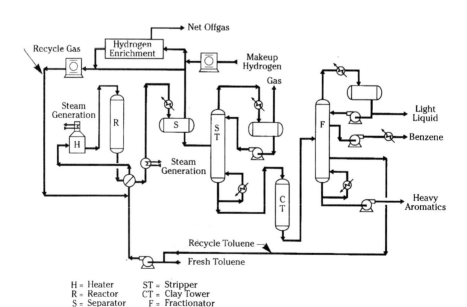

H = Heater ST = Stripper
R = Reactor CT = Clay Tower
S = Separator F = Fractionator

FIGURE 2.3.1 UOP THDA process for benzene production.

is charged to a stripper for the removal of light ends. Stripper bottoms are percolated through a clay treater to the fractionation section, where high-purity benzene is obtained as an upper sidecut from a benzene fractionation column. Unconverted toluene is recycled to the reactor from the lower sidecut of the benzene column. Heavy-aromatic by-product is withdrawn from the bottom of the column to storage.

The reactor-section process flow in a naphthalene THDA unit is similar to that described for the benzene unit. Fresh feed is mixed with unconverted recycle alkyl aromatics and makeup and recycle hydrogen. The mixture is then heated and charged to the reactor. Materials in the feedstock materials that boil close to naphthalene would make the recovery of high-purity product either impossible or uneconomic if they remain unconverted. Process conditions are set to ensure that these materials are hydrocracked or dealkylated, or both, to products easily separated by fractionation.

In the case of naphthalene, the aromatic-splitter bottoms are charged to a naphthalene splitter, where the small amount of heavy-aromatic coproduct is rejected as a bottoms product. Naphthalene-splitter overhead is directed to the naphthalene fractionator, where high-purity naphthalene is recovered as an overhead product. Naphthalene-fractionator bottoms are recycled to the reactor section. In both benzene and naphthalene THDA units, clay treating of the product is generally required to meet the usual acid-wash color specifications.

Several design options are available for optimizing hydrogen usage in both types of units. Coproduct light ends, primarily methane, must be removed from the reaction section to maintain hydrogen purity. When the supply of makeup hydrogen is limited, consideration also must be given to the elimination of C_3 and heavier nonaromatic hydrocarbons from the makeup gas. If present, these materials hydrocrack and substantially increase hydrogen consumption. During the design stage of hydrodealkylation units, careful attention must be given to hydrogen consumption and availability as related to overall refinery operation.

Depending on the application, THDA units can process a wide variety of feedstocks. For the production of benzene, feedstocks could include extracted light alkylbenzene, suitably treated coke-oven light oil, and pyrolysis coproducts. Feedstocks to produce naphthalene could include heavy reformate, extracted cycle oils from the fluid catalytic cracking (FCC) process, and coal-tar-derived materials.

Benzene produced from commercial THDA units typically has a freeze point of 5.5°C, which exceeds nitration-grade benzene specifications.

PROCESS ECONOMICS

Although THDA yields are about 99 percent on a molar basis, they are considerably lower on a weight basis because of the change in molecular weight. Weight yields for the dealkylation of toluene to benzene are shown in Table 2.3.1. Investment and utility requirements are shown in Table 2.3.2.

The economics of benzene manufacture via the THDA process are very sensitive to the relative prices of benzene and toluene. As a general rule, THDA becomes economically viable when the price of benzene (per unit volume) is more than 1.25 times the price of toluene. For this reason, the THDA process has become the process used to meet benzene demand during peak periods. When benzene is in low demand, THDA units are not operated. However, a UOP-designed THDA is easily revamped at low cost to a Tatoray process unit. This flexibility greatly extends the utilization of expensive processing equipment and provides a means of generating a wider product state (for example, benzene and mixed xylenes) during periods of low benzene demand.

TABLE 2.3.1 THDA Yields

Benzene production	Feeds, wt %	Product, wt %
Hydrogen (chemical consumption)	2.3	
Methane		17.7
Ethane		0.6
Benzene		83.6
Toluene	100	
Heavy aromatics		0.4
Total	102.3	102.3

TABLE 2.3.2 THDA Process Investment and Utility Requirements*

Estimated battery-limits erected cost	$5.3 million
Utilities:	
Electric power, kW	69
Fuel, 10^6 kcal/h (10^6 Btu/h)	1.8 (7.2)
Steam, MT/h (klb/h)	0.69 (1.5)
Cooling water, m^3/h (gal/min)	13.7 (60)

*Basis: 50,000 MTA (1095 BPSD) toluene feed

Note: MT/h = metric tons per hour; MTA = metric tons per annum; BPSD = barrels per stream day.

CHAPTER 2.4
BP-UOP CYCLAR PROCESS

John J. Jeanneret
UOP
Des Plaines, Illinois

INTRODUCTION

In recent years, light hydrocarbons have become increasingly attractive as fuels and petrochemical feedstocks, and much effort has been devoted to improving the recovery, processing, and transportation of liquefied petroleum gas (LPG) and natural gas. Because production areas are often located in remote areas that are far removed from established processing plants or consumers, elaborate product-transport infrastructures are required. Although natural gas can be moved economically through pipelines, condensation problems limit the amount of LPG that can be transported in this way. Thus, most LPG is transported by such relatively expensive means as special-purpose tankers or rail cars. The high cost of transporting LPG can often depress its value at the production site. This statement is especially true for propane, which is used much less than butane for gasoline blending and petrochemical applications.

British Petroleum (BP) recognized the problem with transporting LPG and in 1975 began research on a process to convert LPG into higher-value liquid products that could be shipped more economically. This effort led to the development of a catalyst that was capable of converting LPG into petrochemical-grade benzene, toluene, and xylenes (BTX) in a single step. However, BP soon realized that the catalyst had to be regenerated often in this application and turned to UOP* for its well-proven CCR* technology, which continuously regenerates the catalyst. UOP developed a high-strength formulation of the BP catalyst that would work in CCR service and also applied the radial-flow, stacked-reactor design originally developed for the UOP Platforming* process. The result of this outstanding technical collaboration is the BP-UOP Cyclar* process.

---

*Trademark and/or service mark of UOP.

PROCESS CHEMISTRY

The Cyclar process converts LPG directly to a liquid aromatics product in a single operation. The reaction is best described as dehydrocyclodimerization and is thermodynamically favored at temperatures above 425°C (800°F). The dehydrogenation of light paraffins (propane and butanes) to olefins is the rate-limiting step (Fig. 2.4.1). Once formed, the highly reactive olefins oligomerize to form larger intermediates, which then rapidly cyclize to naphthenes. These reactions—dehydrogenation, oligomerization, and cyclization—are all acid-catalyzed. The shape selectivity of the zeolite component of the catalyst also promotes the cyclization reaction and limits the size of the rings formed. The final reaction step is the dehydrogenation of the naphthenes to their corresponding aromatics. This reaction is highly favored at Cyclar operating conditions, and the result is virtually complete conversion of the naphthenes.

The reaction intermediates can also undergo a hydrocracking side reaction to form methane and ethane. This side reaction results in a loss of yield because methane and ethane are inert at Cyclar operating conditions.

Because olefins are a key reaction intermediate, they can of course be included in the feed to the Cyclar unit. Heavier paraffins, such as pentanes, can also be included in the feed. Olefins and pentanes are almost completely converted in the Cyclar unit, but the unit must be designed to handle them because they result in a higher catalyst-coking rate than pure butane and propane feedstocks.

Although the reaction sequence involves some exothermic steps, the preponderance of dehydrogenation reactions results in a highly endothermic overall reaction. Five moles of hydrogen are produced for every mole of aromatic components formed from propane or butane.

Because propane and butanes are relatively unreactive, the Cyclar process requires a catalyst with high activity. At the same time, the production of methane and ethane from unwanted hydrocracking side reactions must be minimized. An extensive joint effort by BP and UOP has resulted in a catalyst that combines several important features to ensure efficient commercial operation:

- At the conditions necessary for high selectivity to aromatics, the conversion performance of the catalyst declines slowly.

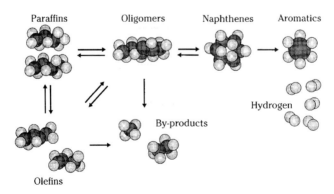

FIGURE 2.4.1 Cyclar reaction mechanism.

- The selectivity to aromatics is nearly constant over the normal range of conversion, resulting in stable product yield and quality. Thus, economic process performance can be maintained despite normal fluctuations in unit operation.
- At normal process conditions, the rate of carbon deposition on the catalyst is slow and steady, amounting to less than 0.02 wt % of the feed processed. Because the carbon levels on spent catalyst are low, regeneration requirements are relatively mild. Mild regeneration conditions extend the life of the catalyst and make it insensitive to process upsets and changes in feedstock composition.
- The catalyst exhibits high thermal stability and is relatively insensitive to common feedstock contaminants. Regeneration fully restores the activity and selectivity of the catalyst to the performance seen with fresh catalyst.
- High mechanical strength and low attrition characteristics make the catalyst well suited for continuous catalyst regeneration.

DESCRIPTION OF THE PROCESS FLOW

A Cyclar unit is divided into three major sections. The reactor section includes the radial-flow reactor stack, combined feed exchanger, charge heater, and interheaters. The regenerator section includes the regenerator stack and catalyst transfer system. The product recovery section includes the product separators, compressors, stripper, and gas recovery equipment.

The flow scheme is similar to the UOP CCR Platforming process, which is used widely throughout the world for reforming petroleum naphtha. A simplified block flow diagram is shown in Fig. 2.4.2. Fresh feed and recycle are combined and heat exchanged against reactor effluent. The combined feed is then raised to reaction temperature in the charge heater and sent to the reactor section. Four adiabatic, radial-flow reactors are arranged in a vertical stack. Catalyst flows vertically by gravity down the stack, and the charge flows radially across the annular catalyst beds. Between each reactor, the vaporized charge is reheated to reaction temperature in an interheater.

The effluent from the last reactor is split into vapor and liquid products in a product separator. The liquid is sent to a stripper, where light saturates are removed from the C_6+ aromatic product. Vapor from the product separator is compressed and sent to a gas recovery section, typically a cryogenic unit, for separation into a 95 percent pure hydrogen product stream, a fuel gas stream of light saturates, and a recycle stream of unconverted LPG. Hydrogen is not recycled.

Because coke builds up on the Cyclar catalyst over time at reaction conditions, partially deactivated catalyst is continually withdrawn from the bottom of the reactor stack for regeneration. Figure 2.4.3 shows additional details of the catalyst regeneration section. A discrete amount of spent catalyst flows into a lock hopper, where it is purged with nitrogen. The purged catalyst is then lifted with nitrogen to the disengaging hopper at the top of the regenerator. The catalyst flows down through the regenerator, where the accumulated carbon is burned off. Regenerated catalyst flows down into the second lock hopper, where it is purged with hydrogen and then lifted with hydrogen to the top of the reactor stack. Because the reactor and regenerator sections are separate, each operates at its own optimal conditions. In addition, the regeneration section can be temporarily shut down for maintenance without affecting the operation of the reactor and product recovery sections.

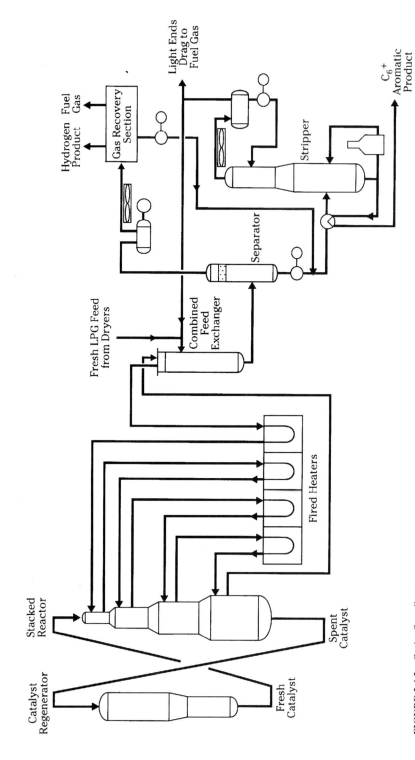

FIGURE 2.4.2 Cyclar flow diagram.

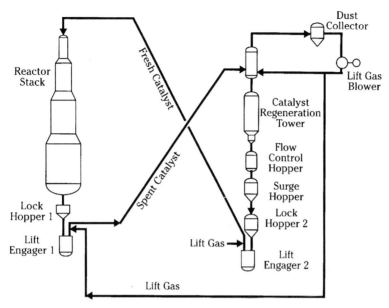

FIGURE 2.4.3 Catalyst regeneration section.

FEEDSTOCK CONSIDERATIONS

Propane and butanes should be the major components in the feedstock to a Cyclar unit. The C_1 and C_2 saturates should be minimized because these components act as inert diluents. Olefins should be limited to less than 10 percent of the feed. Higher concentrations of olefins require hydrogenation of the feed. The C_5 and C_6 components increase the rate of coke formation in the process and should be limited to less than 20 wt % and 2 wt %, respectively, in Cyclar units designed for LPG service. Cyclar units can be designed to process significantly higher amounts of C_5 and C_6 materials if necessary. In general, the feed to a Cyclar unit should meet the specifications outlined in Table 2.4.1.

TABLE 2.4.1 Feedstock Specifications

Contaminant	Limit
Sulfur	<20 mol ppm
Water	No free water
Oxygenates	<10 wt ppm
Basic nitrogen	<1 wt ppm
Fluorides	<0.3 wt ppm
Metals	<50 wt ppb

PROCESS PERFORMANCE

The major liquid products from a Cyclar process unit are BTX and C_9+ aromatics. These products may be separated from one another by conventional fractionation downstream of the Cyclar stripper column.

In general, aromatics yield increases with the carbon number of the feedstock. In a low-pressure operation, the overall aromatics yield increases from 62 wt % of fresh feed with an all-propane feedstock to 66 percent with an all-butane feed. With this yield increase comes a corresponding decrease in fuel gas production. These yield figures can be interpolated linearly for mixed propane and butane feedstocks. The distribution of butane isomers in the feed has no effect on yields.

The distribution of aromatic components in the liquid product is also affected by feedstock composition. Butane feedstocks produce a product that is leaner in benzene and richer in xylenes than that produced from propane (Fig. 2.4.4). With either propane or butane feeds, the liquid product contains about 91 percent BTX and 9 percent heavier aromatics.

The Cyclar unit produces aromatic products with nonaromatic impurities limited to 1500 ppm or less. Thus, marketable, high-quality, petrochemical-grade BTX can be obtained by fractionation alone, without the need for subsequent extraction.

The by-product light ends contain substantial amounts of hydrogen, which may be recovered in several different ways, depending on the purity desired:

- An absorber-stripper system produces a 65 mol % hydrogen product stream.
- A cold box produces 95 mol % hydrogen.
- An absorber-stripper system combined with a pressure-swing absorption (PSA) unit produces 99 mol % hydrogen.
- A cold box combined with a PSA unit is usually more attractive if large quantities of 99+ mol % hydrogen is desired.

EQUIPMENT CONSIDERATIONS

The principal Cyclar operating variables are feedstock composition, pressure, space velocity, and temperature. The temperature must be high enough to ensure nearly complete conversion of reaction intermediates to produce a liquid product that is essentially free of nonaromatic impurities but low enough to minimize nonselective

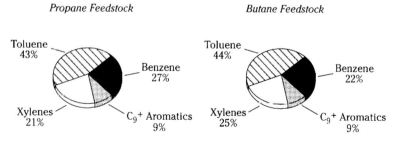

FIGURE 2.4.4 Cyclar aromatic product distribution.

thermal reactions. Space velocity is optimized against conversion within this temperature range to obtain high product yields with minimum operating costs.

Reaction pressure has a big impact on process performance. Higher pressure increases reaction rates, thus reducing catalyst requirements. However, some of this higher reactivity is due to increased hydrocracking, which reduces aromatic product yield. UOP currently offers two alternative Cyclar process designs. The low-pressure design is recommended when maximum aromatics yield is desired. The high-pressure design requires only half the catalyst and is attractive when minimum investment and operating costs are the overriding considerations (Fig. 2.4.5).

Various equipment configurations are possible depending on whether a gas turbine, steam turbine, or electric compressor drive is specified, whether air-cooling or water-cooling equipment is preferred, and whether steam generation is desirable.

CASE STUDY

The overall material balance, investment cost, and utility consumption for a representative Cyclar unit are shown in Table 2.4.2. The basis for this case is a low-pressure Cyclar unit processing 54 metric tons per hour (MT/h) [15,000 barrels per day (BPD)] of a feed consisting of 50 wt % propane and 50 wt % butanes. The investment cost is limited to the Cyclar unit and stripper column and does not include further downstream product fractionation. The estimated erected cost for the Cyclar unit assumes construction on a U.S. Gulf Coast site in 1995. The scope of the estimate includes engineering, procurement, erection of equipment on-site, and the initial load of Cyclar catalyst.

COMMERCIAL EXPERIENCE

The UOP CCR technology, first commercialized in 1971 for the Platforming process has been applied to the Oleflex* and Cyclar process technologies. More than 100 CCR

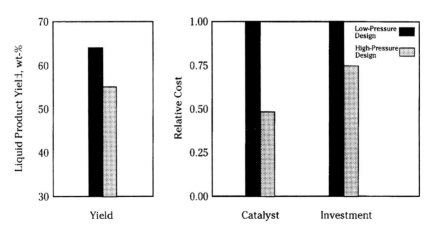

FIGURE 2.4.5 Effect of Cyclar operating pressure.

---

*Trademark and/or service mark of UOP.

TABLE 2.4.2 Material Balance and Investment Cost*

Overall material balance	MTA
LPG feedstock	430,000
Products:	
Benzene	66,700
Toluene	118,800
Mixed xylenes	64,000
C_9+ aromatics	24,600
Hydrogen (95 mol %)	29,400
Fuel gas	126,500
Estimated erected cost, million $U.S.	79.0

Utility consumption:	(MTA)
Electric power, kW	5,500
High-pressure steam, MT/h	27 (credit)
Low-pressure steam, MT/h	7
Boiler feed water, MT/h	33
Cooling water, m^3/h	640
Fuel fired, million kcal/h	70

*Basis: 54 ton/h (15,000 BPD) of LPG feedstock. Feed composition: 50 wt % propane, 50 wt % butanes.

Note: MTA = metric tons per annum; MT/h = metric tons per hour; BPD = barrels per day.

units are currently operating throughout the world. The combination of a radial-flow, stacked reactor and a continuous catalyst regenerator has proven to be extremely reliable. On-stream efficiencies of more than 95 percent are routinely achieved in commercial CCR Platforming units.

BP commissioned the first commercial-scale Cyclar unit at its refinery in Grangemouth, Scotland, in January 1990. This demonstration unit was designed to process 30,000 metric tons per annum (MTA) of propane or butane feedstock at either high or low pressure over a wide range of operating conditions. The demonstration effort was a complete success because it proved all aspects of the Cyclar process on a commercial scale and supplied sufficient data to confidently design and guarantee future commercial units. The Cyclar unit at Grangemouth demonstration unit was dismantled in 1992 after completion of the development program.

In 1995, UOP licensed the first Cyclar-based aromatics complex in the Middle East. This Cyclar unit is a low-pressure design that is capable of converting 1.3 million MTA of LPG into aromatics. The associated aromatics complex is designed to produce 350,000 MTA of benzene, 300,000 MTA of *para*-xylene, and 80,000 MTA of *ortho*-xylene. This new aromatics complex is due on-stream by the end of 1997.

BIBLIOGRAPHY

Doolan, P. C., "Cyclar: LPG to Valuable Aromatics," DeWitt Petrochemical Review, Houston, March 1989.

Doolan, P. C., and P. R. Pujado, "Make Aromatics from LPG," *Hydrocarbon Processing*, September 1989.

Gosling, C. D., F. P. Wilcher, and P. R. Pujado, "LPG Conversion to Aromatics," Gas Processors Association 69th Annual Convention, Phoenix, March 1990.

Gosling, C. D., G. L. Gray, and J. J. Jeanneret, "Produce BTX from LPG with Cyclar," CMAI World Petrochemical Conference, Houston, March 1995.

Gosling, C. D., F. P. Wilcher, and P. R. Pujado, "LPG Conversion to Aromatics," Gas Processors Association 69th Annual Convention, Phoenix, March 1990.

Gosling, C. D., F. P. Wilcher, L. Sullivan, and R. A. Mountford, "Process LPG to BTX Products," *Hydrocarbon Processing,* December 1991.

Martindale, D. C., P. J. Kuchar, and R. K. Olson, "Cyclar: Aromatics from LPG," UOP Technology Conferences, various locations, September 1988.

CHAPTER 2.5
UOP ISOMAR PROCESS

John J. Jeanneret
UOP
Des Plaines, Illinois

INTRODUCTION

The UOP* Isomar* process is used to maximize the recovery of a particular xylene isomer from a mixture of C_8 aromatic isomers. The Isomar process is most often applied to *para*-xylene recovery, but it can also be used to maximize the recovery of *ortho*-xylene or *meta*-xylene. The term *mixed xylenes* is used to describe a mixture of C_8 aromatic isomers containing an equilibrium distribution of *para*-xylene, *ortho*-xylene, *meta*-xylene, and ethylbenzene (EB). In the case of *para*-xylene recovery, a mixed-xylenes feed is charged to a UOP Parex* unit, where the *para*-xylene isomer is preferentially extracted at 99.9 wt % purity and 97 wt % recovery per pass. The Parex raffinate, which is almost entirely depleted of *para*-xylene, is then sent to the Isomar unit (Fig. 2.5.1). The Isomar unit reestablishes an equilibrium distribution of xylene isomers, essentially creating additional *para*-xylene from the remaining *ortho* and *meta* isomers. Effluent from the Isomar unit is then recycled to the Parex unit for recovery of additional *para*-xylene. In this way, the *ortho* and *meta* isomers and EB are recycled to extinction. A complete description of the entire aromatics complex may be found in Chap. 2.1.

PROCESS CHEMISTRY

The two main categories of xylene isomerization catalysts are EB-dealkylation catalysts and EB-isomerization catalysts. The primary function of both catalyst types is to reestablish an equilibrium mixture of xylene isomers; however, they differ in how they handle the EB in the feed. An EB-dealkylation catalyst converts EB to a valuable benzene coproduct. An EB-isomerization catalyst converts EB into additional xylenes.

---

*Trademark and/or service mark of UOP.

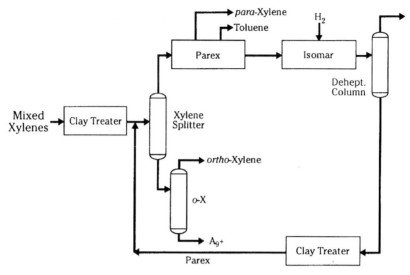

FIGURE 2.5.1 Typical Parex-Isomar loop.

UOP offers both an EB-isomerization catalyst, I-9,* and an EB-dealkylation catalyst, I-100.* Both I-9 and I-100 are bifunctional catalysts that incorporate zeolitic acidic sites and platinum metal sites. The acid function on both catalysts serves the same function: isomerization of xylenes.

The I-9 catalyst system isomerizes EB to xylenes through a naphthene intermediate (Fig. 2.5.2). The metal function first saturates the EB to ethylcyclohexane, then the acid function isomerizes it to dimethylcyclohexane, and finally the metal function dehydrogenates the naphthene to xylene. Because the isomerization of EB is an equilibrium limited reaction, the conversion of EB is usually limited to about 30 wt % per pass. In a typical aromatics complex using the I-9 catalyst, naphthenes are recycled back to the Isomar unit through the xylene column and Parex unit to suppress the for-

---

*Trademark and/or service mark of UOP.

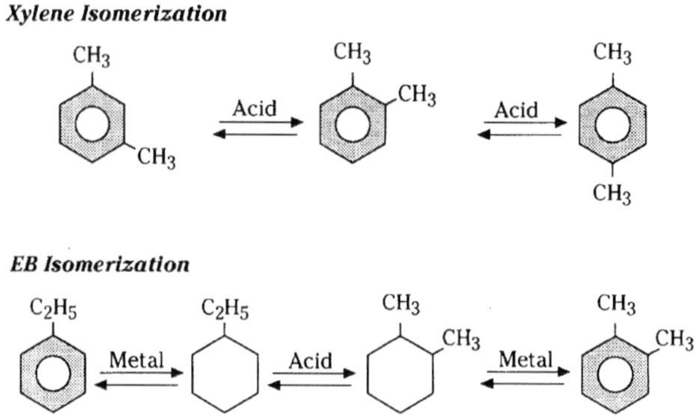

FIGURE 2.5.2 I-9 catalyst chemistry.

mation of naphthenes in the Isomar unit and thereby increase the yield of *para*-xylene from the complex.

The I-100 catalyst system uses an EB-dealkylation mechanism in which the ethyl group is cleaved from the aromatic ring by the acid function of the catalyst (Fig. 2.5.3). This reaction is not equilibrium-limited, thereby allowing EB conversion of up to 70 wt % per pass. Because this reaction does not involve a naphthene intermediate, C_8 naphthenes need not be recycled back through the Parex-Isomar loop.

All xylene isomerization catalysts exhibit some loss of aromatic rings across the reactor. Because a typical Parex-Isomar loop is designed with a recycle-to-feed ratio of 2.5 to 3.0, any ring loss across the isomerization reactor is magnified accordingly. In the Isomar process, the precise level of expected C_8 ring loss varies with catalyst type and operating severity but is normally in the range of 1.5 to 4 wt % per pass. The lower end of this range represents operation with the I-100 catalyst, and the upper end represents the I-9 catalyst. Minimizing ring loss in the Isomar process results in maximum yield of xylenes from the aromatics complex.

The proper selection of isomerization catalyst type depends on the configuration of the aromatics complex, the composition of the feedstocks, and the desired product slate. The choice of isomerization catalyst must be based on an economic analysis of the entire aromatics complex. The C_8 fraction of the reformate from a typical petroleum naphtha contains approximately 15 to 17 wt % EB, but up to 30 wt % EB may be in a similar pyrolysis gasoline (pygas) fraction. Using an EB-isomerization catalyst maximizes the yield of *para*-xylene from an aromatics complex by converting EB to xylenes. An EB-isomerization catalyst is usually chosen when the primary goal of the complex is to maximize the production of *para*-xylene.

Alternatively, an EB-dealkylation catalyst can be used to debottleneck an existing Parex unit or crystallizer by converting more EB per pass through the isomerization unit and eliminating the requirement for naphthene-intermediate circulation around the Parex-Isomar recycle loop. For a new aromatics complex design, using an EB-dealkylation catalyst minimizes the size of the xylene column and Parex and Isomar units required to produce a given amount of *para*-xylene. However, this reduction in

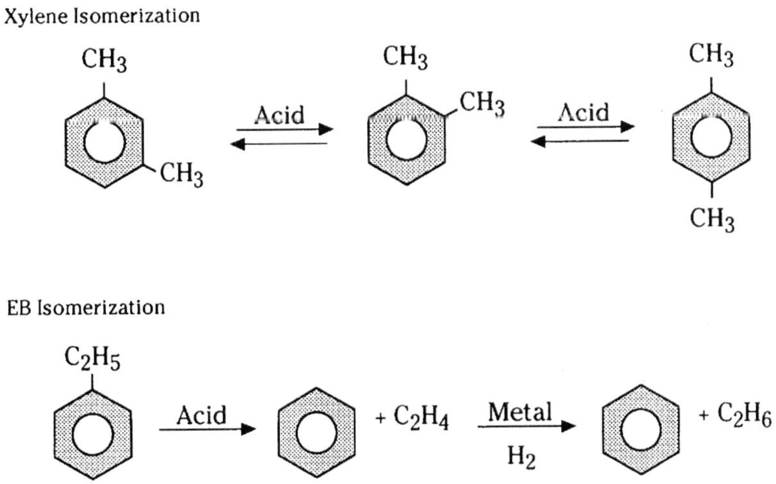

FIGURE 2.5.3 I-100 catalyst chemistry.

size of the Parex-Isomar loop comes at the expense of lower *para*-xylene yields, because all the EB in the feed is being converted to benzene rather than to additional *para*-xylene. Lower *para*-xylene yield means that more feedstock will be required, which increases the size of the CCR* Platforming,* Sulfolane,* and Tatoray units in the front end of the complex as well as most of the fractionators.

DESCRIPTION OF THE PROCESS FLOW

An Isomar unit is always combined with a recovery unit for one or more xylene isomers. Most often, the Isomar process is combined with the UOP Parex process for *para*-xylene recovery (Fig. 2.5.1). Fresh mixed-xylenes feed to the Parex-Isomar loop is first sent to a xylene column, which can be designed either to recover a portion of the *ortho*-xylene in the bottoms or simply reject C_9+ aromatic components to meet feed specifications for the Parex unit. The xylene column overhead is then directed to the Parex unit, where 99.9 wt % pure *para*-xylene is produced at 97 wt % recovery per pass. The Parex raffinate from the Parex unit, which contains less than 1 wt % *para*-xylene, is sent to the Isomar unit.

The feed to the Isomar unit is first combined with hydrogen-rich recycle gas and makeup gas to replace the small amount of hydrogen consumed in the Isomar reactor (Fig. 2.5.4). The combined feed is then preheated by exchange with the reactor effluent, vaporized in a fired heater, and raised to reactor operating temperature. The hot feed vapor is then sent to the reactor, where it is passed radially through a fixed bed of catalyst. The reactor effluent is cooled by exchange with the combined feed and then sent to the product separator. Hydrogen-rich gas is taken off the top of the product separator and recycled back to the reactor. A small portion of the recycle gas is purged to remove accumulated light ends from the recycle gas loop. Liquid from the bottom of the product separator is charged to the deheptanizer column. The C_7- overhead from the deheptanizer is cooled and separated into gas and liquid products. The dehep-

*Trademark and/or service mark of UOP.

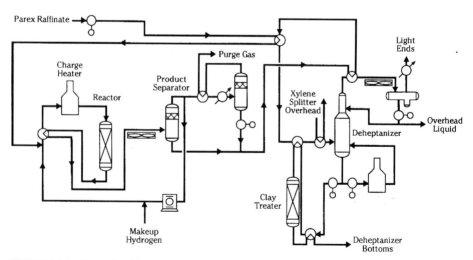

FIGURE 2.5.4 Isomar flow diagram.

tanizer overhead gas is exported to the fuel gas system. The overhead liquid is recycled back to the debutanizer column of the Platforming unit so that any benzene in this stream may be recovered in the Sulfolane unit. The C_8+ fraction from the bottom of the deheptanizer is clay-treated, combined with fresh mixed-xylenes feed, and recycled back to the xylene column.

FEEDSTOCK CONSIDERATIONS

The feedstock to an Isomar unit usually consists of raffinate from a Parex unit. At times, charging the fresh mixed-xylenes feed directly to the Isomar unit may be desirable, or the Isomar unit may be used in conjunction with fractionation to produce only *ortho*-xylene. In any case, the feed to an Isomar unit should meet the specifications outlined in Table 2.5.1.

Any nonaromatic compounds in the feed to the Isomar unit are cracked to light ends and removed from the Parex-Isomar loop. This ability to crack nonaromatic impurities eliminates the need for extracting the mixed xylenes, and consequently the size of the Sulfolane unit can be greatly reduced. In a UOP aromatics complex, the reformate from the CCR Platforming unit is split into C_7- and C_8+ fractions. The C_7- fraction is sent to the Sulfolane unit for recovery of high-purity benzene and toluene. Because modern, low-pressure CCR Platforming units operate at extremely high severity for aromatics production, the C_8+ fraction that is produced contains essentially no nonaromatic impurities and thus can be sent directly to the xylene recovery section of the complex.

PROCESS PERFORMANCE

The performance of the xylene isomerization catalysts can be measured in several specific ways, including the approach to equilibrium in the xylene isomerization reaction itself, the conversion of EB per pass, and the xylene ring loss per pass. Approach to equilibrium is a measure of operating severity for an EB-isomerization catalyst, and EB conversion is a measure of severity for an EB-dealkylation catalyst. For both catalyst types, ring loss increases with operating severity. In a *para*-xylene application, for example, high EB conversion in the Isomar unit is beneficial for the Parex unit but is accompanied by higher ring loss and thus lower overall yield of *para*-xylene from the complex.

Perhaps the best way to compare xylene isomerization catalysts is to measure the overall *para*-xylene yield from the Parex-Isomar loop. Figure 2.5.5 compares the *para*-xylene yield, based on fresh mixed-xylenes feed to the Parex-Isomar loop, for

TABLE 2.5.1 Isomar Feedstock Specifications

Contaminant	Effect	Limit
Water	Deactivates catalyst, promotes corrosion. Irreversible.	200 ppm, max.
Total chloride	Increases acid function, increases cracking. Reversible.	2 ppm, max.
Total nitrogen	Neutralizes acid sites, deactivates catalyst. Irreversible.	1 ppm, max.
Total sulfur	Attenuates Pt activity, increases cracking. Reversible.	1 ppm, max.
Lead	Poisons acid and Pt sites. Irreversible.	20 ppb, max.
Copper	Poisons acid and Pt sites. Irreversible.	20 ppb, max.
Arsenic	Poisons acid and Pt sites. Irreversible.	2 ppb, max.

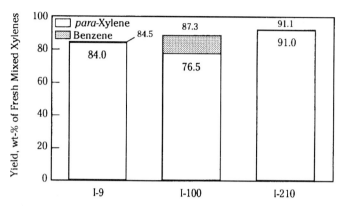

FIGURE 2.5.5 Parex-Isomar yields.

both the I-9 and I-100 catalyst systems. The basis for this comparison is the flow scheme shown in Fig. 2.5.1. The composition of the mixed-xylenes feed is 17 wt % EB, 18 wt % *para*-xylene, 40 wt % *meta*-xylene, and 25 wt % *ortho*-xylene. The operating severity for the I-9 catalyst is 22.1 wt % *para*-xylene in the total xylenes from the Isomar unit. The operating severity for the I-100 catalyst is 65 wt % conversion of EB per pass. With the I-9 catalyst, the overall yield of *para*-xylene is 84 wt % of the fresh mixed-xylenes feed. Because it has lower ring loss per pass, the I-100 catalyst exhibits a higher overall yield of benzene and *para*-xylene, but the yield of *para*-xylene alone is only 76.5 wt %. Thus, more mixed xylenes are required to produce a target amount of *para*-xylene with the I-100 catalyst.

Figure 2.5.5 also shows the yields for a new UOP EB-isomerization catalyst called I-210.* The I-210 catalyst system relies on the same reaction chemistry as the I-9 catalyst but is much more selective and exhibits lower ring loss. The ring loss of the I-210 catalyst is only about 1.5 wt % per pass versus about 4 wt % for the I-9. With the I-210 catalyst, the overall yield of *para*-xylene is 91 wt % of the fresh mixed-xylenes feed, a yield improvement of 7 wt % over that of the I-9 catalyst. At the time this chapter was written, commercial manufacturing trials for the I-210 catalyst were just beginning. UOP expects to have at least one Isomar unit running with the I-210 catalyst by the time this handbook is published.

EQUIPMENT CONSIDERATIONS

The charge heater is normally a radiant convection-type heater. The process stream is heated in the radiant section, and the convection section is used for a hot-oil system or steam generation. The heater can be designed to operate either on fuel gas or fuel oil, and each burner is equipped with a fuel gas pilot. A temperature controller at the heater outlet regulates the flow of fuel to the burners. Radiant-section tubes are constructed of 1.25% Cr–0.5% Mo. Tubes in the convection section are carbon steel.

The Isomar process uses a radial-flow reactor. The vapor from the charge heater enters the top of the reactor and is directed to the sidewall. The vapors then travel radially through a set of scallops, through the fixed catalyst bed, and into a center pipe. The reactor effluent then flows down through the center pipe to the reactor out-

---
*Trademark and/or service mark of UOP.

let. The advantage of the radial-flow reactor is low pressure drop, which is important in the Isomar process because the reaction rates are sensitive to pressure. Low pressure drop also reduces the power consumption of the recycle gas compressor. The reactor is constructed of 1.25% chrome (Cr)–0.5% molybdenum (MO) alloy.

The purpose of the product separator is to split the condensed reactor effluent into liquid product and hydrogen-rich recycle gas. The pressure in the product separator determines the pressure in the reactor. Separator pressure is regulated by controlling the rate of hydrogen makeup flow. Hydrogen purity in the recycle gas is monitored by a hydrogen analyzer at the recycle-gas compressor suction. When hydrogen purity gets too low, a small purge is taken from the recycle gas. The product separator is constructed of killed carbon steel.

The recycle gas compressor is usually of the centrifugal type and may be driven by an electric motor or a steam turbine. The compressor is provided with both seal oil and lube oil circuits and an automatic shutdown system to protect the machine against damage.

The purpose of the deheptanizer column is to remove light by-products from the reactor effluent. The deheptanizer usually contains 40 trays and incorporates a thermosiphon reboiler. Heat is usually supplied by the overhead vapor from the xylene column located upstream of the Parex unit. The deheptanizer column is constructed of carbon steel.

The combined feed-effluent exchanger is constructed of 1.25% Cr–0.5% Mo. Other heat exchangers in the Isomar unit are constructed of carbon steel.

CASE STUDY

A summary of the investment cost and utility consumption for a typical I-9 Isomar unit is shown in Table 2.5.2. The basis for this case is an Isomar unit processing 5600 metric tons per hour (40,000 barrels per day) of raffinate from a Parex unit. This case corresponds to the case study for an integrated UOP aromatics complex presented in Chap. 2.1. The investment cost is limited to the Isomar unit, deheptanizer column, and downstream clay treater. The estimated erected cost for the Tatoray unit assumes construction on a U.S. Gulf Coast site in 1995. The scope of the estimate includes engineering, procurement, erection of equipment on the site, and the initial load of I-9 catalyst.

COMMERCIAL EXPERIENCE

The first UOP Isomar unit went on-stream in 1967. Since that time, UOP has licensed a total of 38 Isomar units throughout the world. Thirty UOP Isomar units have been

TABLE 2.5.2 Investment Cost and Utility Consumption*

Estimated erected cost, million $U.S.	21.5
Utility consumption:	
Electric power, kW	908
High-pressure steam, MT/h	19.6
Cooling water, m^3/h	232
Fuel-fired, million kcal/h	21.0

*Basis: 5600 MT/h (40,000 BPD) Parex raffinate

Note: MT/h = metric tons per hour; BPD = barrels per day.

commissioned, and eight more are in various stages of design or construction. UOP offers a choice of both EB-isomerization and EB-dealkylation catalyst types. This choice of catalyst provides additional flexibility to tailor the distribution of products from an aromatics complex.

BIBLIOGRAPHY

Jeanneret, J. J., "Developments in *p*-Xylene Technology," DeWitt Petrochemical Review, Houston, March 1993.

Jeanneret, J. J., "*para*-Xylene Production in the 1990's," UOP Technology Conferences, various locations, May 1995.

Jeanneret, J. J., C. D. Low, and V. Zukauskas, "New Strategies Maximize *para*-Xylene Production," *Hydrocarbon Processing*, June 1994.

Kuchar, P. J., J. J. Jeanneret, C. D. Low, and J. Swift, "Processes for Maximum BTX Complex Profitability," UOP Technology Conferences, various locations, September 1992.

CHAPTER 2.6
UOP PAREX PROCESS

John J. Jeanneret
UOP
Des Plaines, Illinois

INTRODUCTION

The UOP* Parex* process is an innovative adsorptive separation method for the recovery of *para*-xylene from mixed xylenes. The term *mixed xylenes* refers to a mixture of C_8 aromatic isomers that includes ethylbenzene, *para*-xylene, *meta*-xylene, and *ortho*-xylene. These isomers boil so closely together that separating them by conventional distillation is not practical. The Parex process provides an efficient means of recovering *para*-xylene by using a solid zeolitic adsorbent that is selective for *para*-xylene. Unlike conventional chromatography, the Parex process simulates the countercurrent flow of a liquid feed over a solid bed of adsorbent. Feed and products enter and leave the adsorbent bed continuously at nearly constant compositions. This technique is sometimes referred to as *simulated moving bed* (SMB) separation.

In a modern aromatics complex (Fig. 2.6.1), the Parex unit is located downstream of the xylene column and is integrated with a UOP Isomar* unit. The feed to the xylene column consists of the C_8+ aromatics product from the CCR* Platforming* unit together with the xylenes produced in the Tatoray unit. The C_8 fraction from the overhead of the xylene column is fed to the Parex unit, where high-purity *para*-xylene is recovered in the extract. The Parex raffinate is then sent to the Isomar unit, where the other C_8 aromatic isomers are converted into additional *para*-xylene and recycled back to the xylene column. A complete description of the entire aromatics complex may be found in Chap. 2.1.

UOP Parex units are designed to recover more than 97 wt % of the *para*-xylene from the feed in a single pass at a product purity of 99.9 wt % or better. The Parex design is energy-efficient, mechanically simple, and highly reliable. On-stream factors for Parex units typically exceed 95 percent.

---

*Trademark and/or service mark of UOP.

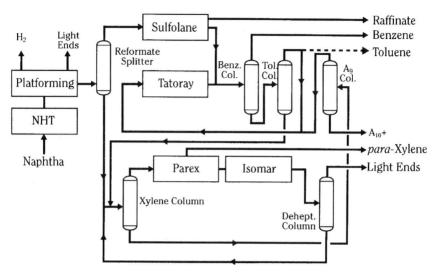

FIGURE 2.6.1 UOP aromatics complex—maximum *para*-xylene.

PAREX VERSUS CRYSTALLIZATION

Before the introduction of the Parex process, *para*-xylene was produced exclusively by fractional crystallization. In crystallization, the mixed-xylenes feed is refrigerated to $-75°C$ ($-100°F$), at which point the *para*-xylene isomer precipitates as a crystalline solid. The solid is then separated from the mother liquor by centrifugation or filtration. Final purification is achieved by washing the *para* xylene crystals with either toluene or a portion of the *para*-xylene product.

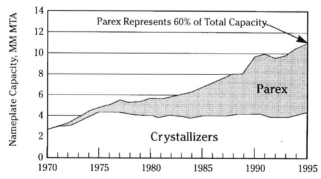

FIGURE 2.6.2 Total installed *para*-xylene capacity.

UOP PAREX PROCESS

Soon after it was introduced in 1971, the UOP Parex process quickly became the world's preferred technology for *para*-xylene recovery. The last new *para*-xylene crystallizer was built in 1975. Since that time, virtually all new *para*-xylene production capacity has been based on the UOP Parex process (Fig. 2.6.2).

The principal advantage of the Parex adsorptive separation process over crystallization technology is the ability of the Parex process to recover more than 97 percent of the *para*-xylene in the feed per pass. Crystallizers must contend with a eutectic composition limit that restricts *para*-xylene recovery to about 65 percent per pass. The implication of this difference is clearly illustrated in Fig. 2.6.3: a Parex complex producing 250,000 metric tons per annum (MTA) of *para*-xylene is compared with a crystallizer complex producing 168,000 MTA. The upper numbers in the figure indicate the flow rates through the Parex complex; the lower numbers indicate the flow rates through a comparable crystallizer complex. A Parex complex can produce about 50 percent more *para*-xylene from a given-size xylene column and isomerization unit than a complex using crystallization. In addition, the yield of *para*-xylene per unit of fresh feed is also improved because a relatively smaller recycle flow means lower losses in the isomerization unit. The technologies could also be compared by keeping the *para*-xylene product rate constant. In this case, a larger xylene column and a larger isomerization unit would be required to produce the same amount of *para*-xylene, thus increasing both the investment cost and utility consumption of the complex.

A higher *para*-xylene recycle rate in the crystallizer complex not only increases the size of the equipment in the recycle loop and the utility consumption within the loop, but also makes inefficient use of the xylene isomerization capacity. Raffinate from a Parex unit is almost completely depleted of *para*-xylene (less than 1 wt %), whereas mother liquor from a typical crystallizer contains about 9.5 wt % *para*-xylene. Because the isomerization unit cannot exceed an equilibrium concentration of *para*-xylene (23 to 24 wt %), any *para*-xylene in the feed to the isomerization unit reduces the amount of *para*-xylene produced in that unit per pass. Thus, the same isomerization unit produces about 60 percent more *para*-xylene per pass when processing Parex raffinate than it does when processing crystallizer mother liquor.

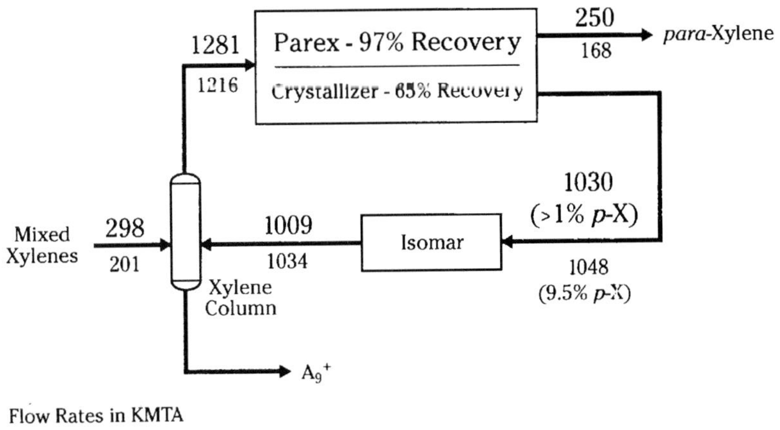

Flow Rates in KMTA

FIGURE 2.6.3 Comparison of Parex with crystallization.

PROCESS PERFORMANCE

The quality of *para*-xylene demanded by the market has increased significantly over the last 20 years. When the Parex process was introduced in 1970, the standard purity for *para*-xylene sold in the market was 99.2 wt %. By 1992, the purity standard had become 99.7 wt %, and the trend toward higher purity continues. All Parex units built after 1991 are designed to produce 99.9 wt % pure *para*-xylene at 97 wt % recovery per pass. Most older Parex units can also be modified to produce 99.9 wt % purity.

FEEDSTOCK CONSIDERATIONS

Most of the mixed xylenes used for *para*-xylene production are produced from petroleum naphtha by catalytic reforming. Modern UOP CCR Platforming units operate at such high severity that the C_8+ fraction of the reformate contains virtually no nonaromatic impurities and may be charged directed to the xylene recovery section of the complex. In many integrated aromatics complexes, up to half of the total mixed xylenes are produced from the conversion of toluene and C_9 aromatics in a UOP Tatoray unit.

Nonaromatic impurities in the feed to a Parex unit increase utility consumption and take up space in the Parex unit, but they do not affect the purity of the *para*-xylene product or the recovery performance of the Parex unit.

Feedstocks for Parex must be prefractionated to isolate the C_8 aromatic fraction and clay-treated to protect the adsorbent. If the Parex unit is integrated with an upstream refinery or ethylene plant, prefractionation and clay treating are designed into the complex. If additional mixed xylenes are purchased and transported to the site, they must first be stripped, clay-treated, and rerun before being charged to the Parex unit. In general, feed to a Parex unit should meet the specifications outlined in Table 2.6.1.

TABLE 2.6.1 Parex Feedstock Specifications

Contaminant	Effect	Limit
Benzene	Binds to adsorption sites, decreases capacity; reversible	500 ppm max.
Methylethyl benzenes	Extracted along with *para*-xylene, contaminates product	100 ppm max.
Other C_9 aromatics	Accumulates in the circulating desorbent; increases utilities	500 ppm max.
Total sulfur	Binds to adsorption sites, decreases capacity; reversible	1 ppm max.
Total nitrogen	Binds to adsorption sites, decreases capacity; reversible	1 ppm max.
Active oxygen	Binds to adsorption sites, decreases capacity; reversible	1 ppm max.
Carbonyl	Binds to adsorption sites, decreases capacity; reversible	2 ppm max.
Total chloride	Breaks down zeolite crystal structure; irreversible	5 ppm max.
Water	Too much causes hydrothermal damage to adsorbent; irreversible	60 ppm max.
Bromine index	Olefins polymerize, deposit on adsorbent, decrease capacity; irreversible	20 max.
Color (Pt-Co)	Specification on the *para*-xylene product	10 max.
Lead	Alters selectivity of the adsorbent; irreversible	5 ppb max.
Arsenic	Alters selectivity of the adsorbent; irreversible	1 ppb max.
Copper	Alters selectivity of the adsorbent; irreversible	5 ppb max.

DESCRIPTION OF THE PROCESS FLOW

The flow diagram for a typical Parex unit is shown in Fig. 2.6.4. The separation takes place in the adsorbent chambers. Each adsorbent chamber is divided into a number of adsorbent beds. Each bed of adsorbent is supported from below by specialized internals, or grids, that are designed to produce highly efficient flow distribution. Each internals assembly is connected to the rotary valve by a "bed line." The internals between each adsorbent bed are used to inject or withdraw liquid from the chamber and simultaneously collect liquid from the bed above and redistribute the liquid over the bed below.

The Parex process is one member of UOP's family of Sorbex* adsorptive separation processes. The basic principles of Sorbex technology are the same regardless of the type of separation being conducted and are discussed in Chap. 10.3. The number of adsorbent beds and bed lines varies with each Sorbex application. A typical Parex unit has 24 adsorbent beds with 24 grids and 24 bed lines connecting the grids to the rotary valve. Because of practical construction considerations, most Parex units consist of two adsorption chambers in series with 12 beds in each chamber.

The Parex process has four major streams that are distributed to the adsorbent chambers by the rotary valve. These *net* streams include:

- *Feed in:* mixed-xylenes feed
- *Dilute extract out: para*-xylene product diluted with desorbent
- *Dilute raffinate out:* ethylbenzene, *meta*-xylene, and *ortho*-xylene diluted with desorbent
- *Desorbent in:* recycle desorbent from the fractionation section

At any given time, only four of the bed lines actively carry the net streams into and out of the adsorbent chamber. The rotary valve is used to periodically switch the positions of the liquid feed and withdrawal points as the composition profile moves down the chamber. A pump provides the liquid circulation from the bottom of the first adsorbent chamber to the top of the second. A second pump provides circulation from the bottom of the second adsorbent chamber to the top of the first. In this way, the two adsorbent chambers function as a single, continuous loop of adsorbent beds.

The dilute extract from the rotary valve is sent to the extract column for separation of the extract from the desorbent. The overhead from the extract column is sent to a finishing column, where the highly pure *para*-xylene product is separated from any toluene that may have been present in the feed.

The dilute raffinate from the rotary valve is sent to the raffinate column for separation of the raffinate from the desorbent. The overhead from the raffinate column contains the unextracted C_8 aromatic components: ethylbenzene, *meta*-xylene, and *ortho*-xylene, together with any nonaromatics that may have been present in the feed. The raffinate product is then sent to an isomerization unit, where additional *para*-xylene is formed, and then recycled back to the Parex unit.

The desorbent from the bottom of both the extract and raffinate columns is recycled back to the adsorbent chambers through the rotary valve. Any heavy contaminants in the feed accumulate in the desorbent. To prevent this accumulation, provision is made to take a slipstream of the recycle desorbent to a small desorbent rerun column, where any heavy contaminants are rejected. During normal operation, mixed xylenes are stripped, clay-treated, and rerun prior to being sent to the Parex unit. Thus,

---

*Trademark and/or service mark of UOP.

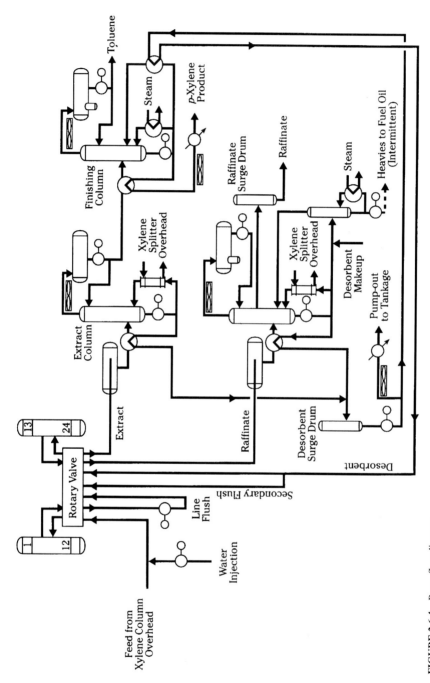

FIGURE 2.6.4 Parex flow diagram.

few heavy contaminants need to be removed from the bottom of the desorbent rerun column.

EQUIPMENT CONSIDERATIONS

UOP supplies a package of specialized equipment that is considered critical for the successful performance of the Parex process. This package includes the rotary valve; the adsorbent chamber internals; and the control system for the rotary valve, pumparound pump, and net flows. The erected cost estimates that UOP provides for the Parex process include the cost of this equipment package.

The rotary valve is a sophisticated, highly engineered piece of process equipment developed by UOP specifically for the Sorbex family of processes. The UOP rotary valve is critical for the purity of the *para*-xylene product and for the unsurpassed reliability of the Parex process. The design of the UOP rotary valve has evolved over 30 years of commercial Sorbex operating experience.

The adsorbent chamber internals are also critical to the performance of the Parex process. These specialized internals are used to support each bed of adsorbent and to prevent leakage of the solid adsorbent into the process streams. Each internals assembly also acts as a flow collector and distributor and is used to inject or withdraw the net flows from the adsorbent chamber or redistribute the internal liquid flow from one adsorbent bed to the next. Proper flow distribution in the adsorbent chambers is essential for purity and recovery performance in the Parex process and is difficult to achieve in larger-diameter vessels. As the size of Parex units has increased over the years, the design of adsorbent chamber internals has evolved to ensure proper flow distribution over increasingly larger-diameter vessels.

The Parex control system supplied by UOP is a specialized system that monitors and controls the flow rates of the net streams and adsorbent chamber circulation and ensures proper operation of the rotary valve.

Because of the mild operating conditions used in the Parex process, the entire plant may be constructed of carbon steel.

The Parex process is normally heat-integrated with the upstream xylene column. The xylene column is used to rerun the feed to the Parex unit. The mixed xylenes are taken overhead, and the heavy aromatics are removed from the bottom of the column. Before the overhead vapor from the xylene column is fed to the adsorption section of the Parex unit, it is used to reboil the extract and raffinate columns of the Parex unit.

UOP offers High Flux* high-performance heat-exchanger tubing for improved heat-exchange efficiency. High Flux tubing is made with a special coating that promotes nucleate boiling and increases the heat-transfer coefficient of conventional tubing by a factor of 10. Specifying UOP High Flux tubing for the reboilers of the Parex fractionators reduces the size of the reboilers and may also allow the xylene column to be designed for lower-pressure operation. Designing the xylene column for lower pressure reduces the erected cost of the column and lowers the utility consumption in that column.

UOP also offers MD* distillation trays for improved fractionation performance. The MD trays are used for large liquid loads and are especially effective when the volumetric ratio between vapor and liquid rates is low. The use of MD trays provides a large total weir length and reduces froth height on the tray. Because the froth height is lower, MD trays can be installed at a smaller tray spacing than conventional distillation trays. The use of MD trays in new column designs results in a smaller required

---

*Trademark and/or service mark of UOP.

diameter and lower column height. Consequently, MD trays are often specified for large xylene columns, especially when the use of MD trays can keep the design of the xylene column in a single shell.

CASE STUDY

A summary of the investment cost and utility consumption for a typical Parex unit is shown in Table 2.6.2. The basis for this case is a Parex unit producing 400,000 MTA of 99.9 wt % pure *para*-xylene product. This case corresponds to the case study for an integrated UOP aromatics complex presented in Chap. 2.1. Because the Parex unit is tightly heat-integrated with the upstream xylene column, the investment cost and utility consumption estimates include both. The estimated erected costs for these units assume construction on a U.S. Gulf Coast site in 1995. The scope of the estimate includes engineering, procurement, erection of equipment on the site, and the initial inventory of Parex adsorbent and desorbent.

COMMERCIAL EXPERIENCE

UOP's experience with adsorptive separations is extensive. Sorbex technology, which was invented by UOP in the 1960s, was the first large-scale commercial application of continuous adsorptive separation. The first commercial Sorbex unit, a Molex* unit for the separation of linear paraffins, came on-stream in 1964. The first commercial Parex unit came on-stream in 1971. UOP has licensed more than 100 Sorbex units throughout the world. This total includes 59 Parex units, of which 47 units are in operation and 12 others are in various stages of design and construction. UOP Parex units range in size from 24,000 MTA of *para*-xylene product to more than 463,000 MTA.

---

*Trademark and/or service mark of UOP.

TABLE 2.6.2 Investment Cost and Utility Consumption*

Estimated erected cost, million $U.S.:	
Xylene column	25.4
Parex unit	73.0
Utility consumption:	
Electric power, kW	3,400
Medium-pressure steam, MT/h (klb/h)	16.3 (credit 35.9)
Cooling water, m^3/h (gal/min)	218 (960)
Fuel fired, million kcal/h (million Btu/h)	109 (433)

*Basis: 400,000 MTA of *para*-xylene product

BIBLIOGRAPHY

Jeanneret, J. J., "Developments in *p*-Xylene Technology," DeWitt Petrochemical Review, Houston, March 1993.

Jeanneret, J. J., "*para*-Xylene Production in the 1990's," UOP Technology Conferences, various locations, May 1995.

Jeanneret, J. J., C. D. Low, and V. Zukauskas, "New Strategies Maximize *para*-Xylene Production," *Hydrocarbon Processing,* June 1994.

Prada, R. E., et al., "Parex Developments for Increased Efficiency," UOP Technology Conferences, various locations, September 1992.

CHAPTER 2.7
UOP TATORAY PROCESS

John J. Jeanneret
UOP
Des Plaines, Illinois

INTRODUCTION

The UOP* Tatoray process is used to selectively convert toluene and C_9 aromatics (A_9) into xylenes and benzene. In a modern aromatics complex, the Tatoray process is integrated between the aromatics extraction and xylene recovery sections of the plant (Fig. 2.7.1). Extracted toluene is fed to the Tatoray unit instead of being blended into

*Trademark and/or service mark of UOP.

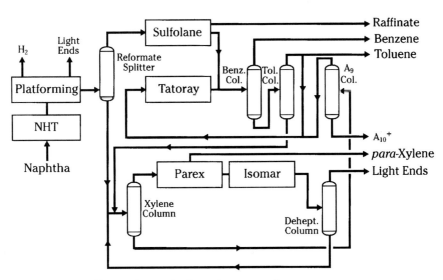

FIGURE 2.7.1 UOP aromatics complex for maximum *para*-xylene.

the gasoline pool or sold for solvent applications. If the goal is to maximize the production of *para*-xylene from the complex, the A_9 by-product can also be fed to the Tatoray unit instead of blending it into the gasoline pool. Processing A_9 in a Tatoray unit shifts the chemical equilibrium in the unit away from benzene production and toward xylene production.

In recent years, the strong demand for *para*-xylene has begun to outstrip the supply of mixed xylenes. The Tatoray process provides an ideal way of producing additional mixed xylenes from low-value toluene and heavy aromatics. The incorporation of a Tatoray unit into an aromatics complex can more than double the yield of *para*-xylene from naphtha feedstock. The entire aromatics complex is discussed in more depth in Chap. 2.1.

PROCESS CHEMISTRY

The two major reactions in the Tatoray process, disproportionation and transalkylation, are illustrated in Fig. 2.7.2. The conversion of toluene alone into an equilibrium mixture of benzene and xylenes is called *disproportionation*. The conversion of a blend of toluene and A_9 into xylenes through the migration of methyl groups between methyl-substituted aromatics is called *transalkylation*. In general, both reactions proceed toward an equilibrium distribution of benzene and alkyl-substituted aromatics. Because methyl groups are stable at reaction conditions, the reaction equilibrium is easy to estimate when the feed consists of all methyl-substituted aromatics (Fig. 2.7.3). The reaction pathways become more complex when other alkyl groups are present in the feed. Propyl and heavier groups either hydrocrack to lighter methyl and ethyl groups or dealkylate completely. Ethyl groups either crack to methyls, dealkylate, or transalkylate.

The transalkylation reaction must be conducted in a hydrogen atmosphere even though no net hydrogen is consumed in the reaction. In practice, a small amount of

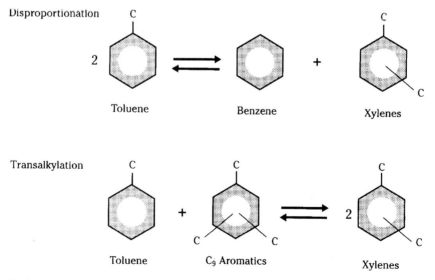

FIGURE 2.7.2 Major Tatoray reactions.

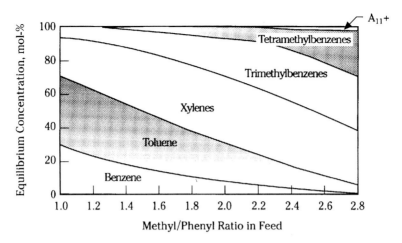

FIGURE 2.7.3 Equilibrium distribution of methyl groups at 700 K.

hydrogen is always consumed because of the dealkylation and hydrocracking side reactions mentioned previously. Hydrogen consumption increases for heavier feedstocks because these generally contain heavier alkyl groups. Any saturates in the feed also crack and contribute to additional hydrogen consumption.

The maximum theoretical conversion per pass is limited by equilibrium and is a function of the feedstock composition. For example, theoretical conversion for a pure toluene feed is approximately 59 wt % per pass. Operating at high conversion minimizes the amount of unconverted material that must be recycled back through the benzene-toluene (BT) fractionation section of the complex. A smaller recycle stream minimizes the size of the benzene and toluene columns, minimizes the size of the Tatoray unit, and minimizes the utility consumption in all of these units. However, as conversion increases, the rate of catalyst deactivation also increases as does the selectivity to saturated and heavy aromatic by-products. The optimum level of conversion in a Tatoray unit is usually in the range of 46 to 48 wt % per pass.

UOP introduced the current generation of Tatoray catalyst, TA-4, in 1988. Since that time, extensive commercial experience has demonstrated that the TA-4 catalyst system displays stable selectivity over many years of continuous operation. Most Tatoray units are designed to run for at least 12 months between catalyst regenerations; however, many Tatoray units have been operated for several years without regeneration. The catalyst is regenerated in situ by a simple carbon-burn procedure.

DESCRIPTION OF THE PROCESS FLOW

The Tatoray process uses a simple flow scheme consisting of a fixed-bed reactor and a product separation section (Fig. 2.7.4). The fresh feed to the Tatoray unit is first combined with hydrogen-rich recycle gas, preheated by exchange with the hot reactor effluent, and then vaporized in a fired heater, where it is raised to reaction temperature. The hot feed vapor is then sent to the reactor, where it is sent down-flow over a fixed bed of catalyst. The reactor effluent is then cooled by exchange with the combined feed, mixed with makeup gas to replace the small amount of hydrogen consumed in the Tatoray reactor, and then sent to a product separator. Hydrogen-rich gas

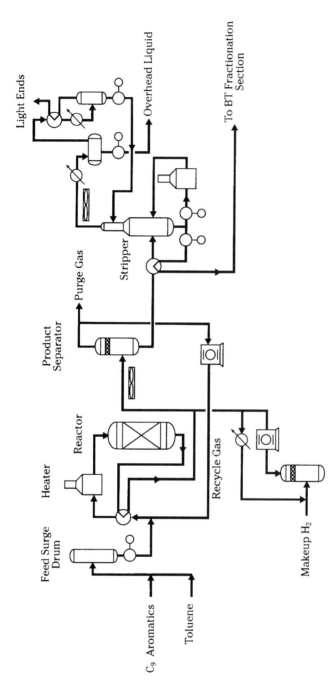

FIGURE 2.7.4 Tatoray flow diagram.

is taken off the top of the product separator and recycled back to the reactor. A small portion of the recycle gas is purged to remove accumulated light ends from the recycle gas loop. Liquid from the bottom of the product separator is sent to a stripper column. The C_5^- overhead from the stripper is cooled and separated into gas and liquid products.

The stripper overhead gas is exported to the fuel gas system. The overhead liquid is recycled back to the debutanizer column of the UOP Platforming* unit so that any benzene in this stream may be recovered in the UOP Sulfolane* unit. The benzene and xylene products, together with the unreacted toluene and A_9, are taken from the bottom of the stripper and recycled back to the BT fractionation section of the aromatics complex.

FEEDSTOCK CONSIDERATIONS

The feed to a Tatoray unit is typically pure toluene or a blend of toluene and A_9. The toluene is usually derived from reformate. The benzene and toluene in the reformate is extracted in a Sulfolane unit, clay-treated, and then fractionated into individual benzene and toluene products. A small portion of the toluene is sometimes taken for sale into solvent applications or blended back into the gasoline pool. The rest of the toluene is sent to the Tatoray unit for conversion into additional benzene and xylenes. The A_9 portion of the feed usually consists of the straight-run A_9 recovered from the reformate together with the A_9 by-products produced in the xylene isomerization unit and in the Tatoray unit itself.

The Tatoray process is capable of processing feedstocks ranging from 100 wt % toluene to 100 wt % A_9. As shown in Fig. 2.7.5, the product composition shifts away from benzene and toward xylenes as the A_9 concentration in the feed increases. Depending on the economics of each individual project, the optimal concentration of A_9 in the feed to the Tatoray unit usually is in the range of 40 to 60 wt %. Any saturates in the feed are generally cracked to light ends. Because these cracking reactions

---

*Trademark and/or service mark of UOP.

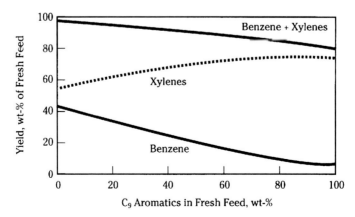

FIGURE 2.7.5 Tatoray yield structure.

adversely affect the activity of the catalyst and lead to accelerated catalyst deactivation, a limitation on saturates in the feed is usually specified. Limitations are also imposed on the allowable concentration of bicyclics and heteromolecules in the feed; both of these materials may contribute to accelerated deactivation of the catalyst. In general, feed to a Tatoray unit should meet the specifications outlined in Table 2.7.1.

PROCESS PERFORMANCE

The ability to process A_9 in a Tatoray unit makes more feedstock available for xylene production and dramatically shifts the selectivity of the unit away from benzene and toward xylenes. Figure 2.7.6 illustrates that a typical aromatics complex without a Tatoray unit can produce approximately 200,000 metric tons per year (MTA) of *para*-xylene from 25,000 barrels per day (BPD) of Arabian Light naphtha, 70 to 150°C (160 to 300°F) cut. If an A_7 Tatoray unit (toluene feed only) is added to the complex, the same 25,000 BPD of naphtha can produce 280,000 MTA of *para*-xylene, an increase of 40 percent. When an A_7-A_9 Tatoray unit is added to the complex, the endpoint of the naphtha is increased from 150 to 170°C (300 to 340°F) to maximize the amount of

TABLE 2.7.1 Tatoray Feedstock Specifications

Contaminant	Effect	Limit
Nonaromatics	Increased cracking, increased H_2 consumption, lower benzene purity	2 wt % max.
Water	Depresses transalkylation activity; reversible	100 ppm max.
Olefins	Promotes deposition of coke on catalyst	20 BI* max.
Total chloride	Promotes cracking of aromatic rings; reversible	1 ppm max.
Total nitrogen	Neutralizes active catalyst sites; irreversible	0.1 ppm max.
Total sulfur	Affects quality of the benzene product	1 ppm max.

*Bromine index.

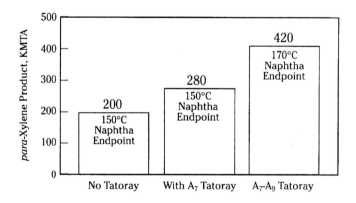

Basis: 985 KMTA (25,000 BPD), Light Arabian Naphtha

FIGURE 2.7.6 Maximum *para*-xylene yield with Tatoray.

A_9 precursors in the feed. About 25,000 BPD of this heavier naphtha produces about 420,000 MTA of *para*-xylene, an increase of 110 percent over the base complex.

The Tatoray process produces petrochemical-grade benzene and xylene products. Benzene purity with a 100 percent toluene feed easily meets the ASTM specifications for refined 545-grade benzene. With a feed of 50 percent toluene and 50 percent C_9 aromatics, the benzene product purity meets the ASTM specifications for refined 535-grade benzene. The xylene product from a Tatoray unit contains an equilibrium distribution of xylene isomers and is low in ethylbenzene. A typical xylene fraction from a Tatoray unit contains 23 to 25 wt % *para*-xylene, 50 to 55 wt % *meta*-xylene, 23 to 25 wt % *ortho*-xylene, and only 1 to 3 wt % ethylbenzene. This low ethylbenzene concentration makes the xylenes produced by Tatoray extremely valuable as feedstock to either a UOP Parex* unit or a *para*-xylene crystallization unit.

EQUIPMENT CONSIDERATIONS

Because the Tatoray process uses relatively mild operating conditions, special construction materials are not required. The simplicity of the process design and the use of conventional metallurgy results in low capital investment and maintenance expenses for the Tatoray process. The simple design of the Tatoray process also makes it ideal for the conversion of existing reformers, hydrodealkylation units, and hydrotreaters to Tatoray service. To date, two idle reforming units, two hydrodealkylation units, and one hydrodesulfurization unit have been successfully converted to service as Tatoray units.

The charge heater is normally a radiant-convection-type heater. The process stream is heated in the radiant section, and the convection section is used for a hot-oil system or for steam generation. The heater can be designed to operate either on fuel gas or fuel oil, and each burner is equipped with a fuel-gas pilot. A temperature controller at the heater outlet regulates the flow of fuel to the burners. Radiant-section tubes are constructed of 1.25 percent Cr and 0.5 percent Mo. Tubes in the convection section are carbon steel.

The Tatoray process uses a simple down-flow, fixed-bed, vapor-phase reactor. The reactor is constructed of 1.25 % Cr–0.5 % Mo.

The purpose of the product separator is to split the condensed reactor effluent into liquid product and hydrogen-rich recycle gas. The pressure in the product separator determines the pressure in the reactor. Product separator pressure is regulated by controlling the rate of hydrogen makeup flow. Hydrogen purity in the recycle gas is monitored by a hydrogen analyzer at the recycle gas compressor suction. When hydrogen purity gets too low, a small purge is taken from the recycle gas. The product separator is constructed of killed carbon steel.

The recycle gas compressor is usually of the centrifugal type and may be driven by an electric motor or a steam turbine. The compressor is provided with both a seal-oil and lube-oil circuit and an automatic shutdown system to protect the machine against damage.

The stripper column is used to remove light by-products from the product separator liquid. The stripper column usually contains 40 trays and incorporates a thermosiphon reboiler. Heat is usually supplied by the overhead vapor from the xylene column located upstream of the Parex unit. The stripper column is constructed of carbon steel.

The combined feed exchanger is constructed of 1.25 % Cr–0.5 % Mo. Other heat exchangers are constructed of carbon steel.

---

*Trademark and/or service mark of UOP.

TABLE 2.7.2 Investment Cost and Utility Consumption*

Estimated erected cost, million $U.S.	17.6
Utility consumption:	
Electric power, kW	395
High-pressure steam, MT/h	7.2 (15.8 klb/h)
Cooling water, m^3/h	158 (696 gal/min)
Fuel fired, million kcal/h	12.6 (50 million Btu/h)

*Basis: 98.5 ton/h (17,000 barrels per day) of feedstock. Feed composition: 60 wt % toluene, 40 wt % C$_9$ aromatics.

Note: MT/h = metric tons per hour.

CASE STUDY

A summary of the investment cost and utility consumption for a typical Tatoray unit is shown in Table 2.7.2. The basis for this case is a Tatoray unit processing 98.5 MT/h (17,000 BPD) of a feed consisting of 60 wt % toluene and 40 wt % A$_9$. This case corresponds to the case study for an integrated UOP aromatics complex presented in Chap. 2.1. The investment cost is limited to the Tatoray unit and stripper column and does not include further downstream product fractionation. The estimated erected cost for the Tatoray unit assumes construction on a U.S. Gulf Coast site in 1995. The scope of the estimate includes engineering, procurement, erection of equipment on the site, and the initial load of TA-4 catalyst.

COMMERCIAL EXPERIENCE

As the demand for *para*-xylene has grown in recent years, the UOP Tatoray process has become increasingly popular. By 1995, UOP had licensed a total of 32 Tatoray units, 25 of which were licensed in the last 10 years. Currently, 24 UOP Tatoray units are in operation, and eight additional units in various stages of design and construction. The feedstocks processed in UOP Tatoray units range from 100 wt % toluene to a mixture of 30 wt % toluene and 70 wt % A$_9$. Design feed rates range from 17 to 146 m^3/h (2600 to 22,000 BPD).

BIBLIOGRAPHY

Jeanneret, J. J., "Developments in *p*-Xylene Technology," DeWitt Petrochemical Review, Houston, March 1993.

Jeanneret, J. J., "*para*-Xylene Production in the 1990s," UOP Technology Conferences, various locations, May 1995.

Jeanneret, J. J., C. D. Low, and V. Zukauskas, "New Strategies Maximize *para*-Xylene Production," *Hydrocarbon Processing,* June 1994.

Kuchar, P. J., J. J. Jeanneret, C. D. Low, and J. Swift, "Processes for Maximum BTX Complex Profitability," UOP Technology Conferences, various locations, September 1992.

P · A · R · T · 3

CATALYTIC CRACKING

CHAPTER 3.1
EXXON FLEXICRACKING IIIR FLUID CATALYTIC CRACKING TECHNOLOGY

Paul K. Ladwig
Exxon Research and Engineering Company
Florham Park, New Jersey

FOREWORD: WHY SELECT FLEXICRACKING IIIR?

This chapter provides general technical information on Flexicracking IIIR* technology, the current Exxon Research and Engineering (ER&E) state-of-the-art fluid catalytic cracking process. It also covers the major technical reasons for selecting ER&E's technology:

The technology is up to date as a result of a continuous improvement program.

Improvements flow from a significant R&D effort, extensive feedback on operating units, and 50 years of design experience.

The main advantages for Flexicracking IIIR are

- Feed and yield flexibility
- Unsurpassed reliability
- Modern cost-effective design
- Applicability to grass roots and upgrades

INTRODUCTION: EVOLUTION OF FLEXICRACKING IIIR

Exxon invented and commercialized the fluid catalytic cracking (FCC) process 50 years ago and continues to be the leader in implementing FCC process improvements.

---
*Trademark of Exxon Corporation.

ER&E's Flexicracking technology was developed from many years of design and operating experience plus an ongoing, aggressive R&D effort. Flexicracking technology continues to improve and evolve. The new ER&E Flexicracking IIIR unit can efficiently handle feeds from hydrotreated vacuum gas oil (VGO) to residual oil (resid). Even highly aromatic lube extracts can be economically processed. The Flexicracking IIIR unit is based on the advanced features in use in the current Flexicracking units and also includes many new features. Several of these new features were suggested by the ER&E proprietary Advanced Process Model (APM), a highly sophisticated fundamental hydrodynamic and kinetic model of the FCC process. Development and commercialization of the new Flexicracking IIIR used proprietary finite-element hydrodynamic models, cold flow models, pilot plant runs, and commercial demonstrations of key new technology applications. Flexicracking IIIR unit process goals are high performance in feed flexibility, product selectivity, conversion, operating flexibility, ease of operation, reliability, and service factor.

2.5 Million Barrel per Day Cracking Capacity

Today, 23 FCC units (FCCUs) are operating in Exxon refineries around the world with a total capacity of 1.2 million barrels per day. In addition, 44 licensed units with a total capacity of 1.3 million barrels per day are in operation. Extensive operating feedback and design experience with these units was incorporated in the Flexicracking IIIR design. This chapter provides information on ER&E-designed FCCUs.

Resids and Lube Extracts Are Routine Feeds

ER&E-designed units process a wide variety of feedstocks ranging from hydrotreated VGO to vacuum resids. Lube extracts, deasphalted oil (DAO), coker gas oil, etc. are also routine feeds. There are obvious strong economic incentives to process marginal feedstocks. Deep-cut vacuum pipestill (VPS) operations, resid feeds, and highly aromatic feeds are, therefore, a fact of life for ER&E-designed units. For example, a well-known refinery performance assessment organization surveyed 50 European FCCUs in 1988. Normalization of the survey, to account for poorer quality feeds and differences in operating conditions, shows that conversion on ER&E-designed units was better than or equal to conversion on competitive units.

Associated Technologies

ER&E offers associated technologies that can be integrated with the Flexicracking IIIR process. For example:

Cat Feed Hydrotreating (GO-fining).* This is a modern and proven process that provides the refiner with an efficient and economical way of upgrading the quality of Flexicracking feeds. Important quality criteria for a cat cracking feedstock are low levels of metals, Conradson carbon residue, multiring aromatics, nitrogen, and sulfur. Feeds that are suitable for GO-fining include thermal and coker gas oils, deasphalted oils, heavy cat cycle oils, and heavy vacuum gas oils.

---

*Trademark of Exxon Corporation.

Wet Gas Scrubbing. While FCCUs are one of the most important units within a modern refining complex, they are also a potential source of atmospheric emissions. As early as 1971, ER&E began development work on a scrubbing system to control atmospheric emissions from FCCUs. The resulting Exxon wet gas scrubbing (WGS) unit is a simple, effective, and economic method of controlling particulates and sulfur oxides to meet current regulations. Since the start-up of the first unit in 1974, a total of 14 units have been installed.

THE FLEXICRACKING IIIR PROCESS

Main Features

The Flexicracking IIIR unit, as shown in Fig. 3.1.1, uses the minimum reactor cyclone vessel elevation consistent with the appropriate reactor-riser length. This minimizes structure height and cost. The proprietary, commercially proven, reactor-riser termination minimizes dilute-phase cracking. A state-of-the-art feed injection system provides excellent feed atomization, dispersion, and catalyst-oil contacting. A hydrodynamically engineered feed injection zone greatly enhances the benefits of the new efficient feed injectors by minimizing undesirable riser-reactor backmixing.

Precise Process Control

The Flexicracking IIIR unit catalyst circulation system uses J-bend transfer lines with throttling slide valves for both the spent and the regenerated catalyst circuits. This configuration provides excellent reactor temperature control via catalyst circulation control and/or feed temperature control. Catalyst circulation is smooth and stable and essentially insensitive to catalyst type or fines content. No potentially troublesome expansion joints in the critical catalyst transfer lines are required with the Flexicracking IIIR unit layout. The overall configuration is flexible enough to allow pressure balance control with a single throttling slide valve, if wanted.

Reliable Regenerator

The Flexicracking IIIR unit uses a high-velocity, high efficiency regenerator with a plate-grid air distributor for the best possible air distribution and ultralow carbon on regenerated catalyst (CRC), for example, 0.03 wt %. The grid uses a reliable, low-thermal-stress design that is inherently resistant to process upsets. This is accomplished by cooling the grid with regeneration air. The regenerator design is also simple and uncluttered for high reliability. No baffles or mechanically complex pipe distributors are needed. A single low-temperature grid seal, not subject to catalyst erosion, is located in the air plenum.

Efficient Stripping

An efficient staged-catalyst stripper is provided to minimize coke production and loss of reactor products to the regenerator. Many laboratory and commercial tests have shown that the catalyst stripper can be significantly improved by some relatively inex-

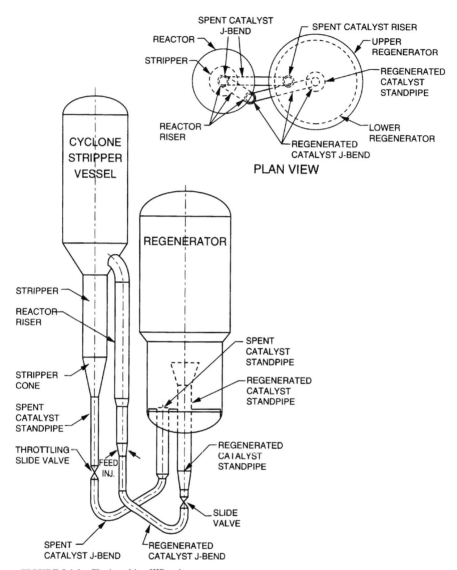

FIGURE 3.1.1 Flexicracking IIIR unit.

pensive design and operational changes. These new design features have been incorporated into the Flexicracking IIIR design.

Good Heat Balance Control

Normally even poor feeds can be accommodated with the inclusion of an ER&E CO combustor, which can be a steam generator or an oil heater. This flexibility is advantageous to many refineries with a steam containment problem. The ER&E proprietary adiabatic combustor design greatly improves the reliability of the combustor as compared to conventional water wall designs. In addition, the ER&E design minimizes the amount of supplemental fuel firing which is necessary to ensure complete combustion of CO. Alternatively, a proven UOP catalyst cooler can be offered as part of the Flexicracking IIIR unit, if the feed quality requires this type of heat balance control equipment.

Proven Safety

Well-proven emergency shutdown systems (ESDs) are included to protect the unit from damage during upset conditions.

Patented Trickle Valves

Patented ER&E trickle valves on the reactor and regenerator cyclones promote smooth catalyst flow in the cyclone diplegs, giving an extra margin to avoid cyclone flooding. The design features lower leakage relative to conventional trickle valves. It is also essentially jam-proof, which is especially important in the potentially coking environment of the reactor cyclone vessel. These innovations greatly reduce wear on the wear-prone secondary trickle valves.

Designed to Be Cost-Effective

The Flexicracking IIIR unit reactor (cyclone vessel) elevation has been reduced to the minimum consistent with the riser-reactor length necessary for short contact time (SCT) catalytic cracking. This, of course, results in reduced structural costs and improved accessibility.

The Flexicracking IIIR unit high-efficiency riser termination is not only more efficient than previous designs, but is lower in capital cost and pressure drop. The lower pressure drop results in lower operating costs.

The Flexicracking IIIR unit stripper is more efficient than previous designs and is also lower in capital cost. The Flexicracking IIIR unit regenerator grid is more thermal-transient resistant than previous designs and is also lower in capital cost.

MAJOR PROCESS FEATURES

Feed Injection

Improved Atomization/Distribution and Reduced Backmixing. Over the last decade, it has become increasingly obvious that improved feed injection systems could result in improved process performance and profitability. Improved feed injection systems became essential with the advent of SCT reactor designs and operations with resids and other marginal feeds. The objectives of improved feed injection systems are

- Optimum feed atomization and distribution of the atomized feed across the riser allows rapid mixing, vaporization, and reaction to occur. Poor atomization or poor distribution can lead to temperature maldistributions and catalyst hot spots that cause increased gas and coke production.
- Reduced catalyst backmixing in the injection zone. An ideal plug flow reaction system minimizes yield debits due to backmixing, which recontacts the reaction products with fresh catalyst and uncracked feed.

New Injectors Reduce Drop Size and Improve Distribution. Feed injector design improvements have been developed through laboratory studies, cold flow testing, and many commercial trials. Continuous improvements have been demonstrated from the original straight pipe injectors to "conventional" nozzles and now the latest design of state-of-the-art injectors. Figure 3.1.2 shows the relative drop size and spray distribution pattern of the latest injectors compared to the older conventional nozzles that they have replaced. The improved atomization and distribution shown are achieved at relatively low steam rates and pressure drops. Riser and nozzle erosion experience has also been excellent. The improvements in atomization and spray distribution are clear and the commercial results have been outstanding.

New Injectors Improve Riser-Reactor Temperature Profile. Poor feed injector performance can lead to poor yields from the riser-reactor. Temperature maldistribution above the injection zone caused by maldistribution of feed and catalyst results in nonuniform mixing and nonuniform reaction conditions. Yield debits are the result. Figure 3.1.3 shows typical commercial riser temperature profiles before and after the installation of new state-of-the-art feed injectors. The temperature profile after the

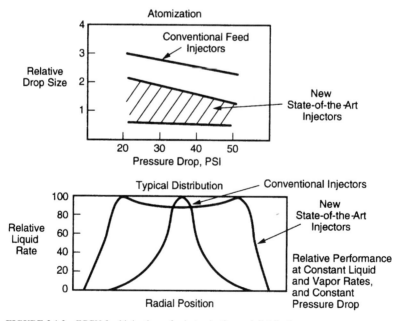

FIGURE 3.1.2 FCCU feed injection—feed atomization and distribution.

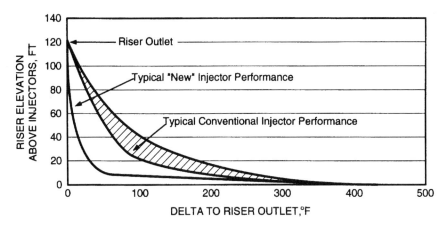

FIGURE 3.1.3 FCCU feed injector performance—riser temperature profiles.

installation of the new injectors clearly indicates improved mixing and increased vaporization lower in the riser. Substantial improvements in both gas and coke selectivities have been demonstrated in dozens of feed injection system upgrades.

New Feed Injection Zone Targets Plug Flow. Recent tracer tests of feed injection zones have shown that backmixing in the injection zone can be drastically reduced by adjusting and optimizing the design and operating conditions (Fig. 3.1.4). Reducing backmixing in the zone is required to provide uniform temperatures and reaction conditions, and to approach the ideal plug flow reaction system performance.

Commercial Results Verify High Profitability. The commercial results from improved feed atomization and dispersion have been very positive. Table 3.1.1 shows typical before and after commercial data from one unit that installed new feed injectors without optimizing the feed injection zone. Further improvements would have

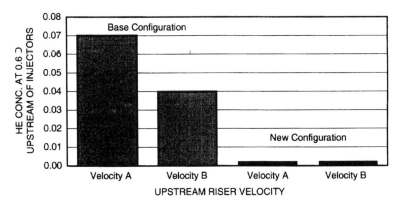

FIGURE 3.1.4 Feed injection zone optimization—backmixing improvement. Helium tracer test results—He added at injectors.

TABLE 3.1.1 FCCU Feed Injector Performance—Typical Before/After Yield Comparison

Constant reactor temperature, feed quality, and catalyst		
	Base case injectors	New state-of-the-art injectors
$C_2 -$ (including H_2S), wt %	3.0	2.9
C_3 total, LV % (wt %)	9.6 (5.4)	9.5 (5.3)
C_4 total, LV % (wt %)	15.4 (10.0)	16.8 (10.9)
C_5/430°F (221°C) naphtha, LV % (wt %)	58.9 (48.8)	63.1 (52.3)
LCO, LV % (wt %)	21.0 (21.3)	17.7 (17.9)
780°F + (415°C +) bottoms, LV % (wt %)	4.3 (5.2)	3.5 (4.3)
Coke, wt %	6.3	6.3
Conversion, LV % (wt %) FF	74.7 (73.5)	78.8 (77.8)

Note: LV = liquid volume; LCO = light cycle oil; FF = fresh feed.

been realized by optimizing the feed injection zone. Refineries have taken advantage of the operating margins allowed by these improved yield selectivities to increase conversion, increase throughput, or run marginal feeds.

Riser and Riser Termination

ER&E Pioneered All-Riser Cracking in 1955. The evolution of cracking catalysts has resulted in a corresponding evolution in the reaction system process configuration (from the old backmixed grid-type reactor configuration to modern plug flow all-riser configurations). Today, the fluid catalytic cracking unit "reactor" vessel is no longer a reactor, but has become a cyclone vessel/catalyst stripper. The riser-reactor and riser termination have become the reactor. The Exxon Baytown 2 FCC unit was a pioneer state-of-the-art unit in this evolution. In 1955, a riser termination consisting of rough-cut cyclones was installed in a cyclone containment vessel and is still in operation. The rough-cut cyclones terminated a long, straight, vertical riser-reactor.

Riser Terminations. The riser-reactor termination device for Flexicracking IIIR units has been the subject of an intensive R&D effort and has been continuously improved. Tools for evaluating the process performance of alternative termination configurations have improved dramatically in recent years. In addition to conventional cold flow modeling, the riser termination has been developed by using advanced finite-element two- and three-dimensional two-phase (gas-solid) hydrodynamic models. Commercial unit tracer tests have been used to validate this modeling. Figure 3.1.5 shows the predicted and actual vapor residence time distribution for one commercial test. This improved modeling prediction technique allows accurate evaluation and prediction of potential process performance of various riser and riser termination configurations that might be considered for an FCCU modernization.

Commercial Results Demonstrate Profitability. The benefits of short contact time cracking are not automatic. The cracking intensity represented by the bed and/or dilute phase must be replaced by another variable or conversion will be reduced. As the catalyst holdup and vapor residence time in the reaction system are reduced by an all-riser type of conversion, other intensity variables such as catalyst/oil, reactor temperature, or catalyst activity are increased to compensate. An 11 to 12% decrease in coke yield

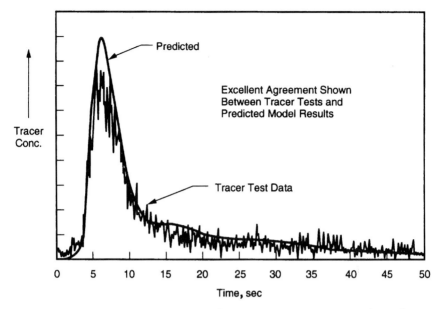

FIGURE 3.1.5 Commercial riser termination performance—gas tracer response at cyclone outlet.

and a 20 to 25% decrease in dry gas yield have been achieved at constant conversion. The decreased coke yield would obviously be used to increase feed rate, add marginal feedstocks, and/or increase conversion, thus further enhancing the credits for SCT. Typical commercial before and after yield comparisons for a conversion to an all-riser SCT reactor configuration are shown in Table 3.1.2.

Commercial Results Demonstrate Operability. The ER&E design has been in commercial operation without any operability problems. There have been no start-up, shutdown, coking, erosion, or any other reported problems with the riser termination.

Dry Gas Reduction Demonstrated. Reductions in dry gas yield have been clearly demonstrated in many commercial tests following modernization of older grid units to all-riser and SCT unit configurations. Figure 3.1.6 shows the dry gas yield as a function of reactor temperature for typical grid and all-riser unit configurations. This figure covers many commercial unit reactor conversions and a wide range of operating temperatures and conditions.

SCT Yields are "Tunable." Process analysis by the ER&E proprietary Advanced Process Model has identified the relative contributions of the intensity parameters to yield selectivity. The effect of microactivity test (MAT), catalyst type, catalyst-to-oil ratio, and reactor temperature on SCT yields is predictable. The insight provided allows tailoring SCT performance via operating parameters.

Stripper

Strippers Can Be Improved. An efficient staged-catalyst stripper is provided to minimize coke production and loss of reactor products to the regenerator. Many commer-

TABLE 3.1.2 Short Contact Time Riser Conversion Commercial Performance Tests

Constant feed rate/quality and constant coke and dry gas yield.

	Base operation	SCT operation
Conversion	Base	+ 3.5 LV % FF
Dry gas	Base	Constant
C_3 unsaturate	Base	+ 23%
C_4 unsaturate	Base	+ 24%
Total LPG saturate	Base	+ 11%
Naphtha	Base	+ 2%
Distillate	Base	− 5%
HCO + bottoms	Base	− 13%
Coke, wt %	Base	Constant
Naphtha $(R + M)/2$	Base	+ 1.5%

Note: LV % = liquid volume percent; FF = fresh feed; R = research octane; M = motor octane.

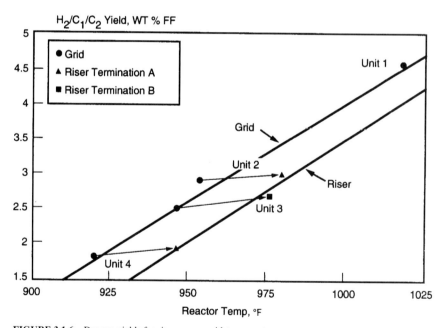

FIGURE 3.1.6 Dry gas yields for riser versus grid-type reactor.

cial tests have shown that the catalyst stripper can be significantly improved by some relatively inexpensive design and operational changes.

Stripper Improvements Are Economic. Large improvements to conventional stripper performance can be gained from operational as well as hardware changes. Laboratory and commercial studies have been conducted to improve the performance of existing hardware, as well as to develop new design features and options for modernizing existing designs.

Operational Changes Enhance Performance. Laboratory tests on spent catalyst samples from many commercial units have shown a significant potential to improve stripper performance and reduce the coke yield from existing strippers (Fig. 3.1.7). This graph shows that 20 to 40 percent reductions in coke yield, relative to the base case, are possible. Commercial tests of operational changes have also shown significant improvements in stripper performance. Figure 3.1.8 shows a roughly 15 percent reduction in coke yield following an adjustment to stripper operating conditions. These and other operational changes can be applied to new units, as well as many existing units, to improve performance.

Hardware Modifications. Hardware modifications are under development to achieve improved stripper performance and commercial trials are scheduled. The potential credits make this an attractive area for continued study and improvement.

Regenerator

High-Velocity Regenerator Minimizes Holdup. The Flexicracking IIIR unit uses a high-velocity, high-efficiency regenerator. A high-velocity dense bed minimizes

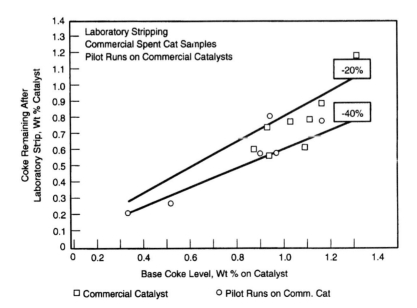

FIGURE 3.1.7 Ultimate stripper performance—potential coke reduction.

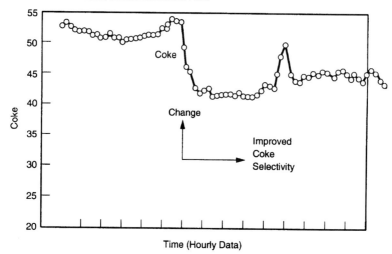

FIGURE 3.1.8 Commercial FCCU stripper optimization test results.

inventory, which is highly desirable in this era of catalyst optimization; i.e., faster catalyst changeouts are possible. The Flexicracking IIIR low-velocity dilute phase minimizes entrainment to the cyclones, which results in lower catalyst losses and longer cyclone life. The high efficiency of the regenerator can be seen from carbon on regenerated catalyst; i.e., most units operate in the 0.03 wt % region. The efficiency of the ER&E design is due, in part, to the use of a flat plate grid for an air distributor. A grid with several thousand grid holes gives a finer, more uniform air distribution than a pipe distributor with several hundred pipes. The grid is also less mechanically complex than a pipe distributor.

Plate Grid Reliability Enhanced. A new design for a plate-type air distributor grid has been developed and commercially tested. The design substantially reduces the thermomechanical stresses that develop during unit operation. The relative deformations experienced by a conventional grid and a modernized grid with reduced stresses and deformations are shown in Fig. 3.1.9. This reduction in deformation and stresses translates directly into improved reliability.

Reliability Improvements Continue. Cyclone lining and trickle valve details have been continuously upgraded to improve service life and reliability. Inspection histories and process performance histories have been reviewed to reevaluate standards for cyclone design and operation.

Control Systems

Tight Reactor Temperature Control Can Be Profitable. ER&E-designed control systems can be tailored to the needs, abilities, and wants of individual refineries. At

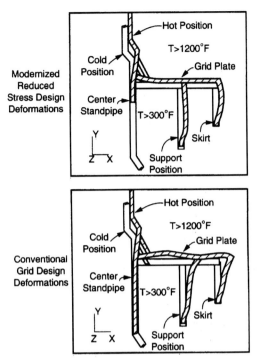

FIGURE 3.1.9 FCCU reliability improvement by reduced thermomechanical regenerator grid stresses.

the simplest level, the instrumentation system handles the heat balance and controls reactor temperature by using catalyst circulation rate, feed preheat, or even feed rate (when other control variables are at their constraint values). Carbon balance would be controlled manually by the operator at this level. At the intermediate level, the supervisory control system handles heat and carbon balance and also controls reactor temperature by using cascades and multiple variables. At the advanced level, commercially available optimization and control algorithms can be obtained to maximize performance and control the heat and carbon balance. This strategy maintains the desired tight control on reactor temperature but allows other controlled variables a greater range of control. The algorithms employed can vary. Several different strategies have been successfully used on Flexicracking units.

Flexicracking IIIR Unit Designed for Tight Temperature Control. The Flexicracking IIIR unit is capable of reactor temperature control via catalyst circulation and/or feed temperature control. Either control mode allows tight reactor temperature control. The dual J-bend design, with throttling slide valves in each circuit, circulates catalyst extremely smoothly. This greatly facilitates tight reactor temperature control. Smooth circulation is maintained over a wide range of catalyst types and particle sizes. These features allow a Flexicracking IIIR unit, using intermediate or advanced controls, to precisely control reactor temperature to ±1°F or better.

Tight reactor temperature control is necessary to consistently attain the desired yields and octane. The modern Flexicracking IIIR unit can provide this tight reactor temperature control.

Advanced Yield Prediction Capability

It is, of course, essential to maximize selectivity for each type of feed and maximize conversion for each operating objective. Process analysis by the ER&E proprietary Advanced Process Model has identified the relative contributions of the various sections of the reaction system to the ultimate process yields. The APM is a sophisticated, fundamental, hydrodynamic, and kinetic-based computer model for predicting FCC yields and product qualities. This model may be the most advanced FCCU model in use anywhere. The model uses easily obtained feed characterization data and unit-specific hydrodynamic information. The model captures the interactions among feedstock composition, catalyst formulation, operating conditions, and unit configuration. The technique for simulating unit configuration is very flexible, which allows the use of the model for new geometry designs as well as for existing units. It therefore allows the unit configuration to be tailored to specific feeds and specific processing objectives.

Example 1—Grid versus All-Riser. The power of this analysis can be seen by comparing the relative effects of an older grid-type configuration to a newer all-riser configuration with an efficient riser termination and an improved stripper. Figure 3.1.10 illustrates the relative gas and coke contributions from the various sections of the reaction system for these old and new system configurations.

Example 2—Smart Engineering Allows Low Cost Resid Operations. Knowledge of catalyst, hydrodynamic, and operational effects can be highly profitable. Figure 3.1.11 shows how one unit was debottlenecked to process resid feed on top of an already highly aromatic lube extracts feed. Cost of the modifications required was insignificant. Effective engineering tools have been developed to implement this type of cost-effective "drop-in" technology application.

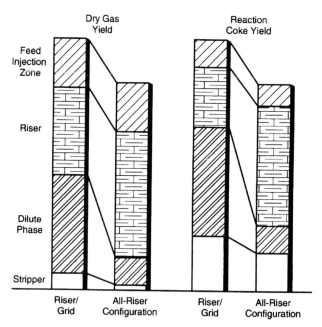

FIGURE 3.1.10 FCCU reaction system performance, dry gas and coke yield improvements—modern all-riser reactor versus riser/grid reactor.

Example 3—Low Emissions Mogas. The ER&E Flexicracking IIIR unit is ideal for low-emissions mogas production. Isobutylene and olefins production for associated processes such as methyl tertiary butyl ether (MTBE) can be easily maximized. Figure 3.1.12 shows the step change in selectivity to light olefins that is possible with Flexicracking IIIR technology.

RELIABILITY

Reliability Is the Key to Profitability

Reliability is measured by time on stream and capacity utilization. Although significant profitability gains can be achieved by process performance improvements, they can be lost very quickly if the unit is not operating. The incentives for maximizing capacity utilization and minimizing downtime are, therefore, obvious. Not so obvious is that production losses due to unscheduled outages will affect profitability more than production losses due to scheduled outages; i.e., unexpected downtime is worse than scheduled downtime.

Systematic Approach to Improved Reliability

Over the last decade, ER&E has conducted a systematic analysis of refinery, process, and equipment reliability to find the sources of both scheduled and unscheduled outages. Experts in different disciplines such as materials, mechanical, machinery, instru-

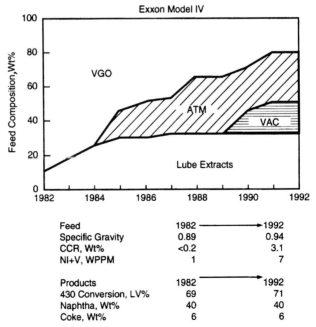

FIGURE 3.1.11 Staged technology program provides effective route to increased-resid catalytic cracking.

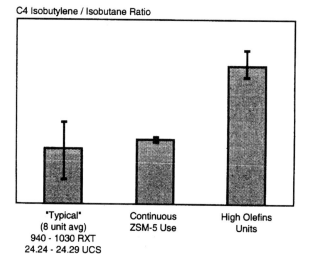

FIGURE 3.1.12 Latest ER&E designs provide high selectivities to light olefins.

mentation, computing, process engineering, process design, operations, and maintenance have participated in the reliability analysis and studies. Key sources of production losses were identified, e.g., component unreliability, electrical/instrument/computer failure, machinery limitations, and process procedures. The sources were then analyzed to eliminate or mitigate problem areas and thereby improve reliability. The benefit was a significant improvement in capacity utilization and increased profitability. These learnings have been incorporated into the latest Flexicracking IIIR unit design, as well as into revamps and improvements on existing Flexicracking units.

Systematic Approach Has Worked

Because of these efforts, over the last decade significant improvements in worldwide FCCU reliability and on-stream service factors have been achieved. Reliability analysis remains a very high priority item. The goal is continuous improvement.

Example 1—ER&E FCCUs Rated Best in Capacity Utilization. A well-known refinery performance assessment organization surveyed 50 European FCCUs in 1988 and concluded that ER&E-designed units had a significantly better capacity utilization than competitive units.

Example 2—Unplanned Shutdowns Reduced. Capacity loss from unplanned shutdowns of Exxon FCCUs has decreased significantly over the 1980–1988 time frame. Figure 3.1.13 shows this decrease.

Example 3—Recent Flexicracking Unit Exceeds Expectations. The thirteenth grassroots Flexicracking unit successfully started up in 1987. The first scheduled turnaround occurred in 1991, 4 years after the initial start-up. During these 4 years, or 1461 days, the unit operated at an average capacity of 31,630 barrels per stream day (BPSD). The capacity utilization factor was 113%, based on the design capacity of 25,000 barrels per day (BPD).

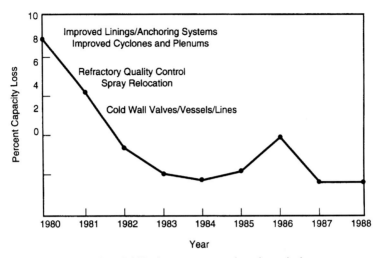

FIGURE 3.1.13 FCCU reliability improvement—unplanned capacity loss.

During the run, the unit throughput was increased step by step up to a maximum capacity of 35,000 BPD. A capacity of 38,500 BPD was demonstrated following the recent turnaround. It was achieved after the installation of new feed injectors designed by ER&E, and represents a 10 percent capacity increase compared to the previous maximum feed capacity of 35,000 BPD.

In the 4-year period following initial startup, the Flexicracking unit had 25 feed outages, most of which were caused by external factors, i.e., 12 power failures, 5 instrument malfunctions, 3 expander problems, and 2 personnel strikes. Only three of the outages were caused by mechanical problems in the fluid solids section. No pressure or high-temperature excursions resulted from any of the outages.

Despite these outages and resulting thermal cycles, the unit was in excellent condition at the time of the first turnaround, after 4 years of operation.

RESID CONSIDERATIONS

Flexicracking IIIR technology incorporates all of the advanced features developed specifically to handle a wide range of marginal feeds. Figure 3.1.14 shows how the feed quality varies among six different ER&E-designed units which have been revamped/modified to process heavier feeds. Notice that although conversion decreases for the poorer feeds, conversion as a percentage of ultimate conversion remains relatively high. Ultimate conversion is calculated from fundamental feed qualities.

Good engineering philosophy for any design is to keep it simple, reliable, and economic. The following guidelines are useful for considering resid design options:

1. *Modernization is first approach.* Modernization of an existing VGO FCCU to handle resid is normally feasible and requires much lower investment than construction of a grass-roots resid cracker.

2. *Single-stage regeneration is best approach.* Single-stage regeneration is normally feasible, even when processing resids. Single-stage regeneration is more cost-effective than two-stage regeneration. Additional equipment is more costly, more

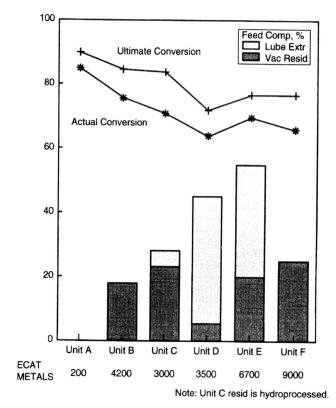

FIGURE 3.1.14 High conversions maintained with poor feeds.

complex to operate and leads to an incremental decrease in reliability. Flexicracking IIIR units do not require two-stage regenerators.

3. *Partial burn regeneration is first approach.* A partial burn regenerator operation can economically export combustion heat from the regenerator to an external CO combustor, thus permitting heat balanced operation with resids and other marginal feeds. The CO combustor also permits decoupling the heat balance from the fluid solids process operations. This allows optimization of the regenerator temperature and the feed injection zone conditions. Significant air blower savings are possible since the CO is burned externally at a much lower pressure than the regenerator pressure. The ER&E adiabatic CO combustor design is ideal for this service. Reliability is significantly better than with conventional water wall designs and the need for supplemental fuel firing is greatly reduced. If necessary to satisfy the heat balance with very high coke make resids, a UOP catalyst cooler design based on commercially demonstrated technology can be offered.

4. *Short contact time is necessary.* A short contact time configuration is normally used for processing resids and other marginal feeds. Most Exxon and licensed reactor-bed units have been converted to configurations with shorter contact time cracking and some to true short contact time riser cracking. The trend to true short contact time cracking with 100 percent plug flow continues. Many of the units converted to shorter contact time cracking will eventually be converted to true SCT.

5. *A modern feed injection system is essential.* Since riser cracking by definition must take place in the riser, feed atomization, uniform cross-sectional feed distribution, and feed penetration to the center of the riser are critical. Therefore, a modernized feed injector zone and proprietary ER&E feed injectors should normally accompany any short contacting upgrade.

   6. *Coke formation need not be a problem.* Coke formation in the cyclone vessel is not a problem for ER&E-designed units processing resids and/or marginal feeds. The key is a modern feed injection system, a properly designed SCT reactor configuration, and correct operating conditions.

UPGRADING WITH FLEXICRACKING IIIR TECHNOLOGY

Modernization Is Profitable

Many engineering challenges are posed when Flexicracking IIIR technology is specified for a FCCU modernization. Grass-roots designs have fewer constraints. The reduced degrees of freedom in implementing a modernization calls for extensive experience and sophisticated design tools. However, essentially any FCCU can benefit from an aggressive modernization program. For example:

- Exxon operates some of the oldest and some of the newest FCCUs. Several of them are vintage 1940s and they are still economical. The reason is a continuous improvement policy. This means that older FCCUs are periodically upgraded and modernized.
- Two Model II FCCUs, designed in 1947, are now operating with high reaction temperature (HRT) and high-temperature regeneration (HTR). Another unit, a Model III FCCU, designed for 25,000 BPD in 1953, is now operating at about 125,000 BPD with HRT and HTR.

Resid Feeds Handled with FCCU Modernization

State-of-the-art resid conversion has been demonstrated with proven, economic, simple, and reliable upgrades of existing FCCUs. ER&E resid processing technology is continuously improving and has reached the point where expensive and complex hardware and process configurations are no longer necessary; i.e., resid operations can almost always be accommodated by modernization.

Model IV Converted to Resid Processing Unit

For example, one existing Model IV unit that was originally constructed in the 1960s has been converted to a state-of-the-art resid processing unit. This Model IV operates with essentially 100 percent North Sea atmospheric resid. Operating data from this unit are shown in Table 3.1.3. The unit operates in a partial burn mode with a CO boiler and does not include a catalyst cooler. Yields and overall process performance of this modernized unit compare favorably with the performance of recent state-of-the-art designs reported in the literature. This VGO unit is now operating very reliably as a resid cracker without the complexity and expense of some grass-roots industry resid crackers.

TABLE 3.1.3 Model IV FCCU Modernized for Resid Service—Typical Yield Performance

Feedstock properties:	
Specific gravity	0.9314
Basic nitrogen, wt ppm	625
Sulfur, wt %	1.0
Conradson carbon, wt %	4.7
Ni on catalyst, wt ppm	3075
V on catalyst, wt ppm	5550
Yields, wt % on FF:	
C_2^- (including H_2S)	2.8
C_3 (total)	3.5
C_4 (total)	6.9
C_5/430°F (221°C) (gasoline)	43.0
430°F (221°C)/700°F (371°C) (LCO)	24.2
700°F + (371°C +) (HCO + slurry)	12.9
Coke	6.7
Conversion, wt %	62.9

Note: FF = fresh feed; LCO = light cycle oil.

Catalyst Cooling

This particular converted Model IV unit did not require catalyst cooling, as it operates with a partial burn regenerator. However, for those operations that require catalyst cooling, ER&E can offer UOP catalyst cooling technology.

ECONOMICS OF EXXON FLEXICRACKING IIIR TECHNOLOGY

The estimated costs for a typical 30 thousand barrels per stream day (kBPSD) Flexicracking IIIR unit processing two different feedstocks are presented in Table 3.1.4. The first case is for a conventional gas oil operation and the second is for a resid FCC processing a 5 Conradson carbon feed. Feeds from both cases are derived from Middle Eastern crudes. Utility requirements for both cases are shown in Table 3.1.5.

ER&E-DESIGNED COMMERCIAL FCC UNITS

ER&E has designed over 2.6 million barrels per day (million BPD) of catalytic cracking capacity of the world total of about 12 million BPD; ER&E-designed units range in capacity from 3 to 140 thousand barrels per day (kBPD). Since 1976, ER&E has designed 19 new Flexicracking units with over 600 kBPD of total capacity. Seventeen of these units are in operation. ER&E has experience with a full spectrum of unit configurations and reactor designs ranging from the early bed cracking reactors of the 1940s to modern all-riser reactors. A list of ER&E-designed fluid catalytic cracking

TABLE 3.1.4 Flexicracking IIIR Investment Estimates

Direct material and labor, 2d quarter 1994, Gulf Coast, million $

	30 kBPSD vacuum gas oil	30 kBPSD resid (5 Conradson carbon feed)
Fluid solids section*	17.1	22.1
Primary fractionator†	13.7	13.8
Light ends‡	6.2	6.2
Total§	37.0	42.1

*Includes reactor, regenerator, air blower, fresh and spent catalyst hoppers, regenerator flue gas tertiary cyclone and waste heat recovery, stack.

†Includes primary fractionator, sidestream strippers, wet gas compressor, and feed preheat train.

‡Includes absorber deethanizer, sponge absorber, de-butanizer, fuel gas scrubber, and C_3-C_4 and naphtha treating facilities.

§Includes all capital facilities within battery limits, common facilities such as pipe racks and utility headers, vendor shop fabrication of all vessels and piping, direct labor wages, and inland freight to site. Items excluded are warehouse spares, offsites/utilities, site preparation, sales tax on materials, escalation, catalyst/chemicals, indirect costs, field labor overheads, basic engineering costs, Exxon Engineering services, royalties/licensing fees, contractor's engineering/construction fees, vendor shop inspection costs, vendor representatives, loss on surplus materials, start-up costs, owner's costs, and project contingency for changes.

units is presented in Table 3.1.6. Figure 3.1.15 is a photograph of the most recent ER&E-designed grass-roots Flexicracking IIIR unit. The unit, located in Southeast Asia, started up in 1995 and has all of ER&E's latest Flexicracking IIIR technology features. Figure 3.1.16 is a photograph of an Exxon Flexicracking IIIR unit located in Canada. The unit was revamped in 1994 to upgrade to the latest ER&E Flexicracking IIIR technology features.

SUMMARY

Exxon invented the fluid catalytic cracking process and has been an industry leader since commercializing the technology in 1942. ER&E has maintained an ongoing, aggressive R&D effort in this technology that has culminated in the current Flexicracking IIIR design. This design allows efficient processing of many feeds previously considered impractical for fluid catalytic cracking.

There are still gains to be made in approaching maximum conversion or volume expansion (Fig. 3.1.17). The ER&E Flexicracking R&D effort is directed at capturing this improved performance.

The Flexicracking unit design will continue to evolve as part of ER&E's continuous improvement program. Changing feedstocks and operating objectives will also be reflected in new designs. The current trends are to even heavier feeds, with higher sulfur and metals, and to reduced-emissions gasoline. These trends are expected to have long-term impact on the Flexicracking process as well as catalyst design and formulation.

TABLE 3.1.5 Flexicracking IIIR Estimated Utility Requirements[a]

	30 kBPSD Vacuum gas oil	30 kBPSD resid (5 Conradson carbon residue feed)
Requirements:		
Electric power,[b] kW		
Air blower	6286	7590
Wet gas compressor	2113	2262
Others	1093	1455
Cooling water,[c] gal/min (m^3/h)	3816 (867)	3819 (868)
Boiler feed water,[d] gal/min (m^3/h)	183 (41)	516 (117)
Steam at 125 lb/in^2 (8.6 bar gage) klb/h (MT/h)	27 (12.2)	29 (13.2)
Instrument air, SCFM (N m^3/min)	300 (8)	300 (8)
Utility air,[e] SCFM (N m^3/min)	605 (16)	605 (16)
Nitrogen, SCFM (N m^3/min)	50 (1)	50 (1)
Steam produced, klb/h (MT/h)		
High pressure, 600 lb/in^2 (41 bar gage)	78 (35.4)	206 (93.4)
Medium pressure, 125 lb/in^2 (8.6 bar gage)	5 (2.3)	29 (13.2)
Catalyst consumption,[f] ton/day (MT/day)	2 (1.8)	5–10 (4.5–9.0)

[a]Includes FCC fluid solids section, regenerator overhead system, fractionation, feed preheat, and light ends facilities.

[b]Air blower and wet gas compressor requirements provided separately to facilitate alternative driver selection studies. Utility estimates shown based on an axial air blower and centrifugal wet gas compressor with electric drivers.

[c]Based on air fin cooling down to 150°F (65°C) and 30°F (−1°C) cooling water temperature rise.

[d]Boiler feed water includes 10% blowdown.

[e]Includes catalyst addition/withdrawal transport requirements.

[f]Catalyst addition rates for resid operation depend on feed metal levels. Addition rates shown are typical values covering a feed metals range of 10–20 wt ppm Ni + V + Na.

Note: SCFM = standard cubic feet per minute; MT/h = metric tons per hour; MT/day = metric tons per day; N m^3/min = normal cubic meter per minute.

TABLE 3.1.6 ER&E-Designed Catalytic Cracking Units[a]

Company	Location	Capacity, total feed, BPSD	Type of unit	Initial operation
Exxon unit[b,c]	United States	91,000	Model II	1943
Exxon unit[b,c]	United States	95,000	Model II	1943
Licensed unit	South America	47,000	Dual reactors	1943
Exxon unit[b,c]	United States	62,700	Model II (T/L)	1944
Exxon unit	Canada	23,000	Model II (T/L)	1948
Licensed unit[b,c,e]	United States	140,000	Model III	1949
Exxon unit	Europe	85,000	Model III	1951
Exxon unit	Canada	29,200	Model IV	1953
Licensed unit	Europe	40,000	Model IV	1953
Exxon unit[b]	Europe	15,400	Model IV	1953
Exxon unit[b]	Europe	20,000	Model IV	1953
Licensed unit[b]	United States	12,000	Model IV	1953
Exxon unit	Canada	12,600	Model IV	1953
Licensed unit	United States	11,000	Model IV	1953
Exxon unit[b]	Europe	9,360	Model IV	1953
Licensed unit	United States	30,000	Model IV	1954
Licensed unit	Africa	9,740	Model IV	1954
Licensed unit	United States	61,000	Model IV	1954
Licensed unit	South America	19,280	Model IV	1954
Licensed unit	Southeast Asia	10,800	Model IV	1954
Licensed unit	Far East	9,360	Model IV	1954
Licensed unit	Far East	10,800	Model IV	1954
Licensed unit	Canada	9,000	Model IV	1954
Licensed unit	United States	8,000	Model IV	1955
Licensed unit	Far East	30,000	Model IV	1955
Licensed unit[b]	Europe	30,000	Model IV	1956
Exxon unit	Canada	34,000	Model IV	1956
Exxon unit	Far East	36,000	Model IV	1956
Licensed unit	Far East	14,700	Model IV	1957
Licensed unit	United States	30,000	Model IV	1957
Exxon unit[b]	Europe	23,150	Model IV	1957
Licensed unit	Southeast Asia	22,500	Model IV	1957
Licensed unit[b]	South America	25,000	Model IV	1957
Exxon unit	Central America	24,000	Model IV	1958
Exxon unit[b,c]	United States	90,800	T/L	1958
Licensed unit	Central America	25,000	Model IV	1959
Licensed unit	United States	55,000	Model IV	1959
Exxon unit[b]	South America	16,000	Model IV	1960
Licensed unit	Central America	51,000	Model IV	1960
Confidential client	N.A.	21,000	Model IV	1960
Licensed unit	United States	22,000	Model IV	1963
Licensed unit	United States	58,000	Model IV	1963
Exxon unit[b]	Europe	30,190	Model IV	1965
Exxon unit[b]	Europe	40,000	Model IV	1967
Exxon unit[a,b]	Europe	35,600	Model IV	1969
Exxon unit[b]	United States	55,000	Model IV	1969
Exxon unit	Far East	81,000	Model IV	1971
Licensed unit[a]	Far East	27,000	Model IV	1973
Exxon unit[a,b]	Canada	55,000	Model IV	1975
Licensed unit[b,d]	United States	45,000	Side/side	1978
Licensed unit[b]	United States	17,500	Model IV	1979

TABLE 3.1.6 ER&E-Designed Catalytic Cracking Units[a] (*Continued*)

Company	Location	Capacity, total feed, BPSD	Type of unit	Initial operation
Licensed unit[a,b]	Far East	18,000	Flexicracking (R/B)	1979
Licensed unit[b]	United States	19,000	Model IV	1979
Licensed unit[a-c,f]	United States	89,200	Flexicracking (R/B)	1979
Licensed unit[a,b]	United States	34,000	Flexicracking (R/B)	1980
Licensed unit[a-c,f]	United States	33,000	Flexicracking (T/L)	1980
Licensed unit[a,b]	United States	17,500	Model IV	1981
Licensed unit[a,b]	South America	95,400	Flexicracking (R/B)	1982
Licensed unit[a,b,e]	Europe	31,200	Flexicracking (R/B)	1982
Licensed unit[a,b]	Far East	28,500	Flexicracking (R/B)	1982
Licensed unit[a,b]	Canada	27,500	Flexicracking (R/B)	1983
Licensed unit[a,b]	Europe	16,500	Flexicracking (R/B)	1983
Licensed unit[a,b,e]	Europe	31,200	Flexicracking (R/B)	1983
Licensed unit[a-c]	United States	42,000	Flexicracking (T/L)	1984
Licensed unit[a,b,e]	Europe	15,000	Flexicracking (T/L)	1992
Licensed unit[a,b]	South America	23,400	Flexicracking (T/L)	1986
Licensed unit[a,b]	South America	34,300	Flexicracking (T/L)	1986
Exxon unit[a,b]	Far East	39,100	Flexicracking (T/L)	1985
Licensed unit[a,b,e]	Europe	30,000	Flexicracking (T/L)	1987
Exxon unit	Europe	30,000	Flexicracking (T/L)	1993
Exxon unit	Southeast Asia	23,000	Flexicracking IIIR	1994

[a]Units on which Flexicracking design features have been incorporated.
[b]Units with high-temperature regeneration facilities installed or under design by ER&E.
[c]Units with regenerator flue gas scrubbing.
[d]Original design by others.
[e]Units with ER&E-licensed flue gas expander installed or planned.
[f]Units for which ER&E flue gas expander design specifications have been prepared, but expander was not installed.

Note: T/L = Transfer line; R/B = Riser/bed; BPSD = barrels per stream day; N.A. = not available.

FIGURE 3.1.15 Latest grass-roots ER&E-designed Flexicracking IIIR unit, located in Thailand.

FIGURE 3.1.16 Exxon Flexicracking IIIR unit located in Canada.

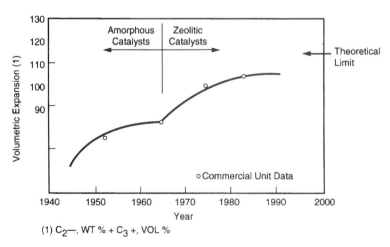

(1) $C_2{-}$, WT % + $C_3{+}$, VOL %

FIGURE 3.1.17 FCC volume expansion, historical and future.

CHAPTER 3.2
THE M.W. KELLOGG COMPANY FLUID CATALYTIC CRACKING PROCESS

Tiffin E. Johnson and Phillip K. Niccum
The M.W. Kellogg Company
Houston, Texas

INTRODUCTION

Fluid catalytic cracking (FCC) is a process for converting higher molecular weight hydrocarbons into lighter, more valuable products through contact with a powdered catalyst at appropriate process conditions. Typically, the FCC process is used to convert excess refinery gas oils and heavier refinery streams into gasoline, C_3 and C_4 olefins, and light cycle oil, bringing refinery output more in-line with product market demands. FCC is often the heart of a modern refinery because of its adaptability to changing feedstocks and product demands and because of the high margins that exist between FCC feedstocks and converted FCC products. As oil refining has evolved over the last 50 years, the FCC process has evolved with it, meeting the challenges of cracking heavier, more contaminated feedstocks, increasing operating flexibility, accommodating environmental legislation, and maximizing reliability. In 1990, The M.W. Kellogg Company and Mobil Oil Corporation joined together to offer the FCC technology discussed here.

FEEDSTOCKS

The modern FCC unit can accept a broad range of feedstocks, a fact which contributes to catalytic cracking's reputation as one of the most flexible refining processes. Examples of common feedstocks for a conventional distillate feed FCC are

- Atmospheric gas oils
- Vacuum gas oils

- Coker gas oils
- Thermally cracked gas oils
- Solvent-deasphalted oils
- Lube extracts
- Hydrocracker bottoms

Residual oil (resid) FCC (RFCC) units charge Conradson carbon residue and metal-contaminated feedstocks, such as atmospheric residues or mixtures of vacuum residue and gas oils. Depending on the level of carbon residue, sulfur, and metallic contaminants (nickel, vanadium, and sodium), these feedstocks may be hydrotreated or deasphalted before being fed to an RFCC unit. Feed hydrotreating or deasphalting reduces the carbon residue and metals of the feed, reducing both the coke-making tendency of the feed and catalyst deactivation.

PRODUCTS

Products from the FCC and RFCC processes are typically:

- Fuel gas (ethane and lighter hydrocarbons)
- Hydrogen sulfide
- C_3 and C_4 liquefied petroleum gas (LPG)
- Gasoline
- Light cycle oil
- Fractionator bottoms product (slurry oil)
- Coke (combusted internally to produce regenerator flue gas)

Although gasoline is typically the desired product from an FCC or RFCC process, design and operating variables can be adjusted to maximize other products. The three principal modes of FCC operation are maximum gasoline production, maximum light cycle oil production, and maximum light olefin production, often referred to as *maximum LPG operation*.

Maximum Gasoline

The maximum gasoline mode is characterized by use of an intermediate cracking temperature, high catalyst activity, and high catalyst-to-oil ratio. Recycle normally is not used since the conversion after a single pass through the riser is already high. Maximization of gasoline yield requires the use of an effective feed injection system, a short contact time vertical riser, and efficient riser effluent separation to maximize the cracking selectivity to gasoline in the riser and to prevent secondary reactions from degrading the gasoline after it exits the riser.

Maximum Middle Distillate

The maximum middle-distillate mode of operation is a low-severity operation in which the first-pass conversion is held to a low level to restrict recracking of light cycle oil formed during initial cracking. Severity is lowered by reducing riser outlet temperature and catalyst-to-oil ratio. The lower catalyst-to-oil ratio often is achieved by the use of a fired feed heater, which significantly increases feed temperature. Additionally, catalyst activity sometimes is lowered by reducing fresh catalyst make-up rate or fresh catalyst activity. Since during low-severity operation a substantial portion of the feed remains unconverted in a single pass through the riser, recycle of heavy cycle oil to the riser is used to reduce the yield of lower-value, heavy streams, such as slurry product.

Middle distillate production also is maximized through the adjustment of fractionation associated with the FCC. When maximizing middle distillate production, crude distillation units are operated to minimize middle distillate components in the FCC feedstock, since these components either degrade in quality or convert to gasoline and lighter products in the FCC. In addition, while maximizing middle distillate production, FCC gasoline endpoint typically would be minimized within middle distillate flash-point constraints, shifting gasoline product into light cycle oil (LCO).

If it is desirable to increase gasoline octane or LPG yield while maximizing LCO production, ZSM-5 catalyst additive can be used. ZSM-5 selectively cracks gasoline boiling-range linear molecules, increasing gasoline research and motor octane ratings, decreasing gasoline yield, and increasing C_3 and C_4 LPG yield. LCO yield is reduced only slightly.

Maximum Light Olefin Yield

The yields of propylene and butylene may be increased above that of the maximum gasoline operation by increasing the riser temperature above 538°C and by use of ZSM-5. At the same time, other operating variables may require adjustment to keep the regenerator temperature in an optimum range.

If the unit is not equipped with a catalyst cooler, catalyst activity may be lowered slightly during maximum light olefin operation to control regenerator temperature. Preferably, activity will be lowered through the use of a lower rare-earth-content fresh catalyst, since this also provides an improvement in coke selectivity and further increases light olefin yields by minimizing hydrogen transfer reactions.

Maximization of light olefin yield also requires the use of an effective feed injection system and efficient riser effluent separation to minimize coke production in the riser and thermal cracking of products exiting the riser at the elevated operating temperatures.

PROCESS DESCRIPTION

The FCC process may be divided into the converter, flue gas, main fractionator, and vapor recovery sections. The number of product streams, the degree of product fractionation, flue gas processing steps, and several other aspects of the process will vary from unit to unit, depending on the requirements of the application.

Converter

The Kellogg Orthoflow FCC converter shown in Fig. 3.2.1 consists of regenerator, stripper, and disengager vessels, with continuous closed-loop catalyst circulation between the regenerator and disengager/stripper. The term *Orthoflow* derives from the

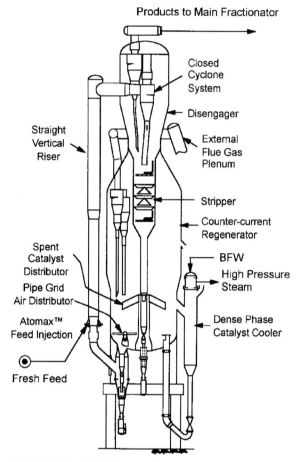

FIGURE 3.2.1 M.W. Kellogg Orthoflow converter.

in-line stacked arrangement of the disengager and stripper over the regenerator. This arrangement has the following operational and cost advantages:

- Essentially all-vertical flow of catalyst in standpipes and risers
- Short regenerated and spent catalyst standpipes allowing robust catalyst circulation
- Uniform distribution of spent catalyst in the stripper and regenerator
- Low overall converter height
- Minimum structural steel and plot area requirements

Preheated fresh feedstock, plus any recycle feed, is charged to the base of the riser reactor. On contact with hot regenerated catalyst, the feedstock is vaporized and converted into lower-boiling fractions (light cycle oil, gasoline, C_3 and C_4 LPG, and dry gas). Product vapors are separated from spent catalyst in the disengager cyclones and flow via the disengager overhead line to the main fractionator and vapor recovery unit for quenching and fractionation. Coke formed during the cracking reactions is deposited on the catalyst, thereby reducing its activity. The coked catalyst separated from the reactor products in the disengager cyclones flows via the stripper and spent catalyst standpipe to the regenerator. The discharge rate from the standpipe is controlled by the spent catalyst plug valve.

In the regenerator, coke is removed from the spent catalyst by combustion with air, which is supplied to the regenerator air distributors from an air blower. Flue gas from the combustion of coke exits the regenerator through two-stage cyclones, which remove all but a trace of catalyst from the flue gas. Flue gas is collected in an external plenum chamber and flows to the flue gas train. Regenerated catalyst, with its activity restored, is returned to the riser via the regenerated catalyst plug valve, completing the cycle.

Atomax* Feed Injection System

The Orthoflow FCC design employs a regenerated catalyst standpipe, a catalyst plug valve, and a short, inclined lateral to transport regenerated catalyst from the regenerator to the riser. From the plug valve, the catalyst flows up the dense-phase lateral, which turns vertical before the oil injection pickup point.

The catalyst then enters a feed injection cone surrounded by multiple flat-spray atomizing feed injection nozzles, as shown in Fig. 3.2.2. The flat, fan-shaped sprays provide uniform coverage and maximum penetration of feedstock into catalyst and prevent catalyst from bypassing feed in the injection zone. Proprietary feed injection nozzles, known as Atomax* nozzles, are used to achieve the desired feed atomization

---

*Trademark of The M.W. Kellogg Company.

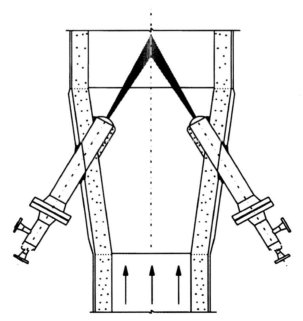

FIGURE 3.2.2 Atomax feed injection nozzle.

and spray pattern while minimizing feed-pressure requirements. The hot regenerated catalyst vaporizes the oil feed, raises it to reaction temperature, and supplies the necessary heat for cracking.

The cracking reaction proceeds as the catalyst and vapor mixture flows up the riser. The riser outlet temperature is controlled by the amount of catalyst admitted to the riser by the catalyst plug valve.

Riser Termination

At the top of the riser, all of the selective cracking reactions have been completed. It is important to minimize product vapor residence time in the disengager to prevent unwanted thermal or catalytic cracking reactions that produce dry gas and coke from more valuable products.

Closed cyclone technology is used to separate product vapors from catalyst with minimum vapor residence time in the disengager. This system (Figure 3.2.3) consists of riser cyclones directly coupled to secondary cyclones housed in the disengager vessel. The riser cyclones effect a quick separation of the spent catalyst and product vapors exiting the riser. The vapors flow directly from the outlet of the riser cyclones into the inlets of the secondary cyclones, then to the main fractionator for rapid quenching.

Closed cyclones almost completely eliminate post-riser thermal cracking with its associated dry gas and butadiene production. Closed cyclone technology is particular-

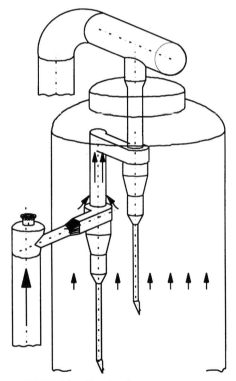

FIGURE 3.2.3 Closed cyclone system.

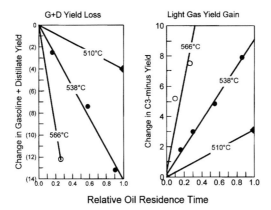

FIGURE 3.2.4 Thermal cracking reaction data from pilot plant.

ly important when the riser is operating at the high temperatures (for example, 538°C or higher) typical of maximum gasoline or maximum light olefin operations. Figure 3.2.4 shows the strong effect of temperature on thermal recracking of gasoline and distillate to produce predominantly dry gas.

Spent Catalyst Stripping

Catalyst separated in the cyclones flows through the respective diplegs and discharges into the stripper bed. In the stripper, hydrocarbon vapors from within and around the catalyst particles are displaced by steam into the disengager dilute phase, minimizing hydrocarbon carryunder with the spent catalyst to the regenerator. Stripping is a very important function because it minimizes regenerator bed temperature and regenerator air requirements, resulting in increased conversion in regenerator-temperature-limited or air-limited operations.

The catalyst entering the stripper is contacted by upflowing steam introduced through two steam distributors. The majority of the hydrocarbon vapors entrained with the catalyst are displaced in the upper stripper bed. The catalyst then flows down through a set of hat and doughnut baffles. In the baffled section, a combination of residence time and steam partial pressure is used to allow the hydrocarbons to diffuse out of the catalyst pores into the steam introduced via the lower distributor.

Stripped catalyst, with essentially all strippable hydrocarbons removed, passes into a standpipe aerated with steam to maintain smooth flow. At the base of the standpipe, a plug valve regulates the flow of catalyst to maintain the spent catalyst level in the stripper. The catalyst then flows into the spent catalyst distributor and into the regenerator.

Regeneration

In the regenerator, coke is burned off the catalyst to supply the heat requirements of the process and to restore the catalyst's activity. The regenerator is operated in either complete CO combustion or partial CO combustion mode. The combustion products from the burning of the coke include CO_2, CO, H_2O, SO_x, and NO_x. In the regenerator cyclones, the combustion gases are separated from the catalyst.

Regeneration is a key part of the FCC process and must be accomplished in an environment that preserves catalyst activity and selectivity so that the reaction system can deliver the desired product yields.

The Kellogg Orthoflow converter uses a countercurrent regeneration system to accomplish this. (The concept is illustrated in Fig. 3.2.5.) The spent catalyst is introduced and distributed uniformly near the top of the dense bed. This is made possible by the spent catalyst distributor. Air is introduced near the bottom of the bed. The design allows coke burning to begin in a low oxygen partial pressure environment that controls the initial burning rate, preventing excessive particle temperatures. The hydrogen in the coke combusts more quickly than the carbon, and most of the water formed is released near the top of the bed. These features together minimize catalyst deactivation during the regeneration process.

With this unique approach, the Kellogg countercurrent regenerator achieves the advantages of multiple regeneration stages, yet accomplishes them with the simplicity, cost efficiency, and reliability of a single regenerator vessel.

Catalyst Cooler

A regenerator heat-removal system may be included to keep the regenerator temperature and catalyst circulation rate at the optimum values for economic processing of the feedstock. The requirement for a catalyst cooler usually occurs in processing resid feedstocks, especially at high conversion.

The Kellogg regenerator heat-removal system is shown in Fig. 3.2.6. It consists of an external catalyst cooler that generates high-pressure steam from heat absorbed from the regenerated catalyst.

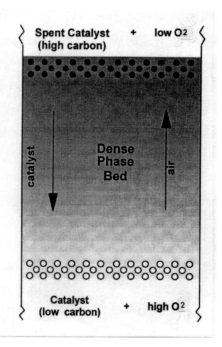

FIGURE 3.2.5 Countercurrent regeneration.

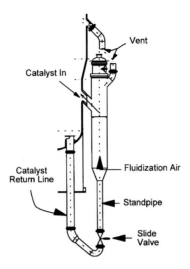

FIGURE 3.2.6 Regenerator dense-phase catalyst cooler.

Catalyst is drawn off the side of the regenerator and flows downward as a dense bed through an exchanger containing bayonet tubes. The catalyst surrounding the bayonet tubes is cooled, then transported back to the regenerator. Air is introduced at the bottom of the cooler to fluidize the catalyst. A slide valve is used to control the catalyst circulation rate and, thus, the heat removed. Varying the catalyst circulation gives control over regenerator temperature for a broad range of feedstocks, catalysts, and operating conditions.

Gravity-circulated boiler feed water flows downward through the inner tubes, while the generated steam flows upward through the annulus between the tubes, as shown in Fig. 3.2.7.

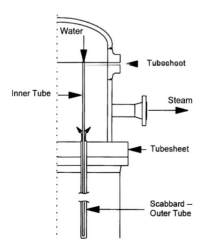

FIGURE 3.2.7 Catalyst cooler water and steam flow.

Flue Gas Section

Flue gas exits the regenerator through two-stage cyclones and an external plenum chamber into the flue gas train as shown in Fig. 3.2.8. Energy from the regenerator flue gas is recovered in the form of mechanical energy by means of a flue gas expander and in the form of heat by the generation of steam in the flue gas cooler.

Power Recovery

A flue gas expander is included to recover energy by reducing flue gas pressure. A third-stage catalyst separator is installed upstream of the expander to protect the expander blades from undue erosion. Overflow from the third-stage separator flows to the expander turbine, where energy is extracted in the form of work. The expander and a butterfly valve located near the expander inlet act in series to maintain the required regenerator pressure. A small quantity of gas with most of the catalyst is taken as underflow from the third-stage catalyst separator and is recombined with the flue gas downstream of the expander.

The expander may be coupled with the main air blower, providing power for blower operation, as shown in Fig. 3.2.8. Or the air blower may be driven by a separate electric motor or steam turbine, with expander output used solely for electrical power generation. If the expander is coupled with the air blower, a motor/generator is required in the train to balance expander output with the air blower power requirement, and a steam turbine is included to assist with start-up. The steam turbine may be designed for continuous operation as an economic outlet for excess steam, or a less expensive turbine exhausting to atmosphere may be used only during start-up.

Flue Gas Heat Recovery

The flue gas from the expander flows to a flue gas cooler, generating superheated steam from the sensible heat of the flue gas stream. If the unit is designed to operate in a partial CO combustion mode, a CO boiler would be installed rather than a flue gas cooler. The gases then pass to the stack. In some cases, an SO_x scrubbing unit or electrostatic precipitator is installed, depending on the governing environmental requirements for SO_x and particulate emissions.

Main Fractionator Section

The process objectives of the main fractionator system are to:

- Condense superheated reaction products from the FCC converter to produce liquid hydrocarbon products.
- Provide some degree of fractionation between liquid sidestream products.
- Recover heat that is available from condensing superheated FCC converter products.

A process flow diagram of the main fractionator section is shown in Fig. 3.2.9.

Superheated FCC converter products are condensed in the main fractionator to produce wet gas and raw gasoline from the overhead reflux drum, light cycle oil from the bottom of the LCO stripper, and fractionator bottoms from the bottom of the main fractionator. Heavy naphtha from the upper section of the main fractionator is utilized as an absorber oil in the secondary absorber in the vapor recovery section (VRS). Any

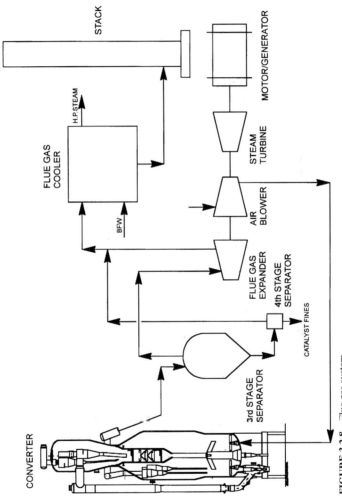

FIGURE 3.2.8 Flue gas system.

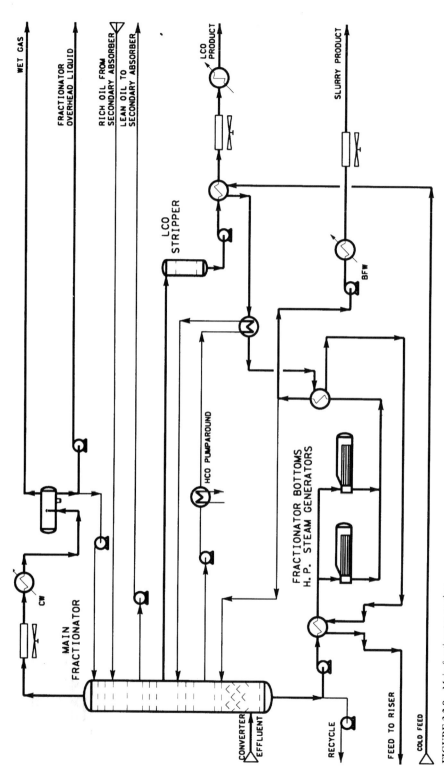

FIGURE 3.2.9 Main fractionator section.

fractionator bottoms recycle and heavy cycle oil recycle also are condensed and sent to the RFCC converter.

Heat recovered from condensation of the converter products is used to preheat fresh converter feed, to reboil the stripper and debutanizer towers in the VRS, and to generate high-pressure steam. Heat that cannot be recovered and utilized at a useful level is rejected to air and finally to cooling water.

Fractionator Overhead

Fractionator overhead vapor flows to the fractionator overhead air cooler, then to the overhead trim cooler. Fractionator overhead products, consisting of wet gas, raw gasoline, and a small amount of reflux and sour water, are condensed in the overhead reflux system.

Net products and reflux are recovered in the fractionator overhead reflux drum. Wet gas flows to the wet gas compressor low-pressure suction drum in the VRS. Raw gasoline is pumped to the top of the primary absorber and serves as primary lean oil.

Heavy Naphtha Pumparound

Fractionation trays are provided between the LCO and heavy naphtha draw in the main fractionator. Desired fractionation between the LCO and raw gasoline is achieved by induced reflux over these trays. Circulating reflux and lean oil are pumped to the pumparound system.

Significant quantities of C_4 and C_5 boiling-range material are recovered in the return rich oil from the secondary absorber. This recovered material is vaporized and leaves the fractionator in the overhead product stream. Lighter components recovered in the secondary absorber are recycled between the fractionator and VRS.

Light Cycle Oil

LCO is withdrawn from the main fractionator and flows by gravity to the top tray of the LCO stripper. Steam is used to strip the light ends from the LCO to improve the flash point. Stripped LCO product is pumped through the fresh feed/LCO exchanger, the LCO air cooler, and the LCO trim cooler, then delivered to the battery limits.

Heavy Cycle Oil Pumparound

Net wet gas, raw gasoline, and LCO products are cooled and HCO reflux is condensed in this section. Total condensed material is collected in a total trap-out tray, which provides suction to the pumparound pump.

Net tray liquid is pumped back to the cleanup trays below. The circulating reflux is cooled by exchanging heat first with the debutanizer in the VRS, then by preheating fresh FCC feed.

Main Fractionator Bottoms Pumparound

FCC converter products, consisting of hydrocarbon gases, steam, inert gases, and a small amount of entrained catalyst fines, flow to the main fractionator tower above the fractionator bottoms steam distributor. The converter products are cooled and washed free of catalyst fines by circulation of a cooled fractionator bottoms material over a baffled tower section above the feed inlet nozzle.

Heat removed by the bottoms pumparound is used to generate steam in parallel kettle-type boilers and to preheat fresh FCC feed, as required. Fractionator bottoms product is withdrawn at a point downstream of the feed preheat exchangers. The bottoms product is cooled through a boiler feedwater preheater and an air cooler, and then is delivered to the battery limits.

Fresh Feed Preheat

The purpose of this system is to achieve required FCC converter feed preheat temperature, often without use of a fired heater. The fresh feed may be combined from several sources in a feed surge drum. The combined feed then is pumped through various exchangers in the main fractionator section to achieve the desired feed temperature.

Vapor Recovery Section

The VRS consists of the wet gas compressor section, primary absorber, stripper, secondary absorber, and debutanizer. The VRS receives wet gas and raw gasoline from the main fractionator overhead drum. The section is required to:

- Reject C_2 and lighter components to the fuel gas system.
- Recover C_3 and C_4 products as liquids with the required purity.
- Produce debutanized gasoline product with the required vapor pressure.

A process flow diagram of a typical VRS is shown in Fig. 3.2.10.

Additional product fractionation towers may be included, depending on the desired number of products and required fractionation efficiency. These optional towers often include a depropanizer to separate C_3 and C_4 LPG, a C_3 splitter to separate propane from propylene, and a gasoline splitter to produce light and heavy gasoline products.

Wet Gas Compression

Wet gas from the fractionator overhead reflux drum flows to a two-stage centrifugal compressor. Hydrocarbon liquid from the low-pressure stage and high-pressure gas from the high-pressure stage are cooled in the air-cooled condenser and are combined with liquid from the primary absorber and vapor from the stripper overhead. This combined two-phase stream is cooled further in the high-pressure trim cooler before flowing into the high-pressure separator drum.

Stripper

Liquid from the high-pressure separator is pumped to the top tray of the stripper, which is required to strip C_2 and lighter components from the debutanizer feed. Thus, the stripper serves to control the C_2 content of the C_3/C_4 LPG product. Stripped C_2 and lighter products are rejected to the primary absorber. Absorbed C_3 and heavier products are recovered in the stripper bottoms.

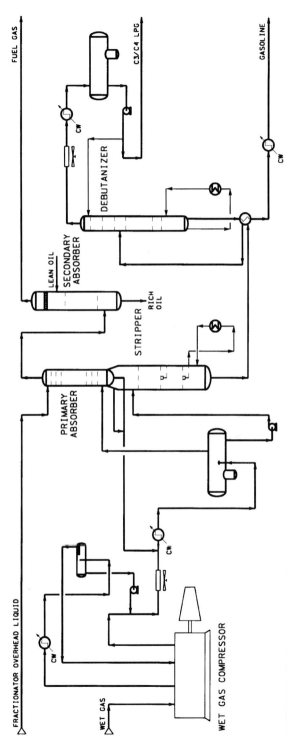

FIGURE 3.2.10 Vapor recovery section.

Primary Absorber

Vapor from the high-pressure separator drum flows to a point below the bottom tray in the absorber. Raw gasoline from the main fractionator and supplemental lean oil from the bottom of the debutanizer combine and flow to the top tray of the absorber. This combined liquid feed serves to absorb C_3 and heavier components from the high-pressure vapor.

Secondary Absorber

Vapor from the primary absorber overhead contains recoverable liquid products. Gasoline-boiling-range components and a smaller quantity of C_4 and C_3 boiling-range material are recovered in the secondary absorber by contacting the primary absorber overhead with heavy naphtha lean oil from the main fractionator.

Rich oil containing recovered material returns to the main fractionator. Sour fuel gas from the top of the secondary absorber flows to the amine treating section and, finally, to the fuel gas system.

Debutanizer

Liquid from the bottom of the stripper exchanges heat with the debutanizer bottoms and flows to the debutanizer. The debutanizer is required to produce a gasoline product of specified vapor pressure, as well as a C_3/C_4 stream containing minimal amounts of C_5 boiling-range materials. The debutanizer reboiler is heated by HCO pumparound. The debutanizer overhead condensing duty is supplied by an air-cooled condenser followed by a trim condenser utilizing cooling water.

The debutanizer overhead liquid product, C_3/C_4 LPG, is pumped to amine and caustic treating sections, then to product storage. The debutanizer bottoms stream, debutanized gasoline, exchanges heat with the debutanizer feed and cooling water prior to caustic treating and delivery to product storage.

PROCESS VARIABLES

There are a large number of variables in the operation and design of an FCC unit that may be used to accommodate different feedstocks and operating objectives. Operational variables are those that may be manipulated while on stream to optimize the FCC performance. Decisions on design variables must be made before the unit is constructed.

Operational Variables

FCC operating variables can be grouped into dependent and independent variables. Many operating variables, such as regenerator temperature and catalyst circulation rate, are considered dependent because operators do not have direct control of them. Independent variables are those over which the operators have direct control, such as riser outlet temperature and recycle rate.

Two dependent operating variables useful in a discussion of other variables are conversion and catalyst-to-oil ratio. Conversion is a measure of the degree to which the feedstock is cracked to lighter products and coke during processing in the FCC. It is defined as 100 percent minus the volume percent yield of LCO and heavier liquid products. In general, as conversion of feedstock increases, the yields of LPG, dry gas, and coke increase, while the yields of LCO and fractionator bottoms decrease. (Gasoline yield increases, decreases, or remains constant depending on the conversion level.) Catalyst-to-oil ratio is the ratio of catalyst circulation rate to charge rate on a weight basis. At constant charge rate, catalyst/oil increases as catalyst circulation increases. At constant riser temperature, conversion increases as catalyst/oil increases because of the increased contact of feed and catalyst.

The following are six important independent operating variables:

- Riser temperature
- Recycle rates
- Feed preheat temperature
- Fresh feed rate
- Catalyst makeup rate
- Gasoline endpoint

Riser Temperature

Increasing the riser temperature set point will signal the regenerated catalyst valve to increase the hot catalyst flow as necessary to achieve the desired riser outlet temperature. The regenerator temperature also will rise because of the increased temperature of the catalyst returned to the regenerator and because of increased coke laydown on the catalyst. When steady state is reached, both the catalyst circulation and regenerator temperature will be higher than they were at the lower riser temperature. The increased riser temperature and catalyst circulation (catalyst/oil) result in increased conversion.

Compared to the other means of increasing conversion, increased riser temperature produces the largest increase in dry gas and C_3 yields, but less increase in coke yield. This makes increasing riser temperature an attractive way to increase conversion when the unit is close to an air limit, but has some spare gas handling capacity.

Increasing riser temperature also significantly improves octane. The octane effect of increased reactor temperature is about $1(R+M)/2$ per $25°F$, where R = research and M = motor. However, beyond a certain temperature, gasoline yield will be negatively affected. The octane effect often will swing the economics to favor high riser temperature operation.

Recycle Rates

HCO and slurry from the main fractionator can be recycled to the riser to increase conversion and/or increase regenerator temperature when spare coke-burning capacity is available. Coke and gas yield will be higher from cracking HCO or slurry than from cracking incremental fresh feed, so regenerator temperature and gas yield will increase significantly when HCO or slurry is recycled to the riser. Therefore recycle of slurry to the riser is an effective way to increase regenerator temperature if this is required. If the FCC is limited on gas handling capacity, the use of HCO or slurry recycle will require a reduction in riser temperature, which will depress octane. Conversion also could fall.

Operation with HCO or slurry recycle together with lower riser temperature sometimes is used when the objective is to maximize LCO yield. This is accomplished because the low riser temperature minimizes cracking of LCO-boiling-range material into gasoline and lighter products, while the recycle of the heavy gas oil provides some conversion of these streams into LCO.

Sometimes, slurry recycle is employed to take entrained catalyst back into the converter. This is done most often when catalyst losses from the reactor are excessive.

Feed Preheat Temperature

Decreasing the temperature of the feed to the riser increases the catalyst circulation rate required to achieve the specified riser outlet temperature. The increase in catalyst circulation rate (catalyst/oil) causes increased conversion of the FCC feedstock. Compared to raising riser outlet temperature, increasing conversion via lower preheat temperature produces a larger increase in coke yield, but smaller increases in C_3 and dry gas yield and octane. Feed preheat temperature has a large effect on coke yield because reducing the heat supplied by the charge to the riser requires an increase in heat from the circulating catalyst to satisfy the riser heat demand.

When the FCC is near a dry gas or C_3 production limit—but has spare coke burning capacity—reducing preheat temperature is often the best way to increase conversion. Conversely, if the FCC is air-limited, but has excess light ends capacity, high preheat (and riser temperature) is often the preferred mode of operation.

In most cases, reducing preheat will lead to a lower regenerator temperature because the initial increase in coke yield from the higher catalyst circulation (catalyst/oil) is not enough to supply the increased reactor heat demand. In other cases, reducing feed preheat temperature may result in an increased regenerator temperature. This is especially likely if the feed preheat temperature is reduced to the point that it hinders feed vaporization in the riser, or if catalyst stripping efficiency falls because of higher catalyst circulation rate.

Fresh Feed Rate

As feed rate to the riser is increased, the other independent operating variables usually must be adjusted to produce a lower conversion so that the unit will stay within controlling unit limitations (i.e., air blower capacity, catalyst circulation capability, gas compressor capacity, downstream C_3 and C_4 olefin processing capacity). The yield and product-quality effects associated with the drop in conversion are chiefly a function of changes in these other independent variables. Economically, feed rate is a very important operating variable because of the profit associated with each barrel processed.

Catalyst Makeup Rate

Each day, several tons of fresh catalyst are added to the FCC catalyst inventory. Periodically, equilibrium catalyst is withdrawn from the FCC to maintain the inventory in the desired range. Increasing the fresh catalyst makeup rate will increase the equilibrium catalyst activity because, in time, it lowers the average age and contaminant (Ni, V, and Na) concentrations of the catalyst in the inventory.

With other independent FCC operating variables held constant, increasing catalyst activity will cause more conversion of feedstock and an increase in the amount of coke deposited on the catalyst during each pass through the riser. To keep the coke burning in balance with the process heat requirements, as the activity increases the regenerator temperature will increase and the catalyst circulation (catalyst/oil) will fall. The conversion usually will increase with increasing activity because the effect of higher catalyst activity outweighs the effects of the lower catalyst circulation rate.

If riser or feed preheat temperatures are adjusted to keep conversion constant as activity is increased, the coke and dry gas yields will decrease. This makes increasing catalyst activity attractive in cases where the air blower or gas compressor are limiting, but where some increase in regenerator temperature can be tolerated. Typically, regenerator temperature will be limited to around 717°C in consideration of catalyst activity maintenance. If the riser temperature must be lowered to stay within a regenerator temperature limitation, the conversion increase will be lost.

Gasoline Endpoint

Gasoline/LCO cut point can be changed to significantly shift product yield between gasoline and LCO while maintaining both products within acceptable specifications. Changing the cut point can significantly alter gasoline octane and sulfur content. A lower cut point results in lower sulfur content and generally higher octane, but, of course, gasoline yield is reduced.

Design Variables

Several FCC process design variables are available to tailor the unit design to the requirements of a specific application:

- Feed dispersion steam rate
- Regenerator combustion mode
- Regenerator heat removal
- Disengager and regenerator pressures
- Feed temperature

Feed Dispersion Steam Rate

Selection of a design feed dispersion steam rate influences the sizing of the feed injection nozzles, so dispersion steam rate is both a design and operating variable. Design dispersion steam design rates are commonly in the range between 2 and 5 wt % of feed, depending on feed quality. The lower values are most appropriate for vacuum gas oil feedstocks, while dispersion steam rates near the upper end of the range are most appropriate for higher-boiling, more-difficult-to-vaporize resid feedstocks. Once the feed nozzle design has been specified, a dispersion steam operating range is recommended for optimizing the unit during operation.

Regenerator Combustion Mode

Oxygen-lean regeneration is most appropriate for use with heavy resids where regenerator heat release and air consumption are high because of high coke yield. In addition, oxygen-lean regeneration offers improved catalyst activity maintenance at high catalyst vanadium levels due to reduced vanadium mobility at lower oxygen levels. In grass-roots applications, therefore, oxygen-lean regeneration is preferred for heavy resid operations with high catalyst vanadium loadings.

On the other hand, for better-quality resids and gas oil feedstocks, complete CO combustion is preferred because of its simplicity of operation. Other factors in the selection of regeneration mode are listed below:

- A unit designed to operate in an oxygen-lean mode of regeneration must include a CO boiler to reduce CO emissions to environmentally safe levels. If a CO boiler is included, the FCC unit also may be operated in a full CO combustion mode, with the CO boiler serving to recover sensible heat from the flue gas.
- In some cases, complete CO combustion will allow the unit to operate with a lower coke yield, thereby increasing the yield of liquid products.
- Unit investment cost is lower for oxygen-lean regeneration because of reduced regenerator, air blower, and flue gas system size.
- SO_x emissions can be controlled to lower levels with complete CO combustion, because of a lower coke burning rate and because SO_x reducing catalyst additives are more effective at the higher regenerator oxygen content.
- Steam production can be maximized by operating in an oxygen-lean mode of regeneration, because of combustion in the CO boiler.
- In some cases, regenerator heat-removal systems (such as catalyst coolers) may be avoided if the unit is operated in an oxygen-lean mode of regeneration.

Regenerator Heat Removal

Depending on the feedstock, desired conversion, and regenerator combustion mode, a regenerator heat-removal system may be required to control regenerator temperature in a range chosen to provide an optimum catalyst-to-oil ratio and minimum catalyst deactivation. Kellogg uses an external dense-phase catalyst cooler for control of heat balance, providing maximum reliability and operating flexibility. The magnitude of the heat-removal requirements may be seen in Fig. 3.2.11, which shows the amount of heat removal required to absorb the heat associated with increased feed Conradson carbon residue content. Direct heat removal from the regenerator is just one of several means available for control of the unit heat balance. Flue gas CO_2/CO ratio also is a variable that may influence the unit heat balance, as shown in Fig. 3.2.12.

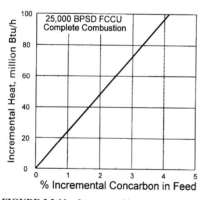

FIGURE 3.2.11 Incremental heat generation.

Disengager and Regenerator Pressures

In the Kellogg Orthoflow converter design, the regenerator pressure is held 7 to 10 lb/in^2 higher than the disengager pressure to provide the desired differential pressures across the spent and regenerated catalyst plug valves. The process designer may still

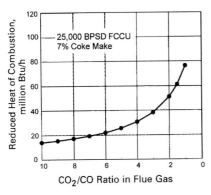

FIGURE 3.2.12 Effect of partial CO combustion relative to complete CO combustion.

specify the overall operating pressure of the system. Lower operating pressures tend to favor product yield selectivity, spent catalyst stripper performance, and air blower horsepower requirements, but these advantages come with increased vessel sizes and thus higher investment cost.

In addition, the economics of flue gas expanders are improved with increased regenerator operating pressure. Economic analyses comparing high- and low-pressure designs have concluded that investment in a lower-pressure unit is the most economical, even if a flue gas expander is included in the analysis.

Feed Temperature

The design feed temperature affects the feed preheat exchanger train configuration and the possible requirement of a fired feed heater. In general, modern FCC designs do not include fired feed heaters, except for those units designed to emphasize the production of middle distillates.

ADVANCED PROCESS CONTROL

FCC units have a large number of interactive variables, making advanced process control (APC) especially beneficial. The benefits of an FCC APC system include the following:

- Operation closer to targets and constraints
- Improved stability and smoother operation
- Enhanced operator information
- Faster response to changes in refinery objectives

These benefits translate into economic gains typically ranging from $0.05 to $0.20 per barrel of feed, not including the less tangible benefits associated with the APC installation.

The Kellogg APC system for the Orthoflow FCC consists of five modules, as shown in Fig. 3.2.13. Although each module is independently implemented, the required interactions are accounted for in the control algorithms. The following is a general description of the functions performed by each module.

Severity Control Module

This module manipulates the riser outlet temperature, feed flow rates, and feed temperature to operate the unit within its constraints while satisfying a specified operating objective. Typical constraints considered by the system include catalyst plug valve differential, regenerator temperature, coke burning rate, wet gas make, and fractiona-

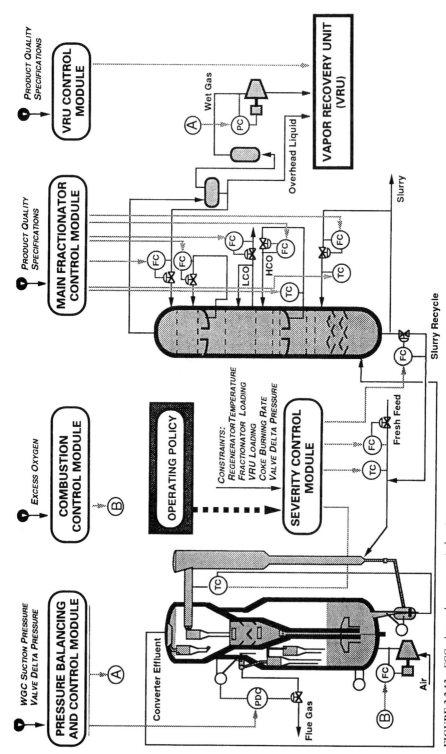

FIGURE 3.2.13 FCC advanced process control system.

tor overhead liquid flow rate. The operating objective is selected by the operator through a menu.

Depending on client requirements and specific refinery objectives, the menu may include options such as

- Maximize reactor temperature while maintaining a feed rate target.
- Maximize feed rate while maintaining a reactor temperature target.
- Minimize feed temperature while maintaining a reactor temperature target.

Combustion Control Module

This module maintains the oxygen composition in the regenerator flue gas at a specified control target by manipulating the airflow rate. This module compensates for changes in the feed rate, recycle flow rate, riser outlet temperature, and feed temperature in a feedforward manner, which assists the system in maintaining the flue gas oxygen concentration close to the target at all times.

Pressure Balancing and Control Module

This module controls the overall converter pressure by manipulating the wet gas compressor suction pressure, which is controlled to maximize the utilization of available air blower and wet gas compressor capacity. It also distributes the available pressure differential across the catalyst valves by manipulating the reactor/regenerator pressure differential, which maximizes converter catalyst circulation capacity.

Fractionator Control Module

This module increases recovery of more valuable products by more closely meeting the quality specifications. It also maximizes high-level heat recovery while observing loading and heat-removal constraints of the unit.

These objectives are accomplished by manipulation of the bottoms pumparound return temperature, the HCO pumparound return temperature, the overhead reflux rate, and the LCO product flow rate. In addition to optimizing the steady-state operation of the main fractionator, the system is configured to react to several different disturbance variables, such as reactor feed rate and riser temperature, in a feedforward manner. This minimizes the transient effects of the disturbances on the fractionator operation.

CATALYST AND CHEMICAL CONSUMPTION

Initial Charge of FCC Catalyst

The initial charge of catalyst to a unit should consist of an equilibrium catalyst with good activity and low metals content. The circulating inventory depends on the coke burning capacity of the unit. Catalyst inventory in the Kellogg Orthoflow design (Fig. 3.2.1) is minimized by the use of a dual-diameter regenerator vessel. This provides a moderately high regenerator bed velocity, which minimizes bed inventory, while the expanded regenerator top section minimizes catalyst losses by reducing catalyst entrainment to the cyclones.

Fresh FCC Catalyst

Catalysts containing rare-earth-exchanged ultrastable Y zeolite are preferred. The ultrastabilization processes provide the zeolite with excellent stability and low coke selectivity, while the rare-earth exchange increases activity and further increases stability. The optimum level of rare earth will depend on the desired tradeoff between gasoline yield, coke selectivity, light olefin yields, and gasoline octane. Depending on the level of bottoms upgrading desired, active matrix materials may be included in the catalyst to increase the ratio of LCO to slurry oil.

When feedstock metals are low, hydrothermal deactivation of catalyst with age is the major factor in catalyst deactivation, setting the fresh catalyst addition rate required to achieve the desired equilibrium catalyst activity with a given fresh catalyst. Proper regenerator design can be used to minimize catalyst makeup requirements, with countercurrent regeneration and low regenerator temperatures minimizing deactivation rate.

The feedstocks for many units contain high levels of nickel and vanadium. In these units, control of equilibrium catalyst metals with fresh catalyst additions is the primary defense against metals contamination and deactivation, as shown in Fig. 3.2.14. At higher feed metal loadings, additional means of controlling the effects of metals become economic. The deleterious effects of nickel contamination can be passivated by the addition of antimony or bismuth. Vanadium effects can be mitigated by employing selective metal traps, either incorporated in the catalyst or as separate particles that selectively bind vanadium and prevent it from reaching and destroying the zeolite. In addition, older/higher-metal catalyst particles can be removed selectively from the unit inventory by magnetic separation using the MagnaCat* process, provid-

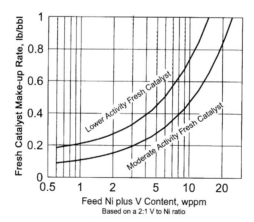

FIGURE 3.2.14 Typical fresh catalyst makeup requirements for constant equilibrium catalyst activity.

ing increased activity and lower equilibrium catalyst metal concentrations for a given fresh catalyst makeup rate.

The FCC catalyst market advances rapidly, and improved products are continuously becoming available. Kellogg continually evaluates the characteristics and performance of commercial fresh and equilibrium catalysts. Several catalyst families from the major vendors have shown the attributes required for effective FCC operation. Within these families, there are variations in activity, rare-earth content, and matrix activity that may be used to optimize the catalyst formulation for a particular application. Although general guidelines help to narrow the choices, the best way to choose the optimum catalyst is through pilot plant testing with a representative feedstock.

Spent Catalyst Disposal

Equilibrium catalyst withdrawn from the regenerator typically is disposed of in a landfill or used in concrete or brick manufacturing.

Passivator

Antimony or bismuth solutions may be required for nickel passivation, especially if the equilibrium catalyst nickel content exceeds 2000 ppm. Passivation reduces the coke and gas make associated with the metals, which translates into increased liquid recovery and reduced compressor requirements. Typically, metals passivation can reduce the coke make by 10 percent and the hydrogen yield by 50 to 70 percent.

Other Chemical Requirements

Diethanolamine (DEA) is required for the amine treating system, a corrosion inhibitor solution is injected into the main fractionator overhead system, and phosphate injection is used in the slurry steam generators and waste heat boiler.

INVESTMENT COSTS AND UTILITIES

The following provides typical investment cost and utilities information for a 50,000 barrel per stream day (BPSD) FCC, including the costs of the converter, flue gas system (without power recovery), main fractionator and vapor recovery sections, and amine treating.

Installed cost, U.S. Gulf Coast—first quarter 1994: 1950–2150 $/BPSD
High-pressure steam production: 40–200 lb/bbl
Electrical power consumption: 0.7–1.0 kWh/bbl

---

*Trademark of Ashland Petroleum Company.

BIBLIOGRAPHY

"New Developments in FCC Feed Injection and Riser Hydrodynamics," R. B. Miller et al., AIChE 1994 Spring National Meeting, Atlanta, 18 April 1994.

"FCC closed cyclone system eliminates post-riser cracking," A. A. Avidan, F. J. Krambeck, H. Owen, and P. H. Schipper, *Oil and Gas Journal,* 16 March 1990.

"Improve regenerator heat removal," T. E. Johnson, *Hydrocarbon Processing,* November 1991.

"Magnetic Separation Enhances FCC Unit Profitability," D. Kowalczyk, R. J. Campagna, W. P. Hettinger, and S. Takase, 1991 NPRA Annual Meeting, San Antonio, 17 March 1991.

"Advances in FCC Vanadium Tolerance," T. J. Dougan, U. Alkemade, B. Lakhanpal, and L. T. Brock, 1994 NPRA Annual Meeting, San Antonio, 20 March 1994.

CHAPTER 3.3
UOP FLUID CATALYTIC CRACKING PROCESS

Charles L. Hemler and Daniel G. Tajbl
UOP
Des Plaines, Illinois

INTRODUCTION

The fluid catalytic cracking (FCC) process is a process for the conversion of straight-run atmospheric gas oils, vacuum gas oils, certain atmospheric residues, and heavy stocks recovered from other refinery operations into high-octane gasoline, light fuel oils, and olefin-rich light gases. The features of the FCC process are relatively low investment, reliable long-run operations, and an operating versatility that enables the refiner to produce a variety of yield patterns by simply adjusting operating parameters. The product gasoline has an excellent front-end octane number and good overall octane characteristics. Further, FCC gasoline is complemented by the alkylate produced from the gaseous olefinic by-products because alkylate has superior midrange octane and excellent sensitivity.

In a typical FCC unit, the cracking reactions are carried out in a vertical reactor riser in which a liquid oil stream contacts hot powdered catalyst. The oil vaporizes and cracks to lighter products as it moves up the riser and carries the catalyst powder along with it. The reactions are rapid, and only a few seconds of contact time are necessary for most applications. Simultaneously with the desired reactions, *coke,* a carbonaceous material having a low ratio of hydrogen to carbon (H/C), deposits on the catalyst and renders it less catalytically active. The spent catalyst and the converted products are then separated, and the catalyst passes to a separate chamber, the regenerator, where the coke is combusted to rejuvenate the catalyst. The rejuvenated catalyst then passes to the bottom of the reactor riser, where the cycle begins again.

DEVELOPMENT HISTORY

Early development of the process took place in the late 1930s. Military requirements pushed widespread commercialization during World War II, when more than 30 units were built and operated.

The first commercial FCC unit was brought on-stream in the United States in May 1942. This design, Model I, was quickly followed with a Model II design (Fig. 3.3.1). A total of 31 units were designed and built. Although engineered by different organizations, these units were similar in concept because the technology came from the same pool, a result of wartime cooperative efforts. Of those first units, several remain in operation today. The principal features of the Model II unit included a reactor vessel near ground level and the catalyst regenerator offset and above it. A rather short transfer line carried both catalyst and hydrocarbon vapor to a dense-bed reactor. Dual slide valves were used at various points in the unit, and this configuration resulted in a low-pressure regenerator with a higher-pressure reactor. Commercial evidence indicated that although conversions were rather low on these early units [40 to 55 liquid volume percent (LV %)], a large portion of the cracking reactions actually took place in the short transfer line carrying both hydrocarbon and catalyst.

After the war, the stacked FCC design (Fig. 3.3.2), which featured a low-pressure reactor stacked directly above a higher-pressure regenerator, was commercialized. This design was a major step toward shifting the cracking reaction from the dense phase of the catalyst bed to the dilute phase of the riser. In the mid-1950s, the straight-riser design, also called the *side-by-side design* (Fig. 3.3.3), was introduced. In this unit, the regenerator was located near ground level, and the reactor was placed to the side in an elevated position. Regenerated catalyst, fresh feed, and recycle were directed to the reactor by means of a long, straight riser located directly below the reactor. Compared with earlier designs, product yields and selectivity were substantially improved.

A major breakthrough in catalyst technology occurred in the mid-1960s with the development of zeolitic catalysts. These sieve catalysts demonstrated vastly superior

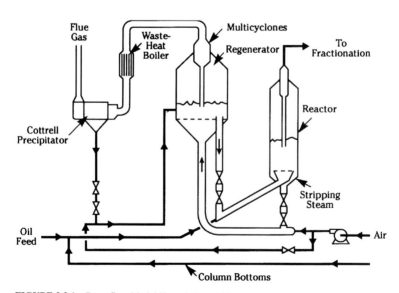

FIGURE 3.3.1 Downflow Model II catalytic cracking unit.

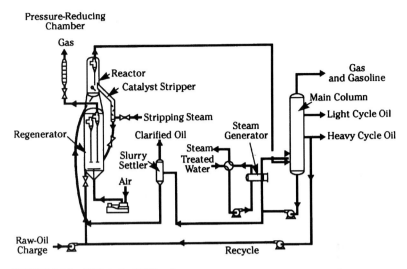

FIGURE 3.3.2 UOP stacked FCC unit.

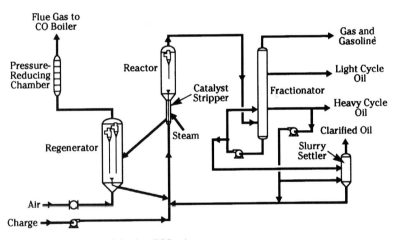

FIGURE 3.3.3 UOP straight-riser FCC unit.

activity, gasoline selectivity, and stability characteristics compared to the amorphous silica-alumina catalysts then in use. The availability of zeolitic catalysts served as the basis for most of the process innovations that have been developed in recent years.

The continuing sequence of advances first in catalyst activity and then process design led to an emphasis on achieving more of the reactions within the dilute phase of the riser, or *riser cracking* as it is commonly called. In 1971, UOP* commercial-

---

*Trademark and/or service mark of UOP.

ized a new design based on this riser-cracking concept, which was then quickly extended to revamps of many of the existing units. Commercial results confirmed the advantages of this system compared to the older designs. Riser cracking provided a higher selectivity to gasoline and reduced gas and coke production that indicated a reduction in secondary cracking to undesirable products.

This trend has continued throughout the years as process designs emphasize greater selectivity to desired primary products and a reduction of secondary by-products. When processing conditions were relatively mild, extended risers and rough-cut cyclones were adequate. As reaction severities were increased, vented risers and direct-connected cyclones were used to terminate the riser. To achieve even higher levels of hydrocarbon containment, further enhancements to prestrip, or displace, hydrocarbons that would otherwise be released from the cyclone diplegs into the reactor vessel now provide an even more selective operation. One example of such selective riser termination designs is the VSS* vortex separation system (Fig. 3.3.4). Such designs have truly approached the concept of all-riser cracking, where almost all of the reaction now takes place within the riser and its termination system.

The emphasis on improved selectivity with all-riser cracking has placed a premium on good initial contact of feed and catalyst within the riser. Thus, much attention over the years has been given to improving the performance of the feed distributor as well as to properly locating it. The quantity of dispersant and the pressure drop required as well as the mechanical characteristics of various feed nozzles have been carefully studied.[1] The feed nozzle, though important, is just one component of a complete feed distribution system. Again the push for higher reaction severities has placed an even greater emphasis on the characteristics of this complete feed distribution system in the design of a modern FCC unit.

Thus far, the discussion has centered on the reactor design; however, significant changes have taken place on the regeneration side. For the first 20 years or so of its his-

---

*Trademark and/or service mark of UOP.

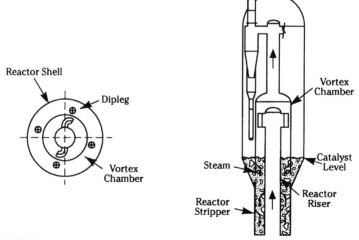

FIGURE 3.3.4 UOP VSS system.

tory, the regenerator of the FCC unit was operated so that the flue gas contained substantial quantities of carbon monoxide (CO) and carbon dioxide (CO_2). In this partial combustion mode, the spent catalyst was regenerated to the point of leaving a few tenths of a percent of carbon still remaining on the regenerated catalyst. A major improvement in FCC technology in the early 1970s was the development of catalysts and hardware to permit complete internal combustion of CO to CO_2. In 1973, an existing FCC unit was revamped to include a new high-efficiency concept in regeneration technology to achieve direct conversion of CO within the unit. This advance was followed by the start-up in early 1974 of a new UOP FCC unit specifically designed to incorporate the new regenerator technology. The development of the new regenerator design and operating technique resulted in reduced coke yields, lower CO emissions (which satisfy environmental standards), and improved product distribution and quality.

A typical FCC unit configuration has a single regenerator to burn the coke from the catalyst. Although the regenerator can be operated in either complete or partial combustion, complete combustion has tended to predominate in new unit designs because an environmentally acceptable flue gas can be produced without the need for additional hardware, such as a CO boiler. This boiler would be required for the partial combustion mode to keep CO emissions low.

With the tightening of crude supplies and refinery economics in the late 1970s, refiners began to look more closely at the conversion of heavier feed components, particularly atmospheric residues. To effectively process highly contaminated residues, Ashland Oil and UOP cooperated to develop a fluidized catalytic cracking approach that would extend the feedstock range. The result of this cooperation, the RCC* process for reduced crude oil conversion, was first commercialized in 1983. Since that time, residue processing has steadily increased to the point that more than half of the new units licensed now process residue or a major residue component.

A sketch of the 1983 version of an RCC unit is shown in Fig. 3.3.5. Among its many innovative features were a two-stage regenerator to better handle the higher coke production that resulted from processing these residues and a new design for a catalyst cooler to help control regeneration temperatures. The two-stage regenerator aided in regulating the unit heat balance because one stage operated in complete combustion and the other operated in partial combustion. The single flue-gas stream that was produced passed to a CO boiler to satisfy flue-gas CO emissions. The RCC* unit also featured a new style of dense-phase catalyst cooler. This cooler aided in not only regulating the regenerator temperature and resulting heat balance but also maintaining catalyst circulation to provide adequate reaction severity.[2]

The process design, catalyst advancements (especially the improvements in metal tolerance), and this additional heat balance control from a reliable catalyst cooler have allowed the RCC unit to economically extend the range of acceptable feedstocks to rather heavy atmospheric residues. Such RCC operations have encouraged refiners to push existing FCC units to process more contaminated feedstocks. Equipment such as the catalyst cooler (Fig. 3.3.6) has been extremely successful in revamps[3] and has found widespread application because of the cooler's ability to vary the level of heat removal in a controlled fashion.

The inventive and innovative spirit that has characterized FCC development from its early days has led to a variety of mechanical and process advancements to further improve the selectivity of the cracking reactions. Thus, improved feed distributors, more effective riser termination devices, and designs that emphasize selective short-time cracking have all been recent process advancements. The pivotal role of catalytic cracking in the refinery almost dictates that even further improvements will be forthcoming.

---

*Service mark of Ashland Oil.

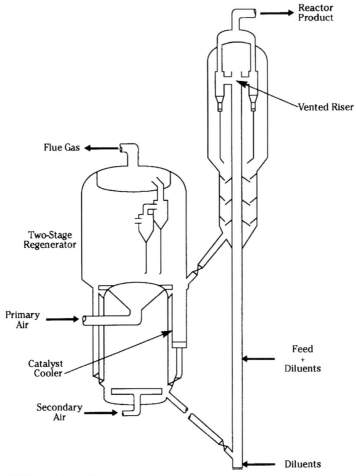

FIGURE 3.3.5 RCC process.

PROCESS CHEMISTRY

Because the chemistry of catalytic cracking is complex, only a broad outline is attempted here. Readers interested in more detailed discussion are referred to an article by Venuto and Habib.[4]

Feedstocks for the FCC process are complex mixtures of hydrocarbons of various types and sizes ranging from small molecules, like gasoline, up to large complex molecules of perhaps 60 carbon atoms. These feedstocks have a relatively small content of contaminant materials, such as organic sulfur, nitrogen compounds, and organometallic compounds. The relative proportions of all these materials vary with the geographic origin of the crude and the particular boiling range of the FCC feedstock. However, feedstocks can be ranked in terms of their *crackability,* or the ease with which they can be converted in an FCC unit. Crackability is a function of the relative proportions of paraffinic, naphthenic, and aromatic species in the feed.

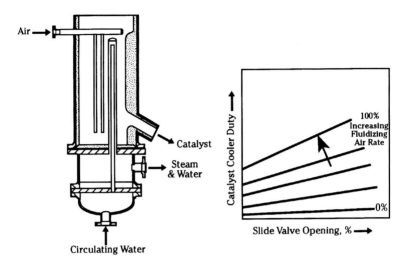

FIGURE 3.3.6 UOP catalyst cooler.

Generally the crackability of FCC feedstocks can be correlated against some simple parameter like feedstock hydrogen content or the UOP characterization factor K:

$$K = \frac{\sqrt[3]{T_B}}{sg}$$

where T_B is an average boiling point of the feedstock, °R, and sg is its specific gravity. A large amount of experimental and commercial data can be classified as shown in Table 3.3.1.

Sulfur compounds do not seriously affect crackability; the cracked sulfur compounds are distributed into the liquid products, thus creating a need for product cleanup before final use. In addition, sulfur also exits from the FCC unit in the form of H_2S and sulfur oxides, the latter posing a potential air-pollution problem.

The organometallic compounds deposit on the circulating catalyst, and after regeneration, almost all the metals in the feedstock remain deposited on the catalysts. These deposited metals have two rather serious deleterious effects: they affect product distribution by causing more light gases, especially hydrogen, to be formed, and they have a serious deactivating effect on the catalyst. To counteract these effects, more fresh catalyst must be added to maintain activity. Heavy polynuclear aromatic-ring compounds are extremely refractory, and these molecules are generally accepted as coke precursors.

TABLE 3.3.1 Feedstock Crackability

Range of characterization factor K	Relative crackability	Feedstock type
>12.0	High	Paraffinic
11.5–11.6	Intermediate	Naphthenic
<11.3	Refractory	Aromatic

In general, the relative amounts of these contaminants in the FCC feedstock increase as the endpoint of the feedstock increases. As endpoints increase into the *nondistillable* range, above about 566°C (1050°F), the increase in these contaminants is dramatic, thus posing a major processing problem. One solution to this problem is to hydrotreat the FCC feedstock. Much of the sulfur and nitrogen leaves the hydrotreater in relatively easily disposable forms of H_2S and NH_3 rather than with the products or as flue gas oxides from the FCC unit. The metals are deposited irreversibly on the hydrotreating catalyst, which is periodically replaced. In addition to removing contaminants, hydrotreating upgrades the crackability of the FCC feed, and hydrotreated feeds do, in fact, crack with better product selectivity because of their increased hydrogen contents.

A carbonium ion mechanism for catalytic cracking reactions is now generally accepted. All cracking catalysts, either the older amorphous silica alumina or modern zeolites, are acidic materials; and reactions of hydrocarbons over these materials are similar to well-known carbonium ion reactions occurring in homogeneous solutions of strong acids. These reactions are fundamentally different from thermal cracking. In thermal cracking, bond rupture is random, but in catalytic cracking, it is ordered and selective.

Various theories have been proposed to explain how the cracking process is initiated, that is, how the first carbonium ions are formed. One theory proposes that the carbonium ion is formed from an olefin, which in turn could be formed by thermal effects on initial catalyst-oil contact, or may be present in the feed. The temperatures involved in catalytic cracking are in the range where thermal cracking can also occur. Alternatively, the carbonium ion could be formed by the interaction of the hydrocarbon molecule with a Brönsted or Lewis acid site on the catalyst. The exact mechanism is not well understood.

Once formed in the feed, the carbonium ions can react in several ways:

- Crack to smaller molecules
- React with other molecules
- Isomerize to a different form
- React with the catalyst to stop the chain

The cracking reaction normally follows the rule of β scission. The C-C bond in the β position relative to the positively charged carbon tends to be cleaved:

$$\text{Carbonium ion} \xrightarrow{\beta \text{ scission}} \text{Olefin} + \text{Carbonium ion}$$

This reaction is most likely because it involves a rearrangement of electrons only. Both of the fragments formed are reactive. The olefin may form a new carbonium ion with the catalyst. The R^+, a primary carbonium ion, can react further, usually first by rearrangement to a secondary carbonium ion and repetition of the β scission.

The relative stability of carbonium ions is shown in the following sequence:

$$\underset{\text{Tertiary}}{C-\overset{C}{\underset{\oplus}{C}}-C} \; > \; \underset{\text{Secondary}}{C-\underset{\oplus}{C}-C} \; > \; \underset{\text{Primary}}{C-C-\underset{\oplus}{C}}$$

Reactions in the system will always proceed toward the formation of the more stable carbonium ion. Thus, isomerizations of secondary to tertiary carbonium ions are common. These reactions proceed by a series of steps including migration of hydride or even alkyl or aryl groups along the carbon chain. Of course, this reaction leads to a product distribution that has a high ratio of branched- to straight-chain isomers:

$$\underset{\substack{\text{Secondary}\\ \text{n-paraffin}}}{\text{C-C-C-C-C-C-R}^{\oplus}} \longrightarrow \underset{\substack{\text{Tertiary}\\ \text{isoparaffin}}}{\text{C-C-C-C(}^{\oplus}\text{)-C-R with C branch}}$$

This equation is an oversimplification of a complicated mechanism.

An example of carbonium ion reaction with other molecules is shown in the following reaction:

$$\text{C-C-C-C-C-C-R}^{\oplus} + \text{C-C(C)-C-R'} \longrightarrow$$

$$\text{C-C-C-C-C-C-R} + {}^{\oplus}\text{C-C(C)-C-R'}$$

Note the formation of the more stable tertiary carbonium ion.

The subject of catalytic coke formation by cracking catalysts, especially its chemical nature and formation, is also a complex topic for which many theories have been proposed. The formation of coke on the catalyst, an unavoidable situation in catalytic cracking, is likely due to dehydrogenation (degradation reactions) and condensation reactions of polynuclear aromatics or olefins on the catalyst surface. As coke is produced through these mechanisms, it eventually blocks the active acid sites and catalyst pores. The only recourse is to regenerate the catalyst to retain its activity by burning the coke to CO and CO_2 in the FCC regenerator. This coke combustion becomes an important factor in the operation of the modern FCC.

THERMODYNAMICS OF CATALYTIC CRACKING

As in the chemistry of cracking, the associated thermodynamics are complex because of the multitude of hydrocarbon species undergoing conversion. The key reaction in cracking, β scission, is not equilibrium-limited, and so thermodynamics are of limited value in either estimating the extent of the reaction or adjusting the operating variables. Cracking of relatively long-chain paraffins and olefins can go to more than 95 percent completion at cracking temperature.

Certain hydrogen-transfer reactions act in the same way. Isomerization, transalkylation, dealkylation, and dehydrogenation reactions are intermediate in the attainment of equilibrium. Condensation reactions, such as olefin polymerization and paraffin alkylation, are less favorable at higher temperatures.

The occurrence of both exothermic and endothermic reactions contributes to the overall heat of reaction, which is a function of feedstock, temperature, and extent of conversion. In general, highly endothermic cracking reactions predominate at low to intermediate conversion levels. At high conversion, some of the exothermic reactions begin to exert an influence. Overall, the reaction is quite endothermic, and heat must be supplied to the system. This heat is provided by the regenerated catalyst.

CATALYST HISTORY

Paralleling the significant improvements in FCC unit design was a corresponding improvement in FCC catalysts. The first catalysts used were ground-up amorphous silica alumina. Whether synthetic or naturally occurring, these catalysts suffered from low activity and poor stability relative to the catalysts available today. Additionally, they had poor fluidization characteristics. Often, fines had to be collected from the flue gas and returned to the unit to assist in maintaining smooth catalyst circulation.

In 1946, spray-dried (microspheroidal) synthetic silica-alumina catalysts were introduced. This type of catalyst, containing 10 to 13 percent alumina, was in general use until a more active and stable catalyst high in alumina (25 wt % alumina) became available in the late 1950s. In addition to improved activity and stability, these spray-dried catalysts also had improved fluidization characteristics.

The most significant catalyst development occurred during the early 1960s, when molecular sieves were introduced into fluid cracking catalysts. The resulting catalysts exhibited significantly higher activity and stability compared with catalysts available at the time. These crystalline catalysts were, and are, ideally suited for the short-contact-time riser cracking concept. Besides being more active, these materials are also more selective toward gasoline production compared to the initial amorphous type.

A wide variety of catalysts can be used in an FCC unit: from low-activity amorphous catalysts to high-activity zeolite-containing catalysts. As an example of relative activities, Table 3.3.2 summarizes pilot-plant results from processing the same feedstock at identical conditions over various catalysts. The present, commercially available high-activity zeolitic catalysts exhibit widely varying matrix compositions, zeolite content, and chemical consistency, yet many can provide the high activity levels required for modern operations. Table 3.3.3 shows the varying characteristics of four commercially available zeolite-containing catalysts.

Many of today's catalysts exhibit a trend toward attrition resistance in response to the concern for reducing particulate emissions. This trend has also affected modern FCC unit design by reducing the amount of catalyst carried to the cyclones.

PROCESS DESCRIPTION

Every FCC complex contains the following sections (Fig. 3.3.7):

- *Reactor and regenerator.* In the reactor, the feedstock is cracked to an effluent containing hydrocarbons ranging from methane through the highest-boiling materi-

TABLE 3.3.2 Effect of Catalyst Activity*

	Amorphous	Low-activity sieve	Moderate-activity sieve	High-activity sieve
Conversion, LV %	63.0	67.9	76.5	78.9
Gasoline, LV %	45.1	51.6	55.4	57.6
RONC	93.3	92.6	92.3	92.3

*Basis: Middle East sour gas oil, 23.7°API gravity (sg = 0.912), 11.84 UOP K factor, 2.48 wt % sulfur.

Note: RONC = research octane number, clear; °API = degrees on American Petroleum Institute scale.

TABLE 3.3.3 Fresh FCC Catalysts

	A	B	C	D
Average bulk density, g/mL	0.56	0.73	0.78	0.42
Surface area, m^2/g	251	127	306	529
Composition, wt %:				
Alumina	28.9	32.7	61.5	21.9
Rare earths	2.7	2.6	0.0	1.9

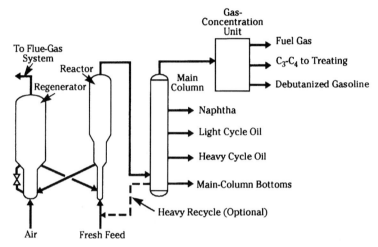

FIGURE 3.3.7 Overall flow diagram for a UOP FCC complex excluding flue gas system option.

al in the feedstock plus hydrogen and hydrogen sulfide. In the regenerator, the circulating spent catalyst is rejuvenated by burning the deposited coke with air at high temperatures.

- *Main fractionator.* Here the reactor effluent is separated into the various products. The overhead includes gasoline and lighter material. The heavier liquid products, heavier naphtha, and cycle oils are separated as sidecuts, and slurry oil is separated as a bottoms product.
- *Gas-concentration unit.* In this section, usually referred to as the *unsaturated-gas plant,* the unstable gasoline and lighter products from the main fractionator overhead are separated into fuel gas, C_3-C_4 for alkylation or polymerization, and debutanized gasoline that is essentially ready for use except for possible chemical treating.

Depending on the objectives of the refiner, some unconverted materials in the feedstock boiling range may be recycled to the reactor. In general, *conversion,* which is typically defined as 100 minus the liquid volume percentage of products heavier than gasoline, is never carried to completion. Some main-column bottoms material, referred to as *clarified oil* or *slurry oil,* is a product usually used for fuel oil blending. Light cycle oil, recovered as a sidecut product, is generally used for home heating, although a fraction might be suitable for diesel fuel blending stock.

The modern FCC unit is likely to have any of a number of optional units associated with the flue gas system. As discussed later, the flue gas contains a significant amount of available energy that can be converted to usable forms. Typically, the flue gas is composed of catalyst fines; nitrogen from the air used for combustion; the products of coke combustion (the oxides of carbon, sulfur, nitrogen, and water vapor); and trace quantities of other compounds. The flue gas exits the regenerator at high temperature, approximately 700 to 780°C (1292 to 1436°F), and at pressures of typically 10 to 40 lb/in^2 gage (0.7 to 2.8 bar gage). The thermal and kinetic energy of the flue gas can be converted to steam or used to drive a turboexpander-generator system for electrical power generation. Unconverted CO in the flue gas can be combusted to CO_2 in a CO boiler that produces high-pressure steam. Catalyst fines may be removed in an electrostatic precipitator.

Reactor-Regenerator Section

The heart of a typical FCC complex (Fig. 3.3.8) is the reactor-regenerator section. In the operation of the FCC unit, fresh feed and, depending on product-distribution objectives, recycled cycle oils are introduced into the riser together with a controlled amount of regenerated catalyst. The charge may be heated, either by heat exchange or, for some applications, by a fired heater.

The hot regenerated catalyst vaporizes the feed, and the resultant vapors carry the catalyst upward through the riser. At the top of the riser, the desired cracking reactions are completed, and the catalyst is quickly separated from the hydrocarbon vapors to minimize secondary reactions. The catalyst-hydrocarbon mixture from the riser is discharged into the reactor vessel through a device that achieves a significant degree of catalyst-gas separation. Final separation of catalyst and product vapor is accomplished by cyclone separation.

The reactor effluent is directed to the FCC main fractionator for resolution into gaseous light olefin coproducts, FCC gasoline, and cycle stocks. The spent catalyst drops from the reactor vessel into the stripping section, where a countercurrent flow of steam removes interstitial and some adsorbed hydrocarbon vapors. Stripped spent catalyst descends through a standpipe and into the regenerator.

During the cracking reaction, a carbonaceous by-product is deposited on the circulating catalyst. This material, called *coke,* is continuously burned off the catalyst in the regenerator. The main purpose of the regenerator is to reactivate the catalyst so that it can continue to perform its cracking function when it is returned to the conversion section. The regenerator serves to gasify the coke from the catalyst particles and, at the same time, to impart sensible heat to the circulating catalyst. The energy carried by the hot regenerated catalyst is used to satisfy the thermal requirements of the cracking section of the unit (the *heat-balance* concept is be discussed in more detail in the next section).

Depending on the specific application, the regenerator may be operated at conditions that achieve complete or partial internal combustion of CO to CO_2, or alternatively, CO may be converted to CO_2 in an external CO boiler. If internal conversion of CO to CO_2 is used, the sensible heat of the flue gas can be recovered in a waste-heat boiler. Flue gas is directed through cyclone separators to minimize catalyst entrainment prior to discharge from the regenerator.

To maintain the activity of the working catalyst inventory at the desired level and to make up for any catalyst lost from the system with the flue gas, fresh catalyst is introduced into the circulating catalyst system from a catalyst storage hopper. An additional storage hopper is provided to hold spent catalyst withdrawn from the circulating system as necessary to maintain the desired working activity and to hold all the catalyst inventory when the FCC unit is shut down for maintenance and repairs.

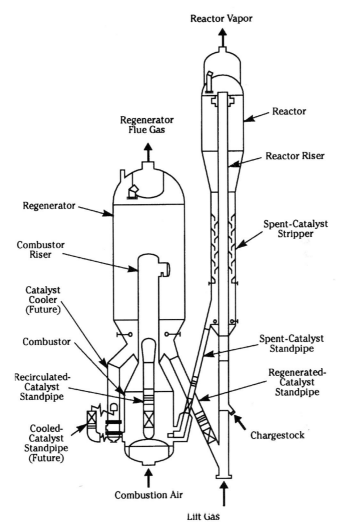

FIGURE 3.3.8 Modern UOP combustor-style FCC unit.

Heat Balance

The schematic diagram of the FCC heat balance in Fig. 3.3.9 shows the close operational coupling of the reactor and regenerator sections. As with other large commercial process units, the FCC unit is essentially adiabatic. The overall energy balance can be written in the following form:

$$\underset{\substack{\text{Heat of} \\ \text{combustion} \\ \text{of coke}}}{Q_{RG}} = \underset{\substack{\text{Enthalpy} \\ \text{difference between} \\ \text{products and feed}}}{(Q_P - Q_{FD})} + \underset{\substack{\text{Enthalpy difference} \\ \text{between flue gas and} \\ \text{regeneration air}}}{(Q_{FG} - Q_A)} + \underset{\substack{\text{Heat of} \\ \text{reaction}}}{Q_{RX}} + \underset{\text{Losses}}{(Q_{L1} + Q_{L2})}$$

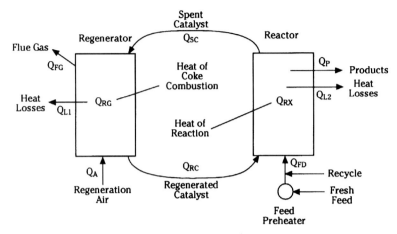

FIGURE 3.3.9 FCC heat balance.

This equation, which has been greatly simplified to present only the major heat terms, describes the basis of the overall reactor-regenerator heat balance. The energy released by burning coke in the regenerator, Q_{RG}, is sufficient to supply all the heat demands for the rest of the reactor and regenerator. Heat is needed to:

- Bring the feed to reaction temperatures
- Vaporize the feed
- Supply the endothermic heat of reaction and various smaller reactor side energy requirements and losses
- Raise the incoming regeneration air temperature to flue gas conditions and to satisfy regenerator losses

The circulating catalyst becomes the mechanism for transferring the needed energy from the regenerator to satisfy the reactor requirements. Thus, all the reactor heat requirements are supplied by the enthalpy difference between regenerated and spent catalyst ($Q_{RC} - Q_{SC}$).

The circulating catalyst rate then becomes a key operating variable because it not only supplies heat but also affects conversion according to its concentration in the reactor relative to oil, expressed in terms of the well-known *catalyst-to-oil ratio*. In practice, the catalyst-to-oil ratio is not a directly controlled variable: changes in the ratio result indirectly from changes in the main operating variables. For instance, an increase in the catalyst-to-oil ratio results from an increase in reactor temperature, a decrease in regenerator temperature, or a decrease in feed preheat temperature. When process conditions are changed so that an increase in the catalyst-to-oil ratio occurs, an increase in conversion is also typically observed.

Fractionation Section

Product vapors from the reactor are directed to the main fractionator, where gasoline and gaseous olefin-rich coproducts and other light ends are taken overhead and routed to the gas concentration unit. Light cycle oil, which is recovered as a sidecut, is stripped for removal of light ends and sent to storage. Net column bottoms are yielded as slurry or clarified oil. Because of the high efficiency of the catalyst-hydrocarbon separation system used in the modern UOP reactor design, catalyst carryover to the fractionator is minimized; the net heavy product yielded from the bottom of the fractionator does not have to be clarified unless the material is to be used in some specific application, such as the production of carbon black, that requires low solids content. In some instances, heavy material can be recycled to the reactor riser.

Maximum usage is made of the heat available at the main column. Typically, light and heavy cycle oils are used in the gas-concentration section for heat-exchange purposes, and steam is generated by a circulating main-column bottoms stream.

Gas-Concentration Section

The gas-concentration section, or unsaturated-gas plant, is an assembly of absorbers and fractionators that separate the main-column overhead into gasoline and other desired light products. Sometimes olefinic gases from other processes such as coking are sent to the FCC gas-concentration section.

A typical four-column gas-concentration plant is shown in Fig. 3.3.10. Gas from the FCC main-column overhead receiver is compressed and directed with primary-absorber bottoms and stripper overhead gas through a cooler to the high-pressure receiver. Gas from this receiver is routed to the primary absorber, where it is contacted by the unstabilized gasoline from the main-column overhead receiver. The net effect of this contacting is a separation between C_3+ and C_2- fractions on the feed to the primary absorber. Primary-absorber offgas is directed to a secondary, or "sponge," absorber, where a circulating stream of light cycle oil from the main column is used to absorb most of the remaining C_5+ material in the sponge-absorber feed. Some C_3 and C_4 material is also absorbed. The sponge-absorber-rich oil is returned to the FCC main column. The sponge-absorber overhead, with most of the valuable C_3+ material removed but including H_2S, is sent to fuel gas or other processing.

Liquid from the high-pressure separator is sent to a stripper column, where most of the C_2- is removed overhead and sent back to the high-pressure separator. The bottoms liquid from the stripper is sent to the debutanizer, where an olefinic C_3-C_4 prod-

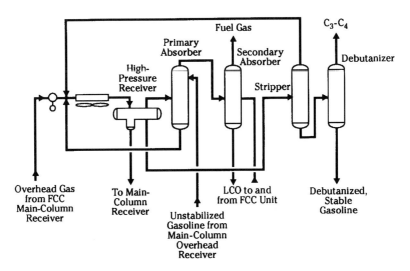

FIGURE 3.3.10 Typical FCC gas concentration plant.

uct is separated and can be sent to either alkylation or catalytic condensation for further gasoline production. The debutanizer bottoms, which is the stabilized gasoline, is sent to treating, if necessary, and then to storage.

This section has described the minimum gas-concentration configuration. Sometimes a gasoline splitter is included to split the gasoline into light and heavy cuts. Any H_2S in the fuel gas or C_3-C_4 product can be removed through absorption in an amine system. Thus, some gas-concentration plants contain six or seven columns.

MODERN UOP FCC UNIT

A modern FCC unit reflects the combination of process and mechanical features probably as well as any process unit in the refinery. Fundamentals of fluidization, fluid flow, heat transfer, mass transfer, reaction kinetics, thermodynamics, and catalysis are applied and combined with the practical experience relating to mechanical design to produce an extremely rugged unit with some sophisticated features. The result is a successful process that combines selective yields with a long run length.

Reactor

The advantages of a reaction system that emphasizes short contact time cracking have led to a modern unit design (Fig. 3.3.8) that is well suited for today's high-activity, superior-selectivity zeolitic catalysts. Great emphasis has been placed on the proper initial contacting of feedstock and catalyst followed by a controlled plug-flow exposure. The reaction products and catalyst are then quickly separated as the hydrocarbons are displaced and stripped from the catalyst before the catalyst passes to the regenerator. This all-riser cracking mode produces and preserves a gasoline-selective yield pattern that is also rich in C_3-C_4 olefins. Higher reaction temperatures have been used to further increase gasoline octanes and yields of the light olefins for downstream alkylation and etherification units.

These individual reaction-side improvements have not been limited to just new unit designs. Many older FCC units have been revamped in one or more of the important areas of feed-catalyst contacting, riser termination, or catalyst stripping. Risers, catalyst standpipes, and slide valves have been replaced as many of these older units have pushed for much higher operating capacities over the years.

Regenerator

A modern UOP FCC unit features a high-efficiency regenerator design, termed a *combustor regenerator*. The combustor-style regenerator was developed to provide a more uniform coke-air distribution and to enhance the ability to burn completely. The regenerator uses a fast fluidized bed as a low-inventory carbon-burning zone followed by a higher-velocity transport-riser heat-exchange zone. The overall combination has excellent catalyst retention and produces flue gas and regenerated catalyst of uniform temperature. Regeneration efficiency and operability are improved, and catalyst inventory is substantially decreased. This reduction in catalyst inventory has economic significance not only from the initial cost of the first catalyst inventory but also from a daily catalyst makeup cost as well.

The combustor configuration was first introduced in the 1970s. Before that, FCC regenerators were operated typically to produce a partial combustion of the coke deposited on the catalyst. Some coke, generally a few tenths of a weight percent, was left on the catalyst after regeneration. The flue gas produced from the coke that was burned in the regenerator often contained about equal proportions of CO and CO_2. As environmental considerations were becoming more significant, a flue gas CO boiler was needed to reduce CO emissions to an acceptable level. If the regenerator can be modified to achieve a more complete combustion step, the capital cost of a CO boiler can be eliminated.

The extra heat of combustion that would be available from burning all the CO to CO_2 also could make a significant change in the heat balance of the FCC unit. The increased heat availability means that less coke needs to be burned to satisfy a fixed reactor heat demand. Because additional burning also produces a higher regenerator temperature, less catalyst is circulated from the regenerator to the reactor.

Another important effect that results from the increased regenerator temperature and the extra oxygen that is added to achieve complete combustion is a reduction in the residual carbon left on the regenerated catalyst. The lower this residual carbon, the higher the effective catalyst activity. From a process viewpoint, complete combustion produces a reduced catalyst circulation rate, but the catalyst has a higher effective activity. Because less coke was needed to satisfy the heat balance, the reduction in coke yield led to a corresponding increase in FCC products.

To assist in the burning of CO, small quantities of noble metal additives are extremely effective when blended with the catalyst. This *promoted catalyst,* as it was called, was widely used in existing units and as an alternative to a complete mechanical modification of the regenerator to a combustor-style configuration. New units were designed with the combustor configuration, which could operate in complete combustion without the more expensive promoted catalyst.

The combustor-style regenerator has proved itself in many varied operations over the years. It has been shown to be an extremely efficient device for burning carbon and burning to low levels of CO. Whether for very small or large units, afterburning has been virtually eliminated, and low levels of carbon on regenerated catalyst are routinely produced.

Yield Versatility

One of the strengths of the FCC process is its versatility to produce a wide variety of yield patterns by adjusting basic operating parameters. Although most units have been designed for gasoline production, UOP has designed units for each of the three major operational modes:

Gasoline Mode. The most common mode of operation of the FCC unit is aimed at the maximum production of gasoline. This mode is better defined as an operation producing a *high* gasoline yield of a specified octane number.

This condition requires careful control of reaction severity, which must be high enough to convert a substantial portion of the feed but not so high as to destroy the gasoline that has been produced. This balance normally is achieved by using an active and selective catalyst and enough reaction temperature to produce the desired octane. The catalyst circulation rate is limited, and reaction time is confined to a short exposure. Because this severity is carefully controlled, no recycle of unconverted components is normally needed.

High-Severity Mode. If additional reaction severity is now added to the system, a high-severity operation producing additional light olefins and a higher-octane gasoline results. This case is sometimes described as a *liquefied petroleum gas (LPG)* mode, or even as a *petrochemical FCC,* because of the increased quantity of light material that is produced and the increased aromatics in the gasoline product. If isobutane is available to alkylate the light olefins or if they are etherified or polymerized into the gasoline boiling range, high total gasoline yields and octanes are produced.

Distillate Mode. If the reaction severity is strictly limited, then the FCC unit can be used for the production of distillates. Changing operating conditions can shift from the normally gasoline-oriented yield distribution to one with a more nearly equal ratio of gasoline to cycle oil. Additional distillates can be produced at the expense of gasoline by reducing the endpoint of the gasoline and dropping the additional material into the light cycle oil product. The usual limitation in this step is reached when the resulting cycle oil reaches a particular flash point specification.

Typical yield patterns for these three modes of operation are shown in Table 3.3.4. The feedstock for these cases was a Middle East vacuum gas oil (VGO). These yields are typical for a particular feedstock. In general, FCC yield patterns are a function of feedstock properties; for instance, a feedstock with a lower UOP K factor and hydrogen content is more difficult to crack and produces a less favorable yield pattern.

The data in Table 3.3.4 show certain trends. As the severity of the FCC unit is increased from low to high, the production of coke and light ends increases, gasoline octane increases, and in general, the liquid products become more hydrogen deficient. Also, the high-severity case overcracks a considerable amount of the gasoline to C_3-C_4 material.

FEEDSTOCK VARIABILITY

The early FCC units were designed primarily to operate on virgin VGOs. These feedstocks, which typically came from high-quality crude oils, were characterized as good cracking feedstocks because of high UOP K factors of 12.0 or more. In the mid-1990s, many refiners are faced with processing less favorable materials. In addition, refiners have been forced to convert more of the nondistillable portion of the barrel to remain competitive. Thus, a greater proportion of FCC feedstock has its origin in the bottom of the barrel. These components may be cracked stocks in the VGO boiling range, or they may be previously virgin nondistillables. Coker and visbreaker gas oils are commonly blended in FCC feed. The next source of heavy FCC feed has traditionally been vacuum-tower residue blended into the feed in proportions consistent with the FCC coke-burning capabilities. Some refiners have chosen to solvent-extract the vacuum residue to provide a nondistillable FCC feed component that has significantly less metal and asphaltene than the vacuum residue itself. Others have gone to the limit and charge certain whole atmospheric residues to their FCC units.

This section briefly discusses two significant FCC operations: the hydrotreating of FCC feeds for yield improvement and environmental concerns and the cracking of various solvent-extracted oils and whole residues.

FCC Feed Hydrotreating

Because the FCC feed can include a substantial amount of sulfur-containing materials, the products, including the flue gas, are typically rich in sulfur compounds. This situation in turn has led to specialized flue gas treating systems and scrubbers for external cleanup or to catalyst modifications and feed hydrotreating as internal process approaches for the reduction of sulfur levels. Of these approaches, only feed

TABLE 3.3.4 Product Yield and Properties for Typical Modes of Operation

	Middle-distillate mode		Gasoline mode	Light-olefin mode
	Full range	Undercut		
	Product yields			
H_2S, wt %	0.7	0.7	1.0	1.0
C_2-, wt %	2.6	2.6	3.2	4.7
C_3, LV %	6.9	6.9	10.7	16.1
C_4, LV %	9.8	9.8	15.4	20.5
C_5+ gasoline, LV %	43.4	33.3	60.0	55.2
Light cycle oil, LV %	37.5	47.6	13.9	10.1
CO, LV %	7.6	7.6	9.2	7.0
Coke, wt %	4.9	4.9	5.0	6.4
	Product properties			
LPG, vol/vol:				
C_3 olefin/saturate	3.4	3.4	3.2	3.6
C_4 olefin/saturate	1.6	1.6	1.8	2.1
Gasoline:				
ASTM 90% point, °C	193	132	193	193
ASTM 90% point, °F	380	270	380	380
RONC	90.5	91.3	93.2	94.8
MONC	78.8	79.3	80.4	82.1
Light cycle oil:				
ASTM 90% point, °C	354	354	316	316
ASTM 90% point, °F	670	670	600	600
Flash point, °C (°F)	97 (207)	55 (131)	97 (207)	97 (207)
Viscosity, cSt @ 50°C (122°F)	3.7	2.4	3.1	3.2
Sulfur, wt %	2.9	2.4	3.4	3.7
Cetane index	34.3	31.8	24.3	20.6
Clarified oil:				
Viscosity, cSt @ 100°C (210°F)	10.9	10.9	9.0	10.1
Sulfur, wt %	5.1	5.1	6.0	6.8

Note: ASTM = American Society for Testing and Materials; RONC = research octane number, clear; MONC = motor octane number, clear.

Source: Reprinted from D. A. Lomas, C. A. Cabrera, D. M. Cepla, C. L. Hemler, and L. L. Upson, "Controlled Catalytic Cracking," UOP 1990 Technology Conference.

hydrotreating provides any significant processing improvement because the addition of hydrogen can dramatically increase the cracking potential of any given feed. This increase can be even more meaningful when the initial feed is poor in quality or when the feed is contaminated. Table 3.3.5 shows the results of hydrotreating poor-quality feed at two different levels of hydrogen addition. As feedstock quality declines and growing emphasis is placed on tighter sulfur regulations, feed hydrotreating will receive even more consideration.

Cracking of High-Boiling Feedstocks

Reference has been made to the cracking of high-boiling fractions of the crude. As refiners seek to extend the range of the feedstocks that are processed in FCC units, the most frequent sources of these heavier feeds are

TABLE 3.3.5 Hydrotreating of FCC Feedstock

	Untreated feed	Mildly desulfurized	Severely hydrotreated
Gravity, °API (specific gravity)	18.4 (0.944)	22.3 (0.920)	26.3 (0.897)
UOP K factor	11.28	11.48	11.67
Distillation D-1160, °C (°F):			
5%	275 (527)	266 (510)	249 (481)
50%	410 (770)	399 (750)	375 (707)
95%	498 (928)	497 (926)	467 (873)
Sulfur, wt %	1.30	0.21	0.04
Nitrogen, wt %	0.43	0.32	0.05
Hydrogen, wt %	11.42	12.07	12.74
Cracking performance at equivalent pilot plant conditions:			
Conversion, LV %	59.0	66.1	82.5
Gasoline, LV %	41.1	46.0	55.6
Coke, wt %	8.8	6.1	5.6

- A deeper cut on a vacuum column
- The extract from solvent extraction of the vacuum-tower bottoms
- The atmospheric residue itself

Regardless of the source of these high-boiling components, a number of problems are typically encountered when these materials are processed in an FCC unit, although the magnitude of the problem can vary substantially:

- *Additional coke production.* Heavy feeds typically have high levels of contaminants, such as Conradson carbon levels. Because much of this material deposits on the catalyst with the normal coke being deposited by the cracking reactions, the overall coke production is substantially higher. Burning this coke requires additional regeneration air. In an existing unit, this coke-burning constraint often limits capacity.
- *Necessity for metal control.* Metals in the heavy feeds deposit almost quantitatively on the catalyst. These metals produce two significant effects. First, they accelerate certain metal-catalyzed dehydrogenation reactions, thereby contributing to light-gas (hydrogen) production and to the formation of additional coke. A second, more damaging effect is the situation in which the presence of the metal contributes to a catalyst activity decline caused partly by limited access to the catalyst's active sites. This latter effect is normally controlled by catalyst makeup practices (adding and withdrawing catalyst).
- *Distribution of sulfur and nitrogen.* The level of sulfur and nitrogen in the products, waste streams, and flue gas generally increases when high-boiling feeds are processed because these feed components typically have higher sulfur and nitrogen contents than their gas oil counterparts. In the case of nitrogen, however, the problem is not just one of higher nitrogen levels in the products. One portion of the feed nitrogen is basic in character, and the presence of this basic nitrogen acts as a temporary catalyst poison to reduce the useful activity of the catalyst.
- *Heat-balance considerations.* Heat-balance control may be the most immediate and troublesome aspect of processing high-boiling feeds. As the contaminant car-

bon increases, the first response is normally an increase in regenerator temperature. Adjustments in operating parameters can be made to assist in this control, but eventually, a point will be reached for heavier feeds when the regenerator temperature is too high for good catalytic performance. At this point, some external heat removal from the regenerator is required and would necessitate a mechanical modification like a catalyst cooler.

Currently, approximately 25 UOP-licensed units have a high-boiling feed as a significant portion of the FCC charge. Interestingly, the product qualities from these operations are not much different from those for similar gas oil operations. In general, the octane levels of the gasoline remain good, the cycle oil qualities are similar, and the heavy fuel oil fraction has a low viscosity and a low metal content and still remains distillable.

Demetallized Oil. For the last several years, demetallized oil (DMO) has been included as a major component of the feed in several UOP-designed FCC units. This DMO results from the extraction of a vacuum-tower bottoms stream using a light paraffinic solvent. Modern solvent-extraction processes provide a higher DMO yield than is possible in the propane-deasphalting process that has been used to prepare FCC feed at a variety of locations for many years. Consequently the DMO is more heavily contaminated. In general, DMOs are still good cracking stocks, but they can be further improved by hydrotreating to reduce contaminant levels and to increase their hydrogen content.

Atmospheric Residue. A number of refiners have chosen to add atmospheric residue as a blend component to their existing FCC units. This choice has resulted from the desire to convert the heaviest portions of the crude. Atmospheric residue has ranged from a relatively low proportion of the total feed all the way to a situation in which it represents the entire feed to the unit. To improve the handling of these high-boiling feeds, several units have been revamped to upgrade them from their original gas-oil designs. Some units have proceeded to increase the amount of residue in a stepwise fashion: modifications to the operating conditions and processing techniques are made as more experience is gained in the processing of high-boiling feeds.

As expected, the properties of the high-boiling feedstocks currently being processed in units originally designed for gas-oil feeds vary across a wide range. Typical of some of this variation are the four feed blends described in Table 3.3.6. They range from clean, sweet residues to more contaminated residues with up to approximately 4 wt % Conradson carbon residue.

TABLE 3.3.6 Typical Residue Cracking Stocks

	A	B	C	D
Gravity, °API (specific gravity)	28.2 (0.886)	24.5 (0.907)	26.4 (0.896)	22.4 (0.919)
UOP K factor	12.1	11.75	12.1	11.95
Sulfur, wt %	0.98	1.58	0.35	0.77
Conradson carbon residue, wt %	1.01	1.25	2.47	3.95
Metals, wt ppm:				
Ni	0.2	1.6	0.7	2.8
V	0.8	2.3	0.5	3.5
Nondistillables at 565°C (1050°F), LV %	10	8	13	23

TABLE 3.3.7 RCC Unit Feedstocks

	A	B	C	D
Gravity, °API	21.3	19.1	21.2	22.4
(specific gravity)	(0.9260)	(0.9396)	(0.9267)	(0.9194)
UOP K factor	11.8	11.7	11.9	12.2
Sulfur, wt %	1.1	2.1	0.55	0.1
Nitrogen, wt %	0.14	—	0.19	0.23
Conradson carbon, wt %	—	—	3.8	5.6
Ramsbottom carbon, wt %	5.0	5.5	—	—
Metals, wt ppm				
Nickel	13	15	2.5	2.2
Vanadium	31	45	3.7	1

RCC Operations When feedstocks with higher levels of contamination are considered for processing, the design modifications and heat management practices from RCC technology come to the forefront. Some examples of the types of feedstock that RCC units have processed are shown in Table 3.3.7. The higher Conradson carbon and metal levels clearly show that an RCC operation can process more difficult feedstocks. High metal levels on the equilibrium catalyst have also been demonstrated in RCC units: one unit reached more than 10,000 ppm of nickel plus vanadium, and another unit operated with more than 15,000 ppm of nickel on the equilibrium catalyst. Even though these values are extremely high, operating or economic limitations will still continue to determine the character of the feedstock that can be processed.

PROCESS COSTS

The following sections present typical process costs for FCC units. These costs are included here for orientation purposes only; specific applications need to be evaluated individually.

Investment

The investment for a new 35,000 barrel per stream day FCC unit operating with 5.5 wt % coke on the basis of fresh feed is shown in Table 3.3.8. In general, costs for other capacities vary according to a ratio of capacities raised to a power of about 0.6.

Operating Costs: Power Recovery

The cost of operating an FCC unit can be reduced significantly with the introduction of power recovery. Refiners have long recognized the significant amount of available energy in the flue gas. After several years of effort, a power-recovery system was developed that converted the dynamic energy of the flue gas into mechanical energy by means of a combination of equipment, the heart of which was an expander turbine.

The recoverable energy is a function of the pressure drop through the expander. The following equation may be used to calculate expander shaft horsepower (shp):

TABLE 3.3.8 Investment Costs

Process equipment included	Estimated erected cost,* million $
Reactor-regeneration, gas concentration, fractionation, and electrostatic precipitator	103
All of the above plus power recovery	114

*Investment accurate within ±40%; current as of 1995, U.S. Gulf Coast erection.

$$\text{shp} = 2.808(EGT)\left(\frac{K}{K-1}\right)\left[1 - \left(\frac{P_E}{P_I}\right)^{(K-1)/K}\right]$$

where E = overall efficiency factor (0.8)
 G = molar flow rate per pound of flue gas, moles/sec
 R = gas constant
 T = inlet temperature, °R
 K = adiabatic exponent (1.313 for flue gas)
 P_E = exhaust pressure
 P_I = inlet pressure

As now typically applied in modern FCC designs, the power recovered and converted into electricity generally meets or exceeds air-blower requirements.

In addition to power recovery, the FCC unit can be designed to produce significant amounts of higher-pressure steam [600 lb/in² gage (42 bar gage)]. The main source of this steam is from exchange with the main-column bottoms and from the flue-gas steam generator.

Typical operating costs for an FCC unit with and without power recovery are shown in Table 3.3.9. The data are for a unit processing a Middle East VGO in a gasoline operation with a five-column gas-concentration unit, electrostatic precipitator, and flue-gas steam generator.

TABLE 3.3.9 Typical Operating Costs

	Power recovery*	Without power recovery*
Utilities:		
Electricity, kWh/1000 bbl FF	−300	1200
Steam, lb/bbl FF:		
600 lb/in² (42 bar) gage	12	32
150 lb/in² (10.5 bar) gage	−14	−14
50 lb/in² (3.2 bar) gage	2.3	2.3
Treated water, lb/bbl FF	34	34
Cooling water, gal/bbl FF	175	175
Materials:		
FCC catalyst, lb/bbl FF	0.16	0.16

*All positive quantities are consumptions (debits); negative quantities are net exported amounts (credits).

Note: FF = fresh feed.

MARKET SITUATION

The FCC process is one of the most widely employed refining processes. More than 450 FCC units have been built worldwide since the process was first commercialized. As of late 1994, about 350 units were operating and about 20 additional new units were in some phase of design or construction. Operating capacity data compiled in late 1994 are listed in Table 3.3.10.

On average, FCC charge capacity can be about one-third the capacity of the crude unit, representing almost exactly the volume fraction of the VGO in the crude. This figure can go as high as one-half the crude capacity if a coker or other process in the refinery converts vacuum residue to additional FCC feed. However, some refiners will send some part of the VGO directly to fuel oil product, and in these cases, the figure is much lower.

The FCC process will clearly be the conversion process of choice in future situations in which gasoline rather than middle distillate is the desired product. The large amount of FCC capacity in North America is shown in Table 3.3.10. The process has the ability to produce middle distillate if desired, but in general, the yields and properties of jet or diesel fuels obtainable from hydrocracking are superior to those from FCC.

REFERENCES

1. Schnaith, M. W., A. T. Gilbert, D. A. Lomas, and D. N. Myers, "Advances in FCC Reactor Technology," paper AM-95-36, 1995 NPRA Annual Meeting, San Francisco, March 19–21, 1995.
2. Hemler, C. L., and A. G. Shaffer, Jr., "The Keys to RCC Unit Success," AIChE Spring 1991 National Meeting, Houston, April 7–11, 1991.
3. Kauff, D. A., and B. W. Hedrick, "FCC Process Technology for the 1990's," paper AM-92-06, 1992 NPRA Annual Meeting, New Orleans, March 22–24, 1992.
4. Venuto, P. B., and E. T. Habib, "Catalyst-Feedstock-Engineering Interactions in Fluid Catalytic Cracking," *Catalysis Reviews: Science and Engineering,* **18**(11), 1–150 (1978).

TABLE 3.3.10 Worldwide FCC Capacity

	Crude capacity, million BPCD*	Catalytic cracking, million BPCD*
North America	18.8	6.0
Asia and Pacific Rim	14.4	2.1
Western Europe	14.2	2.1
Eastern Europe and former Soviet Union	12.9	0.7
South America and Caribbean	5.8	1.0
Middle East	5.3	0.3
Africa	2.8	0.2
Total	74.2	12.4

*BPCD = barrels per calendar day.
Source: *Oil & Gas Journal,* December 1994.

CHAPTER 3.4
STONE & WEBSTER–INSTITUT FRANCAIS DU PÉTROLE RFCC PROCESS

David A. Hunt
Refining Section
Stone & Webster Engineering Corporation
Houston, Texas

HISTORY

Stone & Webster (S&W), in association with Institut Francais du Pétrole (IFP), is the licenser of the S&W-IFP residual fluid catalytic cracking (RFCC) process. The original S&W-IFP RFCC process was developed during the early 1980s by Total Petroleum Inc. at its Arkansas City, Kansas, and Ardmore, Oklahoma, refineries. Because the development of this process saw heavy input from an operating company, unit operability and mechanical durability were incorporated into the design to ensure smooth operation and long run lengths. To process the heavy, viscous residual feedstocks, which can contain metals in high concentrations and produce relatively high amounts of coke, the design incorporates an advanced feed injection system, a unique regeneration strategy, and a catalyst transfer system which produces extremely stable catalyst circulation. Recent technology advances have been made in the areas of riser termination, reactant vapor quench, and mix temperature control (MTC).

Today 23 full-technology S&W-IFP RFCC units have been licensed worldwide (revamp and grass roots), more than all other RFCC licensers combined. Within the Pacific Rim, S&W-IFP's 13 licensed units outnumber the competition 2 to 1. From 1980 to 1995, there were 16 operating S&W-IFP FCC units totaling more than 80 years of commercial operation. A recently licensed RFCC, located in Japan, is shown in Fig. 3.4.1. A listing of all S&W-IFP (full-technology) licensed RFCC units is shown in Table 3.4.1.

While the conception of this technology was based on processing residual feed, the technology has been proven and is widely accepted for processing lighter gas oil feed stocks. Stone & Webster and IFP have ample experience revamping gas oil FCC units to upgrade the feed injection system, combustion air distributor, riser termination device, etc. At present more than 40 FCC units are processing over 1,700,000 barrels per day (BPD) of FCC feed employing the S&W-IFP feed injection technology. In fact, S&W-IFP systems have replaced feed injection systems of virtually every competing licenser, but never have been replaced by other technologies.

FIGURE 3.4.1 S&W-IFP RFCC unit located in Japan. Photograph shows second- and first-stage regenerators and main fractionator. Note the external cyclones on the second-stage regenerator.

TABLE 3.4.1 S&W-IFP Full RFCC Technology Units

Refinery	Location	Capacity, BPSD	Start-up
A	Kansas	20,000	1981
B	Oklahoma	25,000/40,000*	1982
C	Canada	19,000	1985
D	Japan	40,000	1987
E	Australia	25,000	1987
F	Canada	25,000	1987
G	China	23,000	1987
H	China	21,000	1989
I	China	28,000	1990
J	China	21,000	1990
K	China	21,000	1991
L	Japan	30,000	1992
M	Japan	31,600	1994
N	Uruguay	9,000	1994
O	California	31,000	†
P	Singapore	24,000	1995
Q	Korea	50,000	1995
R	Korea	30,000	1995
S	Thailand	37,000	1996
T	Malaysia	55,000	Hold
U	Thailand	14,000	—
V	India	15,000	1997
W	India	60,000	1998

*Design capacity was 40,000 BPSD. Currently operating at 25,000 BPSD.
†On hold waiting permitting.
Note: BPSD = barrels per stream day.

PROCESS DESCRIPTION

RFCC Converter

Two configurations of the grass-roots RFCC unit are offered. The first, and the most common, is the stacked regenerator version shown in Fig. 3.4.2, which minimizes plot space. The second is a side-by-side regenerator design configuration, which is discussed in "FCC Revamp to RFCC" below and is more typical of FCC units which have been revamped to RFCC.

The process flow will be presented by using the stacked version shown in Fig. 3.4.2. The RFCC utilizes a riser-reactor, catalyst stripper, first-stage regeneration vessel, second-stage regeneration vessel, catalyst withdrawal well, and catalyst transfer lines. Process flow for the side-by-side configuration is identical except for the catalyst transfer between the first- and second-stage regenerators.

Fresh feed is finely atomized with dispersion steam and injected into the riser through the feed injection nozzles over a dense catalyst phase. The small droplets of feed contact the freshly regenerated catalyst and instantaneously vaporize. The oil molecules intimately mix with the catalyst particles and crack into lighter, more valuable products.

Mix temperature control nozzles, inject a selected recycle stream which quenches the catalyst and feed vapor. This feature allows control of the critical feed-catalyst mix zone temperature independent of the riser outlet temperature. Riser outlet temperature (ROT) is controlled by the regenerated catalyst slide valve.

CATALYTIC CRACKING

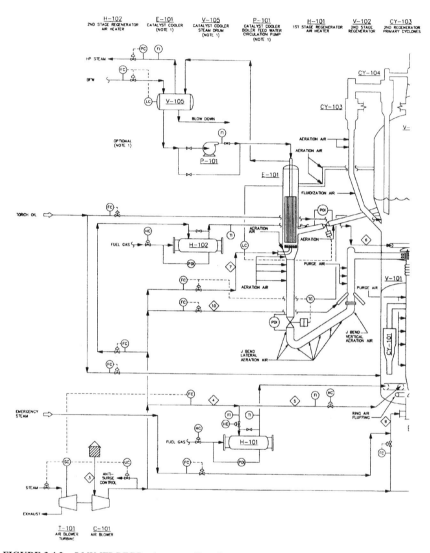

FIGURE 3.4.2 S&W-IFP RFCC unit process flow diagram.

As the reaction mixture travels up the riser, the catalyst, steam, and hydrocarbon product mixture pass through a riser termination device. S&W-IFP currently offers the patented Ramshorn device for this service. This device quickly disengages the catalyst from the steam and product vapors. Reactant vapors are quenched inside the gas outlet tube of the Ramshorn, minimizing thermal product degradation reactions. Reactant vapors are then ducted to the top of the reactor near the reactor cyclone inlets, while catalyst is discharged into the stripper through a pair of catalyst diplegs.

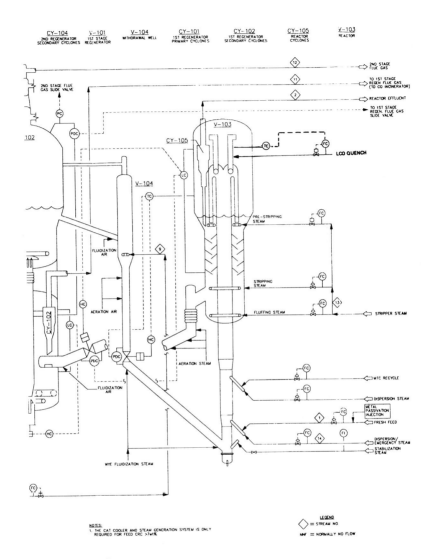

FIGURE 3.4.2 (*Continued*)

This ducting minimizes the vapor residence time and undesirable secondary thermal reactions in the vessel. The vapors and entrained catalyst pass through single-stage high-efficiency cyclones. Reactor products, inerts, steam, and a minute amount of catalyst flow into the base of the main fractionator and are separated into various product streams.

Below each dipleg of the Ramshorn separator, a steam ring quickly strips volatile hydrocarbon material. Without rapid stripping, the volatile hydrocarbon sticking to the catalyst might otherwise react to form polymeric coke as the catalyst travels through the stripper. The stripper portion of this vessel utilizes four baffled stages. Steam from the main steam ring fluidizes the catalyst bed, displaces the entrained hydrocarbons, and strips the adsorbed hydrocarbons from the catalyst before it enters the regeneration system. A steam fluffing ring, located in the bottom head of the stripper, keeps the catalyst properly fluidized and ensures smooth catalyst flow through the spent catalyst transfer line.

Stripped catalyst leaves the stripper through the 45° slanted withdrawal nozzle and then enters a vertical standpipe. The spent catalyst flows down through this standpipe and into a second 45° lateral section that extends into the first-stage regenerator. The spent catalyst slide valve is located near the top of this lower 45° transfer line and controls the catalyst bed level in the stripper. Careful aeration of the catalyst standpipe ensures proper head buildup and smooth catalyst flow. The flow rates from the aeration taps are adjustable to maintain stable standpipe density for different catalyst circulation rates or different catalyst types. The catalyst enters the first-stage regenerator through a catalyst distributor which disperses the catalyst onto the bed surface.

Catalyst and combustion air flow countercurrently within first-stage regenerator vessel. Combustion air is distributed into the regenerator vessel by an air ring. Air rings provide even air distribution across the bed, resulting in proper fluidization and combustion. Partially regenerated catalyst exits near the bottom of the vessel through a hollow stem plug valve which controls the first-stage regenerator bed level. A lift line conveys the partially regenerated catalyst from the first-stage regenerator to the second stage utilizing air injected into the line through the hollow stem of the plug valve. Carbon monoxide–rich flue gases exit the regenerator through two-stage high-efficiency cyclones.

The operational severity of the first-stage regeneration is intentionally mild due to partial combustion. Low temperature results in the catalyst maintaining higher surface area and activity levels. The coke burn percentage can be varied by shifting the burn to the second-stage regenerator, giving the RFCC the operating flexibility for residual as well as gas oil feedstocks. For residual feed, nearly 70 percent of the coke is burnt in the first-stage regenerator while approximately 50 percent is burned during gas oil operation. Essentially all the hydrogen on the coke is burned off the coke in the first-stage regenerator; this step, coupled with low regenerator temperature, minimizes hydrothermal deactivation of the catalyst.

As the catalyst enters the second-stage regeneration vessel, below the combustion air ring, a mushroom grid distributes the catalyst evenly across the bottom head. This grid distributor on the top of the lift line ensures proper distribution of air and catalyst. In the second-stage regenerator, the remaining carbon on the catalyst is completely burned off with excess oxygen, resulting in a higher temperature compared to the first-stage regenerator. An air ring in this regenerator distributes a portion of the combustion air, while the lift air provides the remainder of the air. With most of the hydrogen burnt in the first stage, moisture content in the gases in the second-stage regenerator is low. This allows higher temperatures in the second-stage regenerator without causing hydrothermal catalyst deactivation.

The second-stage regenerator vessel has minimum internals, which increases temperature limitations. Flue gas leaving the regenerator passes through two-stage external cyclones for catalyst removal. The recovered catalyst is returned to the regenerator via diplegs and the flue gas flows to the energy recovery section.

If the feed Conradson carbon residue is greater than 7.0 wt %, a catalyst cooler will be required for the second-stage regenerator (shown as optional in Fig. 3.4.2) to reduce the second-stage regenerator temperature to less than 760°C. A dense-phase catalyst cooler will withdraw catalyst and return it, via an air lift riser, to just beneath the combustion air ring. Heat is recovered from the catalyst by generating saturated high-pressure steam. Large adjustments in the catalyst cooler duty can be made by varying the catalyst circulation rate through the catalyst cooler. Fine catalyst cooler duty corrections can be made by adjusting the fluidization air rate in the cooler.

Hot regenerated catalyst flows into a withdrawal well from the second-stage regenerator. The withdrawal well allows the catalyst to deaerate properly to standpipe density before entering the vertical regenerated catalyst standpipe. This design ensures smooth and even catalyst flow down the standpipe. Aeration taps, located stepwise down the standpipe, serve to reaerate the catalyst and replace gas volume lost by compression. Flow rates for the aeration taps are adjustable to maintain desirable standpipe density, allowing for differences in catalyst circulation rates or catalyst types. The catalyst passes through the regenerated catalyst slide valve, which controls the reactor temperature by regulating the amount of hot regenerated catalyst to the reactor. The catalyst then flows down the 45° slanted wye section to the riser base. Here, stabilization steam nozzles redistribute the catalyst as it travels up toward the feed nozzles. Fluidization in the wye section, and the stabilization nozzles at the riser base, ensure stable and smooth dense-phase catalyst flow to the feed injection zone. A straight vertical section below the feed nozzles stabilizes the catalyst flow before feed injection and serves as a reverse seal preventing oil flow reversal.

Flue Gas Handling

Each RFCC flue gas system is generally unique from one unit to the next because of local environmental requirements and refiner preference. An example of a basic flue gas handling system is shown in Fig. 3.4.3. Each flue gas line will have a flue gas slide valve and orifice chamber. The first-stage regenerator flue gas slide valve (FGSV) controls the pressure differential between the two regenerator vessels, while the second-stage regenerator FGSV directly controls the pressure of the second-stage regenerator. Each orifice chamber consumes the remaining available pressure from the system.

A CO incinerator is located just downstream of the first-stage regenerator orifice chamber and oxidizes all CO gases to CO_2, utilizing fuel gas and combustion air. Exit temperature is typically 980°C with 1 percent excess O_2. Gases from the CO incinerator combine with second-stage regenerator flue gases and enter a flue gas cooler where heat is recovered as high-pressure superheated steam. Flue gases are finally dispersed into the atmosphere through a stack.

Large-capacity RFCC units may employ a power recovery train and tertiary cyclone system on the first-stage regenerator flue gas stream to drive the air blower. Depending on local particulate emission requirements, an electrostatic precipitator (ESP) or other particulate recovery device such as a third-stage cyclone system or flue gas scrubber may be used to recover entrained particulates. Increasing SO_x and NO_x emission requirements may necessitate a flue gas scrubber, SO_x capturing catalyst additive, or similar process for SO_x recovery and/or a selective catalytic reduction (SCR) unit for NO_x recovery.

Catalyst Handling

The RFCC catalyst handling system has three separate and unique functions:

- Spent catalyst storage and withdrawal
- Fresh catalyst storage and addition
- Equilibrium catalyst storage and addition

The spent hopper receives hot catalyst intermittently from the second-stage regenerator to maintain proper catalyst inventory during operation. In addition, the spent cata-

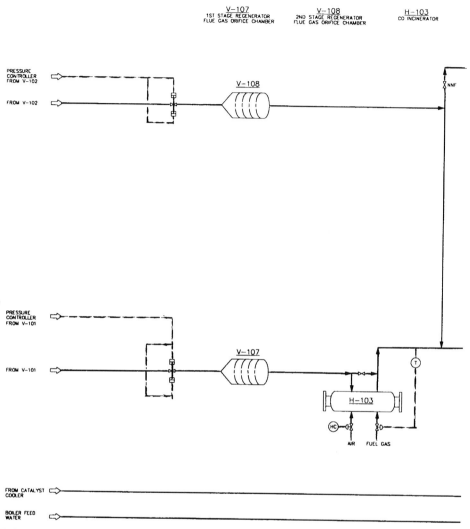

FIGURE 3.4.3 Flue gas handling process flow diagram.

lyst hopper is used to unload, store, and then refill the entire catalyst inventory during RFCC shutdowns.

The fresh catalyst hopper provides storage of catalyst for daily makeup. A fresh catalyst loader, located just beneath the hopper, loads fresh catalyst from the hopper to the first-stage regenerator. Fresh catalyst makeup is based on maintaining optimal unit catalyst activity.

Unique to RFCC designs is a third hopper which is used for equilibrium catalyst. Like the fresh catalyst hopper, the equilibrium catalyst hopper provides storage of catalyst for daily makeup. Equilibrium catalyst serves to flush metals from the unit equi-

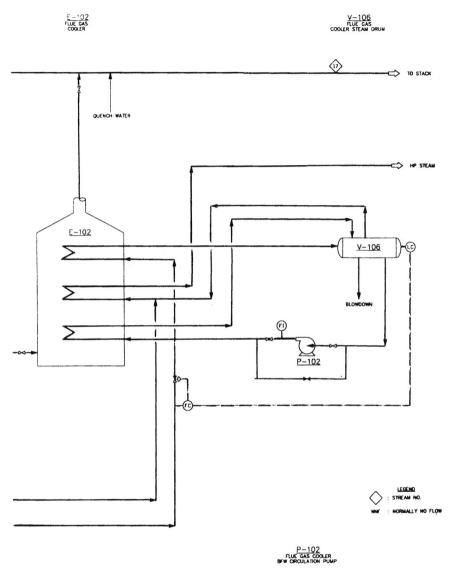

FIGURE 3.4.3 (*Continued*)

librium catalyst in processing of residual feeds with high metal content. However, equilibrium catalyst does not contribute much to cracking activity.[1] As a result, the equilibrium catalyst addition rate is based on targeted metal content on unit equilibrium catalyst, while fresh catalyst makeup rate is based on maintaining unit catalyst activity. An equilibrium catalyst loader is located just beneath the hopper which supplies equilibrium catalyst to the first-stage regenerator.

RFCC FEEDSTOCKS

The most significant advantage of the S&W-IFP RFCC process is the flexibility to process a wide range of feedstocks. Table 3.4.2 lists the range of feedstock properties which have been successfully processed in the S&W-IFP RFCC.

Feedstock to the RFCC can take a variety of forms, from a hydrotreated vacuum gas oil (VGO) to a virgin highly aromatic atmospheric tower bottoms (ATB) such as Arabian light ATB. The RFCC feedstock can also be a blend of various unit streams such as VGO plus coker vacuum gas oil, vacuum tower bottoms (VTB), deasphalted oil (DAO), slop wax, or lube extract. In fact, the number of possible feed constituents to the RFCC is quite large, and it would be impossible to list them all.

What gives the RFCC unit the flexibility to process this wide range of feedstocks is primarily the two-stage regenerator design. A common index which indicates a feedstock's tendency to produce feed-derived coke is the Conradson carbon residue (CCR). As the residual content of a feedstock increases, so does the CCR amount. Table 3.4.3 compares the maximum CCR levels which can be processed in a two-stage regenerator and in a single-stage regenerator.

Recently the increasing need to convert the bottom of the barrel into clean transportation fuels (low sulfur) coupled with the decreasing availability of sweet crudes has ignited an interest in hydrodesulfurization and residual hydrodesulfurization (RDS). Reynolds, Brown, and Silverman showed that it is economically feasible to upgrade VTB using Chevron's vacuum RDS (VRDS) process into feedstock for the S&W-IFP RFCC unit.[2] Processing 100 percent VTB in the RFCC is considerably more attractive than processing it in traditional thermal processors such as delayed and fluid cokers.

OPERATING CONDITIONS

Like traditional FCC units, the S&W-IFP RFCC unit can be operated in maximum distillate, maximum gasoline, or maximum olefin operational modes. Conversion is decreased for maximum distillate operations and increased for the maximum olefin

TABLE 3.4.2 Commercial RFCC Feedstock Operation Experience

Property	Range
Gravity, °API	19–29
Conradson carbon residue, wt %	0–8
Sulfur, wt %	0.2–2.4
Nitrogen, wt %	0.05–0.30
Metals (Ni + V), wt ppm	0–50
540°C + components, LV %	0–58

Note: °API = degrees on the American Petroleum Institute scale; LV = liquid volume.

TABLE 3.4.3 Heavy-Feed Processing Capabilities of Various Heat-Rejection Systems

System	Conradson carbon residue, wt %
Single-stage regenerator	
Full combustion	2.5
Partial combustion	3.5
Partial combustion + MTC	4.0
Catalyst cooler	None
Two-stage regenerator	
Alone	6.0
With MTC	7.0
Catalyst cooler	None

TABLE 3.4.4 Typical RFCC Operating Conditions

Reactor	
Pressure, kg/cm^2 gage	1.1 to 2.1
Temperature, °C	510 to 550
MTC recycle, vol % feed	10 to 25
Feed dispersion steam, wt % feed	3.0 to 7.0
Stripping steam, kg/1000 kg catalyst	3.0 to 5.0
Miscellaneous steam, wt % feed	1.5
First-stage regenerator	
Pressure, kg/cm^2 gage	1.4 to 2.5
Temperature, °C	620 to 690
CO/CO$_2$	0.3 to 1.0
O$_2$, vol %	0.2
Coke burn, wt %	50 to 70
Second-stage regenerator	
Pressure, kg/cm^2 gage	0.7 to 1.4*
Temperature, °C	675 to 760
O$_2$, vol %	2.0
Coke burn, wt %	30 to 50

*Second-stage regenerator pressures reflect a stacked regenerator configuration. For side-by-side regenerator configurations, the second-stage regenerator pressure would be similar to the first-stage regenerator pressure.

TABLE 3.4.5 Commercial RFCC Product Yields

	Unit (year)	
	A (1987)	B (1993)
Feed properties:		
540°C + components, LV %	36	58
CCR, wt %	5.9	4.9
Gravity, °API	22.3	25.1
Yields:		
Dry gas, wt %	4.3	3.2
C$_3$-C$_4$, LV %	24.9	30.5
Gasoline, LV %	60.2	61.5
Light cycle oil, LV %	17.5	14.0
Slurry, LV %	6.6	4.9
Coke, wt %	7.8	8.0
Conversion, LV %	75.9	81.1

operations by adjusting riser outlet temperature and catalyst activity. Typical range of ROTs required for the three operation modes are: maximum distillate, 510°C ROT minimum; maximum gasoline, 510 to 530°C ROT; and maximum olefins, 530 to 550°C ROT. For maximum distillate operation, MTC, discussed in "Mix Temperature Control" below, is critical in order to maintain the required mix temperature to assure vaporization of the heavy residual feed at lower riser outlet temperatures. Likewise, reactant vapor quench technology, discussed in "Amoco Product Vapor Quench" below, is especially critical during maximum olefins operations to reduce postriser thermal cracking at the elevated reactor temperatures.

Other typical operating conditions of the RFCC unit are shown in Table 3.4.4. Examples of observed commercial product yields from an S&W-IFP RFCC unit are shown in Table 3.4.5.

RFCC CATALYST

Catalyst Type

A successful RFCC operation depends not only on the mechanical design of the converter but also on the catalyst selection. In order to maximize the amount of residual content in the RFCC feed, a low-delta-coke catalyst must be employed. Delta coke is defined as

$$\text{Delta coke} = \text{wt \% carbon on spent catalyst} - \text{wt \% CRC}$$

where CRC = carbon on regenerated catalyst, or as

$$\text{Delta coke} = \frac{\text{(coke wt \% feed)}}{\text{(catalyst-to-oil ratio)}}$$

Delta coke is a very popular index and when increased can cause significant rises in regenerator temperature, ultimately reducing the amount of residual feed which can be processed. Commercial delta coke consists of the following components:

- Catalytic coke (deposited slowly as a result of the catalytic reaction)
- Feed-derived coke (deposited quickly and dependent on feed CCR)
- Occluded coke (entrained hydrocarbon)
- Contaminant coke (coke produced as a result of metal contaminant)

Because the feed-derived coke becomes a large contributor to the overall delta coke in processing residual feeds, it is crucial that the overall delta coke be minimized in a RFCC operation.

Stone & Webster–IFP typically recommends a catalyst with the following properties which characterize it as a low-delta-coke catalyst:

- Low rare-earth ultrastable Y (USY) zeolite
- Equilibrium microactivity test (MAT) activity 60 to 65
- Low-delta-coke matrix

At high metal loadings the operator may also consider catalyst with vanadium traps.

Catalyst Addition

Virgin residual feeds may contain large amounts of metals, which ultimately are deposited on the catalyst. Because of the mild two-stage regeneration, the catalyst metal content can be allowed to approach 10,000 wt ppm (Ni + V) before product yields are significantly affected. For an RFCC operation, catalyst addition is based on maintaining catalyst activity as well as metals on catalyst as opposed to maintaining only activity for typical FCC gas oil operations. The most economical way to maintain both activity and metals is to add both fresh catalyst and purchased equilibrium catalyst. Equilibrium catalyst is an effective metal-flushing agent; however, equilibrium catalyst does not contribute much cracking activity.[1] As a result, equilibrium catalyst is added with fresh catalyst in order to economically control both the unit catalyst activity and metal content.

TWO-STAGE REGENERATION

In the S&W-IFP two-stage regeneration process, the catalyst is regenerated in two steps: 50 to 70 percent in the first-stage regenerator and the balance in the second-stage regenerator. The first-stage regeneration is controlled by operating the first stage in an oxygen-deficient environment, producing significant amounts of carbon monoxide. Since the heat of combustion of carbon to carbon monoxide is less than one-third that for combustion to carbon dioxide, much less heat is transferred to the catalyst than a single-stage full-combustion regenerator. For example, a 30,000-BPSD RFCC unit with a feed gravity of 22.5° API and a coke yield of 7.5 wt % at 66 percent coke burn in the first-stage regenerator has reduced the heat transferred to the catalyst by approximately 25×10^6 kcal/h over a full-burn single-stage regenerator.

The remaining carbon on the catalyst is burned in the second-stage regenerator in full combustion mode. Because of the elevated temperature, external cyclones are employed to minimize regenerator internals and allow carbon-steel construction.

Comparison of Two-Stage Regeneration and Single-Stage Regeneration with a Catalyst Cooler

Although both systems operate to control regenerator temperatures, the principles of operation are significantly different. The advantages of the two-stage regeneration system become apparent as the feed becomes heavier and/or its metal content increases.

The benefits of a two-stage regeneration system over a single-stage system with a catalyst cooler are briefly described as follows.

Lower Catalyst Particle Temperature. A catalyst cooler removes heat after it is produced inside the regenerator, while less heat is produced in a two-stage regenerator design. This results in a lower catalyst particle temperature during combustion, reducing overall catalyst deactivation. Since the combustion is occurring in two steps, the combustion severity of each step is low. In the first-stage regenerator, the catalyst enters the bed from the top through the spent catalyst distributor while the combustion air enters the bed at the bottom of the vessel. This countercurrent movement of catalyst and air prevents the contacting of spent catalyst (high carbon) with fresh air containing 21 percent oxygen. All of these factors result in lower catalyst thermal deactivation for the two-stage regeneration system.

Lower Hydrothermal Deactivation. While the catalyst is only partially regenerated in the first stage, most of the water formed by the combustion of the hydrogen in the coke is removed in this vessel. Figure 3.4.4 shows the percent hydrogen on coke burn as a function of carbon burn. Since the temperature of the first-stage regenerator is low, catalyst hydrothermal deactivation is significantly reduced. In the second-stage regenerator, where the bed temperature is high, moisture is minimal and does not pose a significant hydrothermal deactivation risk for the catalyst.

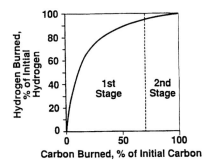

FIGURE 3.4.4 Hydrogen and hydrocarbon burn rates.

Better Metal Resistance. When refiners run high-metal feeds, it is very advantageous to be able to run with high metal levels on the equilibrium catalyst. Studies have clearly shown that high metal levels (particularly vanadium) leads to excessive catalyst deactivation in the presence of steam and oxygen. Since most of the steam in a regenerator comes from the hydrogen in the coke, the moisture content can be calculated in a straightforward manner. For a single-stage regenerator this will usually be more than 10 percent moisture. When steam and vanadium react in the presence of oxygen, vanadic acid is formed, which attacks the alumina in the catalyst zeolite structure. Massive dealumination causes the collapse of the zeolite structure, and the resulting catalyst is left with little activity. The equations

$$2V + \tfrac{5}{2}O_2 \rightarrow V_2O_5$$

and

$$V_2O_5 + 3H_2O \rightarrow 2VO(OH)_3$$
Vanadic acid

describe the generation of vanadic acid. As a result, catalyst in a single-stage regenerator operating in the presence of excess oxygen and steam is prone to vanadic acid attack.

Staging the regeneration can be particularly effective in this situation. In the first-stage regenerator, most of the hydrogen (and subsequent water vapor) is removed at low temperature without the presence of oxygen. This is followed by a full-burn second-stage regenerator where there is excess oxygen but very little moisture. Vanadium destruction of the catalyst structure is minimized, since very little V_2O_5 is present in the first-stage regenerator because of the lack of oxygen and lower temperature, while vanadic acid is minimized in the second-stage regenerator by lack of water. In other words, the reaction

$$2V + \tfrac{5}{2}O_2 \rightarrow V_2O_5$$

proceeds very slowly in the first-stage regenerator because of a lack of oxygen while the reaction

$$V_2O_5 + 3H_2O \rightarrow 2VO(OH)_3$$

proceeds slowly in the second-stage regenerator because of low steam content.

The two-stage regeneration is clearly less severe with regard to catalyst deactivation and this, coupled with the newer generation of catalyst with vanadium traps, will allow refiners to run heavier crudes more efficiently and economically than ever before.

Catalyst Cooler System

The S&W-IFP heavy residual RFCC units (feed CCR greater than 6.0 wt % or 7.0 wt % with MTC) design includes a well-proven catalyst cooler system. The same catalyst cooler design can also be provided for an existing FCC regenerator. This design is operating in more than a dozen units, with several more in the design and construction stages. A few features of the Stone & Webster catalyst cooler system are

- Dense-phase, downward catalyst flow
- Slide-valve-controlled catalyst circulation
- Turndown capability from 0 to 100 percent

- No tube sheet required
- High mechanical reliability
- Cold wall design
- All carbon-steel construction
- High heat transfer—low tube wall temperature
- 100 percent on-stream factor

Catalyst cooler duties can range from as low as 2×10^6 kcal/h up to 35×10^6 kcal/h. In the event that more than a 35×10^6 kcal/h cooler is required, multiple catalyst coolers can be employed on a regenerator.

A schematic diagram for a catalyst cooler coupled to a regenerator (second-stage regenerator in a two-stage regeneration system) is shown in Fig. 3.4.5. Catalyst level inside the cooler is controlled by the inlet catalyst slide valve. Gross temperature control of the regenerator is achieved by the bottom catalyst slide valve and fine temperature control is made by the cooler fluidization air.

S&W-IFP TECHNOLOGY FEATURES

S&W-IFP offers many technology features which improve the product selectivity, unit capacity, and operability of our RFCC designs. These same features are available to refiners who wish to upgrade existing FCC units. In fact, various aspects of the S&W-IFP FCC process have been applied to over 50 FCC revamps.

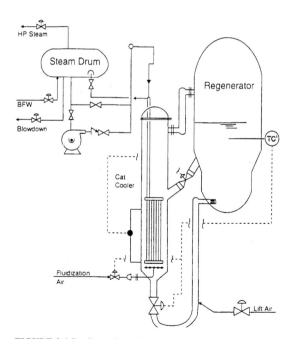

FIGURE 3.4.5 General catalyst cooler arrangement.

Feed Injection System

The feedstock injection system and lower portion of the feed riser are the most critical parts of the RFCC/FCC. The earlier pioneering and patented developments of Total Petroleum Inc. have convinced the refining industry of the value and benefits of advanced feed injection. Basic elements of the S&W-IFP feed injection system are as follows:

- Dense-phase flow of catalyst up to the feed injection point, employing small quantities of steam to stabilize catalyst flow and maintain a uniform catalyst flux across the riser.
- Atomization of the feed external to the riser using steam in a simple but efficient two-fluid nozzle not involving complex internals subject to plugging and erosion.
- Introducing feed into an upward-flowing dense phase of catalyst in a manner which achieves the penetration and turbulence necessary to accomplish rapid heat transfer from the hot catalyst to the fine oil droplets, ensuring rapid vaporization.

Table 3.4.6 lists actual commercial product yield improvements observed after replacing an older feed injection system with the S&W-IFP design.

Basic elements of the S&W-IFP feed injection nozzle are shown in Fig. 3.4.6. This two-fluid nozzle works by injecting oil under pressure against a target plate to break

TABLE 3.4.6 Incremental S&W-IFP Feed Injection System Product Yields

	Delta Yields	
Product	Unit A	Unit B
Dry gas, wt %	+0.0	−1.3
C_3/C_4, LV %	+1.5	+1.5
Gasoline, LV %	+3.4	+6.2
Light cycle oil, LV %	+1.6	−4.5
Slurry, LV %	−6.5	−0.3
Coke, wt %	+0.0	−0.1
Conversion, LV %	+4.9	+4.8

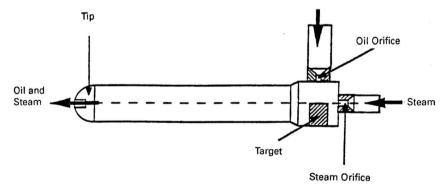

FIGURE 3.4.6 S&W-IFP feed injection nozzle.

the oil into thin sheets that the steam shears as it moves across and through the oil. The oil mist is injected into the riser through a specially designed tip which ensures maximum riser coverage without impinging and damaging the riser wall.

This feed injection system was developed for RFCC operations where the residual feed is highly viscous and difficult to atomize. In order to provide adequate atomization of the residual feedstock, this nozzle design uses oil pressure, steam pressure, and steam rate. For vacuum gas oil feedstocks which are considerably easier to atomize, oil pressure and steam rates can be reduced below those of residual operations.

Mix Temperature Control

An important concern in processing heavy feedstocks with substantial amounts of residual oil is ensuring rapid feed vaporization. This is critical to minimize unnecessary coke deposition due to incomplete vaporization. Unfortunately, in conventional designs, the mix temperature is essentially dependent on the riser outlet temperature. Typically the mix temperature is about 20 to 40°C higher than the riser outlet temperature and can be changed only marginally by catalyst-to-oil ratio.

In many cases, raising the riser outlet temperature to adjust the mix temperature is not desirable since this may result in undesirable nonselective cracking reactions with high production of dry gas. The problem becomes even more critical with less severe operating conditions for maximum distillate production. To address this problem and make the above objectives compatible with each other, riser outlet temperature must be independently adjusted. This is achieved with MTC developed and patented by IFP-Total.

MTC is performed by recycling a selected liquid cut downstream of the fresh feed injection zone. It roughly separates the riser into two reaction zones:

- An upstream zone, characterized by high temperature, high catalyst-to-oil ratio, and very short contact time.
- A downstream zone, where the reaction proceeds under more conventional and milder catalytic cracking conditions.

Creating two separate cracking zones in the riser permits fine tuning of the feed vaporization and cracking to desired products. With MTC, it is possible to raise the mix temperature while maintaining or even lowering the riser outlet temperature. Figure 3.4.7 illustrates the MTC nozzle arrangement and the three temperature zones.

The primary objective of the MTC system is to provide an independent control of the mix temperature. However, as a heat-sink device similar to a catalyst cooler, MTC can be used to increase the amount of residual feed processed in the unit.

Riser Termination Device

Numerous studies have shown that postriser vapor residence time leads to thermal cracking and continued catalyst cracking in the reactor vessel. Unfortunately, these postriser vapor-phase reactions are extremely nonselective and lead to degradation of valuable liquid products, high dry-gas make, and high hydrogen transfer in liquefied petroleum gas (LPG) olefins (low olefin selectivity). The factors that contribute to these phenomena are temperature, time, and surface area. S&W-IFP's riser termination technology is designed to control all three factors.

S&W-IFP's Ramshorn riser termination device, shown in Fig. 3.4.8, quickly separates the catalyst from the vapor. Reactant vapors are ducted to near the reactor

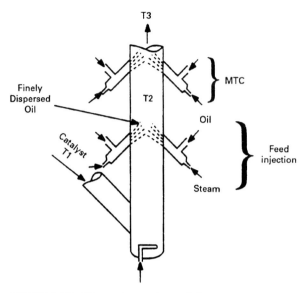

FIGURE 3.4.7 Mix zone temperature control.

cyclones to ensure low vapor residence time in the reactor. By not directly coupling the outlet ducts to the reactor cyclones, catalyst carryover to the fractionator during upsets is significantly reduced. Other problems such as reactor heat-up, differential metal growth, and coking in the reactor are also eliminated.

Amoco Product Vapor Quench

This technology was developed and patented by Amoco and is offered to the industry by Stone & Webster and IFP under an exclusive arrangement. Reactant vapors are quenched, leaving the riser termination system substantially free of catalyst, by injecting a light cycle oil quench. By employing quench technology, nonselective thermal reactions are arrested, resulting in higher gasoline yields and lower dry gas production. In addition, use of the quench technology further preserves the LPG olefins and gasoline octane, minimizes the formation of diolefins, and enhances gasoline stability.

The effectiveness of vapor quench is shown in Table 3.4.7. The data indicate that a reduction in dry gas production is observed even at low riser outlet temperatures. As expected, the impact of quench in terms of dry gas reduction and gasoline yield improvement is more marked at higher temperatures.

The combination of the S&W-IFP riser termination device and Amoco's vapor quench virtually eliminates undesirable postriser reactions.

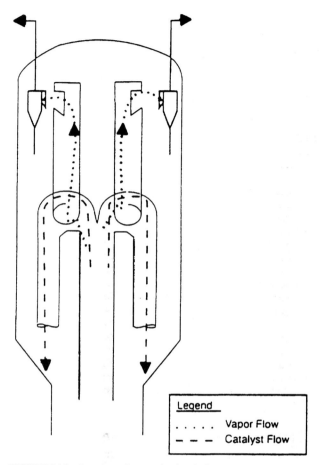

FIGURE 3.4.8 Ramshorn riser termination device.

TABLE 3.4.7 Impact of Reactor Vapor Quench on FCC Yields

	Unit A	Unit B
Temperature, °C:		
Riser outlet	513	549
After quench	484	519
Yield shifts, wt %:		
Dry gas	−0.23	−0.80
Gasoline	+0.43	+1.80

MECHANICAL DESIGN FEATURES

The S&W-IFP RFCC mechanical design philosophy is based on multiple concepts to provide high reliability and maintainability with longer run lengths. Mechanical design efforts have focused on areas of an FCC unit that have historically caused high maintenance costs and increased downtime. These efforts have resulted in an overall mechanical design capable of providing more than 3 years of operation between turnarounds. Some of the features are discussed here.

Cold Wall Design

The cold wall design concept is emphasized throughout the unit in the riser, reactor, regenerators, catalyst cooler, external transfer lines, slide valves, and external cyclones. Internal refractory insulation of vessel pressure parts sufficiently reduces the skin temperatures to permit use of less expensive and easier-to-maintain carbon-steel materials. Lower metal temperatures result in less thermal expansion of the components, minimizing the need for expansion joints to compensate for differential thermal expansion between interconnected components and transfer lines.

External surface areas of the pressure parts are exposed for on-line inspection, thereby reducing inspection and maintenance costs. The internal refractory protects the pressure shell from catalyst erosion, while metal hot spots can be readily detected before they progress to a potentially dangerous level.

Feed Nozzle Fabrication

The S&W-IFP proprietary feed injection nozzles are installed through sleeves in the riser wall. Erosion of the riser wall is avoided by careful selection of the entrance angle of the sleeve and the design of the nozzle spray angle. The nozzle tip and atomizing chamber are made from erosion-resistant material to virtually eliminate wear. In the unlikely event of erosion, those surfaces exposed to erosive conditions are easily replaced and are designed so that normal maintenance can be performed during a scheduled turnaround with removal of the nozzle from the vessel sleeve. Typically, it is only necessary to inspect the nozzles at turnaround and only rarely is any maintenance required.

External Cyclones

Cold wall external cyclones are used on the second-stage regenerator to remove them from the internal, hot environment. The cyclones are attached directly to the cold wall regenerator and the minimal differential thermal expansion is easily accommodated. The size and length/diameter ratio of the external cyclones are not limited by the internal dimensions of the regenerator; therefore, more efficient cyclones can be designed with a shorter, less expensive regenerator. In addition, the external cyclones offer longer turnaround cycles, are insensitive to thermal excursions, and are subject to direct inspection while in operation. The cyclones can be easily monitored for mechanical reliability by using infrared cameras and for process performance by monitoring the dipleg levels with level indicators.

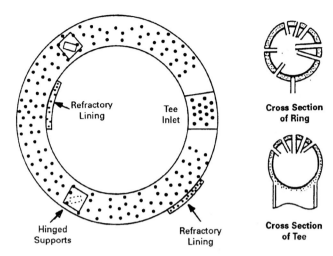

FIGURE 3.4.9 Combustion air ring.

Combustion Air Rings

The S&W-IFP design utilizes proprietary combustion air rings instead of dome or pipe grids. The design provides optimum air distribution and mixing, both vertically and laterally, and overcomes problems of material cracking, distributor erosion, and nozzle erosion experienced with other designs. The use of properly designed nozzles and high-density refractory material on the rings eliminates all damage due to erosion. A combustion air ring is shown in Fig. 3.4.9.

FCC REVAMP TO RFCC (SECOND-STAGE REGENERATOR ADDITION)

Adding a second-stage regenerator is an effective means of converting an existing FCC unit to residual service without losing throughput. To date, three FCC units have been revamped to include a second-stage regenerator and allow the processing of heavy residual feedstocks. These designs retain the existing regenerator as the first-stage regenerator and the reactor/stripper. A new second-stage regenerator, catalyst transfer lines, and CO incinerator; a new or supplemental air blower; and a revamp of the flue gas handling facilities are required. By operating the first-stage regenerator in partial combustion mode, as explained earlier, no additional heat-removal facilities will be required up to a feed CCR of 6.0 wt %. Shown in Fig. 3.4.10 is an FCC unit revamped to include a second-stage regenerator; the figure indicates both new and existing equipment.

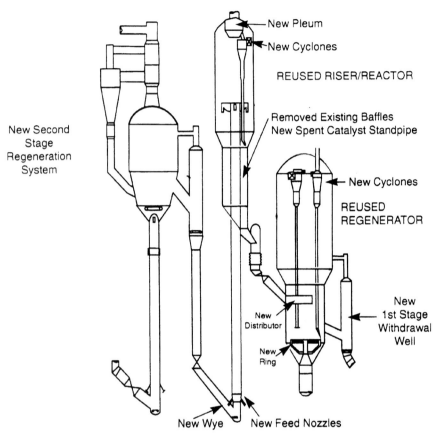

FIGURE 3.4.10 Side-by-side regenerator RFCC revamp design.

REFERENCES

1. Mott, Raymond, "FCC Catalyst Management for Resid Processing," First FCC Forum, Stone & Webster Engineering Corporation, The Woodlands, Tex., May 11–13, 1994.
2. Reynolds, B. E., E. C. Brown, and M. A. Silverman, "Clean Gasoline via VRDS/RFCC," *Hydrocarbon Processing,* April 1992, pp. 43–51.

CHAPTER 3.5
DEEP CATALYTIC CRACKING, THE NEW LIGHT OLEFIN GENERATOR

David A. Hunt
Refining Section
Stone & Webster Engineering Corporation
Houston, Texas

BASIS

The fluid catalytic cracking (FCC) unit is the most important and widely used heavy oil conversion process in the modern refinery. Historically, the FCC unit has operated in maximum gasoline and maximum distillate modes, depending on seasonal product demands and refinery locale. Recently, with the advent of reformulated gasoline requirements, the FCC unit has been increasingly required to operate in the maximum olefin mode. Light isoolefins, isobutylene and isoamylene, from the FCC unit are a necessary feedstock for methyl tertiary butyl ether (MTBE) and tertiary amyl methyl ether (TAME) oxygenated reformulated gasoline blending components. Increased alkylate demand to meet reformulated gasoline requirements also necessitates an increase in light olefins.

At the same time as these changes are occurring in the refining industry, the petrochemical industry is experiencing increased demands for propylene for the manufacture of polypropylene products. Nearly half of the propylene used by the chemical industry is obtained from refineries, and the remainder from steam cracking (SC).[1] As a result, the demand for propylene from both FCC units and SC units is rising.

It is certain that the isoolefin demand for oxygenated gasoline blending components will continue to grow through the end of the century. In addition, the demand for propylene, both as an alkylation feed and for polypropylene production, is also expected to grow. This of course places a considerable strain on the FCC unit and SC in order to meet the demand. Obviously, a need for an economical light olefin generating process is required to meet the demand of these light olefins (C_3 through C_5).

To this end, Stone & Webster has entered into an agreement with the Research Institute of Petroleum Processing (RIPP) and Sinopec International, both located in the People's Republic of China, to exclusively license RIPP's deep catalytic cracking

(DCC) technology outside China. DCC is a newly developed process, similar to FCC, for producing light olefins (C_3-C_5) from vacuum gas oil (VGO) feedstocks. Stone & Webster's proven position in FCC technology and olefins is a natural complement to DCC technology.

A successful commercial trail of the DCC process at the Sinopec Jinan refinery in China has led to three additional DCC units in China and abroad. Table 3.5.1 is a listing of all DCC units operating or under construction to date. Figure 3.5.1 shows a recently constructed DCC unit in China.

PROCESS DESCRIPTION

DCC is a fluidized catalytic process for selectively cracking a variety of feedstocks to light olefins. A traditional reactor/regenerator unit design is employed with a catalyst having physical properties much like those of FCC catalyst. The DCC unit may be operated in one of two operating modes: maximum propylene (Type I) and maximum isoolefins (Type II). Each operational mode employs a unique catalyst and operating conditions. DCC reaction products are light olefins, high octane gasoline, light cycle oil, dry gas, and coke. A small amount of slurry oil is also produced.

DCC maximum propylene operation (Type I) employs both riser and bed cracking at severe reactor conditions. Maximum isoolefin operation (Type II) utilizes riser cracking, like a modern FCC unit, at slightly milder conditions than Type I operation. Figure 3.5.2, a process flow diagram of a Type I DCC process, serves as a basis for the process description. (Note that the only difference between the Type I and Type II designs is an extended riser with riser termination device above the reactor bed level.)

Fresh feed is finely atomized by steam and injected into the riser through Stone & Webster proprietary FCC feed injection nozzles over a dense phase of catalyst. The atomized oil intimately mixes with the catalyst and begins to crack into lighter, more valuable products. A good feed injection system is required for DCC, just as for FCC operation, to ensure rapid oil vaporization and selective catalytic cracking reactions.

Riser steam is injected just above the feed injection point to supplement feed dispersion and stripping steam in order to achieve optimal hydrocarbon partial pressure for the DCC operation. Simple steam injection nozzles are employed for riser steam injection. (Steam requirements for DCC Type II operation are considerably less and may not need additional steam injection nozzles.)

Slurry recycle is injected, if required, just above the riser steam nozzles. This recycle stream is not required to increase overall conversion but rather to optimize the unit heat balance, as a large slurry reaction product is coke.

At the top of the riser, catalyst, steam, and hydrocarbon pass through a riser terminator located below the reactor bed. Conversion of the DCC feedstock can be regulat-

TABLE 3.5.1 DCC Commencement Status

Location	Feed capacity, MTY*	Start-up date
Jinan, China	150,000	1994
Anqing, China	400,000	1995
Daqing, China	120,000	1995
Rayong, Thailand†	725,000	1997

*MTY = Metric tons per year.
†Design by Stone & Webster Engineering Corporation.

FIGURE 3.5.1 Grass-roots DCC unit located in China. From left to right note the regenerator, reactor, and main fractionator.

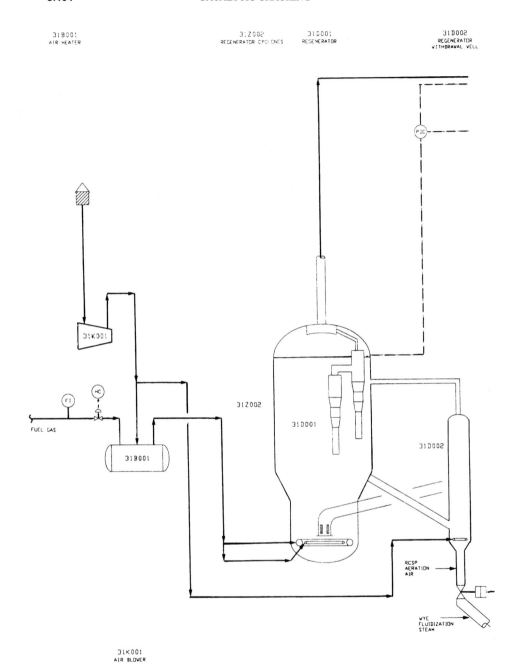

FIGURE 3.5.2 Maximum propylene DCC unit (Type I) process flow diagram.

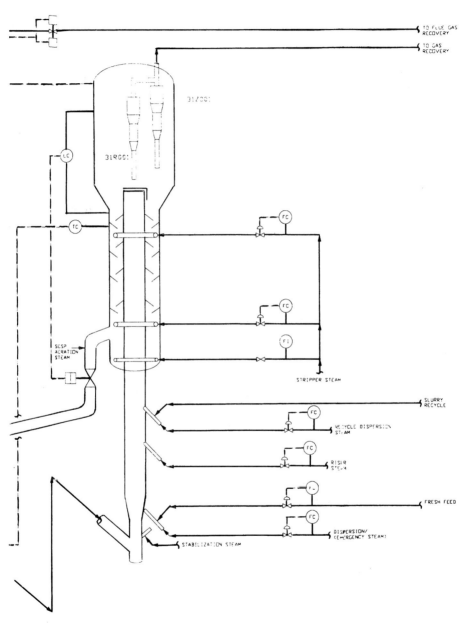

FIGURE 3.5.2 (*Continued*)

ed by adjusting the catalyst bed height (hydrocarbon weight hourly space velocity) above the riser distributor. Two-stage high-efficiency reactor cyclones remove entrained catalyst from the reactor vapors. Products, inerts, steam, and a small amount of catalyst flow from the reactor into the base of the main fractionator to begin product separation.

The regenerated catalyst slide valve controls the reactor bed temperature by regulating the amount of hot regenerated catalyst entering the riser. Nominal reactor temperatures and pressures are listed in "Operating Conditions" below.

The stripper portion of the reactor vessel uses baffled stages and staged stripping. A prestripping steam ring, located beneath the first stripper baffle, strips the volatile hydrocarbon material just as the catalyst enters the stripper. Without this initial stripping, volatile hydrocarbon adsorbed on the catalyst might otherwise react to form polymeric coke as the catalyst travels down the stripper. Steam from the main steam ring fluidizes the catalyst bed, displaces the entrained hydrocarbons, and strips the remaining adsorbed hydrocarbons from the catalyst before it enters the regeneration system. A steam fluffing ring, located in the bottom head of the stripper, keeps the catalyst properly fluidized and ensures smooth catalyst flow into the spent catalyst standpipe.

Spent catalyst leaves the stripper through a slanted standpipe. Aeration taps, located stepwise down the standpipe, serve to reaerate the catalyst and replace gas volume lost by compression. The spent catalyst slide valve, located near the point where the standpipe enters the regenerator, maintains proper bed level in the reactor/stripper. Reactor bed level is optimized with respect to conversion and unit operability.

Spent catalyst is dispersed inside the regenerator by a catalyst distributor just above the combustion air rings. Combustion air rings provide even air distribution across the regenerator bed, resulting in proper fluidization and combustion. The regenerator operates in a full combustion mode with approximately 2 vol % excess oxygen. Regenerator flue gases exit through two-stage high-efficiency regenerator cyclones which remove entrained catalyst from the flue gas. Typical regenerator temperature is near 700°C. Regenerator/reactor differential pressure is controlled by a flue gas slide valve.

Hot regenerated catalyst is withdrawn from the regenerator, just below the regenerator bed level, into a catalyst withdrawal well. The withdrawal well allows the catalyst to deaerate properly to standpipe density before entering the vertical regenerated catalyst standpipe. A small air ring located in the withdrawal well serves to maintain proper catalyst fluidization. Aeration taps, located stepwise down the standpipe, serve to reaerate the catalyst and replace gas volume lost by compression. Catalyst passes through the regenerated catalyst slide valve, which controls the reactor temperature by regulating the amount of hot catalyst entering the riser/reactor section. A straight vertical section below the feed nozzles stabilizes the catalyst flow and serves as a reverse seal, preventing oil reversals into the regenerator. Here, stabilization steam nozzles redistribute the catalyst, maintaining a dense catalyst phase, as it travels up toward the critical feed injection zone.

The DCC gas recovery section employs a low-pressure-drop main fractionator design with warm reflux overhead condensers to condense the large amounts of steam used in the converter. A large wet gas compressor is required, relative to FCC operation, because of the high amounts of dry gas and liquefied petroleum gas (LPG). The absorber and stripper columns, downstream of the wet gas compressor, are specifically designed for enhanced C_3 recovery at relatively low gasoline rates. Following the traditional debutanizer and depropanizer for contaminant removal, a deethanizer and C_3 splitter are required to produce polymer-grade propylene. For DCC units in or near a petrochemical process, a cryogenic ethylene recovery unit utilizing Stone & Webster's Advanced Recovery System (ARS) technology may be of interest for ethylene recovery and essentially complete propylene recovery.

The flue gas handling system, downstream of the DCC regenerator, requires considerations no different than those of a FCC system. It consists of a flue gas slide valve to control the differential pressure between the reactor and regenerator followed by an orifice chamber. Heat is recovered by a flue gas cooler in the form of high-pressure superheated steam. Depending on local particulate emission specifications, the system may contain a third-stage cyclone separator upstream of the flue gas slide valve or an electrostatic precipitator (ESP) upstream of the stack. SO_x or NO_x emission requirements may necessitate a flue gas scrubber or SO_x-capturing catalyst additive to reduce SO_x emissions and/or a selective catalytic reduction (SCR) process for NO_x recovery.

CATALYST

The most critical part of the DCC process is the catalyst. RIPP's research and development efforts have resulted in the development of several proprietary catalysts, each with unique zeolites. All catalysts have physical properties similar to those of FCC catalysts.

The catalyst designated CRP-1 was developed for use in the DCC maximum propylene operation (Type I). CRP has a relatively low activity to ensure high olefin selectivity and low hydrogen transfer reactions. The catalyst also exhibits a high degree of hydrothermal stability and low coke selectivity.

CS-1 and CZ-1 were developed to produce high isobutylene and isoamylene selectivity as well as propylene selectivity. Again, these catalysts are ~ow-hydrogen-transfer catalysts with good hydrothermal and coke-selective properties.

All three types of catalyst are currently manufactured by Qilu Petrochemical Company's catalyst facility in China. Stone & Webster is currently qualifying a second DCC catalyst supplier outside China.

FEEDSTOCKS

The DCC process is applicable to various VGO feedstocks for propylene and isoolefin production. Feedstocks may also include wax, naphtha, and residual oils. Paraffinic feedstocks are preferred; however, successful pilot plant trails have also been performed with naphthenic and hydrotreated aromatic feeds.

OPERATING CONDITIONS

A range of typical operating conditions for both Type I (maximum propylene) and Type II (maximum isoolefins) is shown in Table 3.5.2. Also indicated are typical FCC and SC operating conditions for comparison. A more severe reactor temperature is required for the DCC process than for FCC. Type II DCC reactor temperature is less severe than Type I, to increase isoolefin selectivity, but still more than FCC. Steam usage for DCC operations is more than for FCC, but considerably less than for SC. DCC catalyst circulation rates are higher than FCC operations, while regenerator temperatures are similar or less.

TABLE 3.5.2 DCC, FCC, and SC Operating Conditions

	DCC Type I, max. C_3	DCC Type II, max. isoolefins	FCC	SC
Temperatures:				
Reactor, °C	550–565	525–550	510–550	760–870
Regenerator, °C	670–700	670–700	670–730	—
Reactor pressure, kg/cm² gage	0.7–1.0	1.0–1.4	1.4–2.1	1.0
Reaction time, s	*	2 (riser)	2 (riser)	0.1–0.2
Catalyst/oil, wt/wt	9–15	7–11	5–8	—
Steam injection, wt % feed	20–30	10–15	2–7	30–80

*Riser residence time approximately 2 s plus 2–8 weight hourly space velocity (WHSV) in reactor bed.

DCC PRODUCT YIELDS

DCC Maximum Propylene (Type I)

A typical DCC maximum propylene yield slate for a Daqing (paraffinic) VGO is shown in Table 3.5.3. For comparison purposes, FCC and SC maximum olefin yields for the same feedstock are also shown in Table 3.5.3.

Propylene is abundant in the DCC LPG stream and considerably higher than for FCC. DCC LPG also contains a large amount of butylenes where the isobutylene fraction of the total butylenes is higher than for FCC (38 to 42 wt % versus 17 to 33 wt %).[1] Subsequent MTBE production is enhanced over FCC operations because of the additional available isobutylene. These high olefin yields are achieved by selectively overcracking naphtha.

Large amounts of dry gas are produced by the DCC because of the severe reactor temperature. DCC dry gas is rich in ethylene, which can be recovered for petrochemi-

TABLE 3.5.3 Yields for DCC Type I versus FCC and Steam Cracking

	wt % of feed		
Component	DCC (Type I)	FCC	SC
H_2	0.3	0.1	0.6
Dry gas (C_1-C_2)	12.6	3.8	44.0
LPG (C_3-C_4)	42.3	27.5	25.7
Naphtha (C_5–205°C)	20.2	47.9	19.3
Light cycle oil (205–330°C)	7.9	8.7	4.7
Slurry oil (330°C+)	7.3	5.9	5.7
Coke	9.4	6.1	—
Light olefins:			
C_2	5.7	0.9	28.2
C_3	20.4	8.2	15.0
C_4	15.7	13.1	4.1

Source: Lark Chapin and Warren Letzsch, "Deep Catalytic Cracking, Maximize Olefin Production," NPRA Annual Meeting, AM-94-43, March 20–22, 1994.

cal sales. Nonetheless, the DCC operation produces considerably less dry gas and more LPG than steam cracking. The primary DCC product is propylene, whereas ethylene is the major SC component. (Steam cracking is a thermal reaction whereas DCC is predominantly catalytic.)

Because of high conversion, the DCC C_5+ liquid products are all highly aromatic. Consequently octane values of the DCC naphtha are very high. For this yield slate, an 84.7 motor octane number, clear (MONC) and 99.3 research octane number, clear (RONC) were measured.[2] DCC C_5+ naphtha has greater than 25 wt % benzene, toluene, and xylene (BTX) content and is a good BTX extraction candidate. Because of high diolefin content, selective hydrotreating is usually required. Selective hydrotreating can be achieved without losing octane.

Coke make is considerably higher than in FCC operation. The high heat of reaction required for the conversion of the feed to DCC products demands a high reactor temperature. As a result the coke yield is larger.

The sensitivity of olefin yield for three VGO types is shown in Table 3.5.4. Daqing VGO is highly paraffinic, Arabian light is moderately aromatic, while Iranian is highly aromatic. Propylene and butylene yields are very high for paraffinic feedstocks but only decrease moderately for the most aromatic feeds. The data were generated in RIPP's 2 barrel per day (BPD) DCC pilot unit.

DCC Maximum Isoolefin (Type II)

Pilot plant DCC maximum olefin yields are shown in Table 3.5.5. Large olefin yields are produced by overcracking naphtha at less severe conditions than Type I. The high olefin selectivity is indicative of very low hydrogen transfer rates. Butylene and amylene isomer breakdowns are shown in Table 3.5.6. Note that the isoolefins in the DCC Type II operation approach their respective thermodynamic equilibrium. As a result, isobutylene and isoamylene yields are very large, each over 6.0 wt % of feed.

DCC INTEGRATION

It is possible to incorporate a DCC process in either a petrochemical or refining facility. Idled FCC units in operating facilities are particularly attractive for DCC implementation. A few possible processing scenarios are discussed.

TABLE 3.5.4 DCC Type I Olefin Yields for Various VGO Feedstocks

	Daqing	Arabian Light*	Iranian*
Specific gravity	0.84	0.88	0.91
UOP K factor	12.4	11.9	11.7
Olefin yield, wt % feed:			
C_2	6.1	4.3	3.5
C_3	21.1	16.7	13.6
C_4	14.3	12.7	10.1

*Hydrotreated vacuum gas oil.

TABLE 3.5.5 DCC Maximum Isoolefin Yields (Type II)

Component	Yield, wt % of feed
C_2-	5.59
C_3-C_4	34.49
C_5 + naphtha	39.00
Light cycle oil	9.77
Heavy cycle oil	5.84
Coke	4.31
Loss	1.00
Light olefins:	
$C_2=$	2.26
$C_3=$	14.29
$C_4=$	14.65
$i\text{-}C_4=$	6.13
$C_5=$	9.77
$i\text{-}C_5=$	6.77

Source: Z. T. Li, W. Y. Shi, N. Pan, and F. K. Jaing, "DCC Flexibility for Isoolefins Production," *Advances in Fluid Catalytic Cracking,* ACS, vol. 38, no. 3, pp. 581–583.

TABLE 3.5.6 Olefin Isomer Distribution DCC Type II Operation

Component	Equilibrium value	DCC max isoolefin
Butylene isomers:		
1-butene, wt %	14.7	12.8
t-2-butene, wt %	24.5	26.7
c-2-butene, wt %	16.7	18.6
isobutylene, wt %	44.1	41.9
Amylene isomers:		
1-pentene, wt %	5.2	5.2
t-2-pentene, wt %	12.2	17.6
c-2-pentene, wt %	12.0	7.9
isoamylene, wt %	70.6	69.3

Source: Z. T. Li, W. Y. Shi, N. Pan, and F. K. Jaing, "DCC Flexibility for Isoolefins Production," *Advances in Fluid Catalytic Cracking,* ACS, vol. 38, no. 3, pp. 581–583.

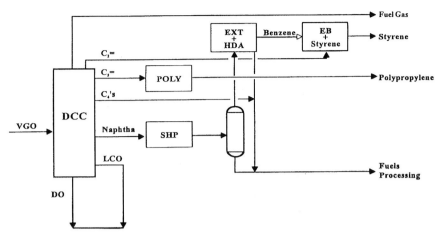

FIGURE 3.5.3 Polypropylene and styrene production scheme (EXT = aromatics extraction, HDA = hydrodealkylation, SHP = selective hydrogenation).

One possible scenario is utilization of a DCC unit to increase propylene production in an ethylene facility. DCC naphtha, ethane, propane, and butane could be sent to the SC for additional ethylene yield. It may be possible to debottleneck the existing product splitter to accommodate the DCC gaseous stream.

A DCC unit could be incorporated into a refining facility for polypropylene and styrene production. An example of such a processing scheme is shown in Fig. 3.5.3.

Another example of DCC integration is for supporting reformulated gasoline production as shown in Fig. 3.5.4. An ethylene recovery unit using Stone & Webster's ARS technology could be incorporated into this scheme for polymer ethylene and propylene sales.

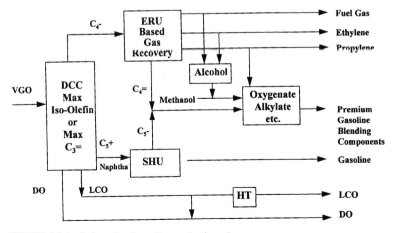

FIGURE 3.5.4 Reformulated gasoline production scheme.

REFERENCES

1. Chapin, Lark and Warren Letzsch, "Deep Catalytic Cracking, Maximize Olefin Production," NPRA Annual Meeting, AM-94-43, March 20–22, 1994.
2. Zaiting, L., J. Fukang, and M. Enze, "DCC—A New Propylene Production Process from Vacuum Gas Oil," NPRA Annual Meeting, AM-90-40, March 25–27, 1990.

PART 4

CATALYTIC REFORMING

CHAPTER 4.1
UOP PLATFORMING PROCESS

**Natasha Dachos, Aaron Kelly,
Don Felch, and Emanuel Reis**

UOP
Des Plaines, Illinois

PROCESS EVOLUTION

The Platforming* process is a UOP*-developed and -engineered catalytic reforming process in widespread use today throughout the petroleum and petrochemical industries. The first UOP Platforming unit went on-stream in 1949. It was the first commercialization of a process that has become a standard feature in refineries worldwide.

In the Platforming process, light petroleum distillate (naphtha) is contacted with a platinum-containing catalyst at elevated temperatures and hydrogen pressures ranging from 345 to 3450 kPa (50 to 500 lb/in^2 gage). Platforming produces a high-octane liquid product that is rich in aromatic compounds. Chemical hydrogen, light gas, and liquefied petroleum gas (LPG) are also produced as reaction by-products.

Originally developed to upgrade low-octane-number straight-run naphtha to high-octane motor fuels, the process has since been applied to the production of LPG and high-purity aromatics. A wide range of specially prepared platinum-based catalysts permits tailored processing schemes for optimum operation. With proper feed preparation, Platforming efficiently handles almost any refinery naphtha.

Since the first Platforming unit was commercialized, UOP has been at the industry forefront in advancing reforming technology. UOP has made innovations and advances on all fronts, including process-variable optimization, catalyst formulation, equipment design, and maximization of liquid and hydrogen yields. Innovations were driven by the need to increase yields and octane and still control coke deposition on the catalyst. The path to higher yields and octane lay in low-pressure, high-severity operations. However, high severity also increased coke production and subsequent coke laydown and deactivation of the catalyst.

The first Platforming units were designed as semiregenerative (SR), or fixed-bed, units employing monometallic catalysts. Semiregenerative Platforming units are periodically shut down to regenerate the catalyst. This regeneration includes burning off catalyst coke and reconditioning the catalyst's active metals. To maximize the length

---
*Trademark and/or service mark of UOP.

of time (cycle) between regenerations, these early units were operated at high pressures in the range of 2760 to 3450 kPa (400 to 500 lb/in^2 gage).

A typical SR Platforming flow diagram is presented in Fig. 4.1.1. In the process flow, feed to the Platforming unit is mixed with recycled hydrogen gas, raised to the reaction temperature first by a feed-effluent combined feed exchanger and then by a fired heater, and then charged to the reactor section. Because most of the reactions that occur in the Platforming process are endothermic, the reactor section is separated into several stages, or reactors. Interheaters are installed between these stages to maintain the desired temperature across all the catalyst in the reactor section. Effluent from the last reactor is cooled by the feed-effluent heat exchanger for maximum heat recovery. Additional cooling to near-ambient temperature is provided by air or water cooling. The effluent is then charged to the separation section, where the liquid and gas products are separated. A portion of the gas from the separator is compressed and recycled back to the reactor section. The net hydrogen produced is sent to hydrogen users in the refinery complex or to the fuel header. The separator liquid is pumped to a product stabilizer, where the more-volatile light hydrocarbons are fractionated from the high-octane liquid product.

UOP initially improved the Platforming process by introducing bimetallic catalysts to SR Platforming units. These catalysts enabled a lower-pressure, higher-severity operation: about 1380 to 2070 kPa (200 to 300 lb/in^2 gage), at 95 to 98 octane with typical cycle lengths of 1 year. The increased coking of the catalyst at the higher severity limited the operating runs and the ability to reduce pressure. Catalyst development alone could not solve these problems. Process innovation was needed. In the 1960s, cyclic reforming was developed to sidestep this barrier. Cyclic reforming still employs fixed-bed reforming, but the reactors can be taken off-line, regenerated, and then put back into service without shutting down the unit and losing production.

UOP recognized the limitations of fixed-bed catalyst stability and so commercialized Platforming with continuous regeneration, the CCR* Platforming process, in 1971. The process employs continuous catalyst regeneration in which catalyst is continuously removed from the last reactor, regenerated in a controlled environment, and then transferred back to the first reactor (Fig. 4.1.2). The CCR Platforming process represents a step change in reforming technology. With continuous regeneration, coke laydown is no longer an issue because the coke is continuously burned off and the catalyst is reconditioned to its original performance. The CCR Platforming process has enabled ultralow-pressure operations at 345 kPa (50 lb/in^2 gage) and produced product octane levels as high as 108. The new approach has been so successful and accepted by the refining and petrochemical industry that more than 95 percent of new catalytic reformers are designed with continuous regeneration. In addition, many units that were originally built as SR Platforming units have been revamped to CCR Platforming units.

In summary, throughout its history, the UOP Platforming process has evolved continuously. In this process, the operating pressure has been lowered by more than 2760 kPa (400 lb/in^2 gage) and hydrogen yields have doubled. Product octane was increased by more than 12 numbers along with a C_5+ yield increase of 2 liquid volume percent (LV %). The evolution of UOP Platforming performance is depicted in Fig. 4.1.3, which shows the increase in both process yields and octane through time and innovation. The performance of the UOP Platforming process is approaching theoretical limits.

---

*Trademark and/or service mark of UOP.

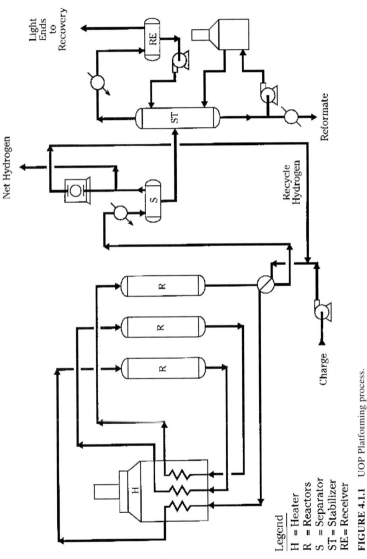

FIGURE 4.1.1 UOP Platforming process.

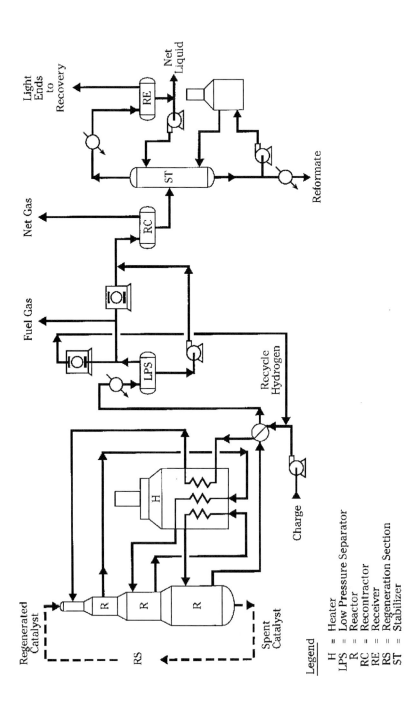

FIGURE 4.1.2 UOP CCR Platforming process.

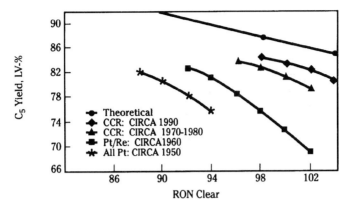

FIGURE 4.1.3 Evolution of UOP Platforming performance.

PROCESS CHEMISTRY

Feed and Product Compositions

The Platforming naphtha charge typically contains C_6 through C_{11} paraffins, naphthenes, and aromatics. The primary purpose of the Platforming process is to produce aromatics from the paraffins and naphthenes. The product stream is a premium-quality gasoline blending component because of the high-octane values of the aromatics. Alternatively, the aromatics-rich product stream can be fed to a petrochemical complex where valuable aromatic products such as benzene, toluene, and xylene (BTX) can be recovered. In motor fuel applications, the feedstock generally contains the full range of C_6 through C_{11} components to maximize gasoline production from the associated crude run. In petrochemical applications, the feedstock may be adjusted to contain a more-select range of hydrocarbons (C_6 to C_7, C_6 to C_8, C_7 to C_8, and so forth) to tailor the composition of the reformate product to the desired aromatics components. For either naphtha application, the basic Platforming reactions are the same. However, for aromatics applications, the more-difficult C_6 and C_7 reactions are emphasized.

Naphthas from different crude sources vary greatly in their hydrocarbon composition and thus in their ease of reforming. The ease with which a particular naphtha feed is processed in a Platforming unit is determined by the mix of paraffins, naphthenes, and aromatics in the feedstock. Aromatic hydrocarbons pass through the unit essentially unchanged. Naphthenes react relatively easily and are highly selective to aromatic compounds. Paraffin compounds are the most difficult to convert, and the relative severity of the Platforming operation is determined by the level of paraffin conversion required. Low-severity (low-octane) operations require little paraffin conversion, but higher-severity operations require a significant degree of conversion.

Naphthas are characterized as lean (high paraffin content) or rich (low paraffin content). Rich naphthas, with a higher proportion of naphthene components, are easier to process in the Platforming unit. Figure 4.1.4 demonstrates the effect of naphtha composition on the relative conversion of the feedstock under constant operating conditions in the Platforming process. A rich naphthenic charge produces a greater volumetric yield of reformate than does a lean charge.

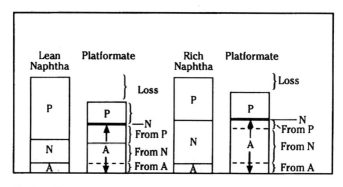

P = Paraffins
N = Naphthenes
A = Aromatics

FIGURE 4.1.4 Typical conversion of lean and rich naphthas.

Reactions

Platforming reactions can generally be classified under four categories: dehydrogenation, isomerization, dehydrocyclization, and cracking. The extent to which each of the reactions occurs for a given Platforming operation depends on the feedstock quality, operating conditions, and catalyst type.

Because the Platforming feed is made up of many paraffin and naphthene isomers, multiple reforming reactions take place simultaneously in the Platforming reactor. The rates of reaction vary considerably with the carbon number of the reactant. Therefore, these multiple reactions occur in series and in parallel to one another. The generalized reaction mechanisms are demonstrated in Fig. 4.1.5 and the generalized reactions are shown in Fig. 4.1.6.

Dehydrogenation of Naphthenes. The principal Platforming reaction in producing an aromatic from a naphthene is the dehydrogenation of an alkylcyclohexane. This reaction takes place rapidly and proceeds essentially to completion. The reaction is

Reactions at Active Sites

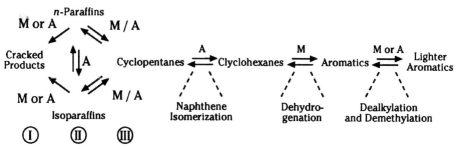

Predominant Active Sites: A = Acid, M = Metal, I = Hydrocracking and Demethylation(M); II = Paraffin Isomerization; III = Dehydrocyclization.

FIGURE 4.1.5 Generalized Platforming reaction mechanism.

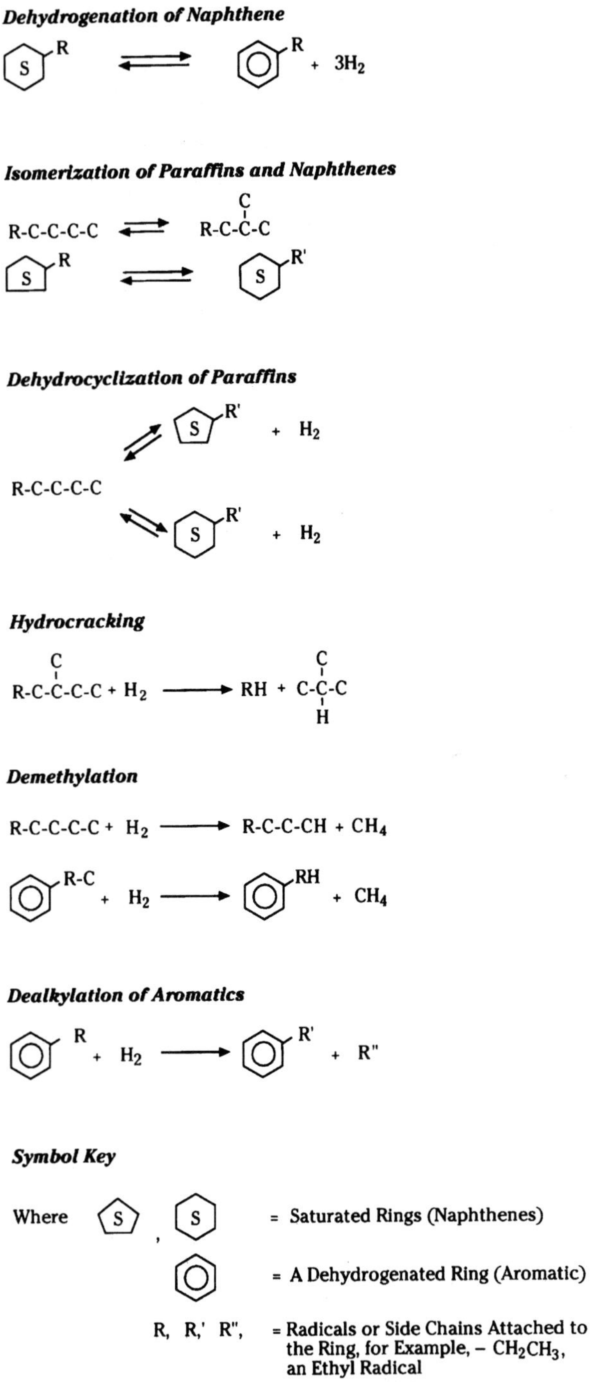

Dehydrogenation of Naphthene

[S-ring]-R ⇌ [aromatic ring]-R + $3H_2$

Isomerization of Paraffins and Naphthenes

R-C-C-C-C ⇌ R-C-C(C)-C

[S-ring]-R ⇌ [S-ring]-R'

Dehydrocyclization of Paraffins

R-C-C-C-C ⇌ [5-ring S]-R' + H_2

⇌ [6-ring S]-R' + H_2

Hydrocracking

R-C-C(C)-C-C + H_2 ⟶ RH + C-C(C)(H)-C

Demethylation

R-C-C-C-C + H_2 ⟶ R-C-C-CH + CH_4

[ring]-R-C + H_2 ⟶ [ring]-RH + CH_4

Dealkylation of Aromatics

[ring]-R + H_2 ⟶ [ring]-R' + R''

Symbol Key

Where (S pentagon), (S hexagon) = Saturated Rings (Naphthenes)

(hexagon) = A Dehydrogenated Ring (Aromatic)

R, R,' R'' = Radicals or Side Chains Attached to the Ring, for Example, – CH_2CH_3, an Ethyl Radical

FIGURE 4.1.6 Generalized Platforming reactions.

highly endothermic, is favored by high reaction temperature and low pressure, and is promoted by the metal catalyst function. Because this reaction proceeds rapidly and produces hydrogen as well as aromatics, naphthenes are the most desirable component in the Platforming feedstock.

Isomerization of Paraffins and Naphthenes. The isomerization of an alkylcyclopentane to an alkylcyclohexane must take place before an alkylcyclopentane can be converted to an aromatic. The reaction involves ring rearrangement, and thus the possibility for ring opening to form a paraffin exists. The paraffin isomerization reaction occurs rapidly at commercial operating temperatures. Thermodynamic equilibrium, however, slightly favors the isomers that are more highly branched. Because branched-chain isomers have a higher octane then straight-chain paraffins, this reaction makes a contribution to product octane improvement. Isomerization reactions are promoted by the acid-catalyst function.

Dehydrocyclization of Paraffins. The most-difficult Platforming reaction to promote is the dehydrocyclization of paraffins. This reaction consists of difficult molecular rearrangements of a paraffin to a naphthene. Paraffin cyclization becomes easier with increasing molecular weight of the paraffin because the probability of ring formation increases. Partially offsetting this effect is the greater likelihood of the heavy paraffins to hydrocrack. Dehydrocyclization is favored by low pressure and high temperature and requires both metal and acid catalyst functions.

Cracking: Hydrocracking and Dealkylation. The difficulty of naphthene isomerization and paraffin cyclization combined with the acid function dictates a high probability for paraffin hydrocracking. Paraffin hydrocracking is favored by high temperature and high pressure. As paraffins crack and disappear from the gasoline boiling range, the remaining aromatics become concentrated in the product, thereby increasing product octane. However, hydrogen is consumed, and the net liquid product is reduced.

The dealkylation of aromatics entails either making the alkyl group (a side chain on the aromatic ring) smaller or removing the alkyl group completely. An example of the latter is converting toluene to benzene. If the alkyl side chain is large enough, the reaction is similar to paraffin cracking. Dealkylation is favored by high temperature and high pressure.

Relative Reaction Rate

The primary reactions for the C_6 and C_7 paraffins proceed at vastly different rates. Because the hydrocracking rate for hexane is at least 3 times greater than the dehydrocyclization rate for hexane, only a small fraction of normal hexane is converted into aromatics. The rate of heptane dehydrocyclization is approximately 4 times that of hexane. Therefore, a substantially greater conversion of normal heptane to aromatics occurs than for hexane.

Reactions of naphthenes in the feedstock show significant differences between the alkylcyclopentanes and the alkylcyclohexanes. The alkylcyclopentanes react slowly and follow two competing paths. The desired reaction is isomerization to an alkylcyclohexane followed by dehydrogenation to aromatics. The competing reaction is decyclization to form paraffins. In contrast, the alkylcyclohexanes dehydrogenate rapidly and nearly completely to aromatics.

The relative ease of isomerization to an alkylcyclohexane increases with increasing carbon number. The ratio of alkylcyclopentane isomerization rate to total alkylcyclopentane reaction rate demonstrates the expected selectivity to form aromatics. At low pressure, this ratio is 0.67 for methylcyclopentane and 0.81 for dimethylcyclopentane.

The conversion of hydrocarbon types as a function of position in the catalyst bed for a moderate-severity Platforming operation is shown in Figs. 4.1.7 to 4.1.10. The feedstock is a rich BTX naphtha with a paraffin, naphthenes, and aromatics (PNA) content of 42, 34, and 24 wt %, respectively. As the naphtha feed passes through the catalyst bed, total aromatics concentration increases and the concentration of naphthenes and paraffins decreases as they undergo conversion (Fig. 4.1.7). The high rate of conversion of cyclohexanes is shown by the rapidly decreasing concentration of naphthenes in the first 30 percent of the catalyst bed. The remaining naphtha conversion occurs at a slower rate and is indicative of cyclopentane conversion and dehydrocyclization of paraffins through a naphthene intermediate. At the reactor outlet, naphthene concentration approaches a low steady-state value, which represents the naphthene intermediary present in the paraffin dehydrocyclization reactions. Paraffin conversion is nearly linear across the reactor bed.

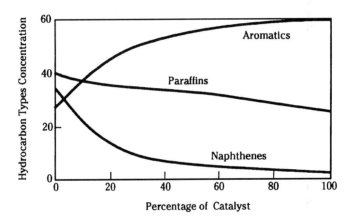

FIGURE 4.1.7 Hydrocarbon-type profiles.

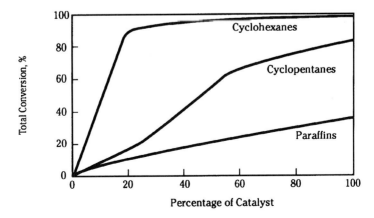

FIGURE 4.1.8 Reactant-type conversion profiles.

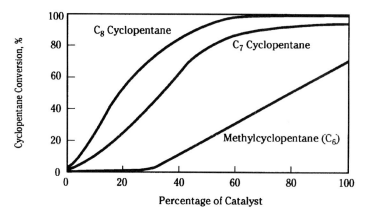

FIGURE 4.1.9 Cyclopentane conversion by carbon number.

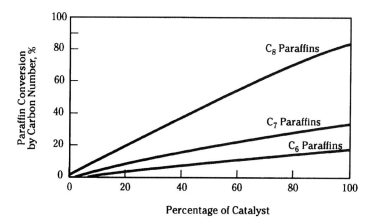

FIGURE 4.1.10 Paraffin conversion by carbon number.

Figure 4.1.8 illustrates the conversion of the three reactive species in the Platforming feedstock. The relative rates of conversion are markedly different. In the first 20 percent of the catalyst, 90 percent of the cyclohexanes are converted, but conversion is only 15 percent for cyclopentanes and 10 percent for paraffins. Cyclopentanes are much less reactive than cyclohexanes.

Figure 4.1.9 shows the relative reaction rate of cyclopentanes by carbon number. Heavier components, which have a greater probability of isomerizing from a five- to six-carbon ring, convert more readily than do the lighter components. The most-difficult reaction, the conversion of paraffins, is characterized by carbon number in Figure 4.1.10. As with the cyclopentanes, the heavier paraffins convert more readily than do the lighter paraffins. The relative ease of conversion associated with increasing carbon number for alkylcyclopentanes and paraffins explains why higher-boiling-range feedstocks are easier to process.

In summary, paraffins have the lowest reactivity and selectivity to aromatics and are the most difficult components to process in a Platforming unit. Although alkylcyclopentanes are more reactive and selective than paraffins, they still produce a significant amount of nonaromatic products. Alkylcyclohexanes are converted rapidly and quantitatively to aromatics and make the best reforming feedstock. As a general rule, heavier components convert more easily and selectively to aromatics than do the lighter components.

Heats of Reaction

Typical heats of reaction for the three broad classes of Platforming reactions are presented in Table 4.1.1. The dehydrocyclization of paraffins and dehydrogenation of naphthenes are endothermic. In commercial Platforming units, the majority of these reactions take place across the first two reactors, as indicated by the large negative-temperature differentials observed. In the final reactor, where a combination of paraffin dehydrocyclization and hydrocracking takes place, the net heat effect in the reactor may be slightly endothermic or exothermic, depending on processing conditions, feed characteristics, and catalyst.

Catalysts

Platforming catalysts are heterogeneous and composed of a base support material (usually Al_2O_3) on which catalytically active metals are placed. The first Platforming catalysts were monometallic and used platinum as the sole metal. These catalysts were capable of producing high octane products; however, because they quickly deactivated as a result of coke formation on the catalyst, they required high-pressure, low-octane Platforming operations.

As refiners needed more activity and stability to move to lower pressure and higher octane, UOP introduced bimetallic catalysts in 1968. These catalysts contained platinum and a second metal, rhenium, to meet increasing severity requirements. Catalyst metals are typically added at levels of less than 1 wt % of the catalyst by using techniques that ensure a high level of metal dispersion over the surface of the catalyst. To ensure the dual functionality of the catalyst, a promoter such as chloride or fluoride is added. The dual function refers to the fact that some Platforming reactions are metal-catalyzed and others are acid-catalyzed as described in the "Reactions" section of this chapter. Most catalyst development for SR Platforming has followed the path of maximizing the efficiency and balance found in the dual-function catalyst system.

The performance of UOP commercial fixed-bed catalysts is shown in Fig. 4.1.11. The R-56* catalyst, which was first commercialized in 1992, is UOP's most stable and most active SR Platforming catalyst. In addition to R-56, UOP has recently introduced the R-72* catalyst, which is a high-yield, bimetallic catalyst. The key feature of the R-72 catalyst in comparison to conventional platinum-rhenium (Pt-Re) catalysts is its

---

*Trademark and/or service mark of UOP.

TABLE 4.1.1 Heats of Reaction ΔH

Reaction	ΔH, kJ/mol H_2
Paraffin to naphthene	+43.93 (endothermic)
Naphthene to aromatic	+70.71 (endothermic)
Hydrocracking	−56.48 (exothermic)

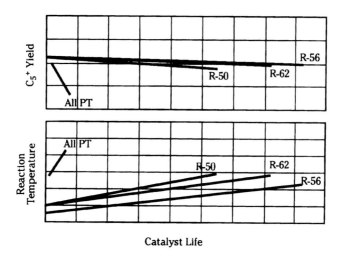

FIGURE 4.1.11 Performance summary of commercial fixed-bed UOP Platforming catalysts.

ability to more efficiently convert paraffins and naphthenes to aromatics with less cracking to light end products. The R-72 catalyst consistently gives a 1 to 2 LV % C_5+ yield advantage compared to Pt-Re catalysts. However, the R-72 catalyst is not so stable as the Pt-Re catalyst. To address the stability requirement, UOP has developed a unique patented staged-loading approach in which the R-72 catalyst is loaded into the front reactors and the R-56 catalyst is loaded into the back reactor. In this manner, the refiner obtains the full yield benefit of the R-72 catalyst and the activity and stability of the R-56 catalyst.

With the introduction of the UOP CCR Platforming process in 1971, Platforming catalyst development began a second parallel track to address the specific needs of the continuous process. The first UOP CCR Platforming unit used a conventional Pt-Re catalyst, but UOP quickly developed the R-30* series catalyst to provide higher yields of gasoline and hydrogen. Catalyst development for the CCR Platforming process has focused on the following broad areas:

- Lower coke make to reduce regenerator investment.
- Higher tolerance to multiple regeneration cycle to maximize catalyst life and minimize catalyst costs. This area includes improvement (reduction) in the rate of surface area decline, which is important because reduced catalyst surface area increases the difficulty of dispersing the metals on the catalyst surface.
- Improved catalyst strength to reduce attrition in the unit as a result of catalyst transfer.
- Metals optimization to reduce the metal content of the catalyst and thus reduce the refinery working-capital requirement.

In 1992, UOP commercialized the R-130* CCR Platforming catalyst series. The R-130 series is characterized by high surface-area stability, activity, and strength. The

---

*Trademark and/or service mark of UOP.

R-130 series catalyst have further optimized the CCR Platforming process. With the R-130 series, new units can be built with a lower inventory of catalyst, and throughput or octane or both can be increased in existing units. Other benefits of the R-130 series catalyst include low attrition, longer life, constant yields, and characteristics that lead to lower utilities and reduced environmental emissions.

PROCESS VARIABLES

This section describes the major process variables and their effect on unit performance: catalyst type, reactor pressure, reactor temperature, space velocity, hydrogen-to-hydrocarbon (H_2/HC) molar ratio, chargestock properties, catalyst selectivity, and catalyst activity. The relationship between the variables and process performance is generally applicable to both SR and continuous regeneration modes of operation.

Catalyst Type

Catalyst selection is usually tailored to the refiner's individual needs. A particular catalyst is typically chosen to meet the yield, activity, and stability requirements of the refiner. This customization is accomplished by varying basic catalyst formulation, chloride level, platinum content, and the choice and quantity of the second metal.

Differences in catalyst types can affect other process variables. For example, the required temperature to produce a given octane is directly related to the type of catalyst.

Reactor Pressure

The average reactor operating pressure is generally referred to as *reactor pressure*. For practical purposes, a close approximation is the last reactor inlet pressure. The reactor pressure affects reformer yields, reactor temperature requirement, and catalyst stability.

Reactor pressure has no theoretical limitations, although practical operating constraints have led to a historical range of operating pressures from 345 to 4830 kPa (50 to 700 lb/in^2 gage). Decreasing the reactor pressure increases hydrogen and reformate yields, decreases the required temperature to achieve product quality, and shortens the catalyst cycle because it increases the catalyst coking rate. The high coking rates associated with lower operating pressures require continuous catalyst regeneration.

Reactor Temperature

The primary control for product quality in the Platforming process is the temperature of the catalyst beds. Platforming catalysts are capable of operating over a wide range of temperatures. By adjusting the heater outlet temperatures, a refiner can change the octane of the reformate and the quantity of aromatics produced.

The reactor temperature can be expressed as the weighted average inlet temperature (WAIT), which is the summation of the product of the fraction of catalyst in each reactor multiplied by the inlet temperature of the reactor, or as the weighted average bed temperature (WABT), which is the summation of the product of the fraction of catalyst in each reactor multiplied by the average of its inlet and outlet temperatures.

Temperatures in this chapter refer to the WAIT calculation. Typically, SR Platforming units have a WAIT range of 490 to 525°C (914 to 922°F). CCR Platforming units operate at a WAIT of 525 to 540°C (977 to 1004°F).

Space Velocity

Space velocity is a measurement of an amount of naphtha processed over a given amount of catalyst over a set length of time. Space velocity is an indication of the residence time of contact between reactants and catalyst. When the hourly volume charge rate of naphtha is divided by the volume of catalyst in the reactors, the resulting quotient, expressed in units of h^{-1}, is the liquid hourly space velocity (LHSV). Alternatively, if the weight charge rate of naphtha is divided by the weight of catalyst, the resulting quotient, also expressed in units of h^{-1}, is the weighted hourly space velocity (WHSV). Although both terms are expressed in the same units, the calculations yield different values. Whether LHSV or WHSV is used depends on the customary way that feed rates are expressed in a given location. Where charge rates are normally expressed in barrels per stream day, LHSV is typically used. Where the rates are expressed in terms of metric tons per day, WHSV is preferred.

Space velocity works directly with reactor temperature to set the octane of the product. The greater the space velocity, the higher the temperature required to produce a given product octane. If refiners wish to increase the severity of a reformer operation, they can either increase the reactor temperature or lower the space velocity by decreasing the reactor charge rate. At constant severity, a change in space velocity with the corresponding change in required WAIT has, at best, a small impact on product yields. Higher space velocities may lead to slightly higher yields as a result of less time available in the reactors for dealkylation reactions to take place. This advantage is partially offset by the higher rate of hydrocracking reactions at higher temperatures.

Hydrogen-to-Hydrocarbon Molar Ratio

The H_2/HC ratio is the ratio of moles of hydrogen in the recycle gas to moles of naphtha charged to the unit. Recycle hydrogen is necessary to maintain catalyst-life stability by sweeping reaction products from the catalyst. The rate of coke formation on the catalyst is a function of the hydrogen partial pressure present.

An increase in the H_2/HC ratio increases the linear velocity of the combined feed and supplies a greater heat sink for the endothermic heat of reaction. Increasing the ratio also increases the hydrogen partial pressure and reduces the coking rate, thereby increasing stability with little effect on product quality or yields. Directionally, lower H_2/HC ratios provide higher C_5+ and hydrogen yields, although this benefit is difficult to measure in commercially operating units.

Chargestock Properties

Chargestock characterization by hydrocarbon type (paraffin, naphthene, and aromatics) and a distillation analysis are required to predict the performance of the Platforming unit in terms of total liquid reformate and hydrogen yield as well as catalytic activity and stability. Chargestocks with a low initial boiling point (IBP), less than 75°C (167°F) measured according to American Society for Testing and Materials specification ASTM D-86, generally contain a significant amount of C_5 material, which is not converted to valuable aromatics products; serve to dilute the final product; and have a negative impact on the design of the unit. For this reason, feedstocks are generally C_6+ naphthas. The endpoint of the chargestock is normally set by the gasoline specifications for the refinery with the realization that a significant rise, typically 15 to 25°C (27 to 45°F), in endpoint takes place between the naphtha charge and reformate product.

The effect of hydrocarbon types in the chargestock on aromatics yield was discussed in the "Process Chemistry" section and can be further illustrated by examining a broad range of chargestock compositions. Licensers typically develop a large database of feedstocks that have been analyzed and tested under controlled conditions to characterize expected reforming yields over a range of octanes. This database allows yields to be predicted for future chargestocks of known composition. Four chargestocks of widely varying compositions were chosen from such a database and are summarized in Fig. 4.1.12.

The chargestock range chosen covers lean through rich feeds. The aromatics-plus-cyclohexanes content is a measure of their ease of conversion, and the paraffins-plus-cyclopentanes content indicates the difficulty of reforming reactions. The effect of feedstock composition on aromatics yield is shown in Fig. 4.1.13. Increasing conversion leads to an increase in the total yield of aromatics for each of the feedstocks. Feeds that are easier to process produce the highest yield of aromatics at any level of conversion.

Catalyst Selectivity

Catalyst selectivity can be easily described as the amount of desired product that can be yielded from a given feedstock. Usually, the selectivity of one catalyst is compared with that of another. At constant operating conditions and feedstock properties, the catalyst that can yield the greatest amount of reformate at a given octane in motor fuel applications or the greatest amount of aromatics in a BTX operation has the greatest selectivity.

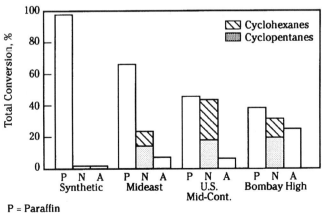

FIGURE 4.1.12 Naphtha characterization by hydrocarbon type.

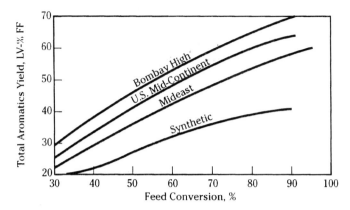

FIGURE 4.1.13 Feedstock conversion and aromatics yield.

Catalyst Activity

Activity is the ability of a catalyst to promote a desired reaction with respect to reaction rate, space velocity, or temperature. Activity is also expressed in a relative sense in that one catalyst is more active than another. In motor fuel applications, activity is generally expressed as the temperature required to produce reformate at a given octane, space velocity, and pressure. A more-active catalyst can produce the desired octane reformate at a lower temperature.

CONTINUOUS PLATFORMING PROCESS

In parallel with bimetallic catalyst improvements and SR Platforming process and regeneration advancement, UOP began to develop the CCR Platforming process. To provide continuous Platforming operation, UOP eliminated the necessity of regeneration in situ by implementing continuous regeneration. In the CCR Platforming unit, partially aged catalyst in the reactors is continuously replaced with catalyst that has been freshly regenerated in an external regenerator (CCR section) to maintain a low average age for the reactor catalyst. Thus, continuous high-selectivity and high-activity characteristics associated with new catalyst can be achieved at significantly higher severities than with the SR Platforming process. For example, a SR Platforming unit operates at a severity that steadily builds coke up on the catalyst surface over the length of a cycle (6 to 18 months), at which point the unit is shut down and the catalyst regenerated. Throughout the cycle, yields decline. In contrast, with a modern CCR Platforming unit, the catalyst is regenerated approximately every 3 days, and yield does not decline.

The CCR Platforming flow scheme incorporates many engineering innovations. Depending on the size of the unit, many SR Platforming units are also built to include some of these innovations. This design allows for an easier transition between SR and CCR Platforming units if the SR Platforming unit is later converted to meet future operating requirements.

Movable-Catalyst-Bed System

In a conventional SR Platforming unit, the reactors are configured side by side. The CCR Platforming unit uses a UOP-patented reactor stack. The reactors are stacked one on top of another to achieve a compact unit that minimizes plot area requirements. The catalyst flows gently by gravity downward from reactor to reactor and this flow simplifies catalyst transfer and minimizes attrition. Catalyst transfer is greatly simplified in comparison to other reforming technologies, which employ side-by-side reactor configurations that require the catalyst to be pneumatically lifted from the bottom of each reactor to the downstream reactor. In contrast, with the reactor stack, catalyst is lifted only twice during each cycle: from the bottom of the reactor stack to the top of the regenerator and then from the bottom of the regenerator back to the top of the reactor stack. The catalyst transfer requires no operator intervention. Catalyst transfer rates have been designed from as low as 91 kg/h (200 lb/h) to as high as 2721 kg/h (6000 lb/h), depending on the capacity and the operating severity of the Platforming unit.

CCR System

The ability to continuously regenerate a controlled quantity of catalyst is the most-significant innovation of the CCR Platforming unit. The catalyst flows by gravity from the last reactor into an integral (to the reactor) catalyst collector vessel. The catalyst is then lifted by either nitrogen or hydrogen lifting gas to a catalyst hopper above the regeneration tower. Catalyst then flows to the regeneration tower, where the catalyst is reconditioned. Regenerated catalyst is then returned to the top of the reactor stack by a transfer system similar to that used in the reactor-regenerator transfer. Thus, the reactors are continuously supplied with freshly regenerated catalyst, and product yields are maintained at fresh catalyst levels.

The regeneration and reactor sections of the unit are easily isolated to permit a shutdown of the regeneration system for normal inspection or maintenance without interrupting the Platforming operation.

Improvements are continuously being made in the CCR regeneration section design. In addition to its atmospheric and pressurized regenerators, UOP introduced the CycleMax* regenerator in 1995. The CycleMax regenerator incorporated the latest innovations and combined them with the best aspects of previous generations at lower cost.

Low-Pressure-Drop Features

Minimum pressure drop in the reactor section is critical for efficient ultralow-pressure operation. Low pressure drop minimizes recycle gas compressor differential pressure and horsepower. The result is lower utility consumption. The cost for even one pound of additional pressure drop across the compressor is high. Minimum pressure drop also permits the operation at the lowest possible average reactor pressure, which increases reformate and hydrogen yields.

UOP employs a variety of special equipment to minimize the pressure drop throughout the plant circuit. Either vertical combined feed-effluent exchangers (VCFEs) or new PACKINOX* welded-plate exchangers introduced in the 1990s are used to maximize thermal efficiency and minimize pressure drop. The patented reac-

---

*Trademark and/or service mark of UOP.

tor-stack design, fired heater design, and plot-plan layout further reduce plant pressure drop to the economic minimum level.

Secondary-Recovery Schemes

Several innovative schemes for increased liquid recovery and separator gas purification have been developed. The need for increasing liquid recovery is more critical with the lower-pressure designs, where the production of hydrogen and C_5+ material is increased as a result of more-selective processing. This advantage can be lost if a recovery system is not installed downstream of the reactor section. At low operating pressures, the flash pressure of the separator has been reduced. The vapor liquid equilibrium thermodynamically allows for more C_4's, C_5's, and C_6+ material to leave with the vapor, resulting in valuable C_5+ product loss and lower-purity hydrogen production. To prevent this phenomenon, different types of recontacting schemes can be employed.

One type of scheme often installed is reactor-effluent vapor-liquid recontacting. Reactor effluent, after being cooled, is physically separated into vapor and liquid portions. Part of the vapor is directed to the recycle-compressor suction for use as recycle gas. The remaining vapor, called the *net separator gas,* is compressed by a booster compressor and discharged into either a drum or an adsorber. The liquid from the separator is also pumped to the drum or absorber to recontact with the net separator gas at elevated pressure to obtain increased liquid recovery and hydrogen purity.

Another method involves chilling the net-separator gas. Depending on downstream pressure requirements, net gas from either the compressor suction or discharge is cooled to approximately 5°C (41°F) by a refrigeration system. Separation of the vapor and liquid improves hydrogen purity and recovers additional liquid, which would be routed to the stabilizer with the liquid from the low-pressure separator.

In addition, proprietary systems have been developed that more efficiently improve the recovery of liquid product. UOP offers a proprietary system, RECOVERY PLUS,* that improves the recovery of the liquid product at minimum operating costs.

Advantages of CCR Platforming

From both economic and technical standpoints, the CCR Platforming process is superior to the SR Platforming process. The following reasons are provided to substantiate that claim:

- The CCR Platforming unit allows for the lowest possible pressure operation and the highest possible yields. At these conditions, the SR Platforming catalyst is completely deactivated after only a few days of operation; in contrast, the high coking rate is managed by the CCR section. Both the hydrogen and C_5+ yields are maximized with the CCR Platforming process. The C_5+ yield advantage is illustrated in Fig. 4.1.14, and the hydrogen yield advantage is shown in Fig. 4.1.15.
- Equally important to high yields in the economics of reforming are constant nondeclining yields. As the catalyst is deactivated by coke deposition in the SR Platforming process, the yields begin to decline. With the CCR Platforming process, the reformate, aromatics, and hydrogen yields remain consistent and constant. This result is particularly important for downstream users because inconsis-

---

*Trademark and/or service mark of UOP.

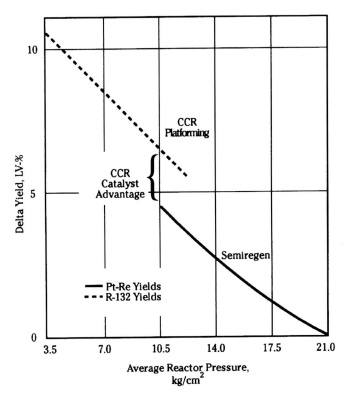

FIGURE 4.1.14 C_5+ yield at decreasing pressure.

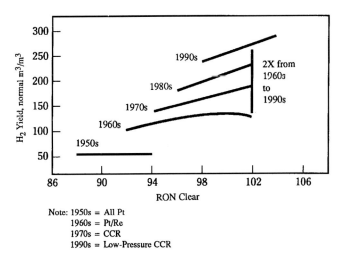

FIGURE 4.1.15 Yield efficiency improvement.

tent quality can lead to their products not meeting specifications. The CCR section ensures proper redispersion of the metals and chloride balance to maintain fresh catalyst activity.
- CCR Platforming units have higher on-stream efficiency and are able to handle upset scenarios without long-term shutdown or significant decline in performance. For example, a compressor-trip or feed-upset scenario can lead to significant problems with the SR Platforming unit because of increased coke levels, which inevitably shorten catalyst cycle length. However, the continuous transfer and regeneration of catalyst in the CCR Platforming unit allow for faster resumption of normal operations. The independent operation of the reactor and catalyst regeneration sections and the mechanical improvements to the design of the CCR Platforming unit lead to greater on-stream availability for the CCR Platforming unit. Recent customer surveys indicate the average time between planned turnarounds is 3.4 years.
- Catalyst is not regenerated in situ. The reactor section operates only in its primary function of providing the catalytic environment for the reforming reactions. It is not exposed to harsh regeneration conditions.

CASE STUDIES

Two cases have been presented to compare the SR Platforming and CCR Platforming processes. The unit capacities are the same for the two modes of operation, but the CCR Platforming unit is run at a higher operating severity, 102 research octane number, clear (RONC) compared to 97 RONC for SR Platforming. The performance advantage of the CCR Platforming process is clearly demonstrated in the case studies. However, UOP continues to license SR Platforming units because gasoline specifications vary in different regions of the world. Some refiners may choose to build a SR Platforming unit to meet current regulations. That unit can later be converted to a CCR Platforming unit to meet future regulations.

Operating Conditions

Table 4.1.2 shows the relative operating severities for the SR and CCR Platforming units. The CCR Platforming unit operates at higher severity and lower reactor catalyst

TABLE 4.1.2 Relative Severities of CCR and SR Platforming Units

Operating mode	SR	CCR
Catalyst type	R-56	R-134
Charge rate, MTD (BPSD)	2351 (20,000)	2351 (20,000)
LHSV, h^{-1}	Base	Base $\times$ 1.8
H$_2$/HC	Base	Base $\times$ 0.5
RONC	97	102
Reactor pressure, kPa (lb/in^2 gage)	Base (Base)	Base $-$ 1035 (Base $-$ 150)
Separator pressure, kPa (lb/in^2 gage)	Base (Base)	Base $-$ 1000 (Base $-$ 145)
Cycle life, months	12	Continuous

Note: MTD = metric tons per day; BPSD = barrels per stream day.

inventory. In addition, the CCR unit runs continuously compared to 12-month SR Platforming cycle lengths.

Product Yields and Properties

Typical product yields for the SR and CCR Platforming units operating at the conditions presented in Table 4.1.2 are shown in Table 4.1.3. Many of the benefits of the CCR Platforming unit are shown in Table 4.1.3. More and higher-purity hydrogen is produced. The higher severity of the CCR Platforming unit results in similar liquid volume for the two units. However, the reformate produced by the CCR Platforming unit is more valuable than that produced by the SR Platforming unit. Taking into account both the higher octane value and the increased on-stream efficiency of the CCR Platforming unit, 80 million more octane-barrels, or 11.4 million more metric octane-tons, are produced per year with the CCR Platforming unit than with the SR Platforming unit. *Octane-yield* is defined as the product of the reformate yield, octane, and operating days.

Economics

The estimated erected cost (EEC) for the two units is presented in Table 4.1.4. The EEC is based on fourth-quarter, 1995, U.S. Gulf Coast, inside-battery-limits erection to UOP standards. The EEC for the CCR Platforming unit is higher than for the SR Platforming unit. The main difference in cost is for the CCR regeneration section. The choice of the Platforming mode of operation depends on capital available and operating severity required. In general, the break point between SR Platforming and CCR Platforming units is an operating severity of 98 RONC. In some regions of the world, 98 RONC or lower severity is sufficient to meet the local gasoline requirements.

TABLE 4.1.3 Yield Comparison of CCR and SR Platforming Units

	SR	CCR	Delta
Hydrogen yield, SCFB	1085	1709.0	+624.0
Hydrogen purity, mol %	80	92.6	+12.6
C_5+ yield, LV %	79.3	79.4	+0.1
C_5+ yield, wt %	85.2	88.2	+ 3
Octane-barrel, 10^6 bbl/year	513	583	+80
Octane-ton yield, 10^6 MTA	64.9	76.3	+11.4

Note: SCFB = standard cubic feet per barrel; MTA = metric tons per annum.

TABLE 4.1.4 Estimated Erected Cost

	Cost, Million $ U.S.	
	SR	CCR
Estimated erected cost	33	48.3
Catalyst base cost	0.9	1.1
Catalyst metals cost	2.5	1.5

CATALYTIC REFORMING

Many of the new SR Platforming units are built with a reactor stack and with the flexibility to be converted to CCR Platforming units at a later date. Thus, the cost of the CCR section is spread out over a longer period, and the profits made from the SR Platforming operation can be used to finance it.

However, to meet the gasoline restrictions in many regions in the world or to produce aromatics, an operating severity higher than 98 RONC is required. For aromatics production, the operating severity is typically 104 to 106 RONC. Therefore, the CCR Platforming process is the only feasible mode of operation. Typically, the increased yields and octane (that is, more and higher-value product), increased on-stream efficiency, and better operating flexibility far outweigh and make up for the incremental cost difference.

The estimated operating requirements for the two units are presented in Table 4.1.5. These estimates are based on the assumption that the units are operated at 100 percent of design capacity at yearly average conditions.

UOP's design philosophy is to minimize consumption of utilities and maximize energy conservation within economic constraints. The operating requirements of the CCR Platforming unit are higher because of the CCR regenerator, lower-pressure operation, and a more intricate recontacting scheme.

The operating revenues and costs expected for the SR and CCR Platforming units are listed in Table 4.1.6 and summarized in Table 4.1.7. The nomenclature used by UOP follows standard definitions. For clarification, the definitions for the economic parameters are listed in Table 4.1.8.

The economics shown in Tables 4.1.6 and 4.1.7 are favorable for either mode of operation. Both modes of operation have a payback time of less than 2 years. However, the economics of the CCR Platforming process are superior as a direct result of the differences in operating severity and flexibility of the two modes of operations. The CCR Platforming unit produces more valuable reformate at 102 RONC ($24.60 per barrel) versus the SR Platforming reformate at 97 RONC ($23.00 per barrel). On-stream efficiency of the CCR Platforming unit is 8640 hours per year compared to 8000 hours per year for the SR Platforming unit. Although the CCR Platforming utility costs are higher than those for the SR Platforming unit, these costs are offset by the increase in both product quantity and value, as demonstrated by pretax profit and return on investment.

UOP COMMERCIAL EXPERIENCE

UOP has designed more than 730 Platforming (both SR and CCR) units around the world. These units have a total feedstock capacity of more than 7.9 million barrels per

TABLE 4.1.5 Operating Requirements

	SR	CCR
Electric power, kWh	246	6,142
Fuel fired, million kcal (10^6 Btu)	44.1 (175)	55.4 (220)
Cooling water, m^3/h (gal/min)	293 (1290)	534 (2350)
High-pressure steam generated,* MT/h (1000 lb/h)	6.3 (14)	9.5 (21)
Boiler feed water, MT/h (1000 lb/h)	16.6 (36.5)	2.16 (47.6)
Condensate return,* MT/h (1000 lb/h)	8.6 (19)	11.1 (24.4)

*Net stream, exported unit.
Note: MT/h = metric tons per hour.

UOP PLATFORMING PROCESS

TABLE 4.1.6 Operating Economics

	$/day	
	SR	CCR
Product value:		
C_5+, SR at $23/bbl and CCR at $24.60/bbl*	358,640*	392,790
H_2, $0.34/lb	38,990	61,585
LPG, $14/bbl	7,155	11,380
Fuel gas, $0.05/lb	30,860	11,220
Total value	435,645	476,975
Total value, million $/year	145	172
Operating days/year	333	360
Operating costs:		
Feedstock cost, $18.50/bbl	370,000	370,000
Utility costs:		
Electric power, 5 cents/kWh	300	7,370
Fuel fired, $2.10/$10^6$ Btu	8820	11,090
Cooling water, 0.10 $/1000 gal	190	340
Boiler feedwater, 0.40 $/1000 lb	350	460
Condensate return, 0.40 $/1000 lb	(+)180	(+)235
Steam make, at $3.45/1000 lb	(+)1140	(+)1740
Total cost	378,340	387,285
Total cost, million $/year	126	139

*At 97 RONC and 102 RONC, respectively.

TABLE 4.1.7 Economic Summary

Description	SR	CCR
Gross key* product value, million $/year	120	141
Raw materials† less by-products,‡ million $/year	98	103
Consumables,§ million $/year	0.3	0.75
Utilities,† million $/yr	2.8	6.2
Total fixed costs,¶ million $/year	5.5	6.5
Capital charges, million $/year	3.5	5.2
Net cost of production, million $/year	110	122
Pretax profit, million $/year	10	20
Pretax return on investment, %	30	41
Payout period (gross), years	1.5	1.3

*The key product is the octane-barrels of reformate.

†Variable costs

‡Defined as the feed cost minus the value of the by-products, which are LPG, hydrogen, and fuel gas.

§Includes catalyst and platinum makeup from attrition and recovery losses.

¶Includes labor, maintenance, overhead, and capital expenses.

TABLE 4.1.8 UOP Nomenclature for Economic Analysis

Term	Definition
Gross margin:	A measure of net receipts exclusive of all capital and operating expenses
Variable costs:	Manufacturing costs that are directly related to the production rate
Fixed costs:	Manufacturing costs that are constant regardless of the production rate
Gross profit:	The total net income prior to considering income tax deductions (key product revenue minus cash cost of production)
Capital charges:	Depreciation and amortization expenses associated with the capital plant investment
Net cost of production:	Total manufacturing costs inclusive of capital charges
Pretax return:	Portion of the gross profit that is subject to income taxes; also termed *taxable income*
Pretax return on investment (ROI):	Simplified approximation of the annual percentage of return that can be expected for each dollar invested. Expressed as the ratio of pretax profits to total plant investment; does not consider compounded interest effects

stream day (BPSD). The feedstocks range from benzene-toluene (BT) cuts to full-range, lean Middle East naphthas and rich U.S. and African naphthas and hydrocracked stocks with capacities ranging from 150 to 60,000 BPSD. Research octane numbers run from 93 to 108 over a wide range of catalysts.

The UOP CCR Platforming process is the most successful reforming process offered by any licenser. As of mid-1995, UOP's unparalleled commercial experience includes:

- 130 UOP CCR Platforming units operating today around the world
- 17 units operating at state-of-the art reactor pressure 50 lb/in^2 gage
- 38 units operating at or below 90-lb/in^2 gage reactor pressure
- 2,876,100 BPSD CCR Platforming unit operating capacity
- More than 1200 operating unit-years of experience in CCR Platforming area
- Every CCR Platforming unit ever started up still operating
- 41 more UOP CCR Platforming units in design and construction.

UOP also has extensive experience in revamping fixed-bed reforming units to CCR Platforming units.

PART · 5

DEHYDROGENATION

CHAPTER 5.1
UOP OLEFLEX PROCESS FOR LIGHT OLEFIN PRODUCTION

Joseph Gregor
UOP
Des Plaines, Illinois

INTRODUCTION

The UOP* Oleflex* process is catalytic dehydrogenation technology for the production of light olefins from their corresponding paraffin. An Oleflex unit can dehydrogenate propane, isobutane, normal butane, or isopentane feedstocks separately or as mixtures spanning two consecutive carbon numbers. This process was commercialized in 1990, and by 1995 more than 300,000 metric tons per year (MTA) of propylene and more than 1,500,000 MTA of isobutylene were produced from Oleflex units located throughout the world.

PROCESS DESCRIPTION

The UOP Oleflex process is best described by separating the technology into three different sections:

- Reactor section
- Product recovery section
- Catalyst regeneration section

Reactor Section

Hydrocarbon feed is mixed with hydrogen-rich recycle gas (Fig. 5.1.1). This combined feed is heated to the desired reactor inlet temperature and converted at high monoolefin selectivity in the reactors.

---

*Trademark and/or service mark of UOP.

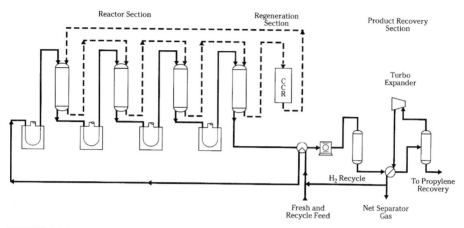

FIGURE 5.1.1 Oleflex process flow.

The reactor section consists of several radial-flow reactors, charge and interstage heaters, and a reactor feed-effluent heat exchanger. The diagram shows a unit with four reactors, which would be typical for a unit processing propane feed. Three reactors are used for butane or isopentane dehydrogenation. Three reactors are also used for blends of C_3-C_4 or C_4-C_5 feeds.

Because the reaction is endothermic, conversion is maintained by supplying heat through interstage heaters. The effluent leaves the last reactor, exchanges heat with the combined feed, and is sent to the product recovery section.

Product Recovery Section

A simplified product recovery section is also shown in Fig. 5.1.1. The reactor effluent is cooled, compressed, dried, and sent to a cryogenic separation system. The dryers serve two functions: (1) to remove trace amounts of water formed from the catalyst regeneration and (2) to remove hydrogen sulfide. The treated effluent is partially condensed in the cold separation system and directed to a separator.

Two products come from the Oleflex product recovery section: separator gas and separator liquid. The gas from the cold high-pressure separator is expanded and divided into two streams: recycle gas and net gas. The net gas is recovered at 85 to 93 mol % hydrogen purity. The impurities in the hydrogen product consist primarily of methane and ethane. The separator liquid, which consists primarily of the olefin product and unconverted paraffin, is sent downstream for processing.

Catalyst Regeneration Section

The regeneration section, shown in Fig. 5.1.2, is similar to the CCR* unit used in the UOP Platforming* process. The CCR unit performs four functions:

- Burns the coke off the catalyst
- Redistributes the platinum

---

*Trademark and/or service mark of UOP.

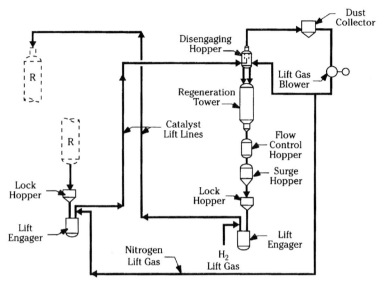

FIGURE 5.1.2 Oleflex regeneration section.

- Removes the excess moisture
- Reduces the catalyst prior to returning to the reactors

The slowly moving bed of catalyst circulates in a loop through the reactors and the regenerator. The cycle time around the loop can be adjusted within broad limits but is typically anywhere from 5 to 10 days, depending on the severity of the Oleflex operation and the need for regeneration. The regeneration section can be stored for a time without interrupting the catalytic dehydrogenation process in the reactor and recovery sections.

DEHYDROGENATION COMPLEXES

Propylene Complex

Oleflex process units typically operate in conjunction with fractionators and other process units within a production complex. In a propylene complex (Figure 5.1.3), a propane-rich liquefied petroleum gas (LPG) feedstock is sent to a depropanizer to reject butanes and heavier hydrocarbons. The depropanizer overhead is then directed to the Oleflex unit. The once-through conversion of propane is approximately 40 percent, which closely approaches the equilibrium value defined by the Oleflex process conditions. More than 90 percent of the propane conversion reactions are selective to propylene and hydrogen; the result is a propylene mass selectivity in excess of 86 wt %. Two product streams are created within the C_3 Oleflex unit: a hydrogen-rich vapor product and a liquid product rich in propane and propylene.

Trace levels of methyl acetylene and propadiene are removed from the Oleflex liquid product by selective hydrogenation. The selective diolefin and acetylene hydrogenation step is accomplished with the Hüls SHP process, which is available for license through UOP. The SHP process selectively saturates diolefins and acetylenes

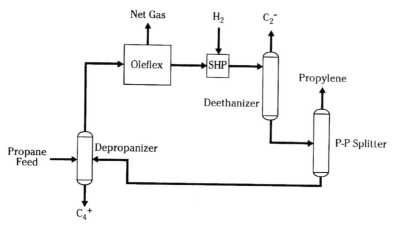

FIGURE 5.1.3 C_3 Oleflex complex.

to monoolefins without saturating propylene. The process consists of a single liquid-phase reactor. The diolefins plus acetylene content of the propylene product is less than 5 wt ppm when an SHP unit is incorporated into the design of the Oleflex propylene complex.

Ethane and lighter material enter the propylene complex in the fresh feed and are also created by nonselective reactions within the Oleflex unit. These light ends are rejected from the complex by a deethanizer column. The deethanizer bottoms are then directed to a propane-propylene (P-P) splitter. The splitter produces high-purity propylene as the overhead product. Typical propylene purity ranges between 99.5 and 99.8 wt %. Unconverted propane from the Oleflex unit concentrates in the splitter bottoms and is returned to the depropanizer for recycle to the Oleflex unit.

Ether Complex

A typical etherification complex configuration is shown in Fig. 5.1.4 for the production of methyl tertiary butyl ether (MTBE) from butanes and methanol. Ethanol can be substituted for methanol to make ethyl tertiary butyl ether (ETBE) with the same process configuration. Furthermore, isopentane may be used in addition to or instead of field butanes to make tertiary amyl methyl ether (TAME) or tertiary amyl ethyl ether (TAEE). The complex configuration for a C_5 dehydrogenation complex varies according to the feedstock composition and processing objectives.

Three primary catalytic processes are used in an MTBE complex:

- Paraffin isomerization to convert normal butane into isobutane
- Dehydrogenation to convert isobutane into isobutylene
- Etherification to react isobutylene with methanol to make MTBE

Field butanes, a mixture of normal butane and isobutane obtained from natural gas condensate, are fed to a deisobutanizer (DIB) column. The DIB column prepares an isobutane overhead product, rejects any pentane or heavier material in the DIB bottoms, and makes a normal butane sidecut for feed to the paraffin isomerization unit.

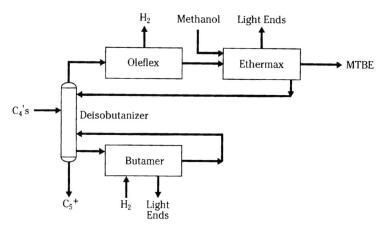

FIGURE 5.1.4 MTBE production facility.

The DIB overhead is directed to the Oleflex unit. The once-through conversion of isobutane is approximately 50 percent. About 93 percent of the isobutane conversion reactions are selective to isobutylene and hydrogen. On a mass basis, the isobutylene selectivity is 90 wt %. Two product streams are created within the C_4 Oleflex unit: a hydrogen-rich vapor product and a liquid product rich in isobutane and isobutylene.

The C_4 Oleflex liquid product is sent to an etherification unit, where methanol reacts with isobutylene to make MTBE. Isobutylene conversion is greater than 99 percent, and the MTBE selectivity is greater than 99.5 percent. Raffinate from the etherification unit is depropanized to remove propane and lighter material. The depropanizer bottoms are then dried, saturated, and returned to the DIB column.

PROPYLENE PRODUCTION ECONOMICS

A complex producing 250,000 MTA of propylene is chosen to illustrate process economics. Given the more favorable C_4 and C_5 olefin equilibrium, butylene and amylene production costs are substantially lower per unit of olefin when adjusted for any differential in feedstock value. The basis used for economic calculations is shown in Table 5.1.1. This basis is typical for U.S. Gulf Coast prices prevailing in mid-1995 and can be used to show that the pretax return on investment for such a complex is approximately 25 percent.

Material Balance

The LPG feedstock is the largest cost component of propylene production. The quantity of propane consumed per unit of propylene product is primarily determined by the selectivity of the Oleflex unit because fractionation losses throughout the propylene complex are small. The Oleflex selectivity to propylene is 92 mol % (86.4 wt %) as discussed previously, and the production of 1.00 metric ton (MT) of propylene requires approximately 1.16 MT of propane.

An overall mass balance for the production of polymer-grade propylene from propane is shown in Table 5.1.2 for a polymer-grade propylene complex producing

TABLE 5.1.1 Utility, Feed, and Product Valuations for Economic Calculations

Utility values		
Fuel gas	$2.10/million Btu	$8.33/million kcal
High-pressure steam	$3.45/klb	$7.60/MT
Low-pressure steam	$3.00/klb	$6.60/MT
Boiler feed water	$0.40/klb	$0.88/MT
Condensate	$0.40/klb	$0.88/MT
Cooling water	$0.08/kgal	$0.02/m^3
Electric power	$0.05/kWh	$0.05/kWh
Feed and product values		
C_3 LPG (95 wt % propane)	$0.34/gal	$177/MT
Hydrogen (99.99 mol %)	$1.08/kSCF	$450/MT
Propylene (99.5 wt %)	$0.22/lb	$485/MT

Note: MT = metric tons; SCF = standard cubic feet.

TABLE 5.1.2 Material Balance for a 250,000-MTA Propylene Complex

	Flow rate, MT/h	Flow rate, MTA
Feed:		
C_3 LPG (95 wt % propane)	38.275	306,200
Products:		
Propylene (99.5 wt %)	31.250	250,000
Hydrogen (99.99 mol %)	1.089	8,712
Fuel by-products	5.936	47,488
Total products	38.275	306,200

Note: MT/h = metric tons per hour; MTA = metric tons per annum.

250,000 MTA, based on 8000 operating hours per year. The fresh LPG feedstock is assumed to be 95 wt % propane with 3 wt % ethane and 2 wt % butane. The native ethane in the feed is rejected in the deethanizer along with light ends produced in the Oleflex unit and used as process fuel. The butanes are rejected from the depropanizer bottoms. This small butane-rich stream could be used as either a by-product or as fuel. In this example, the depropanizer bottoms were used as fuel within the complex.

The Oleflex process coproduces high-quality hydrogen. Project economics benefit when a hydrogen consumer is available in the vicinity of the propylene complex. If chemical hydrogen cannot be exported, then hydrogen is used as process fuel. A pressure-swing adsorption (PSA) unit has been included in the complex to generate a chemical hydrogen coproduct from the net hydrogen stream; this evaluation assumes that hydrogen is sold as a separate product. The PSA technology is available through UOP.

Utility Requirements

Utility requirements for a complex producing 250,000 MTA of propylene are summarized in Table 5.1.3. These estimates are based on the use of an extracting steam turbine to drive the Oleflex reactor effluent compressor. A water-cooled surface condenser is used on the steam turbine exhaust. An electric motor drive was chosen in this example for the propane-propylene splitter heat-pump compressor. Other types of drivers, such as back-pressure steam turbines, condensing steam turbines, or gas turbines, are suitable alternatives for either of these compressors.

Some high-pressure steam is generated within the convection section of the Oleflex heater, and the rest is imported from an off-site boiler. The requirement for the boiler feed water identified in Table 5.1.3 is for the inside battery limit (ISBL) steam generation within the Oleflex heater, and the high-pressure steam value in Table 5.1.3 is the net requirement from the off-site boiler.

No fuel gas credit is taken for the net hydrogen product, but fuel gas credits were applied for the depropanizer bottoms and deethanizer overhead streams. The fuel gas produced in the propylene complex is more than enough to supply the Oleflex heaters. Any excess fuel gas is exported to the off-site boiler or fuel gas system.

Propylene Production Costs

Representative costs for producing 250,000 MTA of polymer-grade propylene using the Oleflex process are shown in Table 5.1.4. These costs are based on feed and product values defined in Table 5.1.1. The fixed expenses in Table 5.1.4 consist of estimated labor costs and maintenance costs and include an allowance for local taxes, insurance, and interest on working capital. The hydrogen product in Table 5.1.4 is applied as a credit when determining the per-ton propylene production cost.

Capital Requirements

The ISBL erected cost for an Oleflex unit producing 250,000 MTA of polymer-grade propylene is approximately $116 million (U.S. Gulf Coast, mid-1995 erected cost). This figure includes the reactor and product recovery sections, a modular CCR unit, a

TABLE 5.1.3 Net Utility Requirements for a 250,000-MTA Propylene Complex

Utility requirements	Consumption	Utility cost $/h	Utility cost $/MTA C_3
Electric power	10,055 kW	503	16.10
High-pressure steam	62.9 MT/h	478	15.30
Boiler feed water	21.7 MT/h	19	0.61
Steam condensate	(83.8 MT/h)	−74	−2.37
Cooling water	3,604 m³/h	72	2.30
Net fuel gas	(24.2 million kcal/h)	−202	−6.46
Net utilities			25.48

Note: Parentheses indicate utility export. MTA = metric tons per annum; MT/h = metric tons per hour.

TABLE 5.1.4 Cost for Producing 250,000 MTA of Polymer-Grade Propylene Using the Oleflex Process

	Revenues, million $/year	Costs	
		million $/year	$/MT C_3
Propylene product	121.2	—	—
Hydrogen product	3.9	—	(15.6)
Propane feedstock	—	54.2	216.8
Net utilities	—	6.4	25.6
Catalyst and chemicals	—	3.1	12.4
Fixed expenses	—	5.0	20.0
Total	125.1	68.7	259.2

Note: Parentheses indicate credit. MTA = metric tons per annum; MT = metric tons.

Hüls SHP unit, and a fractionation section consisting of a depropanizer, deethanizer, and heat-pumped P-P splitter. The costs are based on an extracting steam turbine driver for the reactor effluent compressor and a motor-driven heat pump. Capital costs are highly dependent on many factors, such as location, cost of labor, and the relative workload of equipment suppliers.

Outside battery limit (OSBL) costs for an Oleflex complex vary widely depending on the extent of existing infrastructure. For a grassroots project, the OSBL expenditure could be as much as 60 percent of the ISBL cost.

Total project costs include ISBL and OSBL erected costs and all owner's costs. This example assumes an inclusive mid-1995 total project cost of $220 million including:

- ISBL erected costs for all process units
- OSBL erected costs (off-site utilities, tankage, laboratory, warehouse, for example)
- Initial catalyst loadings
- Capitalized loan interest during construction
- Project development including site procurement and preparation

Overall Economics

Because the feedstock represents such a large portion of the total production cost, the economics for the Oleflex process are largely dependent on the price differential between propane and propylene. Assuming the values of $177/MT for propane and $485/MT for propylene, or a differential price of $308/MT, the pretax return on investment is approximately 25 percent for a complex producing 250,000 MTA of propylene.

CHAPTER 5.2
UOP PACOL DEHYDROGENATION PROCESS

Peter R. Pujadó
UOP
Des Plaines, Illinois

INTRODUCTION

Paraffins can be selectively dehydrogenated to the corresponding monoolefins by using suitable dehydrogenation catalysts. Iron catalysts have long been used for the dehydrogenation of ethylbenzene to styrene, and catalysts made of chromia (chrome oxide) supported on alumina have long been used for the dehydrogenation of light paraffins (for example, n-butane to n-butene) and the deeper dehydrogenation of olefins to diolefins (for example, n-butene to 1,3-butadiene). However, newer commercial processes for the dehydrogenation of light and heavy paraffins are based on the use of noble-metal catalysts because of the superior stability and selectivity of these catalyst systems.

In the late 1940s and through the 1950s, the pioneering work done at UOP* by Vladimir Haensel on platinum catalysis for the catalytic reforming of naphthas for the production of high-octane gasolines and high-purity aromatics showed that platinum catalysts have interesting dehydrogenation functions. This research area was later pursued by Herman Bloch and others also within UOP. In 1963-64, UOP started development work on heterogeneous platinum catalysts supported on an alumina base for the dehydrogenation of heavy n-paraffins. The resulting successful process, known as the Pacol* process (for paraffin conversion to olefins), was first commercialized in 1968. The advent of the UOP Pacol process marked a substantial transformation in the detergent industry and contributed to the widespread use of linear alkylbenzene sulfonate (LAS or LABS) on an economical, cost-effective basis. As of mid-1995, more than 30 Pacol units have been built, and practically all new linear alkylbenzene (LAB) capacity built on a worldwide basis over the last two decades makes use of UOP's Pacol catalytic dehydrogenation process.

---

*Trademark and/or service mark of UOP.

Maintaining technological superiority over some 30-odd years requires continued innovation and improvement, principally of the dehydrogenation catalyst, the reactor design, and operating conditions because these have the greatest impact on the overall process economics. The first commercial Pacol dehydrogenation catalysts, denoted DeH-3 and DeH-4, came on-stream in the mid-1960s. They were soon superseded by a newer generation, DeH-5, that was commercialized in 1971 and dominated the market for several years. In 1983, DeH-7 catalyst was introduced. This new catalyst exhibited about 1.75 times the stability of its predecessor, DeH-5, and soon replaced it as the dominant Pacol catalyst. But, the catalyst development efforts continue, and a new catalyst, DeH-9, is expected to be commercialized in 1996. All these various generations of paraffin dehydrogenation catalysts have resulted in improved yields at higher conversion and higher operating severities, thus allowing for smaller and more economical units for a given production capacity.

Since 1980, UOP has adapted similar catalysts to the selective catalytic dehydrogenation of light olefins (propane to propylene and isobutane to isobutylene) in the Oleflex* process; a number of large-capacity units have been built for this application. Because of the higher severity, light paraffin dehydrogenation units make use of UOP's proprietary CCR* continuous catalyst regeneration technology, which was originally developed and commercialized for the catalytic reforming of naphthas at high severity. Because the Pacol process operates at a lower severity, catalyst runs are significantly longer, and CCR technology is not needed.

PROCESS DESCRIPTION

The catalytic reaction pathways found in the dehydrogenation of n-paraffins to n-monoolefins [linear internal olefins (LIO)] in addition to other thermal cracking reactions are illustrated in Fig. 5.2.1. A selective catalyst is required if only LIO is to be the main product.

In the Pacol reaction mechanism, the conversion of n-paraffins to monoolefins is near equilibrium, and therefore a small but significant amount of diolefins and aromatics is produced. In the alkylation process, the diolefins consume 2 moles of benzene to yield heavier diphenylalkane compounds or form heavier polymers that become part

---

*Trademark and/or service mark of UOP.

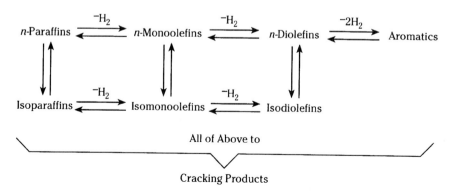

FIGURE 5.2.1 Dehydrogenation reaction pathways.

of the heavy alkylate and the bottoms by-products of the hydrofluoric (HF) acid regenerator. Thus, diolefin formation represents a net loss of alkylate yield. In 1984, UOP developed the DeFine* process, a highly selective catalytic hydrogenation process to convert diolefins back to monoolefins. Detergent complexes licensed prior to 1986 included only Pacol and HF Detergent Alkylate* units. The first DeFine unit came on-stream during the fourth quarter of 1986; all subsequent Pacol process units have also incorporated DeFine hydrogenation reactors, and DeFine reactors have also been retrofitted into a growing number of existing older Pacol units. Both Pacol and DeFine processes are also used in the latest process developed and commercialized by UOP, the Detal* process, for the production of LAB using a heterogeneous solid catalyst instead of the older, traditional HF acid catalyst.

The dehydrogenation of n-paraffins is an endothermic reaction with a heat of reaction of about 125 kJ/g · mol (30 kcal/g · mol; 54,000 Btu/lb · mol). The equilibrium conversion for the dehydrogenation reaction is determined by temperature, pressure, and hydrogen partial pressure. As expected, the equilibrium conversion increases with temperature and decreases with pressure and with increasing hydrogen-to-hydrocarbon ratio. Kinetically, the overall conversion depends on space velocity (feed-to-catalyst ratio): excessively high space velocities do not allow for sufficient conversions, and space velocities that are too low lead to lower selectivities because of the onset of side and competitive reactions.

Figure 5.2.2 illustrates the flow scheme of an integrated complex incorporating Pacol, DeFine, and HF Detergent Alkylate units or Pacol, DeFine, and Detal units. The main differences between the two flow schemes are in the alkylation section as a result of the elimination of the HF acid handling and neutralization facilities; for example, no alumina treater is used in conjunction with a Detal process unit.

In the Pacol process, linear paraffins are dehydrogenated to linear olefins in the presence of hydrogen over a selective platinum dehydrogenation catalyst. An adiabatic radial-flow reactor with feed preheat is normally used to compensate for the endothermic temperature drop and to minimize pressure drop within an efficient reactor volume. Relatively high space velocities are used so that only a modest amount of catalyst is required. Hydrogen and some by-product light ends are separated from the

*Trademark and/or service mark of UOP.

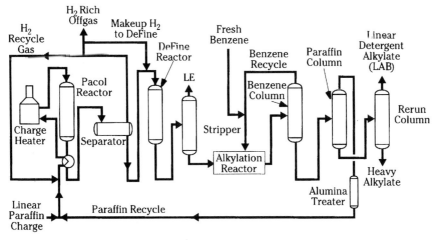

FIGURE 5.2.2 Integrated LAB complex.

dehydrogenation reactor effluent, and a part of this hydrogen gas is recycled back to the dehydrogenation reactor to minimize coking and enhance catalyst stability. The separator liquid is an equilibrium mixture of linear olefins and unconverted n-paraffins, which are charged to a DeFine reactor for the selective conversion of diolefins to monoolefins. A near-stoichiometric amount of hydrogen is also charged to the DeFine reactor. The DeFine reactor effluent is stripped to remove dissolved light hydrocarbons. The stripper bottoms, a mixture of monoolefins and unconverted n-paraffins, is then charged together with benzene to the alkylation unit, where benzene is alkylated with the monoolefins to produce LAB. Small amounts of heavy alkylate and, if HF is used, polymer from the acid regenerator bottoms are also formed. Benzene and n-paraffins are fractionated from the alkylation reactor effluent and then recycled to the alkylation and Pacol reactors, respectively. The final column fractionates the LAB product overhead and recovers heavy alkylate as bottoms product.

A similar process scheme can be used to produce concentrated n-olefins. Figure 5.2.3 illustrates the flow scheme of an integrated complex featuring the Pacol, DeFine, and Olex* processes. In this combination, the Pacol and DeFine processes are the same as described previously. The stripper bottoms stream, which consists of an equilibrium mixture of n-paraffins and n-monoolefins, is now sent to an Olex separation unit. The Olex process uses continuous liquid-phase, simulated countercurrent adsorptive separation technology to recover high-purity n-olefins out of the mixture. The olefinic extract and the paraffinic raffinate streams that leave the adsorption chamber both contain desorbent. These two streams are fractionated for the removal and recovery of the desorbent, which is then recycled back to the adsorption chamber. The paraffin raffinate is recycled to the Pacol dehydrogenation unit for complete conversion of the unconverted n-paraffins to the ultimate n-olefin product. Table 5.2.1 shows the olefins composition of a typical Olex process.

The LIO produced by the Pacol process and recovered in an Olex unit is premium material for the production of detergent alcohols via hydroformylation. Oxo technologies, such as Shell's, Exxon's (formerly Norsolor's and Ugine Kuhlmann's), or Enichem Augusta's can be used. In mid-1995, three integrated Pacol-Olex-Oxo com-

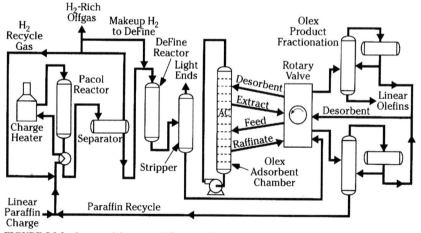

FIGURE 5.2.3 Integrated detergent olefins complex.

*Trademark and/or service mark of UOP.

TABLE 5.2.1 Typical Olex Extract Composition

Composition	With PEP, wt %	Without PEP, wt %
Linear monoolefins	95.0	92.5
Other monoolefins	3.1	3.0
Diolefins	0.5	0.5
Total	98.5	96.0
Olefins	98.6	96.0
Aromatics (see text)	0.2	3.0
Paraffins	1.2	1.0
Total	100.0	100.0

plexes were operating with a total production capacity of approximately 230,000 metric tons (MT) of detergent alcohols per year. Surfactants made from detergent alcohols manufactured according to this combination of technologies show superior properties in terms of detergency and solubility.

PACOL PROCESS IMPROVEMENTS

Repeated successful attempts have been made over the years to increase the per-pass conversion of n-paraffins across the Pacol reactor and still preserve a high selectivity and high overall yield of linear olefins.

The more severe operating conditions used for higher reactor conversions also result in faster deactivation of the dehydrogenation catalyst. The catalyst used in the Pacol process has a direct impact on the reaction kinetics but not on equilibrium conversion, which is governed by thermodynamic principles. Therefore, most of the process improvements have been associated with modifications in reactor design or in operating conditions.

A high-conversion Pacol process was developed partially in response to the significant increase in feedstock and utility costs that occurred between 1974 and 1981. Operating the process at higher per-pass conversions affords several advantages. A smaller combined-feed stream to the dehydrogenation reactor permits a smaller-size unit and results in lower capital investment and utility costs. As the unconverted n-paraffins pass through the alkylation reaction zone and are separated by fractionation for recycle to the Pacol reactor, the reduction in the recycle stream also decreases the capital investment and operating cost of the detergent alkylation unit. All recent units are of the high-conversion type.

Criteria for the high-conversion design were to maintain the same selectivity to linear olefins and increase conversion. This approach required changes in operating conditions. Figure 5.2.4 shows the effect of pressure on olefin selectivity at constant temperature and hydrogen-to-feed mole ratio. At lower pressures, higher n-paraffin conversion can be obtained and selectivity can be maintained because of the more favorable dehydrogenation equilibrium.

A similar effect can be observed when the hydrogen-to-feed ratio is lowered. The latest designs of the Pacol process take advantage of both of these variables. The net result is a 30 percent increase in n-paraffin conversion compared to the earlier designs.

Overall, the Pacol catalyst possesses an attractive catalyst life in terms of metric tons of LAB produced per kilogram of catalyst. A typical run on a single Pacol catalyst load ranges from 30 to 60 days, depending on operating severity. As shown on

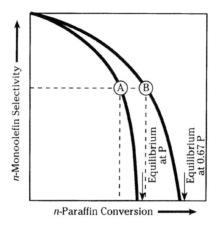

FIGURE 5.2.4 Effect of pressure on conversion.

Fig. 5.2.5, two parallel reactors were used for most units built through 1987. In this design, one reactor operates at any given time and the second reactor is on standby. When the decline in catalyst activity warrants a change, the reactors are switched. To expedite the change and minimize interruption in production, a start-up heater is provided. For safe operation and isolation, each valve shown in the drawing actually represents a double block and bleed valve. Thus, 16 large valves are required on process lines. These valves cycle from cold-to-hot and hot-to-cold service at each change of the reactors and require regular maintenance to control leakage. To minimize maintenance and simplify the operation, a new reactor design (Fig. 5.2.6) is now used commercially in two units. This design provides a catalyst hopper on the top and on the bottom of a single reactor and a hydrogen and a nitrogen purge system. When catalyst activity has declined sufficiently, the catalyst from the reactor is withdrawn to the lower hopper, and fresh catalyst is loaded from the top hopper, thereby eliminating the need for valves in large-diameter process lines. An additional catalyst volume inside the reactor vessel is provided as a preheating zone. A portion of the hydrogen-rich recycle gas passes through a heat exchanger and is used for preheating the catalyst. The hydrogen-rich gas is also used to purge hydrocarbons from the catalyst that leaves the reactor. This design is similar in concept to that used commercially in more than 100 UOP CCR Platforming* units.

Other improvements have been carried out over the years on the overall Pacol design. Some reactor design changes that are in the demonstration stage are expected to make a profound positive impact on future unit designs and operating performance.

Outside the reactor sector, other process design changes made over the past few years have also contributed to enhancing the reliability and economics of the Pacol process. One, for example, reflects the introduction of rotary screw compressors instead of the reciprocating or centrifugal machines used in earlier Pacol units. Rotary screw compressors are especially effective when lube oil contamination of the process gas cannot be tolerated. Nonlubricated screw compressors can deliver gases with the same reliability as a centrifugal compressor, and the positive displacement of screw compressors makes them well suited for applications that require high compression ratios and large changes in gas molecular weights. In addition, screw machines offer

---

*Trademark and/or service mark of UOP.

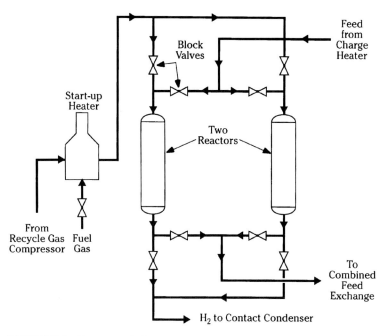

FIGURE 5.2.5 Two-reactor design.

economic benefits over comparable reciprocating machines in terms of lower installation costs and not requiring a spare.

Changes in engineering design also resulted in increased energy efficiency, reduced fractionation losses, and improved operational stability. Some of these design changes concerned the Pacol unit itself, but many were more closely associated with the associated downstream units.

As in the design of other process units, significant energy savings were achieved by relatively small incremental expenditures in increased exchanger area and by rearrangement of the heat exchanger network. For example, a low-pressure-drop contact condenser was advantageously introduced to cool the reactor effluent after the hot combined-feed effluent exchanger. Also, the application of efficient mixing technology in the reaction zone enhanced the quality of the reaction environment and allowed operation with recycle ratios close to their minimums.

YIELD STRUCTURE

If expressed on a weight basis, the yield of linear olefins from n-paraffins in the Pacol process depends on the molecular weight of the feedstock. In the common situation in which linear olefins are produced for the manufacture of LAB, typically from n-paraffins in the C_{10} to C_{13} carbon range, about 1.05 kg of feed is required per 1.00 kg of linear olefins, or about 97 percent of the theoretical stoichiometric yield.

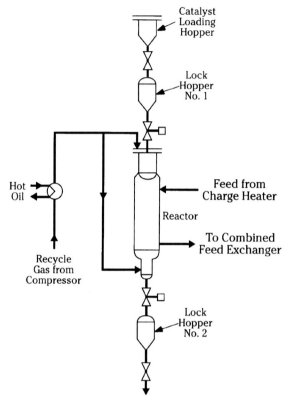

FIGURE 5.2.6 One-reactor design.

COMMERCIAL EXPERIENCE

More than 30 Pacol process units have been built and brought on-stream around the world since the mid-1960s, and practically without exception, all are still operating. A few other units are in various stages of design and construction. Most Pacol units are directly integrated with a benzene alkylation unit for the production of LAB without the need for an intermediate separation or recovery of the LIOs. These units represent an aggregate design capacity in excess of 1.3 million metric tons per year (MTA) of LAB; however, through revamps and expansions, the actual operating capacity is significantly larger. In addition, other Pacol units are associated with Olex units to recover LIO for the production of detergent alcohols.

PROCESS ECONOMICS

Because a Pacol unit is never found by itself, but is instead integrated with a DeFine unit and either a HF Detergent Alkylate unit, a Detal solid-acid alkylation unit, or an Olex LIO separation unit, the economics can be discussed only in conjunction with the associated units. Details on LAB production can be found in Chap. 1.5.

TABLE 5.2.2 Economics for LIO Production Using the UOP Pacol, DeFine, and Olex Processes*

	Unit cost, $	Per MT of LIO	
		Units	$
Raw materials:			
n-Paraffins (98% purity)	400/MT	1.05 MT	420.0
By-product credits		−0.05 MT	−13.1
Catalysts and chemicals			32.2
Utilities:			
Power	0.05/kWh	305 kWh	15.3
Steam	7.1/MT	0.16 MT	1.1
Cooling water	0.01/m^3	17 m^3	0.2
Fuel fired (92% eff.)	2.32/GJ	16.30 GJ	37.8
Labor, maintenance, direct overhead, and supervision			25.2
Overhead, insurance, property taxes, depreciation, amortization			98.6
Total cost of production			617.3

*Estimated erected cost: $65,000,000 (basis: 60,000 MT/year of LIO). MT = metric tons.

As a different example, economics for the production of 60,000 MTA of LIO starting from n-paraffins are shown in Table 5.2.2. The complex includes Pacol, DeFine, and Olex units and reflects typical economic conditions. The resulting production cost of $617/MT of LIO compares favorably with the costs of production of LIO or LAO by other routes. A typical product composition is shown in Table 5.2.1. If desired, the aromatic content can be reduced by adding the UOP proprietary PEP* (Pacol Enhancement Process) for the selective removal of aromatics; introduction of this novel technology has resulted in more than 90 percent reduction in the aromatic content and a 2.5 to 3.0 percent increase in olefin purity.

BIBLIOGRAPHY

Vora, B. V., P. R. Pujadó, and M. A. Allawala, "Petrochemical Route to Detergent Intermediates," 1988 UOP Technology Conference.

---

*Trademark and/or service mark of UOP.

PART · 6

GASIFICATION AND HYDROGEN PRODUCTION

CHAPTER 6.1
KRW FLUIDIZED-BED GASIFICATION PROCESS

W. M. Campbell
The M. W. Kellogg Company
Houston, Texas

INTRODUCTION

In many refinery processing configurations, a petroleum coke by-product is produced that normally has low market value. This coke can be gasified to produce a fuel gas for generating steam and electric power, or to produce hydrogen and synthesis gas for use within the refinery.

The KRW gasification technology is an efficient, nonslagging fluidized-bed process that has been successfully tested on a broad range of bituminous, subbituminous, and lignitic coals and several petroleum cokes. The process was piloted for close to 15 years at a 30-ton/day process development unit (PDU) in Waltz Mill, Pennsylvania. In addition to the gasification testing, hot-gas desulfurization and particulate removal processes also were piloted.

In an air-blown operating mode, using limestone or dolomite as an in-bed sulfur sorbent, the KRW gasifier produces a low-Btu gas that is cleaned of residual sulfur and particulate, then delivered to a combined-cycle power plant as fuel. In an oxygen-blown operating mode, the produced gas—containing hydrogen, CO, CO_2, and some methane and H_2O—can be processed to recover hydrogen or synthesis gas, or can be used as a medium-Btu fuel.

HISTORY

KRW Energy Systems Inc., under a number of contracts with the U.S. Department of Energy, its precursor agencies, and the Gas Research Institute (GRI), engaged in the development of a pressurized, fluidized-bed gasification process for converting many varieties and ranks of coal to clean, environmentally acceptable fuel gas. From 1972 to 1984, the program was conducted by Westinghouse Electric Corporation at its

Waltz Mill facility. In 1984, ownership of the technology and management of the program was taken over by The M. W. Kellogg Company. Operation of the PDU was continued until the facility was decommissioned in 1988. The cost of the development program over that period of time was approximately $150 million (U.S.).

The primary goals of the program were the development of a process to produce low-Btu fuel gas for fueling a combustion turbine combined-cycle power plant and the development of a process to produce a medium-Btu industrial fuel or synthesis gas that could be used as a feedstock for substitute natural gas or liquid synfuels.

Development began in 1972 under the sponsorship of the Office of Coal Research and was directed toward development of a two-stage, air-blown gasification system consisting of a coal devolatilizer and a gasifier-agglomerator. In 1979, GRI became a part of the program, and greater emphasis was placed on the development of a medium-Btu, oxygen-blown process. On the basis of experimental breakthroughs in the process design, it was demonstrated that caking coals, highly reactive coals, and coals with wide ranges of ash content could be processed successfully in a single-stage gasification process.

As part of the development program, a calcium-based sulfur sorbent was added to the coal feed to the gasifier. Both limestone ($CaCO_3$) and dolomite ($CaCO_3 \cdot MgCO_3$) were tested. A hot-gas cleanup (HGC) system also was built and tested at Waltz Mill. The HGC system included a barrier filter to remove particulate from the gas and a fixed-bed desulfurization system to remove hydrogen sulfide (H_2S) and carbonyl sulfide (COS).

A cold-flow model of the lower section of the gasifier was built to assist in the scaleup of the gasifier to commercial size. The model was 3 meters in diameter, about the maximum size of the combustion section of a large commercial gasifier. An off-line hot-gas filtration test cell also was built to permit testing of a number of alternative filter designs. Ceramic and sintered metal candles, as well as Westinghouse-designed ceramic cross-flow filter elements, were tested either in the test cell or in the PDU.

KRW SINGLE-STAGE GASIFICATION PROCESS

The KRW single-stage gasification process was designed to operate in an air- or oxygen-blown mode to produce either low- or medium-Btu gas. Run-of-mine coal is dried to a surface moisture of about 5 percent (which is needed for satisfactory pressurization and transfer into the gasifier) and is crushed in an impact mill to a typical feed size of -3 mm. The dry, sized coal is conveyed to a lockhopper system, where it is pressurized, discharged through a rotary feeder, and pneumatically transported by air or recycle process gas into the gasifier through a central feed jet. The gasifier is a refractory-lined pressure vessel.

As indicated in Fig. 6.1.1, steam and oxidant (air or oxygen) are injected as a mixture into the gasifier through an annulus surrounding the feed tube. The coal rapidly devolatilizes in the jet, and the volatiles, recycle gas, and some of the char combust, supplying the heat for the endothermic gasification. The high jet velocity and the rapid devolatilization and combustion in the jet induce vigorous mixing and dispersal of char in the bed, allowing even highly caking coals to be processed. The high temperature of the combusting jet assures that the products of devolatilization (tars, oils, and light hydrocarbons) are cracked and gasified to methane, carbon monoxide, and hydrogen.

The high velocity of the jet causes the surrounding char bed to be well back-mixed, assuring that the steam present continues the gasification of the char, leaving particles that are rich in ash. The unique fluid dynamic design of the gasifier allows the ash-rich particles to agglomerate as they are heated by the jet and cooled by the reacting fluid bed. The ash-rich particles contain mineral compounds that form eutectics with

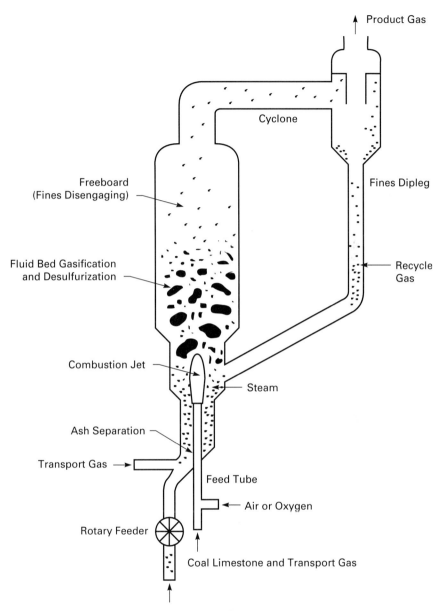

FIGURE 6.1.1 KRW fluid-bed gasifier.

melting points from 540°C to more than 1090°C. The lower-melting compounds melt, wetting the surface of the ash particles and causing them to be sticky and to agglomerate. The agglomerated ash particles are heavier than the char particles in the fluid bed and tend to settle to the bottom of the gasifier, where they are continuously removed through a rotary feeder.

The bottom annulus of the gasifier is designed to cool the ash particles as they leave the gasifier. A cooling gas stream flows upward through the exiting ash, cooling it and helping to reentrain the lighter char particles and to direct them back into the gasifier fluid bed. The melting characteristics of coal and coke ashes can differ widely, and there is a temperature range for each feed within which the gasifier operates best. There can be a significant change in the ash melting temperature range of the naturally occurring ash and the various ash eutectic mixtures that form in the gasifier. Addition of a calcium-based sulfur sorbent with the feed also can have an effect on these ash-melting temperature ranges. When limestone is used in the gasifier, the bottom discharge material is often referred to as *lash,* a combination of spent limestone and ash.

The ash-melting characteristics for candidate feeds are measured on a hot-stage microscope, and the operating temperature range for each is determined. This temperature range can determine the extent of char conversion in the gasifier. Typical operating ranges are shown for various ranks of coal in Fig. 6.1.2.

Product gas from the gasifier enters one or more cyclones, where particulates are separated from the gas. The recovered particulates, which contain unconverted hydrocarbon (char), are returned to the lower section of the gasifier, where they are induced to flow again through the hot jet and bed to convert additional char to gas and to agglomerate the ash. The amount of fines circulated through the gasifier and cyclones differs with various feeds. Fines are contained in the feed, and additional fines are formed by attrition and reaction in the gasifier. An equilibrium quantity of circulating fines builds up, depending on feed friability, operating temperature, and the solids

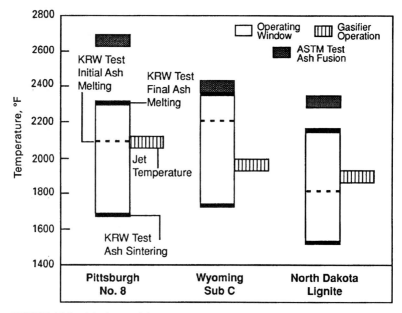

FIGURE 6.1.2 Ash characteristics.

inventory of the gasifier. The superficial gas velocity in the gasifier freeboard is set to limit the fines entrainment and recirculation to a reasonable level.

Raw product gas from the cyclones, containing a small amount of very fine particulate, is cooled in a steam-generating heat exchanger before entering a barrier filter, where essentially all of the fines are removed by back-blowing sections of elements while the filter remains on line. The fines contain unconverted char that may require combustion prior to disposal. A number of different barrier filter designs were tested at Waltz Mill. Sintered metal and ceramic candle filters were tested extensively. Westinghouse-designed ceramic cross-flow filter elements also were tested in a supporting test facility. Candle and cross-flow filters are shown in Fig. 6.1.3.

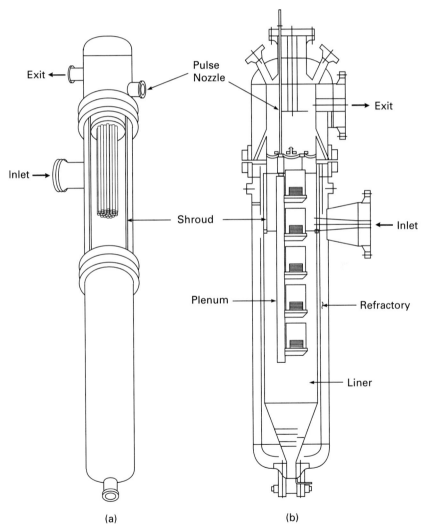

FIGURE 6.1.3 Candle and cross-flow filters. (*a*) Ceramic candle filter; (*b*) Westinghouse cross-flow filter.

The gasifier product gas contains hydrogen sulfide and carbonyl sulfide. With a sulfur sorbent such as limestone or dolomite in the gasifier, the sulfur content of the gas will range from 300 to 600 wt ppm of H_2S and COS. If no sorbent is used, the sulfur content of the gas will be in proportion to the sulfur in the feed. In air-blown gasification, the sulfur content will be about 2000 wt ppm per percent sulfur in the feed. In oxygen-blown gasification, the sulfur content will be about 5000 wt ppm per percent sulfur in the feed.

A fixed-bed desulfurization system using regenerable zinc ferrite as the sorbent was tested in the PDU at Waltz Mill. Sulfur levels of 10 to 20 ppm in the treated gas were achieved in this system. The product gas also contains traces of ammonia (NH_3) and hydrogen cyanide (HCN), derivatives of the nitrogen in the feed. Using limestone in the gasifier tends to reduce the level of these components. If there is any alkali in the feed, some of it may vaporize in the gasifier and may have to be removed by an alkali sorbent or by cooling the gas going to the filter.

Overall Environmental Emissions

The KRW gasification system utilizes in situ (limestone in the gasifier) and external desulfurization systems. The equilibrium H_2S leakage from the gasifier ranges from 300 to 600 vol ppm, depending on operating conditions but independent of the amount of sulfur in the feed as long as sufficient limestone is added. The external desulfurization system can reduce the H_2S to less than 20 to 30 vol ppm with a number of available zinc-based sorbents.

Ammonia is formed in the gasifier from fuel-bound nitrogen. Ammonia data were not specifically recorded during the petcoke tests, but during testing of coal with similar levels of fuel-bound nitrogen, ammonia in the product gas typically contained less than 5 percent of the fuel-bound nitrogen when limestone was present in the gasifier. Ammonia was not removed by the zinc ferrite in the external desulfurization testing.

The particulates in product gas are removed in the barrier filter to a level of less than 5 ppm. Volatile metals and alkali tend to accumulate on the particulate as the gas is cooled. The particulates contain a high percentage of carbon and are usually sent along with the ash to a combustor, where the remaining carbon is burned and the calcium sulfide is oxidized to sulfate. In this hot-gas cleanup system, there is no aqueous condensate produced, although some may be produced in subsequent processing of the gas. The solid waste from the process consists mainly of spent limestone and metals from the petcoke. After processing in the combustor/sulfator, this material should be suitable for landfill disposal.

DESCRIPTION OF THE KRW PROCESS DEVELOPMENT UNIT

The KRW PDU was a nominal 15- to 30-ton/day coal gasification pilot plant. The range of capacity reflects the difference in the air- and oxygen-blown mode of operation. A photograph of the facility is shown in Fig. 6.1.4*a*. Figure 6.1.4*b* to *e* shows several components of the PDU.

A process schematic diagram is shown in Fig. 6.1.5. Raw coal was crushed and dried in an impact mill, then screened through a 6-mesh screen (−3 mm) prior to transfer to storage bins. Metallurgical coke breeze or petroleum coke also was prepared and stored, since a nonvolatile fuel is preferred for start-up of the gasifier. Limestone or dolomite sorbent is similarly dried, sized, and stored as required.

FIGURE 6.1.4a Overall view of KRW PDU.

FIGURE 6.1.4b Coal feed injection line and ash removal annulus.

The feeds were pressurized in a lockhopper system, passed through rotary valves, and pneumatically conveyed by air to a central jet in the bottom of the gasifier. When the oxidant was air, it was preheated to about 565°C prior to injection into the gasifier. Superheated steam was mixed with the oxidant and entered the gasifier via an annulus surrounding the fuel jet. Agglomerated ash and spent sorbent were cooled and discharged through a rotary valve at the bottom of the gasifier to depressurizing lockhoppers. A delumper upstream of the rotary valve was included to crush the occasional large agglomerate formed in the gasifier.

Fines escaping the gasifier with the product gas were separated in three cyclone stages. Solids captured by the first two cyclones were returned to the combustion sec-

FIGURE 6.1.4c L valve used to return secondary cyclone fines to the gasifier.

tion of the gasifier through a sloped dipleg and a nonmechanical L valve, respectively. Fines captured in the third, nonrecycle, cyclone were quenched with water in a venturi, depressurized, and recovered in a filter press.

For the hot-gas cleanup test program, the product gas from the third-stage cyclone was adiabatically humidified and cooled to about 650°C with water injected through an atomizing nozzle. The cooled gas entered a barrier filter where the balance of the particulates was removed from the gas. A small slipstream of recycle gas was quenched with water and was recompressed to use for fluidization and ash cooling in the gasifier.

The particulate-free gas then proceeded to the zinc ferrite desulfurization reactor for H_2S and COS removal. The desulfurized fuel gas then was depressurized and combusted in a thermal oxidizer. The desulfurizer was a single fixed-bed unit, which oper-

FIGURE 6.1.4d Hot filter, desulfurizer, and tertiary cyclone (clockwise from left).

ated either in the absorption or regeneration mode. The reactor was designed to remove all the sulfur from the gas when no sulfur sorbent was used in the gasifier. It was later modified (reduced in size) to operate in cleanup mode following primary desulfurization in the gasifier. The zinc ferrite was regenerated with steam and air. When a sulfur sorbent was used in the gasifier, the spent regeneration gas was returned to the gasifier for reabsorption of the SO_2.

TEST RESULTS OBTAINED IN THE PDU

The PDU tests have provided an extensive database characterizing the effects of feedstock properties and process operating variables on gasification performance. A summary of the characteristics of the feeds tested in the PDU is given in Table 6.1.1. The principle operating variables were coal feed rate, sorbent/coal ratio, recycle gas ratio, bed height, temperature, and pressure. Reports are available in the literature describing each test campaign.[1-6] Test data included all input and output flows and compositions, axial temperature and pressure profiles in the gasifier, and pressures and temperatures at each major piece of equipment. Solid samples were collected and analyzed every hour or, at least, every shift from the gasifier, cyclones, and barrier filter.

The gasifier test program included more than 13,000 hours of hot operation, while the hot-gas cleanup test program completed more than 3000 hours of operation. The test data were reduced, steady-state operating periods were defined, and complete heat and material balances were prepared for each period. The database includes more than 250

FIGURE 6.1.4e Cold-flow scaleup facility.

such balances. Various laboratory and bench-scale studies have contributed to the fundamental understanding of fluidization phenomena, particle attrition and elutriation, particle collection and analysis, materials selection, and reactor mathematical modeling.

The gasifier test program also included tests with in-bed desulfurization. Limestone or dolomite sorbent was added with the coal, and most of the sulfur in the feed was captured on the sorbent. The formed calcium sulfide is rejected with the ash and needs to be oxidized before disposal as landfill. Downstream hot-gas cleanup testing included particulate removal and removal of the balance of the sulfur to low levels.

Initially in the PDU, sintered metal candle filters were used to remove particulate from the hot product gas. The candles were attacked by sulfur, which resulted in premature blinding and, ultimately, mechanical failure of the candles. The sintered metal

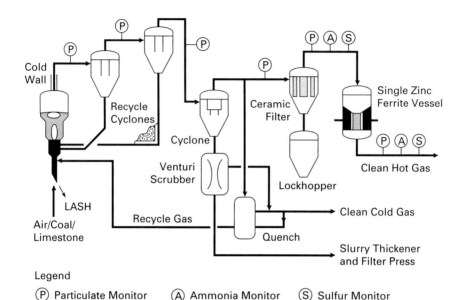

FIGURE 6.1.5 PDU process schematic.

TABLE 6.1.1 Range of KRW PDU Feedstocks

	Bituminous	Subbituminous	Lignites	Cokes	Non-U.S.A.
Ash, %	7–12	8	11–23	0.2–14	18–45
Moisture, %	2–7	18–22	14–18	0.9–10	2–4
Caking index, FSI	2–9	0	0	0	0–0.5
Reactivity	4–9	60–71	28–44	1–2	5–49
Sulfur, %	1–4	0.4–0.6	0.6–0.9	1–2.5	0.5–0.9
Total carbon, %	65–74	54	45–52	80–90	49–62
Ash fusion temp., °C	1260–1650	1270–1320	1170–1370	1390–1430	1390–1420
Volatile matter, %	16–36	33–44	32–36	2–10	23–25

Note: FSI = free-swelling index

candles were replaced with ceramic elements. More than 1000 hours of hot (540 to 650°C) operating experience were logged on the ceramic candles. A number of the ceramic candles cracked, but modifications to shroud the candles from impact by the inlet gas flow minimized this problem.

A fixed-bed desulfurizer using zinc ferrite as the sorbent removed the residual sulfur from the gas down to the 5 to 15 ppm level. The spent sorbent was regenerated with steam and air. Zinc ferrite was one of the earlier sorbents developed for this service and has a number of disadvantages. First, it requires the moisture content of the gas to be about 30 percent to prevent formation of iron carbides in the sorbent particles. The steam lowers the heating value of the product gas substantially. Second, the sorbent tends to lose mechanical strength as it ages, which is thought to be associated with the change in size of the metal oxide as it cycles from the oxide form to the sulfide/sulfate and back to the oxide.

A number of improved sorbents have been developed and tested by Kellogg and others for the hot-gas cleanup service. These newer sorbents are stronger and more attrition-resistant and have been tested in fluidized-bed applications where even lower sulfur leakage is experienced than with the fixed-bed designs. One of the advantages of fluidized-bed designs over fixed-bed designs is the insensitivity to inlet sulfur loading, since the sorbent is continuously regenerated.

Petroleum cokes from both Chevron's El Segundo and Arco's Watson refineries in California were processed in the KRW gasifier. The analysis of the feed cokes and typical gas analyses from oxygen-blown gasification of these cokes is given in Table 6.1.2. Carbon conversion levels achieved during the petroleum coke tests ranged from 77 to 95 wt %. Because of the low ash content of the coke feed, the carbon content of the bottom ash product was greater than 90 wt %. The carbon in this ash can be further converted in a combustor, if necessary. If sorbent is added with the coke feed, the calcium sulfide formed would be oxidized to sulfate in the combustor, making it safe for disposal as landfill. The metals in this material are normally not leachable, but tests should be made for each application.

Petroleum coke contains varying amounts of ash. Low-melting eutectics can form agglomerates in the gasifier at higher temperatures. These conglomerates can grow and plug the ash outlet of the gasifier. One ash component that seems to be troublesome is Fe_2O_3, which may be combined with sulfur in the coke feed. During the high-temperature (1135°C) petroleum coke tests, ash annulus plugging was experienced.

TABLE 6.1.2 Product Gas from Oxygen-Blown KRW Gasification of Petroleum Cokes

Coke analysis	
Ultimate analysis, wt %:	
Carbon	87.1–90.3
Hydrogen	3.8–4.0
Oxygen	1.5–2.0
Nitrogen	1.6–2.5
Sulfur	2.1–2.3
Proximate analysis, wt %:	
Volatiles	9.0–9.7
Fixed C	80.4–89.2
Moisture	0.9–10.2
Ash	0.2–0.4
Product gas analysis	
Gasification at 980–1135°C, vol %:	
Hydrogen	13.5–16.8
CO	34.3–45.6
CO_2	27.3–36.4
Methane	0.1–0.9
Nitrogen	0.4–0.7
Water	8.7–13.9
H_2S	0.3–0.6
Btu/SCF	4.3–5.4

Note: SCF = standard cubic feet.

The agglomerates that formed consisted principally of compounds of iron, aluminum, and silica, all of which tended to concentrate in the agglomerates.

Since the amount of ash in petroleum coke is small, the addition of a calcium-based sorbent to modify the ash-melting characteristics should be considered. Vanadium and nickel are found in most petroleum cokes in concentrations far greater than in coals or lignites. These metals remain in the ash product and do not appear to have a material effect on the gasification or ash-agglomerating process.

The carbon conversion level in a fluidized-bed gasifier is determined by the operating temperature of the gasifier, the volume of the bed, and the reactivity of the coke itself. Coke reactivity is readily determined by a thermogravimetric analyzer (TGA), which subjects the coke to gasification conditions, measuring the reactivity of the coke with increasing conversion level by change in weight with time.

The operating-temperature limits in a fluidized-bed gasifier can be determined by measuring the ash-melting characteristics of any candidate coke feed on a hot-stage microscope. The melting characteristics of the ash in the coke feed and the ash produced in the TGA both should be measured, since the melting points of the various eutectic mixes can be substantially different than in the raw coke.

COMMERCIAL-SCALE DESIGN

Extensive testing was done on a 3-m-diameter cold-flow scaleup facility (CFSF) of the combustion zone of a commercial gasifier. A picture of this facility is shown in Fig. 6.1.6. The CFSF tests confirmed the effectiveness of the central jet design on mixing in a large gasifier, the aeration requirements of the grid area surrounding the jet and the interaction of jet velocity and jet submergence on flow patterns. The combination of bench-scale CFSF and PDU data was used to develop mathematical models for design and prediction of performance of commercial gasifiers and the elutriation of fines from these units.

The capacity of a KRW gasifier depends on the operating pressure of the gasifier, whether it is air- or oxygen-blown, whether sorbent is added with the feed, the amount of fines that are in the feed and are formed in the gasifier, and physical constraints, such as vessel shipping clearances. A typical gasifier is 3.7 m in internal freeboard diameter and 24 to 27 m tall.

In air-blown mode, when using in-bed sorbent desulfurization and operating at 300-lb/in^2 pressure, this size gasifier processes about 800 ton/day of coke. This configuration probably would be used if the product gas were to be used for electric power generation. In oxygen-blown mode, the feed capacity would be 1000 to 1500 ton/day, depending on whether sorbent was used. The addition of a calcium-based sorbent to the gasifier has several effects:

- It increases the reactivity of the carbon, requiring a smaller bed volume to achieve the desired conversion level.
- It increases the density of the bed and permits higher gas velocity and, thus, higher feed capacity.

The choice of hot-gas cleanup technology depends on the application. For electric power production, gasifier in-bed desulfurization followed by hot-gas filtration and hot-gas desulfurization is generally most economic and results in the highest cycle efficiency. For production of hydrogen or synthesis gas, oxygen-blown gasification with cold-gas cleanup (e.g., alkaline wash) is often preferred because of the downstream processing sequence. The decision to use a sulfur sorbent in the gasifier is a

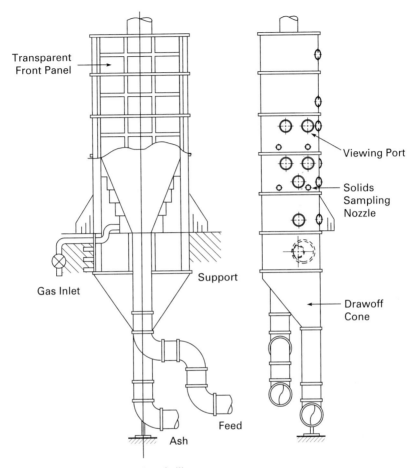

FIGURE 6.1.6 Cold-flow scaleup facility.

separate economic choice affected by the level of sulfur in the feed and whether a separate sulfur product is desired.

Many applications of petroleum coke gasification are expected to include requirements for combinations of hydrogen or synthesis gas and steam and electric power production. Integration of the refinery with the gasifier island and the power island often gives substantial capital and operating cost savings to each.

APPLICATION

A simplified block diagram of a petroleum-coke-based gasification unit associated with a refinery is shown in Fig. 6.1.7. A 15,000 metric ton per day (mtpd) refinery processing a heavy crude or a reduced crude produces approximately 1900 mtpd of petroleum coke.

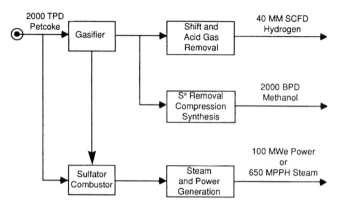

FIGURE 6.1.7 Petcoke commercial schematic.

Sufficient coke is fed to the oxygen-blown gasifiers to produce the gas required for hydrogen and methanol synthesis gas. In this scheme, the unconverted char from the gasifier and a portion of the feed coke is directed to a coke combustor for generation of steam. The products produced are 1.2×10^6 std m^3/day of hydrogen and 260 mtpd of methanol. The methanol is used to make methyl tertiary butyl ether (MTBE), an oxygenate used in the refinery gasoline pool. Approximately 300 metric tons per hour (mtph) of steam is generated that could produce up to 105 MWe of electric power.

A wide variation in the quantity of these streams is possible to match the needs of the refinery. The gasifier can produce fuel for the gas turbine in a combined-cycle configuration, or steam can be produced for a simple cycle turbogenerator arrangement. The choice requires an economic and operating availability study for each particular application. In the case shown in Fig. 6.1.7, one gasifier would process about 1400 ton/day of coke, bypassing the balance to the combustor.

If availability considerations required two gasifiers, the full 2900 mtpd of coke could be processed, with the excess gas going to the gas turbine. If electric power availability is critical, different combinations of gas and steam turbines might be selected to satisfy this criteria. An oxygen-blown gasifier is required for synthesis gas production, but an air-blown gasifier would satisfy the gas turbine and would reduce the size of the oxygen plant required.

CONCLUSIONS

Extensive pilot plant testing of lignitic, subbituminous, and bituminous coals in the KRW fluidized-bed gasifier indicate that a wide variety of carbonaceous feeds can successfully be gasified to produce fuel gas or synthesis gas. In the air-blown mode, a low-Btu (100 to 150 Btu/SCF) fuel gas is produced. In the oxygen-blown mode, a medium-Btu (250 to 300 Btu/SCF) gas can be produced for use as fuel or synthesis gas. Both coke breeze and petroleum coke were successfully fed to the gasifier. In processing petroleum coke, the high hydrogen and carbon monoxide content of the produced gas suggests its subsequent processing into hydrogen for use in the refinery or for production of oxygenates to upgrade the refinery gasoline pool. The economics of producing oxygenates may not be attractive in smaller capacity units.

The first commercial KRW gasifier is currently being designed for use by Sierra Pacific Power Company at its Tracy power station near Reno, Nevada. The project processes 800 mtpd of western bituminous coal in an air-blown gasifier to produce 100 MWe of electric power. It is expected to be in operation in early 1997.

REFERENCES

1. "Cold-Flow Scale-up Facility Experimental Results and Comparison of Performance at Different Bed Configurations," topical report FE-19122-46.
2. "32 Month Gasifier Mechanistic Study, Process Analysis, Fiscal Years 1984, 1985," topical report FE-21063-26.
3. "50-Month Gasifier Mechanistic Study, Fiscal Years 1984 through 1988," final report FE-21063-69.
4. "Advanced Development of a Pressurized Ash Agglomerating Fluidized-Bed Coal Gasification System," final report FE-19122-51.
5. "Assessment of Coal Gasification/Hot-Gas Clean-up Based Advanced Gas Turbine Systems," Southern Company Services, prepared for U.S. Department of Energy, Washington, D.C.
6. "Southern Company Services' Study of an (Oxygen-Blown) KRW-Based GCC Power Plant," EPRI report GS-6876, Southern Company Services, prepared for Electric Power Research Institute, Palo Alto, Calif.

CHAPTER 6.2
FW HYDROGEN PRODUCTION

James D. Fleshman
Foster Wheeler USA Corporation
Clinton, New Jersey

INTRODUCTION

As hydrogen use has become more widespread in refineries, hydrogen production has moved from the status of a high-tech specialty operation to an integral feature of most refineries. This has been made necessary by the increase in hydrotreating and hydrocracking, including the treatment of progressively heavier feedstocks.

Steam reforming is the dominant method for hydrogen production. This is usually combined with pressure-swing adsorption (PSA) to purify the hydrogen to greater than 99 vol %.

As hydrogen production grows, a better understanding of the capabilities and requirements of the modern hydrogen plant becomes ever more useful to the refiner. This will help the refiner get the most from existing or planned units and make the best use of hydrogen supplies in the refinery.

USES OF HYDROGEN

Overview

There has been a continual increase in refinery hydrogen demand over the last several decades. This is a result of two outside forces acting on the refining industry: environmental regulations and feedstock shortages. These are driving the refining industry to convert from distillation to conversion of petroleum. Changes in product slate, particularly outside the United States, are also important. Refiners are left with an oversupply of heavy, high-sulfur oil, and in order to make lighter, cleaner, and more salable products, they need to add hydrogen or reject carbon.

Within this trend there are many individual factors depending on location, complexity of the refinery, etc.

TABLE 6.2.1 Hydrogen Consumption Data
Chemical consumption only

Process	Wt % on feed	SCF/bbl	Wt % on crude
Hydrotreating:			
Straight-run naphtha	0.05	20	0.01
FCC naphtha/coker naphtha	1	500	0.05–0.01
Kerosene	0.1	50	0.1–0.02
Hydrodesulfurization:			
Low-sulfur gas oil to 0.2% S	0.1	60	0.03
High-sulfur gas oil to 0.2% S	0.3	170	0.04
Low-sulfur gas oil to 0.05% S	0.15	80	0.04
High-sulfur gas oil to 0.05% S	0.35	200	0.05
FCC gas oil/coker gas oil	1	600	0.1
Cycle oil hydrogenation	3	1700	0.3
Hydrocracking vacuum gas oil	2–3	1200–1800	0.5–0.8
Deep atmospheric residue conversion	2–3.5	1200–2200	1–2

Note: FCC = fluid catalytic cracker; SCF = standard cubic feet.
Source: Lambert et al.[8]

Hydrogen Demand

The early use of hydrogen was in naphtha hydrotreating, as feed pretreatment for catalytic reforming (which in turn was producing hydrogen as a by-product). As environmental regulations tightened, the technology matured and heavier streams were hydrotreated. These included light and heavy distillates and even vacuum residue.

Hydrotreating has also been used to saturate olefins and make more stable products. For example, the liquids from a coker generally require hydrotreating, to prevent the formation of polymers.

At the same time that demand for cleaner distillates has increased, the demand for heavy fuel oil has dropped. This has led to wider use of hydrocracking, which causes a further large increase in the demand for hydrogen.

Table 6.2.1 shows approximate hydrogen consumption for hydrotreating or hydrocracking of various feedstocks.

HYDROGEN PRODUCTION

Hydrogen has historically been produced in catalytic reforming, as a by-product of the production of the high-octane aromatic compounds used in gasoline. Changes in this process have had a large impact on the refinery hydrogen balance.

As reforming has changed from fixed-bed to cyclic to continuous regeneration, pressures have dropped and hydrogen production per barrel of reformate has increased. Recent changes in gasoline composition, due to environmental concerns, have tended to reduce hydrogen production, however. Besides limits on aromatics, requirements for oxygenates in gasoline have resulted in reduced reforming severity, as the high-octane oxygenates have displaced reformate from the gasoline pool. The only safe statement is that the situation will continue to change.

Where by-product hydrogen production has not been adequate, hydrogen has been manufactured by steam reforming. In some cases partial oxidation has been used, par-

FIGURE 6.2.1 Modern hydrogen plant.

ticularly where heavy oil is available at low cost. However, oxygen is then required, and the capital cost for the oxygen plant makes partial oxidation high in capital cost.

Figure 6.2.1 shows a typical modern hydrogen plant. This unit produces 82 million SCFD (at 60°F and 14.7-lb/in^2 absolute) [92,000 (N) m^3/h (N represents normal temperature and pressure at 0°C and 1.0332 kg/cm^2 absolute)] of hydrogen from natural gas for a Far Eastern refinery, at a purity of 99.9 vol %. The Foster Wheeler Terrace Wall* steam reforming furnace is visible in the background, with the 12 absorbers and two surge drums of the pressure-swing adsorption unit in the foreground.

Table 6.2.2 shows approximate hydrogen production from various processes.

---

*Registered trademark of Foster Wheeler.

TABLE 6.2.2 Hydrogen Production Data

Process	Wt % on feed	SCF/bbl	Wt % on crude
Continuous regeneration reformer	3.5	1600	0.35–0.60
Semiregenerative reformer	1.4–2.0	600–900	0.15–0.30
Residue gasification	20–25	12,000–16,000	1–5
Catalytic cracking	0.05–0.10	30–60	0.01–0.04
Thermal cracking	0.03	20	0.01
Ethylene cracker	0.5–1.2	—	—
Steam (methane) reformer	30	12,000	—

Source: Lambert et al.[8]

Chemistry

Steam Reforming. In steam reforming, light hydrocarbons such as methane are reacted with steam to form hydrogen:

$$CH_4 + H_2O = 3H_2 + CO$$

$$\Delta H = 97{,}400 \text{ Btu/lb} \cdot \text{mol } (227 \text{ kJ/g} \cdot \text{mol})$$

where ΔH is the heat of reaction. The reaction equation can be generalized to:

$$C_nH_m + (n)H_2O + (n + m/2)H_2 + nCO$$

The reaction is typically carried out at approximately 1500°F (815°C) over a nickel catalyst packed into the tubes of a reforming furnace. Because of the high temperature, hydrocarbons also undergo a complex series of cracking reactions, plus the reaction of carbon with steam. These can be summarized as

$$CH_4 = 2H_2 + C$$
$$C + H_2O = CO + H_2$$

Carbon is produced on the catalyst at the same time that hydrocarbon is reformed to hydrogen and CO. With natural gas or similar feedstock, reforming predominates and the carbon can be removed by reaction with steam as fast as it is formed. When heavier feedstocks are used, the carbon is not removed fast enough and builds up. Carbon can also be formed where the reforming reaction does not keep pace with heat input, and a hot spot is formed.

To avoid carbon buildup, alkali materials, usually some form of potash, are added to the catalyst when heavy feeds are to be used. These promote the carbon-steam reaction and help keep the catalyst clean. The reforming furnace is also designed to produce uniform heat input to the catalyst tubes, to avoid coking from local hot spots.

Even with promoted catalyst, cracking of the feedstock limits the process to hydrocarbons with a boiling point of 350°F (180°C) or less: natural gas, propane, butane, and light naphtha. Heavier hydrocarbons result in coke building up on the catalyst. Prereforming, which uses an adiabatic catalyst bed operating at a lower temperature, can be used as a pretreatment to allow heavier feeds to be used without coking. A prereformer will also make the fired reformer more tolerant of variations in heat input.

After reforming, the CO in the gas is reacted with steam to form additional hydrogen, in the water-gas shift reaction:

$$CO + H_2O = CO_2 + H_2$$

$$\Delta H = -16{,}500 \text{ btu/lb} \cdot \text{mol } (-38.4 \text{ kJ/g} \cdot \text{mol})$$

This leaves a mixture consisting primarily of hydrogen and CO_2. After CO_2 removal, which we shall discuss later, many plants use methanation—the reverse of reforming—to remove the remaining traces of carbon oxides:

$$CO + 3H_2 = CH_4 + H_2O$$
$$CO_2 + 4H_2 = CH_4 + 2H_2O$$

Partial Oxidation. Hydrogen can also be produced by partial oxidation of hydrocarbons:

$$CH_4 + \tfrac{1}{2}O_2 = CO + 2H_2$$

$$\Delta H = -10,195 \text{ Btu/lb} \cdot \text{mol} \ (-23.7 \text{ kJ/g} \cdot \text{mol})$$

The shift reaction also participates so the result is a mixture of CO and CO_2 in addition to H_2. Temperature in partial oxidation is not limited by catalyst tube materials, so higher temperature may be used, which results in reduced methane slippage.

Steam Reforming/Wet Scrubbing

Figure 6.2.2 shows the flow sheet for a wet scrubbing plant, based on steam reforming of natural gas. Plants with similar configurations came into widespread use about 1960, when high-pressure steam reforming became economical. They were built until the mid-1980s, when they were generally supplanted by plants using PSA.

Feedstock at 450 lb/in^2 (31 bar) gage is preheated and purified to remove traces of sulfur and halogens in order to protect the reformer catalyst. The most common impurity is H_2S; this is removed by reaction with ZnO. Organic sulfur may also be present; in this case recycled product hydrogen is mixed with the feed and reacted over a hydrogenation catalyst (generally cobalt/molybdenum) to convert the organic sulfur to H_2S. If chlorides are present they are also hydrogenated and then reacted with a chloride adsorbent.

The feed is then mixed with steam, preheated further, and reacted over nickel catalyst in the tubes of the reformer to produce synthesis gas—an equilibrium mixture of H_2, CO, and CO_2. Steam:carbon ratio is a key parameter, since high steam levels aid in methane conversion. Residual methane in the synthesis gas will pass through the plant unchanged (along with any N_2 in the feed). This will reduce the hydrogen purity so it is important to ensure a low methane leakage. A high steam:carbon ratio and high reforming temperatures are used for this reason. Excess steam is also used to prevent

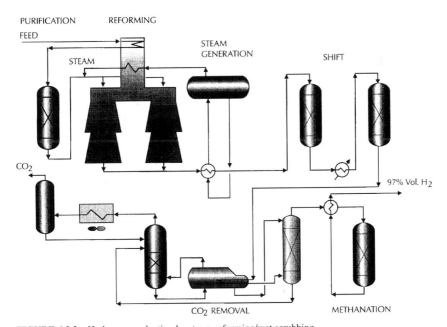

FIGURE 6.2.2 Hydrogen production by steam reforming/wet scrubbing.

coke formation on the catalyst. Typical reformer outlet conditions for hydrogen production are 1500°F and 300 lb/in^2 (815°C and 21 bar) gage.

Much of the design and operation of hydrogen plants involves protecting the reforming catalyst and the catalyst tubes. The extreme temperatures and the sensitivity of the catalyst tend to magnify small upsets. Minor variations in feed composition or operating conditions can have significant effects on the life of the catalyst or the reformer itself. This is particularly true of changes in molecular weight of the feed gas, or poor distribution of heat to the catalyst tubes.

The synthesis gas passes through the reformer waste heat exchanger, which cools the gas and generates steam for use as process steam in the reformer: the surplus is exported. The cooled gas [still at about 650°F (345°C)] is reacted over a fixed bed of iron oxide catalyst in the high-temperature shift converter, where the bulk of the CO is reacted, then cooled again and reacted over a bed of copper zinc low-temperature shift catalyst to convert additional CO.

The raw hydrogen stream is next scrubbed with a solution of a weak base to remove CO_2. The flow scheme in Fig. 6.2.2 is based on use of a potassium carbonate solution in water to react with the CO_2; a similar process uses an ethanolamine solution. CO_2 in the gas reacts reversibly with potassium carbonate to form potassium bicarbonate. The solution is depressed and steam-stripped to release CO_2, with the heat for the regenerator reboiler coming from the hot synthesis gas. The regenerator overhead stream is then cooled to condense water. The CO_2 is available for recovery or can be vented.

The raw hydrogen leaving the CO_2 removal section still contains approximately 0.5 percent CO and 0.1 percent CO_2 by volume. These will act as catalyst poisons to most hydrogen consumers, so they must be removed, down to very low levels. This is done by methanation, the reverse of reforming. As in reforming, a nickel catalyst is used, but as a fixed bed.

Typical final hydrogen purity is 97 vol %, with the remaining impurities consisting mainly of methane and nitrogen. Carbon oxide content is less than 50 vol ppm.

Product hydrogen is delivered from the methanator at approximately 250 lb/in^2 (17 bar) gage, and must generally be compressed before final use. This is done in a reciprocating compressor. Centrifugal compressors are not feasible because of the low molecular weight; the pressure rise per foot of head is too low, and too many stages would be required.

Steam Reforming/PSA

Plants built after the mid-1980s were generally based on pressure-swing adsorption. PSA is a cyclic process which uses beds of solid adsorbent to remove impurities from the gas. The purified hydrogen passes through the adsorbent beds with only a tiny fraction absorbed. The beds are regenerated by depressurization, followed by purging at low pressure.

When the beds are depressed, a waste gas (or "tail gas") stream is produced, consisting of the impurities from the feed (CO_2, CO, CH_4, N_2), plus some hydrogen. This stream is burned in the reformer as fuel. Reformer operating conditions in a PSA-based plant are set so that the tail gas provides no more than about 85 percent of the reformer fuel. This limit is important for good burner control because the tail gas is more difficult to burn than regular fuel gas. The high CO_2 content can make it difficult to produce a stable flame.

As the reformer operating temperature is increased, the reforming equilibrium shifts, resulting in more hydrogen and less methane in the reformer outlet and hence less methane in the tail gas. Actual operating conditions can be further optimized according to the relative cost of feed, fuel, and export steam.

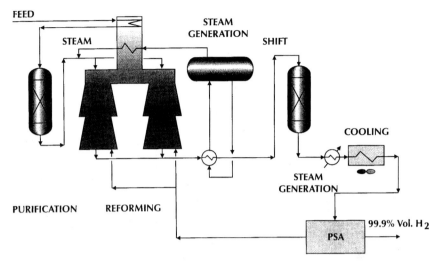

FIGURE 6.2.3 Hydrogen production by steam reforming/PSA.

The flow sheet for a typical PSA-based hydrogen plant is shown in Fig. 6.2.3. As in the wet scrubbing process, the feed is purified and reformed, followed by shift conversion. Only a single stage of shift conversion is used, since a very low CO residual is not required. Any CO remaining in the raw hydrogen will be removed and recovered as reformer fuel. After cooling, the gas is purified in the PSA unit.

The PSA unit is simpler to operate than a wet scrubbing system, since it has no rotating equipment or circulating solutions. In addition, the adsorbent will remove methane and nitrogen, which could not be removed by the wet scrubbing process. Typical hydrogen recoveries in a PSA unit are in the 80 to 90 percent range, with product purity generally 99.9 vol %.

Because of the loss of hydrogen to the PSA tail gas, the reformer and front end of a PSA plant are larger than in a wet scrubbing plant. A PSA plant uses less process steam and does not require heat for the reboiler; this leaves additional steam available for export. Capital cost for the two schemes is generally very similar. The additional export steam can provide a strong utility cost advantage for the PSA plant in addition to its purity and operability advantages.

Product Properties

Hydrogen purity depends primarily on the purification method. This is illustrated in Table 6.2.3. In wet scrubbing, the major impurities are methane and nitrogen. Methane

TABLE 6.2.3 Composition of Product Hydrogen

	Wet scrubbing	PSA
Hydrogen purity, vol %	95–97	99–99.99
Methane	2–4 vol %	100 vol ppm
$CO + CO_2$, vol ppm	10–50	10–50
Nitrogen, vol %	0–2	0.1–1.0

TABLE 6.2.4 Impurities—Ease of Removal by PSA

Easy	Moderate	Difficult	Not removed
C_3H_6	CO	O_2	H_2
C_4H_{10}	CH_4	N_2	He
C_5+	CO_2	Ar	
H_2S	C_2H_6		
NH_3	C_3H_8		
BTX	C_2H_4		
H_2O			

Note: BTX = benzene, toluene, and xylenes.
Source: Miller and Stoecker.[5]

in the product is the residual left after reforming, or is formed in the methanator from residual CO or CO_2. Nitrogen in the feed is carried through the plant unchanged, although there is a dilution effect because of the larger volume of hydrogen compared to the feedstock.

In a PSA plant, most impurities can be removed to any desired level. Table 6.2.4 shows the difficulty of removal of impurities. Removal of a more difficult impurity will generally ensure virtually complete removal of easier impurities. Nitrogen is the most difficult to remove of the common impurities, and removing it completely requires additional adsorbent. Since it acts mainly as a diluent, it is usually left in the product. The exception is where the hydrogen is to be used in a very high pressure system such as a hydrocracker. In that case the extra cost for nitrogen removal is justified by the savings in hydrogen purge losses.

In the case of a nitrogen-free feedstock such as liquid petroleum gas (LPG) or naphtha, a purity of 99.99 percent can be readily achieved. In this case, carbon monoxide is the usual limiting component. Since CO must be removed to ppm levels, the other impurities—CO_2 and H_2O—are removed to virtually undetectable levels. A typical residual of about 100 ppm CH_4 remains because of inefficiencies in the purge system.

Operating Variables

Operating Conditions. The critical variables for steam reforming are temperature, pressure, and steam:hydrocarbon ratio. Picking the operating conditions for a particular plant involves an economic tradeoff among these three factors.

Steam reforming is an equilibrium reaction, and conversion of the hydrocarbon feedstock is favored by high temperature, which in turn carries a fuel penalty. Because of the volume increase in the reaction, conversion is also favored by low pressure, which conflicts with the need to supply the hydrogen at high pressure. In practice, temperature and pressure are limited by the tube materials.

Table 6.2.5 shows the effect of changes in temperature, pressure, and steam:carbon ratio. The degree of conversion is measured by the remaining methane in the reformer outlet, known as the *methane leakage.*

Shift Conversion. In contrast to reforming, shift conversion is favored by low temperature. The gas from the reformer is reacted over iron oxide catalyst at 600 to 700°F (315–370°C), with the limit set by the low-temperature activity of the catalyst. In wet scrubbing plants using a methanator, it is necessary to remove CO to a much lower

TABLE 6.2.5 Effect of Operating Variables on the Reformer

Temperature		Absolute pressure		Steam:carbon ratio	CH_4 in outlet, mol % (dry basis)
°F	°C	lb/in²	bar		
1500	815	350	24	3.0	8.41
1550	845	350	24	3.0	6.17
1600	870	350	24	3.0	4.37
1550	845	300	21	3.0	5.19
1550	845	400	28	3.0	7.09
1550	845	350	24	2.5	8.06
1550	845	350	24	3.5	4.78

TABLE 6.2.6 Effect of Process Variables on Shift Conversion

HTS inlet temperature		LTS inlet temperature		Reformer steam: carbon ratio	CO in outlet, mol % dry basis
°F	°C	°F	°C		
600	315	—	—	3.0	2.95
700	370	—	—	3.0	4.07
600	315	—	—	5.0	1.53
700	370	—	—	5.0	2.33
600	315	400	205	3.0	0.43
600	315	500	260	3.0	0.94
700	370	400	205	3.0	0.49
700	370	500	260	3.0	1.04
600	315	400	205	5.0	0.19
600	315	500	260	5.0	0.46
700	370	400	205	5.0	0.21
700	370	500	260	5.0	0.50

level to avoid excessive temperatures in the methanator. In those plants the gas is cooled again and reacted further over a copper-based catalyst at 400 to 500°F (205 to 260°C). Table 6.2.6 shows the effect of temperature and steam:carbon ratio on the CO remaining after shift conversion.

Alternative Processes

Partial Oxidation. Partial oxidation (POX) reacts hydrocarbon feed with oxygen at high temperatures to produce a mixture of hydrogen and carbon monoxide. Since the high temperature takes the place of a catalyst, POX is not limited to the light, clean feedstocks required for steam reforming. Partial oxidation is high in capital cost, and for light feeds it has been generally replaced by steam reforming. However, for heavier feedstocks it remains the only feasible method.

In the past, POX was considered for hydrogen production because of expected shortages of light feeds. It can also be attractive as a disposal method for heavy, high-sulfur streams, such as asphalt or petroleum coke, which sometimes are difficult to dispose of.

Consuming all of a refinery's asphalt or coke by POX would produce more hydrogen than is likely to be required. Because of this, and the economies of scale

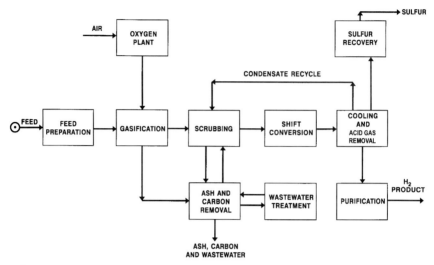

FIGURE 6.2.4 Hydrogen production by partial oxidation.

TABLE 6.2.7 Asphalt Composition—Partial Oxidation Feedstock

Density at 15°C	1.169 kg/L
Carbon	85.05 wt %
Hydrogen	8.10 wt %
Nitrogen	0.80 wt %
Sulfur	6.00 wt %
Ash	0.05 wt %
V	600 ppm
Ni	200 ppm

required to make POX economic, hydrogen may be more attractive if produced as a by-product, with electricity as the primary product.

Figure 6.2.4 is a block flow diagram of a unit to produce electricity from asphalt, with hydrogen as a by-product. Besides being high in carbon, the asphalt contains large amounts of sulfur, nitrogen, nickel, and vanadium (Table 6.2.7). Much of the cost of the plant is associated with dealing with these components.

The asphalt is first gasified with oxygen in an empty refractory-lined chamber to produce a mixture of CO, CO_2, and H_2. Because of the high temperature, methane production is minimal. Gas leaving the gasifier is first quenched in water to remove solids, which include metals (as ash) and soot. Metals are removed by settling and filtration, and the soot is recycled to the gasifier. The gas is further cooled and H_2S is removed by scrubbing with a selective solvent. Sulfur removal is complicated by the fact that a significant amount of carbonyl sulfide (COS) is formed in the gasifier. This must be hydrolyzed to H_2S, or a solvent that can remove COS must be used.

Hydrogen processing in this system depends on how much of the gas is to be recovered as hydrogen, and how much is to be used as fuel. Where hydrogen production is a relatively small part of the total gas stream, a membrane unit can be used to withdraw a hydrogen-rich stream. This is then purified in a PSA unit. In the case where maximum hydrogen is required, the entire gas stream may be shifted to convert CO to H_2, and a PSA unit used on the total stream.

Catalytic Partial Oxidation. Also known as *autothermal reforming,* catalytic partial oxidation reacts oxygen with a light feedstock, passing the resulting hot mixture over a reforming catalyst. Since a catalyst is used, temperatures can be lower than in non-catalytic partial oxidation, which reduces the oxygen demand.

Feedstock requirements are similar to steam reforming: light hydrocarbons from refinery gas to naphtha may be used. The oxygen substitutes for much of the steam in preventing coking, so a lower steam:carbon ratio can be used. Since a large excess of steam is not required, catalytic POX produces more CO and less hydrogen than steam reforming. Because of this it is suited to processes where CO is desired, for example, as synthesis gas for chemical feedstocks. Partial oxidation requires an oxygen plant, which increases costs. In hydrogen plants, it is therefore used mainly in special cases such as debottlenecking steam reforming plants, or where oxygen is already available on site.

By-Product Recovery

Carbon dioxide and steam are the major by-products from hydrogen manufacture.

Carbon Dioxide. Where there is a market for CO_2, recovery can be very attractive. Historically the major use has been in the food industry, with recent growth being for injection in enhanced oil recovery. A substantial amount of CO_2 is available from hydrogen plants: a plant making 10 million SCFD [11,000 (N) m³/h] of hydrogen from natural gas vents 2.5 million SCFD or 145 tons (132 tonnes) per day of CO_2.

Recovery of CO_2 is easiest in older plants using wet scrubbing. These produce a concentrated CO_2 stream which needs only final purification to remove traces of H_2, CO, and CH_4, followed by compression.

More recent plants, using PSA, can use a vacuum-swing adsorption (VSA) system for CO_2 recovery (Fig. 6.2.5). Tail gas from the PSA system is compressed and fed to

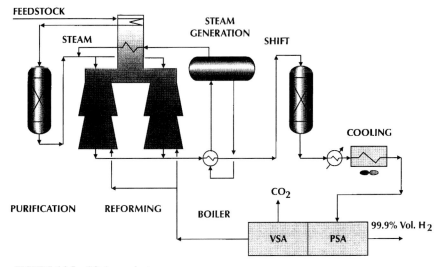

FIGURE 6.2.5 CO_2 by-product recovery.

the VSA system, which uses a separate set of adsorber vessels. By using vacuum regeneration, the system can split the tail gas into a CO_2 product stream, a hydrogen-rich stream which is recycled to the reformer, and a nitrogen-rich reject stream.

Besides recovering CO_2, the VSA system increases overall hydrogen production by recovering hydrogen which would otherwise have been lost in the tail gas.

A wet scrubbing system can also be installed upstream of a PSA unit to recover CO_2. This can also be used in a revamp to increase capacity by reducing the load on the PSA system.

Steam. Most hydrogen plants generate steam, mainly for use as process steam with the excess available for export. A typical 50 million SCFD [56,000 (N) m^3/h] unit based on PSA will export between 70,000 and 160,000 lb/h (between 30 and 70 tonnes/h), depending on configuration. Plants with air preheat are at the lower end of the steam production range, while steam export can be further increased by adding auxiliary burners between the radiant and convection sections.

Catalysts

Hydrogen plants are one of the most extensive users of catalysts in the refinery. Catalytic operations include hydrogenation, steam reforming, shift conversion, and methanation. Sulfur and halogen removal are actually done by reaction with solid adsorbents, but they are included here for completeness.

Reforming. Because of the high temperatures and heat load of the reforming reaction, reforming catalyst is used inside the radiant tubes of a reforming furnace. The catalyst is subject to severe operating conditions: up to 1600°F (870°C), with typical pressure drops of 40 lb/in^2 (2.8 bar). In order to withstand these conditions, the carrier is generally an alumina ceramic, although some older formulations use calcium aluminate.

The active agent in reforming catalyst is nickel, and normally the reaction is controlled by both diffusion and heat transfer. The catalyst is therefore made in rings to provide increased mass and heat transfer at minimum pressure drop. To further increase heat transfer, most catalyst vendors now offer specially shaped catalyst.

Even with a high-strength carrier, catalyst life is limited as much by physical breakdown as by deactivation. Thermal cycling is especially hard on the catalyst; when the tubes are heated they expand and the catalyst tends to settle in the tube, then when the tube cools and the tube contracts the catalyst is crushed. This can cause voids to form in the tubes, leading to hot spots and ultimately to ruptured tubes.

The main poisons are sulfur and chlorides, which are present in small quantities in most feedstocks. Sulfur poisoning is theoretically reversible, and the catalyst can often be restored to near full activity by steaming. However, in practice the deactivation may cause the catalyst to overheat and coke, to the point that it must be replaced.

Chlorides are an irreversible poison: the chlorine combines with the nickel to form nickel chloride, which is volatile. The nickel migrates and recrystallizes, reducing the catalyst activity.

The catalyst is also sensitive to poisoning by heavy metals and arsenic, although these are rarely present in feedstocks.

The catalyst is supplied as nickel oxide. During start-up the catalyst is heated in a stream of inert gas, then steam. When the catalyst is near the normal operating temperature, hydrogen or a light hydrocarbon is added to reduce the nickel oxide to metallic nickel. Steaming the catalyst will oxidize the nickel, but most catalysts can readily be rereduced.

Shift Conversion. The second important reaction in a steam reforming plant is the shift conversion reaction:

$$CO + H_2O = CO_2 + H_2$$

The equilibrium is dependent on temperature, with low temperatures favoring high conversions.

Two basic types of shift catalyst are used in steam reforming plants: iron/chrome high temperature shift catalysts, and copper/zinc low temperature shift catalysts.

High-Temperature Shift. High-temperature shift catalyst operates in the range of 600 to 800°F (315 to 430°C). It consists primarily of magnetite, Fe_3O_4, with chrome oxide, Cr_2O_3 added as a stabilizer. The catalyst is supplied in the form of Fe_2O_3 and CrO_3, and must be reduced. This can be done by the hydrogen and carbon monoxide in the shift feed gas, and occurs naturally as part of the start-up procedure.

If the steam:carbon ratio of the feed is too low, the reducing environment is too strong and the catalyst can be reduced further, to metallic iron. This is a problem, since metallic iron will catalyze Fischer-Tropsch reactions and form hydrocarbons. In older wet scrubbing plants this was rarely a problem, since the steam:carbon ratio of the process gas was in the range of 5 to 6, too high for iron formation. In some newer plants with steam:carbon ratios below 3, the shift catalyst is slowly converted to iron, with the result that significant amounts of hydrocarbons are formed over the high-temperature shift catalyst.

To slow down (but not eliminate) overreduction, the catalyst can be doped with copper, which acts by accelerating the conversion of CO. It increases activity at lower temperatures, but also makes the catalyst sensitive to poisoning by sulfur and chlorides.

High-temperature shift catalyst is very durable. In its basic form it is not sensitive to most poisons, and has high mechanical strength. It is subject to thermal sintering, however, and once it has operated at a particular temperature, it loses its activity at lower temperatures.

Low-Temperature Shift. Low-temperature (LT) shift catalyst operates with a typical inlet temperature of 400 to 450°F (205 to 230°C). Because of the lower temperature, the reaction equilibrium is better and outlet CO is lower.

Low-temperature shift catalyst is economic primarily in wet scrubbing plants, which use a methanator for final purification. The main advantage of the additional conversion is not the extra hydrogen that is produced, but the lower CO residual. This reduces the temperature rise (and hydrogen loss) across the methanator.

PSA-based plants generally do not use LT shift, since any unconverted CO will be recovered as reformer fuel. Since an LT shift increases hydrogen production for a fixed reformer size, it may be used in revamps to increase production.

Low-temperature shift catalyst is sensitive to poisoning by sulfur and chlorides. It is also mechanically fragile and sensitive to liquid water, which can cause softening of the catalyst followed by crusting or plugging.

The catalyst is supplied as copper oxide on a zinc oxide carrier, and the copper must be reduced by heating it in a stream of inert gas with measured quantities of hydrogen. Reduction is strongly exothermic and must be closely monitored.

Methanation. In wet scrubbing plants, final hydrogen purification is by methanation, which converts CO and CO_2 to CH_4. The active agent is nickel, on an alumina carrier.

The catalyst has a long life, as it operates under clean conditions and is not exposed to poisons. The main source of deactivation is plugging from carryover of CO_2 removal solutions.

The most severe hazard is overtemperature, from high levels of CO or CO_2. This can result from breakdown of the CO_2 removal equipment or from exchanger tube leaks which quench the shift reaction. The results of breakthrough can be severe, since the methanation reaction produces a temperature rise of 125°F per 1 percent of CO, or 60°F per 1 percent of CO_2. While the normal operating temperature in a methanator is approximately 600°F (315°C), it is possible to reach 1300°F (700°C) in cases of major breakthrough.

Feed Purification. Long catalyst life in modern hydrogen plants is attributable to a great extent to effective feed purification, particularly sulfur and chloride removal. A typical natural gas or other light hydrocarbon feedstock contains traces of H_2S and organic sulfur. Refinery gas may contain organic chlorides from a catalytic reforming unit.

In order to remove these it is necessary to hydrogenate the feed to convert the organic sulfur to H_2S, which is then reacted with zinc oxide: organic chlorides are converted to HCl and reacted with an alkali metal adsorbent. Purification is done at approximately 700°F (370°C), since this results in best use of the zinc oxide, as well as ensuring complete hydrogenation.

Coking of Reforming Catalyst. Coking of the reformer catalyst is the most characteristic problem in a hydrogen plant. While it may be similar in appearance to the fouling of heater tubes found in other units, additional precautions are necessary here. A major reason for the high reliability of modern units is the reduction in catalyst coking. This is due to advances in catalyst technology and in reformer design.

While light, methane-rich streams such as natural gas or light refinery gas are the most common feeds to hydrogen plants, there is often a requirement to process a variety of heavier feedstocks, including LPG and naphtha, because of seasonal variations in feedstock price, an interruptible supply of natural gas, or turnarounds in a gas-producing unit. Feedstock variations may also be inadvertent, for example, changes in refinery offgas composition from another unit.

When using heavier feedstocks in a hydrogen plant, the primary concern is coking of the reformer catalyst. There will also generally be a small capacity reduction due to the additional carbon in the feedstock and additional steam required. This increases the load on the shift and CO_2 removal section of the plant. The size of this effect will depend on the feedstocks used and on the actual plant. Coking, however, is of more immediate concern, since it can prevent the plant from operating.

Coking is most likely about one-third the way down the tube, where both temperature and hydrocarbon content are high enough. In this region, hydrocarbons can crack and polymerize faster than the coke is removed by reaction with steam or hydrogen. Once the catalyst is deactivated, temperature increases further and coking accelerates. Farther down the tube, where the hydrocarbon-to-hydrogen (HC/H_2) ratio is lower, there is less risk of coking. Coking depends to a large extent on the balance between catalyst activity and heat input; more active catalyst produces more hydrogen at lower temperature, reducing the risk of coking. Uniform heat input is especially important in this region of the catalyst tube, since any catalyst voids or variations in catalyst activity can produce localized hot spots, leading to coke formation or tube failure.

Coke formation results in hot spots in catalyst tubes, and can produce characteristic patterns known as *giraffe necking* or *tiger tailing*. It increases pressure drop, reduces conversion of methane, and can cause tube failure. Coking may be partially alleviated by increasing the steam:hydrocarbon ratio to change the reaction conditions, but the most effective solution is to replace the reformer catalyst with one designed for heavier feeds.

In addition to the reforming and shift reactions over reforming catalyst, a number of side reactions occur. Most of these include the production or removal of carbon. Carbon is continuously formed on the catalyst, but ordinarily reacts with steam faster than it can build up. Heavier feeds produce more carbon. Unless the process conditions or the catalyst is changed, the carbon can accumulate.

Standard methane reforming catalyst uses nickel on an alpha-alumina ceramic carrier. The alumina is acidic, which promotes hydrocarbon cracking and can form coke with heavier feeds. Some catalyst formulations use a magnesia/alumina spinel which is more neutral than alpha-alumina. This reduces cracking on the carrier, and allows somewhat heavier feedstocks to be used: typically into the LPG range. The drawbacks to this approach include difficulty in reducing the catalyst unless there is a supply of hydrogen in the reducing gas, and the possible damage to the catalyst by hydration of the catalyst during start-up.

Further resistance to coking can be achieved by adding an alkali promoter, typically some form of potash (KOH), to the catalyst. Besides reducing the acidity of the carrier, the promoter catalyzes the reaction of steam and carbon. While carbon continues to be formed, it is removed faster than it can build up. This approach can be used with naphtha feedstocks up to a boiling point of 350°F (about 180°C).

Under the conditions in a fired reformer, potash is volatile, and it is incorporated into the catalyst as a more complex compound which slowly hydrolyzes to release KOH. The promoted catalyst is used only in the top half of the catalyst tubes, since this is where the hydrocarbon content, and the possibility of coking, is the highest. In addition, this keeps the potash out of the hottest part of the tube, reducing potash migration.

Alkalized catalyst allows the use of a wide range of feedstocks, but it does have drawbacks. In addition to possible potash migration, which can be minimized by proper design and operation, the catalyst is also somewhat less active than conventional catalyst.

Prereforming. Another option to reduce coking in steam reformers is to use a prereformer. This uses a fixed bed of very active catalyst, operating at a lower temperature, upstream of the fired reformer (Fig. 6.2.6). Inlet temperatures are selected so that there is minimal risk of coking. Gas leaving the prereformer contains only steam, hydrogen, carbon oxides, and methane. This allows a standard methane catalyst to be used in the fired reformer. This approach has been used with feedstocks up to light kerosene. The drawback to this approach is the need for a separate prereformer reactor and a more complicated preheat train.

Since the gas leaving the prereformer poses reduced risk of coking, this also makes the fired reformer more "forgiving." It can compensate to some extent for variations in catalyst activity and heat flux in the primary reformer.

Besides its use for feedstock flexibility, a prereformer can be used to reduce the fuel consumption and steam production of the reformer. Since the prereformer outlet gas does not contain heavier hydrocarbons, it can be reheated to a higher temperature than the original feedstock without the risk of carbon formation. The higher preheat temperature reduces the radiant duty and fuel consumption, as well as steam production.

Reformer Design

Equipment Configuration. Designs for steam reforming furnaces must deal with the problems caused by the extremely high process temperatures. These include thermal expansion, cracking, and overheating. The high temperatures also mean the use of exotic alloys; as an example, a common tube material is HP-45, which contains 25 percent chrome and 35 percent nickel, with the element niobium added to stabilize the grain structure.

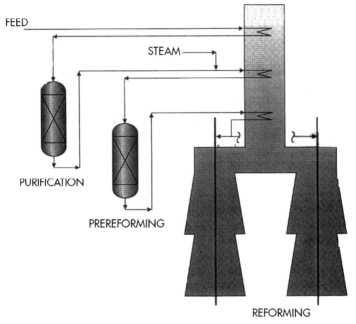

FIGURE 6.2.6 Prereformer flow scheme.

Tube expansion at reforming temperatures is approximately 10 in (250 mm) for a typical 40-ft (12-m) tube. This expansion is taken up at the cold end of the tube by connecting the tubes to the inlet header with long, relatively flexible tubes called pigtails. A counterweight system is used to support the tube and ensure that the tube is kept in constant tension to prevent bowing.

The combination of light components and good thermal conductivity results in a relatively low danger of thermal cracking compared to heavier hydrocarbons. This means that high flux rates, typically above 20,000 Btu/h · ft^2 (63,000 W/m^2), may be used. This in turn requires that heat flux be very uniform to avoid hot spots. In larger furnaces, firing is from both sides of the tube, and measures are taken to ensure that heat flux is relatively uniform over the length of the tube. This may be done by using a radiant wall design such as a terrace wall unit, or positioning the flame next to the coldest part of the tube in down-fired units.

Since catalyst is packed into the tubes, many multiple passes are used to keep pressure drop to a manageable level. There are several hundred parallel passes in a large furnace. Careful packing of the catalyst into the tubes ensures even flow distribution.

Several reformer configurations have evolved to deal with these factors: Terrace Wall, side-fired, down-fired, and bottom-fired furnaces are used. These designs are summarized below.

Terrace Wall. Foster Wheeler's Terrace Wall reformer was developed to handle the high temperatures and high heat fluxes used in steam reforming. This design uses a long, relatively narrow firebox, with the tubes in a single row down the center (Figs. 6.2.7 to 6.2.9). The burners are located in terraces along the sides, and fire upward against sloping, refractory lined walls. Generally two terraces are used. The hot refractory then radiates heat to the tubes, resulting in a very uniform, controlled heat distrib-

FIGURE 6.2.7 This Terrace Wall reformer located in a North American refinery produces over 120 million SCFD of hydrogen from natural gas or light refinery gas.

ution. This helps to avoid localized overheating and carbon laydown. The process gas flow is downward and the flue gas flow upward. The convection section is located above the radiant section. Larger furnaces often use two radiant cells located side by side, and sharing a common convection section.

The radiant wall design provides uniform heat flux and is resistant to localized overheating, even in the event of catalyst coking. The vertical stacking of the furnace, with the convection section located above the radiant section, results in smaller plot area for most sizes.

The updraft arrangement minimizes power required for fans, and the furnace can be designed to operate in natural draft, without fans.

Side-Fired. This design is similar to the terrace wall furnace, with burners located at multiple levels (often six levels). Special burners are used to direct the flames back against the walls. It is possible to get additional control of firing from the larger number of burners.

Down-Fired. This design uses burners located on the roof of the furnace, firing downward (Fig. 6.2.10). Multiple rows of tubes are used, alternating with rows of burners. Special burners are used to ensure the proper flame pattern. This is required in order to get good heat distribution along the length of the tube. Both process gas and flue gas flow is downward.

The multiple rows allow reduced cost for extremely large units, as is required in large methanol or ammonia plants. The convection section is located at grade; this allows good fan access and more stable fan mounting but increases the plot area required. Fewer (but larger) burners are required.

Cylindrical. The furnace is in the shape of a vertical cylinder, with burners located in the center of the floor. The tubes are arranged in a ring around the burners. Ample spacing between the tubes allows radiation to be reflected from the furnace walls and reach the backs of the tubes, in order to provide good heat distribution. Both process gas and flue gas flow is upward.

This design is used for smaller units, with an upper plant size limit of 5 to 10 million SCFD [5500 to 11,000 (N) m^3/h] of hydrogen with a single reformer. Since the hot end of the tubes is at the top, the tubes can be anchored at the top and expand downward. The counterweights or spring hangers used on larger units are not necessary, reducing the cost of the furnace. These units are generally shop-fabricated. Size is therefore limited by shipping restrictions.

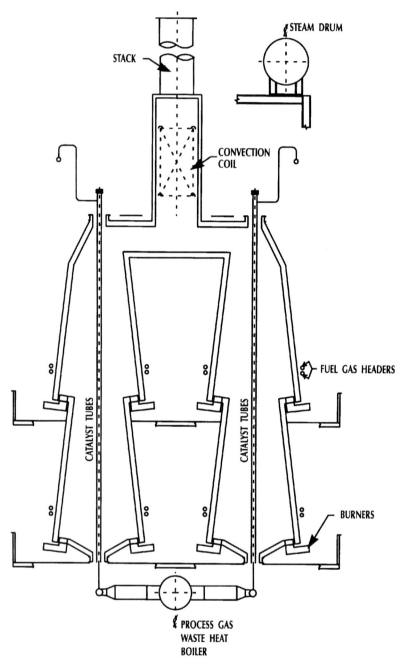

FIGURE 6.2.8 Natural draft reformer.

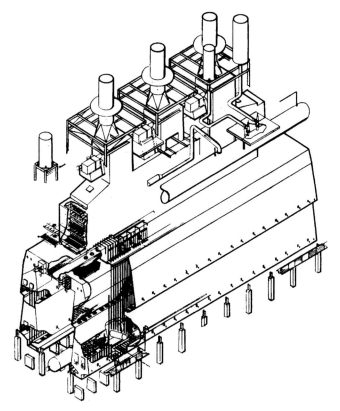

FIGURE 6.2.9 Terrace Wall reformer.

Plant Operation

Several operations are characteristic of hydrogen plants. They include loading catalyst into the reformer tubes, measuring tube metal temperatures, and pinching off of catalyst tubes.

Catalyst Loading. The goal of catalyst loading is to fill the 40-ft-long (12-m) tubes completely, without voids and without fracturing any of the catalyst rings. Early reformers were loaded by filling the tubes with water and dumping the catalyst in. This was discontinued after it was found that on start-up water trapped inside the catalyst turned to steam and fractured the rings.

Traditionally, loading has been done by first loading the catalyst into cloth tubes known as socks, then lowering the socks into the tubes. By manipulating the rope, the catalyst is dumped, falling only a few inches. This is a slow process, requiring vibration of the tubes to eliminate voids and careful measurement of the tube pressure drop and volume loaded into each tube to ensure consistency.

Catalyst Tube Temperature Measurement. As hydrogen plant technology has matured, competitive pressures have made it necessary to operate plants closer to their

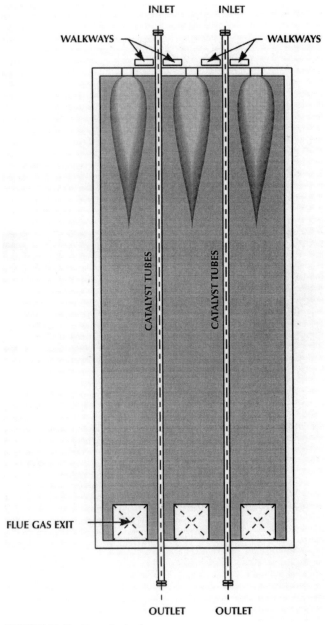

FIGURE 6.2.10 Down-fired reformer.

limits, including temperature limits on the catalyst tubes. To avoid tube failures, many plants now monitor tube metal temperatures each day, or even each shift. Optical (actually infrared) pyrometers are used to measure temperatures, since thermocouples do not survive the 1700 to 1800°F (930 to 980°C) conditions.

Besides measuring metal temperatures, it is important to identify variations in temperature which may indicate catalyst problems. Catalyst deactivation will raise the tube temperature, as it becomes necessary to fire harder to reach the same conversion. Poisoning also often causes variations in catalyst activity, leading to hot spots on the tubes, and distinctive patterns, known as tiger tailing or giraffe necking. Catalyst breakup from thermal cycling can also cause similar patterns, as well as hot tubes from plugging.

Whether one is measuring temperature or identifying patterns, accurate readings require a clear view of the catalyst tubes, preferably from a direction perpendicular to the metal surface. The Terrace Wall or side-fired furnaces provide an advantage in this case, since most furnaces include viewing ports to allow measurement of temperature on virtually all tubes. The multiple tube rows common in down-fired furnaces require viewing from the end of the tube rows, making accurate measurement difficult.

Tube Damage and Pinching. The life of the catalyst tubes depends to a large extent on the condition of the catalyst, which in turn is subject to damage by poisoning or mechanical stress. Poisoning is possible from sulfur or chlorides, either in the feedstock or low-quality steam, while mechanical stress is usually from thermal cycling. The metal tubes have a higher coefficient of thermal expansion than the catalyst. As the tubes heat up they expand and the catalyst settles farther down the tube. When the tube cools, it contracts and the catalyst is fractured. After a number of cycles, the catalyst can break up, plugging the tube or forming voids.

Breakup from thermal cycling can be aggravated by high pressure drop in the catalyst tubes. A smaller tube diameter can reduce furnace cost, since catalyst volume and tube weights are reduced for a given heat input. However, pressure drop is increased at smaller diameters, leading to more stress on the catalyst. During process upsets it becomes easier to exceed the crush strength of the catalyst, and the catalyst fractures.

As the condition of the catalyst worsens, hot spots can develop in tubes and the tube can rupture. Shutting down the furnace at this point to repair the tubes would lead to lost production, as well as extra heating/cooling cycles. Individual tubes can be isolated on line to seal them off and continue operation. This is done by pinching the pigtails shut with a hydraulic clamp while the unit is operating. Many operators shut off hydrocarbon feed while this is done, keeping steam flowing to the tubes.

The tubes themselves are also subject to damage from thermal cycling. As the tubes heat up, the outside and hotter part of the tube wall expands more than the inner portion, leading to high stress levels. The metal will creep in operation, normalizing the stress. The process is then reversed when the tube cools. Continued cycling can lead to cracks.

INTEGRATION INTO THE MODERN REFINERY

Purification

A wide variety of processes are used to purify hydrogen streams. Since the streams are available at a wide variety of compositions, flows, and pressures, the best method of purification will vary.

Factors which must be considered in selecting a purification method are

- Cost (investment and operating)
- Hydrogen recovery
- Product purity
- Pressure profile
- Turndown
- Proven reliability

Wet Scrubbing. Wet scrubbing systems, particularly amine or potassium carbonate systems, are used for removal of acid gases such as H_2S or CO_2. Most depend on chemical reaction and can be designed for a wide range of pressures and capacities. They were once widely used to remove CO_2 in steam reforming plants, but have generally been replaced by PSA units except where CO_2 is to be recovered. They are still used to remove H_2S and CO_2 in partial oxidation plants.

Wet scrubbing systems remove only acid gases or heavy hydrocarbons, but not methane or other light gases, hence have little influence on product purity. Therefore, wet scrubbing systems are most often used as a pretreatment step, or where a hydrogen-rich stream is to be desulfurized for use as fuel gas.

PSA. Pressure-swing adsorption uses beds of solid adsorbent to separate impure hydrogen streams into a very pure high-pressure product stream and a low-pressure tail gas stream containing the impurities and some of the hydrogen. The beds are then regenerated by depressuring and purging (Fig. 6.2.11 and 6.2.12). Part of the hydrogen—typically 10 to 20 percent—is lost into the tail gas.

The cost of the system is relatively insensitive to capacity. This makes PSA more economic at larger capacities, while membrane units tend to be favored for smaller plants.

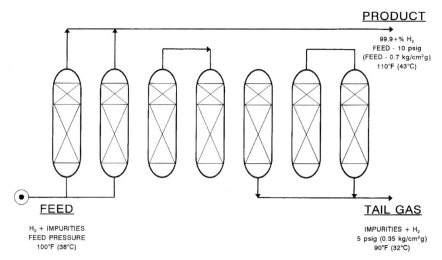

FIGURE 6.2.11 This figure illustrates the flow through a PSA unit for the different steps in the cycle. In the first step impure hydrogen enters the bottom of the bed, with pure hydrogen leaving the top. In the next step pure hydrogen is recovered as the bed is partially depressurized into another bed at lower pressure. These "equalizations" are a key to the high recovery of hydrogen in modern PSA units. The bed is then vented to the tail gas system and purged with pure hydrogen from another bed.

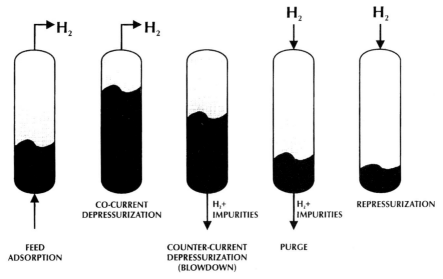

FIGURE 6.2.12 PSA process steps.

PSA is generally the first choice for steam reforming plants because of its combination of high purity, moderate cost, and ease of integration into the hydrogen plant. It is also often used for purification of refinery offgases, where it competes with membrane systems.

Turndown is simple to about 30 percent of flow, where it is limited by the accuracy of flow measurement. Systems can be designed to go somewhat lower by adding low-range transmitters. Reliability is very high.

It is not generally economic to design a PSA unit to process synthesis gas from both steam reforming and hydrogen/hydrocarbon gas. Doing so causes problems with both the fuel balance and the adsorbents. Tail gas from the steam reforming unit consists largely of CO_2 and is returned at low pressure to the reformer furnace as fuel. The plant fuel balance requires that the tail gas from the hydrocarbon PSA be compressed into the fuel system. Combining these two units would result in too much fuel gas to supply the reformer furnace, with much of the CO_2 from the synthesis gas being compressed into the refinery fuel system. In addition, the adsorbents for the two systems are different, and combining them would affect the hydrogen recovery.

Membranes. Membrane units separate gases by taking advantage of the difference in rates of diffusion through membranes. Gases which diffuse faster (including hydrogen) become the permeate stream, and are available at low pressure. The slower gases become the nonpermeate and leave the unit at close to feed pressure.

Membrane units contain no moving parts or switch valves and have potentially very high reliability. The major threat is from components in the gas (such as aromatics) which attack the membranes, or from liquids, which plug them.

Membranes are fabricated in relatively small modules; for larger capacity more modules are added. Cost is therefore virtually linear with capacity, making them more competitive at lower capacities.

Design of membrane systems involves a tradeoff between pressure drop (or diffusion rate) and surface area, as well as between product purity and recovery. As the surface area is increased, the recovery of fast components increases; however more of the slow components are recovered, which lowers the purity. Operating them at turn-down changes the relationship between diffusion rate and surface area: modules may be taken out of service to keep conditions constant.

Cryogenic Separation. Cryogenic separation units operate by cooling the gas and condensing some or all of the gas stream. Depending on the product purity required, separation may be by simple flashing or by distillation. Cryogenic units tend to be more expensive than other processes, especially in smaller sizes. This is partly because of the feed pretreatment required to remove compounds which would freeze, such as water or CO_2. They are therefore used either in very large sizes or where they offer a unique advantage, such as the ability to separate a variety of products from a single feed stream. One example is the separation of light olefins from a hydrogen stream.

Hydrogen recovery is in the range of 95 percent, with purity above 98 percent obtainable. Once the material has been condensed, additional fractionation is relatively cheap.

Feedstocks

The best feedstocks for steam reforming are light, saturated, and low in sulfur; this includes natural gas, refinery gas, LPG, and light naphtha. These feeds can be converted to hydrogen at high thermal efficiency and low capital cost.

Natural Gas. Natural gas is the most common hydrogen plant feed, since it meets all the requirements for reformer feed, and is low in cost. A typical pipeline natural gas (Table 6.2.8) contains over 90 percent C_1 and C_2, with only a few percent of C_3 and heavier hydrocarbons. It may contain traces of CO_2, with often significant amounts of N_2. The N_2 will affect the purity of the product hydrogen: it can be removed in the PSA unit if required, but at increased cost.

Purification of natural gas, before reforming, is usually relatively simple. Traces of sulfur must be removed to avoid poisoning the reformer catalyst, but the sulfur content is low, and generally consists of H_2S plus some mercaptans. Zinc oxide, often in combination with hydrogenation, is usually adequate.

TABLE 6.2.8 Typical Natural Gas Composition

Component	Volume %
CH_4	81.0
C_2H_6	10.0
C_3H_8	1.5
C_4H_{10}	0.5
$C_5H_{12}+$	0.2
N_2	5.8
CO_2	1.0
Sulfur (H_2S, RSH)	5 vol ppm
Total	100.0

Note: RSH = mercaptans.

Refinery Gas. Light refinery gas, containing a substantial amount of hydrogen, can be an attractive steam reformer feedstock. Since it is produced as a by-product, it may be available at low cost. Processing of refinery gas will depend on its composition, particularly the levels of olefins and of propane and heavier hydrocarbons.

Olefins can cause problems by forming coke in the reformer. They are converted to saturated compounds in the hydrogenator, giving off heat. This can be a problem if the olefin concentration is higher than about 5 percent, since the hydrogenator will overheat. A recycle system can be installed to cool the reactor, but this is expensive and wastes heat.

Heavier hydrocarbons in refinery gas can also form coke, either on the primary reformer catalyst or in the preheater. If there is more than a few percent of C_3 and higher compounds, a promoted reformer catalyst should be considered, in order to avoid carbon deposits.

When hydrogen content is greater than 50 vol % and the gas is at adequate pressure, the gas should first be considered for hydrogen recovery, using a membrane or pressure-swing adsoprtion unit. The tail gas or reject gas—which will still contain a substantial amount of hydrogen—can then be used as steam reformer feedstock.

Refinery gas from different sources varies in suitability as hydrogen plant feed. Catalytic reformer offgas, as shown in Table 6.2.9 for example, is saturated, very low in sulfur, and often has a high hydrogen content. This makes excellent steam reformer feedstock. It can contain small amounts of chlorides. These will poison the reformer catalyst and must be removed.

The unsaturated gas from an FCC or coker, on the other hand, is much less desirable. Besides olefins, this gas contains substantial amounts of sulfur, which must be removed before the gas is used as feedstock. These gases are also generally unsuitable for direct hydrogen recovery, since the hydrogen content is usually too low.

Hydrotreater offgas lies in the middle of the range. It is saturated, so it is readily used as hydrogen plant feed. Content of hydrogen and heavier hydrocarbons depends to a large extent on the upstream pressure. Sulfur removal will generally be required.

The process scheme shown in Fig. 6.2.13 uses three different refinery gas streams to produce hydrogen. First, high-pressure hydrocracker purge gas is purified in a membrane unit. Product hydrogen from the membrane is available at medium pressure and is combined with medium pressure offgas, which is first purified in a PSA unit. Finally, low-pressure offgas is compressed, mixed with reject gases from the membrane and PSA units, and used as steam reformer feed. The system also includes a recycle loop to moderate the temperature rise across the hydrogenator from the saturation of olefins.

TABLE 6.2.9 Typical Catalytic Reformer Offgas Composition

Component	Volume %
H_2	75.5
CH_4	9.6
C_2H_6	7.6
C_3H_8	4.5
C_4H_{10}	2.0
$C_5H_{12}+$	0.8
Total	100.0

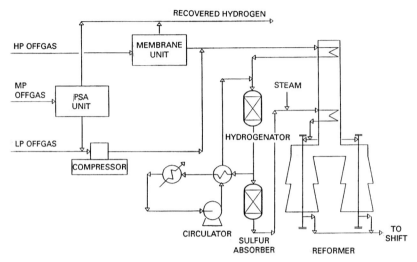

FIGURE 6.2.13 Feed handling and purification with multiple feedstocks.

Liquid Feeds. Liquid feeds, either LPG or naphtha, can be attractive feedstocks where prices are favorable. Naphtha is typically valued as low-octane motor gasoline, but at some locations there is an excess of light straight-run naphtha, and it is available cheaply. Liquid feeds can also provide backup feed, if there is a risk of natural gas curtailments.

The feed handling system needs to include a surge drum, feed pump, and vaporizer, usually steam-heated. This will be followed by further heating, before desulfurization. The sulfur in liquid feeds will be in the form of mercaptans, thiophenes, or heavier compounds. These compounds are stable and will not be removed by zinc oxide, therefore a hydrogenator will be required. As with refinery gas, olefins must also be hydrogenated if they are present.

The reformer will generally use a potash-promoted catalyst to avoid coke buildup from cracking of the heavier feed. If LPG is to be used only occasionally, it is often possible to use a methane-type catalyst at a higher steam:carbon ratio to avoid coking. Naphtha will require a promoted catalyst unless a prereformer is used.

HEAT RECOVERY

In selecting a heat recovery system for a new plant, a number of factors must be balanced: environmental regulations, capital cost, operating cost, and reliability. The relative importance of these will vary from project to project.

The environmental regulations with the most impact on plant design are typically NO_x limitations. Other impacts such as SO_x or water emissions are minimal, because low sulfur fuel is typically used and there are few emissions other than flue gas. The choice of heat recovery system can have a major effect on NO_x production, since both the amount of fuel fired and the flame temperature will be affected. Preheating combustion air will reduce firing, but since NO_x formation is strongly influenced by flame temperature, there will be an overall increase in NO_x formation. Other methods of

TABLE 6.2.10 Economics of Air Preheat versus Steam Generation*

Effect of air preheat	Fuel fired, million Btu	Steam produced, klb	BFW, klb	Total
Reduction per hour	85.4	22.9	22.9 + blowdown ~1%	
Unit cost (low fuel cost):	0.95	4.40	0.888	
Cost per hour, $	−81.17	100.66	−20.53	−1.04
Cost per year, $				−8700
Unit cost (high fuel cost):	3.00	10.00	1.40	
Cost per hour, $	−256.33	228.77	−32.65	−60.22
Cost per year, $				−505,800

*Basis: 45 million SCFD [50,000 (N) m³/h], 8400 h/year.
Note: BFW = boiler feedwater.

reducing firing, such as prereforming or heat exchange reforming, do not affect the flame temperature and will therefore reduce NO_x production. Any of these methods can also be useful if there is a limit on the total amount of fuel fired, such as when a plant is to be expanded under an existing permit.

Capital cost and operability will generally favor steam generation. This is the simplest scheme, and is favored wherever the additional steam can be used (Table 6.2.10). No additional catalysts are necessary, and if a Terrace Wall or a side-fired reformer is used, it is possible to build the reformer as a natural draft unit. This eliminates the forced and induced draft fans and further improves reliability. In cases where steam has little value, air preheat, prereforming, or heat exchange reforming will be favored, although capital cost will be increased with these options.

Prereforming

In prereforming, the reformer feed is processed in an adiabatic fixed bed, with a highly active catalyst because of the lower temperature.

The reformer feedstock is mixed with steam and passed over the prereforming catalyst, as is shown in Fig. 6.2.6. As reforming proceeds, the temperature falls. The gas is then reheated and sent to the primary reformer. For feedstocks heavier than methane, heat of cracking will tend to raise the temperature and can result in a temperature rise for liquid feeds or heavy refinery gases. The technology is well proven, and the catalyst has been used for other applications in naphtha reforming. Other than the reactors, the only additional equipment required is a preheat coil in the reformer.

On the other hand, only a limited amount of heat can be recovered, since the reactor inlet temperature is limited to about 930°F (500°C) to avoid cracking of the feedstock. Much of the savings in energy comes from the ability to reheat the feed to a high temperature. Since the pre-reformer outlet contains no hydrocarbons heavier than methane, there is little risk of cracking.

The high activity catalyst is also sensitive to deactivation, and provision must be made to allow catalyst changeout during operation.

Heat-Exchange Reforming

The process gas leaving the reformer can be used as a heat source for additional reforming. Reforming catalyst is packed in the tubes of a heat exchanger, and the pri-

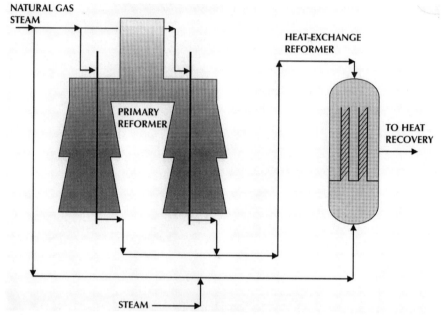

FIGURE 6.2.14 Heat-exchange reforming.

mary reformer outlet gas flows in the shell. Various arrangements are used to cope with tube expansion, such as the one shown in Fig. 6.2.14. Here the hot gas from the primary mixes with the gas leaving the open-ended catalyst tubes, and then flows along the outside of the catalyst tubes. An advantage of the heat exchange reformer is that it can reach higher temperatures and recover more heat than the prereformer, although at higher equipment cost.

The temperature in the heat exchange reformer is lower than that in the primary, and the steam:carbon ratio in the heat exchange reformer is increased to improve equilibrium and reduce the methane leakage down to the same range as in the primary. This also shifts the reforming heat load to a lower temperature, improving the heat balance.

The main effect of the heat-exchange reformer is to reduce the fuel demand and steam generation. Table 6.2.11 shows this reduction: from 159,000 lb/h (72 tonnes/h) with the primary reformer alone, to 77,000 lb/h (35 tonnes/h) with the addition of the heat-exchange reformer. By combining the heat exchange reformer with air preheat, there is a further reduction in the steam generation and fuel demand for the plant: export steam is reduced to 21,000 lb/h.

ECONOMICS

Process Route

Capital costs for hydrogen production are illustrated in Fig. 6.2.15, which compares costs for purification, steam reforming, and partial oxidation. Where hydrogen is

TABLE 6.2.11 Utility Comparison Heat Exchange Reformer

	Heat-exchange reformer		
	Base case	Cold air	Air preheat
Hydrogen, million SCFD	50	50	50
Primary reformer, °F	1500	1550	1600
Natural gas, million SCFD:			
Feed	20.9	18.9	17.6
Fuel	1.7	1.2	0.8
Total	22.6	20.1	18.4
Steam export lb/h	159,000	77,000	21,000

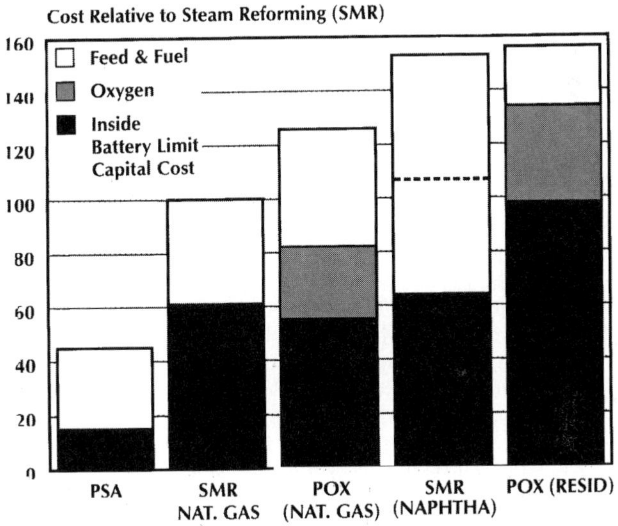

FIGURE 6.2.15 Production cost of different cost-process routes. PSA = pressure-swing adsorption; SMR = steam-methane reforming; POX = partial oxidation.

already available in sufficient quantity, it is cheapest to merely purify it as required. In most cases this is not sufficient, and it is necessary to manufacture it.

Figure 6.2.15 illustrates why steam reforming is favored over partial oxidation. For light feedstocks, capital cost for the inside battery limit (ISBL) plants are similar for steam reforming or partial oxidation. However, when the cost of oxygen is included, the cost for partial oxidation (POX) rises substantially. Naphtha reforming is slightly higher in capital cost than reforming of natural gas. Feedstock cost will depend on the value of the naphtha; where the naphtha is valued as motor gasoline it cannot compete with natural gas. Where there is a surplus of low-octane naphtha, it may be valued at fuel cost or even below; in this case steam reforming of naphtha can be attractive.

For partial oxidation of residual fuel, a substantial amount of equipment is required to handle the soot, ash, and sulfur (Fig. 6.2.4). The cost for this additional equipment, as well as the additional oxygen required, means that heavy oil must be much cheaper than natural gas to justify partial oxidation. Alternatively, partial oxidation may be used as a way to dispose of a stream such as petroleum coke or asphalt, which is considered a waste product.

Capital Cost

Where capacity, feedstock, and method of heat recovery are known for a steam reforming plant, a reasonable estimate may be made of capital cost, typically to an accuracy of ±30%. For a 50 million SCFD [56,000 (N) m^3/h] hydrogen plant, based on natural gas feed, and using steam generation for heat recovery, capital cost is approximately $30,000,000.

This assumes a battery limit process unit, including the equipment shown in Fig. 6.2.3, on the U.S. Gulf Coast constructed in first quarter 1995 through mechanical completion. It also assumes that the site is free of above- and below-ground obstructions. It does not include the cost of land, taxes, permits, warehouse parts, escalation, catalyst, and support facilities.

Make or Buy

In recent years refiners have been presented with a viable alternative to building their own hydrogen plant. It is possible to buy hydrogen like a utility "over the fence" from one of the major industrial gas companies, such as BOC Gases. These companies typically have experience producing and selling many industrial gases such as hydrogen, nitrogen, and oxygen and several have entered alliances with plant manufacturers which allow them to produce hydrogen more efficiently and more cheaply than refiners can do on an individual plant basis.

Advantages of Making Hydrogen (Building and Owning a Plant). Historically most refiners have built their own hydrogen plants as needed. These plants have been fully integrated into the refinery, in many cases as part of the hydrotreating or hydrocracking complex. This approach allows the refiner to maintain complete control over both the project and the ongoing supply. If the hydrogen plant is combined with other investments or is a duplicate of an existing plant, there can be capital savings compared to a stand-alone hydrogen plant. However, most hydrogen plants are of a sufficient scale that they can be efficiently managed as individual construction projects.

There can also be other benefits to owning the plant from an operations standpoint. Typically control of the hydrogen plant would be combined with that of other processing units which can lead to operating labor efficiencies where the same operators are responsible for multiple units. A view can also be taken that operations between hydrogen-producing units and hydrogen-consuming units can be better managed when everything is owned by the refiner. This point has become more difficult to argue as computerized process control has become more commonplace and output signals can be shared between operations.

Advantages of Buying Hydrogen "over the Fence." With a properly structured gas supply agreement it is possible to be guaranteed the necessary security of hydrogen supply while achieving both a lower total hydrogen cost with less total risk to the refiner. Refiners can capitalize on the experience of companies whose business is to produce hydrogen, while concentrating their resources (both financial and technical) on refining, where their value added is greatest. Over-the-fence suppliers can typically achieve savings on both capital and operating costs, which are passed on to the hydrogen purchasers. Additionally, once a supply agreement has been reached, the refiner's hydrogen costs are fixed and known. All potential uncertainties, such as capital cost overruns and operating cost variance become the responsibility of the hydrogen supplier.

Capital savings over what an individual refiner can achieve can come about for several reasons. Some of these are

1. *Multiple plant experience.* A typical refiner would add only one hydrogen plant at a time while a hydrogen supplier would typically build several hydrogen plants a year. This ongoing plant construction effort allows for savings in standardization of design and procurement of major equipment as well as ongoing efforts to minimize plant capital.

2. *Alliances with technology providers.* Several industrial gas companies are now allied with major engineering procurement/construction (EPC) companies to provide over-the-fence hydrogen supply. These alliances help ongoing capital reduction and value engineering efforts to further reduce capital costs.

3. *Multiple-customer, multiple-plant networks.* In those areas where gas companies can supply multiple customers from a single plant or a network, there are significant economy-of-scale benefits that would not be available to a refiner. As an example, the capital costs of one 50 million SCFD plant would be significantly less than that of two 25 million SCFD plants. It is also possible in these situations to provide product sharing between plants as a backup or as additional purchase flexibility.

The cumulative effect of these benefits could be a plant capital reduction of over 10 percent (significantly more if multiple customers can be combined).

In some instances operating costs for hydrogen suppliers can also be lower than for the refiner. Operating experience for a collection of plants can be monitored and individual plant performance can be compared to collective norms. Minor variations in performance can be noted earlier and preventive action can be taken to avoid major outages and problems. Also, purchases for catalyst replacements and spare parts for many units can be pooled, leading to additional savings.

Summary. Unlike in the past, refiners looking to add hydrogen capacity now have a viable alternative to investing their own capital to add capacity. There are advantages and disadvantages to both building a plant and buying hydrogen over the fence, but there are potentially significant savings associated with buying hydrogen as a utility. Any refiner considering an increase in hydrogen requirements should seriously consider hydrogen purchase, if only as a mechanism to minimize its own capital outlays.

UTILITY REQUIREMENTS

Typical utility requirements for a 50 million SCFD hydrogen plant feeding natural gas are as follows (no compression is required).

Feedstock	730 million Btu/h (770 GJ/h)
Fuel	150 million Btu/h (158 GJ/h)
Export steam, 600 lb/in^2 gage/700°F	120,000 lb/h (54 tonnes/h)

BFW   160,000 lb/h (72 tonnes/h)
Cooling water   900 gal/min (200 m³/h)
Electricity   800 kW

REFERENCES

1. B. J. Cromarty, K. Chlapik, and D. J. Ciancio, "The Application of Pre-reforming Technology in the Production of Hydrogen," NPRA Annual Meeting, San Antonio, March 1993.
2. J. D. Fleshman, *Chem. Eng. Prog.*, **89**(10), 20 (1993).
3. A. Fuderer, "Catalytic Steam Reforming of Hydrocarbons," U.S. patent 4,337,170, June 29, 1982.
4. M. H. Hiller, G. Q. Miller, and J. J. Lacatena, "Hydrogen for Hydroprocessing Operations," NPRA Annual Meeting, San Antonio, March 1987.
5. G. Q. Miller and J. Stoecker, "Selection of a Hydrogen Separation Process," NPRA Annual Meeting, San Francisco, March 1989.
6. *Physical and Thermodynamic Properties of Elements and Compounds,* United Catalysts, Inc.
7. M. V. Twigg, *Catalyst Handbook,* 2d ed., Wolfe Publishing, London, 1989.
8. G. J. Lambert, W. J. A. H. Schoeber, and H. J. A. Van Helden, "The Hydrogen Balance in Refineries," Foster Wheeler Heavy Oil Processing and Hydrogen Conference, Noordwijk, The Netherlands, April 1994.

PART · 7

HYDROCRACKING

CHAPTER 7.1
MAK MODERATE-PRESSURE HYDROCRACKING

M. G. Hunter
The M.W. Kellogg Company
Houston, Texas

D. A. Pappal
Mobil Technology Company
Paulsboro, New Jersey

C. L. Pesek
Akzo Nobel Catalysts
Houston, Texas

INTRODUCTION

The increasing market demand for middle distillates and the need for lower-sulfur, cleaner-burning transportation fuels has directed the refining industry's attention to adding conversion capacity by hydrocracking. Hydrocracking is a highly flexible process option that can be used to convert virtually any refinery stream into value-added products. It is particularly well suited to the production of low-sulfur, high-quality middle-distillate fuel components and can be integrated synergistically with other conversion technologies, such as fluid catalytic cracking (FCC) and coking.

In the 1960s and 1970s, the most common application of hydrocracking was the maximum production of naphtha for gasoline reforming. As the growth in demand for middle-distillate products has increased and the demand for gasoline has begun to level off, the application of hydrocracking technology for maximum conversion of vacuum gas oil (VGO) to jet and diesel fuel has gained in importance.

The maximum conversion of heavy vacuum gas oils to diesel and lighter products typically has required that hydrocracking plants be designed for operating pressures in excess of 140 bar gage. The capital costs for both new high-pressure hydrocracking

TABLE 7.1.1 Typical Hydrocracker Operating Conditions

	High-pressure hydrocracking	MAK-MPHC
Conversion, wt %	70–100	20–70
Pressure, bar gage	100–200	70–100
Liquid hourly space velocity	0.5–2.0	0.5–2.0
Average reactor temperature, °C	340–425	340–425
Hydrogen circulation, (N) m^3/m^3	650–1700	350–1200
Hydrogen consumption, (N) m^3/m^3	200–600	70–200

Note: (N) = standard temperature and pressure.

equipment and the incremental hydrogen to feed that equipment has made such facilities difficult to justify within increasingly capital-constrained operating environments. Operating at less than total conversion can open up opportunities to optimize the relationship between pressure, conversion, catalyst life, hydrogen consumption, and product quality, leading to substantially reduced capital investment and highly profitable returns.

Moderate-pressure hydrocracking (MPHC) is a once-through hydrocracking process for the conversion of heavy gas oils to low-sulfur distillates and unconverted oil that is highly upgraded relative to raw feed. Operating at lower pressure significantly reduces capital investment and results in substantially less hydrogen consumption. Furthermore, the process requirements for MPHC are within the range of many existing VGO desulfurization units. The typical range of operating conditions for MPHC is shown in Table 7.1.1.

Mobil Research and Development Corporation, Akzo Nobel Chemicals, and The M.W. Kellogg Company have formed a partnership to offer MAK-MPHC and other full-conversion hydrocracking technologies for license to the refining industry.

Mobil and Akzo Nobel have actively engaged in MPHC research for more than 10 years.[1-4] Mobil's first commercial MPHC installation was successfully started up in 1983. Mobil operates five hydrocrackers, two of which are partial-conversion MPHC designs that process heavy VGO into middle-distillate products. This research and operating experience has led to an advanced capability to apply hydrocracking conversion technology to heavy feedstocks under moderate-pressure conditions.

Akzo Nobel has commercialized a family of hydrotreating and hydrocracking catalysts that are combined to achieve the optimal balance between activity and selectivity for each specific refining application. Akzo Nobel zeolite-based hydrocracking catalysts have been selected for application in nine units around the world, with an excess of 1.1 million kilograms produced.

M.W. Kellogg is a leading technology-based international engineering and construction firm that has experience in the design, engineering, and construction of 22 hydrocrackers.

HYDROCRACKING COSTS

The high installed costs for hydrocracking equipment is related to high-hydrogen partial-pressure requirements and process conditions requiring the use of exotic construction materials. The required operating-pressure level is determined by a complex relationship between feed properties, desired conversion level, catalyst life, and product-quality constraints. To operate at near 100 percent of total conversion with extinction recycle, high hydrogen pressures are required to limit catalyst deactivation to

TABLE 7.1.2 Hydrocracker Cost Comparison

	High-pressure hydrocracker* base		MPHC
Total pressure, bar gage	>140	100	70
Conversion, %	90–100	70	50
Relative installed cost ISBL hydrocracker	100	73	62
Relative installed cost plus H$_2$ plant†	134	94	62
Relative operating cost‡	1	0.7	0.6

*Assumes 140-bar gage cold high-pressure separator, 40% recycle ratio (70% crack per pass) and 100% conversion to diesel and lighter.

†Cost including hydrogen plant for incremental hydrogen demand above that required for a 70 bar gage, 50% conversion MPHC.

‡Includes all utilities, catalyst costs, and hydrogen valued at $0.08 U.S. per normal cubic meter.

acceptable rates. In general, heavier, higher endpoint feedstocks will necessitate higher design pressures for a given conversion level and catalyst life. Also, the aromaticity of hydrocracked products will be directly proportional to hydrogen partial pressure.

While MPHC will yield a high-quality diesel fuel component, the kerosene fraction will, in general, not qualify as a specification turbine fuel. In many refining situations, however, the kerosene cut will be suitable for blending into the jet fuel pool. Designing a hydrocracker to produce high smoke-point specification jet fuel can result in significant quality giveaway in the naphtha and diesel fractions and inefficient use of hydrogen resources.

Designing a hydrocracker for single-pass, partial-conversion operating mitigates to a large extent the need for high operating pressures. The single-pass hydrocracker is not subject to fouling or high catalyst deactivation rates that can result from the buildup of polynuclear aromatics in the recycle oil stream. The kinetic impact of lower hydrogen pressure is compensated for by a decreased conversion level and by lowering of liquid hourly space velocity (LHSV) if necessary. The naphtha, diesel, and unconverted bottoms products are not oversaturated, and hydrogen consumption thereby is minimized. The economic keys to MPHC are the values of the low-sulfur distillate and upgraded bottoms products relative to the untreated feed.

Relative installed costs [inside battery limits (ISBL)] based on curve-type estimates for a full-conversion hydrocracker and two MPHC options are shown in Table 7.1.2. The relative operating costs shown include estimated utility, hydrogen, and catalyst costs. The grass-roots MPHC unit that utilizes available refinery hydrogen can require less than half of the capital investment associated with a high-pressure hydrocracker. As will be illustrated, it is possible to revamp existing moderate-pressure hydroprocessing equipment to achieve conversion levels of 30 to 50 percent. In such situations, the capital investment is substantially below that stated in Table 7.1.2.

TECHNOLOGY DEVELOPMENT

A simplified flow diagram of a typical moderate-pressure hydrocracker is shown in Fig. 7.1.1. Advanced reactor internals design technology allows the application of multibed reactors while maintaining stable operations and maximizing catalyst utilization. In most cases, catalyst requirements are such that only a single reactor vessel is needed. Both high- and low-temperature separators are utilized in a reaction section to enhance operability and heat integration with the fractionation section. Because of the

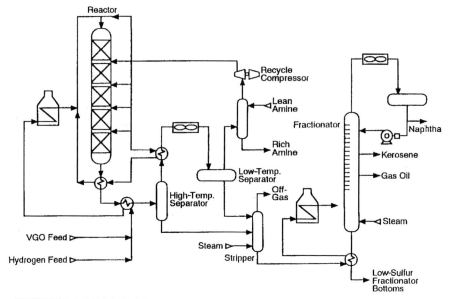

FIGURE 7.1.1 MAK-MPHC flow diagram.

low light-ends make from MPHC, a simple stripper followed by the fractionating column can be specified for product recovery. Lower design pressure and a minimum equipment count result in substantial capital cost savings.

Reactor Internals

Reactor internals performance is extremely important for the safe, reliable, and profitable operation of the hydrocracking process. Mobil's experience in operating hydrocracking reactors dates back to June of 1967 with the start-up of Mobil's first hydrocracker. The unit was designed to hydrocrack the extinction diesel range gas oils to gasoline. The unit could not operate at design conditions because of significant radial temperature maldistribution (see Fig. 7.1.2, and 7.1.3) in the two first-stage reactors. The impact of the radial temperature maldistribution included:

- Reactor runaways
- Loss of cycle length
- Processing limited to low-heat-release feedstocks
- Significant loss of product uplift

The construction of a second new unit was to be completed in October of 1968, and reactor performance similar to that of the first unit would not be acceptable. Therefore, a crash research program was conducted by Mobil beginning in May of that year to define and test quench zone modifications for the unit under construction. Because of the extremely tight schedule, the experimental portion of the program was limited to 1 month. In addition, the modified quench zone had to fit within the existing available reactor height. A full-scale plywood and clear plastic quench zone cold-flow

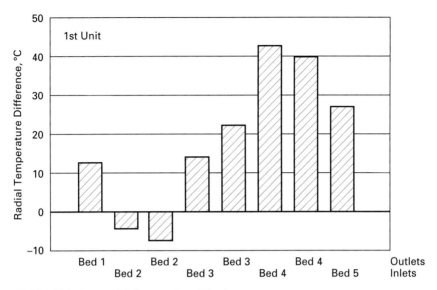

FIGURE 7.1.2 Reactor A before quench modifications.

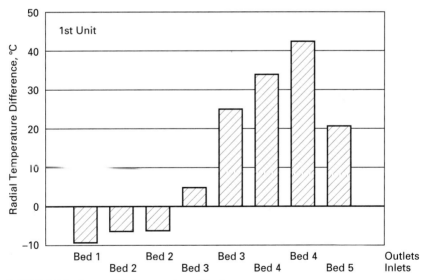

FIGURE 7.1.3 Reactor B before quench modifications.

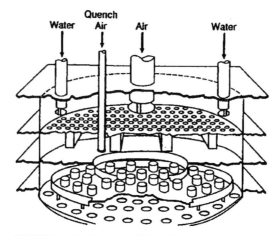

FIGURE 7.1.4 Schematic of full-scale quench zone model.

model (Fig. 7.1.4) was constructed to test candidate designs. The data collected at simulated full-scale process conditions were used to compare the different designs by visual observations, tracer compound concentration profiles in the gas and liquid, and pressure drop. A quantitative index method was used to rate the candidate designs based on tracer concentrations in the liquid and gas phases.

The mixing index was defined as 100 minus the standard deviation for the tracer concentration in samples collected from four quadrants of the quench zone outlet. The following criteria were quantified in order to rate the quench zone designs:

- Quench gas distribution throughout the process vapor
- Transverse liquid mixing
- Transverse vapor mixing

The most effective quench apparatus tested was a design that researchers have named the Spider-Vortex.* The Spider-Vortex surpassed the target mixing index of 80 in all three categories. The Spider-Vortex was fabricated and installed without delaying the October unit completion date.

Commercial data from the second hydrocracker confirmed the excellent results observed in the cold-flow temperatures. Quench zone performance was improved dramatically over that of the first unit. With the original hardware, fully 75 percent of the radial temperature differences exceeded 80°C. The largest radial temperature difference was about 83°C. With the Spider-Vortex system in the new unit, all measured radial temperature differences were less than 8°C (see Fig. 7.1.5). The Spider-Vortex also was retrofitted into the first hydrocracker at the earliest opportunity with equally impressive results (see Fig. 7.1.6). The retrofit resulted in maximum radial differences measured anywhere in the reactor by only about 6°C.

The design of the first Spider-Vortex installation could not be optimized because of the extremely short development time available. The second application utilized newer developments and resulted in improved performance relative to the first. Since its initial development in 1968, the Spider-Vortex design has continued to be improved. The system recently was retrofitted into a third Mobil hydrocracking reac-

---

*Trademark of Mobil.

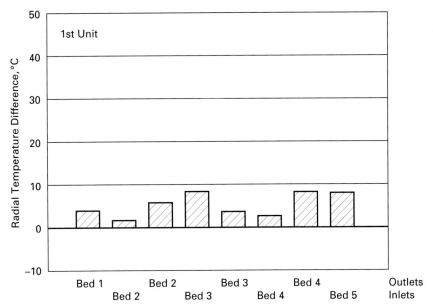

FIGURE 7.1.5 Unit 2 after quench modifications.

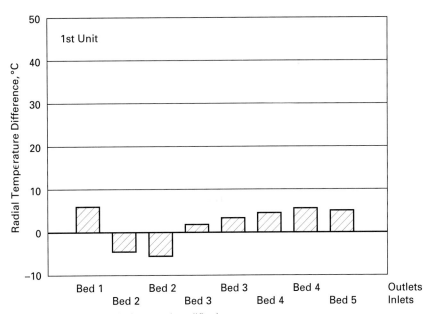

FIGURE 7.1.6 Reactor A after quench modifications.

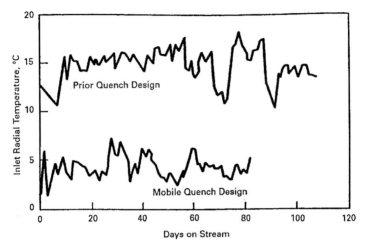

FIGURE 7.1.7 Catalyst bed inlet radial temperatures improved with Mobil quench design.

tor with dramatic results. Radial temperature differences across the top of the catalyst beds dropped from about 8°C to about 3°C (see Fig. 7.1.7). The bed outlet radial temperature differences decreased from 17°C to 19°C with the previous quench zone design to about 3°C (see Fig. 7.1.8). These improvements in temperature distribution and catalyst utilization translate into better yields, longer catalyst life, and more efficient use of limited hydrogen resources. A complete description of the Spider-Vortex development program can be found elsewhere.[5]

A second important feature of the Spider-Vortex is the gas-liquid redistribution equipment. This hardware achieves small-scale contacting of the process gas and liq-

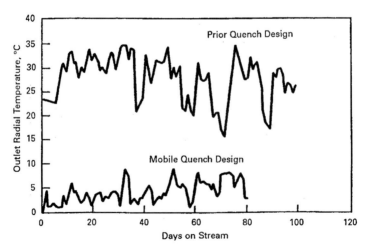

FIGURE 7.1.8 Catalyst bed outlet radial temperatures improved with Mobil quench design.

uid and distributes both phases across the next catalyst bed. Uniform wetting at the top of the packed bed is critical because liquid distribution generally does not improve as the phases proceed downward. Maldistribution at the top of the catalyst beds therefore must be minimized to realize full catalyst utilization in the reactor. Mobil has conducted extensive experimentation and modeling in this area and has developed very effective gas-liquid distribution equipment that has a high degree of flexibility.

The advantage of the Spider-Vortex is clear and striking. This system transfers to the catalyst bed a spreading, highly atomized spray of process liquid scattered within the process gas. At the catalyst surface, the individual gas-liquid sprays actually overlap to achieve complete coverage of the solids surface, resulting in full utilization of the catalyst volume. The alternative hardware creates comparatively little liquid breakup, instead producing discrete liquid systems that cover smaller, nonoverlapping regions of catalyst and leave large areas of catalyst unwetted.

Detailed fluid mechanic and reactor modeling calculations of several cases have shown a catalyst activity advantage of between 3 and 11°C for the Spider-Vortex redistribution system compared to the commonly used device. The advantage varies considerably with gas and liquid flow rates and could be even larger for certain high-severity applications. The Spider-Vortex system and Mobil's engineering design are available as features of the MAK-MPHC process.

Catalyst Selection

Another key feature of the MAK-MPHC process is the specification of a dual catalyst system. The system consists of a pretreat catalyst formulated for high hydrodenitrogenation (HDN) activity followed by a hydrocracking catalyst with both activity and selectivity tailored to meet specific conversion objectives. Since organic nitrogen compounds are poisons to acidic hydrocracking catalysts, nitrogen must be converted to low levels in order to achieve significant hydrocracking activity. As can be seen in Fig. 7.1.9, hydrotreating catalyst formulations optimized for HDN are substantially more active for nitrogen removal than are hydrocracking (MHC) catalysts. Only at very high temperatures does the denitrogenation activity of the MHC catalyst begin to

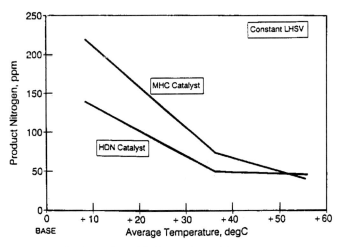

FIGURE 7.1.9 Comparison of denitrogenation activity.

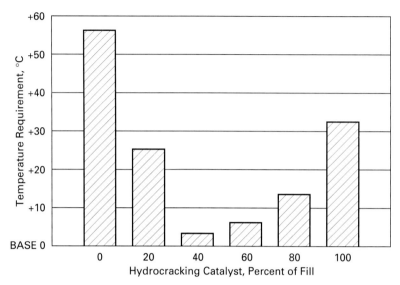

FIGURE 7.1.10 Required temperature for 60 percent conversion.

match that of the HDN material. Hydrotreating catalysts also are significantly less expensive than hydrocracking catalysts and are less sensitive to temperature instability. Therefore, it is best to carry out the high-heat-release desulfurization and denitrogenation reactions in the beds that contain only hydrotreating catalyst.

Figure 7.1.10 is a plot of the required operating temperature versus the percent hydrocracking catalyst employed in the system. The plot shows the pilot plant average reactor temperature adjusted to achieve 60 wt % conversion to 343°C-minus product at constant overall liquid hourly space velocity. For this particular feedstock, the lowest temperature required for 60 wt % conversion was obtained with a catalyst system consisting of 60 percent HDN catalyst followed by 40 percent MHC catalyst. The optimal catalyst distribution will depend on many factors, including feedstock properties and processing objectives. Lower overall catalyst requirements and the use of less expensive hydrotreating catalysts result in substantially reduced cost.

Pilot Plant Yields and Properties

The results of pilot plant tests conducted at both 54-bar and 82-bar hydrogen partial pressure are presented in this section. These data represent the range of expected operations for both grass-roots MPHC and potential revamp applications.

Results at 54-bar Hydrogen Pressure. A series of pilot plant runs was made in 1990 to support the application of Akzo Nobel catalysts in an existing Mobil MPHC unit. The experiments were conducted at 54-bar hydrogen pressure and 500 normal cubic meters [(N) m^3] of recycle gas per cubic meter of feed, with yields, product properties, and catalyst deactivation rates determined at two conversion levels. Properties of the VGO feedstock used in these studies are shown in Table 7.1.3.

The VGO feed was processed at two different temperature levels, resulting in nominal conversion levels of 37 and 46 percent. The catalyst system consisted of Akzo

TABLE 7.1.3 Pilot Plant Feed Properties—54-bar Studies

Specific gravity at 15°C	0.907
Sim. distillation, D-2887, °C	
IBP	295
10%	352
50%	429
90%	518
FBP	635
Sulfur, wt %	2.1
Nitrogen, wt ppm	860
CCR, D-4530, wt %	0.4

Note: IBP = initial boiling point; FBP = final boiling point.

TABLE 7.1.4 MPHC Pilot Plant Yields Catalysts: KF 843/KC 2600

Conversion, LV % (to 343°C−)	37	46
Average reactor temperature, °C	Base	Base+4
Yields, vol % of feed:		
C_5–166°C, naphtha	9.2	13.7
166–227°C, kerosene	7.8	10.4
227–343°C, light gas oil	23.8	26.6
343–388°C, heavy gas oil	18.5	17.1
388°C+, LSFO	44.9	37.1
Hydrogen consumption, (N) m^3/m^3	84	103

Note: LV % = liquid volume percent; LSFO = low-sulfur fuel oil.

Nobel KF-843 in the treating section and KC-2600 in the cracking section. The LHSV was consistent with a greater than 2-year catalyst cycle at the highest conversion level. The yields and hydrogen consumptions are presented in Table 7.1.4.

Total middle-distillate (166–343°C) yields of 31.6 and 37.0 percent were achieved at 37 and 46 percent conversion, respectively. Hydrogen consumptions were only 84 and 103 (N) m^3/m^3, respectively. While these yields do not represent the maximum distillate selectivity for Akzo Nobel catalysts, the high activity and stability of KC-2600 allows for an extended operating cycle and better cumulative yield compared to less active catalyst alternatives.

Middle-distillate and unconverted bottoms product properties at the 46 percent conversion level are shown in Table 7.1.5.

The kerosene and light gas oil fractions are excellent blending components for jet fuel and diesel products in this refinery. The heavy gas oil and unconverted bottoms are excellent low-sulfur, low-viscosity fuel oil components. The low density and nitrogen content of the heavy gas oil (HGO) and bottoms products also would make them excellent feedstocks for FCC.

Results at 82-bar Hydrogen Pressure. A pilot plant program to study a different Akzo Nobel catalyst system, KC-8343/KC-2300, at 82-bar hydrogen pressure has recently been completed. The higher pressure level is indicative of what might be specified in a grass-roots MPHC application designed to achieve in excess of 50 percent conversion. Properties of the VGO feed used in these runs are provided in Table 7.1.6.

TABLE 7.1.5 MPHC Typical Product Properties Catalysts: KF 843/KC 2600 at 46% Conversion

	Kerosene	LGO	HGO	LSFO
TBP cut, °C	166–227	227–343	343–388	388+
Specific gravity at 15°C	0.834	0.871	0.849	0.858
Sulfur, wt ppm		100	100	300
Nitrogen, wt ppm			12	24
Aromatics, wt %	35.8			
Smoke point, mm	16			
Viscosity at 50°C, cSt		2.81	6.85	
Viscosity at 65°C, cSt				14.13
Cetane index		44	56	
Pour point, °C		−21	+15	

Note: LGO = light gas oil; HGO = heavy gas oil.

TABLE 7.1.6 Pilot Plant Feed Properties—82-bar Studies

Specific gravity at 15°C	0.916
Distillation, D-1160, °C:	
IBP	315
10%	383
50%	447
90%	504
FBP	537
Sulfur, wt %	2.25
Nitrogen, wt ppm	800
Viscosity at 50°, cSt	31.2

The VGO feed was processed at three different temperature levels, resulting in nominal conversion levels of 46, 64, and 73 percent. The recycle gas-to-oil ratio was set at 674 (N) m³/m³, and again LHSV was consistent with a greater than 2-year catalyst life at the highest conversion level tested. The yields and hydrogen consumptions for the KF-843/KC-2300 catalyst system are shown in Table 7.1.7.

Table 7.1.7 illustrates the ability of MPHC to deliver high levels of conversion and excellent middle-distillate selectivity at moderate pressure. The hydrogen consumptions are substantially less than for extinction recycle hydrocracking, which minimizes

TABLE 7.1.7 MPHC Pilot Plant Yields Catalysts: KF 843/KC 2300

Conversion, LV % (to 360°C−)	46	64	73
Average reactor temperature, °C	Base	Base+14	Base+20
Estimated cycle, months	>36	32	27
Yields, vol % of feed:			
C₅–149°C, naphtha	11.1	21.0	25.8
149–266°C, kerosene	17.8	29.5	35.1
266–360°C, diesel	23.0	20.7	19.3
360°C–plus, LSGO	53.6	35.7	26.8
Hydrogen consumption, (N) m³/m³	133	176	187

Note: LSGO = low-sulfur gas oil.

TABLE 7.1.8 MPHC Typical Product Properties Catalysts: KF 843/KC 2300 at 64% Conversion

	Kerosene	Heavy diesel	Full-range diesel	LSFO
TBP cut, °C	149–266	266–360	149–360	360+
Specific gravity at 15°C	0.839	0.862	0.848	0.857
Sulfur, wt ppm	10	<100	<50	<100
Nitrogen, wt ppm				20
Aromatics, wt %	28			
Smoke point, mm	15.5			
Freeze point, °C	−50			
Cetane index		51	45	

the need for incremental hydrogen production facilities. A unit can be designed to vary the conversion level over a wide range in response to changing market needs for distillates and low-sulfur gas oil. Typical product properties from this pilot study are shown in Table 7.1.8.

Again, all three products are excellent blending components for use in jet fuel, highway diesel, and other low-sulfur fuels.

COMMERCIAL RESULTS

The Mobil-designed moderate-pressure hydrocracker located at Mobil's joint-venture Kyokuto Petroleum Industries Ltd. (KPI) refinery in Chiba Prefecture, Japan, went on stream in September of 1983. An existing VGO desulfurization unit was converted to an MPHC unit to add conversion capacity to the refinery while maintaining production of low-sulfur gas oil. Increased production of gasoline and distillate was desired and was accomplished by converting lower-valued high-sulfur VGO. The MPHC unit has been meeting or exceeding expectations for more than 10 years. The long-term reliability of this unit is testimony to the operating expertise of Mobil and KPI, the quality of the process design, and the effective performance of Mobil's advanced reactor internals. The design and operation of the KPI MPHC unit has resulted in a leading-edge application of hydrocracking at moderate pressures. The unit design conditions are summarized in Table 7.1.9.

Japanese governmental safety regulations require shutdowns for mandatory yearly inspections of refinery facilities. However, KPI's extremely safe operating history has allowed the relaxation of the mandatory inspection time period to 2 years. This is a major advantage for KPI and significantly improves hydroprocessing unit stream time

TABLE 7.1.9 KPI Chiba MPHC Design Conditions

Feed	343–579°C VGO
	904–934 kg/m^3
	1.8–2.8 wt % sulfur
	186 m^3/h
Conditions	55-bar gage separator pressure
	674 (N) m^3/m^3 feed
	Single stage, single pass
Reactor internals	Mobil Spider-Vortex quench zone and proprietary redistribution system

and profitability. Pilot plant studies (including those previously described in Tables 7.1.3, 7.1.4, and 7.1.5) were initiated to identify a catalyst system capable of operating for two years at a nominal 45 wt % conversion to 343°C − products. On the basis of these pilot studies, a dual-catalyst system comprising Akzo Nobel's KF-843 hydrotreating catalyst and Akzo Nobel's KC-2600 hydrocracking catalyst was chosen for the KPI MPHC unit.

The unit was started up on Akzo Nobel catalyst in August of 1991 and shut down for a scheduled turnaround in July of 1993. At shutdown, the normalized reactor temperature was only about 399°C, which is well below the metallurgical limit of the unit. The average catalyst temperature normalized to constant conversion and feed nitrogen is plotted against days of stream in Fig. 7.1.11. The least squares regression line through the data indicates an aging rate of only 1.2°C per month.

Figure 7.1.12 shows the actual 343°C+ conversion and the net distillate yield normalized to constant conversion and constant 343°C − distillate in the feed. During the first 100 days of operation, the conversion was increased from about 30 wt % in the first week to the target level of 45 wt %. From 100 days through the end of cycle

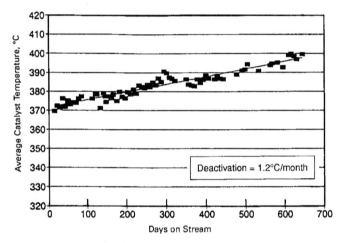

FIGURE 7.1.11 Normalized temperature versus catalyst age.

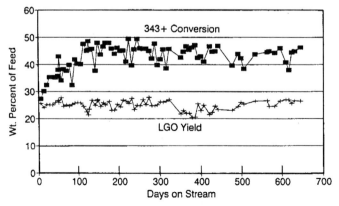

FIGURE 7.1.12 Conversion and yield versus catalyst age.

at approximately 650 days on stream, the conversion was maintained essentially constant at 45 percent. The normalized distillate yield was very stable throughout the cycle, as seen in Fig. 7.1.12. The current cycle was started up with a fresh fill of the same Akzo Nobel catalyst stream and continues to operate successfully.

MPHC GRASSROOTS APPLICATIONS

Many options are available for integrating a grassroots moderate-pressure hydrocracker into basic refinery schemes. The simplest configuration would be the addition of an MPHC unit to a hydroskimming refinery, as shown in Fig. 7.1.13. The MPHC would be a minimum investment alternative for adding VGO conversion and increasing middle-distillate and low-sulfur oil yield. The MPHC unit could be added along with coking or visbreaking in an overall scheme to reduce fuel oil make and increase production of middle distillates.

Most refineries with bottom-of-the-barrel upgrading capability are built around FCC as the primary conversion unit. The shift in product demand from gasoline to middle distillates and toughening restrictions on allowable sulfur emissions will make the addition of MPHC upstream of FCC, as shown in Fig. 7.1.14, an attractive option. The conversion achieved in the MPHC unit will unload the FCC and will provide opportunities to recover incremental gas oils by deep-cut vacuum distillation, coking, visbreaking, and solvent deasphalting. A refinery with existing vacuum and FCC units may find the combination of MPHC and one of these vacuum bottoms upgrading alternatives attractive for increasing distillate and for reducing fuel oil production. The incremental gas oils recovered by vacuum residue conversion methods are normally of poor quality, and substantial product upgrade can be achieved by hydrocracking. Additional FCC feed also can be purchased or the FCC can be converted to run incremental resid.

The yield and selectivity changes for MPHC processing upstream of an FCC unit are illustrated in Fig. 7.1.15. The total C_5+ distillate yield for catalytic cracking untreated VGO is approximately 80 vol %, of which only one-quarter is middle distillate (light cycle oil). When an MPHC unit running at nominally 45 percent conversion

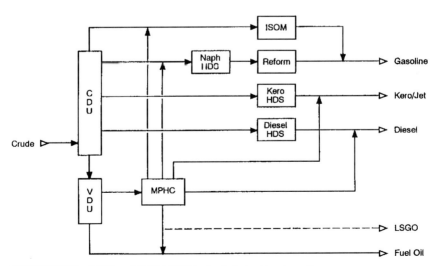

FIGURE 7.1.13 Addition of MAK-MPHC to simple hydroskimming refinery.

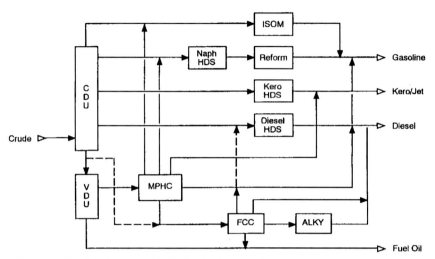

FIGURE 7.1.14 Addition of MAK-MPHC to simple FCC refinery.

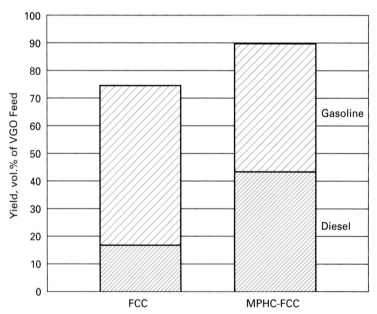

FIGURE 7.1.15 MPHC and FCC yield and selectivity.

is placed upstream of the FCC unit, total C_5+ distillate yield is increased to almost 95 vol %, and the selectivity to middle distillate is increased to 50 percent.

Middle-distillate product quality also is generally improved with MPHC, particularly with respect to sulfur content. The MPHC-FCC combination also will provide greater flexibility to shift product distribution than a stand-alone FCC unit.

REVAMP OF EXISTING HYDROTREATERS

Many refiners have discovered that existing gas oil desulfurization units can be revamped to achieve incremental conversion to more valuable distillate products. In many cases, the catalyst system, together with only minor modifications to the VGO desulfurizer (HDS) product recovery equipment, can result in 30 percent net conversion. The achievable conversion level will depend on the minimum acceptable operating cycle. MAK-MPHC technology can be used to add reactor volume and to achieve as much as 50 percent conversion without sacrificing run length.

A simple comparison of yields and costs for the revamp of an existing 200 m^3/h HDS is presented in Table 7.1.10. Conversion level is increased to 46 percent by the addition of a new reactor vessel, a new hydrogen makeup compressor, a new product fractionator, and modifications to the recycle gas compressor. Total distillate (naphtha plus diesel) yield is increased by 1942 m^3/day with an estimated capital investment of $19.1 million. Operating costs associated with increased utilities, hydrogen consumption, and catalyst usage are estimated at $9.8 million annually.

The project payout will depend on the price differential between distillate product (naphtha plus diesel) and low-sulfur gas oil. The yearly net revenue increase and simple payout of capital are plotted as a function of distillate minus low-sulfur gas oil price differential in Fig. 7.1.16. The actual low-sulfur gas oil price is presented as a parameter at $101 per cubic meter and $126 per cubic meter of product. As the value

TABLE 7.1.10 Simple Economics—Revamp of 200-m^3/h VGO* HDS Unit to MPHC

	Existing VGO HDS	Revamp to MPHC
Hydrogen pressure, bar	54	54
Relative space velocity	1.0	0.4
Yields, m^3/day:†		
Naphtha (C_5–166°C)	95	654
Middle distillate (166–343°C)	382	1765
LSGO (343°C–plus)	4389	2585
Incremental costs:		
Investment,‡ million $U.S.		19.1
Operation,§ million $U.S. per year		9.8

*Based on feedstock from Table 7.1.3.
†Based on 46% conversion and yields from Table 7.1.4.
‡Includes new reactor, new makeup compressor, new fractionator, and recycle gas compressor revamp.
§Includes incremental utilities, hydrogen, and catalyst.
Note: HDS = hydrodesulfurization.

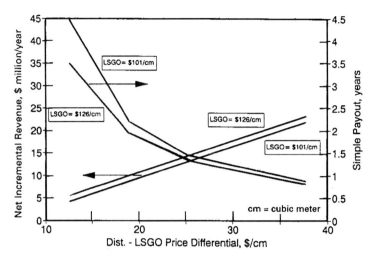

FIGURE 7.1.16 MPHC revamp revenue and payout (U.S. dollars).

of low-sulfur distillate products increases relative to low-sulfur fuel oil, conversion to MPHC can result in quick payouts and high returns on investment.

REFERENCES

1. J. W. M. Sonnemans, F. L. Plantenga, P. H. Desai, and V. J. D'Amico. "Mild Hydrocracking in Heavy Oil in the 80s," NPRA Annual Meeting, AM-84-60, San Antonio, 1984.
2. P. H. Desai et al., "MHC-FCC, an Economic Choice," NPRA Annual Meeting, San Antonio, 1985.
3. W. R. Derr and W. J. Tracy, "Moderate Pressure Hydrocracking for Heavy Vacuum Gas Oils," *Energy Progress,* vol. 6, no. 1, March 1986.
4. R. H. Gilman, M. Y. Asim, and T. A. Reid. "Optimizing FCC Operations Using Pretreatment to Meet Future Market Challenges," NPRA Annual Meeting, AM-91-37, San Antonio, 1991.
5. M. S. Sarli, S. J. McGovern, D. W. Lewis, and P. W. Snyder, "Improved Hydrocracker Temperature Control: Mobil Quench Zone Technology," NPRA Annual Meeting, AM-93-73, San Antonio, 1993.

CHAPTER 7.2
CHEVRON ISOCRACKING—HYDROCRACKING FOR SUPERIOR FUELS AND LUBES PRODUCTION

Alan G. Bridge
Technology Alliances
Chevron Research and Technology Company
Richmond, California

Hydrocracking technology plays the major role in meeting the need for cleaner-burning fuels, effective feedstocks for petrochemical operations, and more effective lubricating oils. Only through hydrocracking can heavy fuel oil components be converted into transportation fuels and lubricating oils whose quality will meet tightening environmental and market demands.

The Chevron Isocracking process, widely licensed for over 30 years, has technological advantages for gasoline, middle-distillate, and lubricating oil production. Optimizing a refinery is always a matter of balance. Every benefit has a cost; every incremental gain in margin trades off against a loss somewhere else. Isocracking helps with this balance by delivering tradeoff advantages with respect to product yield, quality, catalyst choice and run length, capital costs, operating costs, versatility, and flexibility. Through its families of amorphous and zeolitic catalysts, Isocracking provides refiners with essential flexibility in choices of crude to buy, products to sell, specs to meet, configurations to use, and efficiency and profitability to achieve, all with the *tradeoff advantage*. This chapter explains the process chemistry which provides these benefits.

ISOCRACKING CHEMISTRY

Chevron's hydrocracking process was named Isocracking because of the unusually high ratio of isoparaffins to normal paraffins in its light products. A high percentage of isoparaffins increases light naphtha product octane numbers and produces outstanding middle-distillate cold flow properties—kerosene/jet freeze point and diesel pour point. In 1992, Chevron enhanced its process capabilities in heavy paraffin isomeriza-

TABLE 7.2.1 Product Quality from Isocracking

Isocracking removes heavy aromatic compounds and creates isoparaffins to produce middle distillates with outstanding burning and cold flow properties.

- Kerosene with low freeze points and high smoke points
- Diesel fuels with low pour points and high cetane numbers
- Heavy naphthas with a high content of single-ring hydrocarbons
- Light naphthas with a high isoparaffin content
- Heavy products that are hydrogen-rich for feeding FCC units, ethylene plants, or lube oil dewaxing and finishing facilities

---

tion by commercializing the Chevron Isodewaxing* process. When combined with hydrocracking, Isodewaxing is the most efficient way to produce high viscosity index (VI), low pour point lube oil base stocks.

Isocracking provides a unique combination of aromatic saturation and paraffin isomerization which generates an attractive combination of product qualities (see Table 7.2.1). The process removes heavy aromatic compounds and produces middle distillates with outstanding burning qualities—high kerosene/jet fuel smoke points and high diesel cetane numbers. The heavy product is rich in hydrogen, making it a prime candidate for feedstock to lube oil facilities, ethylene crackers, or fluid catalytic cracking (FCC) plants.

THE IMPORTANCE OF HYDROGEN

Hydrocracking removes the undesirable aromatic compounds from petroleum stocks by the addition of hydrogen. The amount of hydrogen required depends on the character of the feedstock.

Isocracking produces cleaner fuels and more effective lubricants from a wide variety of petroleum stocks—different crude oil sources and, in some cases, heavy oils generated by different processing routes. These technical challenges can be illustrated by focusing on feed and product hydrogen contents using a Stangeland diagram[1] as shown in Fig. 7.2.1. This relates the hydrogen content of hydrocarbons to their molecular weight and provides a road map for all hydrocarbons present in petroleum stocks. By comparing the characteristics of feedstocks and products, the processing schemes required to go from one to the other can be represented.

The upper line of Fig. 7.2.1 represents the hydrogen content of pure paraffins, which have the highest hydrogen content of any hydrocarbon series. Aromatic compounds have much lower hydrogen content and fall considerably below the paraffin line. This diagram shows regions which meet the specifications for the most important refined products—motor gasoline, jet/kerosene, diesel, and lubricating oils. The regions for middle distillate and lubes all border the paraffin line. Aromatic compounds hurt the qualities of these products. The motor gasoline region is more complex because both hydrogen-rich isoparaffins and hydrogen-poor aromatics improve octane numbers.

The Isocracking process handles variations in feedstocks easily. Table 7.2.2 shows some of the important properties of the straight-run distillates from four popular crude oils: Arabian light, Sumatran light, Chinese (Shengli), and Russian (Western Siberia). Kerosene smoke point, diesel cetane, and vacuum gas oil VIs reflect the overall aromatic nature of the crude oil. Sumatran light is uniquely paraffinic and has the highest hydrogen content. The sulfur levels of the distillates from these different crude oils are

---

*Trademark of Chevron.

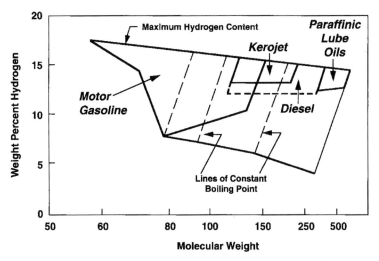

FIGURE 7.2.1 Stangeland diagram showing product hydrogen content. Regions which meet the specifications for jet/kerosene, diesel, and lube products all border the paraffin line. Aromatic compounds hurt the quality of these products.

TABLE 7.2.2 Crude Oil Distillate Qualities

Each crude oil contains distillates of different sulfur levels and burning qualities

			Boiling range					
°F	400–500		500–650		650–800		800–1000	
°C	204–260		260–343		343–427		427–538	
Inspection	Smoke point, mm	Sulfur, wt %	Cetane index	Sulfur, wt %	Viscosity index	Sulfur, wt %	Viscosity index	
Arabian light	22	1.3	51	2.2	65	2.7	55	
Sumatran light	27	0.1	60	0.1	75	0.1	60	
Chinese (Shengli)	20	0.4	52	0.6	40	0.7	26	
Russian (West Siberia)	20	1.1	49	1.8	46	2.3	35	

also shown. Environmental pressures continue to push product sulfur levels down. Although the Stangeland diagram does not include this important contaminant, sulfur occurs primarily in the aromatic component of the feedstock. Isocracking effectively eliminates sulfur as it saturates and cracks the heavy aromatics.

The hydrogen contents of kerosene/jet, diesel, and lube products are shown in Fig. 7.2.2. Again, the paraffinic nature of the Indonesian crude oil is clearly shown. The Russian and Arabian distillates are much more deficient in hydrogen in the vacuum gas oil boiling range. Shengli distillates show less variation in hydrogen content from light to heavy distillates compared to the Russian and Arabian stocks. The refiner's challenge is to upgrade the vacuum gas oils from these and other crude oils into more valuable, hydrogen-rich products. Using Chevron's hydrogen-efficient technology, product specifications can often be exceeded. Consequently, greater blending of lower-quality straight-run or cracked stocks into product pools is possible, thereby increasing the refiner's margin while keeping product prices down.

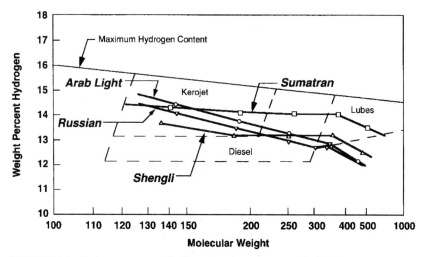

FIGURE 7.2.2 Hydrogen content of distillates from four common crude oils. Some crude oils contain distillates with more hydrogen than others and can be converted to finished products with less effort.

ISOCRACKING CONFIGURATIONS

Two popular configurations are used in the Isocracking process:

- A single-stage once-through plant (SSOT) (see Fig. 7.2.3) is a low-cost facility for partial conversion to light products. This configuration is used when the heavy unconverted oil has value as a lube oil base stock, or as feedstock to an ethylene cracker or FCC unit.
- A two-stage Isocracking unit (see Fig. 7.2.4) is used when maximizing transportation fuel yield is the goal. In this case the unconverted first-stage product is recycle-hydrocracked in a second stage. This configuration can be designed for maximum yield of middle distillates or naphtha, depending on product values. The ratio of kerosene/jet to diesel or middle distillate to naphtha can be varied over a wide range by either changing product fractionator operation or using alternative second-stage catalysts.[2]

ISOCRACKING CATALYSTS

Hydrocracking catalysts for upgrading raw (nonhydrotreated) feedstocks contain a mixture of hydrous oxides for cracking and heavy metal sulfides for hydrogenation.

The simplest method for making hydrocracking catalysts is impregnation of the heavy metals into the pores of the hydrous oxide which has already been formed into the final catalyst shape. The support material can contain a number of components—silica, alumina, magnesia, titania, etc. These are all oxides which can exist in a very high surface area form. The ratio of silica to alumina affects the acidity of the final catalyst and, therefore, its cracking activity. High-silica catalysts have high acidity and high cracking activity; high-alumina catalysts have low acidity and low cracking activity.

Zeolites, crystalline aluminosilicates, are sometimes used in hydrocracking catalysts. Zeolites are very active cracking components which greatly increase the crack-

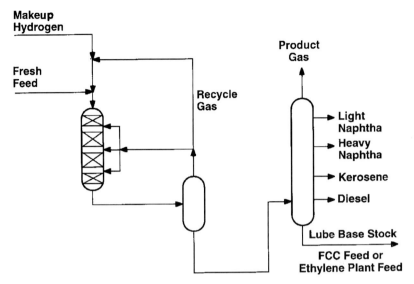

FIGURE 7.2.3 SSOT Isocracking: the simplest, least expensive configuration. Typical configuration for converting heavy oil to lube base stock, FCC feed, or ethylene plant feed.

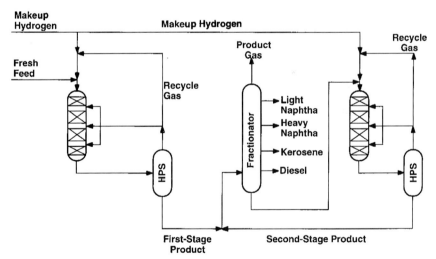

FIGURE 7.2.4 Two-stage isocracking achieves total conversion. Typical configuration for optimizing yields of transportation fuels, middle distillates, and naphtha.

TABLE 7.2.3 Conditions for Testing Isocracking Catalyst on California Gas Oils

High-nitrogen California feedstocks are used in pilot plant studies to differentiate between hydrocracking catalysts.

	Boiling range, °F (°C)	Gravity, °API	Nitrogen, ppm	Pressure, lb/in^2 gage	Catalyst temperature, °F (°C)
Light feed	600–710 (316–377)	23.3	1700	1600	710 (377)
Medium feed	600–900 (316–482)	19.8	2900	1800	732 (389)
Heavy feed	700–980 (371–527)	15.8	5200	2000	763 (406)

Note: °API = degrees on American Petroleum Institute scale.

ing function of dual functional catalysts. This can provide significant improvements in catalyst performance at the cost of a lighter product yield structure. Using zeolites in hydrocracking catalysts introduces a yield/activity tradeoff into the catalyst design and selection process.

Early experience at Chevron with impregnated catalysts showed that the most active hydrocracking catalysts for raw (nonhydrotreated) feedstocks were those with a highly dispersed hydrogenation component, so Chevron developed catalysts designed to optimize dispersion. Instead of impregnating into an already formed support, *cogel* catalysts are made by precipitating all components from solution at the same time into a homogeneous gel. Careful washing, drying, and calcining give the finished catalysts unique and valuable properties and performance.

Isocracking's cogel catalysts have proven to be highly effective with the heaviest part of VGO feeds where the nitrogen compounds are concentrated. Table 7.2.3 shows the pilot plant conditions used to compare the performance of the first cogel catalysts to that of impregnated catalysts. The three straight-run vacuum gas oils (VGOs) from California crude feedstocks used in the tests varied in boiling range from a light (23.3°API gravity) VGO to a heavy (15.8°API) VGO. Nitrogen content ranged from 1700 to 5200 ppm. The impregnated catalysts were a high-silica catalyst and a high-alumina catalyst.

Table 7.2.4 shows that the cogel and high-silica catalysts exhibited the best performance. The high-silica catalysts showed the highest activities on the light feeds, but the cogel catalyst was superior on the heaviest feedstock. The higher denitrification of the cogels is the key to their performance on the heavy ends of vacuum gas oils.

This ability to handle heavy feeds was dramatically demonstrated in long runs designed to measure deactivation (fouling) rates. The cogel fouling rate was an order of magnitude lower than those measured on a variety of high-silica catalysts. This comparison is shown in Fig. 7.2.5, in which the performance is correlated with the active pore volume of the different amorphous catalysts. The active pore volume consists of the volume of pores within the rather narrow pore size range that is needed for optimum conversion of vacuum gas oil feedstocks. The superior stability of the cogel catalyst is a result of the more uniform dispersion of the hydrogenation component and the unique distribution of the pore sizes. This combination is very important for effective processing of heavier feedstocks.

Chevron's Richmond laboratory has created a complete family of amorphous cogel Isocracking catalysts. This consists of catalysts whose exceptional stability is augmented by special capabilities for selective denitrification, conversion to high yields of middle distillates (jet fuel and diesel), and lube base stocks of outstanding quality.

TABLE 7.2.4 Results from Testing Isocracking Catalyst on California Gas Oils

Cogel catalysts are best in converting heavy, high-nitrogen feedstocks.

	Catalyst type		
	Amorphous high-silica impregnated	Amorphous high-alumina impregnated	Amorphous cogel
Light feed	0.27	0.17	0.21
Medium feed	0.21	0.15	0.20
Heavy feed	0.46	0.33	0.46
Light feed	9.4	5.9	7.3
Medium feed	5.3	4.5	5.6
Heavy feed	4.5	4.9	5.5

*Rate constants are first-order, for conversions below 550°F for light and medium feeds, 650°F for heavy feeds.

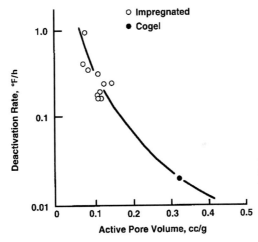

FIGURE 7.2.5 Fouling rate with heavy feeds—cogel versus impregnated catalysts. Cogels show much greater stability and longer run cycles than other amorphous catalysts.

The addition of small amounts of zeolite components to cogel catalysts enhanced the cracking function of the catalysts. Refiners who need to meet seasonal gasoline (mogas) and jet fuel demands rather than maximize diesel have found that cogel catalysts with zeolite components will produce lighter product slates more efficiently. Chevron calls these catalysts *amorphous/zeolite* since both components contribute equally to catalyst performance.

A third category of Chevron hydrocracking catalysts, known as *zeolite*, are noncogel, high-zeolite content catalysts which were introduced by Chevron in the 1980s for naphtha-producing applications. Performance characteristics of these materials are included below in "Product Yields and Qualities."

Depending on the catalyst selected, a complete slate of desired products can be produced from a variety of available feedstocks. Feedstocks have ranged from light naphthas to deasphalted oils. FCC cycle stocks are commonly upgraded by hydrocracking. The desired products vary from country to country, region to region, and refinery to refinery. In many regions, production of good-quality middle distillates and lube oils gives the best margins and hydrocracking is the only process that provides the required feed conversion.

Given the various performance characteristics available from the wide range of high-performance catalysts, Isocracking can deliver the following advantages[3,4]:

- Outstanding activity and resistance to fouling, minimizing hydrocracker capital investment and hydrogen consumption
- Higher yields of desired products
- Product specifications are always met or exceeded
- Long catalyst cycle lengths combined with successful regenerability
- Consistent product yields and qualities through run cycle
- Flexibility to change product mix

PRODUCT YIELDS AND QUALITIES

Meeting the target product yield is the most important property of a hydrocracking catalyst system. Figure 7.2.6 illustrates the different yield structures that Isocracking can provide by judicious choice of catalyst and design parameters.

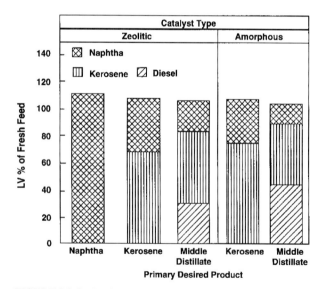

FIGURE 7.2.6 Product yields from alternative catalyst systems. Tailoring Isocracking catalyst systems enables refiners to produce their target product slate from a wide range of crudes.

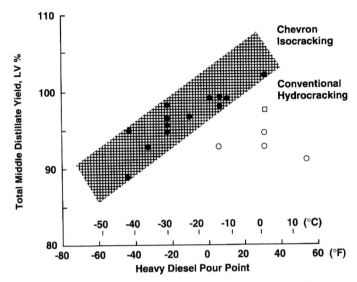

FIGURE 7.2.7 Yield and product quality of middle distillates. Isocracking catalysts increase yields by 5 to 10 percent while maintaining pour points as low as −40°F (−40°C).

Amorphous catalysts like ICR 106 or ICR 120 are used for maximum production of middle distillates. Isocrackers using these amorphous catalysts can achieve a 95 liquid volume percent (LV %) yield of total middle distillate (kerosene plus diesel) while producing only 15 LV % naphtha.

Amorphous Isocracking catalysts give better cold flow properties than other hydrocracking catalysts, but not at the expense of yields (see Fig. 7.2.7). Isocracking catalysts give a 5 to 10 percent higher yield of quality middle distillate with as much as 22°C lower heavy diesel pour point. Isocracking also gives better-quality end-of-run products. With some catalysts, an increase in product aromatic levels occurs as the run cycle progresses. These aromatics cause the burning quality of middle distillates to deteriorate significantly. Isocracking catalysts provide consistent product quality throughout the length of the run. (See Fig. 7.2.8 for variation of jet quality.)

Polynuclear aromatic (PNA) compounds are undesired by-products formed through a complex sequence of chemical reactions occurring during typical hydrocracking conditions.[5] In processing of heavy straight-run feedstocks using zeolitic catalysts in a recycle configuration, PNAs deposit in the cooler parts of the plant. This disrupts hydrocracker operation. To prevent PNA deposits, most hydrocrackers must operate with a heavy product bleed stream. By using amorphous cogel catalysts, which are much less prone to this phenomenon than zeolitic catalysts, and careful unit design, PNA formation can be controlled and unit downtime can be minimized.

Isocracking for Middle-Distillate Production

A two-stage Isocracker using amorphous Isocracking catalysts produces very high yields of kerosene/jet and diesel fuel. The burning qualities of middle distillates produced in the second (recycle) stage are much better than the equivalent stocks pro-

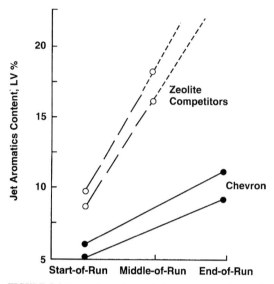

FIGURE 7.2.8 Jet aromatics content over an operating cycle in hydrocracking of Middle Eastern VGO. Chevron catalysts maintain lower jet aromatics and better smoke points throughout a run cycle.

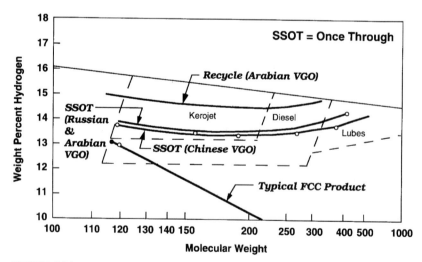

FIGURE 7.2.9 Hydrogen contents for single-stage versus recycle Isocracking products. Both Isocracker configurations add hydrogen where it is needed, but with recycle operations the exceptional product quality enables the refiner to upgrade fuel oil to diesel value.

duced in a once-through unit. Table 7.2.9 shows a typical example for an Arabian vacuum gas oil feedstock. The aromatic contents of both the jet and diesel products are less than 1 percent. This difference is shown dramatically in Fig. 7.2.9, which compares the product hydrogen contents for the yield structures shown on Tables 7.2.5 and 7.2.6 with the corresponding Arabian and Chinese (Table 7.2.9) single-stage operations. For comparison, the very low hydrogen content of FCC products is also shown. In recycle operation, Isocracking produces middle distillates which exceed target specifications for smoke point, cetane, and sulfur. This enables a refiner to blend more lower-value diesel stock into the product pool, thereby upgrading it from fuel oil to diesel value. Figure 7.2.10 shows that diesel blends containing up to 40 percent FCC light cycle oil are possible, depending on the diesel cetane and sulfur specifications.

Isocracking for Naphtha Production

Zeolitic catalysts are generally used in this service since they are more active than amorphous catalysts and produce a higher ratio of naphtha to middle distillate. Chevron's noble metal/zeolite and base metal/zeolite catalysts have different performance characteristics. The noble metal/zeolite catalyst provides higher liquid and jet fuel yields, higher smoke point jet fuel, and longer cycle length. The base metal/zeo-

TABLE 7.2.5 Product Yields and Qualities for Arabian VGO

Recycle Isocracking maximizes middle distillate yields and qualities, producing 5% more heavy diesel than conventional catalysts.

Feed	
Source	Arabian VGO
Gravity, °API	33.8
Sulfur, ppm	8.0
Nitrogen, ppm	0.8
D 2887 Distillation, °C:	
ST/5	363/378
10/30	386/416
50	444
70/90	479/527
95/EP	546/580

	Product yields		Product quality	
Product	wt %	LV %	Characteristic	Value
C_5–82°C	6.8	8.9		
82–121°C	8.8	10.4	P/N/A	58/42/0
121–288°C	49.1	53.8	P/N/A	57/42/1
			Smoke point, mm	41
			Freeze point, °C	≤75
288–372°C	32.5	34.2	Cetane index	>68
			P/N/A	62/37/1
			Cloud point, °C	−18
			Pour point, °C	−39

Note: P/N/A = paraffins/naphthalenes/aromatics; EP = endpoint.

TABLE 7.2.6 Isocracking—Typical Yields and Product Qualities

Isocracking produces high yields of 100 VI lube base stocks.

Feed	
Source	Russian
Gravity, °API	18.5
Sulfur, wt %	2.28
Nitrogen, wt %	0.28
Wax, wt %	6.5
D 2887 distillation, °C:	
ST	435
10/30	460/485
50	505
70/90	525/550
EP	600

| Product | Product yields | | Product quality | |
	wt %	LV %	Characteristic	Value
C_5–180°C	4.8	5.9		
180–290°C	15.4	17.4	Smoke point, mm	22
			Cetane index	56
290–370°C	16.4	18.1	Flash point, °C	145
370–425°C	13.7	15.0		
425–475°C	19.3	21.0	Solvent dewaxed 240N	
			VI	97
			Pour point, °C	−12
475°C+	27.4	29.6	Solvent dewaxed 500N	
			VI	105
			Pour point, °C	−12

lite catalyst provides a lower liquid yield but a higher yield of C_4^- gas and isobutane, a higher-octane light naphtha, and a more aromatic product naphtha. The selection of a noble metal or a base metal catalyst for a hydrocracker depends on the economics of the particular refinery situation.

Chevron has developed and commercialized a number of improved zeolitic Isocracking catalysts (ICR 208, 209, and 210)[9] capable of giving high naphtha yields (see Fig. 7.2.6) and long run lengths in many commercial plants. Figure 7.2.11 shows the very low deactivation rate that is typical of operation with ICR 208 in the Chevron U.S.A. Richmond two-stage Isocracker.

Refiners may take advantage of the high activity and long cycle length of Chevron's zeolitic catalysts by:

- Increasing plant throughput
- Processing more difficult, lower-value feeds
- Decreasing first-stage severity to balance catalyst life in both stages
- Decreasing the hydrogen partial pressure to reduce hydrogen consumption

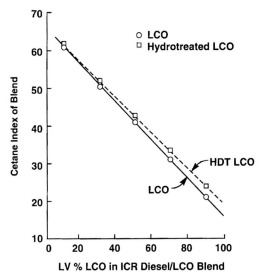

FIGURE 7.2.10 Isocracker diesel upgrades light cycle oils. Using recycle Isocracking, refiners can reduce product costs by adding up to 40 percent FCC light cycle oil to their diesel blends while still meeting 45 cetane index.

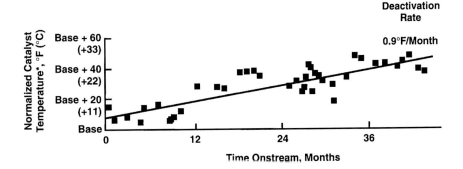

*Normalized to Base Case Conditions

FIGURE 7.2.11 Deactivation rate of zeolitic catalysts. Zeolitic catalyst, ICR 208, has demonstrated long life in Chevron's Richmond Refinery, maintaining product quality throughout the run.

Isocracking for Lube Production

The lube oil industry faces constant change resulting from environmental legislation, new engine designs, consumer demands, competitive pressures, and availability of lube-quality crudes. Lube oil manufacturers must manage these changes to stay competitive. Improved fuel economy and environmental requirements are driving the demand for higher-quality, lower-viscosity multigrade oils. Using conventional mineral oil technology, it is very difficult to meet stringent requirements on volatility. Base

TABLE 7.2.7 Solvent Extraction—Typical Yields and Product Qualities

Furfural extracts contain high levels of heavy aromatics and can be used only in fuel oil or as FCC feed.

Characteristic	Typical inspections
Gravity, °API	13.3
Specific gravity	0.977
Sulfur, wt %	4.3
Nitrogen, ppm	1900
Aromatics, wt %	82
Conradson carbon, wt %	1.4
Aniline point, °F	108
TBP distillation, °F:	
ST	700
10%	788
50%	858
90%	932
EP	986
Carbon, wt %	84.82
Hydrogen, wt %	10.68
Viscosity index	−50

Note: TBP = true boiling point.

oils with very high paraffin content have low volatility for their viscosity and much higher viscosity indexes than more aromatic oils.

A single-stage once-through Isocracker removes heavy aromatics very effectively, thereby producing highly paraffinic lube base stocks. Isocracking has several advantages over the more traditional solvent extraction approach to lube base oil production:

- Solvent extraction upgrades the VI of feed by physical separation; i.e., low-VI components are removed as extract and high-VI components remain in the raffinate. Isocracking upgrades VI by removing low-VI components through aromatics saturation and naphthenic ring opening. By creating higher-VI components, Isocracking allows the use of unconventional crudes for lube production.
- Feedstock (unconventional crudes) and operating costs are lower with Isocracking than with solvent extraction.
- The by-products of Isocracking include valuable high-quality transportation fuels, whereas solvent refining produces a highly aromatic extract which is used in fuel oil or as FCC feed. Typical Isocracking product yields and qualities from a Russian feedstock are shown in Table 7.2.6. For comparison, typical extract qualities are shown in Table 7.2.7.
- Lube Isocrackers are easily adapted to meet other processing objectives. For example, during times of low lube demand, Isocrackers can produce transportation fuels and prepare premium FCC feed. (A corollary of this is that Isocrackers designed for transportation fuels can also be adapted for lube operation.)

Depending on the refiner's processing objective, Chevron amorphous and amorphous/zeolite cogel catalysts are used in Isocrackers operating in base oil production mode.

Isodewaxing

The waxy lube oil produced from hydrocracking must be dewaxed in order to produce lube base stocks which meet quality requirements for finished lubricants. Chevron's Isodewaxing process outperforms traditional solvent or catalytic dewaxing processes in producing high-quality base oils. Traditional dewaxing processes remove wax from lube oils by crystallization (solvent dewaxing) or by cracking the normal paraffin wax molecules to light gas (catalytic dewaxing). In contrast, Chevron's Isodewaxing catalyst isomerizes the normal paraffin wax to desirable isoparaffin lube molecules, resulting in high-VI, low-pour-point base oils, while coproducing small quantities of high-quality middle-distillate transportation fuels.

Operating conditions for Isodewaxing are very similar to conventional lube oil hydrotreating, thus it is generally possible to combine the Isodewaxer/hydrofinisher operations in the same process unit or use an existing hydrotreater for a revamp project.

Generally speaking, the higher the VI, the better the cold flow and thermal stability properties of the lubricant. Isodewaxing economically produces conventional base oils (CBOs) with VIs of 95 to 110 or unconventional base oils (UCBOs) with VIs over 110 from either hydrocracked feedstocks or hydrotreated feedstocks from a solvent-extraction process. In fact, with Isocracking, the higher the wax content in the feedstock, the higher the product VI. UCBOs up to about 130 VI are today typically prepared by severe hydrocracking of vacuum gas oils derived from lube crudes followed by solvent extraction and solvent dewaxing. Isodewaxing can also produce this type of UCBO lubes from hydrocrackate, or from waxy vacuum gas oil, but at a lower cost because solvent dewaxing is not required. Table 7.2.8 shows the capabilities for UCBO manufacture by Isodewaxing on a Sumatran light vacuum gas oil. Isodewaxing produces 2½ times the UCBO yield of conventional dewaxing.

Chevron Isodewaxing catalyst ICR 404 produces mineral-oil-based lubricants that approach the performance of synthetic lubricants, but at a much lower manufacturing cost. Isodewaxed base oils have:

- Better cold flow properties to ensure adequate lubrication during cold engine startups
- Low viscosity (for fuel efficiency) combined with low volatility (to reduce oil consumption and emissions)
- Higher VI for improved lubrication in high-temperature, high-shear conditions
- Greater oxidation stability for longer lubricant life and fewer engine deposits

Isodewaxing is the most cost-effective method to produce a mineral-oil-based lubricant which will meet strict engine performance requirements.

TABLE 7.2.8 Unconventional Base Oil from Sumatran Light VGO

Isodewaxing produces 2½ times the UCBO that is achieved by solvent dewaxing.

VGO → Isocracking → Dewaxing → Hydrofinishing →

Product	Chevron Isodewaxing	Solvent dewaxing
Viscosity at 100°C, cSt	4.5	3.8
VI	130	133
Pour point, °C	−12	−12
Yield, LV % VGO feed	65	25

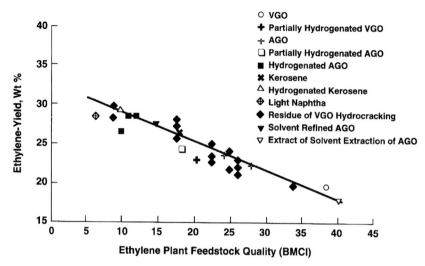

FIGURE 7.2.12 Correlation between ethylene yield and Bureau of Mines correlation index. Ethylene can be produced from many different feedstocks.

Isocracking for Petrochemical Feedstock Production

Both aromatics and olefin users within the petrochemical industry benefit from hydrocracking processes. The aromatics industry takes advantage of the conservation of single-ring compounds in hydrocracked naphthas. These compounds are precursors for the benzene, toluene, and xylenes produced when the naphtha is catalytically reformed. The Isocracking catalysts and configurations used in this application are the same as those used for gasoline production. The olefin industry requires hydrogen-rich feedstocks, since increasing the hydrogen content invariably improves the yield of olefins and decreases the production of heavy, undesirable products. Figure 7.2.12 shows the correlation[10] between the ethylene yield and the Bureau of Mines Correlation Index. This index is closely related to feedstock hydrogen content. Sinopec is operating a Chevron SSOT Isocracker at the Qilu Refinery. The heavy product from that unit is fed to an ethylene cracker. Typical product yields and qualities are shown in Table 7.2.9. Note the excellent quality middle distillates produced in the same operation.

INVESTMENT AND OPERATING EXPENSES

The capital investment required for an Isocracking Unit depends on the type of feedstock to be processed and the quality of the products which are desired. For middle-distillate and lube oil base stock production, the greater the difference in hydrogen content between the feedstock and the desired products, the greater the capital requirement. Feedstock impurities, such as metals, asphaltenes, nitrogen, and sulfur, increase refining difficulty. Care must be taken in the feedstock preparation facilities in order to minimize their effect.

Table 7.2.10 gives a rough idea of typical onplot investment ranges for installing Isocracking units on the U.S. Gulf Coast. Table 7.2.11 shows typical utility requirements for the same plants.

TABLE 7.2.9 Ethylene Plant Feed Production via Isocracking.

SSOT Isocracking upgrades VGO into a high yield of good-quality ethylene plant feed.

Feed	
Source:	Chinese (Shengli)
Gravity, °API	21.4
Sulfur, wt %	1.03
Nitrogen, wt %	0.21
D 2887 Distillation, °C:	
ST/5	314/353
10/30	371/414
50	441
70/90	463/500
95/EP	518/551

	Product yields		Product quality	
Product	wt %	LV %	Characteristic	Value
C_5–129°C	13.31	17.05		
129–280°C	34.63	39.41	Smoke point, mm	26
			Freeze point, °C	−62
280–350°C	12.40	13.64	Cetane index	57
			Pour point, °C	−12
350°C+	37.04	40.45	Sulfur, ppm	7
			BMCI	15

Note: BMCI = Bureau of Mines Correlation Index.

SUMMARY

Chevron's Isocracking configurations and catalyst systems have produced outstanding quality products, from a variety of feedstocks, all around the world. The Isodewaxing technology has heralded in a new era of cost-effective, superior-quality lube oil production.

Chevron is the only major operator of high-pressure hydroprocessing units that also develops refining technology. Chevron's Richmond Refinery contains the largest hydrocracking complex in the world (see Fig. 7.2.13). Within the same plot space, there are 15 high-pressure reactors representing a 45,000-barrels-per-operating-day (BPOD) two-stage Isocracker, two SSOT Isocrackers (with a total feed rate of 30,000 BPOD) producing lube stocks which are then Isodewaxed, and a 65,000-BPOD deasphalted oil hydrotreater. This experience guides Chevron's development of new hydrocracking catalysts and processes. The Isocracking process is offered for license by:

Chevron Products Company, Inc.

Global Lubricants and Technology

Marketing Division

555 Market Street, San Francisco, California, U.S.A. 94120

TABLE 7.2.10 Typical Isocracker Capital Investments*

Operation	$ U.S. per BPOD of feed
Single-stage, once-through for lubes or fuels	1500–2500
Two-stage	
Middle distillate	2000–3000
Naphtha	2500–3500

*U.S. Gulf Coast, mid-1995, on plot only.
Note: BPOD = barrels per operating day.

TABLE 7.2.11 Typical Isocracker Utility Requirements

	Configuration	
	Single-stage	Two-stage
Fuel, million kcal/h	0.5–0.7	0.8–1.2
Power, kW	250–300	300–425
Cooling water, m^3/h	50–70	50–70
Medium-pressure steam, thousand kg/h	0 to 0.2	−0.4 to 0.2
Condensate, m^3/h	−0.4 to −0.7	−0.7 to −0.9

*Basis: consumption per 1000-BPOD capacity.
Note: Each Isocracking plant has its own unique utility requirements depending on the refinery situation and the need to integrate with existing facilities. The above guidelines can be used to give typical operating expenses.

and by

ABB Lummus Global, Inc.

Technology Division

1515 Broad Street

Bloomfield, New Jersey, U.S.A. 07003

REFERENCES

1. A. G. Bridge, G. D. Gould, and J. F. Berkman, "Chevron Hydroprocesses for Upgrading Petroleum Residue," *Oil and Gas Journal,* **85,** January 19, 1981.
2. A. G. Bridge, J. Jaffe, B. E. Powell, and R. F. Sullivan, "Isocracking Heavy Feeds for Maximum Middle Distillate Production," 1983 API Meeting, Los Angeles, May 1993.
3. A. G. Bridge, D. R. Cash, and J. F. Mayer, "Cogels—A Unique Family of Isocracking Catalysts," 1993 NPRA Meeting, San Antonio, March 21–23, 1993.
4. R. L. Howell, R. F. Sullivan, C. Hung, and D. S. Laity, "Chevron Hydrocracking Catalysts Provide Refinery Flexibility," Japan Petroleum Institute, Petroleum Refining Conference, Tokyo, October 19–21, 1988.
5. R. F. Sullivan, M. Boduszynski, and J. C. Fetzer, "Molecular Transformations in Hydrotreating and Hydrocracking," *Journal of Energy and Fuels,* **3,** 603 (1989).

FIGURE 7.2.13 Chevron U.S.A.'s Richmond Refinery contains the largest hydrocracking complex in the world.

6. D. V. Law, "New Catalyst and Process Developments in Residuum Upgrading," The Institute of Petroleum, Economics of Refining Conference, London, October 19, 1993.
7. M. W. Wilson, K. L. Eiden, T. A. Mueller, S. D. Case, and G. W. Kraft, "Commercialization of Isodewaxing—A New Technology for Dewaxing to Manufacture High-Quality Lube Base Stocks," 1994 NPRA Meeting, Houston, November 3–4, 1994.
8. S. J. Miller, "New Molecular Sieve Process for Lube Dewaxing by Wax Isomerization," *Microporous Materials,* **2,** 439–449 (1994).
9. A. J. Dahlberg, M. M. Habib, R. O. Moore, D. V. Law, and L. J. Convery, "Improved Zeolitic Isocracking Catalysts," 1995 NPRA Meeting, San Francisco, March 19–21, 1995.
10. S. Nowak, G. Zummerman, H. Guschel, and K. Anders, "New Routes to Low Olefins From Heavy Crude Oil Fractions," *Catalysts in Petroleum Refining,* pp. 103–127, Elsevier Science Publishers, New York, 1989.

CHAPTER 7.3
UOP UNICRACKING PROCESS FOR HYDROCRACKING

Mark Reno
UOP
Des Plaines, Illinois

INTRODUCTION

Hydrogenation and hydrocracking are among the oldest catalytic processes used in petroleum refining. They were originally employed in Germany in 1927 for converting lignite into gasoline and later used to convert petroleum residues into distillable fractions. The first commercial hydrorefining installation in the United States was at the Standard Oil Company of Louisiana in Baton Rouge in the 1930s. No significant increase in the hydrorefining of petroleum occurred during the next 20 years. In the 1950s, hydrodesulfurization and mild hydrogenation processes experienced a tremendous growth, mostly because large quantities of by-product hydrogen were made available from the catalytic reforming of low-octane naphthas to produce high-octane gasoline.

The first modern hydrocracking operation was placed on-stream in 1959 by the Standard Oil Company of California. The unit was small, producing only 1000 barrels per stream day (BPSD). When a unit was installed to complement an existing fluid catalytic cracking (FCC) unit, refiners quickly recognized that the hydrocracking process had the flexibility to produce varying ratios of gasoline and middle distillate. Thus, the stage was set for rapid growth in U.S. hydrocracking capacity from about 3000 BPSD in 1961 to about 120,000 BPSD in just 5 years. From 1965 to 1983, U.S. capacity had grown eightfold, to about 980,000 BPSD.

Outside the United States, early applications were liquefied petroleum gas (LPG) production by hydrocracking naphtha feedstocks. The excellent-quality distillate fuels produced when hydrocracking gas oils and other heavy stocks led to the choice of hydrocracking as a major conversion step in refinery expansions in locations where diesel and jet fuels were in demand. Interest in high-quality distillate fuels produced by hydrocracking has increased dramatically worldwide. As of 1995, more than 3 million BPSD of hydrocracking capacity is either operating or is in design and construction worldwide.

PROCESS APPLICATIONS

Hydrocracking is one of the most versatile of all petroleum-refining processes. Any fraction from naphtha to nondistillables can be processed to produce almost any desired product with a molecular weight lower than that of the chargestock. At the same time that hydrocracking takes place, sulfur, nitrogen, and oxygen are almost completely removed, and olefins are saturated so that products are a mixture of essentially pure paraffins, naphthenes, and aromatics. Table 7.3.1 illustrates the wide range of applications of hydrocracking by listing typical chargestocks and the usual desired products.

The first eight chargestocks are virgin fractions of petroleum crude and gas condensates. The last four are fractions produced from catalytic cracking and thermal cracking. All these streams are being hydrocracked commercially to produce one or more of the products listed.

PROCESS DESCRIPTION

The UOP* Unicracking* process is carried out at moderate temperatures and pressures over a fixed catalyst bed in which the fresh feed is cracked in a hydrogen atmosphere. Exact process conditions vary widely, depending on the feedstock properties and the products desired. However, pressures usually range between 35 and 219 kg/cm^2 (500 and 3000 lb/in^2 gage), and temperatures between 280 and 475°C (536 and 887°F).

Chemistry

The chemistry of hydrocracking is essentially the carbonium ion chemistry of catalytic cracking coupled with the chemistry of hydrogenation. The result is a highly branched product because of a great tendency to form tert-butylcarbonium ion. The postulated reaction of paraffins is shown in Fig. 7.3.1. The reaction starts with the formation of olefins at metallic sites and the formation of carbonium ions from these olefins at acidic sites. Extensive catalytic cracking followed by hydrogenation to form isoparaf-

---

*Trademark and/or service mark of UOP.

TABLE 7.3.1 Applications of the Unicracking Process

Chargestock	Products
Naphtha	Propane and butane (LPG)
Kerosene	Naphtha
Straight-run diesel	Naphtha and/or jet fuel
Atmospheric gas oil	Naphtha, jet fuel, and/or distillates
Natural gas condensates	Naphtha
Vacuum gas oil	Naphtha, jet fuel, distillates, lubricating oils
Deasphalted oils and demetallized oils	Naphtha, jet fuel, distillates, lubricating oils
Atmospheric crude column bottoms	Naphtha, distillates, vacuum gas oil, and low-sulfur residual fuel
Catalytically cracked light cycle oil	Naphtha
Catalytically cracked heavy cycle oil	Naphtha and/or distillates
Coker distillate	Naphtha
Coker heavy gas oil	Naphtha and/or distillates

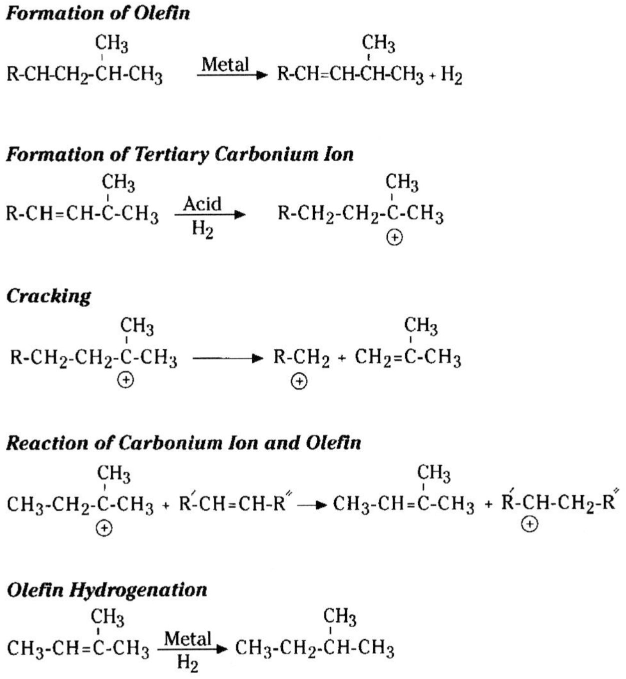

Formation of Olefin

$$\text{R-CH-CH}_2\text{-}\underset{\underset{\text{CH}_3}{|}}{\text{CH}}\text{-CH}_3 \xrightarrow{\text{Metal}} \text{R-CH=CH-}\underset{\underset{\text{CH}_3}{|}}{\text{CH}}\text{-CH}_3 + \text{H}_2$$

Formation of Tertiary Carbonium Ion

$$\text{R-CH=CH-}\underset{\underset{\text{CH}_3}{|}}{\text{C}}\text{-CH}_3 \xrightarrow[\text{H}_2]{\text{Acid}} \text{R-CH}_2\text{-CH}_2\text{-}\underset{\underset{\text{CH}_3}{|}}{\overset{+}{\text{C}}}\text{-CH}_3$$

Cracking

$$\text{R-CH}_2\text{-CH}_2\text{-}\underset{\underset{\text{CH}_3}{|}}{\overset{+}{\text{C}}}\text{-CH}_3 \longrightarrow \overset{+}{\text{R-CH}_2} + \text{CH}_2\text{=}\underset{\underset{\text{CH}_3}{|}}{\text{C}}\text{-CH}_3$$

Reaction of Carbonium Ion and Olefin

$$\text{CH}_3\text{-CH}_2\text{-}\underset{\underset{\text{CH}_3}{|}}{\overset{+}{\text{C}}}\text{-CH}_3 + \text{R}'\text{-CH=CH-R}'' \longrightarrow \text{CH}_3\text{-CH=}\underset{\underset{\text{CH}_3}{|}}{\text{C}}\text{-CH}_3 + \overset{+}{\text{R}'\text{-CH-CH}_2\text{-R}''}$$

Olefin Hydrogenation

$$\text{CH}_3\text{-CH=}\underset{\underset{\text{CH}_3}{|}}{\text{C}}\text{-CH}_3 \xrightarrow[\text{H}_2]{\text{Metal}} \text{CH}_3\text{-CH}_2\text{-}\underset{\underset{\text{CH}_3}{|}}{\text{CH}}\text{-CH}_3$$

FIGURE 7.3.1 Postulated paraffin-cracking mechanism.

fins appears to be the primary reaction. The rate of hydrocracking increases with the molecular weight of the paraffin.

A typical hydrocracking reaction for a cycloparaffin (Fig. 7.3.2) is known as a paring reaction, in which methyl groups are rearranged and then selectively removed from the cycloparaffin without severely affecting the ring itself. Normally the main acyclic product is isobutane. The hydrocracking of multiple-ring naphthene, such as decalin, is more rapid than that of a corresponding paraffin. Naphthenes found in the product contain a ratio of methylcylopentane to methylcyclohexane that is far in excess of thermodynamic equilibrium.

Reactions during the hydrocracking of alkyl aromatics (Fig. 7.3.3) include isomerization, dealkylation, paring, and cyclization. In the case of alkylbenzenes, ring cleavage is almost absent, and methane formation is at a minimum.

Catalyst

Several types of catalyst are used in the hydrocracking process. These catalysts combine cracking and hydrogenation activity in varying proportions to achieve the goal of converting a given feedstock to a desired product. Hydrogenation activity is achieved through the use of metal promoters impregnated on a catalyst support. These promoters can be either base or noble metals, depending on the ultimate application of the catalyst. Cracking activity is achieved by varying catalyst-support acidity. These variations are achieved primarily by using combinations of silica and alumina in both amorphous and crystalline, or zeolitic, support material.

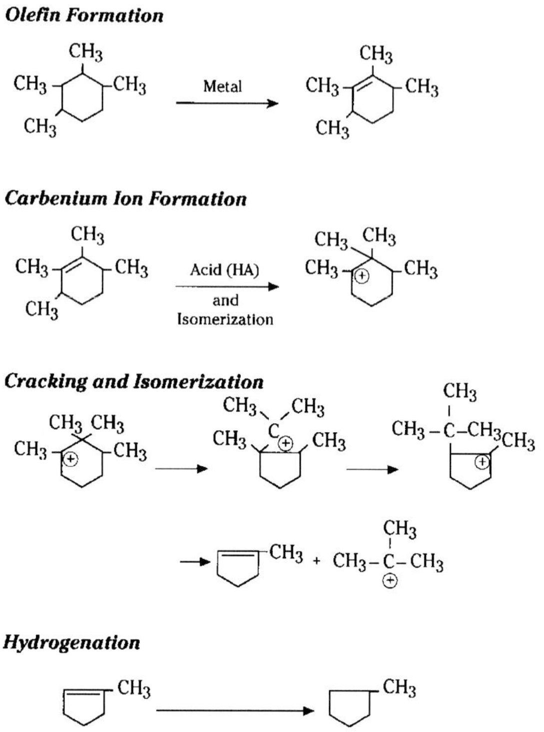

FIGURE 7.3.2 Postulated cracking mechanism for naphthenes.

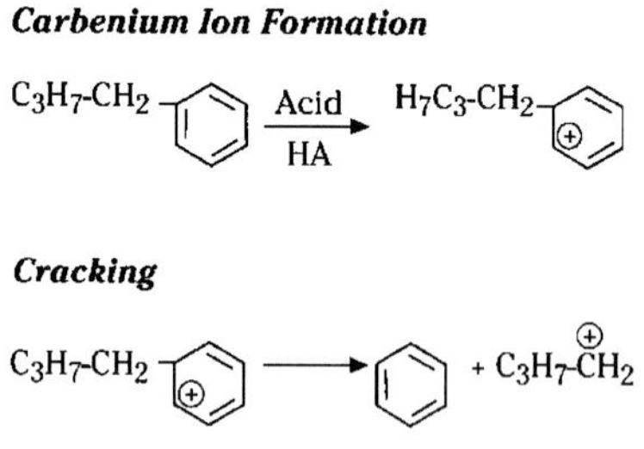

FIGURE 7.3.3 Postulated aromatic-dealkylation mechanism. Isobutane is also formed following butyl carbenium ion isomerization, olefin formation, and hydrogenation.

A postulated network of reactions that occur in a typical hydrocracker processing a heavy petroleum fraction is shown in Fig. 7.3.4. The reactions of the multiring species should be noted. These species, generally coke precursors in nonhydrogenative cracking, can be effectively converted to useful fuel products in a hydrocracker because the aromatic rings can be first hydrogenated and then cracked.

Amorphous silica or alumina was the first catalyst support material to be used extensively in hydrocracking service. When combined with base-metal hydrogenation promoters, these catalysts effectively converted vacuum gas oil (VGO) feedstocks to

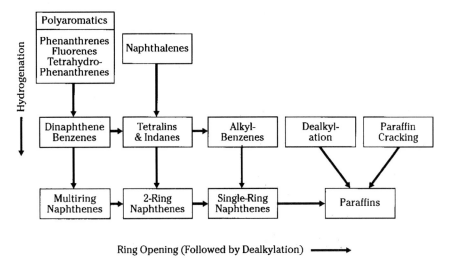

FIGURE 7.3.4 Hydrocracking reactions.

products with lower molecular weight. Over three decades of development, amorphous catalyst systems have been refined to improve their performance, primarily by varying the acid-metal balance. UOP catalysts in this category include DHC*-2, DHC-6, and most recently DHC-8. All catalysts have excellent middle distillate selectivity and activity suitable for full conversion of a wide range of feedstocks.

Crystalline catalyst-support materials, such as zeolites, have been used in hydrocracking catalysts by UOP since the late 1960s. The combination of selective pore geometry and varying acidity has allowed the development of catalysts that convert a wide range of feedstocks to virtually any desired product. UOP now offers catalysts that will selectively produce LPG, naphtha, middle distillates, or lube base oils at high conversion activity using molecular-sieve catalyst support materials. The UOP zeolitic catalysts used in hydrocracking service include HC*-14, 16, 18, 24, 26, 28, 33, and 34. As a result of continued development in this area, catalysts that convert a variety of feeds into virtually any product slate have been discovered. Unlike the amorphous-based catalysts, the zeolites are usually more selective to lighter products and thus more suitable when flexibility in product choice is desired. In addition, zeolitic catalysts typically employ a hydroprocessing catalyst specifically designed to treat the feed to remove contaminants prior to conversion. UOP catalysts such as HC-K and HC-H are used for this service. These materials are specifically designed with high hydrogenation activity to effectively remove feed sulfur and nitrogen. They thus ensure a clean feed to the zeolitic-based catalyst and optimal performance of the conversion catalysts. In recent years, UOP has successfully blended the high activity of zeolite catalysts with the excellent distillate selectivity seen with amorphous supports. These catalysts include DHC-100, HC-22, and DHC-32, all of which have been used commercially. The excellent distillate selectivity of these systems, coupled with superior temperature performance, make them an attractive choice in many situations.

One important consideration of catalyst selection is regenerability. Hydrocracking catalysts typically operate for cycles of 2 years between regenerations but can be operated at virtually any cycle depending on process conditions. When end-of-run conditions are reached, as dictated by either temperature or product performance, the catalyst

---

*Trademark and/or service mark of UOP.

is typically regenerated. Regeneration primarily involves combusting the coke off the catalyst in an oxygen environment to recover fresh catalyst surface area and activity. Regenerations can be performed either with plant equipment if it is properly designed or a vendor regeneration facility. Both amorphous and zeolitic catalyst supplied by UOP are fully regenerable and recover almost full catalyst activity after carbon burn.

Thermodynamics

The reactions that occur in hydrocracking can be grouped into two broad classes:

- Hydrogenation of olefins, aromatic rings, sulfur compounds, nitrogen compounds, and oxygen compounds
- Cracking reactions of C-C bonds

The cracking reactions are common also to fluid catalytic cracking (FCC), and their net effect is a slightly endothermic heat of reaction. However, the hydrogenation reactions are highly exothermic, and when their contribution is added, the net heat of reaction for the hydrocracking process is then exothermic. Thus, a large adiabatic temperature rise occurs in a commercial hydrocracker, and a major part of the engineering involved in the design of a hydrocracker is the control of this temperature rise. Excessively high temperatures can lead to coke deposition and deactivation of the hydrocracking catalyst. Temperature control is accomplished, as will be seen later, by the use of a cold recycle-gas quench injected at various points between a series of catalyst beds.

Hydrocracking Flow Schemes

Over the years, many different flow schemes have been used for hydrocracking service so that various feeds can be processed to a full range of products. UOP has developed four distinct flow configurations for hydrocracking service:

- *Single-stage:* The single-stage flow scheme is the most widely used because of its efficient design, which results in minimum cost for a full-conversion operation. This scheme can employ a combination of hydrotreating and cracking catalysts or simply amorphous cracking catalysts, depending on the final product required.
- *Once-through:* Unlike the single-stage flow scheme, the once-through flow scheme is a partial conversion option that results in some yield of material similar in boiling range to the feed. This material, or unconverted product, is highly saturated and free of feed contaminants but is similar in molecular weight to the feed. If a refinery has a use for this unconverted product, such as FCC feed or high-quality lube base oil, this flow scheme is preferred.
- *Two-stage:* In the two-stage flow scheme, feedstock is treated and partially converted once-through across a first reactor section. Products from this section are then separated by fractionation. The bottoms from the fractionation step are sent to a second reactor stage for complete conversion. This flow scheme is most widely used for large units where the conversion in the once-through first stage allows high feed rates without parallel reactor trains and the added expense of duplicate equipment.
- *Separate-hydrotreat:* Like two-stage flow, the separate-hydrotreat flow scheme uses two reactor stages. In this scheme, however, the recycle gas loops for the two stages are independent, thus allowing the processing of feedstocks with very high contaminant levels or the use of contaminant-sensitive catalyst in the second stage if dictated by product demands.

The single-stage flow scheme is the most widely used hydrocracking flow scheme in commercial service. The flow scheme allows the complete conversion of a wide range of feedstocks and product recovery designed to maximize virtually any desired product. The design of this unit configuration has been optimized to reduce capital cost and improve operating performance. Greater than 95 percent on-stream efficiency is typical.

Figure 7.3.5 illustrates a typical single-stage flow scheme. As shown, feedstock, recycle oil, and recycle gas are exchanged against reactor effluent to recover process heat and are then sent through a final charge heater and into the reactor section. The reactor section contains catalysts that allow maximum production of the desired product slate. In virtually all hydrocracking systems, the reactions are highly exothermic and require cold hydrogen quench injection into the reactors to control reactor temperatures. This injection is accomplished at quench injection points with sophisticated reactor internals that both mix reactants and quench and redistribute the mixture. Proper mixing and redistribution are critical to ensure good temperature control in the reactor and good catalyst utilization through acceptable vapor or liquid distribution.

Reactor effluent is sent through exchange to a hot separator, where conversion products are flashed overhead and heavy unconverted products are taken as hot liquid bottoms. The use of a hot separator improves the energy efficiency of the process by allowing hot liquid to go to the fractionation train and prevents polynuclear aromatic (PNA) fouling of cold parts of the plant. The overhead from the hot separator goes to a cold separator, where recycle gas is separated from the product. The product is then sent to fractionation, and recycle gas is returned to the reactor via the recycle compressor.

The fractionation train typically starts with a stripper column to remove hydrogen sulfide, which is in solution with the products. The removal ensures a relatively clean product in the main fractionator column, thus reducing column costs and metallurgy requirements. The stripper is followed by a main fractionating column with appropriate stages and side draws to remove the desired products. The bottoms from this main column is recycled back to the reactor section for complete feed conversion.

To allow complete conversion without PNA fouling or excessive catalyst coking, UOP has developed several techniques to selectively remove PNAs from the recycle oil stream. Some PNA removal is critical for successful operation at complete conversion. In earlier designs, the unit was simply purged of PNAs by taking a bottoms drag stream. In newer units, PNAs are selectively removed by either fractionation or adsorption, and the result is an increased yield of valuable liquid product.

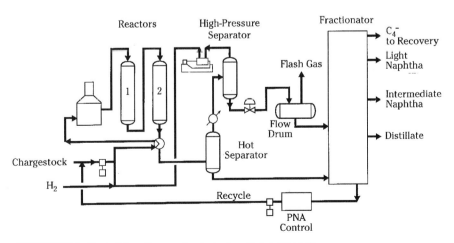

FIGURE 7.3.5 Typical flow diagram of a single-stage Unicracking unit.

YIELD PATTERNS

The versatility of the hydrocracking process is clearly shown by the yield structures in Table 7.3.2. These data are typical yields expected when processing a Middle East VGO for maximum distillate and for maximum naphtha, the two extremes of hydrocracker operation.

INVESTMENT AND OPERATING EXPENSES

Capital investment and operating expenses for a hydrocracker are sensitive to:

- The processibility of the feedstock
- The desired product slate
- The desired product specifications

The desired product slate has a profound effect on the arrangement of equipment, as discussed in the previous section. If the feed has demetallized oil or is more difficult to process for some other reason, operating conditions can be more severe than in hydrocracking a VGO. This additional severity can be manifested in equipment, hydrogen consumption, utilities, and additional catalyst. In general, a jet-fuel operation is more severe than an operation producing a full-range diesel product. And naphtha production requires a higher hydrogen consumption than either jet-fuel or diesel production.

TABLE 7.3.2 Typical Hydrocracker Yields*

	Distillate	Naphtha
Yield:		
NH_3, wt %	0.1	0.1
H_2S, wt %	2.6	2.6
C_2-, wt %	0.8	0.8
C_3, wt %	1.0	3.3
C_4, vol %	2.9	21.4
Light naphtha, vol %	7.3	39.1
Heavy naphtha, vol %	7.7	68.9
Distillate, vol %	94.0	—
Product properties:		
Jet-fuel cut:		
Smoke point, mm	27	—
Freeze point, °C (°F)	−59 (−74)	—
Aromatics, vol %	9	—
Diesel-fuel cut:		
Cetane no.	56	—
Naphtha:		
P/N/A, vol	—	33/55/12
Research octane no.	—	68

*Basis: Feedstock, Middle East VGO; density, 22.2 °API; sulfur, 2.5 wt %.

Note: P/N/A = paraffins/naphthenes/aromatics; °API = degrees on American Petroleum Institute scale.

TABLE 7.3.3 Hydrocracker Capital Investment*

Operation	Distillate	Naphtha
Estimated erected cost, $/BPSD CF	2000–3000	1600–2500

*As of January 1, 1995, based on combined-feed (CF) rate; includes 20% of material+labor as design engineering+construction engineering cost; does not include hydrogen plant; BPSD = barrels per stream day.

TABLE 7.3.4 Typical Hydrocracker Utilities

Power, kW	100–320
Fired fuel, 10^6 Btu/h	2–6
Cooling water, gal/min	20–80
Medium-pressure steam, MT/h (klb/h)	0.11–0.22 (0.25–0.50)
Condensate, MT/h (klb/h)	0.18 (0.4)

Note: Based on 1000 BPSD fresh feed; MT/h = metric tons per hour.

Only typical examples can be given; not every case can be covered. The figures in the accompanying tables are for illustrations only; variation may be expected for specific cases. Typical capital-investment guidelines are given in Table 7.3.3. Typical utility guidelines are given in Table 7.3.4.

PART 8

HYDROTREATING

CHAPTER 8.1
CHEVRON RDS/VRDS HYDROTREATING— TRANSPORTATION FUELS FROM THE BOTTOM OF THE BARREL

David N. Brossard
Chevron Research and Technology Company
Richmond, California

INTRODUCTION

The Chevron Residuum Desulfurization (RDS) and Vacuum Residuum Desulfurization (VRDS) Hydrotreating processes are used by refiners to produce low-sulfur fuel oils, and to prepare feeds for vacuum gas oil (VGO) fluid catalytic crackers (FCCs), residuum FCCs (RFCCs), visbreakers, and delayed cokers. Over half of the fixed-bed residuum hydrotreaters in operation use Chevron's RDS/VRDS Hydrotreating technology.

RDS/VRDS Hydrotreaters upgrade residual oils by removing impurities and cracking heavy molecules in the feed to produce lighter product oils. Early applications of Chevron's residuum hydroprocessing technology were used to remove sulfur from atmospheric residues (ARs) and vacuum residues (VRs), hence the term *desulfurization.* Today, RDS/VRDS Hydrotreaters perform equally well removing nitrogen, carbon residue (see "Process Chemistry" section), nickel, and vanadium from the oil and cracking heavy VR molecules to VGO, distillates, and naphtha products. The amount of impurities removed depends on the feed and on the product specifications desired by the refiner. Sulfur removal greater than 95 percent, metal removal (primarily nickel and vanadium) greater than 98 percent, nitrogen removal greater than 70 percent, carbon residue reduction greater than 70 percent, and cracking of vacuum residue (538°C+ material converted to 538°C−) as high as 60 liquid volume percent (LV%) have been commercially demonstrated. RDS/VRDS Hydrotreating uses fixed beds of catalyst that typically operate at moderately high pressures [150 to 200 atm (2133 to

2850 lb/in^2)] and temperatures [350 to 425°C (662–797°F)] in a hydrogen-rich atmosphere (80 to 95 mol % hydrogen at the reactor inlet) to process the oil feed. The feed to a VRDS Hydrotreater is generally the VR from a crude unit vacuum column with a typical starting true boiling point (TBP) cut point of 538°C (1000°F), although cut points of 575°C (1067°F) and higher are feasible. The feed to an RDS Hydrotreater is generally AR from a crude unit atmospheric column with a typical starting TBP cut point of 370°C (698°F). Other feeds (such as solvent deasphalted oil, solvent deasphalter pitch, vacuum gas oil, and cracked gas oils from visbreakers, FCCs, RFCCs, and cokers) can also be processed in either RDS or VRDS Hydrotreaters.

Residua from many crudes have been successfully processed in RDS and VRDS Hydrotreaters. Table 8.1.1 shows a partial list of crudes that have been commercially processed in Chevron RDS/VRDS Hydrotreaters.

The range of feeds which can be economically processed in RDS/VRDS Hydrotreaters expands significantly when On-Stream Catalyst Replacement (OCR) technology is added to the unit. OCR technology allows spent catalyst to be removed from a guard reactor and be replaced by fresh catalyst while the reactor remains in service. This enables the refiner to process heavy, high-metal feeds or to achieve deeper desulfurization from a fixed-bed residuum hydrotreater (see Chap. 10.1).

HISTORY

Hydrotreating of residual oils was a natural extension of hydrotreating distillate oils and VGOs to remove sulfur.[1,2,3] Chevron's first commercial RDS Hydrotreater was commissioned in 1969. Typical of many early residuum hydrotreaters, Chevron's first RDS Hydrotreater was designed to remove sulfur to produce low-sulfur fuel oil (LSFO). Chevron's first VRDS Hydrotreater, commissioned in 1977, was also designed to produce LSFO.

TABLE 8.1.1 RDS/VRDS Hydrotreater Feedstocks That Have Been Commercially Processed

A.A. Bu Khoosh	Kuwait
Alaskan North Slope	Laguna
Algerian	Margham C.
Arabian Berri	Maya
Arabian Heavy	Mina Saud
Arabian Light	Minas
Arabian Medium	Murban
Basrah Light	Oguendjo
Cabinda	Oman
Colombian Limon	Qatar Land
Dubai	Qatar Marine
Duri	Russian
El Chaure	Shengli No. 2
Gipsland	Statfjord
Indonesian	Suez
Iranian Heavy	Tia Juana Pesado
Iranian Light	Umm Shaif
Isthmus	West Texas Intermediate
Khafji	West Texas Sour
Kirkuk	Zakum

In 1984 Okinawa Sekiyu Seisei, a Japanese refiner, first reported[4] the operation of a Chevron RDS Hydrotreater in "conversion mode." In this operation, the reactor temperature was raised fairly high early in the run—much higher than required to simply produce low-sulfur fuel oil—and held high until the end of run. This operation hydrocracked as much VR as possible to lighter boiling products (VR was "converted" to light products.) It also shortened the run length because of higher catalyst deactivation from coke deposited on the catalyst through more of the run. Conversion mode operation has been favored by many RDS/VRDS Hydrotreater operators in recent years to minimize the production of fuel oil.

An alternative to destroying low-value fuel oil has been to convert it to higher value fuel oil. During the late 1970s and through the 1980s and 1990s, the demand for and value of high-sulfur (3 percent) fuel oil and low-sulfur (1 percent) fuel oil dropped. In some cases, power plant operators have been willing to pay higher prices for fuel oils with much lower sulfur content (0.1 to 0.5 wt %). RDS/VRDS Hydrotreating enabled refiners to produce these lower-sulfur fuel oils. The lowest-sulfur fuel oil commercially produced from sour crudes (about 3 wt % sulfur in the AR) was 0.1 wt %. This fuel oil was produced by the Chevron RDS Hydrotreater at Idemitsu Kosan's refinery in Aichi, Japan (see Table 8.1.2).

The ability of residuum hydrotreaters to improve the economics of conversion units by pretreating their feeds has been understood for many years. The most noticeable economic impact of feed pretreatment is to lower the sulfur content of the feed to the conversion unit. For example, pretreatment of RFCC feed to reduce its sulfur to less than 0.5 wt % eliminates the need to install costly flue gas desulfurization facilities. Addition of hydrogen to the feed by the hydrotreater also improves the product yields and product qualities of the downstream conversion unit. In 1983, at Phillips' Borger, Texas, refinery, the first Chevron RDS Hydrotreater was commissioned to pretreat residuum to feed an existing RFCC unit. Prior to the hydrotreater project, the RFCC had been feeding sweet domestic crude. The 50,000 barrel per day (BPD) RDS Hydrotreater was designed to achieve 92 percent hydrodesulfurization (HDS) and 91 percent hydrodemetallization (HDM) from a mixed domestic and Arabian Heavy AR for a 1-year cycle. In addition to its contribution toward meeting environmental

TABLE 8.1.2 Production of Premium Low-Sulfur Fuel Oil

	RDS feed	RDS 343°C+ product
Sulfur, wt %	3.75	0.09
Viscosity, cSt at 50°C	248	84
Specific gravity, $d^{15}/4°C$	0.9590	0.9105
Carbon residue, %	7.95	2.28
Ni/V, ppm	13/40	<.1/<.1
Nitrogen, ppm	2060	644
Distillation, °C		
IBP	257	272
5%	325	330
10%	353	358
20%	402	393
30%	435	425
40%	467	450
50%	502	480
60%	537	515
70%	—	555

Commercial data from IKC Aichi RDS unit.

requirements and reducing catalyst usage, the feed pretreatment significantly increased the gasoline yield from the RFCC.[5] Since then, all of the residuum hydrotreaters built to pretreat residuum to feed an RFCC unit have been Chevron RDS Hydrotreaters.

An RDS Hydrotreater is used to pretreat residuum to feed a delayed coker unit at Chevron's refinery in Pascagoula, Mississippi.[6] The hydrotreater was originally designed to remove sulfur and metals from the feed to the coker so that the coke would have less sulfur and metals and be easier to sell. Since its commissioning in 1983, the RDS unit has provided significant economic benefit to the refinery. Coke production has been reduced and the proportion of light products is higher than it would have been without the RDS. This includes converting VR (which would otherwise be fed to the coker) to VGO, diesel, and naphtha in the RDS Hydrotreater. In addition, the hydrotreated VR from the RDS produces lower weight percent coke in the coker than the straight-run VR. Both of these effects lead to lower coke production and more light products from the refinery. The RDS Hydrotreater at Pascagoula remains the largest residuum hydrotreater in the world at 96,000 BPD.

Refiners have been hydrotreating residuum for over 25 years. In that time, residuum hydrotreating has changed with the needs of refiners from its initial function of removing sulfur from fuel oil to converting residuum directly and to improving the economics of downstream conversion units. Fixed-bed residuum hydrotreating continues to be a popular route to residuum conversion. Figure 8.1.1 shows the growth trend of fixed-bed residuum hydrotreaters though 1994.

PROCESS DESCRIPTION

A simplified flow diagram for a Chevron RDS/VRDS Hydrotreater is shown in Fig. 8.1.2.

Oil feed to the RDS/VRDS Hydrotreater (AR or VR primarily, but may also include VGO, solvent deasphalted oil, solvent deasphalter pitch, and others) is combined with makeup hydrogen and recycle hydrogen and heated to the reactor inlet temperature. Heat is provided from heat exchange with the reactor effluent and by a reactor charge heater.

The reaction of hydrogen and oil occurs in the reactors in the presence of the catalyst. Hydrotreating reactions on the catalyst remove sulfur, nitrogen, vanadium, nickel, carbon residue, and other impurities from the residuum; hydrogenate the molecules; and crack the residue to lighter products. The required catalyst average temperature (CAT) is initially low, but is gradually increased by 45°C (81°F) or more as the catalyst ages. The net hydrotreating reactions (such as sulfur and metals removal) are exothermic (see "Process Chemistry"). To prevent reactor temperatures from getting too high, quench gas—cold recycled hydrogen gas—is added between reactors and between catalyst beds of multiple-bed reactors to maintain reactor temperatures in the desired range.

Reactors in hydrotreating service have carefully designed internals to assure good distribution of gas and liquid. In multiple-bed reactors, quench spargers disperse the quench gas evenly across the reactor to maintain even reactor temperatures. Chevron provides both single-bed and multiple-bed RDS/VRDS reactors, depending on the needs of the refiner. Single-bed reactors are relatively small, typically 400,000 to 900,000 kg in weight, and therefore single-bed reactors are easier to install and to unload catalyst from than multiple-bed reactors. Multiple-bed reactors tend to be larger, 600,000 to 1,200,000 kg, but take up less plot space in a refinery compared to several single-bed reactors. This is very important in refineries where space is limited.

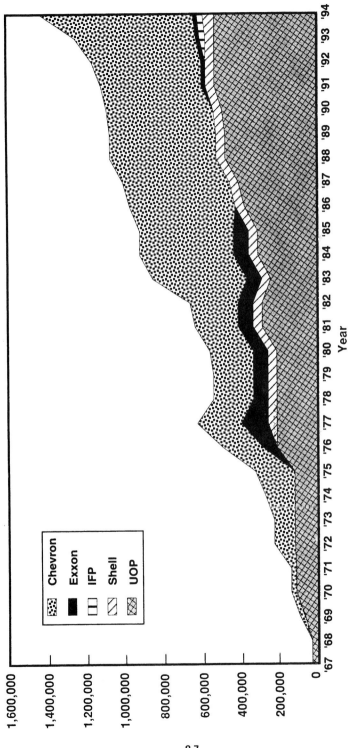

FIGURE 8.1.1 Growth of fixed-bed residuum hycroprocessing. Fixed-bed hydroprocessing is a rapidly growing technology. (*Data from published sources.*)

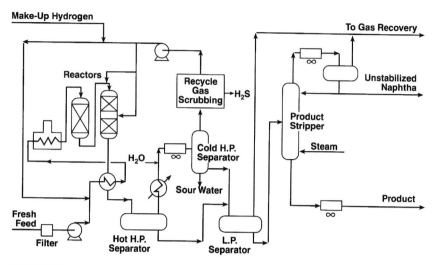

FIGURE 8.1.2 Simplified RDS/VRDS flow scheme.

The reactor effluent is cooled (by heat exchange with the reactor feed) to recover the heat released from the hydrotreating reactions. This heat exchange helps to reduce the fuel required in the feed heater. After cooling, the reactor effluent is flashed in the hot, high-pressure separator (HHPS) to recover hydrogen and to make a rough split between light and heavy reaction products. The reactor effluent heat exchange maintains the HHPS at a constant temperature, which is important in protecting the reaction products. If the HHPS temperature is too high, thermal cracking and coking reactions might take place in the HHPS (in the absence of catalyst) and downstream (in the absence of hydrogen and catalyst) and might degrade the oil. The liquid from the HHPS is let down in pressure, sent to the low-pressure separators, and then on to the product fractionator.

The HHPS vapor is cooled and water is injected to absorb hydrogen sulfide (H_2S) and ammonia (NH_3) produced in the reactors by the hydrotreating reactions. The mixture is further cooled to condense the product naphtha and gas oil and is flashed in the cold, high-pressure separator (CHPS). The CHPS separates the vapor, liquid water, and the liquid light hydrocarbons. The hydrocarbon liquid is let down in pressure and sent to the low-pressure separators. The water is sent to a sour water recovery unit for removal of the hydrogen sulfide and ammonia.

The hydrogen-rich gas from the CHPS flows to the H_2S absorber. There the H_2S that was not removed by the injected water is removed through contact with a lean amine solution. The purified gas flows to the recycle compressor where it is increased in pressure so that it can be used as quench gas and recombined with the feed oil.

Hydrogen from the reactors is purified and recycled to conserve this expensive raw material. Recycling the hydrogen is also important to provide high gas flow rates. High gas-to-feed-oil ratios provide a desirable excess of hydrogen in the reactors (see "Process Chemistry" section) and ensure good gas and liquid flow distribution in the reactors. The recycle hydrogen gas is also used for reactor quench.

Liquid from the low-pressure separators is fed to the atmospheric fractionator, which splits the hydroprocessed oil from the reactors into the desired final products.

PROCESS CHEMISTRY

Heteroatoms

Any atom in a crude oil molecule which is neither hydrogen nor carbon is called a *heteroatom*. Heteroatoms include sulfur, nitrogen, oxygen, nickel, vanadium, iron, sodium, calcium, and other less common atoms.

Carbon Residue

Carbon residue is a measurement of the tendency of a hydrocarbon to form coke. Expressed in weight percent, carbon residue is measured by microcarbon residue (MCR; American Society for Testing and Materials specification ASTM D4530), by Conradson carbon residue (CCR; ASTM D189), a considerably older test, or by Ramsbottom carbon residue (RCR; ASTM D524). MCR is the preferred measurement technique because it is more accurate than the other methods and requires a smaller sample. Instruments that measure MCR are very inexpensive. MCR is roughly equivalent to CCR and both correlate well to RCR. Carbon residue is useful in predicting the performance of a hydrocarbon in a coker or FCC unit. While carbon residue is not a direct measure of, it does correlate well with, the amount of coke formed when the oil is processed in cokers or FCCs.

Asphaltenes

Residual oil is composed of a broad spectrum of molecules. The number of specific molecules in residual oil is too large to classify, and therefore researchers have developed analytical techniques for separating these molecules for better understanding. The most common separation of residual oils is into asphaltenes and maltenes. This is done by diluting the residue with large quantities of normal paraffins such as *n*-heptane or *n*-pentane. The maltene fraction will remain in solution with the paraffin phase while the asphaltene fraction will form a separate phase. This is the principle behind the refinery process called solvent deasphalting (SDA).

The molecules in the maltene fraction can be further separated into fractions of varying polarity by being passed over columns packed with different adsorbents. A full description of these separation techniques is provided by Speight.[7]

There is considerable disagreement about what constitutes an asphaltene molecule beyond its insolubility in a paraffinic solvent. Still, the subject of asphaltenes is important. The high concentration of heteroatoms in the asphaltenes requires that at least some of the asphaltene molecules be hydrotreated to have high removals of the heteroatoms. In addition, the hydrogen content of asphaltene molecules must be increased if they are to be transformed to transportation fuels.

Converting asphaltene molecules to nonasphaltene molecules is a major challenge for refiners. As processing VR or AR becomes more severe, coke is formed (in FCCs or cokers, for example) or the asphaltenes become insoluble in the processed residuum and precipitate out in a sticky, equipment-plugging material commonly referred to as *dry sludge*. Of course, the asphaltenes were soluble in the maltenes in the original residuum, so the processing must have caused some change to the maltenes, or to the asphaltenes, or to both. Dry sludge formation usually limits the practical severity in which residuum can be processed in many conversion units including residuum hydroprocessing units (see "Dry Sludge Formation" below).

Hydroprocessing Reactions

Reactions in an RDS or VRDS Hydrotreater take place in the liquid phase, since much of the residual feed and product molecules do not vaporize at reactor pressure and temperature. The oil in the reactor is saturated with hydrogen gas because the partial pressure of hydrogen is very high and hydrogen is available in great excess (typically 10 to 30 moles of hydrogen for each mole of oil feed). The oil and hydrogen reactant molecules diffuse through the liquid oil filling the catalyst pores and adsorb onto the catalyst surface where the hydrotreating reactions take place. Larger molecules tend to adsorb more strongly onto the catalyst surface than smaller molecules. This means that the large VR molecules tend to dominate the reactions on the catalyst when they can successfully diffuse into the catalyst pores. The product molecules must then desorb from the catalyst surface and diffuse out through the liquid that fills the catalyst pores.

On the catalyst surface, sulfur, nitrogen, nickel, and vanadium atoms are removed from the residual molecules, and carbon-to-carbon bonds are broken. These reactions generally lead to cracking the original oil molecules to smaller molecules, which boil at a lower temperature. As a result the viscosity of the oil is also reduced. When the product is used as fuel oil, less volume of expensive cutter stock (such as jet or diesel) is required to meet a given viscosity specification.

Hydrotreating is very exothermic. The heat produced by the reactions causes the gas and oil to increase in temperature as they pass down through the catalyst beds. The temperature in the reactors is controlled by the addition of hydrogen quench gas between reactors and between catalyst beds within a reactor. The heat produced by the reactions is recovered in the reactor effluent heat exchangers and used to preheat the feed upstream of the feed furnace.

There are fundamental differences between the removal of the different impurities, largely because of the structure of the molecules in the residuum. Sulfur atoms tend to be bound in the oil as "sulfur bridges" between two carbon atoms or to be contained in a saturated ring structure (see Fig. 8.1.3). Removal of these sulfur atoms usually

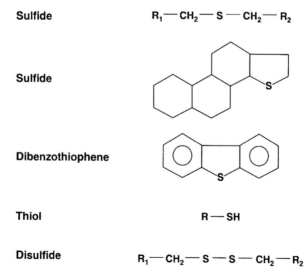

FIGURE 8.1.3 Typical petroleum molecules that contain sulfur atoms. Sulfur atoms usually have simple chemical bonds in petroleum.

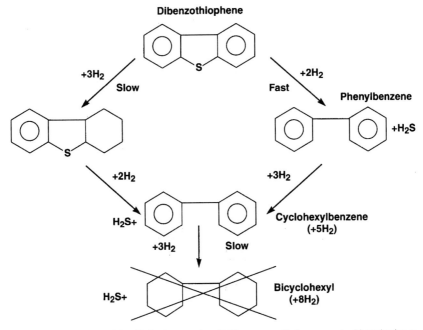

FIGURE 8.1.4 Typical desulfurization reaction. Sulfur can usually be removed without having to saturate aromatic rings.

requires only the breaking of the two sulfur-carbon bonds per sulfur atom and the subsequent addition of four atoms of hydrogen to cap the ends of the bonds that were broken. When the part of the molecule that contains the sulfur can access the catalyst surface, sulfur removal is relatively easy. Figure 8.1.4 shows the hydroprocessing reactions of dibenzothiophene as an example of a sulfur-bearing petroleum molecule. The reaction pathway to produce phenylbenzene is favored because it does not require the saturation of an aromatic ring structure.

Nitrogen atoms tend to be bound in the aromatic rings in the residual molecules (see Fig. 8.1.5). It is usually necessary to saturate the aromatic ring that contains the nitrogen atom with hydrogen before the nitrogen-carbon bonds can be broken and the nitrogen removed. This requirement to saturate aromatic rings makes the removal of nitrogen much more difficult than the removal of sulfur. Figure 8.1.6 shows the hydroprocessing reactions of quinoline as an example of a nitrogen-bearing petroleum molecule. The first step along any reaction pathway toward removal of the nitrogen atom is the saturation of an aromatic ring structure. The amount of nitrogen removed is almost always lower than that of sulfur because of the relative difficulty of the reactions. Also, high levels of removal of nitrogen require high hydrogen partial pressure and catalysts with very high hydrogenation activity.

Nickel and vanadium atoms are generally bound into a porphyrin structure in the residue. Figure 8.1.7 shows a typical vanadyl-porphyrin molecule. These structures are quite flat, and the metals are relatively easy to remove if the catalyst has sufficiently large pores to accommodate the large molecules that contain them. Vanadium tends to be much easier to remove than nickel.

The removed sulfur and nitrogen are converted into hydrogen sulfide and ammonia gases. The hydrogen sulfide and ammonia diffuse out of the catalyst pore with the other

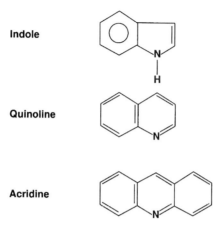

FIGURE 8.1.5 Typical petroleum molecules that contain nitrogen atoms. Nitrogen atoms usually have complex, aromatic bonds in petroleum.

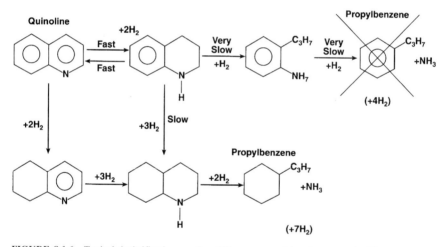

FIGURE 8.1.6 Typical denitrification reaction. Nitrogen can seldom be removed without saturating aromatic rings.

reactants. The removed nickel and vanadium are bound up with sulfur and remain on the catalyst surface. Fresh hydrotreating catalyst undergoes a very rapid fouling as its fresh active metals are covered with a layer of nickel and vanadium from the crude. Fortunately, nickel and vanadium are themselves catalytic metals (although much less active than the original catalytic metals), therefore the catalyst surface retains some activity, though considerably less than the fresh catalyst. Eventually, however, the nickel and vanadium sulfide molecules fill up the catalyst pores and reduce the ability of the large residuum molecules to diffuse through the liquid filling the pores. When access of the residuum molecules to the catalyst surface becomes severely restricted, the catalyst has lost its hydrotreating activity (see "Catalysts" below).

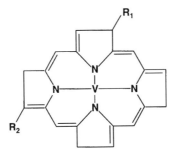

FIGURE 8.1.7 Typical vanadyl-porphyrin molecule. Porphyrin structures are very flat and vanadium is easily removed—if the molecules can diffuse through the catalysts' pores.

Other undesirable side effects of the hydroprocessing reactions occur if some of the high molecular weight residuum molecules that adsorb onto the catalyst surface react with other oil molecules instead of with hydrogen. This is a particular problem when the hydrogen partial pressure in the reactor is low. In this case, the molecules grow larger. If they grow large enough they may not readily desorb from the catalyst surface, but remain on the catalyst surface as coke. The coke formed in this fashion leads to a severe deactivation of the catalyst.

Dry Sludge Formation

One other undesirable effect of hydroprocessing reactions is that the solubility of the asphaltenes usually decreases with increased processing of the residue. This occurs even though the quantity of asphaltenes is reduced during the hydroprocessing. Unfortunately, while the asphaltenes are destroyed by being hydrogenated and cracked, the maltene fraction of the residue is also being hydrogenated and cracked—usually more severely than the asphaltenes. Since the maltenes are generally smaller molecules, it is easy for them to diffuse into the catalyst pores and be hydrotreated. In this hydrotreatment, aromatic rings are hydrogenated and aliphatic side chains are removed by cracking. These reactions reduce the ability of the maltene fraction to solubilize the asphaltenes. Usually, the loss of solubility of the maltenes for the asphaltenes occurs faster than the asphaltenes can be converted and the asphaltenes drop out of solution. The precipitated asphaltenes create dry sludge, which plugs up equipment, and, at its worst, can deposit in the catalyst and eventually form coke. This rapidly deactivates the catalyst. Even when the dry sludge does not deactivate the catalyst or cause operating problems by plugging equipment in the residuum hydrotreater, it can cause problems in the downstream processing units or make the product fuel oil unsalable.

Conversion

Hydrocracking is the transformation of larger, high-boiling-point hydrocarbons into smaller, lower-boiling-point hydrocarbons in the presence of hydrogen. In residuum hydrotreating, this transformation can take place because of the breaking of a carbon-to-carbon bond or because of the removal of a heteroatom that was bonding to two otherwise unconnected pieces of hydrocarbon. Since many of the carbon atoms in VR are in aromatic rings, it is necessary to hydrogenate the molecules and saturate the rings before bonds can be broken and the molecules cracked.

In residuum hydroprocessing, it is common to refer to the hydrocracking of the residue as conversion. In this usage, conversion is defined as the destruction of residue boiling higher than a certain true boiling point temperature [usually 538°C (1000°F)] to product boiling lower than that temperature. Conversion can be calculated as $(F_{T+} - P_{T+})/F_{T+}$, where F_{T+} is the volume fraction of the feed boiling above temperature T and P_{T+} is the volume fraction of the product boiling above temperature T. The hydroprocessing of the residue and conversion of the residue are linked; it is not possible to do one without doing the other. For most conventional residuum hydrotreating catalysts, conversion is primarily a function of the catalyst temperature and the space velocity.

It has been noted[4] that the formation of dry sludge is related to the level of conversion of the residuum hydrotreater. This is certainly true when the catalyst and feed are not changed significantly. The presence of VGO in the feed can lower the conversion at which sludge formation becomes a problem (because of the poor ability of hydroprocessed VGO to solubilize asphaltenes). Small-pore, high-surface-area catalysts can also have a deleterious effect on product stability because they can selectively hydroprocess maltene molecules without destroying any asphaltene molecules. Generally, fixed-bed residuum hydrotreaters achieve conversions between 20 and 60 LV % at normal operating conditions and the onset of dry sludge occurs generally between 45 and 60 LV % conversion depending on the feed, processing conditions, and catalyst system.

Note that the VGO in the feed to an RDS Hydrotreater is a very poor solvent for asphaltenes—particularly after it becomes highly hydrogenated. VRDS Hydrotreaters, therefore, can operate at 5 to 10 percent higher cracking conversion than RDS Hydrotreaters before the onset of dry sludge formation.

CATALYSTS

Designing residuum hydroprocessing catalyst for high activity is a compromise. Early catalysts, which had been developed to hydrotreat light oils, had pore sizes that were too small to hydrotreat residuum; The catalyst pores became plugged with metals and coke and were very quickly deactivated. Catalysts were modified for AR and VR hydroprocessing by increasing their pore sizes, but this led to less active surface area and much lower activity for hydroprocessing reactions.

Fixed-bed residuum hydrotreating catalysts are generally small, extruded pellets made from an alumina base. The pellets are impregnated with catalytic metals—often called *active metals*—that have good activity for hydrogen addition reactions. Active metals that are used for RDS/VRDS Hydrotreater catalysts include cobalt, nickel, molybdenum, and other more proprietary materials. The catalyst pellets are usually small, 0.8 to 1.3 mm in diameter, because the reaction kinetics are usually diffusion-limited; a small catalyst pellet with high surface-to-volume ratio has better diffusion for the relatively large residuum molecules, and this leads to better reactivity. Different shapes of extruded pellets are often used to take advantage of the high surface-to-volume ratio of some shaped pellets while still maintaining reasonable reactor pressure drop.

The pore diameters of residuum hydrotreating catalysts need to be quite large, relative to catalysts found in other refinery processes, to accommodate the large residuum molecules that need to be treated. Unfortunately, as the size of the pores increases, the surface area decreases and so does the catalyst activity. Another complication is that, as the nickel and vanadium atoms are removed from the residuum, they form nickel and vanadium sulfides that deposit on the catalyst surface. The metal sulfides build up on the active surface and fill up the catalyst pores. The metal sulfides tend to deposit near the openings of the catalyst pores and plug these pore openings. This is because the residuum molecules that contain nickel and vanadium are quite large and do not diffuse far into the catalyst pores before they are removed. The diffusion of the large residuum molecules is reduced even further by the plugging of the pore openings by the metal sulfides. If the large molecules cannot enter the pores and thus have no access to the active catalyst surface, they cannot be hydrotreated.

To overcome the limitations of the small pore versus large pore tradeoff, Chevron designs systems of catalyst that are layered such that catalyst activity increases as the residuum moves through the reactor. The catalysts that the residue first contacts have large pores to successfully remove the vanadium and nickel from the large molecules that contain these impurities and to resist deactivation (due to pore plugging) from these removed metals. Later catalysts, which see lower nickel- and vanadium-content oil, can have smaller pores and higher surface activity to perform more hydrotreating reactions.

Chevron generally calls its most metal-tolerant metal-removal catalysts *hydrodemetallization* catalysts. However, Chevron's HDM catalysts also promote hydrodesulfurization, hydrodenitrification (HDN), and other conversion reactions. Catalysts which have higher activity for HDS reactions and carbon residue reduction, and less tolerance for metals, are called *hydrodesulfurization catalysts*. HDS catalysts also have good reactivity for HDM and HDN reactions and cracking conversion in addition to being somewhat metal-tolerant (although not as metal-tolerant as HDM catalysts). Finally, Chevron has a few catalysts that have very high activity for HDN reactions. They are the most difficult reactions to achieve in an RDS/VRDS Hydrotreating unit. HDN catalysts are also very active for HDS, carbon residue reduction, and cracking conversion. They tend to have little demetallization activity and are not very tolerant to metal poisoning.

Selecting the amounts and types of catalysts for an RDS/VRDS Hydrotreater requires extensive pilot plant data, commercial plant data, and a good reactor kinetics model.

VRDS HYDROTREATING

In many countries the demand for gasoline relative to middistillate is much lower than the typical product slate from a refinery that relies on an FCC for its conversion capacity. Many projects have relied on VGO hydrocracking to give high yields of top-quality middle distillates (see Chap. 7.2).

Given the attractiveness of using the VGO component as hydrocracker feed, it has become an important consideration that the residuum hydrotreater be capable of efficiently processing 100 percent VR. Chevron VRDS Hydrotreating has been successfully processing this difficult feedstock since 1977.

The characteristics of an acceptable RFCC feedstock are as shown in Table 8.1.3. These requirements are readily obtained by Chevron RDS units and can be met in a

TABLE 8.1.3 RFCC Feed Targets

Sulfur	0.5% max. to avoid flue gas desulfurization in the RFCC
Carbon residue	7–10% max. to limit catalyst cooling requirements
Nickel + vanadium	5–25 ppm to limit RFCC catalyst consumption
Crackability	A combination of hydrogen content, boiling range, and viscosity that promotes vaporization and cracking at the injection point

Chevron VRDS for VRs derived from Arabian Heavy, Arabian Light, and Kuwait as well as most other popular crude oils.

VR is more difficult to hydrotreat than AR because there is no easily processed VGO in the feed. Therefore, VRDS relies heavily on the ability of catalysts to upgrade the heavy compounds found in the VR. Chevron has tailored catalyst for use in either RDS or VRDS service. Catalysts designed for RDS operation are not necessarily effective for VRDS service and vice versa.

The effectiveness of the catalyst for upgrading VR is indicated by the amount of demetallization, asphaltene removal, and desulfurization achieved while still producing a stable product (no solid or asphaltene precipitation). Table 8.1.4 shows an example of VRDS pilot plant processing of an Arabian Heavy/Kuwait VR mixture to 99 percent HDM, 97 percent HDS, and 82 percent carbon residue reduction. While such high severity is seldom required, it clearly illustrates the capability of the VRDS process.

VRDS reactors must handle very viscous feeds relative to RDS reactors. The unusual two-phase flow reactor hydrodynamics encountered with VR feeds dictate special design considerations to avoid unreasonably high and unstable reactor pressure drops. This was the subject of a considerable amount of pilot plant and scaleup testing prior to the start-up of the first VRDS Hydrotreater in 1977 at Chevron's El Segundo refinery. Experience gained on processing 100 percent VR at El Segundo, together with data from the earlier laboratory study, were the basis for the design of a VRDS Hydrotreater for the Nippon Petroleum Refining Company that was started up at their Muroran, Japan, refinery in 1982.[8] Extensive experience with 100 percent VR has enabled Chevron to produce a trouble-free technology.

Converting RDS to VRDS

The RDS Hydrotreater at the Mizushima refinery of the Japan Energy Company (formerly the Nippon Mining Company) operated with AR for several years and was successfully converted to process VR in 1981.[9] For the next few catalyst cycles following the conversion to VRDS, the unit gradually increased the fraction of VR in its feed until 100 percent VR was processed over the entire run.

TABLE 8.1.4 Chevron VRDS Pilot Plant Performance

	VRDS Feed*	VRDS Product
Boiling range, °C	538+	343+
Gravity, °API	4.6	18.1
CCR, wt %	23.1	5.7
Sulfur, wt %	5.3	0.24
Nitrogen, wt %	0.42	0.15
Nickel+ vanadium, ppm	195	3
Viscosity, cSt at 100°C	5500	32
538°C+ conversion, LV %	54.2	
343°C+ yield, LV %	81.4	
H_2 consumption, SCFB	1650	

*Arabian Heavy/Kuwait in 50/50 volume ratio.

Note: °API = degrees on American Petroleum Institute scale; SCFB = standard cubic feet per barrel.

FEED PROCESSING CAPABILITY

Handling Impurities

In addition to the high concentrations of impurities that have already been discussed (sulfur, nitrogen, carbon residue, nickel, and vanadium), residues also contain high concentrations of particles such as iron sulfide scales and reservoir mud. If fed directly to a fixed-bed RDS/VRDS Hydrotreater, these particles would be filtered out by the small catalyst pellets and would form a crust. This crust would cause the pressure drop across the catalyst bed to increase dramatically, disturb the even distribution of oil and gas in the reactor, and eventually force a plant shutdown because of excessive pressure drop.

The first line of defense against particles in an RDS/VRDS Hydrotreater is the feed filter. These filters have small openings, commonly 25 microns, which allow the filtered oil to pass but retain the larger particulates. When the pressure drop across the filters becomes excessive, they are automatically removed from service and briefly backwashed with oil. The particles that the feed filters remove would cause a very severe increase in pressure drop across the reactors, and therefore feed filtration is required for all residuum hydrotreaters. Even so, small particles pass on to the catalyst beds.

Some of the particles that pass through the feed filters are small enough to pass through the catalyst in the reactors as well. These particles do not affect the operation of the RDS/VRDS Hydrotreater. However, some particles are small enough to pass through the feed filters, but large enough to be trapped by the catalyst beds. These particles fill up the spaces between the catalyst and lead to increasing reactor pressure drop. This increased pressure drop is especially severe when the particles are all removed at one point in the reactor system. A large pressure drop increase can cause the refiner to decrease the feed rate or shut down the plant before the catalytic activity of the catalyst has been expended.

In addition to particles, high-reactivity metals such as iron can cause considerable operating difficulty for a fixed-bed residuum hydrotreater. Oil-soluble iron is highly reactive and is easily removed right on the outside of very active catalyst pellets. The removed iron quickly fills up the area between the catalyst pellets and leads to pressure drop increase. Oil-soluble calcium is also present in some crudes and can cause pressure drop increases and catalyst poisoning.

Catalyst Grading to Prevent Pressure Drop Increase

The high levels of insoluble and soluble iron in some California crudes caused Chevron to develop top bed grading technology[10] to minimize problems associated with these metals. Chevron was the first company to use a catalyst grading system in a commercial hydrotreater, in 1965. Special calcium removal catalysts have been applied in a VRDS Hydrotreater where the feed contained more than 50 wt ppm calcium. Effective catalyst grading combines physical grading of the catalyst by size and shape and grading of the catalyst by activity.

The goal of physical grading is to filter out particles in the grading catalyst that would otherwise plug the top of the smaller active catalyst pellets. By carefully changing the sizes and shapes of the physical grading catalyst, one can remove particles over several layers and therefore reduce their tendency to cause the pressure drop to increase.

Grading catalyst by activity means to gradually increase the surface activity of the catalyst down the reactor system. The oil is then exposed first to catalyst with very low activity that forces the reactive metals (iron and calcium, for example) to penetrate into the catalyst. There the removed metals do not fill up the space between pellets and pressure drop increase is avoided. The catalyst activity is increased in subsequent layers until all of the reactive metals have been removed.

Catalyst grading for preventing pressure drop increase cannot be accurately simulated on a small (pilot plant) scale because the flow regimes are much different. Extensive refinery experience forms the basis for Chevron's catalyst grading techniques. This experience includes data from a commercial unit feeding deasphalted oil that contained particulate and soluble iron, as well as very reactive nickel and vanadium. Proper catalyst grading techniques allow RDS/VRDS Hydrotreaters to run until catalyst activity is used up rather than shut down prematurely due to excessive pressure drop.

Feed Flexibility

Many types of feeds have been processed in RDS/VRDS Hydrotreaters. Successful processing of VR feeds as viscous as 6000 centistokes at 100°C have been commercially demonstrated.[6] Feeds containing up to 500 wt ppm Ni+V have also been commercially processed[11] in a fixed-bed RDS Hydrotreater.

Chevron catalyst systems can tolerate average feed metals over 200 wt ppm Ni+V while maintaining a 1-year run length. Feeds with considerably higher than 200 wt ppm Ni+V can be processed with at least a 1-year run length before catalyst replacement is required if the feed is pretreated in an On-Stream Catalyst Replacement reactor. By lowering the metals in the feed to processable levels, OCR increases the refiner's flexibility to run less expensive high-metal feeds (see Chap. 10.1).

COMMERCIAL APPLICATION

The use of residuum hydrotreatment to produce LSFO for power plants still continues as countries adopt stricter environmental regulations. More commonly, however, the refiner wishes to reduce fuel oil yield and have the flexibility to prepare feedstock for a downstream conversion unit.

Figure 8.1.8 shows a simple scheme for converting residuum to motor gasoline (mogas) using an RFCC. In this scheme, the RDS Hydrotreater significantly upgrades the RFCC feed. Pretreating the residuum feed increases its hydrogen content and reduces its impurities. For many crudes this upgrade is necessary for the RFCC to be operable. For other crudes, whose AR could be fed directly to the RFCC, RDS Hydrotreating improves the economics of the conversion project and increases the yield of market-ready light products.

In some residuum upgrading projects, middle distillates are more desirable products than mogas. In these projects, Chevron's Isocracking process is the preferred processing route to produce high-quality middle distillates from the VGO. Figure 8.1.9 shows a scheme in which the AR is sent to a vacuum tower to prepare VGO feed for an Isocracker. The remaining VR is then sent to a VRDS Hydrotreater to be pretreated for an RFCC. High-severity VRDS Hydrotreating has been shown[12] to prepare suitable feedstock for an RFCC. This scheme provides the optimal usage of hydrogen for upgrading the residuum. The hydrogen required by the Isocracker is just the amount necessary to convert the VGO to middle distillates. The hydrogen required by the VRDS Hydrotreater is just enough to improve the volatility and hydrogen content of the VR to be satisfactory for RFCC feed.

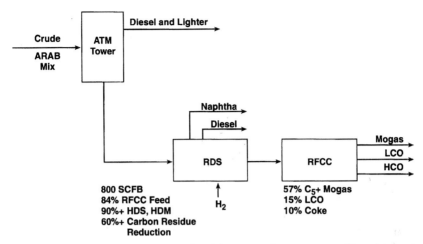

FIGURE 8.1.8 Simple residuum conversion. A low cost project to convert residuum into maximum mogas.

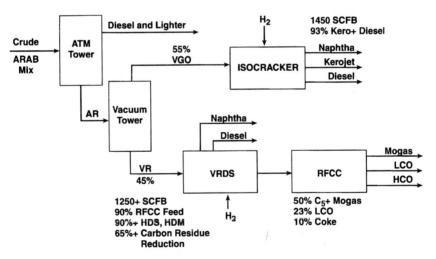

FIGURE 8.1.9 Residuum conversion to middle distillate and mogas. Converting the RDS to accept vacuum residuum and adding an Isocracker enables refiners to produce market-ready middle distillates.

Many residue upgrading projects need to vary the relative production of gasoline and middle distillates with market demands. Figure 8.1.10 shows a scheme where the vacuum column cut point is varied between 425°C (797°F) to produce maximum mogas and 565°C (1049°F) to produce maximum middle distillates. Again, the addition of hydrogen is adjusted to just satisfy the hydrogen upgrading requirements of the product slate.

Finally, some projects need to be installed in phases. Figure 8.1.11 shows the hypothetical transition from a simple upgrading project to a complete and flexible

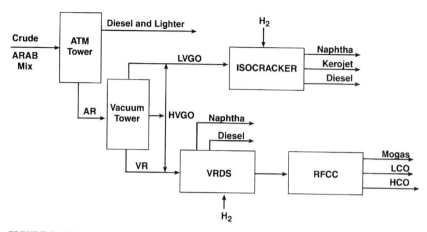

FIGURE 8.1.10 Middle distillate to mogas flexibility with VRDS/RFCC. Refiners can respond to changing demands for mogas and middle distillates by changing VGO cut point in the vacuum tower.

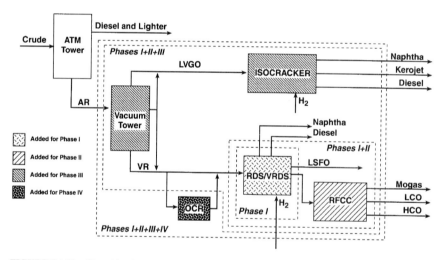

FIGURE 8.1.11 Phased implementation of a residuum conversion project. Chevron's hydroprocessing technologies enable refiners to phase in residuum conversion projects as market demands change.

upgrading project in four phases. Phase 1 consists of an RDS Hydrotreater to reduce the quantity and improve the quality of fuel oil produced. Phase 2 sees the installation of an RFCC to completely destroy the residuum. Phase 3 further extends the project by adding a vacuum column with a variable VGO cut point and an Isocracker to make high-quality middle distillates from the VGO. The RDS Hydrotreater processes either AR that has a higher starting cut point than its original design or pure VR (in the case that all of the VGO is routed to the Isocracker). It is important that the residuum hydrotreater in phase 1 be designed with the flexibility to process either AR or VR. Finally, phase 4 adds an OCR onto the RDS/VRDS Hydrotreater to provide greater flexibility to process inexpensive, high-metal crudes.

Example Yields and Product Properties

Table 8.1.5 shows the yields and product properties from a sample RDS Hydrotreater preparing high-quality RFCC feed (0.4 wt % sulfur) from Arabian Heavy AR (650°F+)

Investment and Utility Consumption Information

Table 8.1.6 shows the estimated investment costs for the RDS Hydrotreater whose yields and product properties are given in Table 8.1.5. These estimated costs are based on similar projects executed by Chevron in its refineries.

Table 8.1.7 summarizes typical running costs (per stream day and per barrel processed) for the 70,000-BPD RDS Hydrotreater whose yield and product properties are shown in Table 8.1.5. Utility estimates are based on Chevron's operating experience. The costs in Table 8.1.7 include no capital charges, either for the RDS Hydrotreater or a hydrogen plant (in the event one is required).

Table 8.1.8 summarizes typical total processing costs for the same 70,000 BPD RDS Hydrotreater. This estimate includes charges for capital for the RDS Hydrotreater at 25 percent of the estimated on-plot and off-plot charges. The charge of $2.50 per thousand cubic feet of hydrogen includes the capital charges for a new hydrogen plant as well as the operating and raw material costs of producing hydrogen.

THE FUTURE

Interest in RDS/VRDS Hydrotreating will continue to expand as environmental restrictions tighten. This is particularly true in developing countries where energy requirements are growing rapidly and shifting away from fuel oil to transportation fuels.

Continuously improving catalysts and process technology have enabled RDS/VRDS Hydrotreating to adapt to refiners' changing requirements. Future demands will be placed on RDS/VRDS Hydrotreating to yield products with lower levels of impurities in the face of increasing impurities in feedstocks. New technologies, such as OCR, are expected to make major contributions to these residuum hydrotreaters of the future.

TABLE 8.1.5 Sample RDS Hydrotreater Yields—RFCC Feed Preparation

	Feed				Products			
	Feedstock	H_2	H_2S	NH_3	C_1-C_4	C_5-280°F	280–650°F	650°F+
BPSD	70,000					1346	10,145	61,058
wt % of feed	100.00	1.43	4.28	0.22	0.23	1.38	12.51	82.81
		(940 SCFB)						
LV % of feed	100.00					1.92	14.49	87.23
Density, °API	11.8					68.2	34.5	19.4
Sulfur, wt %	4.37					0.004	0.034	0.40
Nitrogen, wt %	0.30					0.003	0.016	0.14
Carbon residue, wt %	13.6							5.5
Viscosity, cSt at 50°C	3240							160
Nickel, wt ppm	34							5
Vanadium, wt ppm	97							5

Feed is Arabian Heavy 650°F+ AR

Note: BPSD = barrels per stream day.

TABLE 8.1.6 Estimated Investment Summary for RDS Hydrotreater to Prepare RFCC Feed

Feed Rate, BPSD	70,000
Run length, days	335
Operating factor	0.92
On-plot investment, million $U.S.:	
Major materials	107.2
Reactors	73.3
Other reactor loop	22.5
Fractionation	4.8
Makeup compression	6.6
Installation cost	79.3
Engineering cost	17.9
Indirect cost	29.8
Total on-plot cost	234.2
Total off-plot cost (30% of on-plot), million $U.S.	70.3
Catalyst cost per charge, million $U.S.	8.8

Basis: second quarter 1995, U.S. Gulf Coast.

TABLE 8.1.7 Utility and Running Cost Summary for 70,000-BPD RDS Hydrotreater to Prepare RFCC Feed*

Item	Unit cost†	Rate‡	$/Stream day	$/Bbl FF
Utilities:				
Fuel	$20.00/EFO-bbl (6 million Btu)	272 EFO-BPD	5,440	0.078
Power	$0.05/kWh	27,000 kW	32,400	0.463
400 lb/in² gage steam	$2.00/klb	−22 klb/h	(1,056)	(0.015)
150 lb/in² gage steam	$3.25/klb	116 klb/h	9,048	0.129
Cooling water	$0.05/kgal	8,200 gal/min	590	0.008
Process injection water	$5.40/kgal	66 gal/min	513	0.007
Boiler feed water	$6.30/kgal	85 gal/min	771	0.011
Condensate	$5.40/kgal	(176) gal/min	(1,369)	(0.020)
Total Utilities			46,337	0.661
Hydrogen	$0.85/kSCF	71.7 million SCFD	60,945	0.871
Catalyst	$8.80 million/year		26,206	0.374
Operating labor	$0.20 million/year/shift	2 shift positions	1,191	0.017
Supervision+support labor		(50% of operating labor)	596	0.009
Maintenance	$6.09 million/year	2% of (on plot+off plot)	18,135	0.259
Taxes+insurance	$3.13 million/year	1% of (on plot+off plot+catalyst)	9,330	0.133
Total running cost			162,740	2.324

*Feed is Arabian Heavy 650°F+ AR.

†Typical costs based on Chevron's operating experience.

‡Positive number is consumption or cost, negative (in parentheses) is production or credit.

Note: EFO = equivalent fuel oil; SCF = standard cubic feet; SCFD = standard cubic feet per day; FF = fresh feed.

TABLE 8.1.8 Utility and Total Cost Summary for 70,000-BPD RDS Hydrotreater to Prepare RFCC Feed*

	Unit cost†	Rate‡	$/Stream day	$/bbl FF
Utilities:				
Fuel	$20.00/EFO-bbl (6 million Btu)	272 EFO-BPD	5,440	0.078
Power	$0.05/kWh	27,000 kW	32,400	0.463
400 lb/in² gage steam	$2.00/klb	− 22 klb/h	(1,056)	(0.015)
150 lb/in² gage steam	$3.25/klb	116 klb/h	9,048	0.129
Cooling water	$0.05/kgal	8,200 gal/min	590	0.008
Process injection water	$5.40/kgal	66 gal/min	513	0.007
Boiler feed water	$6.30/kgal	85 gal/min	771	0.011
Condensate	$5.40/kgal	(176) gal/min	(1,369)	(0.020)
Total Utilities			46,337	0.661
Hydrogen	$2.50/kSCF	71.7 million SCFD	179,250	2.561
Catalyst	$8.80 million/year		26,206	0.374
Operating labor	$0.20 million/year/shift	2 shift positions	1,191	0.017
Supervision+support labor		(50% of operating labor)	596	0.009
Maintenance	$6.09 million/year	2% of (on plot+off plot)	18,135	0.259
Taxes+insurance	$3.13 million/year	1% of (on plot+off plot+catalyst)	9,330	0.133
Capital charge	$76.12 million/year	25% of (on plot+off plot)	226,691	3.238
Total processing cost			507,736	7.252

*Feed is Arabian Heavy 650°F+ AR.

†Typical costs based on Chevron's operating experience.

‡Positive number is consumption or cost, negative (in parentheses) is production or credit.

REFERENCES

1. H. A. Frumkin and G. D. Gould, "Isomax Takes Sulfur Out of Fuel Oil," AIChE Meeting, New Orleans, March 16–20, 1969.
2. A. G. Bridge, E. M. Reed, P. W. Tamm, and D. R. Cash, "Chevron Isomax Processes Desulfurize Arabian Heavy Residua," 74th National AIChE Meeting, New Orleans, March 11–15, 1973.
3. A. G. Bridge, G. D. Gould, and J. F. Berkman, "Residua Processes Proven," *Oil and Gas Journal,* **85,** January 19, 1981.
4. K. Saito, S. Shinuzym, F. Fukui, and H. Hashimoto, "Experience in Operating High Conversion Residual HDS Process," AIChE Meeting, San Francisco, November 1984.
5. J. B. Rush and P. V. Steed, "Refinery Experience With Hydroprocessing Resid for FCC Feed," 49th Midyear Refinery Meeting, API, New Orleans, May 16, 1984.
6. B. E. Reynolds, D. V. Law, and J. R. Wilson, "Chevron's Pascagoula Residuum Hydrotreater Demonstrates Versatility," NPRA Annual Meeting, San Antonio, March 24–26, 1985.
7. J. G. Speight, *The Chemistry and Technology of Petroleum,* 2d ed., Marcel Dekker, New York, 1991.
8. H. Kanazawa and B. E. Reynolds, "NPRC's Success With Chevron VRDS," NPRA Annual Meeting, San Antonio, March 25–27, 1984.
9. N. E. Kaparakos, J. S. Lasher, S. Sato, and N. Seno, "Nippon Mining Company Upgrades Vacuum Tower Bottoms in Gulf Resid HDS Unit," Japan Petroleum Institute Petroleum Refining Conf., Tokyo, October 1984.
10. C. Hung, H. C. Olbrich, R. L. Howell, and J. V. Heyse, "Chevron's New HDM Catalyst System for a Deasphalted Oil Hydrocracker," AIChE 1986 Spring National Meeting, April 10, 1986, paper no. 12b.
11. B. E. Reynolds, D. R. Johnson, J. S. Lasher, and C. Hung, "Heavy Oil Upgrading for the Future: The New Chevron Hydrotreating Process Increases Flexibility," NPRA Annual Meeting, San Francisco, March 19–21, 1989.
12. B. E. Reynolds and M. A. Silverman, "VRDS/RFCC Provides Efficient Conversion of Vacuum Bottoms Into Gasoline," Japan Petroleum Institute Petroleum Refining Conf., Tokyo, October 18–19, 1990.

CHAPTER 8.2
HÜLS SELECTIVE HYDROGENATION PROCESS

Scott Davis
UOP
Des Plaines, Illinois

PROCESS DESCRIPTION

Small amounts of acetylenes and dienes are contaminants to many light olefin processing units. For example, both sulfuric and HF alkylation units suffer from increased acid losses as a result of dienes contained in the feedstock. Certain reactive C_5 dienes contained in the C_5 feed to a unit producing tertiary amyl methyl ether (TAME) must be removed, or they will cause fouling of the resin catalyst. Also, polymer-grade propylene typically requires less than 5 wt ppm of dienes and acetylenes to meet specifications.

The removal of acetylenes and dienes in streams rich in C_3 to C_6 olefins is accomplished with the Hüls* Selective Hydrogenation process (SHP). This process is a highly selective catalytic process that can hydrogenate acetylenes and dienes to the corresponding monolefin without affecting the valuable olefin content of the feedstock. In addition, if needed, the Hüls SHP unit can be designed to provide hydroisomerization of some of the olefins. For instance, a Hüls SHP unit processing a C_4 olefin stream derived from fluid catalytic cracking (FCC) not only converts the dienes to monolefins but also converts much of the butene-1 to butene-2. Butene-2 will generate a higher-octane alkylate in HF alkylation units. Typical product diene levels from a Hüls SHP unit range from 25 to 1 wt ppm. The Hüls SHP unit for C_3 to C_6 olefin streams is available for license through Hüls A.G. in Marl, Germany, or from UOP in Des Plaines, Illinois, U.S.A.

---

*Service mark of Chemische Werke Hüls A.G.

PROCESS FLOW

The flow diagram for this process is shown in Fig. 8.2.1. The feed is combined with hydrogen at near stoichiometric ratios to the diene and acetylene content of the feed. Hydrogenation then takes place in a liquid-phase fixed-bed reactor. Unless the hydrogen purity is low, no separation step is required for the removal of light ends from the product. Thus, the reactor effluent can be charged directly to downstream units.

The design of the Hüls SHP unit is simple and requires low capital and nominal operating costs. Low-temperature, liquid-phase operation means that no utilities are required in most cases if the feed and hydrogen are both available at suitable conditions. Heating or cooling duties are limited to cases with relatively high diene concentrations in the feed.

The catalyst represents a low cost for the Hüls SHP unit. Commercial operation for a period of more than 2 years without the need to regenerate or replace the catalyst is typical.

COMMERCIAL EXPERIENCE

A total of 19 units have been licensed. Fifteen of these units are currently on-stream. The feedstocks to these units vary from C_3's to C_5's. Most of these units are designed for diene reduction to less than 5 wt ppm in the product.

INVESTMENT AND OPERATING REQUIREMENTS

In most applications, the Hüls SHP unit consumes virtually no utilities. The capital investment, including catalyst, for the Hüls SHP unit is small, usually within the range of $500,000 to $1,000,000. The flexibility and minimal cost of this technology make the Hüls SHP unit a valuable processing tool.

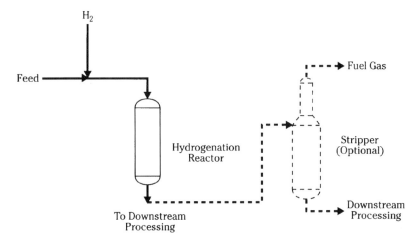

FIGURE 8.2.1 Hüls selective hydrogenation flow diagram.

CHAPTER 8.3
UOP UNIONFINING TECHNOLOGY

James E. Kennedy
UOP
Des Plaines, Illinois

INTRODUCTION

Hydrotreating is one of the most mature technologies found in the 1990s refinery, rivaling the history and longevity of the thermal process. In 1952, UOP and Union Oil Co. of California began licensing hydrotreating under the name of the Unifining process. The partnerships and the development of this technology have gone through a series of changes over the years, and in 1995 the acquisition of the Unocal Process Technology and Licensing group by UOP resulted in the merger of two premier hydroprocessing companies and the combination of their expertise under the UOP* Unionfining* banner.

Generally speaking, the hydrotreating process removes objectionable materials from petroleum distillates by selectively reacting these materials with hydrogen in a catalyst bed at elevated temperature. These objectionable materials include sulfur, nitrogen, olefins, and aromatics. Lighter materials such as naphtha are generally treated for subsequent processing in catalytic reforming units, and the heavier distillates, ranging from jet fuel to heavy vacuum gas oils, are treated to meet strict product-quality specifications or for use as feedstocks elsewhere in the refinery. Many of the product-quality specifications are driven by environmental regulations that are becoming more stringent each year. This push toward more environmentally friendly products is resulting in the addition of hydroprocessing units in refineries throughout the world.

PROCESS CHEMISTRY

The chemistry behind the hydrotreating process can be divided into a number of reaction categories: (hydro)desulfurization, (hydro)denitrification, saturation of olefins,

---

*Trademark and/or service mark of UOP.

8.30 HYDROTREATING

and saturation of aromatics. For each of these reactions, hydrogen is used to improve the quality of the petroleum fraction.

Desulfurization

Desulfurization is by far the most common of the hydrotreating reactions. Sulfur-containing hydrocarbons come in a number of forms, and the ability to remove sulfur from the different types of hydrocarbons varies from one type to the next. The degree to which sulfur can be removed from the hydrocarbon varies from near-complete desulfurization for light straight-run naphthas to 50 to 70 percent for heavier residual materials. Figure 8.3.1 lists several sulfur-containing compounds in order of the difficulty in removing the sulfur.

The reaction of thiophenol, which is at the top of the list in Fig. 8.3.1, proceeds quite rapidly; the reaction is shown schematically in Fig. 8.3.2. Multiring thiophene-type sulfurs are more difficult to treat because the ring structure, which is attached to the sulfur on two sides, must be broken. Figure 8.3.3 is a schematic representation of the reaction for the desulfurization of dibenzothiophene.

In each case, the desulfurization reaction results in the production of hydrogen sulfide (H_2S) in the reactor section of the plant. To complete the desulfurization reaction,

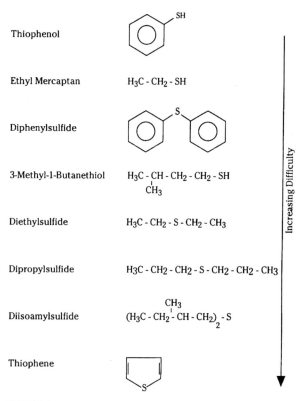

FIGURE 8.3.1 Relative desulfurization reactivities.

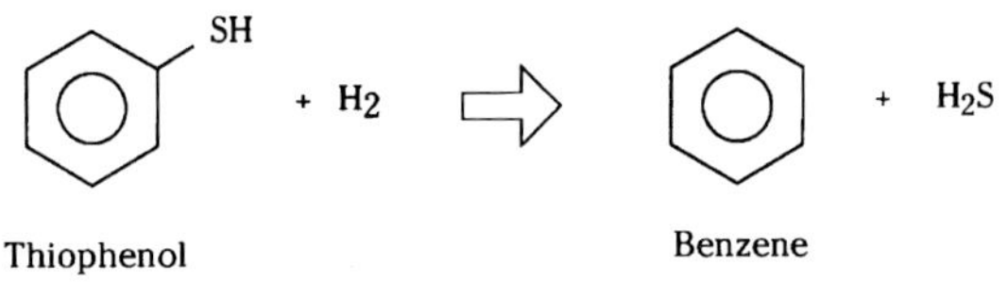

FIGURE 8.3.2 Desulfurization of thiophenol.

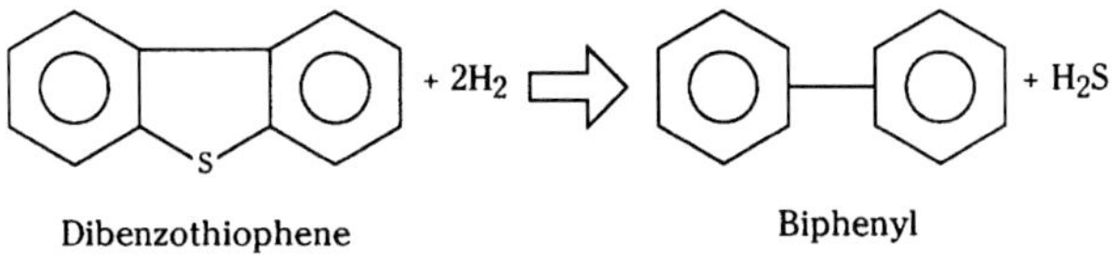

FIGURE 8.3.3 Desulfurization of dibenzothiophene.

the H$_2$S must be removed in downstream fractionation.

Denitrification

The nitrogen compounds that occur naturally in crude oils and that would normally be found in the feed to a hydrotreater can be classified into two categories: *basic* nitrogen, which is generally associated with a six-member ring, and *neutral* nitrogen, which is generally associated with a five-member ring. Examples of these two types of

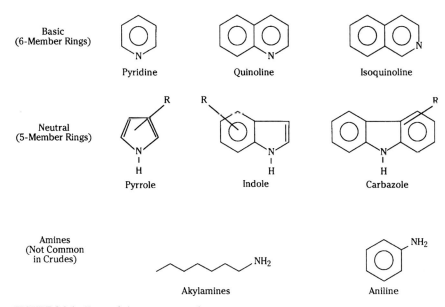

FIGURE 8.3.4 Types of nitrogen compounds.

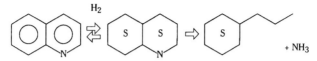

FIGURE 8.3.5 Denitrification of quinoline.

nitrogen are shown in Fig. 8.3.4. The complexity of the nitrogen compounds makes denitrification more difficult than desulfurization.

The denitrification reaction first proceeds through a step that saturates the aromatic ring. This saturation is an equilibrium reaction and normally sets the rate at which the denitrification reaction can occur. Figure 8.3.5 is a schematic representation of a denitrification reaction. The combination of aromatic saturation followed by denitrification results in an increase in the amount of hydrogen required compared to desulfurization. This increased hydrogen consumption also translates into an increase in the amount of heat generated.

The denitrification reaction results in the generation of ammonia (NH_3). To complete the processing, this NH_3 must be removed in downstream fractionation.

Olefin Saturation

Although desulfurization is the most common of the reactions, olefin saturation also proceeds quite rapidly. As shown in Fig. 8.3.6, hydrogen is added to an olefin and the corresponding saturated compound is the product. This reaction is quite fast and highly exothermic. If a significant quantity of olefins is present in the feed, the resulting heat release must be accounted for in the unit design. The ease with which this reaction takes place allows for operation at lower temperatures than the other hydrotreating reactions discussed in this section.

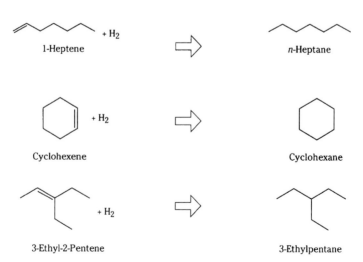

FIGURE 8.3.6 Typical olefin saturation reactions.

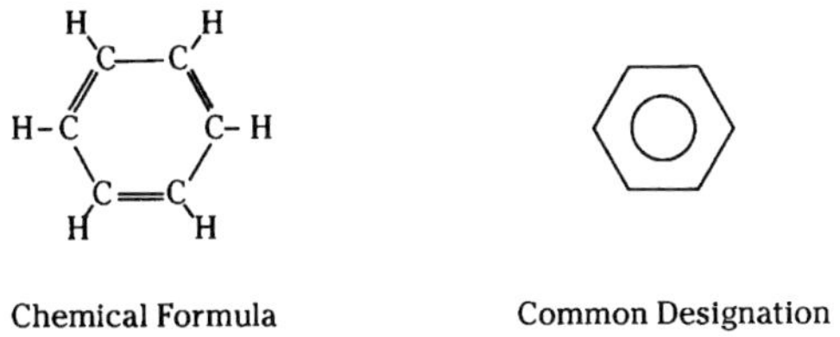

Chemical Formula   Common Designation

FIGURE 8.3.7 Benzene ring.

Aromatic Saturation

Aromatic saturation occurs according to the same principles as olefin saturation in that hydrogen is added to saturate the double bonds in the aromatic or benzene ring. The aromatic or benzene ring is a six-carbon atom ring that contains three double bonds (Fig. 8.3.7). Because this ring structure is quite prominent in many of the materials found in the refinery, the symbol for this benzene ring is simplified and indicated as a hexagon with a circle inside.

Figure 8.3.8 schematically shows three typical aromatic saturation reactions. The S inside the ring represents a six-member carbon ring that has had all of the double bonds saturated. Because these aromatic-saturation reactions are highly exothermic, maintaining a proper temperature profile in the reactor is important. As the catalyst deactivates, the temperatures are raised to maintain conversion until end-of-run (EOR) conditions are approached. In the case of aromatic saturation, EOR occurs when the equilibrium no longer favors aromatic saturation.

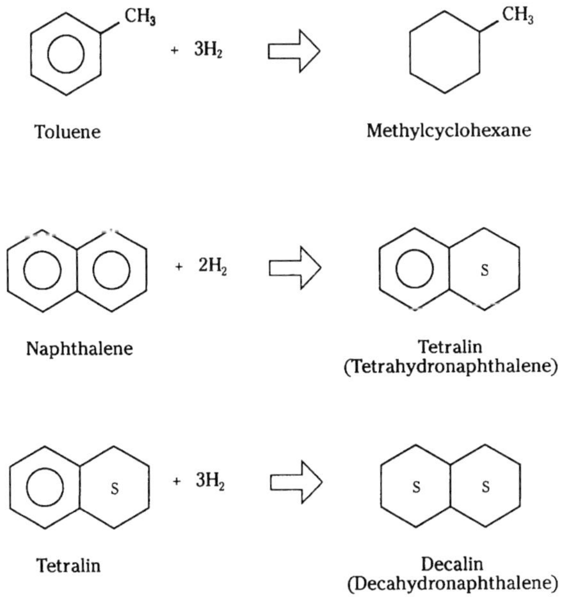

Toluene + 3H$_2$ ⇒ Methylcyclohexane

Naphthalene + 2H$_2$ ⇒ Tetralin (Tetrahydronaphthalene)

Tetralin + 3H$_2$ ⇒ Decalin (Decahydronaphthalene)

FIGURE 8.3.8 Typical aromatic saturation reactions.

Metals Removal

In addition to the previously mentioned typical hydroprocessing functions, the Unionfining unit may also be designed to remove low levels of metals from the feed. The metals to be removed include nickel and vanadium, which are native to the crude oil, as well as silicon and lead-containing materials that are added elsewhere in the refinery. These metals are poisons to downstream processing units and can pose environmental problems if they are contained in a fuel product that will eventually combust. In the past, refiners would operate their hydrotreating unit until the hydrotreating catalyst had no more capacity to absorb metals. In a 1990s hydrotreating unit, the reactor is loaded with a catalyst that is designed specifically to have a high capacity for metals removal if the feed metals are anticipated to be high.

CATALYST

The primary function of the catalyst used in the hydrotreating reaction is to change the rate of reactions. The suitability of a catalyst depends on a variety of factors related to the feed quality and processing objectives. The catalysts used in the UOP Unionfining processes are typically a high surface area base loaded with highly dispersed active metals.

For hydrodesulfurization operations, the catalyst type is typically a *cobalt molybdenum* catalyst. The typical composition for this catalyst is shown in Table 8.3.1. A nickel molybdenum catalyst ($NiO/MoO_3/Al_2O_3$) is sometimes used as an alternative.

In denitrification operations, a catalyst with a different hydrogen function is required to allow operation at normal temperatures. In these instances, the *nickel molybdenum* catalyst is more common. These catalysts are also good desulfurization catalysts; however, their hydrogen consumption could be higher because of their better denitrification activity.

Either of these catalysts provides adequate activity for the saturation of olefins. As previously mentioned, these reactions are fast and occur at temperatures lower than those required for desulfurization or denitrification.

For the saturation of aromatics, the selection of the proper catalyst is quite dependent on the processing objectives. In many cases, a nickel molybdenum catalyst provides the required level of aromatic saturation. In cases where the feed aromatics content is high or the product aromatics specification is low, UOP might suggest a catalyst that has some level of noble metal (such as platinum or palladium) to be used after the nickel molybdenum catalyst.

The metals-removal catalysts are designed specifically for the purpose of removing metals from the feed so that they do not affect the hydrotreating capability of the

TABLE 8.3.1 Typical Composition of Cobalt Molybdenum Catalyst

Species	Range, wt %	Typical, wt %
CoO	1–5	3
MoO_3	6–25	12
Al_2O_3	Balance	Balance

hydroprocessing catalyst. These catalysts typically have a different shape or pore structure or both than the normal hydrotreating catalyst and are often designed to have some reduced level of desulfurization or denitrification activity.

PROCESS FLOW

The actual flow scheme of the UOP Unionfining process varies, depending on the application. Figure 8.3.9 provides a generic look at the flow scheme of a UOP Unionfining unit. The feed is exchanged with the reactor effluent, mixed with recycle hydrogen, and then heated to reaction temperature in a fired heater. The combined feed then flows through the reactor, which contains the catalyst that will accelerate the reaction. The reactor effluent is cooled in exchange with the feed and then in a series of coolers before being separated in a vapor-liquid separator. The vapor portion is recompressed, combined with fresh hydrogen, and returned to the reactor feed. The liquid portion is fed to a fractionator, where it is stripped of light ends, H_2S, and NH_3.

UNIONFINING APPLICATIONS

Generally speaking, the most common way to categorize hydrotreating applications is by feed type. This section provides general information on a limited number of Unionfining applications.

Naphtha Unionfining

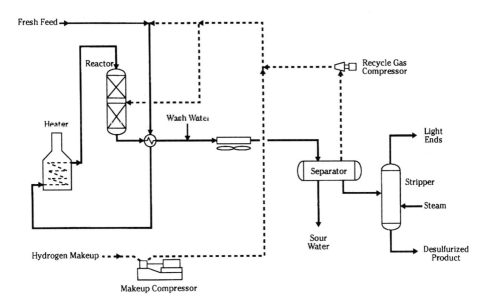

FIGURE 8.3.9 Typical Unionfining process flow.

The main use of the hydrotreating process in naphtha applications is in the preparation of feedstocks for the naphtha reforming unit. The reforming process requires low levels of sulfur, nitrogen, and metals in the feed. The Unionfining process reduces the sulfur and nitrogen to less than 0.5 wt ppm and the metals to nondetectable levels. For olefinic feeds, the Unionfining process is also used to stabilize the naphtha by completely saturating the olefins.

A comparison of the typical processing conditions of the various hydroprocessing operations indicates that naphtha feeds are typically the easiest to hydrotreat. Table 8.3.2 provides a list of typical operating conditions for the applications discussed in this section.

Distillate Unionfining

A distillate Unionfining process is typically installed to improve the quality of kerosene and diesel oils. Although the primary objective is generally desulfurization, a middle-distillate hydrotreater is often installed to improve the combustion properties, stability, color, or odor of the product.

With the development of more stringent environmental criteria around the world, many distillate hydrotreaters are being designed in the 1990s to improve the quality of jet fuel and road diesel. In addition to reducing the sulfur and nitrogen content of the distillates, operating conditions can be adjusted to saturate the aromatics in the kerosene and diesel. The result is an improvement in the smoke point of the jet fuel and the cetane number of the diesel. The role of hydrotreating will continue to expand for the foreseeable future, and the refining industry expects no turning back from the trend of increasing environmental challenges.

Vacuum Gas Oil Unionfining

A vacuum gas oil (VGO) Unionfining process is typically designed to either upgrade the feed quality for further processing or improve the VGO quality so that it can be used as an environmentally friendly fuel oil. Typically, further processing of the VGO occurs in a fluid catalytic cracking (FCC) unit or in a hydrocracking unit.

As can be seen in Table 8.3.2, the conditions required to hydrotreat a VGO stream are more severe than those required to hydrotreat feedstocks with lower molecular

TABLE 8.3.2 Typical Hydrotreating Operating Conditions

Operating conditions	Naphtha	Middle distillate	Light gas oil	Heavy gas oil
LHSV	1.0–5.0	1.0–4.0	1.0–3.0	0.75–2.0
H_2/HC ratio, Nm^3/m^3 (SCFB)	50 (300)	135 (800)	170 (1000)	337 (2000)
H_2 partial pressure, kg/cm^2 (lb/in^2 abs)	14 (200)	28 (400)	35 (500)	55 (800)
SOR reactor temperature, °C (°F)	290 (555)	330 (625)	345 (653)	355 (671)

Note: LHSV = liquid hourly space velocity; N = standard temperature and pressure; SCFB = standard cubic feet per barrel.

weight. As a result, some low level (10 to 30 percent) conversion can take place in a VGO Unionfining unit. This conversion requires that the product fractionation be designed to recover lighter products for use elsewhere in the refinery or for blending with the refinery product streams.

RCD Unionfining Process

The RCD Unionfining process for hydrotreating residual hydrocarbons is not discussed in this chapter; however, the principles involved are the same, but the processing conditions are more severe (Table 8.3.2).

INVESTMENT

The investment associated with the installation of a hydrotreating unit depends on the feed characteristics and the product specifications. Generally speaking, as the feed gets heavier or the individual product specifications are reduced, the processing requirements are increased. These more severe processing conditions can result in more pieces of equipment, larger equipment, and higher operating pressure, all of which increase the cost of the unit. The required capital investment for a hydrotreating unit can vary from $500 to $2000 U.S. per barrel per stream day of capacity.

UOP HYDROPROCESSING EXPERIENCE

The Unionfining process is really a broad family of fixed-bed hydrotreating processes. Naphtha, distillate, VGO, and RCD. Unionfining units are in operation throughout the world. UOP's Unionfining experience is broken down by application in Fig. 8.3.10. More than 500 commercial units have been designed, and these units process literally hundreds of different feed streams.

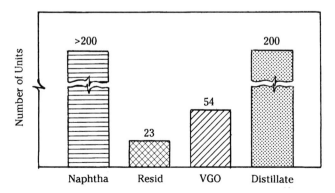

FIGURE 8.3.10   Unionfining experience.

BIBLIOGRAPHY

Ackelson, Donald B., David A. Lindsay, Robert E. Miller, and Milan Skripek, "Hydrotreating-Hydrocracking," Annual Conf. on International Refining and Petrochemicals, Singapore, May 9–10, 1994.

Baron, Ken, Robert E. Milner, Alice Tang, and Laurie Palmer, "Hydrotreating of Light Cycle Oil," Annual Meeting of National Petroleum Refiners Association, New Orleans, March 22–24, 1992.

Nguyen, Tuan A., and Milan Skripek, "Reducing Sulfur in FCC Gasoline via Hydrotreating," AIChE Spring National Meeting, April 17–21, 1994.

Nguyen, Tuan, and Milan Skripek, "VGO Unionfining: Technical Case Studies," *Hydrocarbon Technology International,* Sterling Publications Ltd., London, 1993.

CHAPTER 8.4
UOP RCD UNIONFINING PROCESS

Gregory J. Thompson
UOP
Des Plaines, Illinois

INTRODUCTION

The UOP* RCD Unionfining* (reduced-crude desulfurization) process represents the merger of three of the world's leaders in residual oil processing and catalyst technology. UOP's acquisition of the Unocal PTL Division in January 1995 resulted in the merging of UOP and Unocal's catalyst technology, commercial know-how, and design experience to create a new, improved resid hydrotreating process: RCD Unionfining. At the time of the acquisition, the 25 units licensed worldwide by UOP and Unocal represented more than 130,000 m^3/h [832,000 barrels per day (BPD)] of capacity. Prior to this acquisition in 1993, UOP entered into an alliance with Catalyst & Chemicals Ind. Co. Ltd. (CCIC) in Japan that enabled UOP to offer CCIC's commercially proven portfolio of residual hydrotreating catalysts.

The RCD Unionfining process provides desulfurization, denitrification, and demetallization of reduced crude, vacuum-tower bottoms, or demetallized oil (DMO). Contaminant removal is accompanied by partial conversion of nondistillables. The process employs a fixed bed of catalyst, operates at moderately high pressure, consumes hydrogen, and is capable of greater than 90 percent removal of sulfur and metals. In addition to its role of providing low-sulfur fuel oil, the process is frequently used to improve feedstocks for downstream conversion units, such as cokers, fluid catalytic crackers (FCCs), and hydrocrackers.

MARKET DRIVERS FOR RCD UNIONFINING

The first commercial reduced-crude desulfurization unit, which came on stream in 1967, was a licensed design from UOP. The resid hydrotreating units produced a low-sulfur fuel oil product that was in increasing demand as a result of the stringent laws

---
*Trademark and/or service mark of UOP.

relating to air pollution that were being enacted in the industrialized countries. Units were designed in the 1970s to produce fuel oil with a sulfur level as low as 0.3 wt %. The trend toward low-sulfur fuel oil has now extended around the world.

Other drivers have added to the need for the RCD Unionfining process. Fuel oil demand has been declining at a rate of about 0.2 percent per year since the 1980s, and this decline has been coupled with a growth of about 1.4 percent in the demand for refined products. As the demand for heavy fuel oil has fallen, the price differential between light and heavy crude oil has increased. This price differential has given the refiner an economic incentive to process heavy crude. However, heavy crude not only produces a disproportionate share of residual fuel, but it is also usually high in sulfur content. Because heavy, high-sulfur crude is a growing portion of the worldwide crude oil reserves, refiners looking for future flexibility have an incentive to install substantial conversion and desulfurization capacity to produce the required product slate. In addition to providing low-sulfur fuel oil, the RCD Unionfining process provides excellent feedstock for downstream conversion processes producing more valuable transportation fuels.

CATALYST

Catalysts having special surface properties are required to provide the necessary activity and stability to cope with reduced-crude components. The cycle life of the catalyst used in the RCD Unionfining process is generally set by one of three mechanisms:

> Excessive buildup of impurities, such as scale or coke, that lead to unacceptable pressure drop in the reactor
>
> Coke formation from the decomposition and condensation of heavy asphaltic molecules
>
> Metal deposition in catalyst pores from the hydrocracking of organometallic compounds in the feed

UOP and CCIC provide a complete portfolio of catalysts to handle each of these three mechanisms (Table 8.4.1).

Feed filtration for removing scale particulates is a standard part of the RCD Unionfining design. In most cases, this filtration satisfactorily prevents buildup of scale on the catalyst bed, and the resulting pressure drop does not limit cycle life. However, some feeds contain unfilterable components. Under these circumstances, the catalyst bed itself acts as a filter, and the impurities build up in the top section of catalyst to create unacceptable pressure drop. The CDS-NP series of catalyst helps prevent this problem by increasing the amount of void space in the top of the reactor for the impurities to collect. The CDS-NP catalysts are designed as a macaroni shape in ¼-in and ⅙-in sizes. The macaroni shape maximizes the amount of void space and catalyst surface area available for deposition of the impurities. The catalysts are also loaded from the larger to smaller size. This kind of loading, which is called *grading* the catalyst bed, helps to maximize the void space available for deposition.

Coke formation reduces catalyst effectiveness by decreasing the activity of the reactive surface and decreasing the catalyst-pore volume needed for metals accumulation. For a given catalyst and chargestock operating under steady-state conditions, the amount of coke on the catalyst is a function of temperature and pressure. Successful hydrodesulfurization of reduced crudes requires that temperatures and pressures be selected to limit coke formation. When coke formation is limited, ultimate catalyst life is determined by the rate of metals removal from organometallic compounds in reduced crude and by the catalyst-pore volume available for the accumulation of metals.

TABLE 8.4.1 CCIC Catalyst Portfolio

Catalyst name	Application	Size and shape
CDS-NP1	ΔP relaxation and HDM	¼ in shaped
CDS-NP5		
CDS-NPS1	ΔP relaxation and HDM	⅙ in shaped
CDS-NPS5		
CDS-DM1	HDM	
CDS-DM5	HDM	
CDS-R9	HDS and HDM	⅛, 1/12, 1/16, 1/22 in cylindrical or shaped
CDS-R95	New version of CDS-R9	
CDS-R2	HDS	
CDS-R25H	New version of CDS-R2	

HDS = hydrodesulfurization.

Note: HDM = hydrodemetallization.

The deposition rate of metals from organometallic compounds correlates with the level of desulfurization for a given catalyst. Thus, the rate of catalyst deactivation by metals deposition increases with the increasing level of desulfurization. Metals deposition results from the conversion of sulfur-bearing asphaltenes. Conversion exposes catalyst pores to the portion of the feed that is highest in organometallics.

The catalyst deactivation rate is also a function of feedstock properties. Heavier reduced crudes with high viscosities, molecular weights, and asphaltene contents tend to be more susceptible to coke formation. Hence, higher pressures and lower space velocities are required for processing these materials. Correlations developed on the basis of commercial data and pilot plant evaluation of many different reduced crudes can predict the relationship between the hydrodesulfurization reaction rate and the deactivation rate and reduced-crude properties.

The UOP-CCIC catalyst portfolio has been developed to maximize both the removal of sulfur, metals, and other impurities such as nitrogen and Conradson carbon and the life of the catalyst. The CDS-DM series catalysts are typically loaded downstream of the CDS-NP catalysts and are designed for maximum metal-holding capacity. Although their metals-removal activity is high, they maximize the removal of the resin-phase metals and minimize the removal of asphaltene-phase metals. (For an explanation of the terms *resin phase* and *asphaltene phase,* see the following "Process Chemistry" section.) The removal of asphaltene-phase metals can lead to excessive formation of coke precursors, which ultimately reduce the life of downstream catalysts. The CDS-R9 series catalysts are typically located downstream of the CDS-DM catalysts. These transition catalysts have intermediate activity for demetallization and desulfurization and are used to gradually move from maximum demetallization to maximum desulfurization. Once again, the gradual transition helps to minimize the formation of coke precursors, which could lead to shortened catalyst life. The final catalysts are the CDS-R2 series, which have maximum desulfurization activity. By the time the residual oil reaches this series of catalysts, the metals level is sufficiently low to prevent metals deactivation.

In addition to this portfolio of conventional desulfurization and demetallization catalysts, several custom catalysts are available for the RCD Unionfining process. The

R-HAC1 catalyst is a resid, mild-hydrocracking catalyst intended for use with lighter feedstocks. Although it has the same hydrodesulfurization activity as conventional HDS catalysts, it produces 3 to 4 vol % more diesel fuel without an increase in naphtha or gas yields. The CAT-X catalyst is designed as an FCC feed pretreatment catalyst. The FCC microactivity testing (MAT) of feeds processed over the CAT-X catalyst has shown an increase in the gasoline yield of as much as 5 percent.

PROCESS CHEMISTRY

Resid is a complex mixture of heavy petroleum compounds that are rich in aromatic structures, heavy paraffins, sulfur, nitrogen, and metals. An *atmospheric resid* (*AR*) is a material that has been produced in an atmospheric-pressure fractionation column as a bottoms product (ATB) when the boiling endpoint of the heaviest distilled product is at or near 343°C (650°F). The bottoms is then said to be a 343°C+ (650°F+) atmospheric resid. A *vacuum resid* (*VR*) is produced as bottoms product from a column running under a vacuum when the boiling endpoint of the heaviest distilled product is at or near 566°C (1050°F). The bottoms is then said to be a 566°C+ (1050°F+) vacuum resid.

Resid components can be characterized in terms of their solubility:

Saturates: Fully soluble in pentane; this fraction contains all the saturates.

Aromatics: Soluble in pentane and separated by chromatography; this fraction contains neutral aromatics.

Resins: Soluble in pentane and absorb on clay; this fraction contains polar aromatics, acids, and bases.

Asphaltenes: Those that are insoluble in pentane (*pentane insolubles*) and those that are insoluble in heptane (*heptane insolubles*); the weight percent of pentane insolubles is always greater than the weight percent of heptane insolubles.

Typically, the conversion reaction path in the RCD Unionfining process is from asphaltenes to resins, resins to aromatics, and aromatics to saturates.

With the exception of the lightest fractions of crude oil, impurities can be found throughout the petroleum boiling range. Impurity concentrations of each fraction increase with the boiling point of the fraction. Examples of this situation for sulfur and nitrogen are shown in Figs. 8.4.1 and 8.4.2.

Of all the components in the resid, the asphaltene components are the most difficult to work with. Asphaltene molecules are large and are rich in sulfur, nitrogen, metals (Fe, Ni, V), and polynuclear aromatic compounds. These components are primarily the ones that deactivate the catalyst through metals contamination or coke production. Characterization of some typical atmospheric resids along with their respective asphaltene components is shown in Table 8.4.2. An example of an asphaltene structure can be seen in Fig. 8.4.3. Typically, the impurities are buried deep inside the asphaltene molecule, and so severe operating conditions are required to remove them.

PROCESS DESCRIPTION

Operating Variables

For a specific feedstock and catalyst package, the degree of demetallization, desulfurization, and conversion increases with the increasing severity of the RCD Unionfining

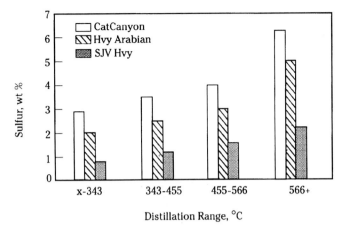

FIGURE 8.4.1 Sulfur distribution.

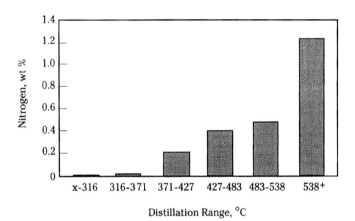

FIGURE 8.4.2 Nitrogen distribution in Hondo California crude.

TABLE 8.4.2 AR Characterization

	Crude source		
	Arabian Heavy	Hondo	Maya
AR properties:			
Sulfur, wt %	4.29	5.9	4.4
NI+V, wt ppm	108	372	500
Asphaltenes, wt %	12.6	13.9	25.2
Asphaltene properties:			
Sulfur, wt %	6.5	7.7	6.4
Ni + V, wt ppm	498	1,130	1,570

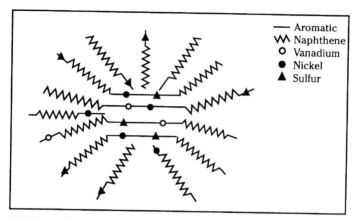

FIGURE 8.4.3 Asphaltene structure.

operation. The operating variables are pressure, recycle-gas rate, space velocity, and temperature.

Pressure. Increasing hydrogen partial pressure decreases the catalyst deactivation rate at constant reactor temperature because the formation of carbonaceous deposits, which deactivate the catalyst, is thereby retarded. The increased pressure also increases the activity for desulfurization, demetallization, and denitrification. Increased hydrogen partial pressure can be obtained by increasing total pressure or by increasing the hydrogen purity of the makeup gas, which together with a recycle-gas scrubber to remove H_2S maximizes hydrogen partial pressure.

Recycle-Gas Rate. Increasing the recycle-gas rate increases the hydrogen-to-hydrocarbon ratio in the reactor. This increased ratio acts in much the same manner as increased hydrogen partial pressure.

Space Velocity. Increasing the space velocity (higher feed rate for a given amount of catalyst) requires a higher reactor temperature to maintain the same impurity removal level and results in an increase in the deactivation rate.

Temperature. Increasing temperature increases the degree of impurity removal at a constant feed rate. Operating at an increased temperature level increases the catalyst deactivation rate. As the catalyst operating cycle proceeds, reactor temperature is usually increased because of the disappearance of active catalyst sites.

Process Flow

A simplified flow diagram of the UOP RCD Unionfining process is presented in Fig. 8.4.4. The filtered liquid feed is combined with makeup hydrogen and recycled separator offgas and sent first to a feed-effluent exchanger and then to a direct-fired heater. In this flow scheme, the direct-fired heater is shown as a two-phase heater, but the alternative of separate feed and gas heaters is also an option. The mixed-phase heater effluent is charged to a guard bed and then to the reactor or reactors. As indicated earlier, the guard bed is loaded with a *graded bed* of catalyst to guard against

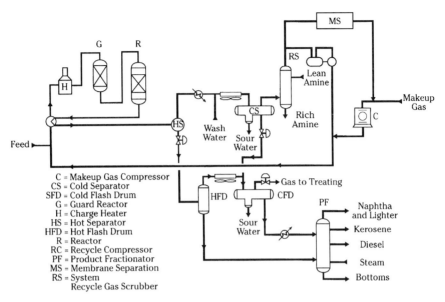

FIGURE 8.4.4 Simplified RCD Unionfining flow scheme.

unacceptable pressure drop, but this catalyst also performs some impurities removal. Removal of the remaining impurity occurs in the reactor.

The RCD Unionfining reactors use a simple downflow design, which precludes problems of catalyst carryover and consequent plugging and erosion of downstream equipment. Because this reactor system has three phases, uniform flow distribution is crucial. UOP provides special reactor internals to ensure proper flow distribution. The reactor-effluent stream flows to a hot separator to allow a rough separation of heavy liquid products, recycle gas, and lighter liquid products. The hot separator overhead is cooled and separated again to produce cold separator liquid and recycle gas, which is scrubbed to remove H_2S before being recycled. A portion of the scrubbed recycle gas is sent to membrane separation to reject light components, mainly methane, that are formed in the reactor. If these components are not removed, they could adversely affect the hydrogen partial pressure in the reactor. Hot separator liquid is fed to a hot flash drum, where the overhead is cooled and mixed with cold separator liquid, and the mixture is charged to the cold flash drum. Bottoms from both the hot and cold flash drums are charged to the unit's fractionation system, which can be set up to either yield low-sulfur fuel oil or match feed specifications for downstream processing.

Process Applications

As trends toward heavier crudes and lower fuel-oil demand have become evident, UOP has devoted increased attention to bottom-of-the-barrel processing. As a result, several flow schemes that offer a variety of advantages have been developed. The most common of these flow schemes is shown in Fig. 8.4.5. Atmospheric residual oil is directly hydrotreated to provide FCC feed. Hydrotreating allows a high percentage of the crude to be catalytically cracked to gasoline while maintaining reasonable FCC catalyst consumption rates and regenerator SO_x emissions. Hydrotreating can also help

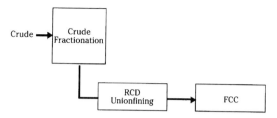

FIGURE 8.4.5 Maximum gasoline production.

refiners meet some of the newly emerging gasoline sulfur specifications (100 wt ppm maximum) in selected parts of the world, such as the United States, Sweden, and Finland.

For upgrading residual oils high in metals, the best processing route may be a combination of solvent extraction (UOP Demex* process) and the RCD Unionfining process. The Demex process separates vacuum residual oil with a high metal content into a DMO of relatively low metal content and a pitch of high metal content. The pitch has several uses, including fuel-oil blending, solid-fuel production, and feed to a partial-oxidation unit for hydrogen production. If the metal and Conradson carbon content of the DMO are sufficiently low, it may be used directly as an FCC or hydrocracker feed component. In some cases, however, hydrotreating the DMO prior to cracking is desirable, as shown in Fig. 8.4.6. This combination of processes shows better economics than either process alone. The arrangement provides an extremely flexible processing route, because a change in feedstock can be compensated for by adjusting the ratio of DMO to pitch in the Demex unit to maintain DMO quality. In some cases, the treated material can be blended with virgin vacuum gas oil (VGO) and fed directly to the conversion unit.

When the RCD Unionfining process is used to pretreat coker feed (fig 8.4.7), it reduces the yield of coke and increases its quality and produces a higher-quality cracking feedstock.

*Trademark and/or service mark of UOP.

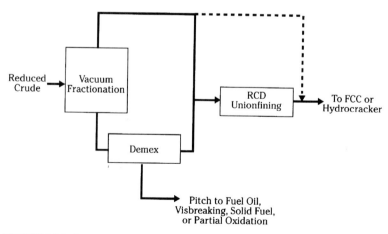

FIGURE 8.4.6 Maximum flexibility flow scheme.

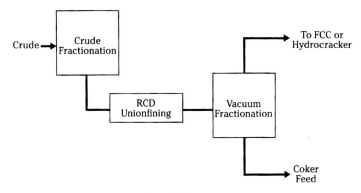

FIGURE 8.4.7 High-quality coke production.

Of course, these examples are just a few of the bottom-of-the-barrel upgrading flow schemes involving the RCD Unionfining process. The correct selection of flow scheme is typically specific to a given refiner's needs and crude type.

OPERATING DATA

The yield and product properties for processing a blended Middle Eastern reduced crude in an RCD Unionfining unit are shown in Table 8.4.3. Utilities required to operate a 132.5-m^3/h [20,000 barrels per stream day (BPSD)] RCD Unionfining unit are shown in Table 8.4.4. The estimated erected cost for this unit is $40 million.

TABLE 8.4.3 Yields and Product Properties of a Middle East Blend Reduced Crude

	Yields on Reduced Crude						
	wt %	vol %	Specific gravity	Sulfur, wt %	Nitrogen, vol %	Viscosity, cSt at 50°C	V + N, ppm
Charge:							
Raw oil	100.00	100.00	12.1	4.1	0.31	2,259	141
Chem. H$_2$ consump.	1.29 (140 m^3/m^3)	—	—	—	—	—	—
Products:							
NH$_3$	0.19	—	—	—	—	—	—
H$_2$S	3.91	—	—	—	—	—	—
C$_2^-$	0.67	—	—	—	—	—	—
C$_3$	0.36	—	—	—	—	—	—
C$_4$	0.36	—	—	—	—	—	—
C$_5$–154°C	1.10	1.50	0.720	0.004	0.004	—	—
154–360°C	14.70	16.70	0.868	0.02	0.02	2-3	—
360°C+	80.00	84.20	0.935	0.47	0.17	151	18
Total	101.29	102.40	—	—	—	—	—

TABLE 8.44 Typical Utilities Required for a RCD Unionfining Unit*

Utility	Reactor	Fractionation	Total
Fired duty, million kcal/h (million Btu/h)	10.16	1.49	11.65(46.27)
Electricity, kW	4937	171	5108
Steam consumed, kg/h (lb/h)			
High-pressure steam	4042	—	4042 (8911)
Low-pressure steam	—	666	666 (1468)
Cooling water, m^3/h (gal/h)	59	86.3	145.3 (38,400)

*Basis: 132.5-m^3/h (20,000-BPSD) unit.

COMMERCIAL INSTALLATIONS

The first commercial direct reduced-crude desulfurization unit was a UOP unit that went on-stream in 1967 at the Chiba, Japan, refinery of Idemitsu Kosan. The first commercial direct vacuum-resid conversion unit was a UOP unit that went on-stream in 1972 at the Natref, South Africa, refinery. As of early 1996, more than 130,000 m^3/h (830,000 BPSD) of RCD Unionfining capacity has been licensed. These units process a variety of feeds, including DMOs and vacuum resids and atmospheric resids. Applications for this process include conventional desulfurization, downstream conversion unit pretreatment, and resid nondistillable conversion.

CHAPTER 8.5
UOP CATALYTIC DEWAXING PROCESS

Orhan Genis
*UOP Ltd.
Guildford, Surrey, England*

INTRODUCTION

The UOP Catalytic Dewaxing process, formerly known as the Unicracking*/DW* process, is a fixed-bed catalytic dewaxing (DW) process for improving the flow properties of various feedstocks to produce fuel or lube products that can be used in extremely cold conditions. Feedstocks to the process may be deasphalted oil, vacuum gas oil, waxy atmospheric gas oils, or various lubestock cuts. The process applications may be classified in two major areas:

Production of various grades of conventional lube oils for extremely low temperature service

Production of middle-distillate components that may be blended into products for extreme winter conditions

The Catalytic Dewaxing process operates across a rather narrow range of design parameters. The primary roles are lube oil dewaxing and middle-distillate flow property improvement. At the same time, the process deep-hydrotreats kerosene or diesel fuel to remove sulfur and nitrogen, and also saturates aromatic compounds. Key process features of the Catalytic Dewaxing process include:

- Excellent product stability
- Excellent product color
- Constant product quality throughout a catalyst cycle
- Minimum viscosity reduction
- Long catalyst cycle life
- Flexibility to produce lubestocks and process distillates in the same unit

---

*Trademark and/or service mark of UOP.

8.50 HYDROTREATING

PROCESS CHEMISTRY

The Catalytic Dewaxing process uses a dual-function, non-noble-metal zeolite catalyst to selectively hydrocrack the paraffinic (waxy) components in the feedstock. The first stage of the process involves hydrotreatment of the incoming feedstock through olefin saturation, desulfurization, and denitrification reactions. Pretreating protects the dewaxing catalyst and provides a feed with a low organic sulfur and nitrogen content, which improves the hydrocracking performance. The zeolite catalyst support selectively absorbs and cracks the normal and near-to-normal paraffins, which are the main constituents of wax. The cracking of paraffins is achieved by using both the acid and metal functions of the catalyst. Figure 8.5.1 shows the postulated paraffin-cracking mechanism.

CATALYST

The process uses two kinds of catalysts. The first is a high-activity desulfurization and dentrification catalyst, which gives an optimum balance between process objectives and cost. The second selectively cracks straight-chain paraffins. This flexible catalyst system enables a refiner to vary the feedstocks and contaminants without affecting product quality or run length.

UOP has several highly active, long-lived Catalytic Dewaxing catalysts. Process objectives determine the type of catalyst used in a particular unit. Catalytic Dewaxing

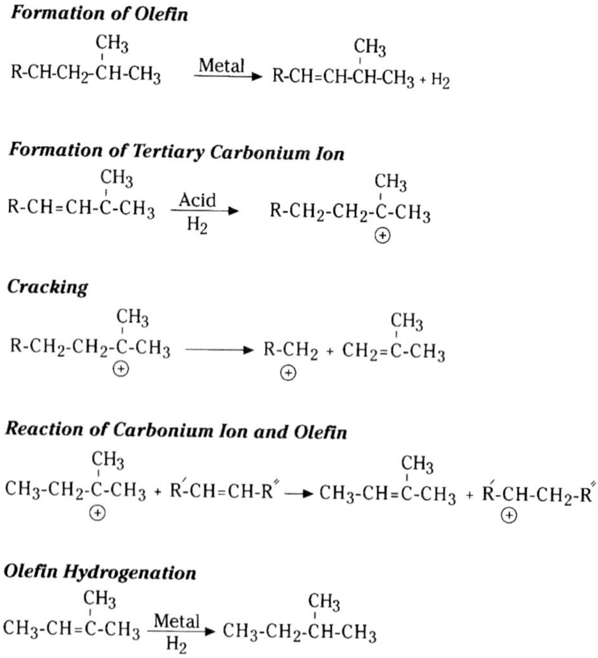

Formation of Olefin

$$R\text{-}CH\text{-}CH_2\text{-}\underset{|}{C}H\text{-}CH_3 \quad \xrightarrow{\text{Metal}} \quad R\text{-}CH\text{=}CH\text{-}\underset{|}{C}H\text{-}CH_3 + H_2$$
$$\phantom{R\text{-}CH\text{-}CH_2\text{-}}CH_3 \phantom{\xrightarrow{\text{Metal}}} \phantom{R\text{-}CH\text{=}CH\text{-}}CH_3$$

Formation of Tertiary Carbonium Ion

$$R\text{-}CH\text{=}CH\text{-}\underset{|}{\overset{CH_3}{C}}\text{-}CH_3 \quad \xrightarrow[H_2]{\text{Acid}} \quad R\text{-}CH_2\text{-}CH_2\text{-}\underset{\oplus}{\overset{CH_3}{C}}\text{-}CH_3$$

Cracking

$$R\text{-}CH_2\text{-}CH_2\text{-}\underset{\oplus}{\overset{CH_3}{C}}\text{-}CH_3 \quad \longrightarrow \quad R\text{-}\underset{\oplus}{C}H_2 + CH_2\text{=}\underset{|}{\overset{CH_3}{C}}\text{-}CH_3$$

Reaction of Carbonium Ion and Olefin

$$CH_3\text{-}CH_2\text{-}\underset{\oplus}{\overset{CH_3}{C}}\text{-}CH_3 + R'\text{-}CH\text{=}CH\text{-}R'' \quad \longrightarrow \quad CH_3\text{-}CH\text{=}\underset{|}{\overset{CH_3}{C}}\text{-}CH_3 + R'\text{-}\underset{\oplus}{C}H\text{-}CH_2\text{-}R''$$

Olefin Hydrogenation

$$CH_3\text{-}CH\text{=}\underset{|}{\overset{CH_3}{C}}\text{-}CH_3 \quad \xrightarrow[H_2]{\text{Metal}} \quad CH_3\text{-}CH_2\text{-}\underset{|}{\overset{CH_3}{C}}H\text{-}CH_3$$

FIGURE 8.5.1 Catalytic Dewaxing postulated paraffin-cracking mechanism.

catalysts typically last 6 to 8 years. During that time, they are regenerated as needed. Typical cycles last 2 to 4 years between regenerations.

PROCESS FLOW

Figure 8.5.2 shows a simplified process flow for a typical Catalytic Dewaxing unit. Fresh feed is preheated and combined with hot recycle gas. The mixture enters the first reactor for treating by a high-activity denitrogenation-desulfurization catalyst, which converts organic nitrogen and sulfur to ammonia and hydrogen sulfide. The reactions are exothermic and cause a temperature rise in the reactor. The reactions are maintained at as low a temperature as possible to maximize catalyst life. The use of a separate reactor for hydrotreating makes the Catalytic Dewaxing process less susceptible to feedstock-induced upsets than single-reactor or swing-reactor processes.

The effluent from the first reactor is cooled with cold quench gas before entering the second reactor, which contains one of UOP's highly selective Catalytic Dewaxing catalysts. These active catalysts function well in the presence of ammonia and hydrogen sulfide. As the feed flows over the dewaxing catalyst, normal paraffins are cracked into smaller molecules, thereby reducing pour point. The average temperature in the second reactor is adjusted to obtain the desired pour point.

Dewaxing reactions are exothermic and must be closely controlled because the dewaxing catalyst is sensitive to temperature. As in the hydrotreater section, reactor temperatures are maintained as low as possible. In the dewaxing reactor, this low temperature not only prolongs catalyst life, but also maximizes liquid yields and helps maintain control.

In both reactors, temperature is controlled by the injection of cold, hydrogen-rich recycle gas at predetermined points. A unique combination of patented internals allows for sufficient mixing of recycle gas with the hot reactants emerging from the previous catalyst bed and the effective distribution of the quenched mixture to the top of the next catalyst bed.

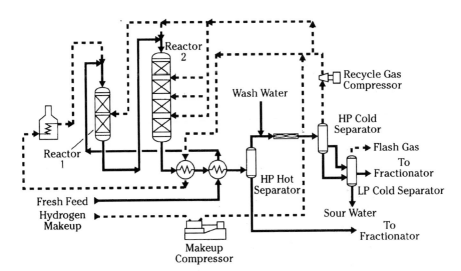

FIGURE 8.5.2 Catalytic Dewaxing process.

Effluent from the second reactor is cooled by exchange with several process streams, including feed and recycle gas. The effluent is then flashed into a hot high-pressure separator, where liquid products are separated from hydrogen-rich vapors. The liquid fraction from this separator is directed to the fractionation section, and vapors are sent to a cold high-pressure separator after being cooled in reactor effluent coolers. Steam condensate or deaerated boiler feed water is injected upstream of the reactor effluent air condensers to minimize corrosion and prevent deposition of ammonia salts. The cold-separator vapor is joined by hydrogen makeup gas to become recycle gas. Liquid hydrocarbon flows into a low-pressure separator.

Flash gas from the low-pressure separator is routed to a light-ends recovery unit or to sour fuel gas. Liquid hydrocarbon from the low-pressure separator exchanges heat with the reactor effluent before flowing into the stripper in the fractionation section.

YIELD PATTERNS

The Catalytic Dewaxing process can be applied to a wide range of feedstocks for dewaxing purposes. Table 8.5.1 shows typical yields and properties for VGO and diesel applications.

INVESTMENT AND OPERATING EXPENSES

The capital investment associated with a Catalytic Dewaxing unit is closely related to the feedstock type and quality as well as the desired level of dewaxing for final products. The flexibility required between processing lubestocks or distillates in the same unit may result in oversizing some of the equipment. The capital investment for a typical Catalytic Dewaxing unit is given in Table 8.5.2, and utilities are listed in Table 8.5.3.

TABLE 8.5.1 Typical Yield and Property Patterns for Catalytic Dewaxing Process

	VGO feed	Diesel feed
Yields, wt %:		
C_1-C_3	0.50	2.50
C_4-260°C (500°F) naphtha	24.50	24.50
Dewaxed product	75.00	73.00
Feed properties:		
°API gravity	27.7	35.1
Sulfur, wt ppm	9500	1.7
Nitrogen, wt ppm	690	1.0
Viscosity, cSt at 100°C (212°F)	4.25	—
Pour point, °C (°F)	30 (86)	21 (70)
Dewaxed product properties:		
°API gravity	27.4	37.5
Sulfur, wt ppm	20	1.0
Nitrogen, wt ppm	20	1.0
Viscosity, cSt at 100°C (212°F)	3.63	—
Pour point, °C (°F)	−20.5 (−5.0)	−12 (+10)

Note: °API = degrees on American Petroleum Institute scale.

TABLE 8.5.2 Capital Investment

Unit feed rate, m³/h (BPSD)	165.7 (25,000)
Feedstock:	
°API gravity	28.3
Specific gravity	0.8855
Sulfur, wt %	0.64
Nitrogen, wt ppm	380
Estimated erected cost, million $	34.5

Note: BPSD = barrels per stream day.

TABLE 8.5.3 Utilities

Power, kW	5100
Steam (tracing only)	Minimal
Cooling water, m³/h (gal/min)	80 (352)
Condensate, m³/h (gal/min)	4 (17.6)
Fuel absorbed, million kcal/h (million Btu/h)	20.5 (81.3)

COMMERCIAL EXPERIENCE

Two Catalytic Dewaxing plants have been put into operation. The first unit was a vacuum gas oil unit processing 10,000 barrels per day of shale oil to remove paraffins at Unocal's shale oil plant in Parachute, Colorado. Since 1988, a Catalytic Dewaxing unit processing 12,000 barrels per day has been producing spindle oil from waxy vacuum gas oil and has also been producing a low-pour diesel product at the OMV refinery in Schwechat, Austria.

CHAPTER 8.6
UOP UNISAR PROCESS FOR SATURATION OF AROMATICS

H. W. Gowdy
UOP
Des Plaines, Illinois

INTRODUCTION

The UOP* Unisar* process saturates the aromatics in naphtha, kerosene, and diesel feedstocks. The use of highly active noble-metal catalysts permits the reactions to take place at mild conditions. Because of the mild conditions and the very selective catalyst, the yields are high, and hydrogen consumption is largely limited to just the desired reactions. A total of 18 Unisar units have been licensed worldwide.

Among the applications of the Unisar process are smoke-point improvement in aircraft turbine fuels, reduction of the aromatic content of solvent stocks to meet requirements for air pollution control, production of cyclohexane from benzene, and cetane number improvement in diesel fuels. The Unisar process also produces low-aromatics diesel with excellent color and color stability.

This process was first applied to upgrading solvent naphthas and turbine fuel. The first commercial Unisar plant, a unit processing 250,000 metric tons per year (MTA) [6000 barrels per stream day (BPSD)] was built in Beaumont, Texas, and went onstream early in 1969. It was designed to saturate the aromatics in untreated straight-run solvent naphtha containing 100 wt ppm sulfur. The aromatics were reduced from 15 to 1.0 vol %. The first catalyst cycle lasted more than 8 years. Another of the early plants was started up at Unocal's San Francisco Refinery in 1971. This unit, which processes 600,000 MTA (14,500 BPSD), reduces the aromatics in hydrocracked turbine stock from 30 wt % to less than 4 wt %. During a test at the latter unit, the aromatics were reduced from 29 wt % to less than 0.1 wt %.

---

*Trademark and/or service mark of UOP.

APPLICATION TO DIESEL FUELS

In the 1990s, the need to increase the cetane number of diesel fuel has grown. Increasing the cetane number improves engine performance and decreases emissions. Cetane number is a strong function of hydrocarbon type and number of carbon atoms. Figure 8.6.1 plots cetane number versus number of carbon atoms for compounds in the diesel boiling range. The graph shows that normal paraffins have the highest cetane number, which increases with chain length. Isoparaffins and mononaphthenes with side chains are intermediate in cetane number, and polynaphthenes and polyaromatics have the lowest cetane numbers.

The saturation of aromatics in diesel-range feeds leads to an increase in cetane number (Fig. 8.6.2). However, whether this reaction alone is sufficient to reach cetane numbers near 50, as required in some European countries, depends on the overall compound distribution in each feed. One version of the advanced Unisar catalyst described later in this chapter has built into it some hydrocracking activity to promote naphthenic ring opening and upgrade these low-cetane feedstocks.

These diesel-range feeds typically have substantially higher boiling points and much higher levels of nitrogen and sulfur than the lighter kerosene and solvent feedstocks for which the original Unisar process was developed.

The original noble-metal catalysts used for the Unisar process have limited tolerance to the nitrogen and sulfur contaminants in diesel-range feeds. Thus, new catalysts have been developed to effectively treat these more difficult feeds. In addition, these feeds must be substantially hydrotreated to remove sulfur and nitrogen before they can be treated with the noble-metal Unisar catalysts. A flow scheme that integrates the

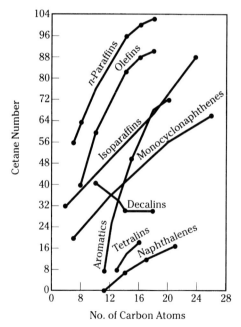

FIGURE 8.6.1 Cetane number versus hydrocarbon type.

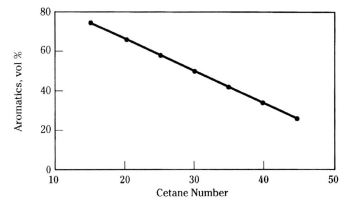

FIGURE 8.6.2 Typical aromatics-cetane relationship.

hydrotreating and aromatics-saturation stages has been developed so that low-sulfur, high-cetane diesel fuels can be efficiently produced.

PROCESS DESCRIPTION

The UOP Unisar process is carried out at moderate temperatures and pressures over a fixed catalyst bed in which aromatics are saturated in a hydrogen atmosphere. Exact process conditions vary, depending on the feedstock properties and the level of aromatics desired in the product.

Chemistry

The primary reaction in the Unisar process is the hydrogenation of aromatics. Other reactions that occur are hydrogenation of olefins, naphthenic ring opening, and removal of sulfur and nitrogen. At conditions that result in significant aromatics hydrogenation, olefins in the feed are completely hydrogenated. When the concentrations of sulfur and nitrogen in the feed are relatively high, the hydrogenation of aromatics is severely limited until the concentrations of heterocompounds have been greatly reduced.

The overall aromatics-saturation reaction rate increases with increases in aromatics concentration, hydrogen partial pressure, and temperature. The reaction rate decreases with increases in the concentration of sulfur and nitrogen compounds and the approach to equilibrium. At low temperatures, aromatics in the product are reduced by increasing the temperature. At these low temperatures, reaction kinetics control the aromatics conversion. However, as temperatures are further increased, a point is reached at which additional temperature increases actually increase aromatics in the product (Fig. 8.6.3). Above this temperature, the reverse dehydrogenation reaction has become dominant, and aromatics conversion is controlled by chemical reaction equilibrium. Thus, using highly active catalysts is important so that lower temperatures—that is, temperatures that are farther from the equilibrium limitation—can be used.

Naphthalene and tetralin have been used as a model to study the reaction mechanism for saturation of diaromatics with the Unisar noble-metal catalyst. The saturation

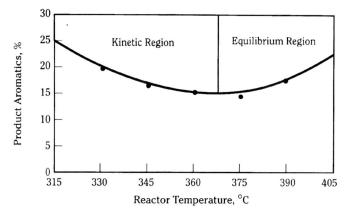

FIGURE 8.6.3 Temperature effect on diesel aromatics reduction.

of naphthalene to tetralin and of tetralin to decalin both fit first-order kinetics quite well. However, when these rate constants were used together in a sequential mechanism in which naphthalene is saturated to tetralin and the tetralin is then saturated to decalin, the resultant calculated product distribution did not fit the distribution observed experimentally.

When the simultaneous saturation of both the naphthalene rings to yield decalin directly is considered along with the saturation of each ring sequentially, the experimental yield distribution can be reproduced satisfactorily. Furthermore, the reaction leading to the simultaneous saturation of both rings of naphthalene to yield decalin directly without the intermediate formation of tetralin is significantly faster than the saturation of only one ring to yield the intermediate tetralin.

The Unisar catalyst study also showed that the overall rate of naphthalene saturation—that is, the sum of the rates of both saturation reactions—is approximately twice as fast as the saturation of tetralin. This result conforms to the generally accepted concept that diaromatics undergo saturation more readily than monoaromatics.

Catalysts

The Unisar catalysts are composed of noble metals on either an amorphous or molecular-sieve support.

The original AS-100* catalyst was developed for kerosene and naphtha solvent hydrogenation. It is highly active and stable in this service. For example, the Unisar plant at the Unocal San Francisco refinery saturates aromatics in the kerosene cut from the hydrocracker to increase the smoke point. This catalyst was loaded in 1971. The original load is still in the reactors, and it has not been regenerated.

The new AS-250 catalyst was specifically developed to treat diesel feedstocks. The catalyst has greatly improved tolerance to the organic sulfur and nitrogen compounds present in diesel-range feeds and much higher activity and stability when treating these feeds. The AS-250 catalyst is 65°C (150°F) more active that its predecessor. This higher activity allows for a more economic Unionfining* design as a result of lower design pressure and higher space velocity.

---

*Trademark and/or service mark of UOP.

The nitrogen and sulfur tolerance of the AS-250 catalyst was demonstrated in a 200-day pilot-plant stability text. The base feed for the study was a 382°C (720°F) endpoint heavy diesel hydrotreated to 50 wt ppm sulfur and 20 wt ppm nitrogen. After more than a month on-stream, a feed containing 450 wt pm sulfur and 135 wt ppm nitrogen was fed to the unit for 24 hours, and then the base feed was returned to the unit. The AS-250 catalyst showed no permanent activity loss or increase in deactivation rate.

Some hydrocracking activity has been built into the AS-250 catalyst to allow naphthenic ring opening to upgrade low-cetane feedstocks. This catalyst delivers high distillate yields, and the converted material is essentially all naphtha.

Typical Process Conditions

The Unisar reactor conditions depend on the feed properties and on the level of aromatics saturation required. Typical operating conditions for commercial Unisar units are

Space velocity: 1.0 to 5.0 vol/vol · h
Pressure range: 3500 to 8275 kPa (500 to 1200 lb/in^2 gage)
Recycle gas H_2 purity: 70 to 90 mol %
Recycle gas rate: 3000 to 6000 standard cubic feet per barrel (SCFB)
Temperature range: 205 to 370°C (400 to 700°F)

Unisar Process Flow

A typical Unisar unit can be represented by the flow diagram shown in Fig. 8.6.4. Fresh feed to the unit is combined with recycle gas from the separator and with make-

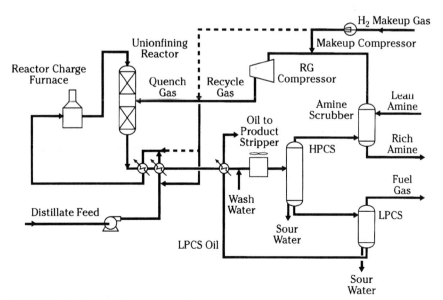

FIGURE 8.6.4 Unisar or Unionfining unit.

up hydrogen. The mixture of gas and feed is heated by exchange with reactor effluent and by a fired heater before entering the reactor. In the reactor, aromatic compounds are hydrogenated to the corresponding naphthenes, olefins are hydrogenated to paraffins, and any organic sulfur compounds are converted to hydrogen sulfide.

Because these catalytic reactions are exothermic, the reactor is divided into multiple catalyst beds that have high-efficiency quench sections in between. In these quench sections, the gas and liquid reactants flowing from the top bed are thoroughly mixed with cold recycle hydrogen to reduce the temperature of the reacting mixture. Then this mixture is distributed over the top of the bed below the quench section. In this way, the temperature is kept in the range necessary for reaction but below the level at which thermodynamic limitations on the reaction rate would be significant.

The reactor effluent stream is initially cooled by heat exchange with the reactor feed and then by air before it enters the gas-liquid separator. The separator gas stream may be scrubbed with an amine solution to remove hydrogen sulfide before the gas is recompressed to the reactor. The need for this scrubbing step depends on the amount of sulfur in the feed.

The separator liquid flows to a stripping column, where any light ends are removed. The finished Unisar product is withdrawn from the bottom of the stripper.

Integrated Unionfining-Unisar Process Flow

When the Unisar process is used to saturate aromatics in feedstocks containing substantial amounts of sulfur and nitrogen, the UOP Unionfining process is used first to remove organic sulfur and nitrogen compounds. Then the Unisar process is used to saturate the aromatics in the hydrotreated feed.

The generalized flow diagram in Fig. 8.6.4 also represents the Unionfining process. Because the Unionfining and Unisar flow diagrams are so similar, the total capital cost would be double that of the Unionfining plant if Unionfining and Unisar steps were done in separate plants. For this reason, the two steps have been combined into the integrated Unionfining-Unisar process (Fig. 8.6.5). The integrated unit has no pressure letdown between the Unionfining and Unisar reactors. Instead, the Unionfining effluent flows to a stripper, where hydrogen sulfide and ammonia are stripped from the hydrotreated feed by hydrogen. This stripped material is then processed in the Unisar reactor.

The integration of these process steps minimizes required equipment, maximizes heat integration, and optimizes utilities. The total cost of this integrated design is just 30 percent more than that of the original Unionfining plant.

Some refiners are faced with lower-sulfur regulations in the immediate future, but new cetane and aromatics specifications are not expected to go into effect until a few years later. Taking into consideration the return on investment over time and the tight availability of capital, the optimal solution for these refiners is to build the Unionfining portion of the complex first and add the Unisar section later. Consequently, this design was also done so that the hydrogen stripper and Unisar stage can be easily added later. The Unionfining reactor, fired heater, and all the heat exchangers and separators in the initial Unionfining unit have the same size requirements for both the Unionfining unit alone and the integrated Unionfining-Unisar unit.

PROCESS APPLICATIONS

A total of 18 Unisar units have been licensed worldwide. Some typical commercial applications of the Unisar process are shown in Table 8.6.1. The aromatics in the dis-

UOP UNISAR PROCESS FOR SATURATION OF AROMATICS 8.61

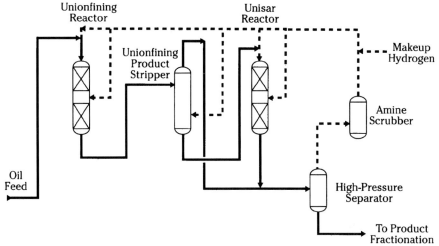

FIGURE 8.6.5 Integrated Unionfining and Unisar unit.

TABLE 8.6.1 Typical Applications

Feedstock type	Distillate	Kerosene	Solvent stock
Feedstock properties:			
°API gravity	37.8	41.3	63.7
Specific gravity	0.8358	0.8189	0.7249
Boiling range, °F (°C)	501–595	301–567	208–277
	(261–313)	(149–297)	(98–136)
Sulfur, wt ppm	3150	340	<2
Aromatics, vol %	24.6	28.2	10.0
C_5 + yields, vol %	101.7	102.1	101.7
Product properties:			
°API gravity	39.0	43.0	65.8
Specific gravity	0.8299	0.8072	0.7173
Sulfur, wt ppm	<2	Nil	Nil
Total aromatics, vol %	<1.0	3.0	<0.5
Hydrogen consumption, SCFB	760	745	330

Note: °API = degrees on American Petroleum Institute scale.

tillate feed are reduced from 24.6 to less than 1 vol %. The aromatics in the kerosene are reduced from 28.2 to 3.0 vol %, and those in the solvent stock are reduced from 10 to less than 0.5 vol %.

An example of upgrading a diesel stock by using the Unisar process is shown in Table 8.6.2. The feedstock properties are shown in the first column. In this example,

TABLE 8.6.2 LCO Upgrading with the Unionfining/Unisar Process

	LCO Feed	Product
°API gravity	0.942 (18.7)	0.852 (34.6)
Sulfur, wt %	1.39	0
Nitrogen, ppm	1107	0
Hydrocarbon types:		
Paraffins	6.6	11.7
Naphthenes	6.1	84.4
Monoaromatics	26.7	3.9
Diaromatics	36.1	0.1
Triaromatics	7.1	0.0
Heterocompounds	14.8	0
Olefins	2.5	0
Cetane number	<21	44.4
Cetane index, D-976	26.6	45.3

the feed is light cycle oil (LCO) from a fluid catalytic cracking unit (FCCU). The LCO has 69.9 wt % aromatics and a cetane number of less than 21. The second column shows the results of using the integrated Unionfining-Unisar process to reduce the aromatics content down to low levels. The aromatics have been reduced to 4 wt %, and the cetane number has increased to 44.4.

CHAPTER 8.7
EXXON DIESEL OIL DEEP DESULFURIZATION (DODD)

Sam Zaczepinski

Exxon Research and Engineering Company
Florham Park, New Jersey

TECHNICAL BACKGROUND OF DODD

Hydrotreating on a variety of feedstocks has been practiced for many years, and Exxon Research and Engineering (ER&E) has generated several technologies which have been applied commercially on a very wide scale. Among these, ER&E developed Hydrofining for treating naphthas and distillate boiling range materials and, recently, has extended its technology to remove sulfur to very low levels. This technology is named Diesel Oil Deep Desulfurization (DODD).

This chapter reviews Distillate Hydrofining and deep desulfurization, then summarizes the application opportunities for Distillate Hydrofining and presents some of the unique features of Exxon Research and Engineering's technology. Distillate Hydrofining is usually practiced to accomplish the desulfurization of heating and diesel oils to meet product sulfur specifications. Other reasons are to improve color and sediment stability. In special applications, the technology can be practiced to saturate aromatics in jet fuels, thereby producing a very high quality jet fuel, or to improve the cetane number of distillate materials.

HYDROFINING CHARACTERISTICS

One feature of Exxon Research and Engineering's technology is its ability to hydrotreat processed (e.g., catalytically or thermally cracked) feedstocks. On the basis of experience with this technology, ER&E is able to practice Hydrofining at low pressures. Thus, efficient hydrogen utilization can be achieved within an overall refinery context. ER&E also has the know-how to manage the catalyst systems to achieve long run lengths with high catalyst activity. This contributes to the exceptionally high service factors of ER&E-designed Hydrofining units.

The following summarizes the commercial experience of Exxon Research and Engineering's Distillate Hydrofining technology. Over 100 units have been constructed. The oldest units have been in service for over 40 years. ER&E-designed units have over 2 million barrels per stream day (BPSD) of installed capacity, with licensed units representing over 25 percent of the total capacity. Units range in size from just over 1 kBPSD to 83 kBPSD. Of particular note, over 30 units process feeds containing thermally or catalytically cracked material (up to 100 percent cracked feed in some cases). Besides distillate hydrotreaters, ER&E has 29 applications of gas oil and residuum desulfurization (Go-Fining and Residfining) representing over 1 million B/D installed capacity.

Figure 8.7.1 is a typical flow diagram of distillate Hydrofining and/or DODD. While the heart of the process is the reactor system, also of note are the preheat furnace, the treat gas recycle gas circuit, and the product fractionation system. Product separation is typically carried out in a staged manner with hot and cold separation ahead of the product stripper. Naphtha is produced from the stripper overhead, while the main product is hydrotreated distillate material from the bottom of the stripper.

Table 8.7.1 provides process parameters (operating conditions, yields, utilities, investments) for a 15 kBPSD DODD unit. The investment is the direct material and labor costs for the equipment shown in Fig. 8.7.1.

From the preceding discussion, it is apparent that Distillate Hydrofining technology is established technology. Recently there have been increased restrictions on the quality of distillate/diesel products. In the United States, regulations are now in effect limiting the sulfur level of diesel fuel used on U.S. highways to 500 wt ppm or 0.05 wt %. There are also limits on the amount of aromatics that can be contained in diesel. Alternatively, this regulatory objective can be achieved by requiring a minimum cetane index. In particular, California has legislation limiting diesel aromatics to 10 percent.

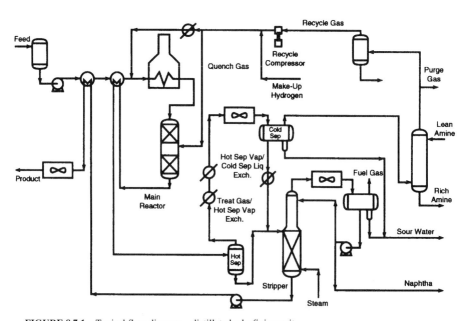

FIGURE 8.7.1 Typical flow diagram—distillate hydrofining unit.

TABLE 8.7.1 Process Parameters for DODD

	Cycle average
Feed rate, of 1.3 wt % S gas oil 75/25 virgin/cracked stock	15 kB/SD (99 m^3/h)
Operating conditions:	
Average reactor temperature °F (°C)	<650 (343)
Average reactor pressure, lb/in^2 (bar) gage	<600 (41.4)
Treat gas system	Recycle
Makeup gas, SCF/B [(N) m^3/m^3]	330 [56]
Chem. H$_2$ consump., SCF/B [(N)m^3/m^3]	230 [39]
Run length, years	>2
Yields:	
Wild naphtha, kBPSD (m^3/h)	0.9 (6)
Sulfur, wt ppm	10
Distillate, kBPSD (m^3/h)	14.1 (93.4)
Sulfur, wt ppm	<500
Utilities	
Steam at 125 lb/in^2 (8.6 bar) gage, klb/h (metric tons/h)	6.3 (2.9)
Power, kW	610
Cooling and process water, gal/min (m^3/h)	450 (102)
Fuel fired, 10^6 Btu/h (MW)	25 (7.3)
Catalyst cost, 1000 $/year	140
Investment, direct material and labor,* million $	9.3

*4th Quarter 1992, U.S. Gulf Coast.
Note: SCF/B = standard cubic feet per barrel; (N) = standard temperature and pressure.

The tight regulations are not unique to the United States. In Europe, the sulfur level in automotive diesel oil (ADO) will be limited to 500 wt ppm in the 1995–1996 time frame. Japan has regulations in place to limit ADO sulfur to 500 wt ppm by 1997. Canada is expected to adopt U.S. regulations soon. Further, there are indications of extending regulations to off-road diesel fuels as well as further restricting sulfur levels and diesel aromatics.

DODD TECHNOLOGY

With tight regulations getting tighter, ER&E's response has been to develop Diesel Oil Deep Desulfurization. The technology is aimed at achieving 0.05 wt % product sulfur easily. The technology has been commercially demonstrated at a number of locations, including the United States, Europe, and Japan. The technology is based on extensive pilot plant work that backs up commercial designs. One particular aspect of Diesel Oil Deep Desulfurization is the importance of getting a good characterization of the feed. ER&E's research characterizes the treatability of the various sulfur species in middle distillates. For the units designed and started up by ER&E, all product specifications have been achieved. In addition to product sulfur level, tight restrictions on product color have been met. Because of its extensive research and commercial experience, ER&E is often able to apply Diesel Oil Deep Desulfurization technology without pilot plant work.

8.66 SULFUR COMPOUND EXTRACTION AND SWEETENING

DODD TECHNOLOGY DATABASE

The sulfur of diesel fuels is receiving increased attention since sulfur has been identified as a contributor to particulate emissions for diesel engines. At the lower sulfur levels (0.05 wt % sulfur in the United States versus 0.1 to 0.5 wt % previously), a new realm of desulfurization chemistry is encountered in which the hardest-to-react sulfur species are controlling. Figure 8.7.2 shows the relative reaction rates (or ease of removal) of various kinds of sulfur-containing molecules. At the 500 ppm sulfur level, the most difficult species on the right will have to be removed.

ER&E's research provides data on the different sulfur species that exist and their distribution in distillate material. With this knowledge, ER&E has taken advantage of new higher-performing catalysts to achieve desired product sulfur levels.

Early correlations of product sulfur level versus residence time, a key process parameter, were developed when desired product sulfur levels were typically in the range of 0.1 to 0.5 wt % (1000 to 5000 wt ppm). These correlations broke down when extended to very low (e.g., 500 wt ppm) product sulfur levels.

As shown in Fig. 8.7.3, the early correlations underpredicted the residence time required to achieve sulfur targets that are currently mandated. Data have now been obtained and are used in engineering work that properly reflect the residence time that is required to achieve very low product sulfur levels.

Exxon Research and Engineering has developed an extensive database on deep desulfurization of distillate materials. Distillates have been tested at over 400 conditions. A wide range of feedstocks has been treated, both light and heavy virgin materials as well as catalytically and thermally cracked stocks. The sulfur types in the feeds vary as do the nitrogen and aromatic contents. Extensive process parameter studies have been conducted. The ranges of conditions considered are 75 to 900 lb/in^2 abs (5.2 to 62.1 bar abs) H_2, 600 to 800°F (316 to 427°C), and 0.2 to 6.0 liquid hourly space velocity (LHSV).

In addition, treat gas effects and nitrogen effects have been evaluated. Finally, long-term pilot plant work has been done so that catalyst deactivation can be characterized. The database provides improved parameters for engineering work required for a commercial unit. Finally, the database has been regressed to provide a fundamental model for deep desulfurization.

Table 8.7.2 summarizes the experience of seventeen Exxon affiliate hydrodesulfurization (HDS) units designed by ER&E to operate in the DODD mode. As shown in

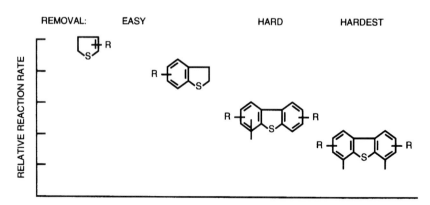

FIGURE 8.7.2 Diesel fuel sulfur reduction.

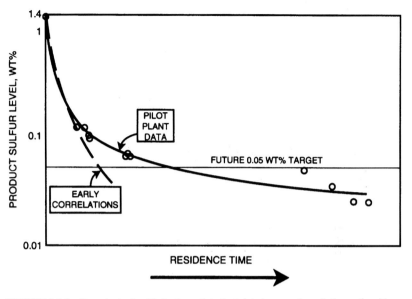

FIGURE 8.7.3 Deep hydrodesulfurization pilot plant data improved predictions of residence time.

TABLE 8.7.2 Commercial HDS Units Designed by ER&E to Run in DODD Mode

Unit	Location	Feed rate kBPSD	Feed rate m³/h	% cracked feed	wt % S in feed	wt % S in product
1	Europe	32.0	212	0	0.80	0.05
2	South America	7.5	50	80	0.11	<0.01
3	North America	3.0	20	40	1.83	0.03
4	Europe	16.6	110	55	0.80	0.04
5	Far East	30.0	199	10	0.70	0.05
6	Far East	23.0	152	10	0.67	0.05
7	North America	10.0	66	100	1.65	0.04
8	North America	35.0	232	46	1.05	0.05
9	North America	38.0	252	0	1.18	0.05
10	North America	12.5	83	0	0.20	0.02
11	North America	6.0	40	100	2.00	0.05
12	North America	12.0	80	0	1.00	0.04
13	North America	7.4	49	0	0.82	0.04
14	North America	6.7	44	100	2.2	0.05
15	North America	15.0	99	15	0.50	0.04
16	North America	28.0	186	50	1.43	0.05
17	Far East*	33.0	146	0	1.40	0.05

*Under construction.

the table, these units are at locations around the world. The feed rates are up to 38 kBPSD and the feeds handled have contained up to 100 percent cracked material. As evident, the product sulfur level has always been at or less than 0.05 wt % sulfur and, in some cases, significantly below that level.

In addition to DODD units at Exxon affiliate locations, a number of licensee units have been designed. Table 8.7.3 indicates the feed rate and percent cracked feed for eight of these units. Both grass-roots units and revamps have been engineered.

Figure 8.7.4 is a photograph of an Exxon-designed 17-kBPD Hydrofiner treating mixed virgin/cracked stock to stringent sulfur specifications in Southeast Asia. The preheat furnace is shown on the left and the reactor is in the center of the photograph.

SUMMARY

Exxon Research and Engineering has Diesel Oil Deep Desulfurization technology that meets current environmental regulations and can, when necessary, be extended to even higher desulfurization levels. The technology is available under license.

TABLE 8.7.3 Hydrofining for Low-Sulfur Diesel—Recent DODD Licensing Units

Licensee	Feed rate		% cracked feed	Unit design
	kBPSD	m^3/h		
1	45	298	25	Revamp
2	60	398	13	Grass roots
3	30	199	45	Revamp
4*	10	66	0	Grass roots
5*	20	133	20	Grass roots
6†	12	80	0	Revamp
7†	15	99	30	Revamp
8†	10	66	0	Grass roots

*Under construction.
†In basic design.

FIGURE 8.7.4 Exxon-designed Hydrofiner treating mixed virgin/cracked stock to stringent sulfur specifications in Southeast Asia.

PART · 9

ISOMERIZATION

CHAPTER 9.1
UOP BENSAT PROCESS

Dana K. Sullivan
UOP
Des Plaines, Illinois

The introduction of reformulated gasoline with mandated limits on benzene content has caused many refiners to take steps to reduce the benzene in their gasoline products. The major source of benzene in most refineries is the catalytic reformer. Reformate typically contributes 50 to 75 percent of the benzene in the gasoline pool.

The two basic approaches to benzene reduction involve prefractionation of the benzene and benzene precursors in a naphtha splitter before reforming, postfractionation in a reformate splitter of the benzene after it is formed, or a combination of the two (Fig. 9.1.1). The benzene-rich stream must then be treated to eliminate the benzene by using extraction, alkylation, isomerization, or saturation (Figs. 9.1.2 and 9.1.3).

If the refiner has an available benzene market, the benzene-rich stream can be sent to an extraction unit to produce petrochemical-grade benzene. Alkylation of the benzene may also be an attractive option if propylene is available, as in a fluid catalytic cracking (FCC) refinery. An isomerization unit saturates the benzene and also increases the octane of the stream by isomerizing the paraffins to a higher-octane mixture. Saturation in a stand-alone unit is a simple, low-cost option.

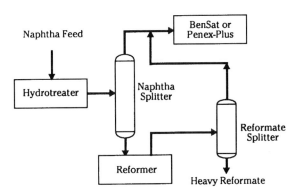

FIGURE 9.1.1 Fractionation for benzene reduction.

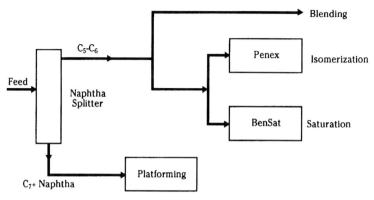

FIGURE 9.1.2 Prefractionation options.

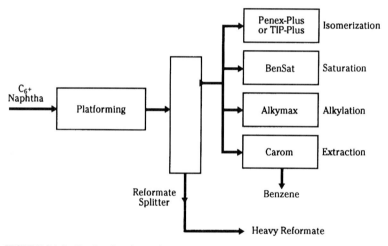

FIGURE 9.1.3 Postfractionation options.

PROCESS DISCUSSION

The UOP* BenSat* process was developed to treat C_5-C_6 feedstocks with high benzene levels. Because almost all the benzene is saturated to cyclohexane over a fixed bed of catalyst, no measurable side reactions take place. Process conditions are moderate, and only a slight excess of hydrogen above the stoichiometric level is required. The high heat of reaction associated with benzene saturation is carefully managed to control the temperature rise across the reactor. Product yield is greater than 100 liquid volume percent (LV %), given the volumetric expansion associated with saturating benzene and the lack of any yield losses from cracking to light ends.

The product has a lower octane than the feed as a result of the conversion of the high-octane benzene into lower-octane cyclohexane. However, the octane can be

---
*Trademark and/or service mark of UOP.

increased by further processing the BenSat product in an isomerization unit, such as a UOP Penex unit (see Chap. 9.3).

PROCESS FLOW

The BenSat process flow is shown in Fig. 9.1.4. The liquid feed stream is pumped to the feed-effluent exchanger and to a preheater, which is used only during start-up. Once the unit is on-line, the heat of reaction provides the required heat input to the feed via the feed-effluent exchanger. Makeup hydrogen is combined with the liquid feed, and flow continues into the reactor. The reactor effluent is exchanged against fresh feed and then sent to a stabilizer for removal of light ends.

CATALYST AND CHEMISTRY

Saturating benzene with hydrogen is a common practice in the chemical industry for the production of cyclohexane. Three moles of hydrogen are required for each mole of benzene saturated. The saturation reaction is highly exothermic: the heat of reaction is 1100 Btu per pound of benzene saturated. Because the benzene-cyclohexane equilibrium is strongly influenced by temperature and pressure, reaction conditions must be chosen carefully.

The UOP BenSat process uses a commercially proven noble metal catalyst, which has been used for many years for the production of petrochemical-grade cyclohexane. The catalyst is selective and has no measurable side reactions. Because no cracking occurs, no appreciable coke forms on the catalyst to reduce activity. Sulfur contamination in the feed reduces catalyst activity, but the effect is not permanent. Catalyst activity recovers when the sulfur is removed from the system.

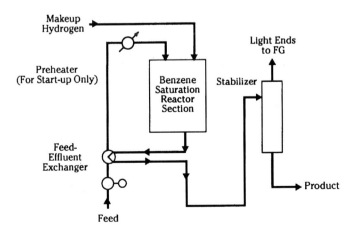

FIGURE 9.1.4 BenSat process flow.

FEEDSTOCK REQUIREMENTS

Light straight-run naphthas must be hydrotreated to remove sulfur. Light reformates usually have very low sulfur contents, and so hydrotreating may not be required. Any olefins and any heavier aromatics, such as toluene, in the feed are also saturated. Table 9.1.1 shows typical feedstock sources and compositions. The makeup hydrogen can be of any reasonable purity and is usually provided by a catalytic reformer.

COMMERCIAL EXPERIENCE

The estimated erected cost (EEC) for a light reformate, fresh-feed capacity of 10,000 barrels per stream day (BPSD) at a feed benzene level of 20 percent by volume is $4.9 million.

Estimated erected costs include inside battery limits, U.S. Gulf Coast open-shop construction, third quarter 1995. The EEC consists of a materials and labor estimate; design, engineering, and contractor's fees; overheads; and expense allowance. The quoted EEC does not include such off-site expenses as cost and site preparation of land, power generation, electrical substations, off-site tankage, or marine terminals. The off-site costs vary widely with the location and existing infrastructure at the specific site. In addition, off-site cost depends on the process unit. A summary of utility requirements is shown in Table 9.1.2.

The first BenSat unit started up in 1994. An additional unit is in design and construction.

TABLE 9.1.1 Typical Feed Compositions, LV %

		Light reformate	
Component	LSR	Light cut	Heartcut
C_5 paraffins	28	29	0
C_5 naphthenes	4	0	0
C_6 paraffins	35	34	47
C_6 naphthenes	17	3	3
C_7+	8	16	24
Benzene	8	18	26
Total	100	100	100

Note: LSR = light straight run.

TABLE 9.1.2 Utilities

Electric, kW	184
Medium-pressure steam, kg/h (klb/h)	7400 (16.3)
Condensate,* kg/h (klb/h)	7400 (16.3)
Cooling water, m³/h (gal/min)	119.5 (526)

*Quantity exported.

CHAPTER 9.2
UOP BUTAMER PROCESS

Nelson A. Cusher
UOP
Des Plaines, Illinois

INTRODUCTION

The first successful efforts in the research and development of catalytic systems for the isomerization of normal paraffins came in the early 1930s. The requirement for high-octane aviation gasoline during World War II accelerated the application of early isomerization research. Light olefinic hydrocarbons were available from the newly developed fluid catalytic cracking (FCC) process and from other mainly thermal operations. These olefinic hydrocarbons could be alkylated with isobutane (iC_4) to produce a high-octane gasoline blending component. However, the supply of isobutane from straight-run sources and other refinery processing was insufficient, and a new source of supply had to be found. Isobutane produced from the new normal paraffin isomerization process met that need.

The first commercial butane isomerization unit went on-stream in late 1941. By the end of the war, 38 plants were in operation in the United States and 5 in allied countries for a total capacity of approximately 50,000 barrels per stream day (BPSD). Five principal isomerization processes, including one developed by UOP, were used in the United States. All were based on Friedel-Crafts chemistry and used aluminum chloride in some form.

The wartime units fulfilled the needs of the time. However, despite many improvements, the units remained difficult and costly to operate. Corrosion rates were excessive, plugging of catalyst beds and equipment was common, and catalyst consumption was high. The units were characterized by high maintenance and operating costs and low on-stream efficiency.

The introduction of the UOP* Platforming* process in 1949 and the rapid spread of such catalytic reforming over dual-functional catalysts in the 1950s served to focus attention on the development of similar catalysts for paraffin isomerization. The term *dual functional* refers to the hydrogenation and controlled-acidity components of a catalyst. Isomerization was known to be one of several reactions that occurred during

---
*Trademark and/or service mark of UOP.

catalytic reforming, and so isolating this reaction for use with feeds that did not require any of the other reactions was a natural next step.

Although the earlier of these dual-functional catalysts eliminated many of the shortcomings of the wartime aluminum chloride catalyst systems, they required relatively high operating temperatures. At these temperatures, unfavorable equilibriums limited per-pass conversion. Further research was conducted, and in 1959, UOP made available to the industry a butane isomerization process, the UOP Butamer* process, that used a highly active, low-temperature hydroisomerization catalyst capable of achieving butane conversion at temperature levels equivalent to the wartime Friedel-Crafts systems without the attendant corrosion or sludge formations. Industry acceptance of the UOP process was rapid, and in late 1959, the first UOP Butamer unit, the first commercial butane isomerization unit to use a low-temperature, dual-functional catalyst system, was placed on-stream on the United States West Coast.

PROCESS DESCRIPTION

The Butamer process is a fixed-bed, vapor-phase process promoted by the injection of trace amounts of organic chloride. The reaction is conducted in the presence of a minor amount of hydrogen, which suppresses the polymerization of olefins formed as intermediates in the isomerization reaction. Even though the chloride is converted to hydrogen chloride, carbon steel construction is used successfully because of the dry environment. The process uses a high-activity, selective catalyst that promotes the desired conversion of normal butane (nC_4) to isobutane at low temperature and, hence, at favorable equilibrium conditions.

Regardless of the iC_4 content of the feed, the butane fraction leaving the unit contains approximately 60 percent by volume of iC_4. Therefore, to obtain optimum plant performance, the refiner wants to charge a butane cut containing the highest practical content of nC_4.

The catalytic reaction is highly selective and efficient and results in a minimum of hydrocracking to light ends or the formation of heavy coproduct. Volumetric yield of iC_4 product based on an nC_4 feed typically approximates slightly more than 100 percent.

PROCESS CHEMISTRY

Isomerization by dual-functional catalysts is thought to operate through an olefin intermediate. The formation of this intermediate is catalyzed by the metallic component, which is assumed for this discussion to be platinum:

$$CH_3-CH_2-CH_2-CH_3 \xrightarrow{Pt} CH_3-CH_2-CH=CH_2 + H_2 \quad (9.2.1)$$

This reaction is, of course, reversible, and because these catalysts are used under substantial hydrogen pressure, the equilibrium is far to the left. However, the acid function (H^+A^-) of the catalyst consumes the olefin to form a carbonium ion and thus permits more olefin to form despite the unfavorable equilibrium:

$$CH_3-CH_2-CH=CH_2 + H^+A^- \rightarrow CH_3-CH_2-\overset{+}{C}H-CH_3 + A^- \quad (9.2.2)$$

---

*Trademark and/or service mark of UOP.

The usual rearrangement ensues:

$$CH_3-CH_2-CH-CH_3 \;+\; \longrightarrow \;\; CH_3-\underset{}{\overset{CH_3}{\underset{+}{C}}}-CH_3 \quad (9.2.3)$$

The isoolefin is then formed by the reverse analog of Eq. (9.2.2):

$$CH_3-\overset{CH_3}{\underset{|}{C}}-CH_3 \;+\; A^- \;\longrightarrow\; CH_3-\overset{CH_3}{\underset{+}{C}}=CH_2 \;+\; H^+A^- \quad (9.2.4)$$

The isoparaffin is finally created by hydrogenation:

$$CH_3-\overset{CH_3}{\underset{|}{C}}=CH_2 \;+\; H_2 \;\xrightarrow{Pt}\; CH_3-\overset{CH_3}{\underset{|}{CH}}-CH_3 \quad (9.2.5)$$

PROCESS VARIABLES

The degree of isomerization that occurs in the Butamer process is influenced by the following process variables.

Reactor Temperature

The reactor temperature is the main process control for the Butamer unit. An increase in temperature increases the iC_4 content of the product toward its equilibrium value and slightly increases cracking of the feed to propane and lighter.

Liquid Hourly Space Velocity (LHSV)

An increase in LHSV tends to decrease the iC_4 in the product at a constant temperature when other conditions remain the same.

Hydrogen-to-Hydrocarbon Ratio (H_2/HC)

The conversion of nC_4 to iC_4 is increased by reducing the H_2/HC ratio; however, the hydrogen effect is slight over the usual operating range. Significant capital savings do result when the H_2/HC ratio is low enough to eliminate the recycle hydrogen compressor and product separator. UOP's standard (and patented) design calls for a H_2/HC ratio of 0.03 molar and allows operation with once-through hydrogen.

Pressure

Pressure has no effect on equilibrium and only a minor influence on the conversion of normal butane to isobutane.

PROCESS CONTAMINANTS

Water poisons the Butamer catalyst. A simple but effective molecular-sieve drying system is used on unit hydrocarbon and gas feeds. Sulfur is a temporary poison that inhibits catalyst activity. The effect of sulfur entering the reaction system is to lower the conversion per pass of normal butane to isobutane.

Butamer catalyst exposed to sulfur essentially recovers its original activity when the sulfur is eliminated from the feed. Also, the effect of sulfur on the Butamer system is minimal because the molecular-sieve feed dryers are also capable of economically removing this material from typical butane fractions. Should a potential feed of relatively high sulfur content be encountered, the bulk of this content would be mercaptan sulfur that is readily removed by simple caustic extraction, such as the UOP Merox* process. The residual sulfur remaining after extraction would then be removed by the feed-drying system of the molecular sieve. Another catalyst poison is fluoride, which also degrades the molecular sieves used for drying. Butamer feeds derived from an HF alkylation unit contain such fluorides, which are removed by passing them over a hot bed of alumina.

The proper design of simple feed-pretreatment facilities effectively controls contaminants and minimizes catalyst consumption.

ISOMERIZATION REACTORS

One characteristic of the Butamer process is that catalyst deactivation begins at the inlet of the first reactor and proceeds slowly as a rather sharp front downward through the bed. The adverse effect that such deactivation can have on unit on-stream efficiency is avoided by installing two reactors in series. Each reactor contains 50 percent of the total required catalyst. Piping and valving are arranged to permit isolation of the reactor containing the spent catalyst while the second reactor remains in operation. After the spent catalyst has been replaced, the relative processing positions of the two reactors are reversed. During the short time when one reactor is off-line for catalyst replacement, the second reactor is fully capable of maintaining continuous operation at design throughput, yield, and conversion. Thus, run length is contingent only on the scheduling of shutdown for normal inspection and maintenance.

In addition to the advantages associated with maximizing on-stream efficiency, a two-reactor system effectively reduces catalyst consumption. Reactors are typically sized so that by the time approximately 75 percent of the total catalyst bed is spent, isomerization decreases to an unacceptable level, and some catalyst replacement is needed. In a single-reactor unit, 25 percent of the original catalyst load, although still active, is discarded when the reactor is unloaded. In a two-reactor system, no active catalyst need be discarded because 50 percent replacement is made when catalyst in the first reactor has been spent. Catalyst utilization is thus 100 percent.

---

*Trademark and/or service mark of UOP.

The choice of a single-reactor or a two-reactor system depends on the particular situation and must be made by evaluating the advantages of essentially continuous operation and increased catalyst utilization against the expense of the somewhat more costly two-reactor installation. Both systems are commercially viable, and Butamer plants of both types are in operation.

PROCESS FLOW SCHEME

The overall process-flow scheme for the Butamer system depends on the specific application. Feed streams of about 30 percent or more iC_4 are advantageously enriched in nC_4 by charging the total feed to a deisobutanizer column. Feeds that are already rich in nC_4 are charged directly to the reactor section. A simplified flow scheme is depicted in Fig. 9.2.1. An nC_4 concentrate, recovered as a deisobutanizer sidecut, is directed to the reactor section, where it is combined with makeup hydrogen, heated, and charged to the Butamer reactor. Reactor effluent is cooled and flows to a stabilizer for removal of the small amount of light gas coproduct. Neither a recycle gas compressor nor a product separator is required because only a slight excess of hydrogen is used over that required to support the conversion reaction. Stabilizer bottoms is returned to the deisobutanizer, where any iC_4 present in the total feed or produced in the isomerization reactor is recovered overhead. Unconverted nC_4 is recycled to the reactor section by way of the deisobutanizer sidecut. The system is purged of pentane and heavier hydrocarbons, which may be present in the feed, by withdrawing a small drag stream from the deisobutanizer bottoms.

The Butamer process may also be incorporated into the design of new alkylation plants or into the operation of existing alkylation units. For this type of application, the inherent capabilities of the iC_4 fractionation facilities in the alkylation unit may be used to prepare a suitable Butamer feed with a high nC_4 content and to recover unconverted nC_4 for recycle.

The major historical use of the Butamer process has been the production of iC_4 for the conversion of C_3 and C_4 refinery olefins to high-octane alkylate. A more recent

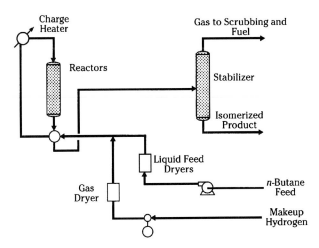

FIGURE 9.2.1 UOP Butamer process.

and larger demand for iC_4 has developed in conjunction with the manufacture of methyl tertiary butyl ether (MTBE), which is a high-octane gasoline blending component particularly useful in reformulated gasolines. Isobutane is dehydrogenated to isobutylene and then made into MTBE. Unconverted butenes and nC_4 are recycled as appropriate to achieve essentially 100 percent conversion of the feed butanes to MTBE.

COMMERCIAL EXPERIENCE

More than 50 UOP Butamer units have been commissioned to date, and 15 others are in engineering design or construction. Product design capacities range from 800 to more than 30,000 BPSD.

Typical yields and investment and operating costs are shown in Tables 9.2.1, 9.2.2, and 9.2.3.

TABLE 9.2.1 Estimated Yields*

	MTA	m³/h	BPSD	SCF/day	wt % on feed
	Feedstock				
Propane	978		37		0.85
Isobutane	29,325		996		25.50
n-butane	82,282		2693		71.55
Isopentane	1,805		56		1.57
n-pentane	610		18		0.53
Total	115,000		3800		100.00
Chemical hydrogen (100% H_2 purity)		65.6		55,600	0.04
	Products				
Isobutane:					
Propane	978		37		0.85
Isobutane	104,190		3540		90.60
n-butane	3,922		128		3.41
Total	109,089		3705		94.86
Heavy-end by-product:					
Isobutane	69		2		0.06
n-butane	2,702		89		2.35
Isopentane	1,058		32		0.92
n-pentane	978		30		0.85
Total	4,807		153		4.18
Light gas:					
Methane	252			39,300	0.22
Ethane	357			29,800	0.31
Propane	541			31,300	0.47
Total	1,150			100,400	1.00

*Basis: Feedstock type: field butane. Hydrogen requirement: does not include that dissolved in the separator liquid. Isobutane purity: 96.5 vol %.

Note: MTA = metric tons per annum; BPSD = barrels per stream day; SCF = standard cubic feet.

TABLE 9.2.2 Estimated Investment Requirements for Butamer Unit with Deisobutanizer Column*

	$ U.S.
Materials and labor	9,100,000
Design, engineering, and contractors' fees	3,900,000
Estimated erected cost	13,000,000
Allowance for catalyst, chemicals, and noble metal on catalyst	380,000

*Basis: Feed rate: 115,000 MTA (3,800 BPSD). U.S. Gulf Coast erection to UOP standards exclusive of off-site costs, fourth quarter 1994. Allowance for catalyst and chemicals reflects current prices FOB point of manufacture.

TABLE 9.2.3 Estimated Operating Requirements*

Utility requirements	Deisobutanizer	Butamer unit	$ U.S. per stream day
Power, kW	200	300	600
Medium-pressure steam:			
At 14.1 kg/cm^2, 1000 kg/h		5.0	
At 200 lb/in^2 gage, 1000 lb/h		11.1	936
Low-pressure steam:			
At 3.5 kg/cm^2, 1000 kg/h	16.3		
At 50 lb/in^2 gage, 1000 lb/h	35.9		2153
Cooling water, m^3/h (gal/min)	35 (155)	89 (390)	77
Total			3766
Catalyst and chemical consumption, $ U.S. per stream day		523	523
Labor allowance/shift:			
Operator			0.50
Helper			0.50

*Basis: Feed rate 115,000 MTA (3800 BPSD). Utility cost basis: electric power $0.05/kW; medium-pressure steam $3.50/klb; low-pressure steam $2.50/klb. Maintenance allowance 3% of erected cost.

CHAPTER 9.3
UOP PENEX PROCESS

Nelson A. Cusher
UOP
Des Plaines, Illinois

INTRODUCTION

A component of refinery gasoline pools that frequently offers the best opportunity for quality improvement is the pentane-hexane fraction, or light straight-run (LSR) naphtha. The LSR is characterized by a low octane number, ordinarily 60 to 70 research octane number (RON), clear. Historically, this fraction, which constitutes approximately 10 percent of a typical gasoline pool in the United States and often a higher percentage in Europe, has been blended directly into gasoline without additional processing except perhaps treating for mercaptan removal. The low octane rating could be increased by approximately 16 to 18 numbers because of its excellent lead susceptibility. The low octane placed the C_5-C_6 straight run in the position of being that segment of the gasoline pool helped most by the addition of lead and least in need of upgrading by processing.

As the petroleum industry moved toward the marketing of motor fuels with reduced or zero lead levels, accommodating the LSR in the gasoline pool became increasingly difficult. The conversion of these C_5 and C_6 paraffins to the corresponding branched isomers to increase their RON, clear, octane number was recognized as a logical and necessary step. One option that UOP* offers to accomplish this upgrading is the Penex* process, which uses a highly active, low-temperature hydroisomerization catalyst. The reliability of this catalyst has been commercially demonstrated since 1959 in butane isomerization (UOP Butamer* process) and since 1969 in C_5-C_6 isomerization.

As a result of U.S. reformulated gasoline legislation for benzene reduction during the 1990s, a variation of the UOP Penex process is being used to saturate all the benzene in the LSR cut and boost the octane of this gasoline fraction.

---

*Trademark and/or service mark of UOP.

PROCESS DISCUSSION

The UOP Penex process is specifically designed for the catalytic isomerization of pentane, hexanes, and mixtures thereof. The reactions take place in the presence of hydrogen, over a fixed bed of catalyst, and at operating conditions that promote isomerization and minimize hydrocracking. Operating conditions are not severe, as reflected by moderate operating pressure, low temperature, and low hydrogen partial pressure requirements.

Ideally, this isomerization catalyst would convert all the feed paraffins to the high-octane-number branched structures: normal pentane (nC_5) to isopentane (iC_5) and normal hexane (nC_6) to 2,2- and 2,3-dimethylbutane. The reaction is controlled by a thermodynamic equilibrium that is more favorable at low temperature.

Table 9.3.1 shows typical charge and product compositions for a C_5-C_6 Penex unit. The compositions of both the C_5 and C_6 fractions correspond to a close approach to equilibrium at the operating temperature.

With C_5 paraffins, interconversion of normal pentane and isopentane occurs. The C_6-paraffin isomerization is somewhat more complex. Because the formation of 2- and 3-methylpentane and 2,3-dimethylbutane is limited by equilibrium, the net reaction involves mainly the conversion of normal hexane to 2,2-dimethylbutane. All the feed benzene is hydrogenated to cyclohexane, and a thermodynamic equilibrium is established between methylcyclopentane and cyclohexane. The octane rating shows an appreciation of some 14 numbers.

PROCESS FLOW

As shown in Fig. 9.3.1, light naphtha feed is charged to one of the two dryer vessels. These vessels are filled with molecular sieves, which remove water and protect the catalyst. The feed is heat-exchanged against reactor effluent before entering a charge heater. It then mixes with makeup hydrogen before entering the reactors. Two reactors normally operate in series.

TABLE 9.3.1 Typical C_5-C_6 Chargestock and Product Compositions

	Percent of total	Chargestock	Product
C_5 paraffins, wt %:	47.5		
Isopentane		42.0	77.0
n-C_5		58.0	23.0
C_6 paraffins, wt %:	45.2		
2,2-dimethylbutane		0.9	31.6
2,3-dimethylbutane		5.0	10.4
Methylpentanes		48.2	46.9
n-C_6		45.9	11.1
C_6 cyclic, wt %:	7.3		
Methylcyclopentane		57.0	52.0
Cyclohexane		17.0	48.0
Benzene		26.0	0
Unleaded octane numbers:			
Research		70.1	83.8
Motor		66.8	81.1

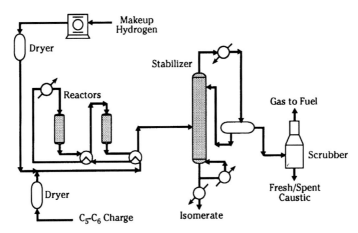

FIGURE 9.3.1 UOP Penex process.

The reactor effluent is cooled before entering the product stabilizer. In newer Penex designs, both the recycle gas compressor and the product separator have been eliminated. Only a slight excess of hydrogen above chemical consumption is used. The makeup hydrogen, which can be of any reasonable purity, is typically provided by a catalytic reformer. The stabilizer overhead vapors are caustic scrubbed for removal of the HCl formed from organic chloride added to the reactor feed to maintain catalyst activity. After scrubbing, the overhead gas then flows to fuel. The stabilized, isomerized liquid product from the bottom of the column then passes to gasoline blending.

Alternatively, the stabilizer bottoms can be separated into normal and isoparaffin components by fractionation or molecular-sieve separation or a combination of the two methods to obtain recycle of the normal paraffins and low-octane methylpentanes (MeC$_5$). Product octanes in the range of 87 to 92 RON, clear, can be achieved by selecting one of the various possible schemes.

The least capital-intensive recycle flow scheme is achieved by combining the Penex process with a deisohexanizer column. The deisohexanizer column concentrates the low-octane methylpentanes into the sidecut stream. This sidecut stream combines with the fresh feed before entering the Penex reactor. The deisohexanizer column overhead, which is primarily isopentane, 2,2-dimethylbutane, and 2,3-dimethylbutane, is recovered for gasoline blending. A small bottoms drag stream, consisting of C_6 naphthenes and C_7's, is also removed from the deisohexanizer column and used for gasoline blending or as reformer feed.

An efficient recycle operation is obtained by combining the Penex process with the UOP Molex* process, which uses molecular sieves to separate the stabilized Penex product into a high-octane isoparaffin stream and a low-octane normal paraffin stream. In this system, fresh feed together with the recovered low-octane normal paraffin stream is charged to the Penex unit. The isomerized product is denormalized in the Molex unit and recovered for gasoline blending.

Many configurations of separation equipment are possible, as shown in Fig. 9.3.2. The optimum arrangement depends on the specific chargestock composition and the required product octane number.

---

*Trademark and/or service mark of UOP.

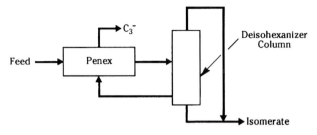

MeC$_5$ and nC$_6$ Recycle Option

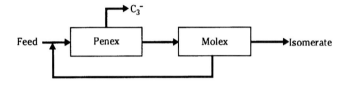

nC$_5$ and nC$_6$ Recycle via UOP Molex Option

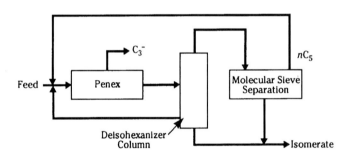

nC$_5$, nC$_6$, and MeC$_5$ Recycle Option

FIGURE 9.3.2 Penex standard flow options.

In addition to increasing octane, another benefit of all Penex-based flow schemes is the saturation of all benzene to cyclohexane. This aspect is particularly important to refiners who want to reduce the level of benzene in their gasoline pool.

Some feedstocks, such as light reformate, can contain high levels of benzene. The performance of the Penex process can be compromised when processing these feedstocks because benzene hydrogenation is a highly exothermic reaction. The heat generated by the benzene hydrogenation reaction can cause the reactors to operate at conditions that are less favorable for octane upgrading. For these applications, UOP offers the Penex-Plus* process, which includes two reactor sections. The first section is designed to saturate the benzene to cyclohexane. The second section is designed to

---

*Trademark and/or service mark of UOP.

isomerize the feed for an overall octane increase. Each reactor is operated at conditions that favor the intended reactions for maximum conversion.

UOP also offers the BenSat* process. This process is similar to the first reactor section of a Penex-Plus unit. Benzene is saturated to cyclohexane with no side reactions. A significant volumetric increase occurs with the BenSat process.

PROCESS APPLICATIONS

As mentioned earlier, the primary purpose of the Penex process is to improve the octane of LSR naphtha. The octane levels for a typical straight-run C_5-C_6 stock are characteristic of the various operating modes (Table 9.3.2).

If the required octane number can be met by recycle of the methylpentanes, the refiner probably would choose fractionation for capital reasons. Where the cost of utilities is high, the refiner might choose a Molex unit, which would separate both nC_5 and nC_6 for recycle. The utility cost would be lower for separating both of these in a Molex unit than it would be for separating the methylpentanes by fractionation, and the refiner would achieve a higher RON.

Separation and recycle during paraffin isomerization are not new. Such options have been installed on many of the isomerization units in operation since the late 1980s. This change is a response to lead phaseout and benzene reduction in gasoline.

The effect of lead elimination on the LSR portion of gasoline can be seen in Table 9.3.3. The octane improvement brought about by modern isomerization techniques can be broken down further. The C_6 portion of the straight run is about 55 RON, clear, and this number is increased to 80 and 93 by once-through and recycle isomerization, respectively. The corresponding figures for the C_5 fraction are 75, 86, and 93.

TABLE 9.3.2 Typical Feed and Product Octane

	RON, clear
Charge	69
Product	
Option 1: no recycle	83
Option 2: recycle of 2- and 3-MeC$_5$+nC_6	88
Option 3: recycle of nC_5+nC_6	89
Option 4: recycle of nC_5+nC_6+2- and 3-MeC$_5$	92

Note: RON = research octane number.

TABLE 9.3.3 Lead Susceptibilities

	Octane number	
	RON, clear	RON + 0.6 g tetraethyl lead/L
U.S. gasoline pool	89	96
Straight-run pentane-hexane:		
Without isomerization	68–70	86–87
Once-through isomerization	83–84	96–97
Isomerized with maximum recycle	92–93	101–103

The important figures, however, are the lead susceptibilities, or the difference between leaded and unleaded octane numbers. As shown in Table 9.3.3, the susceptibility of the entire pool is 7 RONs and that of the C_5-C_6 fraction is 17 to 18. These figures show the principal reason why no one was interested in C_5-C_6 isomerization prior to the worldwide movement toward lead elimination.

The data show that once-through isomerization almost compensates for lead elimination in the LSR fraction and recycle isomerization more than makes up for it. To look at the figures another way, in a typical gasoline pool containing 10 percent LSR naphtha, isomerization provides a way of increasing the pool RON by 2 or more numbers with essentially no yield loss.

Reformulated gasoline legislation in the United States is limiting aromatics concentrations in gasoline. Similar legislation is being enacted or is under consideration in other parts of the world. This limitation on the aromaticity of gasoline further enhances the importance of high-octane aliphatic components such as alkylate and isomerized C_5-C_6.

THERMODYNAMIC EQUILIBRIUM CONSIDERATIONS, CATALYSTS, AND CHEMISTRY

Paraffin-isomerization catalysts fall mainly into two principal categories: those based on Friedel-Crafts catalysts as classically typified by aluminum chloride and hydrogen chloride and dual-functional hydroisomerization catalysts.

The Friedel-Crafts catalysts represented a first-generation system. Although they permitted operation at low temperature, and thus a more favorable isomerization equilibrium, they lost favor because these systems were uneconomical and difficult to operate. High catalyst consumption and a relatively short life resulted in high maintenance costs and a low on-stream efficiency.

These problems were solved with the development of second-generation dual-functional hydroisomerization catalysts. These catalysts included a metallic hydrogenation component in the catalyst and operated in a hydrogen environment. However, they had the drawback of requiring a higher operating temperature than the Friedel-Crafts systems.

The desire to operate at lower temperatures, at which the thermodynamic equilibrium is more favorable, dictated the development of third-generation catalysts. The advantage of these low-temperature [below 200°C (392°F)] catalysts contributed to the relative nonuse of the high-temperature versions. Typically, these noble-metal, fixed-bed catalysts contain a component to provide high catalytic activity. They operate in a hydrogen environment and employ a promoter. Because hydrocracking of light gases is slight, liquid yields are high. The first of these catalysts was commercialized in 1959 in the UOP Butamer process for butane isomerization.

An improved version of these third-generation catalysts is used in the Penex process. Paraffin isomerization is most effectively catalyzed by a dual-function catalyst containing a noble metal and an acid function. The reaction is believed to proceed through an olefin intermediate that is formed by the dehydrogenation of the paraffin on the metal site. The following reactions use butane for simplicity:

$$CH_3-CH_2-CH_2-CH_3 \xrightarrow{Pt} CH_3-CH_2-CH=CH_2 + H_2 \quad (9.3.1)$$

The equilibrium conversion of paraffin is low at paraffin isomerization conditions. However, sufficient olefin must be present to convert a carbonium ion by the strong acid site:

$$CH_3-CH_2-CH=CH_2 + [H^+][A^-] \rightarrow CH_3-CH_2-\overset{+}{CH}-CH_3 + A \quad (9.3.2)$$

Through the formation of the carbonium ion, the olefin product is removed, and equilibrium is allowed to proceed. The carbonium ion in the second reaction undergoes skeletal isomerization, probably through a cycloalkyl intermediate:

$$CH_3-CH_2-\overset{+}{CH}-CH_3 \rightarrow \overset{C\quad CH_3}{\underset{C-C}{/H^+\backslash}} \rightarrow CH_3-\underset{+}{\overset{CH_3}{\underset{|}{C}}}-CH_3 \quad (9.3.3)$$

This reaction proceeds with difficulty because it requires the formation of a primary carbonium ion at some point in the reaction. Nevertheless, the strong acidity of the isomerization catalyst provides enough driving force for the reaction to proceed at high rates. The isoparaffinic carbonium ion is then converted to an olefin through loss of a proton to the catalyst site:

$$CH_3-\underset{+}{\overset{CH_3}{\underset{|}{C}}}-CH_3 + A^- \rightarrow CH_3=\underset{+}{\overset{CH_3}{\underset{|}{C}}}-CH_2 + [H^+][A^-] \quad (9.3.4)$$

In the last step, the isoolefin intermediate is hydrogenated rapidly back to the analogous isoparaffin:

$$CH_3-\overset{CH_3}{\underset{|}{C}}=CH_2 + H_2 \rightarrow CH_3-\overset{CH_3}{\underset{|}{CH}}-CH_3 \quad (9.3.5)$$

Equilibrium limits the maximum conversion possible at any given set of conditions. This maximum is a strong function of the temperature at which the conversion takes place. A more favorable equilibrium exists at lower temperatures.

Figure 9.3.3 shows the equilibrium plot for the pentane system. The maximum isopentane content increases from 64 mol % at 260°C to 82 mol % at 120°C (248°F). Neopentane and cyclopentane have been ignored because they seem to occur only in small quantities and are not formed under isomerization conditions.

The hexane equilibrium curve shown in Fig. 9.3.4 is somewhat more complex than that shown in Fig. 9.3.3. The methylpentanes have been combined because they have nearly the same octane rating. The methylpentane content in the C_6-paraffin fraction remains nearly constant over the entire temperature range. Similarly, the fraction of 2,3-dimethylbutane is almost constant at about 9 mol % of the C_6 paraffins. Theoretically, as the temperature is reduced, 2,2-dimethylbutane can be formed at the expense of normal hexane. This reaction is highly desirable because nC_6 has a RON of 30. The RON of 2,2-dimethylbutane is 93.

Of course, the petroleum refiner is more interested in octane ratings than isomer distributions. Figure 9.3.5 shows the unleaded research octane ratings of equilibrium mixtures plotted against the temperature characteristic of that equilibrium for a typical chargestock. Both the C_5 and the C_6 paraffins show an increase in octane ratings as the temperature is reduced.

Because equilibrium imposes a definite upper limit on the amount of desirable branched isomers that can exist in the reactor product, operating temperatures are thought to provide a simple basis for catalyst comparison or classification. However, temperature is only an approximate comparison that at best can discard a catalyst whose activity is so low that it might be operated at an unfavorably high temperature. Further, two catalysts that operate in the same general low-temperature range may dif-

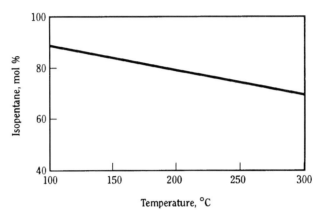

FIGURE 9.3.3 C_5 paraffin equilibrium plot.

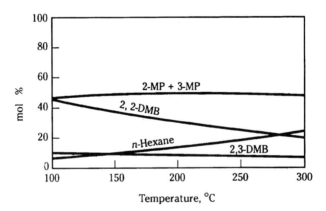

FIGURE 9.3.4 C_6 paraffin equilibrium plot.

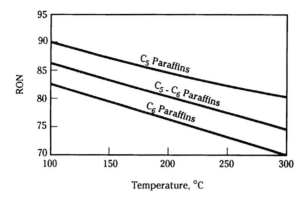

FIGURE 9.3.5 Unleaded RON ratings of equilibrium fractions.

fer in the closeness with which they can approach equilibrium in the presence of reasonable amounts of catalyst.

FEEDSTOCK REQUIREMENTS

To maintain the high activity of the Penex catalyst, the feedstock must be hydrotreated. However, costly prefractionation to sharply limit the levels of C_6 cyclic and C_7 compounds is not required. In fact, the Penex process affords the refiner with remarkably good flexibility in the choice of feedstocks, both at the time of design and even after the unit has been constructed. The latter is important because changes in the overall refinery processing scheme may occur in response to changing market situations. These changes could require that the composition of the isomerization feed be modified to achieve optimal results for the entire refinery.

The Penex system can be applied to the processing of feeds containing up to 15 percent C_7 with minimal or no effect on design requirements or operating performance. Generally, the best choice is to operate with lower levels of C_7+ material because these compounds are better suited for upgrading in a reforming process. Charge containing about 5.0 percent or even higher amounts of benzene is completely acceptable in the Penex chargestock and will not produce carbon on the catalyst. When the feed has extremely high levels of benzene, a Penex-Plus unit is recommended. (The "Plus" section can be retrofitted to an existing Penex unit should the refiner want to process high-benzene feedstocks in an existing Penex unit.) The low-octane C_6- cut recovered from raffinate derived from aromatic-extraction operations typically contains a few percent of olefins and is completely acceptable as Penex feed without prehydrogenation.

Sulfur is an undesirable constituent of the Penex feed. However, it is easily removed by conventional hydrotreating. Sulfur reduces the rate of isomerization and, therefore, the product octane number. Its effect is only temporary, however, and once it has been removed from the plant, the catalyst regains its normal activity.

Water, other oxygen-containing compounds, and nitrogen compounds are the only impurities normally found in the feedstock that will irreversibly poison the Penex catalyst and shorten its life. Fresh feed and makeup hydrogen are dried by a simple, commercially proven desiccant system.

COMMERCIAL EXPERIENCE

Industry acceptance of the UOP Penex process has been widespread. The first Penex unit was placed on-stream in 1958. By mid-1995, more than 80 UOP Penex units had been commissioned, and more than 20 others were in engineering design or construction.

A summary of typical commercial Penex unit yields, product properties, capital costs, utility requirements, and overall operating costs is presented in Tables 9.3.4 through 9.3.9.

TABLE 9.3.4 Typical Estimated Yields for Once-through Processing

	Reactor feed	Reactor product
C_4+ streams, BPD		
iC_4	10	109
nC_4	170	159
iC_5	1,700	3,215
nC_5	2,369	940
Cyclo-C_5	172	121
2,2-dimethylbutane	100	1,565
2,3-dimethylbutane	197	473
2-methylpentane	1,234	1,502
3-methylpentane	899	761
nC_6	2,076	477
Methylcyclopentane	328	290
Cyclo-C_6	278	279
Benzene	277	0
C_7	190	164
Total	10,000	10,136
C_4+ properties		
Specific gravity	0.662	0.651
Reid vapor pressure, kg/cm² (lb/in²)	0.77 (10.9)	0.96 (13.7)
Octane number		
RON, clear	69.3	83.9
RON+3 cm³ tetraethyl lead/U.S. gal	89.1	98.1
MON, clear	67.4	81.9
MON+3 cm³ tetraethyl lead/U.S. gal	87.9	99.6
Hydrogen consumption, SCF/day		1,953,000
Light-gas yields, SCF/day		
C_1		15,000
C_2		7,600
C_3		156,700

Note: BPD = barrels per day; RON = research octane number; MON = motor octane number; SCF = standard cubic feet; i and n indicate iso and normal forms, respectively.

TABLE 9.3.5 Typical Estimated Yields: Penex with Molex Recycle*

Component	Fresh feed to reactor	From Molex to reactor	Stabilizer bottoms	Isomerate product from Molex
C_4+ streams, BPD				
iC_4	10	0	210	210
nC_4	170	0	163	163
iC_5	1,700	102	4,195	4,093
nC_5	2,369	1,253	1,319	66
Cyclo-C_5	172	3	123	120
2,2-dimethylbutane	100	40	1,653	1,613
2,3-dimethylbutane	197	13	544	531
2-methylpentane	1,234	43	1,776	1,733
3-methylpentane	899	23	931	908
nC_6	2,076	555	585	30
Methylcyclopentane	328	7	268	261
Cyclo-C_6	278	6	261	255
Benzene	277	0	0	0
C_7	190	4	176	172
Total	10,000	2,049	12,204	10,155
C_4+ properties				
Specific gravity	0.662	0.643	0.648	0.649
Reid vapor pressure, kg/cm² (lb/in²)	0.77 (10.9)	0.82 (11.7)	0.98 (13.9)	1.01 (14.4)
Octane number				
RON, clear	69.3	56.6	83.4	88.8
RON+3 cm³ tetraethyl lead/U.S. gal	89.1	81.4	97.8	101.1
MON, clear	67.4	55.8	81.4	86.6
MON + 3 cm³ tetraethyl lead/U.S. gal	87.9	80.6	99.3	103.1
Hydrogen consumption, SCF/day				2,039,000
Light-gas yields, SCF/day:				
C_1				17,300
C_2				8,700
C_3				173,400

*Basis: 10,000 BPD.

TABLE 9.3.6 Typical Estimated Yields of Penex with Deisohexanizer Sidecut Recycle

Component	Fresh feed to reactor	From deisohexanizer to reactor	Stabilizer bottoms	Isomerate product from deisohexanizer	Deisohexanizer drag
		C$_4$+ streams, bbl/day			
iC_4	2	0	315	315	0
nC_4	49	0	94	94	0
iC_5	2,433	0	3,381	3,381	0
nC_5	1,885	0	1,033	1,033	0
Cyclo-C$_5$	100	0	70	70	0
2,2-dimethylbutane	57	59	2,813	2,754	0
2,3-dimethylbutane	222	369	898	527	2
2-methylpentane	1,532	1,743	2,906	1,142	20
3-methylpentane	992	1,282	1,506	190	35
nC_6	1,487	856	940	3	82
Methylcyclopentane	561	443	518	0	76
Cyclo-C$_6$	179	285	501	0	216
Benzene	195	0	0	0	0
C$_7$	306	177	345	0	168
Total	10,000	5,214	15,320	9,509	599
		C$_4$+ properties			
Specific gravity	0.661	0.678	0.656	0.640	0.724
Reid vapor pressure, kg/cm^2 (lb/in^2)	0.80 (11.4)	0.40 (5.7)	0.89 (12.6)	1.17 (16.7)	0.25 (3.6)
Octane number					
RON, clear	73.2	72.5	82.6	88.5	77.0
RON+3 cm^3 tetraethyl lead/U.S. gal	91.4	90.5	97.1	101.2	90.8
MON, clear	71.1	71.0	81.0	87.2	69.9
MON+3 cm^3 tetraethyl lead/U.S. gal	90.5	88.7	98.7	105.1	85.3

TABLE 9.3.7 Typical Penex Estimated Investment Costs

	Once-through, million $ U.S.	Penex deisohexanizer, million $ U.S.	Penex-Molex, million $ U.S.
Material and labor	5.7	10.5	15.9
Design, engineering, and contractor's expenses	2.4	3.8	5.5
Total estimated erected cost of ISBL unit	8.1	14.3	21.4

Note: ISBL = inside battery limits; basis = 10,000 BPD.

TABLE 9.3.8 Typical Penex Estimated Utility Requirements*

	Options		
	Once-through	Penex deisohexanizer	Penex-Molex
Electric power, kW	375	975	830
Medium-pressure steam usage (to condensate), 1000 kg/h (klb/h)	9.4 (20.8)	12.0 (26.4)	9.6 (21.2)
Low-pressure steam usage (to condensate), 1000 kg/h (klb/h)	—	24.2 (53.4)	13.4 (29.6)
Cooling water, m^3/h (gal/min)	136 (600)	262 (1153)	277 (1220)

*Basis: 10,000 BPD.

TABLE 9.3.9 Typical Penex Estimated Operating Requirements*

	Once-through, million $ U.S.	Penex-deisohexanizer, million $ U.S.	Penex-Molex, million $ U.S.
Initial catalyst, adsorbent, and noble metal inventory	4.12	4.48	4.79
Annual catalyst and adsorbent costs	0.54	0.59	0.60
Annual chemical cost	0.08	0.09	0.09
Catalyst and chemical operating cost, $/bbl	0.19	0.21	0.21
Number of operators	1.5	2.5	2.5

*Basis: 10,000 BPD and 1995 prices.

CHAPTER 9.4
UOP TIP AND ONCE-THROUGH ZEOLITIC ISOMERIZATION PROCESSES

Nelson A. Cusher
UOP
Des Plaines, Illinois

INTRODUCTION

Light straight-run (LSR) naphtha fractions made in the refinery are predominantly C_5's and C_6's. Some C_7's are also present. They are highly paraffinic and have clear research octane numbers (RONC) usually in the 60s. The nonnormal components have higher octanes than normal paraffins (Table 9.4.1) and are excellent gasoline-blending feedstocks. For the refiner who wants to upgrade the octane of a gasoline pool and has use for a high-purity normal paraffin product, the UOP* IsoSiv* separation technology is a good fit. However, if octane improvement is of primary importance, isomerization technology is the best choice.

Paraffin isomerization to upgrade the octane of light naphtha streams has been known to the refining industry for many years and has gained importance since the onset of the worldwide reduction in the use of lead antiknock compounds. This technology continues to be important in view of current U.S. legislation on reformulated gasoline.

The most cost-effective means to upgrade an LSR feedstock in a grassroots situation is the UOP Penex* process, which is discussed further in Chap. 9.3. However, refiners with idle hydroprocessing equipment, such as old catalytic reformers or hydrodesulfurization units, can consider converting this equipment to the UOP Once-Through (O-T) Zeolitic Isomerization process (formerly known as the Shell Hysomer process). The process scheme is similar to that of a simple hydrotreater, as shown in Fig. 9.4.1, and conversions can be accomplished quickly and at low cost. With O-T Zeolitic Isomerization, a 10 to 12 octane-number increase for the C_5–71°C (160°F) light naphtha can be achieved.

---

*Trademark and/or service mark of UOP.

TABLE 9.4.1 Properties of Common Gasoline Components

	Molecular weight	Boiling point,* °F	Density, * lb/gal	RONC
Isobutane	58.1	10.9	4.69	100+
n-butane	58.1	31.1	4.86	93.6
Neopentane	72.1	49.0	4.97	116
Isopentane	72.1	82.2	5.20	92.3
n-pentane	72.1	96.9	5.25	61.7
Cyclopentane	70.0	120.7	6.25	100
2,2-dimethylbutane	86.2	121.5	5.54	91.8
2,3-dimethylbutane	86.2	136.4	5.54	101.7
2-methylpentane	86.2	140.5	5.57	73.4
3-methylpentane	86.2	145.9	5.44	74.5
n-hexane	86.2	155.7	5.48	94.8
Methylcyclopentane	84.2	161.3	6.28	91.3
2,2-dimethylpentane	100.2	174.6	5.64	92.8
Benzene	78.1	176.2	7.36	100+
2,4-dimethylpentane	100.2	176.9	5.64	83.1
Cyclohexane	84.2	177.3	6.53	83
2,2,3-trimethylbutane	100.2	177.6	5.78	112
3,3-dimethylpentane	100.2	186.9	5.81	98
2,3-dimethylpentane	100.2	193.6	5.83	88.5
2,4-dimethylpentane	100.2	194.1	5.68	55
3-methylhexane	100.2	197.5	5.76	65
Toluene	92.1	231.1	7.26	100+
Ethylbenzene	106.2	277.1	7.26	100+
Cumene	120.2	306.3	7.21	100+
1-methyl-2-ethylbenzene	120.2	329.2	7.35	100+
n-decane	142.3	345.2	6.11	−53

*The values for °C and kg/m^3 can be found in Table 10.5.1.

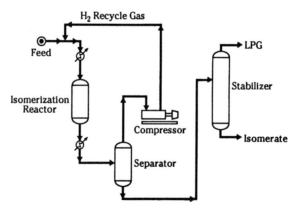

FIGURE 9.4.1 UOP O-T Zeolitic Isomerization process.

For those refiners who need more octane than can be achieved from the once-through operation, an additional 8 to 10 RONC can be gained by adding molecular sieve adsorption to the O-T Zeolitic Isomerization process. Molecular sieve adsorption is used to extract the unreacted normal paraffins so they can be recycled to extinction. This approach of complete isomerization is referred to as the UOP TIP* process. Because O-T Zeolitic Isomerization is an integral part of the TIP process, the ensuing discussion begins with the once-through operation and concludes with a discussion of TIP.

O-T ZEOLITIC ISOMERIZATION PROCESS

Process Chemistry

Thermodynamically, low temperatures are preferred for obtaining maximum amounts of branched paraffins in the reaction product. Operation below 204°C (400°F) for maximum activity requires a catalyst that uses a halide activator. For these catalysts, feed drying is required to eliminate any corrosion or catalyst stability concerns.

The catalyst used in the O-T Zeolitic Isomerization process, however, is based on a strongly acidic zeolite with a recoverable noble-metal component. No external acid activators are used and the catalyst does not produce a corrosive environment. Therefore, feed drying is not necessary.

The catalyst base behaves as an acid of the Brönsted type because it has a high activity for normal-pentane isomerization in the absence of a metal component. At a relatively low hydrogen partial pressure, the carbonium ion concentration generated by the activated low-sodium zeolite is apparently higher than it would have been if the paraffin-olefin equilibrium had been established. This excessive carbonium ion concentration leads to not only high initial conversion but also unstable operation and low selectivity under preferred operating conditions (Fig. 9.4.2). This figure also shows that incorporation of the metal function stabilizes the conversion and lowers the initial activity. These results are to be ascribed to the lower olefin and carbonium ion concentration in the presence of the dual-function catalyst as a result of the paraffin-olefin equilibrium.

The reaction mechanism on the new catalyst is shown in Fig. 9.4.3. Carbonium ions and isoparaffins are generated from normal paraffins by a combination of hydride-ion abstraction and hydride-ion transfer reactions. In the adsorbed state, skeletal rearrangement reactions occur. This reaction is the horizontal path shown in Fig. 9.4.3. Alternatively, while the normal pentane is in the carbonium ion state (nP+ or iP+), it may surrender a proton to form an olefin, which in turn is hydrogenated to form a paraffin (these two paths are vertical).

Even a minute amount of the noble metal stabilizes the conversion to isopentane, provided that the noble metal is well dispersed and distributed throughout the zeolite (Table 9.4.2). However, in commercial applications, more than the minimum amount of noble metal is required. Normally the catalyst contains a few tenths of a percent of precious metal. Proper catalyst preparation methods and start-up procedures are essential for optimal results.

---

*Trademark and/or service mark of UOP.

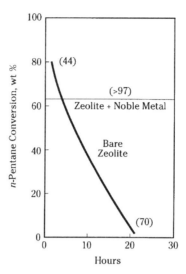

FIGURE 9.4.2 Effect of noble-metal addition on n-pentane isomerization. (Selectivity for isopentane overcracking is indicated in parentheses.)

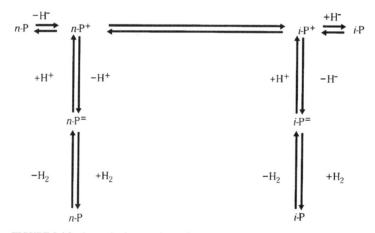

FIGURE 9.4.3 Isomerization reaction path.

Process Description

The O-T Zeolitic Isomerization process is a fixed-bed, vapor-phase process for the catalytic isomerization of low-octane normal pentane or normal hexane or both to high-octane isoparaffins. The isomerization reaction is carried out at 245 to 270°C (470 to 520°F) and 21 to 35 kg/cm^2 (300 to 500 lb/in^2 gage) in the presence of hydrogen. Equipment requirements are a reactor vessel, heater, recycle hydrogen compressor, feed-product heat exchanger, product cooler, phase separator drum, and product stabilizer section.

TABLE 9.4.2 Influence of Metal Load on Zeolite on Hydroisomerization of n-Pentane.

mol metal/100 g zeolite	First-order rate constant
0.000	0.12
0.025	1.5
0.25	1.6
2.00	2.0
5.4	2.3

A comparison of catalytic reforming and O-T Zeolitic Isomerization appears in Table 9.4.3. A brief discussion about the required equipment from the perspective of converting an existing hydrotreater follows.

Reactors. With catalytic reformers that were originally designed for a weight hourly space velocity (WHSV) comparable to that of the O-T Zeolitic Isomerization process, no major modifications to the reactors are required except to eliminate interstage heating. Because of the difference in feed densities, the O-T Zeolitic Isomerization catalyst requirement is typically about 20 percent less than the reformer catalyst requirement.

If the O-T Zeolitic Isomerization unit is to be designed for a lower WHSV or if recycle of normal paraffins to obtain the maximum octane increase is desired, converting from internal to external insulation can achieve about a 25 to 30 percent increase in reactor volume. This increase is possible because of the relatively low operating temperature for the O-T Zeolitic Isomerization process; however, the material used to construct the reactor shell should be checked for pressure or temperature limitations.

Compressors. The recycle-compressor capacity for a reformer is usually more than adequate for the O-T Zeolitic Isomerization process. A 25 kg/cm^2 (350 lb/in^2 gage) reformer will have about twice the capacity required for the O-T Zeolitic Isomerization process. In plants containing two compressors, each with a 50 percent capacity, one compressor can be shut down.

TABLE 9.4.3 Comparison of Catalytic Reforming and O-T Zeolitic Isomerization

	Catalytic reforming	O-T Zeolitic Isomerization
Feed composition	C$_7$+	C$_5$-C$_6$
Feed gravity, °API	52–62	88–90
Operating pressure, kg/cm^2 (lb/in^2 gage)	14–35 (200–500)	21–35 (300–500)
Operating temperature, °C (°F)	About 510 (950)	About 260 (500)
Feed, WHSV	1–5	1–3
H$_2$/HC ratio, mol/mol	5–10	1–4
H$_2$, SCFB	500–1700 produced	About 70 required
Heat of reaction	Highly endothermic	Nearly isothermal
Reid vapor pressure, kg/cm^2 (lb/in^2 gage)		
Feed	0.05–0.07 (0.7–1.0)	0.8–1.0 (12–14)
Product	0.2–0.4 (3–6)	0.9–1.1 (13–16)
Catalyst regeneration	Continuous to about 1 year periodic	Every 2 to 3 years

Note: WHSV = weight hourly space velocity; HC = hydrocarbon; SCFB = standard cubic feet per barrel.

Makeup hydrogen for the O-T Zeolitic Isomerization process can be reformer net gas. If the reformer supplying the hydrogen is a low-pressure unit, a small makeup compressor is required. For a O-T Zeolitic Isomerization unit processing 5000 barrels per day (BPD) of feed, hydrogen makeup is typically about 500,000 standard cubic feet per day (SCF/day).

Heaters and Heat Exchangers. Heat exchange equipment and heaters are usually more than adequate. Interstage reheaters between reactors are not required because the isomerization reaction is mildly exothermic.

Feed Pump. Because of differences in feed gravity, feed rate, vapor pressure, and possible net positive suction head (NPSH), a new feed pump may be required.

Stabilizer System. In the O-T Zeolitic Isomerization process, the amount of light ends produced is substantially less than in the reforming process. In any case, where a reformer has been converted to an O-T Zeolitic Isomerization unit, the stabilizer feed rate is higher even though the stabilizer overhead product is lower than in the reforming operation. The small amount of light ends plus a bottoms product with a higher vapor pressure may dictate an increased reflux rate or a column retray or both.

Commercial Information

The need for a high-octane product to replace the octane lost with lead phaseout and benzene reduction in the gasoline pool has placed more emphasis on isomerization. As previously noted, the attractiveness of the O-T Zeolitic Isomerization process is that it can be adapted to an existing idle hydrotreater, catalytic reformer, or other hydroprocessing unit with minimal investment. The actual time to modify a unit ranges from a few days to a few weeks.

Commercial Installations. As of mid-1995, more than 30 O-T Zeolitic Isomerization units have been commissioned to process 1000 to 13,500 BPD of feed. About half of these are catalytic-reformer or hydrotreater conversions. One unit was assembled from assorted surplus refinery equipment. Of the conversions, one unit is arranged so that it can be operated as either a reformer or a O-T Zeolitic Isomerization unit by switching a few spool pieces.

The oldest of the converted units started up in 1970 in La Spezia, Italy. This unit was integrated with a catalytic reformer so that both units have a common recycle-gas compressor system, product-cooling train, and stabilizer section. Combinations of this sort often result in capital savings of 20 to 40 percent compared to stand-alone isomerization and reforming units. In 10 years of operation, the catalyst in the La Spezia unit was regenerated in situ four times. Typical cycle lengths for O-T Zeolitic Isomerization units are 3 to 4 years.

Typical Performance. Paraffin isomerization is limited by thermodynamic equilibrium so that a once-through, or single-pass, isomerization reactor provides only partial conversion of the normal paraffins. In the reactor, C_5-C_6 paraffins are isomerized to a near-equilibrium mixture, and aromatics become saturated to naphthenes, which, in turn, are partially converted into paraffins. Olefins in the feed are saturated, and C_7+ paraffins are mostly hydrocracked to C_3 to C_6 paraffins.

Tables 9.4.4 and 9.4.5 provide a summary of typical O-T Zeolitic Isomerization yields, product properties, conversion costs, utility requirements, and overall operating costs. Typical C_5+ isomerate yield is 97 to 98 liquid volume percent (LV %) on

TABLE 9.4.4 Typical Estimated Performance, O-T Zeolitic Isomerization Unit, 10,000 BPD

Component	Fresh feed to reactor	Product
Hydrogen consumption, m³/h (1000 SCF/day)	2018 (1710)	—
Light gas yield, m³/h (1000 SCF/day):		
C_1	—	333 (283)
C_2	—	180 (152)
C_3	—	292 (248)
C_4 + streams, LV % on feed:		
iC_4	0.10	2.50
nC_4	0.58	1.41
iC_5	16.84	30.39
nC_5	29.07	16.17
Cyclo-C_5	1.69	1.24
2,2-dimethylbutane	0.51	8.26
2,3-dimethylbutane	1.93	3.74
2-methylpentane	12.08	14.43
3-methylpentane	8.80	9.21
nC_6	19.35	8.24
Methylcyclopentane	1.95	3.35
Cyclo-C_6	3.41	0.96
Benzene	1.75	0.0
C_7	1.94	0.97
Total	100.00	100.87
C_4+ properties:		
Specific gravity	0.659	0.648
Reid vapor pressure, kg/cm³ (lb/in²)	0.8 (10.8)	1.0 (14.2)
Octane number:		
RON, clear	68.1	79.5
RON + 3 cm³ TEL/U.S. gal	88.4	95.5
MON, clear	66.4	77.6
MON + 3 cm³ TEL/U.S. gal	87.3	96.3

Note: BPD = barrels per day; SCF = standard cubic feet; RON = research octane number; MON = motor octane number; TEL = tetraethyl lead; i = iso; n = normal.

feed and the octane number is increased by about 10 to 12, resulting in an isomerate quality of 77 to 80 RONC.

Usually no new major equipment is required when a reformer is converted to an O-T Zeolitic Isomerization unit of the same feed capacity. Thus, the only costs are for new piping and instrumentation, engineering, and a charge of O-T Zeolitic Isomerization catalyst. For a unit with a feed rate of 5000 BPD, capital costs will total $2.7 to $4.0 million. This amount is only about half of the cost of a grassroots installation. Expected catalyst life is 10 to 15 years.

TIP PROCESS

General Description

Some refiners need more octane from the LSR naphtha fraction than is possible from the O-T Zeolitic Isomerization process. As previously noted, the TIP process com-

TABLE 9.4.5 O-T Zeolitic Isomerization Conversion Economics and Performance*

Total capital required, $/BPSD	670
Utilities, per BPSD feed:	
Fuel consumed (90% efficiency), million kcal/h (million Btu/h)	0.0006 (0.0025)
Water at 17°C rise, m^3/day (gal/min)	0.33 (0.06)
Power, kWh	0.05
Steam at 10.5 kg/cm^2 (150 lb/in^2 gage), saturated, kg/h (lb/h)	0.5 (1.1)
Hydrogen consumption, m^3/day (SCF/h)	2.7–6.1 (4–9)
Typical performance:	
Isomerate, RONC	77–80
C_5+ isomerate yield, LV %	97–98
Catalyst expected life, years	10–15

*Basis: Battery limits; U.S. Gulf Coast, 1995, 4000–6000 BPSD, including new stabilizer, new piping and instrumentation, engineering, and catalyst.

bines the O-T Zeolitic Isomerization process with UOP's naphtha IsoSiv process to yield an 87 to 90 RONC product, an improvement of approximately 20 numbers. The TIP unit can be built grassroots, or a UOP IsoSiv unit can be added to an existing O-T Zeolitic Isomerization unit to convert it to a TIP unit. In this type of revamp, generally all existing equipment can be used.

The TIP process uses adsorption technology to remove and recycle the unconverted normal paraffins. During the adsorption step, a shape-selective molecular sieve removes all the unconverted normal paraffins from the isomerate to allow the branched-chain isomers to pass through. These adsorbed normals are then desorbed by stripping with recycle hydrogen and passed directly into the isomerization reactor. Because the entire process is carried out in the vapor phase, utility requirements are low. The entire process operates at a constant low pressure. The presence of hydrogen during the desorption step prevents the buildup of coke on the adsorbent. Like the catalyst, the adsorbent can be regenerated in situ if an upset condition causes coking.

Process Description of TIP

The TIP process is a constant-pressure vapor-phase process operating at a moderate pressure, 14 to 35 kg/cm^2 (200 to 500 lb/in^2 gage) range, and moderate temperatures, 245 to 370°C (475 to 700°F). Hydrogen at a sufficient partial pressure must be present during isomerization to prevent coking and deactivation of the catalyst. A simplified schematic flow sheet is shown in Fig. 9.4.4.

Hydrotreated fresh feed is mixed with the hot recycle stream of hydrogen and C_5-C_6 normal paraffins prior to entering the isomerization reactor. A small stream of makeup hydrogen is also added to the feed of the reactor. The reactor effluent, at near-equilibrium isomerization composition, is cooled and flashed in a separator drum. The liquid product, which contains some unconverted low-octane normal paraffins, is vaporized and passed into a bed of molecular-sieve adsorbent, where the straight-chain normals are adsorbed for recycle back to the isomerization reactor. The branched-chain isomers and cyclic hydrocarbons, which have molecular diameters greater than the diameter of pores in the molecular-sieve adsorbent, cannot be adsorbed and exit from the absorbent bed essentially free of normal paraffins. This isomerate product is stabilized as required to remove any excess hydrogen, 1 to 2 percent cracked products, and any propane or butane introduced with the makeup hydrogen. The hydrogen purge gas from the separator is circulated by means of a recycle

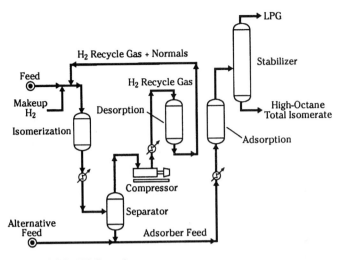

FIGURE 9.4.4 TIP flow scheme.

compressor through a heater and is then used as a purge gas to strip the normal paraffins previously adsorbed on the molecular-sieve adsorbent bed. The hydrogen plus desorbed normals is then mixed with the fresh feed upstream of the isomerization reactor. The isomerization section and the adsorption section of a TIP unit share a common recycle hydrogen loop.

Feedstocks that contain an appreciable amount of heptanes or nonnormal components use an alternative feed point (Fig. 9.4.4). The fresh feed enters the system just upstream of the adsorbers rather than at the isomerization reactor. This feed-entry point allows the nonnormal components and isoheptanes to pass into the final isomerate product without first passing through the isomerization reactor, where some of the heptanes are hydrocracked to liquefied petroleum gas (LPG). With feedstocks having a low normal-paraffin content, it is also more efficient to have the fresh feed enter the system just upstream of the adsorbers to recover the nonnormal components. Only the adsorbed normal paraffins are then sent to the resulting smaller isomerization reactor.

Feeds with high levels of benzene can be processed initially in either the reactor section or the adsorption section. Benzene is saturated completely to cyclohexane in the reactor section, thereby producing a benzene-free isomerate product. For feeds with high levels of benzene, presaturation in a separate reactor at a high space velocity is used to remove the heat of saturation from the TIP reactor. This technology is known as TIP-Plus.* Sending the feed to the adsorption section allows the high-octane benzene to pass into the isomerate product. For feeds that are best processed in the absorber section first but need to minimize benzene in the product, the saturation-section effluent can be sent to the adsorption section of the TIP-Plus process. The refiner needs to evaluate both octane and benzene target levels to determine the proper feed point.

---

*Trademark and/or service mark of UOP.

The TIP unit is normally designed with the capability for an in situ oxidative regeneration of the catalyst and the adsorbent to minimize downtime in the event of an unexpected upset that might coke the catalyst or the adsorbent.

Commercial Information

As of mid-1995, more than 30 TIP units were in operation worldwide. Another four units were in design or construction. Tables 9.4.6 and 9.4.7 provide a summary of typical TIP process yields, product properties, capital costs, utility requirements, and overall operating costs. A 0.6 power factor applied to the ratio of fresh-feed rates can be used with the cost given in Table 9.4.7 for a quick estimate of the investment costs for different-size TIP units. Utilities and catalyst-adsorbent requirements tend to increase in direct proportion to an increase in fresh feed rate.

TABLE 9.4.6 Typical Estimated Yields for the TIP Process, 10,000 BPD

Component	Fresh feed to reactor	Adsorber feed	Recycle paraffins	Isomerate product
H_2 consumption, m^3/h (1000 SCF/day)	2175 (1844)	—	—	—
Light gas yield, m^3/h (1000 SCF/day):				
C_1	—	190 (161)	—	—
C_2	—	81 (69)	—	—
C_3	—	311 (264)	—	—
C_4+ streams, BPSD:				
iC_4	10	337	194	288
nC_4	58	1,035	1,247	136
iC_5	1,684	5,254	1,446	4,523
nC_5	2,907	3,188	3,411	142
Cyclo-C_5	169	153	33	132
2,2-dimethylbutane	51	1,052	215	910
2,3-dimethylbutane	193	528	98	458
2-methylpentane	1,208	2,042	368	1,771
3-methylpentane	880	1,307	230	1,134
nC_6	1,935	1,272	1,301	22
Methylcyclopentane	195	397	68	344
Cyclo-C_6	341	113	19	98
Benzene	175	0	0	0
C_7	194	103	15	89
Total	10,000	16,781	8,645	10,047
C_4+ properties:				
Specific gravity	0.659	0.642	0.632	0.640
Reid vapor pressure, kg/cm^2 (lb/in^2)	0.8 (10.8)	1.2 (16.7)	1.4 (20.6)	1.3 (19.2)
Octane number:				
RON, clear	68.1	79.7	70.7	88.3
RON+3 cm^3 TEL/U.S. gal	88.4	95.6	90.1	100.9
MON, clear	66.4	77.7	69.4	85.8
MON+3 cm^3 TEL/U.S. gal	87.3	96.4	90.4	102.5

TABLE 9.4.7 TIP Process: Economics and Performance

Economics:	
Investment,* $/BPSD	2800–3500
Catalyst and adsorbent inventory, $/BPSD	240
Utilities:	
Fuel consumed (90% furnace efficiency), million kcal/h (million Btu/h)	7.8 (31)
Water at 17°C rise (31°F), m^3/day (gal/min)	2159 (396)
Power, kWh	1455
Steam at 10.5 kg/cm^2 (150 lb/in^2 gage) kg/h (lb/h)	2.8 (6.2)
Hydrogen consumption (70% hydrogen purity), 1000 m^3/day (1000 SCF/h)	17.7 (26)

*Battery limits, U.S. Gulf Coast, 1995, feed rate 4000–6000 BPSD.

Wastes and Emissions

No wastes or emissions are created by the O-T Zeolitic Isomerization or TIP processes. Product stabilization, however, does result in small amounts of LPG ($C_3 + C_4$, rich in iC_4) and in stabilizer vent ($H_2 + C_1 + C_2$) products. The stabilizer vent products are usually used as fuel. The LPG is a valuable by-product that is blended elsewhere in the refinery.

PART · 10

SEPARATION PROCESSES

CHAPTER 10.1
CHEVRON'S ON-STREAM CATALYST REPLACEMENT TECHNOLOGY FOR PROCESSING HIGH-METAL FEEDS

David E. Earls
Chevron Research and Technology Company
Richmond, California

INTRODUCTION

In processing less expensive, high-metal feeds, the need for frequent catalyst change-outs can make conventional fixed-bed residuum hydrotreating technology uneconomical. Chevron developed on-stream catalyst replacement (OCR) to remove metals from feed before it is hydrotreated in fixed-bed residuum desulfurization (RDS) units. The ability to add and withdraw catalyst from the high-pressure, moving-bed OCR reactor while it is onstream gives refiners the opportunity to process heavier, high-metal feeds or to achieve deeper desulfurization while maintaining fixed-bed run lengths and improving product properties.

DEVELOPMENT HISTORY

Development of the OCR process started in 1979 as part of the research on new reactor concepts which might be applied to synthetic fuels and heavy oil upgrading. Most of these alternative fuels are difficult to upgrade to transportation fuels. Typically high in nitrogen, sulfur, and metals, they tend to deactivate catalyst very rapidly. Consequently, conventional fixed-bed hydrotreating processes can not upgrade these feedstocks economically. Chevron determined that if fresh catalyst could be continually moved through a reactor, then catalytic activity could be maintained without shutting down the unit. Theoretically, the metal capacity of the catalyst would be fully utilized in the OCR unit, thus reducing the necessary size of the downstream RDS unit and lowering total operating costs.

Using cold model and catalytic testing, Chevron Research and Technology Co. showed that with the proper equipment design these process improvements could be obtained. The feasibility of the process was proven during the operation of a 200 barrel per day (BPD) demonstration unit at Chevron's Richmond Refinery in 1985. Critical to the design's success was the proof that the valves could operate reliably at the high temperatures and pressures required for resid upgrading.

PROCESS DESCRIPTION

Chevron's OCR process is a countercurrent, moving-bed technology that removes metals and other contaminants from feedstocks prior to processing in fixed-bed residuum hydrotreating reactors. In the OCR reactor, residuum and hydrogen flow upward through the reactor and the catalyst flows downward. This process removes the metals and carbon residue that cause plugging and catalyst deactivation in conventional fixed-bed RDS units. Figure 10.1.1 shows the OCR reactor system, including the equipment used to transfer catalyst into and out of the reactor. OCR reactors can be paired and serve as pretreatment beds for two parallel trains of fixed-bed reactors. In this case, only one set of catalyst transfer vessels is needed to move catalyst for two OCR reactors.

Catalyst Transfer System

In a parallel train system, catalyst transfer is done, on average, once a week to and from each OCR reactor. The amount of catalyst transferred varies from 1.5 to 8 percent of the OCR reactor catalyst capacity. The quantity removed is dictated by the nickel and vanadium content of the feed and the metals concentration on the removed catalyst. The transfer rate is adjusted to allow for changes in operating requirements.

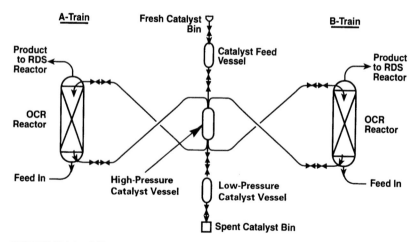

FIGURE 10.1.1 OCR reactor system. Chevron's OCR technology adds and removes catalysts from a high-pressure reactor while it is operating, thus providing refiners the opportunity to process less expensive, high-metal feeds.

Once the requirements are defined, the catalyst is added and withdrawn batchwise on a regular schedule to maintain the required OCR activity.

All the steps required to transfer the catalyst are controlled by a computer-driven, automatic sequencer. Use of the automatic sequencer minimizes the need for operator attention and assures consistent adherence to all necessary procedures. Operation of the OCR catalyst transfer system is easily monitored by the existing RDS board operator.

Catalyst transfer in and out of the OCR reactor is accomplished in a series of steps which do not interrupt the operation of the unit:

1. Fresh catalyst is transferred by gravity into the low-pressure catalyst feed vessel.
2. There, flush oil (usually a heavy gas oil) is added and the mixture is transferred as a slurry to the high-pressure catalyst vessel (HPCV).
3. The low-pressure catalyst feed vessel is then isolated, and the pressure in the HPCV is equalized with the top of the OCR reactor.
4. The fresh catalyst is then transferred as a slurry to the top of the OCR reactor.
5. Once the transfer is complete, as indicated by the level in the HPCV, the double isolation valves are flushed to remove catalyst, and the HPCV is isolated from the system.

Spent catalyst is removed from the bottom of the reactor in a similar manner:

1. The HPCV is pressure-equalized with the bottom of the OCR reactor.
2. The spent catalyst is moved as a slurry in the feed resid from the bottom of the reactor. Once the desired amount of catalyst has been transferred, as indicated by the level in the HPCV, the transfer is stopped and the valves and lines are flushed with oil.
3. The double isolation valves are then closed and the HPCV is isolated from the OCR reactor and depressurized. The spent catalyst is washed of resid and cooled.
4. The catalyst is then transferred as a slurry to the low-pressure catalyst vessel, where the flush oil is drained.
5. The spent catalyst then flows by gravity into the spent catalyst bin for disposal.

Since the catalyst is transferred in a low-velocity oil slurry, catalyst attrition is prevented and the system's lines and valves are protected from erosion. OCR line sizes are smaller than main process lines and special full port valves are used in the catalyst transfer lines. These valves are flushed clear of catalyst before closing to minimize valve wear.

OCR Reactor

A schematic drawing of the OCR reactor is shown in Fig. 10.1.2. The catalyst bed in the OCR unit is essentially a fixed bed, which intermittently moves down the reactor. The catalyst level in the OCR reactor is monitored by a level detector at the top of the reactor. As fresh catalyst is added at the top of the reactor, residuum is fed into the bottom. Both move through the reactor in a countercurrent flow, causing the dirtiest, most reactive residuum to contact the oldest catalyst first. The upflow of the residuum through the OCR reactor slightly expands the catalyst bed. This slight expansion enhances residuum/catalyst contact, minimizes reactor plugging, and creates a consistent pressure drop thus providing for optimum flow patterns through the reactor. Meanwhile, the fully spent OCR catalyst is removed at the bottom of the reactor.

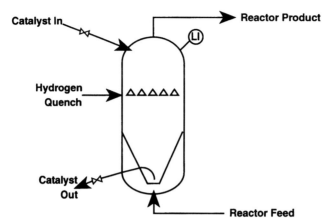

FIGURE 10.1.2 OCR reactor details. The countercurrent flow of reactants and catalyst through the OCR reactor ensures that only the most nearly spent catalyst is removed, thereby minimizing catalyst usage and cost.

The specially designed cone at the bottom of the reactor allows for plug flow of the catalyst to the removal port at the bottom of the reactor. This plug flow ensures that the most metal-loaded, least active catalyst is removed from the reactor. Consequently, catalyst activity is maximized and cost is minimized.

Bed plugging is one of the most common causes for premature shutdown of fixed-bed hydroprocessing units. Particulates and reactive metals depositing on the top layers of the reactor cause an increase in pressure drop and maldistribution of the liquid and gas flow. This, in turn, can lead to localized hot spots and rapid coke formation. Two features of the OCR process dramatically reduce the severity of this problem in downstream fixed-bed units:

1. The most reactive feed metals are deposited on the OCR catalyst and do not enter the fixed-bed unit.
2. Particulate material in the feed is not retained in the OCR bed, but passes through to the fixed-bed unit.

Separating the problems of particulates and reactive metals allows the refiner to optimize catalyst grading for the removal of particulate material in the downstream fixed-bed units. As a result, the problem of metal or coke fouling in the RDS unit is largely neutralized by the high hydrodemetallization (HDM) catalyst activity in the OCR unit.

OCR units operate at the same temperature (approximately 730°F) and pressure (approximately 2000 lb/in^2) as their downstream RDS counterparts. Consequently, integrating OCR into the processing scheme is easy and efficient because it can use the same recycle hydrogen supply, feed pumps, and feed furnace as the fixed-bed RDS reactor.

Catalyst

A special spherical catalyst developed by Chevron Research and Technology Co. was designed to fit the requirements of the OCR process:

- High hydrodemetallization activity and metals capacity to minimize downstream reactor volume and catalyst usage.
- Moderate hydrodesulfurization (HDS) and Conradson carbon removal (HDCCR) activity to reduce coking.
- Strength and hardness to minimize breakage in handling.
- Consistent shape and size to facilitate catalyst transfer and stable bed operation.

Since its introduction in 1992, the OCR catalyst has exhibited low attrition, high crush strength, and exceptional selectivity for residue demetallization. The Ni+V conversion typically exceeds 60 percent for the first 60 days of operation and then gradually trends toward the expected equilibrium conversion level, 50 to 70 percent (see Fig. 10.1.3). In commercial operation catalyst is withdrawn and routinely analyzed for nickel and vanadium content. This analysis confirms that only the most spent OCR catalyst is being withdrawn—a critical factor in minimizing catalyst consumption and cost.

While the main objective of the OCR reactor is to extend catalyst life in downstream fixed-bed reactors by maintaining high HDM performance, the OCR catalyst also achieves high HDS/HDM and HDCCR/HDM activity ratios. As shown in Fig. 10.1.4, the OCR reactors' sulfur and CCR conversion has been excellent. Typically, the HDS conversion stabilizes at the HDS equilibrium objective target of 50 percent while the HDCCR conversion stabilizes at the equilibrium objective target of 30 percent. The level of HDS and HDCCR activity in the OCR reactors greatly improves the overall performance of the OCR/RDS units and significantly extends the run life of the fixed-bed catalyst.

COMMERCIAL OPERATION

Chevron's OCR process has been in commercial operation since 1992. The first unit was installed as a retrofit to a Chevron-licensed RDS unit at the Idemitsu Kosan Company, Ltd. (IKC) Aichi refinery (see Fig. 10.1.5). Chiyoda Corporation provided

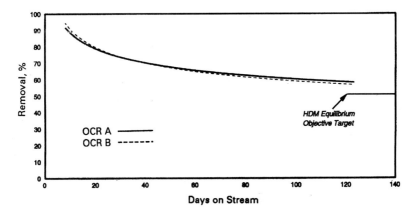

FIGURE 10.1.3 OCR hydrodemetallization performance. By maintaining consistently high HDM performance throughout the run, OCR reactors minimize catalyst consumption and optimize catalytic performance.

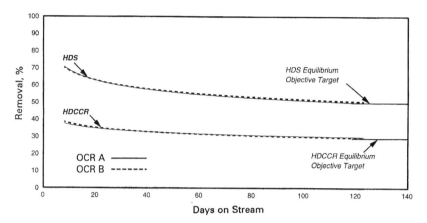

FIGURE 10.1.4 OCR hydrodesulfurization and Conradson carbon removal. The OCR reactors' excellent sulfur and Conradson carbon residue conversion enables downstream RDS units to optimize catalyst usage.

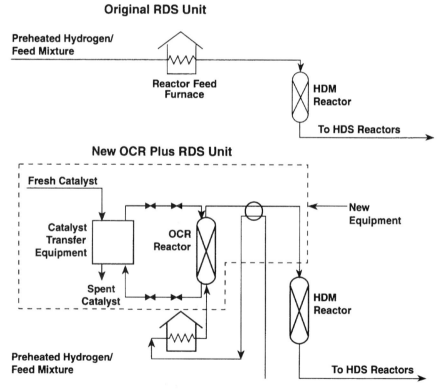

FIGURE 10.1.5 OCR retrofit of Aichi RDS hydrotreating unit. The Aichi retrofit was completed in less than a month of down time, and consisted of adding a new OCR reactor and an OCR reactor feed/effluent exchanger to each of the two reactor trains, as well as common catalyst transfer equipment.

the detailed engineering for the project. The RDS unit at the Aichi Refinery consists of two parallel reactor trains which process a total of 50,000 barrels per stream day (BPSD) of atmospheric residuum (AR) that is fed to a residual fluid catalytic cracking (RFCC) unit. Prior to adding OCR, atmospheric residuum from Arabian light was the required feed. Upgrading the RDS unit with an OCR reactor enabled IKC to switch feeds from 100 percent Arabian Light to a less expensive blend of 50 percent Arabian Light and 50 percent Arabian Heavy without sacrificing RFCC feed quality and fixed-bed catalyst life. Table 10.1.1 shows how the feed rate was increased, yields of naphtha and gas oil increased, and RFCC feedstock properties improved with the addition of the OCR reactors. The OCR unit also improved the activity and fouling rate of the RDS catalyst (see Fig. 10.1.6)

OCR APPLICATIONS

The driving force behind the decision to add OCR technology to an RDS processing scheme is the desire to run heavier, higher-metal feeds, because of crude oil changes or the need to cut deeper into the barrel. As the metal content of the feed rises above 100 to 150 ppm, the catalyst life cycle decreases to the point of being uneconomical. Figure 10.1.7 shows how much the relative catalyst life for a fixed-bed unit decreases as the total feed metals in the residuum increase.

When OCR is added to the processing scheme, total catalyst consumption is less than that for processing with a fixed-bed unit alone. Figure 10.1.8 shows the catalyst consumption required for a fixed-bed RDS unit operating alone versus a combined OCR/RDS processing scheme. The economy of the OCR reactor is especially apparent as the feed metals approach 200 ppm. Total catalyst consumption is lower because only the most heavily loaded catalysts are removed from an OCR. In a fixed-bed reactor, catalyst with low metal loading must be discarded with the spent catalyst at the end of the run. Since only spent catalysts are removed from an OCR reactor, the cata-

TABLE 10.1.1 OCR Improves RDS Operation at the Aichi Refinery

	OCR/RDS, typical RDS run after OCR	RDS, typical RDS run before OCR
Atmospheric resid feed rate, BPSD	50,000	45,000
Properties	OCR feed	RDS feed
Gravity, °API	13.6*	15.1
Sulfur, wt %	3.5	3.1
Conradson carbon residue, wt %	11	10
Ni + V, wt ppm	75	52
RFCC feed properties		
Sulfur, wt %	0.29	0.34
Conradson carbon residue, wt %	4.6	4.7
Ni + V, wt ppm	10	10
Cracking to naphtha and gas oil, LV %	20	15.5
Conradson carbon residue removal, wt %	67	61
Run cycle	1 year	1 year

*Crude oil was 2° API lower with OCR for a substantial savings in crude oil costs; °API = degrees on American Petroleum Institute scale; LV % = liquid volume percent.

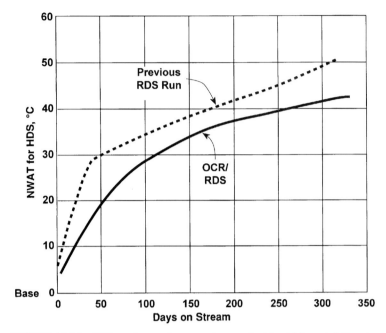

FIGURE 10.1.6 RDS catalyst performance improved with addition of OCR unit. Adding an OCR unit enabled Aichi to switch to a less expensive heavy feed, besides improving the activity of the RDS catalyst.

lyst is fully utilized, thus reducing the total catalyst cost per barrel of feed processed. Coincidentally, this also minimizes the amount of spent catalyst generated per pound of metals removed. The higher metals on the spent catalyst allows for more economic reclamation of the metals from the spent OCR catalysts.

Adding an OCR in front of an RDS is also cost-effective when the goal is to maximize production of lighter, cleaner-burning transportation fuels. The OCR allows less demetallization catalyst to be used in the fixed beds, thus providing more reactor volume for high-activity desulfurization catalyst. Achieving deeper desulfurization in the RDS unit enables refiners to produce ultralow-sulfur fuel oil as well as exceptionally clean feed for an RFCC unit.

Using an RDS unit to prepare feed for an RFCC reactor has gained wide acceptance because high-quality mogas and middle distillates can be produced with little or no low-value by-products. To maximize operating profitability, RFCCs require feeds that are very low in contaminant metals, carbon residue, and sulfur concentration, in addition to having feed volatility sufficiently high to fully vaporize at the feed nozzle. Metals reduce catalyst selectivity and activity, resulting in increased RFCC catalyst consumption. Carbon residue contributes to high coke yields and heat balance problems. Sulfur forces refiners to invest in expensive flue gas desulfurization equipment. Additionally, sulfur in the feed appears in the finished products.

In summary, the yield, product quality, and operating efficiency produced from an RFCC unit are directly related to the quality of the feed. The OCR/RDS technology has been used to process feed for RFCC units from a variety of heavy AR feeds, including Arabian Heavy and Ratawi. Pretreating the residuum in an OCR unit enables the refiner to use less expensive feeds, achieve higher product yields, and pro-

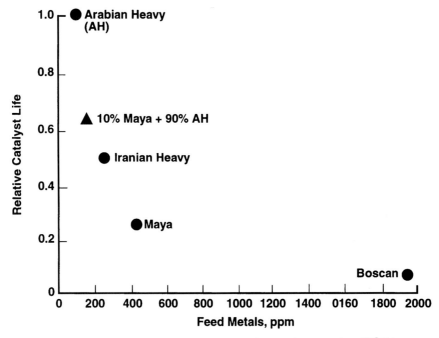

FIGURE 10.1.7 Relative RDS/VRDS catalyst life versus feed metal concentration. (VRDS is vacuum residuum desulfurization.) Conventional fixed-bed hydrotreating cannot economically process high-metal feeds.

duce better product quality while experiencing fewer feed-related operating problems. OCR can also be combined with Chevron's vacuum residuum desulfurization (VRDS) technology to upgrade vacuum residuum from heavy crudes into a synthetic AR with superior RFCC feed qualities. (For a more complete discussion, see Chap. 8.1.)

ECONOMIC BENEFITS OF OCR

The most apparent economic benefit of adding OCR to the processing scheme is the ability to run heavier, high-metal, less expensive crudes. Figure 10.1.9 shows how gains in gross margin can be made by improving unit capacity while maintaining the same fixed-bed investment. Similarly, when retrofitting an existing RDS unit with OCR technology, savings are achieved by extending catalyst life and run lengths. Each catalyst changeout in a fixed-bed unit takes approximately 4 weeks. The shorter run length means a costly reduction in the on-stream factor. Thus, the penalties for processing a high-metal feed in a fixed-bed unit are twofold—higher catalyst cost and reduced on-stream factor. OCR eliminates these limitations by providing maximum catalyst utilization and increasing the on-stream operating factor to approximately 0.96.

The economics of the OCR process are greatly dependent on the difference in price between light and heavy crudes, and each refiner's operating constraints. At a differential of U.S. $1.80/bbl between Arab Light and Arab Heavy crude, switching from

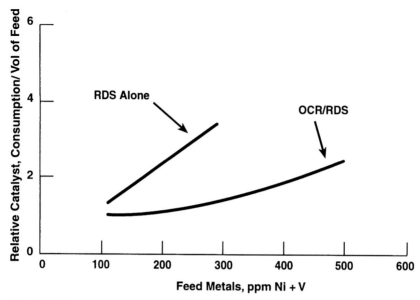

FIGURE 10.1.8 Comparison of catalyst consumption. Total catalyst cost is reduced when OCR technology is added to processing schemes designed to treat heavy, high-metal feeds. (Note: Based on the same reactor volume in both systems.)

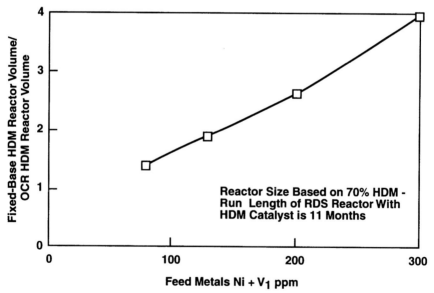

FIGURE 10.1.9 Savings in HDM reactor size with OCR. Throughput capacity of RDS units increases significantly when they are operated in conjunction with an OCR unit.

100 percent Arabian Light to a 50/50 blend of Arabian Light and Arabian Heavy will pay back the OCR investment in less than 2 years. Alternatively, the unit throughput can be increased. In this scenario, with a product upgrade from heavy feed to low-sulfur fuel oil and to middle distillate, an increase in feed rate of 10,000 BPD will pay out in less than 2 years.

OCR is a valuable technology for refiners trying to meet tough environmental guidelines within tight budgetary constraints. The benefits of the OCR process are summarized below:

- Ability to process less expensive, heavy, high-metal feedstocks.
- No interruption in operations to remove spent catalyst or add fresh catalyst to the OCR reactor.
- Prevention of guard-bed plugging problems.
- Longer life for downstream residuum-hydrotreating fixed-bed catalysts.
- Reduced downtime for fixed-bed catalyst changeouts.
- Savings in HDM reactor size.
- Additional throughput capacity with no increase in furnace capacity or NO_x emissions.
- Lower overall catalyst costs.
- Minimized waste from spent catalyst.
- Economical recovery of metals from spent catalyst.

CHAPTER 10.2
FW SOLVENT DEASPHALTING

F. M. Van Tine
Howard M. Feintuch
Foster Wheeler USA Corporation
Clinton, New Jersey

The changing pattern of petroleum-products demand has resulted in a sharp increase in the quantity of distillate products that must be produced from each barrel of crude oil processed. On the other hand, crudes have become heavier and higher in sulfur content. This has resulted in a reduced yield of distillate products that can be produced from the crude by using conventional distillation processes. This imbalance has created the need for processes that can increase the amount of distillates yield per barrel of crude.[1,2]

Vacuum distillation is an efficient means to separate distillates according to their volatility. Nevertheless, the recovery of distillates with a true boiling point (TBP) cut point over 570°C (1060°F) would require distillation temperatures at which thermal cracking reactions are too fast and make distillation impractical.

Solvent deasphalting (SDA) provides an extension of vacuum distillation. It permits practical recovery of much heavier oil, at relatively low temperatures, without cracking or degradation of these heavy hydrocarbons. SDA separates hydrocarbons according to their solubility in a liquid solvent, as opposed to volatility in distillation. Lower-molecular-weight and most paraffinic components are preferentially extracted. The least soluble materials are high-molecular-weight and most aromatic components.[3] This makes the deasphalted oil (DAO) extract light and paraffinic and the asphalt raffinate heavy and aromatic.

Since SDA is a relatively low temperature process, there is essentially no corrosion. Fouling of equipment is extremely low. SDA is a very reliable process with typical stream factors of about 98 percent and run lengths of up to 3 years.[4]

For more than 40 years SDA, generally using propane as the solvent, has been employed in the refining of lubricating oils to separate high-boiling fractions of crude oils beyond the range of economical commercial vacuum distillation. This means that the feed to deasphalting usually is a vacuum residue above the 510°C TBP cut point. More recently SDA has been adapted for the preparation of catalytic cracking feeds, hydrocracking feeds, hydrodesulfurizer feeds, and hard asphalt. For these purposes, heavier-than-propane solvents are used together with higher operating temperatures. This results in maximum yield of valuable DAO and minimum yield of hard asphalt having a softening point usually over 150°C. The term *deep deasphalting* has been frequently used to describe SDA operating under these conditions.

SDA is not a new process but rather a well-proven process with which most refiners are already familiar.[5,6,7]

10.16   SEPARATION PROCESSES

Abbreviations used in this chapter are defined in Table 10.2.15 at the end of the chapter.

PROCESS DESCRIPTION

Figure 10.2.1 is a process-flow diagram for a typical SDA unit utilizing a double-effect evaporative solvent recovery system for recovering solvent from the DAO. Figure 10.2.2 shows an SDA unit.

Extraction Section

Vacuum-residue fresh feed is pumped into the SDA unit and combined with a small quantity of predilution solvent to reduce its viscosity. The combined vacuum residue and predilution solvent is cooled to the desired extraction temperature and then flows into the middle of the extraction tower. Solvent streams from the high-pressure and low-pressure solvent receivers are combined, and a small portion of the combined solvent is used as predilution solvent. The major portion of the solvent is adjusted to the desired temperature by the solvent heater-cooler and then flows into the bottom section of the extraction tower. The solvent flows upward, extracting the more paraffinic hydrocarbons from the vacuum residue, which is flowing down through the tower. The multiring hydrocarbons (asphaltenes) are not dissolved by the solvent and exit from the bottom of the tower. A temperature gradient is maintained across the extraction tower by steam coils located near the top of the tower. The highest temperature is at the top of the tower, and the lowest temperature is at the solvent inlet. In the top section of the tower, less soluble, heavier, and more aromatic hydrocarbons separate from the solution of DAO and solvent. These heavier hydrocarbons flow back down the tower, providing internal reflux and thus improving the separation between the DAO and asphalt.

DAO-Recovery System (Double-Effect Evaporation)

The extract or DAO mix containing DAO plus most of the solvent leaves the top of the extraction tower and flows to the DAO-recovery system. The pressure at the top of the tower is maintained above the vapor pressure of the solvent at the extraction temperature by a back-pressure control valve located in the DAO-mix outlet line. The DAO mix flows to an evaporator, where it is heated to vaporize a portion of the solvent, and then flows into the high-pressure flash tower, where the solvent vapor is separated from the remaining DAO mix. The DAO mix from the bottom of the high-pressure flash tower flows to the pressure vapor heat exchanger (PVHE), where additional solvent is vaporized from the DAO mix by condensing solvent from the high-pressure flash tower. The high-pressure solvent condensed in the PVHE is collected in the high-pressure solvent receiver. The partially vaporized DAO mix flows from the PVHE to the low-pressure flash tower, where the solvent vapors are separated from the remaining DAO-mix liquid. This liquid flows down the tower, where it is further heated by rising solvent vapors from the reboiler. Approximately 90 to 95 percent of the solvent contained in the DAO mix from the extraction tower is recovered by the double-effect evaporation process. From the low-pressure flash tower the DAO mix flows to the DAO stripper, where the remaining solvent is stripped from the DAO by superheated steam. The steam and stripped-solvent vapors are removed from the top of the stripper. The solvent-free DAO product is recovered at the stripper bottom and is then cooled before flowing to battery limits.

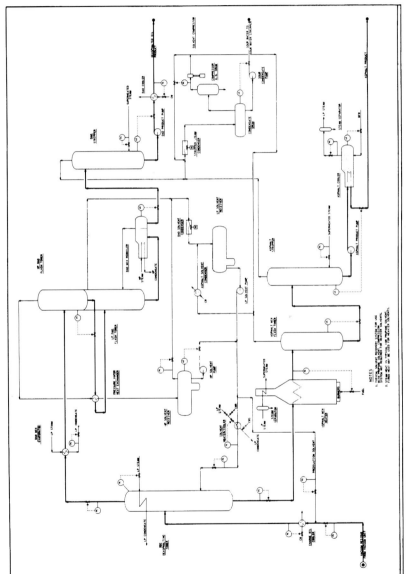

FIGURE 10.2.1 Process-flow diagram for a typical solvent-deasphalting unit.

FIGURE 10.2.2 A solvent-deasphalting unit. On the left side are two rotating-disk contactor (RDC) towers, the charge air cooler, and the high-pressure and low-pressure solvent receivers. On the right side are the DAO stripper, the DAO flash towers, and the asphalt flash tower (only partially visible). In the middle are the heat exchangers. The kettles are the asphalt-steam generator exchangers.

Asphalt-Recovery System

The raffinate phase containing the asphalt and a small amount of solvent (asphalt mix) flows at a controlled rate from the bottom of the extraction tower to the asphalt-mix heater. The hot two-phase asphalt mix from the heater is flashed in the asphalt-mix flash tower, where solvent vapor is taken overhead. The remaining asphalt mix flows to the asphalt stripper, where superheated steam entering below the bottom tray is used to strip the remaining solvent from the asphalt. The wet solvent-vapor overhead combines with the overhead from the DAO stripper. The asphalt product is pumped from the stripper bottom and cooled by generating low-pressure steam.

Solvent-Recovery System

Solvent evaporated in the low-pressure flash tower is condensed and collected in the low-pressure solvent receiver. Solvent from the asphalt flash tower, which operates at the same pressure as the low-pressure flash tower, is condensed in a separate condenser and is also collected in the low-pressure solvent receiver. The reason for segregating the two solvent streams is the potential for accidental fouling by asphalt entrained with solvent vapors from the asphalt flash tower.

The steam and solvent vapors from the DAO and asphalt strippers are cooled, and most of the steam present is condensed and collected in the stripper-condensate drum. This water is pumped from the unit to a sour-water stripper for removal of any dissolved sour gases and hydrocarbons. The noncondensed solvent vapor from this drum is compressed by the solvent compressor and joins the vapor from the asphalt flash tower upstream of the asphalt-solvent condenser.

Makeup solvent is pumped into the low-pressure solvent receiver from the off-site solvent storage drum as required.

This solvent-recovery system is typical of an SDA unit operating with propane solvent. In units using heavier solvents such as butane or pentane, the solvent vapors are completely condensed, separated from the water, and then pumped from the low-pressure solvent receiver. No solvent compressor is required.

TYPICAL FEEDSTOCKS

The SDA process (normally using propane or a propane-butane mixture as the solvent) has been in commercial use for the preparation of lubricant-bright-stock feeds from asphalt-bearing crude residue for many years.[8,9] Many commercial SDA units have also been used for preparing paving and specialty asphalts from suitable vacuum residues.

The increasing use of the fluid catalytic cracking (FCC) process together with the increasing price of crude oil resulted in the need to maximize the quantity of FCC feedstock obtained from each barrel of crude. These conditions led to the extension of the SDA process to the preparation of cracking feedstocks from vacuum residues. The current trend for maximizing distillate-oil production has also prompted the increased use of the SDA process to prepare hydrocracking feedstocks from vacuum residues.

SDA supplements vacuum distillation by recovering additional high-quality paraffinic oil from vacuum residues beyond the range of practical distillation. Although atmospheric residues have been commercially solvent-deasphalted, typical SDA feedstocks are 570°C+ (1060°F+) TBP cut-point vacuum residues. These vacuum residues often contain high levels of metals (primarily nickel and vanadium), carbon residue, nitrogen, sulfur, and asphaltenes. Table 10.2.1 gives three examples of vacuum-residue feedstocks, covering a wide range of properties, that can be processed in an SDA unit.

It is evident from Table 10.2.1 that, in contrast to many catalytic refining processes, a high content of feedstock impurities does not limit the applicability of SDA for processing residues. For a fixed DAO quality, high impurity levels in the feedstock will reduce DAO yield and increase asphalt production, but technically sound extraction is still possible. Table 10.2.2 shows estimated DAO yields for the three feedstocks listed in Table 10.2.1 when producing a DAO of 10 wt ppm total nickel-plus-vanadium content. Table 10.2.3 shows the corresponding estimated asphalt yields and properties.

TABLE 10.2.1 Typical SDA Feedstocks

	Vacuum-residue TBP cut point, °C	Gravity, °API	Conradson carbon residue, wt %	Sulfur, wt %	Ni + V, wt ppm
Heavy Arabian	570	3.6	25.1	5.5	193
Heavy Canadian	570	8.1	17.4	2.7	110
Canadian	570	11.7	15.0	1.5	50

TABLE 10.2.2 Estimated DAO Yields

	wt ppm metals in feedstock	vol % DAO at 10 wt ppm (Ni + V) DAO
Heavy Arabian	193	51
Heavy Canadian	110	73
Canadian	50	84

TABLE 10.2.3 Estimated Asphalt Properties for 10 wt ppm (Ni+V) DAO Production

	Specific gravity at 60/60°F	R&B softening point, °C	Penetration at 25°C	vol % asphalt yield
Heavy Arabian	1.119	128	0	49
Heavy Canadian	1.136	152	0	27
Canadian	1.151	170	0	16

In the future, crude-oil supplies are likely to become heavier, and vacuum residues obtained from processing these crudes will also become heavier and higher in impurity levels. Even under these conditions, SDA will continue to provide an economically and technically attractive route for upgrading these vacuum residues into lighter, more valuable products.

EXTRACTION SYSTEMS

The efficiency of the SDA process is highly dependent on the performance of the liquid-liquid extraction device. Proper design of the extraction device is necessary to overcome the mass-transfer limitations inherent in processing heavy, viscous oils to assure that the maximum yield of a specified quality of DAO is obtained. There are two major categories of extraction devices used for solvent deasphalting: mixer-settlers (a single stage or several stages in series) and countercurrent (multistage) vertical towers.

Mixer-Settler Extraction

Mixer-settlers were the first SDA device used commercially, and this is the simplest continuous-extraction system.[10] It consists of a mixing device (usually an in-line static mixer or a valve) for intimately mixing the feedstock and the solvent before this mixture flows to a settling vessel. The settling vessel has sufficient residence time to allow the heavy asphalt (raffinate) phase to settle by gravity from the lighter solvent-oil phase (extract). A single-stage mixer-settler results in, at best, one equilibrium extraction stage, and therefore the separation between the DAO and asphalt is poorer than that obtainable with a countercurrent multistage extraction tower. This poorer separation is evidenced by the higher nickel and vanadium content of the DAO produced by the single-stage system compared to the multistage system. Table 10.2.4

TABLE 10.2.4 Solvent-Deasphalting Kuwait Vacuum Residue

Alphalt product, vol % on crude	% of feed (Ni + V) in DAO	
	Single-stage	Multistage countercurrent
8	22	8
10	17	4.5
12	13	2

Source: C. G. Hartnett, "Some Aspects of Heavy Oil Processing," API 37th Midyear Meeting, New York, May 1982.

gives data comparing the DAO obtained from Kuwait vacuum residue by using one equilibrium extraction stage versus that obtained from a countercurrent multistage extraction.[11] These data were obtained at a solvent-to-feed ratio of 6:1.

Single-stage mixer-settler extraction devices were gradually replaced by vertical countercurrent towers as the advantage of multistage countercurrent extraction became evident. The economic incentive for obtaining the maximum yield of high-quality DAO for lubricant production has resulted in the use of multistage countercurrent extraction towers in virtually all lubricating-oil refineries.

Recently, some SDA designers have advocated a return to the mixer-settler extraction system for processing vacuum residues to obtain cracking feedstock, a considerably lower-value product than lubricating-oil bright stock. This position is based on the theory that the lower installed cost of the mixer-settler system offsets the product-value loss due to the lower DAO yield. This is true only for low marginal values of the DAO cracking stock over the vacuum-residue feedstock and for small yield losses. The latter assumption is true at very high (in general, greater than 90 vol %) DAO yields. With the heavier crudes being processed today, this is not always a realistic assumption.

Countercurrent Extraction

As shown in Table 10.2.4, countercurrent extraction provides a much more effective means of separation between the DAO and the asphalt than does single-stage mixer-settler extraction. This subsection will discuss the major factors affecting the design of a commercial countercurrent SDA extraction tower.

Countercurrent contact of feedstock and extraction solvent is provided in an extraction vessel called a *contractor* or *extractor* tower. Liquid solvent (light phase) enters the bottom of the extraction tower and flows upward as the continuous phase. The vacuum-residue feedstock enters the upper section of the extraction tower and is dispersed by a series of fixed or rotating baffles into droplets which flow downward by gravity through the rising continuous solvent phase. As the droplets descend, oil from the droplets dissolves into the solvent, leaving insoluble asphalt or resin, saturated with solvent, in the droplets. These droplets collect and coalesce in the bottom of the tower and are continuously withdrawn as the asphalt phase (heavy phase, or raffinate).

As the continuous solvent phase, containing the dissolved DAO, reaches the top of the tower, it is heated, causing some of the heavier, more aromatic dissolved oil to separate from the solution. These heavier liquid droplets flow downward through the ascending continuous solvent-DAO solution and act as a reflux to improve the sharpness of the separation between the DAO and the asphalt. This type of extraction system is analogous to the conventional distillation process.

The most common extractor towers used commercially are the rotating-disk contactor (RDC) developed by Shell and the fixed-element, or slat, towers. RDC contactors have proved to be superior to slat towers because of the increased flexibility inherent in their operation as well as the improved DAO quality obtained by using the RDC.[12] A 3 to 5 percent DAO-yield advantage has been found for the RDC at constant DAO quality.[10,12]

Figure 10.2.3 shows a schematic of a rotating-disk contactor. The RDC contactor consists of a vertical vessel divided into a series of compartments by annular baffles (stator rings) fixed to the vessel shell. A rotating disk, supported by a rotating shaft, is centered in each compartment. The rotating shaft is driven by a variable-speed drive mechanism through either the top or the bottom head of the tower. Steam coils are provided in the upper section of the tower to generate an internal reflux. Calming grids are provided at the top and bottom sections of the tower. The number of com-

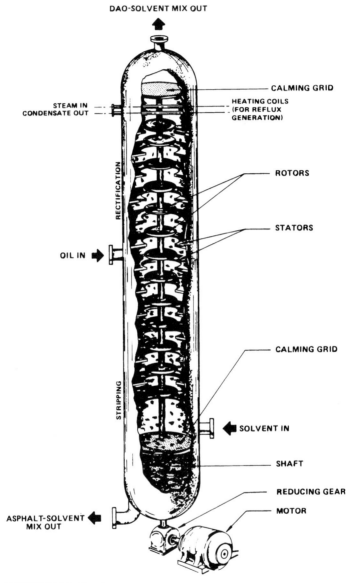

FIGURE 10.2.3 Rotating-disk contactor.

partments, compartment dimensions, location of the feed nozzle, and rotor speed range are all selected to provide optimal performance for a given set of operations.

RDC Capacity

The conditions under which flooding occurs in an RDC or slat tower represent the capacity limit at which the contractor can be operated. Flooding is evidenced by a loss of the interface level between the solvent and the asphalt phases in the bottom of the tower as well as by a deterioration in DAO quality. Usually this condition will appear quite suddenly, and if it is not properly corrected, asphalt may be entrained into the DAO-recovery system.

The maximum capacity of an RDC tower is a function of the energy input of the rotating disk. This energy input is given by the following equation.[12,13]

$$E = \frac{N^3 R^5}{HD^2}$$

where D = tower diameter, ft
E = energy input factor, ft^2/s^3
H = compartment height, ft
N = rotor speed, r/s
R = rotor-disk diameter, ft

The tower capacity is given by the quantity

$$T = \frac{V_D + V_C}{C_R}$$

where V_C = superficial velocity of solvent (continuous phase), ft/h
V_D = superficial velocity of residue (dispersed phase), ft/h
C_R = factor, defined by RDC internal geometry[14]; it can be taken as the smaller value of O^2/D^2 or $(D^2 - R^2)/D^2$
O = diameter of opening in stator, ft
T = tower capacity, ft/h

For a fixed RDC internal geometry and for a given system (at constant solvent/feed ratio) the quantity $(V_D + V_C)$ at flooding (maximum tower capacity) is a smooth function of energy-input quantity E. This function is illustrated by Fig. 10.2.4 for propane deasphalting in lubricating oil manufacture. This type of correlation permits the scaling up of pilot-plant data to a commercial-size unit or recalculation of the capacity of an existing tower for the same system at different conditions.

SOLVENT-RECOVERY SYSTEMS

Compared with the solvent-extraction section, which requires special design expertise, recovery of the solvent from a deasphalting operation is a relatively straightforward exercise in process design. The DAO-solvent and asphalt-solvent streams from the extraction section are heated to cause phase separation, the solvent-vapor phases are removed from the liquid DAO and asphalt phases, the solvent is accumulated for recy-

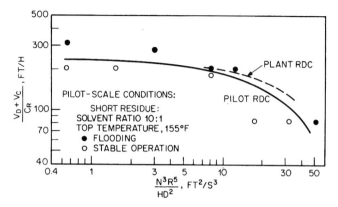

FIGURE 10.2.4 RDC capacity for propane deasphalting. [*Courtesy of Pennwell Publishing Company, publishers of the* Oil and Gas Journal, *59, 90–94 (May 8, 1961).*]

cle back to the extraction section, and the solvent-free DAO and asphalt are recovered as products.

The largest share of investment and operating costs incurred in deasphalting, as in most other solvent-extraction processes, is associated with the DAO solvent-recovery section. Most older SDA units rely on a single stage of solvent evaporation, using large amounts of low-pressure steam to supply the heat required. These units could operate with low-pressure steam because they use propane, which has a relatively low evaporation temperature, as the solvent. Single-stage evaporation represents the minimum capital cost but is also the least energy-efficient design. However, in the past 30 years many units using two stages of solvent evaporation have been installed. This type of solvent-recovery system was described in the section "Process Description" and illustrated in Fig. 10.2.1. When energy costs increase, there is a greater incentive to condense the lower-pressure solvent from the second stage by evaporating additional solvent at still lower pressure and to repeat the process as many times as may be economically justified. This is an application of multiple-effect evaporation, which has been used for more than 100 years in such industries as sugar refining and salt crystallization. It has also been used for more than 45 years in petroleum refining to recover solvents such as furfural, phenol, n-methyl-2-pyrrolidone (NMP), methyl ethyl ketone, toluene mixtures, and other lubricating-oil extraction and dewaxing solvents. A deasphalting unit using double-effect evaporation requires only about 62 percent of the utilities required to operate a unit with single-effect evaporation. A unit using triple-effect evaporation requires only about 54 percent of the utilities required by a unit using single-effect evaporation.[15]

Figure 10.2.5 illustrates a triple-effect solvent-recovery system. This figure, compared with Fig. 10.2.1 (double-effect solvent recovery), shows that the added stage of evaporation requires one additional heat exchanger, one medium-pressure flash tower, one medium-pressure solvent receiver, and one additional solvent pump. Although the cost of installing this additional equipment is appreciable, the return on investment from this improvement is in general less than 3 years.[15] Table 10.2.5 compares estimated investment cost and utility savings for converting an existing single-effect solvent-recovery unit to a double- or triple-effect recovery unit.

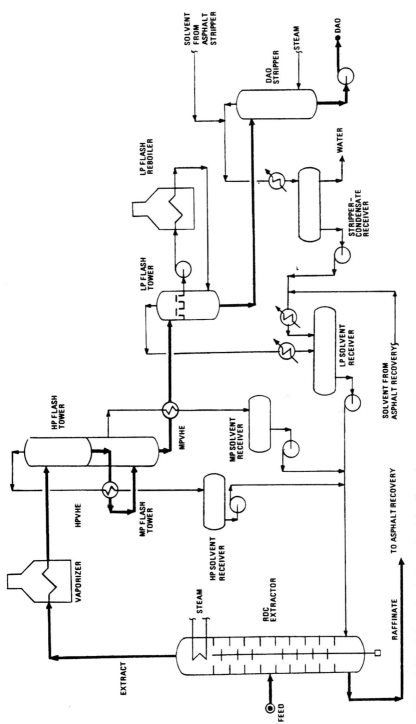

FIGURE 10.2.5 Solvent-deasphalting unit: triple-effect solvent recovery system.

TABLE 10.2.5 Deasphalting-Unit Energy-Reduction Economics*

Deasphalted-oil solvent recovery	Single-effect	Double-effect	Triple-effect
Conversion investment, D&E†	0	$1,300,000	$1,900,000
Annual utility costs‡			
Medium-pressure steam, $3.65/1000 lb	$91,100	$91,100	$91,100
Low-pressure steam, $3.10/1000 lb	931,600	465,200	377,500
Power, $0.04/kWh	218,400	218,400	218,400
Cooling water, $0.068/1000 gal	87,200	47,600	39,200
Fuel, $2.02/$10^6$ net Btu	44,100	44,100	44,100
Total annual utility costs	$1,372,400	$866,400	$770,300
Annual utility reduction	0	$506,000	$602,100
Simple payout before taxes, years	....	2.6	3.2

*J. W. Gleitsmann and J. S. Lambert, "Conserve Energy: Modernize Your Solvent Deasphalting Unit," Industrial Energy Conservation Conference, Houston, April 1983. Charge rate: 1700 BPSD of vacuum residue, 93 vol % DAO yield, 350 stream days/year.
†Estimated; basis: United States Gulf Coast, third quarter 1995.
‡Based on unit costs provided by SRI International, Menlo Park, Calif., December 1994.

The quantity of energy savings obtainable by using multiple-effect evaporative-solvent recovery is directly proportional to the quantity of solvent circulated. SDA units employing relatively large solvent-to-feed ratios, such as units designed to produce lubricating-oil bright stock, will show larger energy savings than units operating at very low solvent-to-feed ratios. The major portion of the energy savings occurs in the heat supplied to the DAO-mix evaporator (or vaporizer).

On units using propane as the solvent, the energy savings will result in reduced consumption of low-pressure steam, while on units using heavier solvents (butane or pentane), the energy savings will be in reduced consumption of fuel.

Figure 10.2.6 illustrates an alternative solvent-recovery system which utilizes the principle that the solvent when above its critical point loses most of its ability to dissolve DAO. The DAO mix from the extractor is heated above the solvent critical temperature in the DAO-mix heater. Two phases are formed in the clarifier; the lighter phase is essentially oil-free, while the heavier phase contains the DAO with a small amount of dissolved solvent. The light phase (solvent) is used to heat the DAO mix; then it is cooled and returned to the extraction tower. The heavy phase is flashed into the DAO stripper, where most of the remaining solvent is released and solvent-free DAO is obtained by stripping with low-pressure superheated steam.

The supercritical solvent-recovery process can be adapted to produce a resin product by including a resin-separation stage between the light-phase–DAO-mix exchanger and the DAO-mix heater. As the extract mix is heated, the heavier, more aromatic resins separate from the solvent and can be recovered by gravity settling from the lighter solvent and DAO phase. Since there is a very limited commercial market for the resins, they are normally recovered with the asphalt rather than as a separate product. If a mixer-settler extraction system is used, the resins may be separated and recycled to the mixer-settler. This resin recycle improves the selectivity of the extraction process. When an RDC or a slat tower with a temperature gradient is used for the extraction, the resin recycle is not implemented, since the internal reflux in the tower already provides for high-efficiency, high-selectivity extraction.

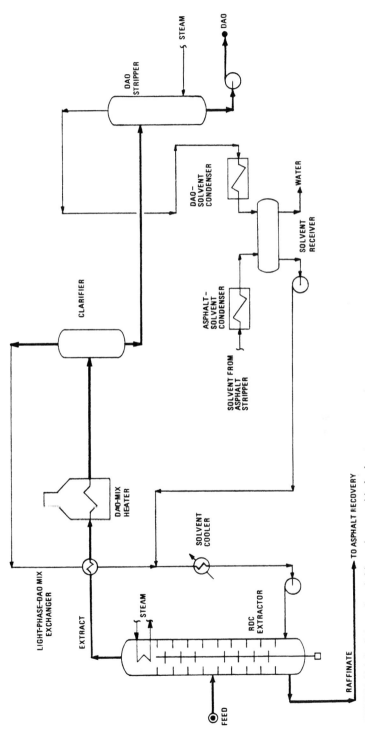

FIGURE 10.2.6 Solvent-deasphalting unit: supercritical solvent-recovery system.

DAO YIELDS AND PROPERTIES

By adjusting such variables as the solvent-to-feed ratio, type of solvent, and extraction conditions, it is possible to adjust the DAO yield to any value between 0 and 100 percent of the vacuum-residue feedstock. At very low DAO yields, DAO quality is at a maximum [low values for metals, nitrogen, sulfur, and Conradson carbon residue (CCR)]; at very high yields DAO quality approaches feed quality.[16] DAO quality does not vary linearly with the yield of DAO owing to the selectivity of the solvent for dissolving the more paraffinic and lower-molecular-weight molecules; these molecules contain fewer impurities than the more aromatic and higher-molecular-weight molecules. This selectivity is shown qualitatively in Fig. 10.2.11.

For design purposes, optimal DAO yield should be established by an economic analysis that not only includes the SDA unit but also considers the impact of DAO quality on the investment and operating cost of the downstream processing units.

DAO is an exceptionally paraffinic material[17] which is characterized by a high pour point, a Watson K factor* over 12, and a low specific gravity. As a virgin material, DAO is essentially free of olefinic hydrocarbons and does not exhibit the stability problems present in "cracked" distillates such as coker or visbreaker distillates.

Physical Properties

DAO physical properties are affected as follows as DAO yield increases:

1. *Specific gravity.* Specific gravity increases as DAO yield increases (DAO becomes heavier). See Table 10.2.6.
2. *Viscosity.* Viscosity increases as DAO yield increases (which corresponds to a heavier DAO). See Table 10.2.6.
3. *Heptane insolubles.* Content of heptane insolubles (asphaltenes) remains very low as DAO yield increases. Nevertheless, the asphaltenes content of the DAO will increase, approaching the feedstock asphaltene content as DAO yield approaches 100 percent. See Table 10.2.6.

---

*The Watson K factor is defined as $K = (TB)^{1/3}/\text{sp gr}$, where TB is the mean average boiling point in degrees Rankine and sp gr is the specific gravity at 60°F/60°F. This factor is an approximate index of paraffinicity, and it ranges from 12.5 for paraffinic products to 10.0 for aromatic products.

TABLE 10.2.6 Solvent-Deasphalting Heavy Arabian Vacuum Residue: DAO Properties

	Vacuum residue	DAO yield, vol % on feed			
		15.1	47.4	65.3	73.8
Gravity, °API	3.6	20.3	14.6	10.8	9.4
Viscosity at 100°C, SSU	70,900	183	599	1590	2540
Viscosity at 150°C, SSU	3,650	82.5	132	263	432
Pour point, °C	74	54	32	38	41
Heptane insolubles, wt %	16.2	0.01	0.01	0.01	0.03

Source: J. C. Dunmyer, "Flexibility for the Refining Industry," *Heat Eng.*, 53–59 (October–December 1977).

4. *Pour point.* At low DAO yields the pour point is high, consistent with the paraffinic character of the DAO. As DAO yield increases, less paraffinic material is dissolved, which in many cases is reflected in a decreasing pour point. As DAO yield continues to increase, the pour point will ultimately near the feed pour point for DAO yields, approaching 100 percent. See Table 10.2.6.

Sulfur

The sulfur distribution between the DAO and the asphalt is a function of DAO yield.[1] Figure 10.2.7 shows a typical relationship between the ratio of sulfur concentration in the DAO to sulfur concentration in the feed as a function of DAO yield. This figure shows an average sulfur-distribution trend and also maximum and minimum ranges expected for a wide number of vacuum-residue feedstocks. For a specific feedstock, the sulfur-distribution relationship is close to linear, especially as DAO yield increases above 50 vol %.[18,19]

Tables 10.2.7 and 10.2.8 give pilot-plant data on sulfur distribution for deasphalting heavy Arabian and heavy Canadian vacuum residues.

The ability of a solvent to reject the feedstock sulfur into the asphalt selectively is not as pronounced as its ability to reject metal contaminants such as nickel and vanadium selectively.[16] This is illustrated in Fig. 10.2.11. The sulfur atoms are more evenly distributed between the paraffinic and aromatic molecules than the metal contaminants, which are heavily concentrated in the aromatic molecules. In many cases, the fact that the metal content in the DAO is low makes hydrodesulfurization of high-yield DAO technically feasible and economically attractive.

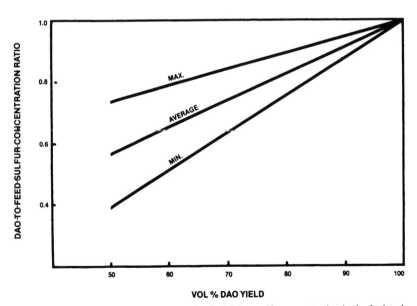

FIGURE 10.2.7 Ratio of sulfur concentration in DAO to sulfur concentration in the feedstock versus DAO yield.

TABLE 10.2.7 Solvent-Deasphalting Heavy Arabian Vacuum Residue

	Vacuum-residue feed	DAO yield, vol % on feed			
		47.4	56.0	65.3	73.8
Gravity, °API	3.6	20.3	14.6	12.4	10.8
Sulfur, wt %	5.5	3.8	3.9	4.0	4.4
Nitrogen, wt %	0.46	0.13	0.22	0.23	0.25
Nickel, wt ppm	58	2	5	8	14
Vanadium, wt ppm	135	5	12	22	43
CCR, wt %	25.1	6.6	...	12.3	14.2

Source: J. C. Dunmyer, "Flexibility for the Refining Industry," *Heat Eng.*, 53–59 (October–December 1977).

TABLE 10.2.8 Solvent-Deasphalting Heavy Canadian Vacuum Residue

	Vacuum-residue feed	DAO yield, vol % on feed			
		65.1	72.8	80.0	81.8
Gravity, °API	8.1	16.3	14.8	13.2	13.0
Sulfur, wt %	2.7	1.6	1.8	2.0	2.1
Nitrogen, wt %	0.45	0.21	0.26	0.32	0.33
Nickel, wt ppm	45	1	5	9	11
Vanadium, wt ppm	65	4	8	17	19
CCR, wt %	17.4	6.0	7.3	9.6	10.0

Source: J. G. Ditman, "Solvent Deasphalting—A Versatile Tool for the Preparation of Lube Hydrotreating Feed Stocks," 38th Midyear API Meeting, Philadelphia, May 17, 1973.

Nitrogen

Figure 10.2.8 shows the ratio of the nitrogen in the DAO to the nitrogen in the feed as a function of DAO yield. It shows the average nitrogen-distribution trend and the maximum and minimum expected for a wide variety of vacuum-residue feedstocks. As shown by a straight line on a semilog plot, this relationship is exponential. Figure 10.2.8 shows that there is little difference among various vacuum residues in the solvent's ability to reject nitrogen into the asphalt selectively. The difference between the maximum and minimum expected values is significantly lower than in the sulfur-distribution plot (Fig. 10.2.7).

Tables 10.2.7 and 10.2.8 give pilot-plant data on nitrogen distribution for heavy Arabian and heavy Canadian vacuum residues.[18,20] SDA exhibits a better ability to reject selectively nitrogen-containing compounds than sulfur-containing compounds.[1,16] (See Fig. 10.2.11 below.)

Metals

The ratio of DAO metal content (Ni + V) to feedstock metal content as a function of DAO yield is shown in Fig. 10.2.9. The straight lines in the figure show that DAO metals content is an exponential function of DAO yield.

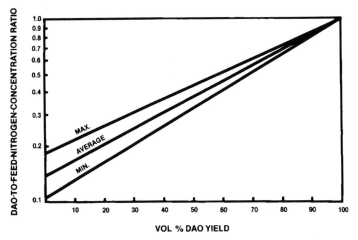

FIGURE 10.2.8 Ratio of nitrogen concentration in DAO to nitrogen concentration in the feedstock versus DAO yield.

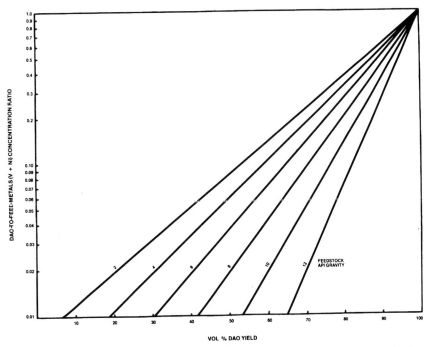

FIGURE 10.2.9 Ratio of metal (Ni+V) concentration in DAO to metal (Ni+V) concentration in the feedstock versus DAO yield.

This trend has been previously reported.[1,16] Figure 10.2.9 also shows that metal distribution is a strong function of the feedstock API gravity. The data in the figure illustrate an average relationship; however, some feedstocks such as Canadian sour and Tia Juana vacuum residues deviate substantially from the average trend. Pilot plant data are normally required to determine the exact DAO yield-quality relationship for a previously untested feedstock.

The nickel and vanadium distributions between the DAO and asphalt are similar but not equal.[16] (See Fig. 10.2.11.)

Tables 10.2.7 and 10.2.8 give pilot plant data on metal distribution for heavy Arabian and heavy Canadian vacuum-residue deasphalting.[18,20] These tables and Fig. 10.2.9 show that metals are rejected from DAO to a much greater extent than sulfur and nitrogen. For example, in deasphalting heavy Arabian vacuum residue at a 65 vol % DAO yield, the following are the ratios of the contaminant level in the DAO to the contaminant level in the feedstock:

Sulfur	72.7%
Nitrogen	50.0%
Nickel	13.8%
Vanadium	16.3%
CCR	49.0%

The high rejection of metals from DAO is of extreme importance in the catalytic processing of DAO. It is possible catalytically to hydroprocess DAO economically owing to the low metals content of DAO obtained even from a high-metal-content vacuum residue.

Conradson Carbon Residue

The deasphalting solvent exhibits a moderate selectivity for carbon rejection from DAO; the selectivity is similar to that of nitrogen rejection but significantly higher than that of sulfur rejection.

Conradson carbon residue* in DAO has a less detrimental effect on the cracking characteristics of DAO than it has in the case of distillate stocks.[4] DAO with 2 wt % CCR is an excellent FCC feedstock; it actually produces less coke and more gasoline than coker distillates.

Figure 10.2.10 shows that the ratio of CCR in DAO to CCR in the feed is an exponential function of DAO yield. As in the case of metals concentrations, the relationship is also a strong function of feedstock API gravity. The data in Fig. 10.2.10 illustrate an average relationship for a number of feedstocks and should not be considered a design correlation.

As in the case of metals, some feedstocks, such as Canadian sour and Tia Juana, deviate substantially from the average trend.

Tables 10.2.7 and 10.2.8 give SDA pilot plant data on CCR distribution for heavy Arabian and heavy Canadian vacuum residues.

See also Fig. 10.2.11.

---

*Conradson carbon residue is a standard test (ASTM D 189) used to determine the amount of residue left after evaporation and pyrolysis of an oil sample under specified conditions. The CCR is reported as a weight percent. It provides an indication of the relative coke-forming propensities of an oil sample.

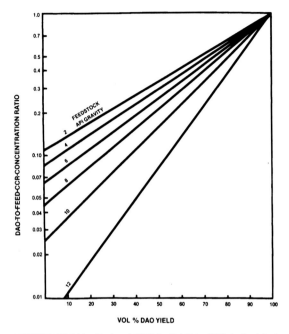

FIGURE 10.2.10 Ratio of CCR in DAO to CCR in feedstock versus DAO yield.

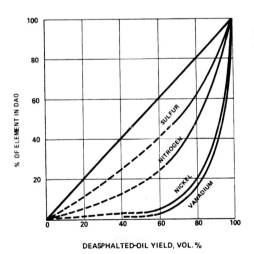

FIGURE 10.2.11 Selectivity in solvent deasphalting. [*Courtesy of the Gulf Publishing Company, publishers of* Hydrocarbon Processing, *52(5), 110–113 (1973).*]

OPERATING VARIABLES

The SDA process uses a light-hydrocarbon solvent to extract valuable paraffinic oils from asphaltene-rich vacuum residues. The SDA unit is usually required to be flexible enough to extract the desired quantity and quality of DAO from varying-quality feedstocks. Four basic operating variables are available to adjust the SDA process to optimal operating conditions:[1]

- Solvent composition
- Solvent-to-feed ratio
- Extraction temperature
- Extraction pressure

When the SDA unit utilizes an RDC, there are available additional operating variables which provide improved flexibility to the process. These variables are

- Temperature gradient. A temperature gradient also can be obtained in any countercurrent extraction tower.
- Rotor speed.

Solvent Composition

Modern SDA units normally use blends of light-hydrocarbon solvents (propane to hexane) to allow maximum operating flexibility. The solubility of oil in the solvent at fixed conditions of temperature and pressure increases as the specific gravity of the solvent increases. Moderate variations in the vacuum-residue-feed quality or DAO yield can be handled by adjusting the temperature level and the temperature gradient across the RDC, but larger variations require adjustments to the solvent composition. For a given operation, there is a fairly narrow range of solvent compositions that can be economically utilized. Proprietary correlations have been developed from pilot plant data so that the proper solvent composition can be selected to obtain an economical commercial deasphalting operation with maximum flexibility for future operating adjustments.

Solvent-to-Feed Ratio

At the normal extraction operating conditions used in commercial SDA units, an increase in the solvent-to-feed ratio results in a larger DAO yield. The molecular weight, density, viscosity, and contaminant level of the DAO are also increased. Since considerable energy is required to recover the solvent, there is an optimal solvent-to-feed ratio for each operation. Pilot plant data and commercial experience are utilized to select this optimal ratio.

Selectivity is the term used to describe the ability of the solvent to recover a paraffinic oil with a low level of contaminants from the vacuum residue. Selectivity can be improved by increasing the solvent-to-feed ratio at constant DAO yield (which requires slightly increasing the extraction temperature).

Table 10.2.9 illustrates the effect of the solvent-to-feed ratio on DAO yield and properties.[21]

Note that as the solvent-to-feed ratio is increased there is a diminishing effect of this ratio on DAO yield. Increasing the solvent-to-feed ratio above some limiting

TABLE 10.2.9 Effect of Solvent-to-Feed Ratio on DAO

Solvent-to-feed volume ratio	DAO, vol % yield on feed	DAO properties	
		Viscosity, SSU at 100°C	Gravity, API
2.8	37.5	51.4	30.1
3.9	50.6	52.4	29.9
11.4	76.0	59.0	28.8

Source: E. E. Smith and C. E. Fleming, *Pet. Refiner*, **36**, 141–144 (1957).

value will result in only a very small (if any) increase in DAO yield. At the same time, the energy required for solvent recovery increases in direct proportion to the quantity of solvent circulated.

Temperature

In general, factors which tend to increase the density of the solvent will increase the capacity of the solvent to dissolve the oil. This capacity is generally referred to as solubility. Consequently, an increase in temperature, which results in a lower-density solvent, reduces the DAO-product yield. As the solvent-extraction temperature is increased, the solvent approaches its critical temperature and its density and solubility are considerably reduced.

From the point of view of DAO-product quality, there is a certain incentive to increase the operating temperature.[1] The corresponding DAO-yield reduction can be compensated for by an increase in the solvent-to-feed ratio. The net result would be an improved selectivity or separation between the DAO and the asphalt fractions and an improved DAO quality (i.e., less metals, nitrogen, and CCR).

The SDA unit design basis is generally established by the ability of downstream units to process the DAO. The solvent-to-feed ratio, the temperature, and other variables should be adjusted to produce the deasphalting operation that results in optimal overall refinery economics, considering not only the deasphalter itself but upstream and downstream processing units and the refinery-product slate. Extraction temperature and extraction-temperature gradient are the variables which provide the usual method of day to day operational control of DAO yield and quality.

Table 10.2.10 illustrates the effect of extraction temperature on DAO yield and properties.

TABLE 10.2.10 Effect of Extraction Temperature on DAO Yield and Quality

Extraction temperature, °C	DAO yield on feed, vol %	DAO properties	
		Viscosity, SSU at 100°C	Gravity, °API
90	58.1	54.4	29.1
93	41.8	52.0	30.0
98	23.9	46.0	31.4
100	19.8	44.5	31.8

Source: E. E. Smith and C. E. Fleming, *Pet. Refiner*, **36**, 141–144 (1957).

Pressure

The solvent-extraction pressure has a direct effect on the density of the solvent. Higher pressure results in a denser solvent with more capacity to dissolve DAO and consequently a larger DAO yield. The process dependence on pressure becomes more evident as operating conditions approach the solvent critical point because the magnitude of the density change relative to the pressure change increases substantially. The extraction pressure is not normally used for day-to-day operational control.

RDC Temperature Gradient

It is possible to improve the quality of the DAO product at a constant DAO yield by maintaining a temperature gradient across the extraction tower. A higher temperature at the top of the RDC as compared with the bottom generates an internal reflux because of the lower solubility of oil in the solvent at the top compared with the bottom. This internal reflux supplies part of the energy for mixing and increases the selectivity of the extraction process in a manner analogous to reflux in a distillation tower.

Table 10.2.11 illustrates the effect of the RDC temperature gradient on the extraction process. Note that the RDC top temperature has been held constant and that the DAO yield is essentially unchanged.

RDC Rotor Speed

The RDC rotor speed has a significant effect on the yield and properties of the DAO and asphalt products. With all other variables held constant, an increase in rotor revolutions per minute within a certain speed range can result in an increased DAO yield. This yield increase is the direct result of higher mass-transfer rates when rotor speed is increased.

The effect of rotor speed on product yields and product properties is more evident at low throughputs and low rotor rates. At high throughputs much of the energy of mixing is obtained from the counterflowing phases themselves; in this case low rotor rates are sufficient to bring the extraction system up to optimal efficiency.

Table 10.2.12 illustrates the effect of rotor speed on a low-throughput operation. Note that the DAO yield is increased with little deterioration of DAO quality.

TABLE 10.2.11 Effect of RDC Temperature Gradient on DAO Quality

RDC temperature gradient, °C	DAO yield on feed, vol %	DAO properties		
		°API	Ni, wt ppm	V, wt ppm
14	83.0	22.3	0.75	0.55
23	83.3	23.4	0.50	0.40

Source: R. J. Thegze, R. J. Wall, K. E. Train, and R. B. Olney, *Oil Gas J.,* **59,** 90–94 (May 8, 1961).

TABLE 10.2.12 Effect of RDC Rotor Speed on Extraction Process

RDC rotor speed, r/min	DAO yield of feed, vol %	DAO properties			Asphalt penetration, 0.1 mm at 25°C
		Viscosity, SSU at 100°C	Gravity, °API	CCR, wt %	
20	76.8	194	23.2	1.4	38
35	80.3	198	23.0	1.5	8
50	83.3	203	22.3	1.5	1

Source: R. J. Thegze, R. J. Wall, K. E. Train, and R. B. Olney, *Oil Gas J.*, **59,** 90–94 (May 8, 1961).

ASPHALT PROPERTIES AND USES

The asphalt yield decreases with increasing DAO yield, and the properties of the asphalt are affected as follows:[22]

- Specific gravity increases, corresponding to a heavier material.
- Softening point increases, and penetration decreases.
- Sulfur content increases.
- Nitrogen content increases.

Table 10.2.13 gives pilot-plant data which illustrates the trend of asphalt properties with decreasing asphalt yield.

Since SDA preferentially extracts light and paraffinic hydrocarbons,[3,23] the resulting asphalt is more aromatic than the original feed. Further, it should be noted that high-softening-point (greater than 105 to 120°C) asphaltenes are free of wax even when precipitated from very waxy residues.[24]

Except for SDA units specifically designed to produce roofing or paving asphalt, the asphalt product is normally considered a low-value by-product. Since there is a very limited commercial market for these by-product asphalts, the refiner must usually find some method of disposing of the asphalt by-product other than by direct sale.

The following are the main uses of the asphalt fraction:

TABLE 10.2.13 Solvent-Deasphalting Heavy Arabian Vacuum-Residue Asphalt Fraction

	Vacuum residue feed	Asphalt yields, vol %				
		84.9	52.6	44	34.7	26.2
Specific gravity, 60°F/60°F	1.0474	1.0679	1.1185	1.1290	1.1470	1.1690
Softening point (R&B), °C	62	79	128	139	164	
Penetration at 25°C, 0.1 mm	24	9	0	0	0	0
Sulfur, wt %	5.5	5.9	6.6	7.3	7.9	8.2
Nitrogen, wt %	0.46	0.53	0.65	0.73	0.79	0.97
Heptane, insoluble, wt %	14.1	71.8	26.8	....	45.1	80.2

Source: *J. C. Dunmyer, "Flexibility for the Refining Industry," *Heat Eng.*, 53–59 (October–December 1977).

Fuel

In some cases, asphalt can be cut back with distillate materials to make No. 6 fuel oil. Catalytic cycle oils and clarified oils are excellent cutter stocks. When low-sulfur-content fuels are required and when the original deasphalter feedstock is high in sulfur, direct blending of the asphalt to make No. 6 fuel oil generally is not possible.

Relatively low softening point asphaltenes can be burned directly as refinery fuel, thereby avoiding the need to blend the asphalt with higher-value cutter stocks.

Direct asphalt burning has been practiced in a number of refineries. However, the high-sulfur-content crudes currently being processed by many refineries result in a high-sulfur-content asphalt which cannot be burned directly as refinery fuel unless a stack-gas sulfur oxide removal process is used to meet United States environmental regulations.

It is possible to use solid (flaked or extruded) asphaltenes as fuel for public utility power plants in conventional boilers with stack-gas cleanup or in modern fluidized-bed boilers.[25] These boilers use fluidized limestone beds directly to capture metals and sulfur oxides from the combustion gases.

Commercial Asphalts

Commercial penetration-grade asphalts can be produced by simply blending SDA asphaltenes with suitable aromatic flux oils. In many cases, this can eliminate the need for air-oxidizing asphalts and thus present obvious economic and environmental advantages. When SDA asphaltenes (which are wax free) are blended with a non-paraffinic flux oil, asphalts having satisfactory ductility can be made even from waxy crudes.[3] This eliminates the need to buy special crudes for asphalt manufacture.

Partial Oxidation

Asphaltenes can be used as a feedstock for synthesis-gas manufacture in partial-oxidation units. This synthesis gas can be used to produce hydrogen for the refinery hydroprocessing units, thereby eliminating the need to steam–re-form more valuable distillate oils or natural gas to produce hydrogen.

INTEGRATION OF SDA IN MODERN REFINERIES

Selection of the optimum residue-upgrading route depends on many factors such as:

- Available feedstock characteristics
- Required flexibility for processing different feedstock
- Feedstock cost
- Product markets
- Product values
- Existing refinery configuration and possibility for process-unit integrations
- Operating costs
- Unit capital-investment costs
- Unit stream factors

Typically, optimization studies involve large computers using linear programming techniques. This optimization is performed during the initial refinery-expansion study phase to determine the most economical conversion route.

For the purpose of illustrating the integration of SDA units in bottom-of-the barrel upgrading, a refinery processing 50,000 barrels per stream day (BPSD) of Kuwait atmospheric residue was selected. The following processing routes are considered:

Base Refinery. (See Fig. 10.2.12.) The basic processing route uses a conventional vacuum-flasher scheme together with vacuum gas oil (VGO) hydrotreating (hydrodesulfurization, or HDS) followed by fluid catalytic cracking. This basic refinery scheme does not provide any vacuum-residue upgrading. The block flow diagram given in Fig. 10.2.12 summarizes the expected product yields when processing 50,000 BPSD of Kuwait atmospheric residue. The products include 20,000 BPSD of heavy, high-sulfur vacuum residue. The main products are summarized in Table 10.2.14.

Maximum-Naphtha Case. (See Fig. 10.2.13.) This processing route is similar to the base refinery, but an SDA unit, which produces additional FCC-unit feedstock from the vacuum residue, has been included. The major change is that instead of the base-case 20,000-BPSD vacuum-residue production, 5400 BPSD of asphalt are produced. Table 10.2.14 summarizes the main products and shows that naphtha production has been increased by 49 percent with respect to the base case. For this illustration FCC was used for the VGO-DAO conversion, although hydrocracking also can be an economically viable route.

Maximum-Distillate Case. In this processing scheme the DAO together with the VGO is cracked in a hydrocracking unit. Figure 10.2.14 shows the flow scheme for this processing route and Table 10.2.14 summarizes the main products. This table shows that the naphtha yield was reduced by 50 percent and the distillate yields (jet fuel plus diesel) increased by 400 percent relative to the base case.

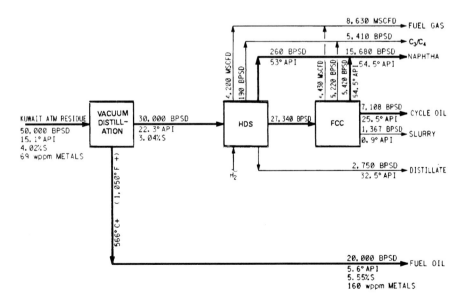

FIGURE 10.2.12 Integration of SDA in modern refineries: base refinery (no SDA unit provided).

TABLE 10.2.14 Integration of SDA in Refineries

	Base refinery	SDA unit application		
		Maximum naphtha	Maximum distillates	Maximum low-sulfur fuel oil
Products, BPSD:				
C_3-C_4 LPG	5,410	8,054	1,383	289
Naphtha	15,680	23,315	8,563	388
Distillates	9,858	14,659	40,407	4,090
Fuel Oil	20,000*	....	....	46,051
Asphalt	....	5,400*	5,400*	
Fuel-oil quality				
°API	5.6	....	....	19.4
wt % sulfur	5.55	....	....	1.55

*Outside No. 6 fuel-oil specifications.

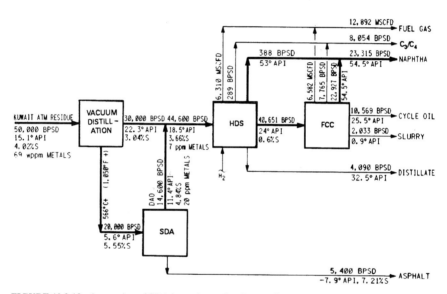

FIGURE 10.2.13 Integration of SDA in modern refineries: maximum-naphtha case.

Maximum Low-Sulfur Fuel Oil. Maximum fuel-oil production is not the general trend in the refinery industry but could be economically attractive under certain market conditions. This processing route is shown in Fig. 10.2.15. In this case the DAO together with the VGO is hydrotreated (HDS) and blended with the asphalt to produce a 1.55 percent sulfur fuel oil. This product corresponds to a 60 percent desulfurization of the atmospheric residue. Compared with direct desulfurization of the atmospheric residue, this route can be economically attractive in many cases. Desulfurization of the DAO plus the VGO blend is a simpler, less expensive process than direct atmospheric-residue hydrotreating.

Lubricating Oil Production. For many years SDA has been used in the manufacture of lubricating oils. In this case SDA produces a short DAO cut, which is further treat-

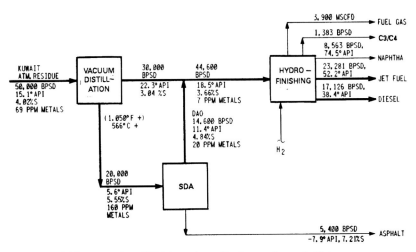

FIGURE 10.2.14 Integration of SDA in modern refineries: maximum-distillate case.

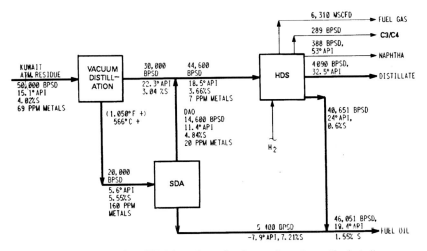

FIGURE 10.2.15 Integration of SDA in modern refineries: maximum low-sulfur fuel oil case.

ed (typically by furfural and then dewaxed) to produce high-quality lubricating oil base stocks. Figure 10.2.16 illustrates the application of SDA in the manufacture of lubricating oils.

TYPICAL UTILITY REQUIREMENTS

For many processing units it is possible to develop utility consumptions based on the feed capacity of the unit or on the amount of one or several of the unit products. For SDA units this is not possible since most of the unit utilities are related to the recovery and circulation of the solvent. SDA utility consumptions consequently depend

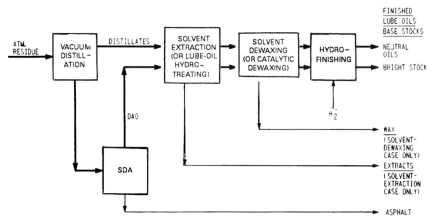

FIGURE 10.2.16 Integration of SDA in modern refineries: lubricating oil production case.

more on the solvent-to-feed ratio than on the DAO yield and unit feed capacity. An accurate estimate of an SDA unit's utility requirements requires definition of these three parameters.

As an example, the figures below give the utility requirements of an SDA unit used for preparation of catalytic-cracker feedstock, with a yield of 73 vol % DAO from Kuwait vacuum residue:

Fuel	76,000 Btu/bbl
Steam	60 lb/bbl
Cooling water	Nil (maximum air cooling)
Power	2 kWh/bbl

These figures are given per barrel of SDA unit feedstock and correspond to a double-effect solvent-recovery system. Utility consumption for a supercritical solvent-recovery system is virtually identical to that for a double-effect solvent recovery. Fuel consumption for the same unit using triple-effect solvent recovery can be expected to be 10 to 20 percent lower than for double-effect solvent recovery.

ESTIMATED INVESTMENT COST

SDA-unit capital-investment costs depend not only on unit feed capacity but also on the solvent-to-feed ratio used in the extraction. SDA units used for the preparation of lubricant-bright-stock feeds are designed for relatively high solvent-to-feed ratios, and investment cost is higher than the corresponding investment for SA units for the preparation of cracking stocks, which operate at a much lower solvent-to-feed ratio.

An accurate capital-investment cost for an SDA unit can be determined only by a detailed front-end design and a definitive estimate. Nevertheless, it is often necessary, when carrying out economic evaluations, to develop a budget-type estimate. Such estimates typically have an accuracy of ±30 percent and require definition of unit capacity and product quality. For preliminary studies, investment rates as low as the following may be used:

SDA for lubricating oils	$2000/bbl of feedstock
SDA for cracking stock	$1000/bbl of feedstock

These investment costs are based on:

- United States Gulf Coast location
- Time basis in third quarter 1995; no forward escalation included
- Exclusions: cost of land, taxes and owner's insurance, licenses, permits, duties, spare parts, start-up cost, interest, forward escalation, support facilities, interconnecting piping, feed, solvent, and product storage

Symbols used in the chapter are listed in Table 10.2.15.

REFERENCES

1. J. A. Bonilla, "Delayed Coking and Solvent Deasphalting: Options for Residue Upgrading," AIChE National Meeting, Anaheim, Calif., June 1982.
2. W. J. Rossi, B. S. Deighton, and A. J. MacDonald, *Hydrocarb. Process.*, **56**(5), 105–110 (1977).
3. J. G. Ditman and J. P. Van Hook, "Upgrading of Residual Oils by Solvent Deasphalting and Delayed Coking," ACS Meeting, Atlanta, April 1981.
4. P. T. Atteridg, *Oil Gas J.*, **61**, 72–77 (Dec. 9, 1963).
5. J. G. Ditman and R. L. Godino, *Hydrocarb. Process.*, **44**(9), 175–178 (1965).
6. J. C. Dunmyer, R. L. Godino, and A. A. Kutler, "Propane Extraction: A Way to Handle Residue," *Heat Eng.* (November–December 1966).

TABLE 10.2.15 Abbreviations

°API	Degrees on American Petroleum Institute scale; API = (141.5/sp gr) − 131.5	LP	Low pressure
		LPG	Liquefied petroleum gas
		MP	Medium pressure
bbl	Barrel (42 U.S. gal)	N	Rotor speed, r/s
BPSD	Barrels per stream day	Ni	Nickel
CCR	Conradson carbon residue	O	Diameter of stator opening, ft
C_R	Factor defined by tower internal geometry	PVHE	Pressure vapor heat exchanger
		R	Rotor-disk diameter, ft
°C	Degrees Celsius	R&B	Ring and ball (softening point)
CWR	Cooling-water return	S	Sulfur
CWS	Cooling-water supply	SCFD	Standard cubic feet per day
DAO	Deasphalted oil	SDA	Solvent deasphalting
D	Tower diameter, ft	sp gr	Specific gravity at 60°F/60°F
D&E	Delivered and erected (cost)	SSU	Seconds Saybolt universal (viscosity)
E	Energy input factor, ft^2/s^3		
°F	Degrees Fahrenheit	TBP	True boiling point
FC	Flow controller	TC	Temperature controller
FCC	Fluid catalytic cracker	V	Vanadium
H	Compartment height, ft	V_C	Solvent superficial velocity, ft/h
HDS	Hydrodesulfurization	V_D	Residue superficial velocity, ft/h
HP	High pressure		
LC	Level controller		

7. R. L. Godino, "Propane Extraction," *Heat Eng.* (March–April 1963).
8. J. G. Ditman and F. T. Mertens, *Pet. Process.* (November 1952).
9. A. Rhoe, "Meeting the Refiner's Upgrading Needs," NPRA Annual Meeting, San Francisco, March 1983.
10. S. Marple, Jr., K. E. Train, and F. D. Foster, *Chem. Eng. Prog.,* **57**(12), 44–48 (1961).
11. C. G. Hartnett, "Some Aspects of Heavy Oil Processing," API 37th Midyear Meeting, New York, May 1982.
12. R. J. Thegze, R. J. Wall, K. E. Train, and R. B. Olney, *Oil Gas J.,* **59,** 90–94 (May 8, 1961).
13. G. H. Reman and J. G. van de Vusse, *Pet. Refiner,* **34**(9), 129–134 (1955).
14. G. H. Reman, *Pet. Refiner,* **36**(9), 269–270 (1957).
15. J. W. Gleitsmann and J. S. Lambert, "Conserve Energy: Modernize Your Solvent Deasphalting Unit," Industrial Energy Conservation Technology Conference, Houston, April 1983.
16. J. G. Ditman, *Hydrocarb. Process.,* **52**(5), 110–113 (1973).
17. S. R. Sinkar, *Oil Gas J.,* **72,** 56–64 (Sept. 30, 1974).
18. J. G. Ditman, "Solvent Deasphalting—A Versatile Tool for the Preparation of Lube Hydrotreating Feed Stocks," API 38th Midyear Meeting, Philadelphia, May 17, 1973.
19. D. A. Viloria, J. H. Krasuk, O. Rodriguez, H. Buenafama, and J. Lubkowitz, *Hydrocarb. Process.,* **56**(3), 109–113 (1977).
20. J. C. Dunmyer, "Flexibility for the Refining Industry," *Heat Eng.,* 53–59 (October–December 1977).
21. E. E. Smith and C. E. Fleming, *Pet. Refiner,* **36,** 141–144 (1957).
22. H. N. Dunning and J. W. Moore, *Pet. Refiner,* **36,** 247–250 (1957).
23. J. G. Ditman and J. C. Dunmyer, *Pet. Refiner,* **39,** 187–192 (1960).
24. J. G. Ditman, "Solvent Deasphalting for the Production of Catalytic Cracking—Hydrocracking Feed & Asphalt," NPRA Annual Meeting, San Francisco, March 1971.
25. R. L. Nagy, R. G. Broeker, and R. L. Gamble, "Firing Delayed Coke in a Fluidized Bed Steam Generator," NPRA Annual Meeting, San Francisco, March 1983.

CHAPTER 10.3
UOP SORBEX FAMILY OF TECHNOLOGIES

John J. Jeanneret and John R. Mowry
UOP
Des Plaines, Illinois

INTRODUCTION

The Sorbex* name is applied to a technique, developed by UOP,* that is used to separate a component or group of components from a mixture by selective adsorption on a solid adsorbent. The Sorbex technology is a continuous process in which feed and products enter and leave the adsorbent bed at substantially constant composition. This technology simulates the countercurrent flow of a liquid feed over a solid bed of adsorbent without physically moving the solid. The principles of Sorbex technology are the same regardless of the type of separation being conducted. The following are examples of commercially proven UOP technologies based on the Sorbex principle; each makes use of a specific adsorbent-desorbent combination uniquely tailored to the specific separation:

- Parex*: separation of *para*-xylene from mixed C_8 aromatic isomers
- MX Sorbex*: *meta*-xylene from mixed C_8 aromatic isomers
- Molex*: linear paraffins from branched and cyclic hydrocarbons
- Olex*: olefins from paraffins
- Cresex*: *para*-cresol or *meta*-cresol from other cresol isomers
- Cymex*: *para*-cymene or *meta*-cymene from other cymene isomers
- Sarex*: fructose from mixed sugars

In addition to these applications, numerous other commercially interesting separations have been identified and demonstrated using the Sorbex process. These applications include ethylbenzene, 1-butene, ethyl toluenes, toluidines, terpenes, chloro and nitro aromatics, alpha and beta naphthols, alkyl naphthalenes, alpha olefins, and tall oil.

---

*Trademark and/or service mark of UOP.

Some of these separations have been commercialized under tolling agreements at a large-scale Sorbex plant that UOP operates in Shreveport, Louisiana.

The general principles of Sorbex technology are described in this chapter. Specific details on some of the Sorbex applications may be found in Chaps. 2.6 and 10.7.

PRINCIPLES OF ADSORPTIVE SEPARATION

Adsorbents can be visualized as porous solids. When the adsorbent is immersed in a liquid mixture, the pores fill with liquid, but the equilibrium distribution of components inside the pore is different from the distribution in the surrounding bulk liquid. The component distributions inside and outside the pores can be related to one another by enrichment factors analogous to relative volatilities in distillation. The adsorbent is said to be selective for any components that are more concentrated inside the pores than in the surrounding bulk liquid.

Adsorption has long been used for the removal of contaminants present at low concentrations in process streams. In some instances, the objective is removal of specific compounds. In other cases, the objective is improvement of general properties, such as color, taste, odor, or storage stability. Common adsorbents are generally classified as polar or nonpolar. Polar, or hydrophilic, adsorbents include silica gel, activated alumina, molecular sieves, and various clays. Nonpolar adsorbents include activated carbons and other types of coal-derived carbons. Polar adsorbents are used when the components to be removed are more polar than the bulk process liquid; nonpolar adsorbents are used when the target components are less polar. Particularly useful are those adsorbents based on synthetic crystalline zeolites, which are generically referred to as *molecular sieves*. A wide variety of selectivities can be obtained in molecular sieves by varying the silica/alumina ratio, crystalline structure, and nature of the replaceable cations in the crystal lattice.

In one commercial separation, linear paraffins are separated from branched-chain and cyclic hydrocarbons by adsorption on 5A molecular sieves. The diameter of the pores is such that only the linear molecules may enter, and branched or cyclic molecules are completely excluded. In this case, the selectivity for linear hydrocarbons is infinite, and the adsorbent acts as a true molecular sieve. Adsorbents that completely exclude unwanted components are rare. In most applications, the pores are large enough to admit molecules of all the components present, and selectivity is the result of electronic interactions between the surface of the adsorbent pores and the individual components.

Adsorption is more efficient than conventional techniques such as liquid-liquid extraction or extractive distillation for many commercially important separations. Considerable development work has identified many adsorbents that are much more selective for specific components than any known solvents. In addition, adsorptive separation exhibits much higher mass-transfer efficiency than conventional extraction or extractive distillation. For example, laboratory chromatographs commonly achieve separation efficiencies equivalent to many thousands of theoretical equilibrium stages in columns of modest length. Such high mass-transfer efficiency stems from the use of small particles of adsorbent with high interfacial area and the absence of significant axial mixing.

In contrast, the trays of conventional liquid-liquid extractors and distillation columns are designed to obtain almost complete axial mixing in each physical stage. Thus, the number of theoretical equilibrium stages is essentially limited to the number of physical stages installed. In theory, this limitation can be partly overcome by the use of packed columns. However, if the packing is small enough to provide interfacial area comparable to that of an adsorbent, maintaining uniform countercurrent flow of the vapor and liquid phases becomes difficult. This flow limitation is less troublesome in an adsorptive system because only one fluid phase is involved.

THE SORBEX CONCEPT

In spite of the potential advantages of adsorptive separation, it did not achieve wide commercial acceptance until the introduction of the UOP Sorbex process in the late 1960s. Prior to the Sorbex process, adsorptive processes were designed much like laboratory chromatographs. Feed was introduced in pulses, and the composition of products varied with time. Integrating such an intermittent process with continuous processes operating both upstream and downstream was difficult. The Sorbex process, for the first time, offered a truly continuous adsorptive separation process that produced products with essentially constant compositions.

The easiest way to understand the Sorbex process is to think of it as a countercurrent flow of liquid feed and solid adsorbent (Fig. 10.3.1). For simplicity, assume the feed is a binary mixture of components A and B, and the adsorbent has a selective attraction for component A. In practice, the feed to a Sorbex unit may contain a multitude of components from which one or more components would be selectively recovered.

The positions of injection and withdrawal of the four net streams divide the adsorbent bed in four zones:

- *Zone 1: adsorption of component A.* This zone is between the point of feed injection and raffinate withdrawal. As the feed flows down through zone 1, countercurrent to the solid adsorbent flowing upward, component A is selectively adsorbed

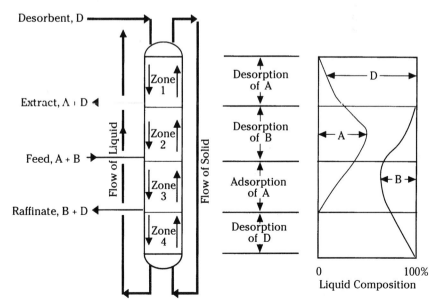

FIGURE 10.3.1 Moving-bed analogy.

from the feed into the pores of the adsorbent. At the same time, the desorbent (component D) is desorbed from the pores of the adsorbent to the liquid stream to make room for A in the pores.

- *Zone 2: desorption of component B.* This zone is between the point of feed injection and extract withdrawal. At the fresh-feed point, the upward-flowing solid adsorbent contains the quantity of component A that was adsorbed in zone 1. However, the pores will also contain a large amount of component B, because the adsorbent has just been in contact with fresh feed. The liquid entering the top of zone 2 contains no B, only A and D. Thus, B is gradually displaced from the pores by A and D as the adsorbent moves up through zone 2. At the top of zone 2, the pores of the adsorbent contain only A and D.
- *Zone 3: desorption of component A.* This zone is between the point of desorbent injection and extract withdrawal. The adsorbent entering zone 3 carries only components A and D. The liquid entering the top of the zone consists of pure D. As the liquid stream flows downward, component A in the pores is displaced by D. A portion of the liquid leaving the bottom of zone 3 is withdrawn as extract; the remainder flows downstream into zone 2 as reflux.
- *Zone 4: isolation zone.* The main purpose of zone 4 is to segregate the feed components in zone 1 from the extract in zone 3. At the top of zone 3, the adsorbent pores are completely filled with component D. The liquid entering the top of zone 4 consists of B and D. Properly regulating the flow rate of zone 4 prevents the flow of component B into zone 3 and avoids contamination of the extract.

The desorbent liquid must have a boiling point significantly different from those of the feed components. In addition, the desorbent must be capable of displacing the feed components from the pores of the adsorbent. Conversely, the feed components must be able to displace the desorbent from the adsorbent pores. Thus, the chosen desorbent must be able to compete with the feed components for any available active pore space in the solid adsorbent solely on the basis of concentration gradients.

DESCRIPTION OF THE PROCESS FLOW

In practice, actually moving a solid bed of adsorbent is difficult. The biggest problem in commercial-size units is ensuring uniform plug flow across large-diameter vessels while minimizing axial mixing. In the Sorbex process, the countercurrent flow of liquid feed and solid adsorbent is accomplished without physical movement of the solid. Instead, countercurrent flow is *simulated* by periodically changing the points of liquid injection and withdrawal along a stationary bed of solid adsorbent. In this *simulated moving-bed* technique, the concentration profile shown in Fig. 10.3.1 actually moves down the adsorbent chamber. As the concentration profile moves, the points of injection and withdrawal of the net streams to the adsorbent chamber are moved along with it.

A simplified flow diagram for a typical Sorbex unit is shown in Fig. 10.3.2. The separation takes place in the adsorbent chamber, which is divided into a number of adsorbent beds. Each bed of adsorbent is supported from below by a specialized grid that also contains a highly engineered flow distributor. Each flow distributor is connected to the rotary valve by a "bed line." The flow distributors between each adsorbent bed are used to inject or withdraw liquid from the chamber or to simply redistribute the liquid over the cross-sectional area of the adsorbent chamber. The number of adsorbent beds and bed lines vary with the Sorbex application.

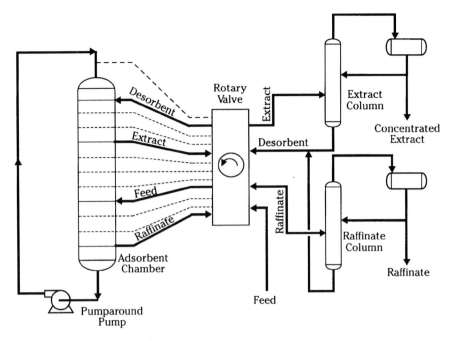

FIGURE 10.3.2 Sorbex flow diagram.

In the Sorbex process, four major streams are distributed to and from the adsorbent chamber by the rotary valve. These *net* streams include:

- *Feed in:* raw mixture of all feed components
- *Dilute extract out:* selectively adsorbed component or components diluted with desorbent
- *Dilute raffinate out:* rejected components diluted with desorbent
- *Desorbent in:* recycle desorbent from the fractionation section

At any given time, only four of the bed lines are actively carrying the net streams into and out of the adsorbent chamber. The movement of the net streams along the adsorbent chamber is effected by a unique rotary valve, specifically developed by UOP for the Sorbex process. Although, in principle, this switching action could be duplicated with a large number of separate on-off control valves, the UOP rotary valve simplifies the operation of the Sorbex unit and improves reliability.

Functionally, the adsorbent chamber has no top or bottom. A pumparound pump is used to circulate process liquid from the last adsorbent bed at the bottom of the adsorbent chamber to the first bed at the top of the chamber. The concentration profile in the adsorbent chamber moves smoothly down past the last bed, through the pump, and back into the first bed. The actual liquid flow rate within each of the four zones is different because the rate of addition or withdrawal of each net stream is different. As the concentration profile moves down the adsorbent chamber, the zones also move down the chamber. The overall liquid circulation rate is controlled by the pumparound pump. This pump operates at four different flow rates, depending on which zone is passing through the pump.

The dilute extract from the rotary valve is sent to the extract column for separation of the extract from the desorbent. The dilute raffinate from the rotary valve is sent to the raffinate column for separation of the raffinate from the desorbent. The desorbent from the bottom of both the extract and raffinate columns is recycled back to the adsorbent chamber through the rotary valve.

COMPARISON WITH FIXED-BED ADSORPTION

Comparing the characteristics of continuous Sorbex operation with the batch operation of conventional liquid chromatography is interesting. In a conventional chromatographic separation (Fig. 10.3.3), pulses of feed and desorbent are alternately charged to a fixed bed of adsorbent. Once again, assume that the feed is a binary mixture of components A and B. As the feed components move through the adsorbent bed, they gradually separate as the less strongly adsorbed component B moves faster than the more strongly adsorbed component A. A second pulse of feed must be delayed long enough to ensure that the fast-moving band of component B from the second pulse does not overtake the slow-moving band of component A from the first pulse.

A mathematical comparison of the Sorbex process with batch chromatography has shown that the batch operation requires three to four times more adsorbent inventory than the Sorbex process does and twice as much circulation of desorbent. This large difference in adsorbent requirement can be explained in physical terms without going into the details of the mathematical analysis. In the Sorbex process, every portion of the adsorbent bed is performing a useful function at all times. In batch chromatography, portions of the adsorbent bed at various times perform no useful function. This situation is most clearly seen near the entrance of the batch chromatograph. As feed enters the adsorbent bed, the adsorbent near the entrance rapidly comes to complete equilibrium with the feed. As feed continues to enter, this section of the adsorbent serves no purpose other than to convey feed further down into the bed. A similar situation occurs when desorbent is introduced. Other nonproductive zones exist within the

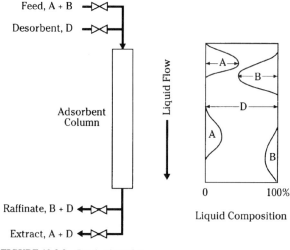

FIGURE 10.3.3 Batch adsorption.

adsorbent bed, between pulses of feed, where excess desorbent is required to keep the bands of component B from overtaking the bands of component A.

COMMERCIAL EXPERIENCE

Invented by UOP in the 1960s, the Sorbex technique was the first large-scale commercial application of continuous adsorptive separation. The first commercial Sorbex unit, a Molex unit for the separation of linear paraffins, came on-stream in 1964. The first commercial Parex unit came on-stream in 1971. UOP has licensed more than 100 Sorbex units throughout the world, including 59 Parex units, 1 MX Sorbex unit, 32 Molex units, 5 Olex units, 3 Sarex units, 1 Cresex unit, and 1 Cymex unit.

BIBLIOGRAPHY

Johnson, J. A. "Sorbex: Continuing Innovation in Liquid Phase Adsorption," Advanced Study Institute on Adsorption, Vimeiro, Portugal, July 1988.

Johnson, J. A., and A. R. Oroskar, "Sorbex Technology for Industrial-Scale Separations," International Symposium on Zeolites as Catalysts, Sorbents, and Detergent Builders, Wurzburg, Germany, September 1988.

Johnson, J. A. and H. A. Zinnen, "Sorbex: A Commercially Proven Route to High Purity Chemicals," *Proc. Royal Swedish Academy of Engineering Sciences Symposium,* Stockholm, March 1987.

Millard, M. T., J. A. Johnson, and R. G. Kabza, "Sorbex: A Versatile Tool for Novel Separations," UOP Technology Conferences, various locations, September 1988.

CHAPTER 10.4
UOP DEMEX PROCESS

E. J. Houde
UOP
Des Plaines, Illinois

INTRODUCTION

The Demex* process is a solvent extraction process developed jointly by UOP* and the Instituto Mexicano del Petroleo (IMP) for the processing of vacuum residue (VR) feedstocks. Developed as an extension of the widely used propane deasphalting technology, the process employs a unique combination of features to separate VR into components whose uses range from incremental feedstock for downstream conversion units to the production of lube basestock and asphalts. Because the Demex process provides the refiner with increased flexibility regarding future processing decisions, including crude selection, refinery debottlenecking, and the potential to reduce crude runs and fuel oil yields, it represents an important element in the refiner's overall bottom-of-the-barrel processing strategy.

PROCESS DESCRIPTION

The Demex process typically separates VR into two components: a relatively contaminant-free, nondistillable demetallized oil (DMO) and a highly viscous pitch. Like propane deasphalting, the Demex process is based on the ability of light paraffinic hydrocarbons to separate the residue's heavier asphaltenic components. Associated with these heavier materials are the majority of the crude's contaminants. Consequently, the lower contaminant content of the recovered DMO allows this material to be used in many refining applications, probably the most important of which is as incremental feedstock to catalytic processes such as fluid catalytic cracking (FCC) or hydrocracking for conversion into transportation fuel products.

Because the pitch recovered from the Demex unit contains most of the contaminants present in the crude, it typically has a high viscosity and a relatively low penetration value. Commercially, Demex pitch has been used in the manufacturing of asphalts and cement and as a blending component in refinery fuel oil pools. Other

---
*Trademark and/or service mark of UOP.

potential uses include the production of hydrogen, synthesis gas, or low-Btu fuel gas and as a solid-fuel blending component.

Unlike conventional propane deasphalting, the Demex process uses a unique combination of heavier solvents, supercritical solvent-recovery techniques, and patented extractor internals to efficiently recover high-quality DMO at high yield. A schematic flow scheme of a modern Demex design is shown in Fig. 10.4.1. This design, which has evolved from experience gained from both pilot plant and commercial operations as well as detailed engineering analyses of its various components, minimizes operating and capital costs and efficiently recovers the desired product yields at the required product qualities.

Incoming VR is mixed with solvent and fed to the vertical extractor vessel. At the appropriate extractor conditions, the VR-solvent blend is separated into its DMO and pitch components. The yield and quality of these components is dependent on the amount of contaminants in the feedstock, the composition and quantity of solvent used, and the operating conditions of the extractor.

Within the extractor, the downflowing asphaltene-rich pitch component and the upflowing DMO-solvent mixture are separated by patented extractor internals. The extractor design also includes a unique liquid flow distribution system to minimize the possibility of fouling the internals. Compared to previous designs, the increased separation efficiency achieved by these two features significantly reduces the size of the extractor vessel and the overall cost of the Demex unit.

The combination of heat exchange with recovered solvent and a direct-fired heater heats the DMO-solvent mixture leaving the top of the extractor to its critical temperature. The separation of the DMO and solvent components of this mixture is accomplished at supercritical conditions within the DMO separator. Recovered solvent is recycled back to the extractor. Because most of the solvent is recovered supercritically, this material can be effectively used for process heat exchange. Consequently, compared to earlier subcritical solvent-recovery designs, supercritical solvent recovery can reduce utilities requirements by more than one-third.

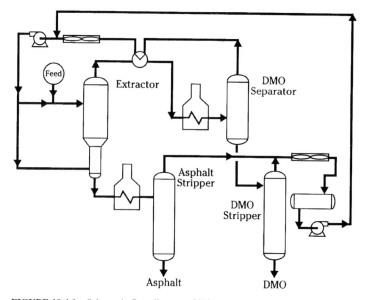

FIGURE 10.4.1 Schematic flow diagram of UOP Demex process.

To minimize solvent loss, any traces of solvent remaining in both the DMO exiting the DMO separator and the pitch from the extractor are recovered in the DMO and pitch strippers, respectively. This recovered solvent is also recycled back to the extractor. If the recovery of an intermediate-quality resin stream is desired—for instance, when specialty asphalts are produced or when independent control of DMO and pitch quality is desired—a resin settler may be added between the unit's extractor and DMO separator.

PRODUCT YIELDS AND QUALITY

The VR fraction of a crude is the usual feedstock for the Demex process. Typical properties of both the vacuum gas oil (VGO) and VR fractions of two common Middle Eastern crudes are presented in Table 10.4.1. As this table illustrates, the VR fraction contains virtually all of the crude's asphaltenic (C_7 insolubles) and organometallic (V+Ni) contaminants and most of the crude's Conradson carbon residue. Each of these contaminants can significantly influence the choice of processing conditions and catalysts used in fixed-bed processing units.

The Demex process can be used to selectively reject the majority of these contaminants. Examples of DMO properties obtained at various extraction levels when processing the two Arabian-based VRs described in Table 10.4.1 are summarized in Tables 10.4.2 and 10.4.3. The selectivity of the process for contaminant rejection is illustrated by the absence of asphaltenes and the significantly reduced amounts of organometallics and Conradson carbon in the recovered DMO. These tables also illustrate that DMO quality decreases with increasing DMO yield. For the Arabian Light case, this decrease results in a variation in demetallization ranging from roughly 98 percent organometallic rejection at 40 percent DMO yield to approximately 80 percent

TABLE 10.4.1 Feedstock Properties

Feedstock	Reduced crude	VGO	Vacuum residue
	Arabian Light:		
Cutpoint, °C (°F)	343+ (650+)	343–566 (650–1050)	566+ (1050+)
Crude, LV %	38.8	26.3	12.5
Specific gravity	0.9535	0.9206	1.0224
Sulfur, wt %	3.0	2.48	4.0
Nitrogen, wt %	0.16	0.08	0.31
Conradson carbon residue, wt %	8.2	0.64	20.8
Metals (V + Ni), wt ppm	34	0	98
UOP K factor	11.7	11.8	11.4
C_7 insolubles, wt %	3.5	0	10
	Arabian Heavy:		
Cutpoint, °C (°F)	343+ (650+)	343–566 (650–1050)	565+ (1050+)
Crude, LV %	53.8	30.6	23.2
Specific gravity	0.9816	0.9283	1.052
Sulfur, wt %	4.34	2.92	6.0
Nitrogen, wt%	0.27	0.09	0.48
Conradson carbon residue, wt %	13.3	0.99	27.7
Metals (V + Ni), wt ppm	125	0	269
UOP K factor	11.5	11.7	11.3
C_7 insolubles, wt %	6.9	0	15

TABLE 10.4.2 DMO Properties of Arabian Light

	DMO yield, LV % of vacuum bottoms		
	40	60	78
Specific gravity	0.9406	0.9638	0.9861
Sulfur, wt %	2.34	2.83	3.25
Nitrogen, wt %	0.1	0.15	0.21
Metals (V + Ni), wt ppm	3	7	19
Conradson carbon residue, wt %	2.85	6.36	10.7
C_7 insolubles, wt %	—	—	0.05
UOP K factor	11.9	11.7	11.6

TABLE 10.4.3 DMO Properties of Arabian Heavy

	DMO yield, LV % of vacuum bottoms	
	30	55
Specific gravity	0.9576	0.9861
Sulfur, wt %	3.53	4.29
Nitrogen, wt %	0.14	0.2
Metals (V + Ni), wt ppm	16	38
Conradson carbon residue, wt %	4.79	10.1
C_7 insolubles, wt %	—	<0.05
UOP K factor	12.0	11.8

rejection at 78 percent DMO yield. The same deterioration in DMO quality with increasing DMO yield is observed for the Arabian Heavy feed case.

Estimated properties of the Demex pitches recovered from the two Arabian feedstock cases are presented in Tables 10.4.4 and 10.4.5. At the higher DMO recovery rates, these materials have zero penetration and can be blended with softer VRs to produce acceptable penetration-grade asphalts.

PROCESS VARIABLES

Several process variables affect the yield and quality of the various Demex products. These variables include extraction pressure and temperature, solvent composition, and extraction efficiency.

TABLE 10.4.4 Pitch Properties of Arabian Light

	Demex extraction level, LV % of vacuum residue		
	40	60	78
Yield, LV % of reduced crude	19.3	12.9	7.0
Specific gravity	1.0769	1.11	1.154
Sulfur, wt %	4.96	5.52	6.31
Metals (V + Ni), wt ppm	154	216	341
Softening point, °C (°F)	88 (190)	102 (215)	177 (368)

TABLE 10.4.5 Pitch Properties of Arabian Heavy

	Demex extraction level, LV % of vacuum residue	
	30	55
Yield, LV % of reduced crude	30.2	19.4
Specific gravity	1.0925	1.1328
Sulfur, wt %	6.93	7.82
Metals (V + Ni), wt ppm	364	515
Softening point, °C (°F)	104 (219)	149 (300)

Extraction Pressure and Temperature

Extraction pressure, which is chosen to ensure that the Demex extractor's solvent-residue mixture is maintained in a liquid state, is related to the critical pressure of the solvent used. During normal operation, when both the extraction pressure and solvent composition are fixed, Demex product yields and qualities are controlled by adjusting the extractor temperature. This adjustment is achieved by varying the temperature of the recycled solvent stream. Increasing the temperature of this stream reduces the solubility of the residue's heavier components and improves DMO quality at the expense of reduced DMO yield. Extraction temperature must be maintained below the critical temperature of the solvent, however, because at higher temperatures no portion of the residue is soluble in the solvent and no separation occurs.

Solvent Composition

Solvents typically used in the Demex process include components such as propane, butanes and pentanes, and various mixtures of these components. Because these materials are generally readily available within a refinery, their use is relatively inexpensive. In addition, because the majority of the solvent is recirculated within the unit, solvent makeup rates are relatively small.

Increasing the solvent's molecular weight increases the yield of recovered DMO by allowing more of the heavier, more-resinous components of the feedstock to remain in the DMO. At the same time, however, the quality of the DMO decreases because these heavier materials have higher contaminant levels. Consequently, proper solvent selection involves balancing increased product yield and decreased product quality. Generally, light solvents, such as propane, are specified when the highest DMO quality is desired. However, light solvents typically produce low DMO yields. Intermediate solvents, such as butanes, are used when a reasonably high yield of high-quality DMO is desired. Finally, heavier solvents, such as pentanes, are used when the maximum yield of DMO is desired, for instance, when the DMO is to be hydrotreated before further processing.

Extraction Efficiency

The separation efficiency of the DMO and pitch products is significantly influenced by the amount of solvent that is mixed with the incoming feed to the Demex extractor. Increasing the amount of solvent improves the separation and produces a higher-quality DMO.

Figure 10.4.2 illustrates the impact of solvent rate on DMO quality. In this example, a DMO containing 40 wt ppm organometallics is recovered at a 3:1 solvent-to-oil

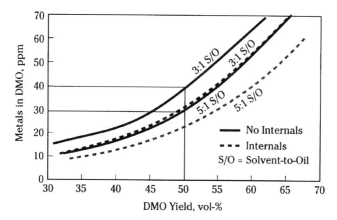

FIGURE 10.4.2 Effect of solvent rate and extractor internals.

(S:O) ratio for 50 vol % DMO yield. When the same feedstock is processed at a higher 5:1 S:O, the organometallic content of the DMO recovered at the same 50 vol % DMO yield is reduced to 30 wt ppm.

Unfortunately, because the quantity of solvent recirculated within the unit is significantly greater than the amount of feedstock being processed, the improved DMO quality achievable at higher solvent rates must be balanced against the additional operating costs associated with the higher solvent recirculation and solvent-recovery requirements and the increased capital costs associated with the larger equipment sizes. In Fig. 10.4.2, the improvement in DMO quality must be balanced against the roughly 50 percent higher operating and capital costs associated with the higher solvent recirculation rate.

The addition of patented Demex extractor internals, however, modifies the relationship between DMO yield and DMO quality by improving the extractor's separation efficiency. As shown in Fig. 10.4.2, the internals may be used to offset higher solvent recirculation rates by allowing either higher-quality DMO to be recovered at the same DMO yield or, conversely, more DMO to be recovered at the same DMO quality. Also, the additional operating and capital costs associated with higher solvent-recirculation rates are eliminated when the intervals are employed.

DMO PROCESSING

Because the most common application of the Demex process involves recovering additional feedstock for catalytic processes such as FCC or hydrocracking, the amount of DMO recovered in the Demex unit can have a significant impact on the quantity and quality of the feedstock used in the conversion unit. Figures 10.4.3 and 10.4.4 summarize the Conradson carbon and organometallic contents of the VGO-DMO blends produced at various DMO recovery rates when processing the Arabian Light and Arabian Heavy feedstocks, respectively.

Figure 10.4.3 indicates that processing the Arabian Light feedstock at DMO recovery rates as high as 78 percent produces VGO-DMO blends with contaminant levels within typical FCC and hydrocracking feedstock specifications. Consequently, the inclusion of the Demex unit increased the amount of feedstock used by the conversion

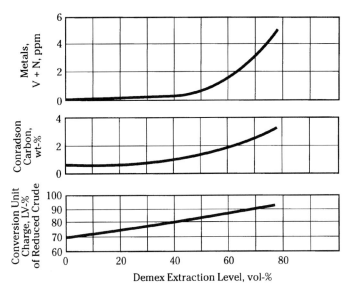

FIGURE 10.4.3 VGO-DMO–blend quality (Arabian Light case).

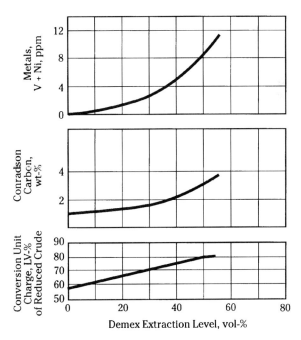

FIGURE 10.4.4 VGO-DMO–blend quality (Arabian Heavy case).

unit by approximately 35 percent. Figure 10.4.4 indicates that a similar percentage increase in conversion unit feedstock is obtained from the Arabian Heavy feedstock when producing a comparable VGO-DMO quality. Because of the higher contaminant content of the Arabian Heavy crude, however, this VGO-DMO quality limit is reached at a lower DMO recovery rate. Thus, hydrotreating DMO recovered from highly contaminated crudes may be an economically feasible bottom-of-the-barrel processing strategy.

PROCESS ECONOMICS

The estimated battery-limits cost for a nominal 20,000 barrels per stream day (BPSD) two-product Demex unit constructed to UOP standards at a U.S. Gulf Coast location is approximately $20 million. Sample utility consumptions for this unit are summarized in Table 10.4.6.

DEMEX PROCESS STATUS

UOP's first Demex unit was licensed in 1977. As of mid-1995, a total of 13 Demex units with capacities ranging from 500 to 42,000 BPSD have been licensed. This number represents a combined licensed capacity of roughly 260,000 BPSD. Approximately 150,000 BPSD of this capacity (nine units) has been placed in commercial service. These commercial applications have ranged from the recovery of incremental feedstock for downstream FCC and hydrocracking units to the production of road asphalt (or solid pitch) and have included both two-product and three-product Demex process configurations.

TABLE 10.4.6 Typical Demex Utilities*

Electric power, kW	1100
Fuel fired, 10^6 Btu/h	33
Steam, MT/h (10^3 lb/h):	
High pressure	31.7 (70)
Medium pressure	3.6 (8)
Low pressure	0.5 (1)
Cooling water, m^3/h (10^3 gal/h)	1,060 (280)

*Basis: 20,000-BPSD capacity.
Note: MT/h = metric tons per hour.

CHAPTER 10.5
UOP ISOSIV PROCESS

Nelson A. Cusher
UOP
Des Plaines, Illinois

INTRODUCTION

Light straight-run (LSR) naphtha fractions made in the refinery are predominantly C_5's through C_7's, with traces of C_8's. They are highly paraffinic and contain moderate amounts of naphthenes, low aromatics, and no olefins. The average clear research octane number (RONC) is usually in the 60s.

The paraffinicity of light naphtha is what makes it a desirable petrochemical cracking stock. The aromatic rings are too thermally stable for cracking, and the naphthenes produce more liquid products. The straight-chain normal paraffins produce more ethylene and less pyrolysis gasoline than the branched-chain paraffins.

Figure 10.5.1 compares pyrolysis unit yields from a normal paraffin feed with yields from a mixed natural gasoline feed. The yields are based on a single-pass pyrolysis operation at equivalent high furnace severities for both feeds. The normal paraffin feed was extracted from a C_5 through C_9 natural gasoline stream. The natural gasoline feed contained 54.4 percent straight-chain paraffins and 45.6 percent branched and cyclic hydrocarbons. The ethylene yield is about 30 percent higher for the all-normals fractions. Propylene, butene, and light-gas yields decrease slightly. The pyrolysis gasoline yield is considerably reduced.

As the endpoint of naphtha is decreased, the paraffinicity of the stream increases; as a result, ethylene production increases and the production of pyrolysis gasoline and fuel oil decreases. The LSR naphtha—especially the 70°C (C_5–160°F) portion, which is about 95 percent paraffinic—is therefore a prime substitute for natural gas liquids as an ethylene plant feed. The nonnormal components of the LSR naphtha fraction have higher octanes than the normal paraffins (Table 10.5.1) and are excellent gasoline blending components.

The UOP* IsoSiv* process uses molecular sieves to physically remove normal paraffins from the LSR feedstock. In the past, gasoline-range IsoSiv units were primarily used to produce specialty chemicals. The normal paraffin product having a 95 to 98 percent purity was cut into single-carbon-number fractions for special solvents. The normal-paraffin–free fraction was usually sent to the gasoline pool as an octane

---

*Trademark and/or service mark of UOP.

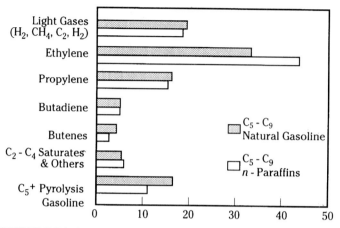

FIGURE 10.5.1 Pyrolysis yield data.

TABLE 10.5.1 Properties of Common Gasoline Components

Component	Molecular weight	Boiling point, °C (°F)	Density, kg/m^3 (lb/gal)	RONC
Isobutane	58.1	−11.7 (10.9)	562 (4.69)	100+
n-butane	58.1	−0.5 (31.1)	582 (4.86)	93.6
Neopentane	72.1	9.4 (49.0)	596 (4.97)	116
Isopentane	72.1	27.9 (82.2)	623 (5.20)	92.3
n-pentane	72.1	36.1 (96.9)	629 (5.25)	61.7
Cyclopentane	70.0	49.3 (120.7)	749 (6.25)	100
2,2-dimethlybutane	86.2	49.7 (121.5)	664 (5.54)	91.8
2,3-dimethylbutane	86.2	58.0 (136.4)	664 (5.54)	101.7
2-methylpentane	86.2	60.3 (140.5)	667 (5.57)	73.4
3-methylpentane	86.2	66.3 (145.9)	652 (5.44)	74.5
n-hexane	86.2	68.7 (155.7)	657 (5.48)	24.8
Methylcyclopentane	84.2	71.8 (161.3)	753 (6.28)	91.3
2,2-dimethylpentane	100.2	79.2 (174.6)	676 (5.64)	92.8
Benzene	78.1	80.1 (176.2)	882 (7.36)	100+
2,4-dimethylpentane	100.2	80.5 (176.9)	676 (5.64)	83.1
Cyclohexane	84.2	80.7 (177.3)	782 (6.53)	83
2,2,3-trimethylbutane	100.2	80.9 (177.6)	693 (5.78)	112
3,3-dimethylpentane	100.2	86.1 (186.9)	696 (5.81)	98
2,3-dimethylpentane	100.2	89.8 (193.6)	699 (5.83)	88.5
2,4-dimethylpentane	100.2	90.1 (194.1)	681 (5.68)	55
3-methylhexane	100.2	91.9 (197.5)	690 (5.76)	65
Toluene	92.1	110.6 (231.1)	870 (7.26)	100+
Ethylbenzene	106.2	136.2 (277.1)	870 (7.26)	100+
Cumene	120.2	152.4 (306.3)	864 (7.21)	100+
1-methyl-2-ethylbenzene	120.2	165.1 (329.2)	881 (7.35)	100+
n-decane	142.3	174.0 (345.2)	732 (6.11)	−53

booster. The more recent IsoSiv units were built to produce high-octane gasoline components; the normal paraffin by-product was sold as petrochemical feedstock or sent to an isomerization reactor.

GENERAL PROCESS DESCRIPTION

The LSR naphtha fractions usually contain 40 to 50 percent normal paraffins. The IsoSiv process (Fig. 10.5.2) separates the normal paraffins from a hydrocarbon mixture by selective adsorption on a molecular sieve material. This material is a crystalline zeolite having uniform pore dimensions of the same order of magnitude as the size of individual hydrocarbon molecules. The molecular sieve used for normal paraffin separation has pore openings in the crystalline structure that are sized to allow molecules of normal paraffin to pass through the pore openings into the internal crystal cavity, where they are retained. Nonnormal hydrocarbons, such as isoparaffins, naphthenes, and aromatics, have larger molecular diameters and are, therefore, excluded from entering the crystal cavity through the pore opening.

The heart of the IsoSiv process is the adsorber section, which consists of vessels filled with molecular sieve adsorbent. The LSR feedstock is fed into one end of an adsorber vessel. The normal paraffins in the feedstock remain in the vessel by being adsorbed into the molecular sieve, and the remainder of the feedstock passes out the other end of the vessel as a nonnormal product. In a subsequent process step, the normal paraffins are recovered from the adsorber vessel as a separate product by use of a purge material. All process hardware in an IsoSiv unit is conventional refinery equipment, such as pumps, furnaces, heat exchangers, and compressors, that is designed to deliver the feedstock and the purge material to the adsorber section.

Typical performance (Table 10.5.2) results in an isomer product that is 98 to 99 percent free of normal paraffins and a normal paraffin product of 95 to 98 percent purity. The high-octane isomer product can have a RON approximately 15 numbers higher than the feed, depending on feed composition. The IsoSiv-grade molecular sieve adsorbent is fully regenerable and has an expected life of 10 to 15 years.

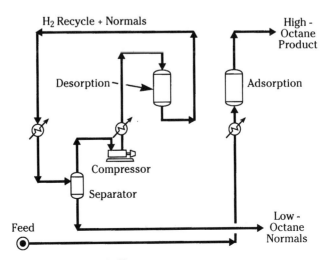

FIGURE 10.5.2 UOP IsoSiv process.

TABLE 10.5.2 Typical Performance

Isomer product purity	98–99%
Isomer research octane	~15 higher than feed RONC
Normal-paraffin product purity	95–98%
Adsorbent expected life	10–15 years

PROCESS PERSPECTIVE

The UOP IsoSiv process gained early acceptance and has maintained a leading position to the present day. The technology of normal paraffin separation by adsorption had its start in the late 1950s in the separation of normal paraffins from gasoline for octane improvement purposes. The first commercial application was an IsoSiv unit installed by the South Hampton Company of Silsbee, Texas. In 1995, more than 45 IsoSiv units were operating as stand-alone units or as part of UOP's TIP* technology in the United States, Australia, Europe, Asia, and South America. These units range in size from 1000 to 35,000 barrels per stream day (BPSD) of feed capacity.

DETAILED PROCESS DESCRIPTION

The naphtha-range IsoSiv process makes use of the highly selective adsorption capability of a molecular sieve. The process is run at a constant, elevated temperature and pressure.

Vapor-phase operation is used to provide straightforward processing. Continuous processing is accomplished through cyclic operation that uses valves actuated by standard, fully automatic sequencing controls to switch adsorption beds. A steady flow of feed and products and constant product purity are maintained. All operating conditions are within the temperature and pressure ranges common to refinery and petrochemical operation. The basic IsoSiv cycle consists of an adsorption and a desorption step.

Adsorption

The feed stream is pumped through the heat exchanger, where it is heated by the non-normal product, and then passes through a feed heater to an adsorption bed. It is then passed upward through one adsorber vessel, where the normal paraffins are selectively adsorbed in the bed. As the normal paraffins are adsorbed, the liberated heat of adsorption creates a temperature front that travels through the bed. This front closely coincides with the mass-transfer front and gives an indication of when the adsorption step should be terminated to prevent the normal paraffins from breaking through the effluent end of the bed. This temperature front is used in the field to set the cycle timer to prevent the front from reaching the bed exit. The unadsorbed isomers and cyclic hydrocarbons that pass through the beds are heat-exchanged against the feed stream to recover heat. This stream is then cooled and condensed, and the high-octane

---

*Trademark and/or service mark of UOP.

liquid is taken as product. The uncondensed vapors are reused as part of the nonadsorbable purge.

Desorption

After the adsorption step, the beds are countercurrently purged with a nonadsorbable medium. This countercurrent purging desorbs the normal paraffins and sweeps these desorbed vapors from the bed, thus maintaining the average partial pressure of the desorbate below the value in equilibrium with the loading on the bed. The continuous removal of the desorbate vapor and the simultaneous transfer of the absorbed phase to the purge gas in an attempt to establish equilibrium drive the desorption stage to completion. A complete removal of the normal paraffin adsorbate is not achieved on each desorption. An economic balance between the bed size, as determined by the fraction of normal adsorbate removed (delta loading), and the purge required determines the degree of normals removed. This stream is then cooled and condensed, and the liquid is taken as normal product. The uncondensed vapors are reused as part of the purge medium.

The elevated temperatures used for vapor-phase adsorption can cause a gradual formation of coke on the beds. To remove any accumulation, a burn-off procedure is incorporated to reactivate the adsorbent at required times. This burn-off capability provides a built-in safeguard against permanent loss of bed capacity as a result of operating upsets. An in situ regeneration procedure is used to burn off the coke deposits and restore full adsorbent capacity.

PRODUCT AND BY-PRODUCT SPECIFICATIONS

The normal product purity is typically 95 to 98 percent. The purity of the isomer product is typically 98 to 99 percent. The high-octane isomer product can have a RON approximately 15 numbers higher than the feed, depending on feed composition.

WASTE AND EMISSIONS

No waste streams or emissions are created by the IsoSiv process. Isomer and normal products are usually stabilized, however. The result is a liquefied petroleum gas product ($C_3 + C_4$, rich in isobutane) and a stabilizer vent ($H_2 + C_1 + C_2$).

PROCESS ECONOMICS

Many factors influence the cost of separating isoparaffins and normal paraffins. These factors include feedstock composition, product purity, and the capacity and location of the unit. Location affects costs of labor, utilities, storage, and transportation. With this in mind, Table 10.5.3 presents investment and utility requirements.

In summary, commercially proven large-scale production technology is available today for the economic production of high-quality isoparaffins and normal paraffins.

TABLE 10.5.3 UOP IsoSiv Process Economics and Performance*

Investment, $/BPSD of normal paraffins in feed:	
Erected capital cost	1900–2500
Adsorbent inventory	205
Utilities, per BPSD:	
Fuel consumed at 90% efficiency, million kcal/h (million Btu/h) per BPSD of total feed	0.0006 (0.0022)
Water at 17°C (31°F) rise, m³/day BPSD (gal/min) per BPSD of normal paraffins in feed	0.82 (0.15)
Power, kWh per BPSD of normal paraffins in feed	0.40
Hydrogen makeup at 70% H_2 purity (solution loss), m³/day (SCF/h) per BPSD of total feed	0.75 (1.1)

*Basis: Battery-limits Gulf Coast location, 1995, excluding product stabilization. Normal-paraffin feed rates of 3000 to 8000 BPSD.

Note: SCF = standard cubic feet.

CHAPTER 10.6
KEROSENE ISOSIV PROCESS FOR PRODUCTION OF NORMAL PARAFFINS

Stephen W. Sohn

UOP
Des Plaines, Illinois

The straight-chain normal paraffins in the kerosene range (C_{12} to C_{18}) have their principal uses in detergent manufacture, chlorinated fire retardants, plasticizers, alcohols, fatty acids, and synthetic proteins. The separation of these straight-chain normal paraffins from other classes of hydrocarbons, such as branched-chain isoparaffins, naphthenes, and aromatics, was a virtual impossibility prior to the advent of the synthetic zeolites known as *molecular sieves*. These uniform, molecular-pore-sized adsorbents, developed by Union Carbide in the early 1950s, opened the way for refiners and petrochemical producers to add adsorption as a means of separating hydrocarbon classes to those already known, such as distillation and liquid-liquid extraction. To date, eight kerosene IsoSiv systems have been started up (Table 10.6.1). The IsoSiv* process is licensed by UOP* subsequent to the joint venture ownership of UOP by Union Carbide and AlliedSignal in 1988.

*Trademark and/or service mark of UOP.

TABLE 10.6.1 Kerosene IsoSiv Commercial Applications

Unit	Feed type	Start-up	Location	Normal paraffin capacity, BPSD
1	Kerosene	1964	United States	2300
2	Kerosene	1971	West Germany	650
3	Kerosene–gas oil	1972	Italy	2600
4	Kerosene	1973	Italy	2600
5	Kerosene–gas oil	1974	Italy	5800
6	Kerosene–gas oil	1976	Italy	5800
7	Kerosene	1983	Brazil	2600
8	Kerosene	1992	People's Republic of China	950

Note: BPSD = barrels per stream day.

GENERAL PROCESS DESCRIPTION

The IsoSiv process separates normal paraffins from a hydrocarbon mixture, such as kerosene or gas oil, by selective adsorption on a molecular-sieve adsorbent material. This material is a crystalline zeolite having uniform pore dimensions of the same order of magnitude as the size of individual hydrocarbon molecules. The molecular sieve used for normal paraffin separation has openings in the crystalline structure that are sized to allow normal paraffin molecules to pass through the pore openings into the internal crystal cavity, where they are retained. Nonnormal hydrocarbons, such as isoparaffins, naphthenes, and aromatics, have larger molecular diameters and are therefore excluded from entering the crystal cavity through the pore opening.

The heart of the IsoSiv process is the adsorber section, consisting of vessels filled with molecular-sieve adsorbent. The kerosene or gas-oil feedstock is fed into one end of an adsorber vessel, the normal paraffins in the feedstock remain in the vessel by being adsorbed in the molecular sieve, and the remainder of the feedstock passes out the other end of the vessel as a denormalized kerosene gas oil. The normal paraffins are recovered from the adsorber vessel as a separate product by using a purged material. All process hardware in an IsoSiv unit is conventional refinery equipment, such as pumps, furnaces, heat exchangers, and compressors, designed to deliver the feedstock and the purge material to the adsorber section and to remove the products from the adsorber section. The kerosene IsoSiv process typically recovers 95 wt % of the normal paraffins in the feedstock and produces a normal paraffin product of 98.5 wt % purity.

PROCESS PERSPECTIVE

During the early 1960s, the appeal of molecular-sieve adsorption led to widespread efforts at innovating new adsorption technology. Many of these efforts were successful, in that they resulted in molecular-sieve processes capable of separating long-chain normal paraffins from kerosene-range feedstocks at just the time when the detergent industry decided to switch to linear alkylbenzene sulfonates as a basis for its formulations of "soft" detergents. The consequent demand for long-chain normal paraffins led to a worldwide wave of construction: at least 12 adsorption plants were built to process kerosene-range feedstocks and use processes developed by Union Carbide, UOP, Esso, British Petroleum, Shell, and Texaco. Among the first units was the South Hampton Company's naphtha IsoSiv unit, which was converted to the kerosene range in 1961. In 1964 Union Carbide Corporation installed at its Texas City, Texas, petrochemical complex an IsoSiv unit producing 100,000 metric tons/year (MTA) (220,000 lb/year) of normal paraffins from kerosene. This unit was to remain the world's largest normal paraffin-producing plant for almost 10 years.

At the beginning of the 1970s, a further extension of adsorption technology was required. The normal paraffins used as substrates for protein production extend into the gas oil feedstock range. Suitable modifications can and have been made to existing adsorption technology to allow successful application to the new requirements. In 1972, Liquichimica S.p.A., now Condea Augusta S.p.A. but then a subsidiary of the Liquigas Group of Italy, installed and started up in Augusta, Sicily, a modified IsoSiv unit to produce 110,000 MTA (242,000 lb/year) of normal paraffins from both kerosene and gas-oil feedstocks. Plant expansions put on-stream in 1973 brought normal paraffin production capacity at Augusta up to approximately 250,000 MTA (551,000 lb/year), making it by a wide margin the largest single normal paraffin–producing installation in the world. A second unit that came on-stream in December 1974 almost doubled previous capacity. A third IsoSiv unit of more than 200,000 MTA (440,000 lb/year) came on-stream in 1976. Total installed capacity is more than 650,000 MTA (4,862,000 lb/year) of normal paraffin production. These units have used feedstocks ranging from kerosene to gas oil and intermediate mixtures of both.

A seventh kerosene IsoSiv unit came on-line in Brazil in 1983. An eighth came on-line in China in 1992.

DETAILED PROCESS DESCRIPTION

The kerosene IsoSiv process employs the highly selective adsorption capability of molecular sieves. The simplified process flow scheme is shown in Fig. 10.6.1. The basic cycle consists of three steps: adsorption, copurge, and desorption. This section describes each in detail.

Adsorption Step

Hydrocarbon feed at elevated temperature and slightly above atmospheric pressure is passed upward through an adsorber vessel, where the normal paraffins are selectively adsorbed in the bed. In processing gas oil feedstock, hexane is added to the gas oil feed to dilute it and prevent capillary condensation from occurring on the adsorbent bed. As the normal paraffins are adsorbed, the liberated heat of adsorption creates a temperature front that travels through the bed. This front closely coincides with the mass-transfer front and gives an indication of when the adsorption step should be terminated to prevent the normal paraffins from breaking through the effluent end of the

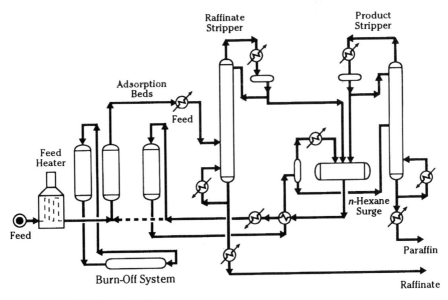

FIGURE 10.6.1 Kerosene IsoSiv process.

bed. This temperature front is used in pilot plant work to determine optimum design conditions and can be employed in commercial units to set the cycle timer to prevent the front from reaching the bed exit. The unadsorbed isomer and cyclic hydrocarbons and some purge hexane that pass through the beds combine with the copurge effluent and are heat-exchanged against the feed stream to recover heat. This stream is then sent to a distillation system, where the hexane purge material is recovered as a distillate product and the heavier isomers are taken as bottoms products.

Copurge Step

After the adsorption step, the normal paraffin–loaded beds are purged in the cocurrent direction with just enough vaporized hexane to displace the nonadsorbed feed and isomeric hydrocarbons from the void spaces in the adsorber vessel. This step is important especially in the production of protein substrates because it ensures that a high-purity product will be recovered from the desorption step. The effluent from the cocurrent purge step is combined with the adsorption effluent stream, as mentioned previously.

Desorption-Purge Step

After the copurge step, the beds are purged countercurrently with hexane. This countercurrent purging desorbs the normal paraffins and sweeps these desorbed vapors from the bed, thus maintaining the average desorbate partial pressure below the value in equilibrium with the loading on the bed. The continuous removal of the desorbate vapor and the simultaneous transfer of the adsorbed phase to the purge gas in an attempt to establish equilibrium drives the desorption toward completion. In addition to this stripping effect, the normal hexane itself becomes adsorbed on the bed and helps displace the heavier normal paraffin desorbate. A complete removal of the heavy normal paraffin adsorbate is not achieved on each desorption. An economic balance between the bed size, as determined by the fraction of heavy normal adsorbate removed (or *delta loading*), and the hexane purge required determines the degree of removal of the heavy normals obtained. As the purge quantity is decreased, the delta loading is decreased; and larger adsorbers are required for a given hydrocarbon feed throughput and cycle time. This decreased delta loading increases the rate of adsorbent deactivation and consequently the required burn-off frequency because the higher residual loading increases the rate of coke formation. Conversely, increasing the purge quantity increases delta loading until the hexane-handling equipment and operating costs become significant factors.

The desorption effluent containing heavy normal paraffins and hexane is partially condensed by heat exchange with the cold hexane purge. The vapor fraction and the condensate are transferred to the normal dehexanizer system, where the normal paraffins are separated from the hexane by standard fractionating techniques. The normal paraffin product from the bottom of the column is cooled and removed from the process. This separation is relatively easy because of the wide difference in boiling point between hexane and the lightest heavy normal paraffin. The recovered hexane from this column is also condensed and circulated back to the hexane accumulator without fractionation. Small additions of fresh hexane are required to make up losses of hexane carried out in both product streams.

The foregoing operation sequence is integrated into continuous processing by the cyclic use of several adsorber vessels. Automatic valves are operated by a sequencing control system. The flow of both feed and products is uninterrupted. Suitable interlocks and alarms are provided so that the plant can operate with a minimum of operator attendance.

Oxidative-Regeneration Description

As the adsorber beds are cycled at the elevated operating temperatures, a carbonaceous deposit gradually accumulates. This deposit reduces the capacity of the adsorbent, and this reduction ultimately results in a breakthrough of normal paraffins into the isomer product stream and decreased normal paraffin recovery. The rate at which this deposit accumulates depends on factors such as temperature, feed impurities, feed properties, cycle time, and residual paraffin loadings. This type of adsorbent deactivation is not permanent, and the original bed capacity can be restored by burning off this deposit under controlled conditions. For a kerosene-type feedstock, a bed can be cycled for 15 to 30 days before oxidative regeneration is necessary. For a gas oil feedstock, the period is reduced to about 6 to 10 days.

When a bed has been cycled to the point at which oxidative regeneration is required, it is removed from the processing operation, and another adsorber vessel is put into operation. This change is made without any interruption in the cycling sequence. The coked bed is removed from cycling after the desorption step and is given an additional long desorption purge to remove as much of the residual normal paraffins as possible. The bed is then completely isolated from the cycling system, and a downflow circulation of nitrogen is pumped by means of a compressor or blower and then passed through a heater to the adsorber vessel. The circulation of hot nitrogen has two purposes: to purge the hexane from the bed and to raise the temperature of the bed to above the coke-ignition point prior to the introduction of oxygen into the system. The effluent gas from the bed is cooled to condense the hydrocarbons and water that desorb.

When the bed is up to temperature, air is introduced into the circulating stream at a controlled rate. The oxygen in the gas combusts with coke in the top of the bed. The heat released from combustion is carried out of the burning zone as a preheat front traveling ahead of the burning front. This preheat front raises the bed temperature even further. This temperature is controlled by regulating the amount of oxygen in the entering gas. Because excessive internal adsorbent temperatures permanently destroy the molecular-sieve crystal, the gas-phase temperature is critical. As the burning front passes through the bed, the temperature drops back to the gas inlet temperature. Because the coke deposit contains hydrogen, water is formed during combustion in addition to carbon oxides. This water must be removed from the system because the molecular-sieve crystal is permanently damaged by repeated exposure to water at high temperatures. To minimize this damage, a dryer is used to prevent the water from accumulating. The proper design of the regeneration process and the rugged nature of the molecular sieve ensure that the adsorbent has a long operating life.

After the regeneration is complete, the bed is cooled down to the process operating temperature and purged of any remaining oxygen by circulating nitrogen. The bed is now ready to go on-stream to replace one of the adsorbers in use so that it in turn can be reactivated.

WASTE AND EMISSIONS

During normal operation of the kerosene IsoSiv unit, the vent gas is not expected to contain more than 5,000 vol ppm of total sulfur on the average. The maximum peak sulfur level in the vent gas stream is not expected to exceed 5 vol % when the unit is operating with feedstocks containing up to 500 wt ppm total sulfur. A second vent stream contains approximately 1000 vol ppm of sulfur during the burn-off of an adsorber bed. The peak concentration is not expected to exceed 5 percent. This vent will also contain approximately 2 vol % carbon monoxide.

Proper handling of these vent gas streams depends on many factors. One suggested method of handling these streams is to feed them to the hexane-heater firebox, provided acceptable stack sulfur levels can still be maintained.

ECONOMICS

Many factors affect the cost of extracting normal paraffins. They include the nature of the feedstock from which the normal paraffins are to be extracted, the specifications of the product normal paraffins, the production capacity or size of the plant, and the location. The last factor includes such items as climatic conditions and availability and cost of labor, utilities, storage, and transportation.

The feedstock is of primary importance. The normal paraffin content of gas oils ranges from 10 to 40 percent, depending on the crude oil source. The higher the normal paraffin content, the more amenable it is for normal paraffin processing. Refiners also find this feed the least attractive for fuel oil or diesel fuel because of its high pour or freeze points. Extracting the normal paraffins reduces the freeze point considerably, thus making the isomer product more salable. Impurities such as the amount of sulfur must also be considered.

Normal paraffin specifications as required by the selected fermentation process are also important. The hydrocarbon range, normal paraffin content, and types of impurities bear directly on whether prefractionation of the feedstock before normal paraffin extraction or postfractionation after extraction is required and on whether and to what degree some form of posttreatment is required to remove trace sulfur and aromatic compounds. The IsoSiv process produces normal paraffins at 98.5 wt % purity.

Plant size is important because large plants tend to be more economical. For normal paraffin extraction, plants producing less than 100,000 MTA (220,000 lb/year) are considered to be relatively small from an economic point of view. However, plants with capacities larger than 500,000 MTA (1,102,000 lb/year) offer little economic incentive. Location is also important.

All these economic considerations, plus an uncertain and rapidly changing economic climate, make estimates of capital-investment and operating costs for extracting normal paraffins extremely tenuous. However, to give some insight into these figures, estimates are presented for kerosene feedstocks of three different molecular weights. These figures appear in Table 10.6.2. Total investment ranges from $12 to $22 million.

Typical performance appears in Table 10.6.3. For all feedstocks, 98.5 wt % normal paraffins in the product are achieved. Normal paraffin recovery is 95 percent. Expected adsorbent life is 5 years; for the highest-molecular-weight feedstock, adsorbent life falls to 3 years.

In summary, commercially proven large-scale production technology is available for the economic production of high-quality normal paraffins in the kerosene range.

TABLE 10.6.2 Kerosene-Range IsoSiv Economics*

	Average molecular weight		
	160	200	240
Capital investment:			
Installed equipment cost	$10,800,000	$15,800,000	$19,800,000
Adsorbent	1,500,000	2,200,000	2,600,000
Total investment	$12,300,000	$18,000,000	$22,400,000
Utilities:			
Fuel (90% furnace efficiency), 10^{10} J/h (10^6 Btu/h)	4.64 (44)	11.4 (108)	16.1 (153)
Power, kW	410	1000	1420
Cooling water [17°C (30.6°F) rise] m^3/min (gal/min)	0.095 (25)	0.208 (55)	0.322 (85)
Steam (50 lb/in^2 gage saturated), kg/h (lb/h)	149 (330)	363 (800)	521 (1150)
Nitrogen, m^3/h (SCF/h)	41.3 (1460)	101 (3580)	144 (5090)
Chemicals (85% purity nC_6 at $1/gal), $/calendar day	1700	4200	5800

*Basis: 100,000 MTA (220,000 lb/year) of normal paraffin product, United States Gulf Coast 1994–95.
Note: SCF = standard cubic feet; MTA = metric tons per year.

TABLE 10.6.3 Kerosene-Range IsoSiv Performance

	Average molecular weight		
	160	200	240
Normal product purity, wt %	98.5	98.5	98.5
Normal product recovery, wt %	95	95	95
Expected adsorbent life, years	5	5	3

BIBLIOGRAPHY

LaPlante, L. J., and M. F. Symoniak, "Here's One Way of Economically Producing Long-Chain Paraffins," NPRA meeting, San Antonio, 1970.

Reber, R. A., and M. F. Symoniak, "IsoSiv: A Separation Process to Product n-Paraffins for Single Cell Protein," American Chemical Society meeting, Philadelphia, April 1975.

CHAPTER 10.7
UOP MOLEX PROCESS FOR PRODUCTION OF NORMAL PARAFFINS

Stephen W. Sohn

UOP
Des Plaines, Illinois

DISCUSSION

The separation of normal paraffins from isoparaffins is done commercially for a number of reasons. In the lighter hydrocarbon range, isoparaffins are often more desirable because of their higher octane values and their superior gasoline alkylation characteristics. In the heavier range, normal paraffins are typically the desired product because of the benefits derived from their linearity in the production of plasticizers, linear alkylbenzene sulfonates, detergent alcohols, and ethoxylates.

This chapter discusses the specific application of the UOP* Molex* process to the separation of normal paraffins from isoparaffins. Although not limited in its application to a particular processing mode or carbon number, the Molex process is most often used in the recovery of normal paraffins for plasticizer and detergent-range applications. Typical carbon numbers are C_6 to C_{10} for plasticizers, C_{10} to C_{14} for linear alkylbenzenes, and C_{13} to $C_{22}+$ (usually heavier than C_{16}) for detergent alcohols.

The UOP Molex process is an established, commercially proven method for the liquid-phase adsorptive separation of normal paraffins from isoparaffins and cycloparaffins using the UOP Sorbex* separations technology (see Chap. 10.3), which uses zeolitic adsorbents. Isothermal liquid-phase operation facilitates the processing of heavy and broad-range feedstocks. Vapor-phase operations, in addition to having considerable heating and cooling requirements, require large variations of temperature or pressure or both through the adsorption-desorption cycle to make an effective separation. Vapor-phase operations also tend to leave a certain residual level of coke on the adsorbent, which must then be regenerated on a cyclic basis. Operation in the liquid

---

*Trademark and/or service mark of UOP.

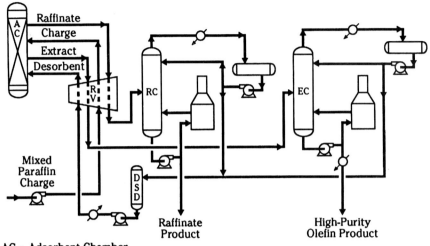

FIGURE 10.7.1 UOP Molex process design characteristics.

phase allows for uninterrupted continuous operation over many years without regenerations.

Refer to Chap. 10.3 for details of the operation of this separations technology. Figure 10.7.1 illustrates the general design characteristics of such units.

YIELD STRUCTURE

Typically, a UOP Molex process unit produces normal paraffins at about 98 to 99 wt % purity and at about 96 to 98 wt % recovery, depending on the amount of adsorbent used relative to the volume of feed.

ECONOMICS

To a certain extent, the economics of the UOP Molex unit is dependent on the feed quality, because some prefractionation and hydrotreating may be required to control the level of contaminants that might otherwise affect unit performance or adsorbent life. If the feed is assumed to have been properly treated, the estimated erected cost of a UOP Molex unit, feeding 353,000 metric tons per year (MTA) (778 million lb/year) of a paraffinic kerosene in the C_{10} to C_{15} range with about 34 percent normal paraffins, was about $15 million in 1995. This unit was designed for the recovery of 120,000 MTA (265 million lb/year) of normal paraffins at 98 percent purity. This $15 million cost represents the fully erected cost within battery limits for a particular UOP Molex unit built on the United States Gulf Coast.

The utility requirements for such a unit per 1000 lb of feed are as follows:

Electric power, kWh	7.9
Hot-oil heat, J (10^6 Btu)	527 (0.5)
Cooling water circulated [15°C (27°F) rise], m^3 (gal)	0.246 (65)

COMMERCIAL EXPERIENCE

As of early 1995, a total of 22 UOP Molex process units had been commissioned. Another three were in various stages of design or construction. Product capacities ranged from 2500 MTA (5.5 million lb/year) to 155,000 MTA (340 million lb/year).

CHAPTER 10.8
UOP OLEX PROCESS FOR OLEFIN RECOVERY

Stephen W. Sohn
UOP
Des Plaines, Illinois

DISCUSSION

The recovery of olefins from mixtures of olefins and paraffins is desirable in a number of areas: recovery of propylene from propane-propylene streams, recovery of C_4 olefins, and especially recovery of heavier olefins for the manufacture of oxoalcohols for plasticizer and detergent applications.

The UOP* Olex* process, a method of separating olefins from paraffins, is another of the many applications of the UOP Sorbex* separations technologies (see Chap. 10.3). This technique involves the selective adsorption of a desired component from a liquid-phase mixture by continuous contacting with a fixed bed of adsorbent. Other commercial methods of separation include extraction and extractive distillation. These methods, however, are less efficient in terms of product recovery and purity, much more energy-intensive, limited in application to lighter molecular weights, and limited as to the range of carbon numbers and of the molecular weights over which they apply.

In contrast, adsorbents have been developed that demonstrate the desirable characteristic of preferential relative adsorptivity for olefins as compared with paraffins. This characteristic permits ready separation of olefins and paraffins even with feedstocks that have a wide boiling range.

Much higher mass-transfer efficiency can be achieved in adsorptive operations than with the conventional equipment used for extraction and extractive distillation. As an example, laboratory-scale chromatographic columns commonly show separative efficiencies equivalent to many thousands of theoretical trays in columns of modest length. This high efficiency results from the use of small particles to give high interfacial areas and form the absence of significant axial mixing of either phase.

---

*Trademark and/or service mark of UOP.

In contrast, trayed fractionating columns and liquid-liquid extractors are designed to provide practically complete axial mixing in each physical element to create interfacial areas for mass transfer. The number of theoretical equilibrium stages is thus limited substantially to the number of physical mixing stages installed. This limitation could be avoided, in theory, by the use of packaged columns. However, if the packing-particle size is small enough to provide interfacial areas comparable with those obtained in adsorptive beds, the capacity to accommodate the counterflow of two fluid phases becomes low. Thus, great difficulty is encountered in obtaining uniform unchanneled flow of both fluid phases. In an adsorptive bed, these limitations are much less severe because only one fluid phase is involved.

In the past, processes employing solid adsorbents for treatment of liquids have not gained wide acceptance except where the quantity of material to be removed was small and frequent regeneration of the adsorbent was therefore not required. One reason for this slow acceptance was the absence of a design that would permit continuous operation. In the usual fixed-bed adsorptive process, the feed stream is discontinuous, and the product streams vary in composition. Thus, integrating the operation of any intermittent process with continuous processes operating upstream and downstream from it is difficult.

The unique process configuration used in Sorbex units eliminates these problems and facilitates continuous adsorptive separation. The Sorbex flow scheme simulates the continuous countercurrent flow of adsorbent and liquid without actual movement of the adsorbent. This system design makes adsorptive separation a continuous process and eliminates the inherent problems of moving-bed operation.

Essentially, the UOP Olex process is based on the selective adsorptive separation of olefins from paraffins in a liquid-phase operation. The adsorbed olefins are recovered from the adsorbent by displacement with a desorbent liquid of a different boiling point.

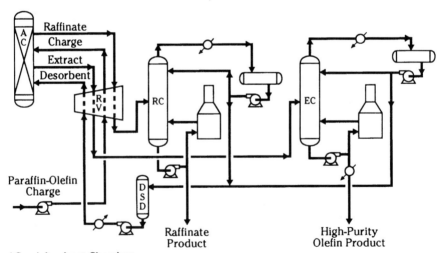

AC = Adsorbent Chamber
DSD = Desorbent Surge Drum
EC = Extract Column
RC = Raffinate Column
RV = Rotary Valve

FIGURE 10.8.1 UOP Olex process.

The flow diagram for the Olex process is shown in Fig. 10.8.1. See Chap. 10.3 for a more detailed description of Sorbex separations technologies.

COMMERCIAL EXPERIENCE

Six UOP Olex process units have been commissioned since the first one came on stream in 1972. In early 1995, another two units were at various stages of design or construction. Five commercial units process heavy (C_{10-13} up to C_{15-18}) olefin feeds and one unit processes a light (C_4) olefin feed. The heavy-feed olefin content ranges from 10 to 13 wt %, and the light feed is approximately 80 wt % olefins.

ECONOMICS

In early 1995, the estimated erected cost of a UOP Olex unit for the production of 52,000 metric tons/year (115 million lb/yr) of olefins in the C_{11} to C_{14} carbon range from a feed stream containing only 10 wt % olefins was about $17 million. This amount was the fully erected cost within battery limits of a unit built on the United States Gulf Coast.

The utility requirements for such a unit per 1,000 lb of feed would be approximately as follows:

Electric Power, kWh	3.0
Hot-oil heat, J (10^6 Btu)	316 (0.3)
Cooling water circulated [15°C (27°F) rise], m³ (gal)	0.132 (35)

P · A · R · T · 11

SULFUR COMPOUND EXTRACTION AND SWEETENING

CHAPTER 11.1
THE M.W. KELLOGG COMPANY REFINERY SULFUR MANAGEMENT

W. W. Kensell
M. P. Quinlan
The M.W. Kellogg Company, USA

INTRODUCTION

Raw crude oil contains sulfur and nitrogen. During processing, the sulfur and nitrogen are converted principally to H_2S and NH_3 and, to a lesser degree, organic sulfur (COS and CS_2) and mercaptans (RSH).

More stringent environmental standards on the emissions of sulfur and nitrogen compounds, together with the low sulfur specifications for petroleum products, have resulted in making sulfur management critical within today's refinery. The importance of sulfur management cannot be overstressed. Today's refineries are processing crudes with higher sulfur contents and are doing more bottom-of-the-barrel conversion. The need for new or revamped sulfur management facilities is expected to grow as demands for cleaner fuels and environment increase and crude oil slates change.

As illustrated in Fig. 11.1.1, sulfur management within a refinery consists of four basic processes. Amine treating units (ATUs) remove H_2S from recycle gas streams in hydroprocessing operations and from fuel gas/liquefied petroleum gas (LPG) recovery units. The amine is regenerated in one or more amine regeneration units (ARUs). Sour

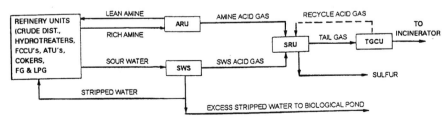

FIGURE 11.1.1 Sulfur removal/recovery.

water strippers (SWSs) remove H$_2$S and NH$_3$ from the sour water streams. Sour water is the result of refinery operations using steam in distillation or steam as a means to reduce hydrocarbon partial pressure or where water injection is used to combat potential corrosion or salt buildup. The sulfur in the acid gas from the ARU and the SWS is removed first by a Claus sulfur recovery unit (SRU) that achieves 92 to 96 percent of the overall sulfur recovery and then by a tail gas cleanup unit (TGCU) that can boost overall sulfur recovery to 99.9 percent. Most refineries now degas the molten sulfur produced. The amine, SWS, SRU, and TGCU processes are discussed in the following chapters.

AMINE

Introduction

The sulfur in crude oil that is converted to H$_2$S during processing typically is removed by a suitable amine. Two broad classifications of refinery amine treating applications are recycle gas treating and fuel gas/LPG recovery. In recycle gas treating, shown schematically in Fig. 11.1.2, the product oil from a hydroprocessing unit has an upper specification limit on its sulfur content. The sulfur in the feed oil reacts with H$_2$ at elevated pressure (typically 35 to 150 bar gage) to form H$_2$S. The reactor product stream is flashed, and a recycle gas stream, containing H$_2$, H$_2$S, and some hydrocarbons, is sent to an amine absorber, where the H$_2$S is removed by the circulating amine stream.

In fuel gas and LPG recovery units, the off-gases and stabilizer overheads from cracking, coking, and reforming units are sent to gas recovery units. The sour fuel gas has H$_2$S removed at low pressure (3.5 to 14 bar gage typically) by the circulating amine. The LPG stream has the bulk H$_2$S removed by amine at 14 to 21 bar gage, then the remaining H$_2$S plus mercaptans is treated by a caustic solution and a proprietary solvent wash that converts the mercaptans to mercapticides. These washes typically achieve a Copper Strip 1A specification. Figure 11.1.3 shows a typical block flow design illustrating the processing steps.

Process Description

Many refineries have multiple amine absorbers served by a common amine regeneration unit. Other refineries have two separate amine regeneration systems, with one system typically dedicated to "clean" users (such as hydrotreaters) and the other dedi-

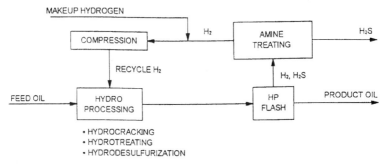

FIGURE 11.1.2 Recycle gas amine treating.

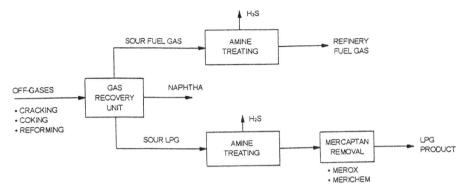

FIGURE 11.1.3 Fuel gas/LPG amine treating.

FIGURE 11.1.4 Amine regeneration unit.

cated to "dirty" users (such as FCC units or cokers). A dual amine regeneration system is illustrated in Fig. 11.1.4. Figure 11.1.5 shows the rich amine streams from the absorbers being combined and sent to the rich amine flash drum to flash off light hydrocarbons and to separate entrained hydrocarbons from the amine. This is necessary to minimize hydrocarbon carryover to the Claus SRUs. The flashed gas is treated with a slipstream of lean amine in the flash gas scrubber prior to routing to the fuel gas system. The flash drum is most often operated at 50 to 75 lb/in^2 gage so that the flashed amine can be delivered to the top of the regenerator without a pump. Steam from the reboilers using 50-lb/in^2 gage saturated steam strips the acid gas (H_2S and CO_2) from the amine. The overhead is cooled to 38 to 49°C to minimize water carryover to the SRU. Provision is made at the reflux accumulator and at the bottom of the

11.6   SULFUR COMPOUND EXTRACTION AND SWEETENING

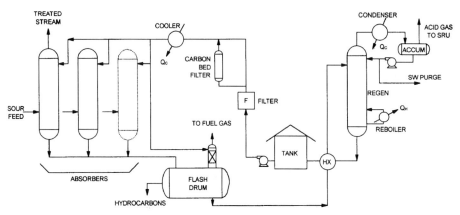

FIGURE 11.1.5   Amine treating unit.

regenerator to skim off light and heavy hydrocarbons respectively. The regenerated amine is filtered and cooled, then distributed to the various absorbers.

Process Variables

Amine selection is normally between monoethanol amine (MEA, 15 to 20 wt %), diethanol amine (DEA, 25 to 33 wt %), and methyl diethanol amine (MDEA, 45 to 50 wt %). MEA, being a primary amine, is highly reactive, but is degraded by COS, CS_2, and even CO_2. These nonregenerable degradation products require that MEA units employ a semibatch reclaimer. DEA is not as reactive as MEA, but still easily achieves treated product specification. Compared to MEA, DEA is more resistant to degradation from COS, CS_2, and CO_2, but DEA cannot easily be reclaimed. Generic MDEA reactivity is low and may not meet treated product specification at low pressures. Its increasing use is a result of its selectivity for H_2S over CO_2 and its lower energy requirements. Formulated MDEA can achieve greater reactivity and still lower energy requirements, but its cost is high. In refineries, DEA is used most often typically as a 25 to 33 wt % solution in water.

Sour feeds from cokers and catalytic crackers typically contain acids (acetic, formic, etc.) and oxygen. These contaminants react with the amine to form heat-stable salts (HSS) and to increase the foaming and corrosivity potential of the amine solution. A water wash ahead of the amine absorbers is recommended to minimize acid carryover with the sour feeds. In extreme cases, if the concentration of the HSS exceeds 10 percent of the amine concentration, a slipstream of the amine will need to be reclaimed.

Ammonia (from the nitrogen in the crude) can concentrate at the top of the regenerator and cause severe corrosion there. A purge on the reflux return line to the SWS keeps the NH_3 at more tolerable levels.

For economy, most refineries will employ a common regenerator for the amine treating associated with the main refinery units. TGCUs typically use a selective amine such as MDEA. The size and operation of the MDEA unit is such that it is nearly always kept separate from other refining amine units.

The required lean amine acid gas residual is a function of the specifications for the treated products. Typically, recycle gas is treated to about 10 vol ppm H_2S, fuel gas

H_2S is 160 vol ppm or lower, and the treated LPG H_2S should not exceed 50 wt ppm. Since the lean amine is in equilibrium with the treated product at the top of the absorber, the required residual at the pressure and temperature conditions can be calculated.

Allowable rich amine loadings (moles of acid gas per mole of amine) vary with the chosen amine and are higher for H_2S than CO_2. At high pressures, high loadings can be employed without exceeding a 70 percent approach to equilibrium at the absorber bottom. However, high loadings need to be weighed against the increased corrosiveness of the rich amine solution when it is depressurized at the separation drum and beyond. Acid gas loadings typically vary from 0.2 to 0.5 mol/mol. In LPG liquid treaters, lower loadings may be necessary because of enhanced LPG-amine contact and tower hydraulics.

Operating Considerations

The major operating considerations for amine units are maintaining the condition of the amine solution, minimizing losses and preventing hydrocarbon carryover to the sulfur plant. Solution cleanliness is achieved by 100 percent particulate filtration and a 10 to 20 percent slipstream filtration through a carbon bed absorber to remove hydrocarbons, foaming, and heat-stable salt precursors. Amine temperatures at the bottom of the regenerator should not exceed 126°C. If high back-pressure from the Claus and TGCUs makes this difficult, the possibility of lowering the amine concentration or of a pumparound regenerator cooling system should be investigated.

While the carbon bed absorber may remove some of the precursors that lead to heat-stable salt formation, the HSS in the amine solution should not be allowed to exceed 10 percent of the amine concentration.

Water washes at the top of the absorbers are an effective way to reduce amine losses, and excess water can be bled off at the reflux purge to the SWS. The rich amine separator drum is a three-phase separator with 20 to 30 minutes' residence time provided to separate the hydrocarbons. Additional hydrocarbon skims also may be provided at the reflux accumulator and at the regenerator tower bottom surge chamber.

Economics

The cost of an ARU is strongly dependent on the circulation and, to a lesser degree, the stripping steam (reboiler size) requirements. Full-flow particulate filtration and large carbon bed adsorbers increase capital cost, but are justified by significantly reduced operating costs and downtime.

SOUR WATER STRIPPING

Sour water in a refinery originates from using steam as a stripping medium in distillation or from reducing the hydrocarbon partial pressure in thermal or catalytic cracking. Also, some refinery units inject wash water to absorb corrosive compounds or salts that might cause plugging. This steam or water comes in contact with hydrocarbons containing H_2S; sour water is the result. The NH_3 present in sour water comes from the nitrogen in the crude oil or from ammonia injected into the crude fractionator to combat corrosion. In addition to H_2S and NH_3, sour water may contain phenols, cyanide, CO_2, and even salts and acids.

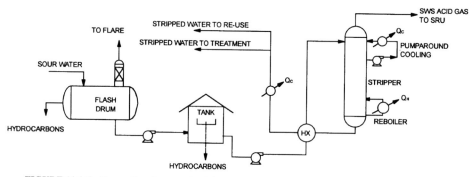

FIGURE 11.1.6 Conventional sour water stripper.

Process Description

A conventional SWS design is illustrated in Fig. 11.1.6. The sour water passes through a flash/separation drum and/or tank to flash off dissolved gases and to remove hydrocarbon oils and solids. The stripper feed is then heated by exchange with the stripper bottoms water. Steam is provided to the bottom of the stripper through a reboiler or by direct steam injection if the reboiler is out of service.

The stripped H_2S and NH_3 vapors pass through a cooling/dehumidification section at the top of the stripper. A pumparound cooler removes the heat. The acid gases, plus the uncondensed water vapor, flow to the sulfur plant at a temperature of 82 to 93°C.

The stripped water is cooled by exchange with the feed and is further cooled by air or water, if necessary, before being reused or sent to a biological treating unit.

Process Chemistry

The chemistry assumes that NH_3 and H_2S are present in the aqueous solution as ammonium hydrosulfide (NH_4HS), which is the salt of a weak acid (H_2S) and a weak base (NH_4OH). The salt hydrolyzes in water to from free NH_3 and H_2S, which then exert a partial pressure and can be stripped. The aqueous phase equilibrium is

$$NH_4^+ + HS^- \rightleftarrows H_2S + NH_3$$

Increasing the temperature shifts the equilibrium to the right, and makes it easier to strip out H_2S and NH_3. H_2S is much less soluble and is therefore more easily stripped. When acidic components such as CO_2 or CN^- are present, they replace HS^- in the above equations, and the NH_3 becomes bound in solution as a salt such as $(NH_4)_2CO_3$. The free NH_3 formed by hydrolysis is small. Thus, the H_2S removal is higher than predicted, while the NH_3 removal is lower.

Process Variables

Steam, fuel gas, and air are all possible media to strip the sour water. To meet stripped water specifications, steam normally is required and is almost exclusively used in refinery sour water treatment.

A typical stripped water specification limits H_2S to 1 to 10 wt ppm and NH_3 to 30 to 200 wt ppm. Normally, it is the NH_3 specification that governs the stripper design, since it is much more difficult to strip than H_2S. Some stripper designs use caustic to free the bound ammonia, particularly when the feed has appreciable CO_2 or cyanides.

The presence of phenols and cyanides in the sour water also can have an impact on the number of strippers. Nonphenolic sour water strippers process sour water with H_2S and NH_3 only. The stripped water is usually suitable for recycle to process units as injection wash water. Phenolic sour water contains phenols and other pollutants from catalytic crackers and cokers, and stripped water from phenolic sour water strippers is corrosive and may poison catalysts if used as injection wash water.

In conventional single-stage strippers, an acid gas containing H_2S and NH_3 is produced. This means that the SRU must be designed for NH_3 burning. An alternative is to use a two-stage stripper (such as Chevron's WWT) that produces separate NH_3 and H_2S product streams.

It is desirable to recycle as much of the stripped water as possible. Stripper water may be reused in the crude desalter, as makeup water for coker/cruder units, as wash water for the hydrotreaters, and occasionally, as cooling tower makeup water. The use of segregated strippers and the specifications of the stripped water determine the extent by which the stripped water can be reused.

Operating Considerations

Major operating considerations for sour water strippers are the foul service and corrosive environment. Some reboilers may last only 6 months to a year without cleaning, and provision for direct steam injection is advisable. The use of pumparound cooling instead of overhead condensing reduces corrosion. Extreme care is needed in metallurgy selection.

Economics

The cost of sour water strippers is strongly dependent on the sour water flow. As would be expected, stripped water specifications and installed tankage capacity also affect the capital costs.

SULFUR RECOVERY

SRUs convert the H_2S in the acid gas streams from the amine regeneration and SWS units into molten sulfur. Typically, a two- or three-stage Claus straight-through process recovers more than 92 percent of the H_2S as elemental sulfur. Most refineries require sulfur recoveries greater than 98.5 percent, so the third Claus stage is operated below the sulfur dew point, it is replaced with a selective oxidation catalyst, such as Superclaus,* or a TGCU follows the Claus unit. It is becoming increasingly popular to degas the produced molten sulfur. Shell, Elf Aquitaine, and others offer proprietary processes that degas the molten sulfur to 10 to 20 wt ppm H_2S.

---

*Trademark of Stork Comprimo.

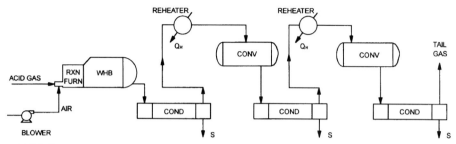

FIGURE 11.1.7 Two-stage Claus SRU.

Process Description

The Claus process, illustrated in Fig. 11.1.7 and photographed with a TGCU in Fig. 11.1.8, consists of a thermal recovery stage followed by two or three stages of catalytic recovery. In the thermal recovery zone, the acid gas is burned in a reaction furnace with the appropriate amount of air to combust approximately one-third the H_2S plus all the hydrocarbons and ammonia in the acid gas feed. The SO_2 from the combustion reacts with the uncombusted H_2S to form elemental sulfur. The products of combustion are cooled in the waste heat boiler and thermal sulfur condenser. Steam is raised at the steam drum associated with the waste heat boiler. Typically, 60 percent or more of the sulfur is recovered in the thermal recovery section of the Claus unit.

Following the thermal stage are two or three catalytic stages, each consisting of reheat (reheater), catalytic conversion (converter), and cooling with sulfur condensation. The sulfur is run down from each of the condensers into a sulfur pit, where optionally the sulfur is degassed. If the overall sulfur recovery requirement is between 96 and 99 percent, the last stage of the three-stage Claus unit can be replaced by a selective oxidation catalyst (such as Superclaus) or by a sub-dew-point reactor [such as Sulfreen* (Elf Aquitaine), CBA (Amoco), or MCRC (Delta-Catalytic)].

Process Chemistry

$$H_2S + \tfrac{3}{2}O_2 \rightarrow SO_2 + H_2O \quad \text{(thermal)}$$

$$H_2S + \tfrac{1}{2}SO_2 \rightleftarrows \tfrac{3}{2}S + H_2O \quad \text{(thermal and catalytic)}$$

Process Variables

Refineries generally require two or more Claus units to assure continued refinery unit operation during upsets, maintenance, or loss of one of the SRU. The choice between two or three is largely one of economics versus flexibility. Some Claus units can now be designed to use oxygen or enriched air when the other Claus unit is down so that only two Claus units are required.

SWS acid gas contains ammonia unless a two-stage SWS is employed. This ammonia can significantly increase the size of the Claus unit and can cause Claus operating

---

*Trademark of Elf Aquitaine.

FIGURE 11.1.8 Claus unit and TGCU.

problems if the ammonia is not fully destroyed in the thermal reactor zone. The design of burners and the reactor furnace configuration are strongly dependent on whether the Claus unit must have ammonia-burning capabilities. If all of the acid gas is not sent to the burner, the amine acid gas should be water-washed to remove traces of ammonia.

Replacing air with enriched air or oxygen significantly enhances the capacity of a Claus unit. This can be particularly attractive when a Claus unit is down or when an existing refinery needs to be revamped to handle higher sulfur capacity.

Reheat may be accomplished by in-line burners (using amine acid gas or fuel gas), hot gas bypass, external heating by steam, etc. These methods vary in cost, reliability, and maintenance requirements. External heating is usually the preferred method, but often the available heat source may not be hot enough to achieve the required reheat temperatures, particularly during catalyst rejuvenation periods.

The purpose of the Claus unit is to assist in achieving the environmentally mandated sulfur recovery requirement. Since the Claus unit often cannot do this alone, the design of the Claus unit (number of stages, selection of last stage between Claus, sub-dew-point and selective oxidation) has to be coupled with TGCU design when sub-dew-point or selective oxidation catalysis in the final SRU stage cannot meet the overall sulfur recovery requirements.

Operating Considerations

Best operating results are achieved when feed flows and compositions are maintained constant. Additionally, hydrocarbon carryover to the Claus unit must be minimized. These objectives are met by designing features into the amine and SWS units, such as large rich amine flash drums, and by providing sour water tankage.

When a unit is operated in the pure Claus mode, it is vital to keep the H_2S/SO_2 ratio in the tail gas at 2/1, since slight deviations cause significant loss in recovery. Superclaus units ahead of the Superclaus reactor should have a H_2S/SO_2 ratio of 10/1 or greater. Running Claus units at low turndown should be avoided because of instrumentation limits and greater corrosion potential.

Economics

The cost of an SRU is strongly dependent on the sulfur capacity and the number of catalytic stages. Ammonia-burning capabilities and low H_2S feed concentrations can significantly increase costs. The H_2S/CO_2 ratio in the feed also affects costs, although most refineries have a relatively rich aggregate acid gas feed. Degassing costs are almost totally dependent on sulfur capacity.

TAIL GAS CLEANUP

Overall sulfur recovery requirements at most refineries in the United States, Germany, etc., are higher than 99 percent, requiring that a TGCU follow the SRUs. The tail gas from the Claus unit contains H_2S, SO_2, CS_2, S vapor and entrained S liquid. Most tail gas cleanup processes hydrogenate/hydrolyze the sulfur compounds to H_2S, and then either recover or convert the H_2S. The H_2S recovery is usually by a selective amine. The H_2S conversion may use a liquid redox or catalytic process. The most popular TGCU processes are the Shell Claus Offgas Treating/Beavon Sulfur Reduction-MDEA (SCOT/BSR-MDEA) units and their clones. These are representative of the

H_2S recovery processes and are capable of achieving overall recoveries of 99.9 percent of the sulfur in the acid gas to the SRUs.

Process Description

Figure 11.1.9 illustrates a SCOT-type TGCU. (Some of the major equipment items also are visible in Fig. 11.1.8.) The tail gas from the Claus unit is heated in the hydrogenerator reactor to the hydrogenation bed inlet temperature by an in-line burner. Fuel gas is combusted substoichiometrically with steam to generate a reducing gas (H_2, CO) and to heat the tail gas. In the reactor, all the sulfur compounds are converted to H_2S according to the process chemistry described below. The reactor products are cooled to generate steam, then further cooled to 38 to 49°C by a circulating quench water system. A bleed stream from the circulating quench water is sent to the SWS.

The gas from the quench tower overhead is then sent to an amine unit. (The absorber and regenerator of the amine section of the TGCU can be seen in Fig. 11.1.8.) The amine is selective but otherwise the flowsheet is almost identical to that described in Chap. 2.2. In SCOT, the absorber operates at low pressure, and there are no hydrocarbons in the tail gas. Thus, a rich amine flash drum is not needed. The filtration of the amine is usually upstream of the regenerator.

Process Chemistry

$$SO_2 + 3H_2 \rightarrow H_2S + 2H_2O$$
$$COS + H_2O \rightleftarrows H_2S + CO_2$$
$$CS_2 + 2H_2O \rightarrow 2H_2S + CO_2$$
$$S_{vap} + H_2 \rightarrow H_2S$$

Process Variables

In the Claus unit burner, typically 5 to 6 percent of the H_2S dissociates into H_2 and sulfur. Depending on the Claus sulfur recovery, it may not be necessary to generate

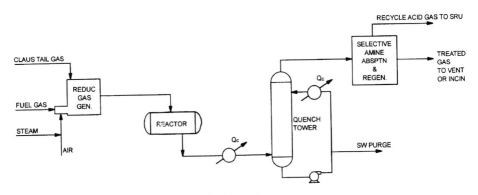

FIGURE 11.1.9 SCOT/BSR-MDEA (or clone) TGCU.

additional reducing gas, enabling the tail gas to be heated externally to hydrogenation bed inlet temperature requirements. Alternatively, a makeup H_2 stream, available elsewhere in the refinery, may negate the need for reducing gas.

The amine is usually a selective amine. Its selection depends on the H_2S specification from the absorber. If the H_2S specification is 10 vol ppm, the absorber vent gas can be vented, thereby saving considerable fuel gas at the incinerator. However, achieving low H_2S levels requires a proprietary formulated MDEA, since generic MDEA will reduce H_2S to only 150 to 250 vol ppm depending on MDEA temperature.

More recently, some refineries have had to meet total sulfur content in the absorber treated gas. This is not usually a problem when the CO_2 in the Claus tail gas is low, but equilibrium constraints can cause COS levels from the hydrogenation reactor to be a problem when CO_2 levels are high. In such cases, a COS hydrolysis reactor downstream of the reactor effluent cooler may be warranted.

Operating Considerations

When the hydrogenation/hydrolysis catalyst loses activity, there is a danger of SO_2 breakthrough. This can cause corrosion in the circulating quench water circuit, and the SO_2 poisons the amine. Catalyst activity and pH levels of the circulating water should be carefully monitored. Maintenance of the MDEA solution is imperative. It is best to filter the MDEA upstream of the regenerator.

Economics

The cost of a SCOT or BSR/MDEA or equivalent clone is usually 75 to 100 percent of the parent Claus unit without degassing.

CHAPTER 11.2
EXXON WET GAS SCRUBBING TECHNOLOGY: BEST DEMONSTRATED TECHNOLOGY* FOR FCCU EMISSION CONTROL

John D. Cunic
Exxon Research and Engineering Company
Florham Park, New Jersey

INTRODUCTION

Within a modern refining complex, one of the major sources of potential atmospheric emissions is the fluidized catalytic cracking unit (FCCU). Potential emissions from an FCCU regenerator are of two general classifications: particulate and gaseous.

The dry particulate emissions are the fine catalyst particles which have passed through the FCCU's cyclone system. The emitted catalyst is fine; in most cases, the majority of this material is in the submicrometer range. In the absence of moisture or sulfuric acid condensation, the normal cause of FCCU stack plume opacity is the presence of fine catalyst particles. Stack opacity is roughly proportional to exit catalyst loading, but is affected by other factors such as particle size distribution and stack diameter. As early as 1974, a New Source Performance Standard (NSPS) had been promulgated by the U.S. Environmental Protection Agency (EPA) regulating the amount of particulate which could be emitted from an FCCU.

With respect to the gaseous pollutants, the two compounds requiring control are carbon monoxide (CO) and sulfur dioxide (SO_2). The use of a CO boiler or high-temperature regeneration (HTR) technology can effectively meet the CO regulations along with controlling other pollutants such as hydrocarbons and ammonia. An NSPS for CO emission was also promulgated in 1974. In 1989, an NSPS for sulfur dioxide

---

*As defined in *Federal Register*, **54**(158): 34009 (1989).

was promulgated. The degree of sulfur dioxide control depends on the technology used to control these emissions.

Concurrent with the passage of the initial Clean Air Act, Exxon began development work on a process to control atmospheric emissions of particulate and SO_2 from the FCCU. The resulting Wet Gas Scrubbing (WGS) process is a simple, effective, and economic method of meeting current and proposed environmental regulations. Since start-up of the first of these units in 1974, in excess of 145 years (as of January 1, 1995) of combined operations have been gained from 14 operating units.

OPERATION

Figure 11.2.1 is a WGS flow plan. Tests have shown that the scrubbers can work equally well either upstream or downstream of a CO or waste heat boiler when the FCCU is combusting all the carbon monoxide within the unit. However, if it is upstream of a CO boiler [i.e., FCCU flue gas contains greater than 500 cm^3/m^3 (vol ppm) CO], the cooled saturated gas leaving the scrubber must be reheated to the high temperatures required for CO combustion. Primarily for this reason, and also to save on quench and ducting materials cost, it is recommended that the WGS be placed downstream of a CO or waste heat boiler.

The WGS removes particulate by washing it from the flue gas stream with droplets of a buffered scrubber liquid while SO_2 is removed by reaction with the buffered solution. Thus, the WGS is designed to accomplish the following five functions:

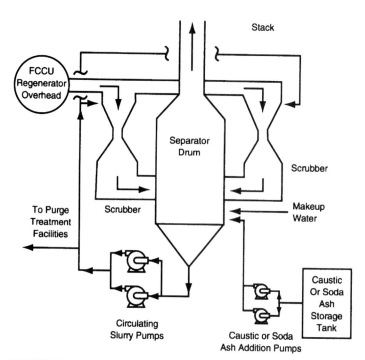

FIGURE 11.2.1 Schematic of scrubbing system.

- Introduction of the flue gas and the scrubber liquid into the unit
- Intimate mixing of the flue gas and scrubber liquid to achieve particulate and SO_2 removal
- Separation of the scrubber liquid from the clean flue gas
- Emission of the clean flue gas
- Disposal of the liquid purge stream in an environmentally acceptable manner

Depending on the configuration of the FCCU, whether it is a grassroots or retrofit application, and the local refinery circumstances, several options exist for carrying out these functions.

FLUE GAS AND SCRUBBER LIQUID

Transporting the flue gas through the scrubber and mixing it with the scrubber liquid requires energy, which can be supplied as flue gas pressure drop. If the flue gas is available at the WGS inlet at approximately 10.34 kPa (1.5 lb/in² gage), as would be the case in an FCCU operated under "full burn" conditions, then a high energy venturi scrubber can be used. Commercial experience has also shown that even if an energy recovery system (expander) has been installed in the FCCU flue gas circuit, a high-energy venturi system can still be used because of its low back-pressure requirement. A schematic of the high-energy venturi is shown in Fig. 11.2.2. In a high-energy venturi scrubber, the pressure drop of the flue gas passing through the throat of the venturi is used to atomize scrubber liquid, which is fed above the venturi throat at low pressures, 6.89 to 34.47 kPa (1 to 5 lb/in² gage), and low flow rates, 6.68×10^{-4} to 2.67×10^{-3} m³/m³ (5 to 20 gal/kft³).

If the WGS is downstream of a CO boiler, and the flue gas pressure is insufficient to use a high-energy venturi, a jet ejector venturi can be used. The jet ejector venturi is shown schematically in Fig. 11.2.3. Here, the scrubber liquid is atomized by pumping it through a spray nozzle. The draft induced by the high-pressure, 413.6 to 827.3 kPa (60

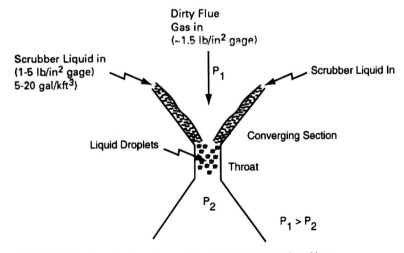

FIGURE 11.2.2 Liquid and gas introduction—high-energy venturi scrubbers.

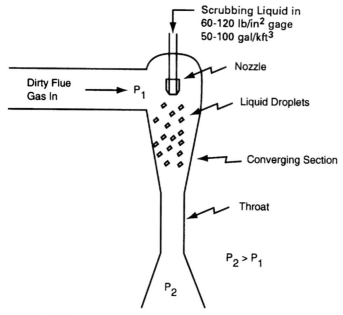

FIGURE 11.2.3 Liquid and gas introduction—jet ejector venturi scrubbers.

to 120 lb/in² gage), and high flow rate, 6.68×10^{-3} to 1.34×10^{-2} m³/m³ (50 to 100 gal/kft³), of the scrubber liquid moves the flue gas into and through the scrubber.

Because the high-energy and jet ejector venturi systems have no moving parts in the flue gas stream, they are superior to conventional venturi scrubbers, which use fans and must operate either in the hot, dusty environment upstream of the scrubber or the cold, wet environment downstream of the scrubber. Both types of WGS systems have been successfully applied commercially.

PARTICULATE AND SO_2 REMOVAL

With either the jet ejector or high-energy venturi, particulate and SO_2 removal takes place in the turbulent section of venturi. Particulate scrubbing occurs by inertial impaction of the liquid droplets with the particles in the gas stream caused by relative velocity differences. To restate this in simplified terms, particulate removal occurs in the same way a bullet strikes a target. In the high-energy venturi, the bullets are the particles and the liquid droplets are the targets; while in the jet ejector venturi the reverse is true. Operational factors which affect the degree of particulate removal include the throat velocity of the scrubber, the liquid-to-gas ratio, the inlet particle size distribution, and the inlet loading of the particles.

Concurrent with the particulate removal, sulfur oxide removal also takes place. The high surface area presented by the liquid droplets and the intimate gas-liquid contact provide ample opportunity for the reaction between the buffered scrubber liquid and the sulfur oxides to take place. Operating parameters which affect sulfur oxide removal include the inlet concentration of the sulfur oxides, the pH of the scrubbing liquid, the liquid-to-gas ratio of the scrubber, and the throat velocity of the gas.

Caustic or soda ash is added to the separated liquid to adjust its pH to the desired level, and the vast majority is circulated back to the venturi. However, a small purge stream is removed to maintain an equilibrium level of suspended solids and dissolved salts.

SEPARATION OF THE SCRUBBER LIQUID FROM THE CLEAN FLUE GAS

Once the pollutants have been transferred from the flue gas to the liquid, the two phases must be separated in a disengaging drum or separator vessel and the cleaned gas emitted to the atmosphere. Separation of the liquid from the gas involves three steps, the first of which is the coalescence of the liquid droplets. Coalescence is the result of relative velocity differences between the variously sized droplets as the gas is decelerated from the scrubber throat. As a result of this phenomenon, droplet growth occurs and few fine droplets enter the separator vessel. The absence of fine droplets permits the separation of the two phases by inertial forces and demisting devices.

Inertial forces, which are the result of a tangential entry design, begin to separate the two phases. However, this step produces only *minimal* separation since the inlet velocities are kept low to minimize erosion in this portion of the scrubbing system. A demisting device provides the final separation of the gas and liquid. The demisting device is selected for its high efficiency, low plugging tendency, and low pressure drop.

CLEAN GAS EMISSION

The separated, clean gas is emitted to the atmosphere through a stack mounted atop the separator vessel. Since the gas is saturated with water, reheat can be added to reduce the length and frequency of the visible steam plume. The length and frequency of the visible steam plume can be estimated from data on WGS operation and local meteorological conditions. While reheat facilities were included in all of the initially installed WGS units, it is only periodically used at one northern location.

PURGE LIQUID RECEIVES TREATMENT

Removing pollutants from the air just to transform them into water pollutants provides little benefit from an environmental viewpoint. Therefore, the purge stream from the WGS system undergoes further treatment to ensure that this stream is discharged in an environmentally acceptable manner. The primary pollutants dealt with are the collected catalyst (suspended solids) and the dissolved salts with high chemical oxygen demand (COD). These functions are carried out in the purge treatment unit (PTU).

PTU DESIGNS

The design of the PTU is highly dependent on local circumstances such as the amount of plot space available, local water table, meteorological conditions, and owner preference. Thus, unlike the standardized designs for the WGS, the PTU designs have been varied. The design of a PTU is a tradeoff between real estate or plot space and investment. Nevertheless, all PTUs have been designed to produce an effluent which conforms to local environmental regulations.

The PTU for the original scrubber system consisted of a large pond. This pond was divided into three basins, one approximately 12.2 by 12.2 m (40 by 40 ft), one approximately 1.2 ha (3 acres), and one approximately 0.81 ha (2 acres). Weirs were incorporated into each basin to allow for adjustment of holdup. The purge flowed into the smallest basin, where most of the catalyst settled, then overflowed to the largest basin where there was sufficient retention time to ensure almost complete separation of the catalyst from the liquid. Also within this basin, a significant portion of the oxidation of the products of SO_2 removal took place. This was accomplished by natural oxygen uptake in the liquid. Finally the last basin was used to ensure complete oxidation of the salts. In this design, no aids were used to assist with either the sedimentation or oxidation process. Because of the extensive plot space requirements for this type of PTU design, no other PTUs have been designed in this manner, nor are any expected.

Obviously not all refineries could expend this amount of real estate on the PTU. Subsequent PTUs were ponds, but included polymers to accelerate the sedimentation process. In addition, aeration devices such as surface and/or static tube aerators were used to reduce the time and plot space required to complete the oxidation of the products of SO_2 removal. Of course, this reduction in plot space requirement was accompanied by an increase in investment.

In the latest generation of PTUs the emphasis has been on minimizing plot space requirements. This has evolved because recent scrubber systems have been part of projects which are being incorporated into existing refineries. In addition, in several instances, a combination of high water tables and local government environmental requirements have restricted the use of ponding, and thus the PTU had to be placed above ground. Thus, a combination of mechanical settling and oxidation devices are used, which results in a compact unit. As has been previously stated, the reduction in plot space is accompanied by an increase in investment. The following table shows the relative plot space and investment for a ponding (includes polymer injection and aeration devices) and an aboveground PTU.

	Plot space	Investment
Ponding	14 × base	Base
Above ground	Base	1.8 × base

THE ABOVEGROUND PTU: THE LATEST GENERATION

A flow plan for an aboveground PTU is shown in Fig. 11.2.4. The purge from the scrubber system is first fed into a back-mixing system where caustic and polymer are injected. Caustic is added to adjust the stream's pH to prevent air stripping of the captured SO_2 in the subsequent oxidation step. Polymer is added to assist in the sedimentation process. After the back-mix system, the purge is then fed into the reactor clarifier where the solids are separated from the process stream. Clarification was selected as the first treatment step since the catalyst is very erosive. Removal of the solids at this point allows for a downgrading of materials in the downstream equipment. Effluent from the clarifier contains typically less than 100 mg/kg (wt ppm) suspended solids which is sufficient to meet most discharge requirements.

Once the solids have been removed from the purge stream, it is pumped to the oxi-

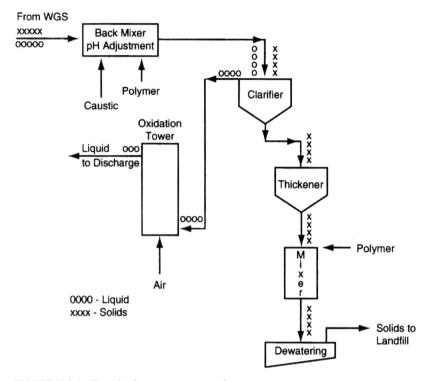

FIGURE 11.2.4 Flow plan for purge treatment unit.

dation tower where it is mixed with compressed air. The type of tower selected has a high internal circulation rate which ensures repeated, intimate mixing of the liquid and air streams. On leaving the oxidation tower, the water can be discharged, since its COD is less than 50 mg/L; however, some refineries have requested that the stream be cooled prior to its discharge in order to meet local requirements.

Meanwhile, the solids which have been collected and concentrated in the clarifier receive further treatment. There is periodic blowdown from clarifier to the agitated clarifier sump. The solids collected in the sump are then pumped to the thickener where they are further concentrated prior to being stored in the bottom of the vessel with the liquid decanted back to the clarifier.

Periodically, a final processing step is performed on the solids. Thickener bottoms pumps transfer a batch of the solids to a mixer where another polymer is added. The solids then go to a large granular-bed vacuum filter where all free water is removed and returned back to the thickener. The moist solids are then loaded into trucks for disposal in a sanitary landfill. Currently, landfill disposal of the solids is acceptable since several high-pressure leachate tests have shown that the leachate contains substantially less than 100 times the drinking water standard for all priority pollutants.

Figure 11.2.5 is a view of a commercial installation of this type of PTU. Total plot space allocated for this unit is approximately 279 m^2 (3000 ft^2). The reactor clarifier is in the left foreground, the thickener is located directly above it, and the granular bed filter is in the left background. Just to the right of the filter is the polymer storage area. The oxidation tower is in the center foreground and the air compressors are in

FIGURE 11.2.5 Typical commercial aboveground PTU system.

the right foreground. Back-mixing equipment is located just under the pipeway. This commercial installation has been in operation since 1984 and has had minimal operational problems.

MEETING ENVIRONMENTAL GOALS

Extensive testing has been conducted on all the WGS facilities by Exxon, its licensees, and various environmental control agencies. A summary of these results is shown in Table 11.2.1. It should be noted that, as designed, all of the WGS facilities are in compliance with their permitted values. In particular, all of the licensee units are in compliance with the current NSPS for FCCUs, which require that particulate emissions be reduced to less than 1 kg (lb) of particulates per 1000 kg (lb) of coke burned. These units are required to meet this standard since they were constructed or modified after issuance of the NSPS. It should also be noted that these units are also in compliance with the NSPS for SO_2 emissions, which require that SO_2 emission be reduced by 90 percent or 50 cm^3/m^3 (vol ppm), whichever results in the higher SO_2 emission.

EPA TESTING

Although one could explore the performance of any of the scrubbing systems in detail, the best choice would be to select the unit which was rigorously tested by the EPA as part of its program to develop the sulfur oxides NSPS. EPA conducted a test program

TABLE 11.2.1 Summary of Typical Performance of Exxon and Licensee FCCU Scrubbing Systems*

Unit	Pollutant†	Inlet rate or concentration‡	Outlet rate or concentration	Regulation§	Collection efficiency, %
Unit A	Part	NM	56.81 lb/h	90.4 lb/h	
	SO_2	915 vol ppm	37 vol ppm	NA	95.6
Unit B	Part	NM	61.9 lb/h	124.2 lb/h	
	SO_2	1011 vol ppm	61.5 vol ppm	NA	93.9
Unit C	Part	NM	63.0 lb/h	190 lb/h	
	SO_2	NM	5.3 vol ppm	NA	
Unit D	Part	560 lb/h	24.8 lb/h	30 lb/h	95.6
	SO_2	771 vol ppm	20 vol ppm	180 vol ppm	97.4
Unit E	Part	NM	0.16 lb/klb (6.2 lb/h)	1 lb/klb¶	
	SO_2	444 vol ppm	16.8 vol ppm	NA	96.2
Unit F	Part	NM	0.26 lb/klb (5.0 lb/h)	1.0 lb/klb¶	
	SO_2	150 vol ppm	7.5 vol ppm	NA	95.0
Unit G	Part	NM	0.82 lb/klb (17.6 lb/h)	1.0 lb/klb¶	
	SO_2	NM	4.7 vol ppm	245 vol ppm	
Unit H	Part	NM		0.62 lb/klb	
	SO_2	NM			

*Conversions: 1 lb/h = 0.454 kg/h; 1 vol ppm = 1 cm³/m³; 1 lb/klb = 1kg/1000 kg.
†Part = particulate (catalyst) emissions; SO_2 = sulfur dioxide emissions.
‡NM = not measured
§NA = no applicable source regulation, allowable emissions rate set by ground level concentration.
¶EPA New Source Performance Standards for FCCU: less than 1.0 pound catalyst emissions per 1000 pounds of coke burned equivalent to 1 kg of catalyst emissions per 1000 kg of coke burned.

on a unit from May 4, 1981, through June 2, 1981. The primary purpose of these tests was to conduct continuous emission monitoring (CEM) of sulfur oxides. However, during the course of this program, emissions of particulates, nitrogen oxides, and hydrocarbons were also measured. It should be noted that, during this entire program, neither Exxon personnel (who were not on site) nor the refinery people made any attempts to optimize WGS performance. Thus, the unit was tested without optimization. Results of these tests are shown in Table 11.2.2.

As can be seen from Table 11.2.2, 12 days of CEM testing showed that the scrubbing system averaged 93 percent sulfur dioxide removal. Subsequent manual testing

TABLE 11.2.2 EPA Testing of Exxon Scrubbing System

SO_2 removal:	
By CEM tests (12 days)	93% average
By Method 8 (3 days)	95% average
Particulate removal by modified Method 5	85.2% average
SO_2 emission rate:	
Inlet (Method 8)	396.8 vol ppm (average)
Outlet (Method 8)	19.6 vol ppm (average)
Particulate emission rate:	
Inlet (modified Method 5)	6.17 lb/1000 lb coke
Outlet (modified Method 5)	0.84 lb/1000 lb coke
NO_x emission rate (Method 7)	93 vol ppm
Hydrocarbon emission rate	23 vol ppm

Note: 1 vol ppm = 1 cm³/m³; 1 lb/1000 lb = 1 kg/1000 kg.

11.24 SULFUR COMPOUND EXTRACTION AND SWEETENING

using a modified EPA Method 8 test confirmed these results, indicating 95 percent sulfur oxide removal across the scrubber [396.8 cm^3/m^3 (vol ppm) in and 19.9 cm^3/m^3 (vol ppm) out]. Nitrogen oxide emissions averaged 93 cm^3/m^3 (vol ppm) and hydrocarbon emissons averaged 23 cm^3/m^3 (vol ppm). Since neither is controlled by the scrubbing system, these values are solely dependent on the operation of the FCCU.

Particulate testing showed that 85.2 percent of the solids entering the WGS were removed and an outlet emission rate of 0.84 kg (lb)/1000 kg (lb) of coke burned. In conducting these tests, EPA's contractor substantially deviated from EPA Method 5 sampling and analytical procedures to provide specific developmental information to EPA. On the basis of these deviations, the outlet particulate loading is believed to be higher than would be expected if "standard" Method 5 practices were followed.

WGS BACKGROUND

Exxon has designed and installed 14 commercial WGS facilities. These are listed in Table 11.2.3. These systems were installed as part of grass-roots FCCUs or as retrofits on existing FCCUs. Scrubber type at each location depended on pressure availability. A typical jet ejector venturi system is shown in Fig. 11.2.6. The flue gas is scrubbed by three jet ejector venturis. Scrubber liquid is circulated by three pumps capable of delivering up to 5400 gal/min at 150 lb/in^2 differential pressure. Normal operating conditions are considerably less severe. The scrubbers are connected to the separator vessel and the clean gas exits through a 10-ft-diameter stack. Total on-site plot space is approximately 3000 ft^2. A typical high-energy venturi is shown in Fig. 11.2.7. In this system, scrubbing is carried out in up to four high-energy venturis. Scrubbing liquid is circulated by a single pump capable of delivering up to 3500 gal/min at 35 lb/in^2 differential pressure. The scrubbers are connected to the separator vessel and the cleaned gas exits through an 8-ft-diameter stack. Total on-site plot space for the unit is about 1800 ft^2.

TABLE 11.2.3 Exxon-Designed Wet Gas Scrubbing Facilities

Company	Type	Grassroots (GR) or retrofit (R)	Capacity, kACFM	Start-up date
Exxon Unit	JEV	R	300	March 1974
Exxon Unit	JEV	R	455	May 1975
Exxon Unit	JEV	R	865	January 1976
Exxon Unit	JEV	R	730	May 1976
Licensee Unit 1	HEV	GR	200	December 1979
Licensee Unit 2	HEV	GR	110	November 1980
Licensee Unit 3	HEV	GR	110	December 1984
Licensee Unit 4	HEV	R	145	April 1985
Licensee Unit 5	HEV	GR	300	July 1991
Licensee Unit 6	HEV	GR	760	December 1992
Licensee Unit 7	HEV	R	245	May 1991
Licensee Unit 8	HEV	GR	337	October 1993
Licensee Unit 9	HEV	R	150	August 1991
Licensee Unit 10	HEV	R	715	Deferred
Licensee Unit 11	HEV	R	356	December 1994
Licensee Unit 12	HEV	GR	312	1999

Note: JEV = jet ejector venturi; HEV = high-energy venturi; kACFM = thousand actual cubic feet per minute.

FIGURE 11.2.6 Typical jet ejector venturi wet gas scrubber system.

ADVANTAGES

On the basis of Exxon's design and operation of the WGS systems, the following advantages have been confirmed.

- *Single-step pollutant removal:* The venturi is the only pollution control system which can remove both particulate and SO_2 pollutants and achieve compliance with expected environmental regulations.

FIGURE 11.2.7 Typical high-energy venturi wet gas scrubber system.

- *Flexible performance:* Day-to-day operating changes (e.g., changes in flue gas rates, composition solids loading, temperature) can be readily handled, if necessary, by small changes in the WGS operating conditions. Even long-term changes, such as changes in FCCU catalyst type, have been handled with little or no adjustment in WGS operation. In addition, the Exxon WGS system has experienced almost every upset that can occur in FCCU operation, including reverse flow, and has not required extensive attention during these upsets.
- *Compact system:* Typical on-site plot area requirements for the Exxon WGS system range from 93 to 465 m^2 (1000 to 5000 ft^2) for FCCUs ranging from 55.2 to

276.0 dm³/s [30 to 150 thousand barrel per day (KBPD)] feed rate. These low plot space requirements arise from the use of multiple high-capacity venturis mounted on a single separator and the ability to locate supplemental equipment, such as the PTU, off site. The lower on-site plot plan requirements also offer advantages with grass-roots plants in that more space can be dedicated to process rather than pollution control units.

- *Reliability:* Exxon's experience with various types of emission control facilities has shown that the WGS system has higher service factors and lower maintenance costs than alternative emission control approaches. In fact, the WGS service factor is equal to or greater than the FCCU itself, and no FCCU has been shut down because of failure of the WGS system. Maintenance costs have proven to be lower than those for either electrostatic precipitators or conventional venturis.
- *Low cost:* In most cases studied, the WGS system was more economical than a combination of feed desulfurization and electrostatic precipitators. Of course, the choice of an emission control system depends on feedstock quality, processing requirements, environmental regulations, and location. However, the attractiveness of FCCU WGS systems increases as either the sulfur content of the FCCU feed is increased (by adding atmospheric resid, for example) or regulations are made more severe.

For example, a study for a Gulf coast grass-roots 73.6-dm³/s [40 thousand barrels per stream day (kBPSD)] FCCU processing either virgin gas oil or a mixture of virgin and coker gas oils was recently considered. A WGS system, with a distillate hydrodesulfurization unit to maintain equivalent distillate quality, was compared to a combination of total feed desulfurization and electrostatic precipitators. All control equipment was designed to meet the current particulate and SO_2 NSPS. For either case, the WGS system/distillate hydrodesulfurization combination showed an economic incentive of about $6 per cubic meter ($1 per barrel) even when credit was taken for FCCU yield improvements due to feed desulfurization.

SUMMARY

The Exxon WGS system offers a unique combination of FCCU particulate and SO_2 control capability with economic and operating advantages when compared to the combination of particulate control via electrostatic precipitators and SO_2 control via feed desulfurization. The WGS system concept can be used with either of the gaseous emission control approaches now in use: CO boilers or high-temperature regeneration. Fourteen commercial units, with over 145 years of combined operation, are now in service. The commercial units have demonstrated over 90 percent particulate removal and over 95 percent SO_2 removal.

CHAPTER 11.3
UOP MEROX PROCESS

D. L. Holbrook
UOP
Des Plaines, Illinois

INTRODUCTION

The UOP* Merox* process is an efficient and economical catalytic process developed for the chemical treatment of petroleum fractions to remove sulfur present as mercaptans (Merox extraction) or to directly convert mercaptan sulfur to less-objectionable disulfides (Merox sweetening). This process is used for liquid-phase treating of liquefied petroleum gases (LPG), natural-gas liquids (NGL), naphthas, gasolines, kerosenes, jet fuels, and heating oils. It also can be used to sweeten natural gas, refinery gas, and synthetic gas in conjunction with conventional pretreatment and posttreatment processes.

Merox treatment can, in general, be used in the following ways:

- To improve lead susceptibility of light gasolines (extraction)
- To improve the response of gasoline stocks to oxidation inhibitors added to prevent gum formation during storage (extraction and sweetening)
- To improve odor on all stocks (extraction or sweetening or both)
- To reduce the mercaptan content to meet product specifications requiring a negative doctor test or low mercaptan content (sweetening)
- To reduce the sulfur content of LPG products to meet specifications (extraction)
- To reduce the sulfur content of coker or fluid catalytic cracking (FCC) C_3-C_4 olefins to save on acid consumption in alkylation operations using these materials as feedstocks or to meet the low-sulfur requirements of sensitive catalysts used in various chemical synthesis processes (extraction)

---

*Trademark and/or service mark of UOP.

PROCESS DESCRIPTION

The UOP Merox process accomplishes mercaptan extraction and mercaptan conversion at normal refinery rundown temperatures and pressures. Depending on the application, extraction and sweetening can be used either singly or in combination. The process is based on the ability of an organometallic catalyst to promote the oxidation of mercaptans to disulfides in an alkaline environment by using air as the source of oxygen. For light hydrocarbons, operating pressure is controlled slightly above the bubble point to ensure liquid-phase operation; for heavier stocks, operating pressure is normally set to keep air dissolved in the reaction section. Gases are usually treated at their prevailing system pressures.

Merox Extraction

Low-molecular-weight mercaptans are soluble in caustic soda solution. Therefore, when treating gases, LPG, and light-gasoline fractions, the Merox process can be used to extract mercaptans, thus reducing the sulfur content of the treated product. In the extraction unit (Fig. 11.3.1), the sulfur reduction attainable is directly related to the extractable-mercaptan content of the fresh feed.

In mercaptan-extraction units, fresh feed is charged to an extraction column, where mercaptans are extracted by a countercurrent caustic stream. The treated product passes overhead to storage or downstream processing.

The mercaptan-rich caustic solution containing Merox catalyst flows from the bottom of the extraction column to the regeneration section through a steam heater, which is used to maintain a suitable temperature in the oxidizer. Air is injected into this stream, and the mixture flows upward through the oxidizer, where the caustic is regenerated by converting mercaptans to disulfides. The oxidizer effluent flows into the disulfide separator, where spent air, disulfide oil, and the regenerated caustic solution are separated. Spent air is vented to a safe place, and disulfide oil is decanted and sent to appropriate disposal. For example, the disulfide oil can be injected into the charge

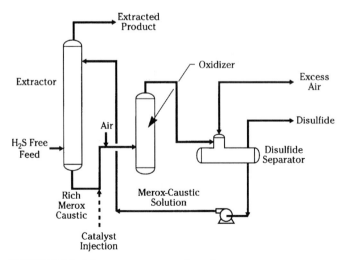

FIGURE 11.3.1 Merox mercaptan-extraction unit.

to a hydrotreating unit or sold as a specialty product. The regenerated-caustic stream is returned to the extraction column. A small amount of Merox catalyst is added periodically to maintain the required activity.

Merox Sweetening

In sweetening units, the mercaptans are converted directly to disulfides, which remain in the product; the total sulfur content of the treated stock is not reduced.

Merox sweetening can be accomplished in four ways:

- Fixed-bed processing with intermittent circulation of caustic solution (Fig. 11.3.2)
- Minimum-alkali fixed-bed (Minalk*) processing, which uses small amounts of caustic solution injected continuously (Fig. 11.3.3)
- Caustic-Free Merox* treatment for gasoline (Fig. 11.3.4) and kerosene (Fig. 11.3.5)
- Liquid-liquid sweetening (Fig. 11.3.6)

Fixed-Bed Sweetening (Conventional). Fixed-bed sweetening (Fig. 11.3.2) is normally employed for virgin or thermally cracked chargestocks having endpoints above about 120°C (248°F). The higher-molecular-weight and more branched mercaptan types associated with these higher-endpoint feedstocks are only slightly soluble in caustic solution and are more difficult to sweeten. The use of a fixed-bed reactor facilitates the conversion of these types of mercaptans to disulfides.

Fixed-bed sweetening uses a reactor that contains a bed of specially selected activated charcoal impregnated with nondispersible Merox catalyst and wetted with caustic solution. Air is injected into the feed hydrocarbon steam ahead of the reactor, and in passing through the catalyst bed, the mercaptans in the feed are oxidized to disulfides. The reactor is followed by a settler for separation of caustic and treated hydrocarbon. The settler also serves as a caustic reservoir. Separated caustic is circulated intermittently to keep the catalyst bed wet. The frequency of caustic circulation over the bed depends on the difficulty of the feedstock being treated and the activity of the catalyst.

An important application of this fixed-bed Merox sweetening is the production of jet fuels and kerosenes. As a result of the development of the Merox fixed-bed system,

---

*Trademarks and/or service marks of UOP.

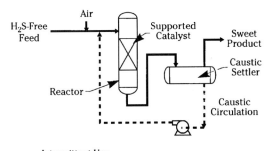

FIGURE 11.3.2 Fixed-bed Merox sweetening unit.

jet fuels and kerosenes (also diesel and heating oils) can be sweetened at costs that are incomparably lower than those of the simplest hydrotreater. The same basic process flow just described is used. However, because of other particular jet-fuel quality requirements, some pretreatment and posttreatment are needed whenever any chemical sweetening process is used.

Fixed-Bed Sweetening (Minalk). This Merox sweetening version is applied to feedstocks that are relatively easy to sweeten, such as catalytically cracked naphthas and light virgin naphthas. This sweetening design achieves the same high efficiency as conventional fixed-bed sweetening but with less equipment and lower capital and operating costs.

The UOP Merox Minalk process (Fig. 11.3.3) relies on a small, controlled, continuous injection of an appropriately weak alkali solution rather than the gross, intermittent alkali saturation of the catalyst bed as in conventional fixed-bed Merox sweetening. This controlled, continuous small injection of alkali provides the needed alkalinity so that mercaptans are oxidized to disulfides and do not enter into peroxidation reaction, which would result if the alkalinity were insufficient.

Caustic-Free Merox. The most recent version of the Merox family is the Caustic-Free Merox process for sweetening gasoline and kerosene (Figs. 11.3.4 and 11.3.5). This technology development uses the same basic principles of sweetening in which the mercaptans are catalytically converted to disulfides, which remain in the treated hydrocarbon product.

The Caustic-Free Merox catalyst system consists of preimpregnated fixed-bed catalysts, Merox No. 21* catalyst for gasoline and Merox No. 31* catalyst for kerosene, and a liquid activator, Merox CF.* This system provides an active, selective, and stable sweetening environment in the reactor. The high activity allows the use of a weak base, ammonia, to provide the needed reaction alkalinity. No caustic (NaOH) is required, and fresh-caustic costs and the costs for handling and disposing of spent caustic are thus eliminated.

A major benefit to the Caustic-Free Merox version for kerosene is that a caustic prewash for removing naphthenic acid is no longer required. The acids are removed simultaneously with mercaptan conversion in the reactor. This procedure not only eliminates a major caustic consumer, but also eliminates hardware and so reduces capital investment.

---

*Trademark and/or service mark of UOP.

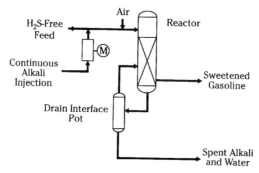

FIGURE 11.3.3 Fixed-bed minimum-alkali Merox sweetening unit.

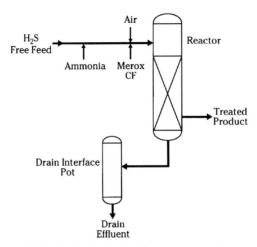

FIGURE 11.3.4 Caustic-Free Merox sweetening for gasoline.

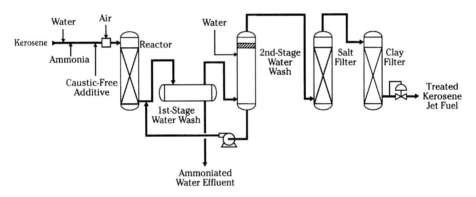

FIGURE 11.3.5 Caustic-Free Merox sweetening for kerosene jet fuel.

The actual design of the Caustic-Free Merox unit depends on whether it is used on gasoline or kerosene. The reactor section is similar to the previously mentioned fixed-bed systems, conventional and Minalk, except for the substitution of a different catalyst, the addition of facilities for continuous injection of the Merox CF activator, and replacement of the caustic injection facilities with ammonia injection facilities, anhydrous or aqueous. For kerosene or jet fuel production, the downstream water-wash system is modified to improve efficiency and to ensure that no ammonia remains in the finished product. Other posttreatment facilities for jet fuel production remain unchanged.

Liquid-Liquid Sweetening. The liquid-liquid sweetening version (Fig. 11.3.6) of the Merox process is not generally used today for new units as refiners switch to the more active fixed-bed systems. Hydrocarbon feed, air, and aqueous caustic soda containing dispersed Merox catalyst are simultaneously contacted in a mixing device, where mercaptans are converted disulfides. Mixer effluent is directed to a settler, from which the

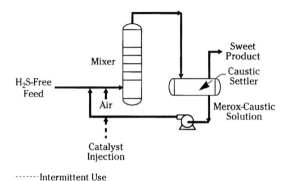

FIGURE 11.3.6 Liquid-liquid Merox sweetening unit.

treated hydrocarbon stream is sent to storage or further processing. Separated caustic solution from the settler is recirculated to the mixer. A small amount of Merox catalyst is added periodically to maintain the catalytic activity.

In general, liquid-liquid sweetening is applicable to virgin light, thermally cracked gasolines and to components having endpoints up to about 120°C (248°F). The mercaptan types associated with catalytically cracked naphthas are easier to oxidize than those contained in light virgin or thermal naphthas, and therefore liquid-liquid sweetening has been successfully applied to catalytically cracked gasolines having endpoints as high as 230°C (446°F).

The various applications of the Merox process on different hydrocarbon streams are summarized in Table 11.3.1.

Merox Process Features

Relative to other treating processes, the Merox process has the following advantages.

Low Operating Cost and Investment Requirement.   The noncorrosive environment in which the process operates requires no alloys or other special materials, thus mini-

TABLE 11.3.1 Merox Process Applications

Hydrocarbon stream	Merox type
Gas	Extraction
LPG	Extraction
Natural gas liquids	Extraction, extraction plus sweetening
Light naphtha	Liquid-liquid sweetening, Minalk sweetening, caustic-free sweetening
Medium or heavy naphtha	Liquid-liquid sweetening
	Caustic-free sweetening
Full-boiling-range naphtha	Extraction plus sweetening, Minalk sweetening, fixed-bed sweetening, caustic-free sweetening
Kerosene or jet fuel	Fixed-bed sweetening
	Caustic-free sweetening
Diesel	Fixed-bed sweetening

mizing investment. In many applications, investment is essentially nil because of the ease of converting existing equipment to Merox treating.

Ease of Operation. Merox process units are extremely easy to operate; usually, the air-injection rate is the only adjustment necessary to accommodate wide variations in feed rate or mercaptan content. Labor requirements for operation are minimal.

Proven Reliability. The Merox process has been widely accepted by the petroleum industry; many units of all kinds (extraction, liquid-liquid, and fixed-bed sweetening) have been placed in operation. By early 1995, more than 1500 of these UOP Merox units had been licensed.

Minimal Chemical-Disposal Requirements. Caustic consumption by atmospheric CO_2, excessive acid in the feedstock, and accumulation of contaminants are the only reasons for the occasional replenishment of the caustic inventory.

Proven Ability to Produce Specification Products. Product deterioration as a result of side reactions does not occur nor does any addition of undesirable materials to the treated product. This fact is especially important for jet-fuel treating. In the Merox process, sweetening is carried out in the presence of only air, caustic soda solution, and a catalyst that is insoluble in both hydrocarbon and caustic solutions and cannot therefore have a detrimental effect on other properties that are important to fuel specifications.

High-Efficiency Design. The Merox process ensures high catalyst activity by using a high-surface-area fixed catalyst bed to provide intimate contact of feed, reactants, and catalyst for complete mercaptan conversion. The technology does not rely on mechanical mixing devices for the critical contact. State-of-the-art Merox technology has no requirement for continuous, high-volume caustic circulation that increases chemical consumption, utility costs, and entrainment concerns.

High-Activity Catalyst and Activators. Active and selective catalysts are important in promoting the proper mercaptan reactions even when the most difficult feedstocks are processed. For the extraction version of the process, UOP offers a high-activity, water-soluble catalyst, Merox WS,* which accomplishes efficient caustic regeneration. As a result, chemical and utility consumption is minimized, and mercaptans are completely converted. For the sweetening version of the Merox process, UOP offers a series of catalysts and promoters that provide the maximum flexibility for treating varying feedstocks and allow refiners to select which catalyst system is best for their situation.

PROCESS CHEMISTRY

The Merox process in all its applications is based on the ability of an organometallic catalyst to accelerate the oxidation of mercaptans to disulfides at or near ambient temperature and pressure. Oxygen is supplied from the atmosphere. The reaction proceeds only in an alkaline environment. The basic overall reactor can be written:

$$4RSH + O_2 \xrightarrow[\text{Alkalinity}]{\text{Merox catalyst}} 2RSSR + 2H_2O \qquad (11.3.1)$$

---

*Trademark and/or service mark of UOP.

where R is a hydrocarbon chain that may be straight, branched, or cyclic and saturated or unsaturated. Mercaptan oxidation, even though slow, reportedly occurs whenever petroleum fractions containing mercaptans are exposed to atmospheric oxygen. In effect, the Merox catalyst speeds up this reaction, directs the products to disulfides, and minimizes undesirable side reactions.

In Merox extraction, in which mercaptans in the liquid or gaseous feedstocks are highly soluble in the caustic soda solution as solvent, the mercaptan oxidation is done outside the extraction environment. Therefore, a mercaptan-extraction step is followed by oxidation of the extracted mercaptan. These steps are:

$$\underset{\text{Oil phase}}{\text{RSH}} + \underset{\text{Aqueous phase}}{\text{NaOH}} \longrightarrow \underset{\text{Aqueous phase}}{\text{NaSR}} + H_2O \qquad (11.3.2)$$

$$\underset{\text{Aqueous phase}}{4\text{NaSR}} + O_2 + 2H_2O \xrightarrow{\text{Merox catalyst}} \underset{\text{Aqueous phase}}{4\text{NaOH}} + \underset{\text{Oil phase (insoluble)}}{2\text{RSSR}} \qquad (11.3.3)$$

According to these treating steps, the treated product has reduced sulfur content corresponding to the amount of mercaptan extracted.

In the case of Merox sweetening, in which the types of mercaptans in the feedstocks are difficult to extract, the sweetening process is performed in situ in the presence of Merox catalyst and oxygen from the air in an alkaline environment. UOP studies have shown that the mercaptan, or at least the thiol (—SH) functional group, first transfers to the aqueous alkaline phase (Fig. 11.3.7) and there combines with the catalyst. The simultaneous presence of oxygen causes this mercaptan-catalyst complex to oxidize, yielding a disulfide molecule and water. This reaction at the oil-aqueous interface is the basis for both liquid-liquid and fixed-bed sweetening by the Merox process and can be written:

$$\underset{\text{Oil phase}}{4\text{RSH}} + O_2 \xrightarrow[\text{Alkalinity}]{\text{Merox catalyst}} \underset{\text{Oil phase}}{2\text{RSSR}} + 2H_2O \qquad (11.3.4)$$

$$\underset{\text{Oil phase}}{2\text{R}'\text{SR} + 2\text{RSH}} + O_2 \xrightarrow[\text{Alkalinity}]{\text{Merox catalyst}} \underset{\text{Oil phase}}{2\text{R}'\text{SSR}} + 2H_2O \qquad (11.3.5)$$

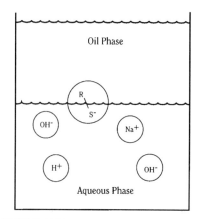

FIGURE 11.3.7 Mercaptide at interface.

Equation (11.3.5) represents the case in which two different mercaptans may enter into this reaction. Petroleum fractions have a mixture of mercaptans so that the R chain may have any number of carbon atoms consistent with the boiling range of the hydrocarbon feed.

Because the process is catalytic, essentially catalyst and caustic soda are consumed. This fact is borne out by commercial experience, in which actual catalyst consumptions are low. Consumption is due mainly to fouling by certain substances and loss through an occasional purge of dirty or diluted caustic solution and a corresponding makeup of fresh caustic to maintain effective caustic concentration.

PRODUCT SPECIFICATIONS

The only product specification applicable to Merox treating is the mercaptan sulfur content of the product because the Merox process per se has no effect on the other properties of the feedstock being treated. Generally, therefore, the Merox process is used to reduce the mercaptan sulfur content, and thereby the total sulfur content, when the process is applied to gases and light stocks in the extraction mode of operation. In the case of heavier chargestocks that require the sweetening mode of operation, the only product specification applied is the mercaptan sulfur content (or sometimes also the doctor test); the total sulfur contents of the untreated feed and the treated product are the same.

Merox-treated products may be finished products sent directly to storage without any further processing or intermediate products that may require either blending into finished stocks or additional processing for making other products.

Table 11.3.2 lists typical quality specifications for treating applications of the Merox process.

PROCESS ECONOMICS

Sample economics of the UOP Merox process in 1995 dollars on the basis of 10,000 barrels per stream day (BPSD) capacity for various applications are given in Table 11.3.3. The capital costs are for modular design, fabrication, and erection of Merox plants. The estimated modular cost is inside battery limits, U.S. Gulf coast, FOB point of manufacturer. The estimated operating costs include catalysts, chemicals, utilities, and labor.

PROCESS STATUS AND OUTLOOK

Continued research has resulted in more than 100 additional patents. The first Merox process unit was put on-stream October 20, 1958. In October 1993, the 1500th Merox process unit was commissioned. Design capacities of these Merox units range from as small as 40 BPSD for special application to as large as 140,000 BPSD and total more than 12 million BPSD.

The application of the operating Merox units is distributed approximately as follows:

- 25 percent LPG and gases
- 30 percent straight-run naphthas

TABLE 11.3.2 Quality Specifications for the Merox Process

Characteristics	Feed Type							
	Gases, LPG, NGL	NGL, LN	MN, HN	FBR gasoline	Jet fuels	Kerosene	Diesels	Heating oils
Feed:								
Mercaptan sulfur, wt ppm	50–10,000	50–2,000	50–5,000	50–5,000	30–1,000	30–1,000	50–800	50–800
H$_2$S, wt ppm*	<10	<10	<10	<10	<1	<1	<1	<1
Acid oil, wt %	.....	.....	<0.01	<0.01	<0.01	<0.01	<0.01	<0.01
Products:								
Mercaptan sulfur, wt ppm	<5–10	<5–10	<5–10	<5–10	<10	<10	<30	<30
Mercaptan sulfur + disulfide sulfur, wt ppm	10–20	<50						

*After caustic prewash, if any, before Merox process. LPG = liquefied petroleum gas; NGL = natural gas liquid; gas = natural gas, refinery gas, or synthesis gas; LN = light naphthas; MN and HN = medium and heavy naphthas; FBR = full boiling range.

TABLE 11.3.3 Merox Process Economics

Product	Merox type	Est. capital, million $	Est. operating costs, cents/bbl
LPG	Extraction*	2.0	0.7
Light naphtha	Minalk	1.0	0.5
	Caustic-free	1.0	2.2
Heavy naphtha and kerosene	Conventional fixed-bed	2.3	2.5
	Caustic-free	2.3	5.8

*Includes pretreating and posttreating facilities.

- 30 percent FCC, thermal, and polymerization gasolines
- 15 percent kerosene, jet fuel, diesel, and heating oils

The Merox process has been thoroughly proved and well-established commercially. Its popular acceptance by the petroleum industry is based on its simplicity and efficiency, low capital and operation costs, and proven reliability. Many refiners have two or more Merox units. Even though the process is approaching 38 years of use, its technology is by no means stagnant, thanks to continuing research and development efforts to ensure an excellent outlook for this remarkably successful process.

P · A · R · T · 12

VISBREAKING AND COKING

CHAPTER 12.1
EXXON FLEXICOKING INCLUDING FLUID COKING

Eugene M. Roundtree
Exxon Research and Engineering Company
Florham Park, New Jersey

INTRODUCTION

This chapter provides general technical information on Flexicoking* technology and covers the major objectives and benefits of Flexicoking technology. Also included is a review of Fluid Coking technology.
    Flexicoking:

- Is residuum fluidized coking, with integrated gasification of up to 97 percent of gross coke production.
- Meets strict environmental regulations.
- Has five units in operation with total capacity greater than 180,000 barrels per day (B/D).
- Benefits from significant advances in utilizing low-Btu gas. Purge coke is readily accepted by the cement industry.

Flexicoking advantages are

- *Process flexibility.* Flexicoking can handle a wide range of heavy crudes. Because coking is a thermal process, it is generally insensitive to feedstock quality. It can handle virtually any pumpable hydrocarbon (e.g., vacuum resid, atmospheric resid, tar sands bitumen, heavy whole crudes, shale oil, catalytic plant bottoms).
- *Reliability.* Studies of Flexicoking unit operating performance have shown that an average mature unit service factor of 90 percent has been obtained.

---

*Trademark of Exxon Corporation.

- *Process improvement activities.* Process alternatives described later in this chapter are the results of Exxon's continuing research and development work aimed at extending and improving Flexicoking technology.
- *Hydrocarbon/hydrogen management.* Coking is a noncatalytic thermal cracking process based on the concept of *carbon rejection*. The heaviest, hydrogen-deficient portions of the feed (i.e., asphaltenes and resins) are rejected as coke, which contains essentially all the feed metals and ash and a substantial portion of the feed sulfur and nitrogen. The carbon rejection route results in lower hydrogen consumption than alternative hydrogen addition–type processes.

Flexicoking also maximizes refinery yield of hydrocarbons, since the coke yield is converted to a clean fuel gas. This gas is used as plant fuel, allowing hydrocarbon fuels to be sent to product sales.

PROCESS DESCRIPTION

Flexicoking is a continuous fluidized-bed thermal cracking process integrated with coke gasification. Flexicoking is a versatile process which is applicable to a wide range of heavy feedstocks. The process can handle virtually any pumpable hydrocarbon stream, including atmospheric and vacuum residua of all types. Processing costs are relatively insensitive to feed contaminants such as organic metals, ash, sulfur, and nitrogen. The Flexicoking process typically converts 99 percent of the vacuum residuum to gaseous and liquid products. The remaining 1 percent is a solid purge which contains more than 99 percent of the metals in the feed. The coker naphtha and gas oils are typically upgraded to salable products by a combination of refining processes such as hydrotreating and catalytic cracking. About 95 percent of the total sulfur in the residuum feed can be recovered from the products as elemental sulfur via commercially available processes.

The Flexicoking unit converts about 97 percent of the gross coke to gas with a lower heating value of 120 to 140 Btu standard cubic foot (SCF). This low-heating-value gas [low-Btu gas (LBG)] can be burned in process heaters and boilers. The coke fines from the Flexicoking unit contain most of the metals in the feedstock and may be suitable for metal recovery.

A simplified flow diagram of the Flexicoking process is shown in Fig. 12.1.1. The process consists of a fluid-bed reactor, a liquid product scrubber on top of the reactor, a heater vessel where circulating coke from the reactor is heated by gas and hot coke from the gasifier, a gasifier vessel, a heater overhead gas cooling system, and a fines removal system.

Residuum feed at 500 to 700°F is injected into the coker reactor, where it is thermally cracked to a full range of vapor products and a coke product which is deposited on the fluidized coke particles. The sensible heat, heat of vaporization, and endothermic heat of cracking of the residuum is provided by a circulating stream of hot coke from the heater. Cracked vapor products are quenched in the scrubber tower. The heavier fractions are condensed in the scrubber and, if desired, may be recycled back to the coking reactor. The lighter fractions proceed overhead from the scrubber into a conventional fractionator where they are split into the desired cut ranges for further downstream processing.

Reactor coke is circulated to the heater vessel where it is heated by coke and gas from the gasifier. A circulating coke stream is sent from the heater to the gasifier where it is reacted at an elevated temperature (1500 to 1800°F) with air and steam to form a mixture of H_2, CO, N_2, CO_2, H_2O, and H_2S, which also contains a small quanti-

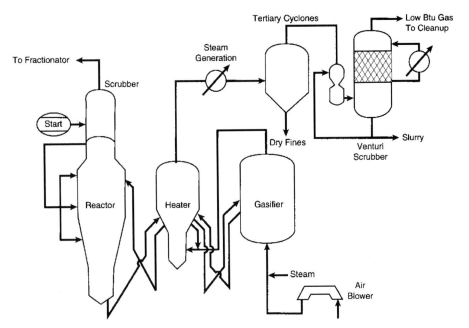

FIGURE 12.1.1 Simplified Flexicoking flow plan.

ty of COS. The gasifier product gas, referred to as coke gas, plus entrained coke particles are returned to the heater and are cooled by cold coke from the reactor to provide a portion of the reactor heat requirement. A return stream of coke sent from the gasifier to the heater provides the remainder of the heat requirement.

The hot coke gas leaving the heater is used to generate high-pressure steam before passing through the tertiary cyclones for removal of entrained coke particles. The remaining coke fines are removed in a venturi scrubber. The solids-free coke gas is then sent to a gas cleanup unit for removal of H_2S.

TYPICAL YIELDS AND PRODUCT DISPOSITIONS

A typical Flexicoking yield pattern of product qualities and disposition is shown in Fig. 12.1.2.

TWO SPECIFIC PROCESS ESTIMATES (YIELDS, QUALITIES, UTILITIES, AND INVESTMENTS)

Flexicoking process estimates for two cases are shown in Tables 12.1.1 to 12.1.5 (Case 1) and Tables 12.1.6 to 12.1.10 (Case 2).

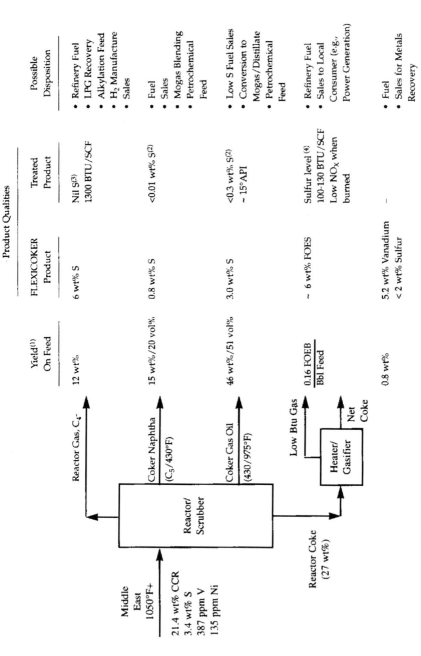

FIGURE 12.1.2 Typical Flexicoking yield pattern showing product qualities and disposition. [*Notes:* (1) Typical yield for Middle East residuum. Correlations are available based on data from a wide range of fields including Jobo, Tia Juana, Bachaquero, Athabasca, Cold Lake, and Lloydminster. (2) Typical hydrotreating severity for low-sulfur products. Design can be varied to provide high-quality diesel and/or jet fuel as well as high-quality catalyst feed. (3) Sulfur removal by MEA scrubbing and Merox treating of C_3's and C_4's. (4) Sulfur content of gas will meet environmental standards using commercially available processing.]

TABLE 12.1.1 Flexicoking Estimates—Case 1: Feed Rate and Properties

Feed type	Maya/Isthmus 52/48 blend
Nominal cut point, °F	1050+
Feed rate, kB/SD	26.3
Feed properties:	
Density, °API	2.9
Sulfur, wt %	5.1
Nitrogen, wt %	0.6
Conradson carbon, wt %	29.7
Nickel, wt ppm	99
Vanadium, wt ppm	481
LV % below 1000°F	5
Feed temperature, °F	500

Note: B/SD = barrels per stream day; °API = degrees on American Petroleum Institute scale; LV % = liquid volume percent.

TABLE 12.1.2 Flexicoking Estimates—Case 1: Estimated Product Yields

Reactor yields	wt % FF	klb/h	LV% FF	kB/SD
H_2S	1.04	4.20		
COS	<0.01	0.01		
CSH	0.04	0.16		
C_2SH	0.04	0.15		
NH_3	0.05	0.20		
CO	0.15	0.61		
CO_2	0.10	0.40		
Total	1.42	5.73		
H_2	0.15	0.61		
Methane	2.69	10.86		
Ethylene	0.61	2.46		
Ethane	2.29	9.24		
Propylene	1.11	4.48		
Propane	1.67	6.74		
Total C_3-	8.52	34.39		
Butadiene	0.05	0.20	0.08	0.20
Butenes	1.43	5.77	2.47	5.77
iC_4	0.19	0.77	0.36	0.77
nC_4	1.04	4.20	1.87	4.20
Total C_4	2.71	10.94	4.78	10.94
$C_5/350°F$	11.51	46.45	16.36	4.30
350/650°F	12.64	51.01	15.05	3.96
650/800°F	9.24	37.29	10.10	3.96
800/975°F	17.10	69.01	17.73	2.66
Total C_5+	50.49	203.76	59.24	15.58
Gross coke*	36.86	148.75		
Total	100.00	403.57	64.02	26.52

*97.3% of gross coke would be gasified, the rest being entrained as fines.
Note: FF = fresh feed.

TABLE 12.1.3 Flexicoking Estimates—Case 1: Estimated Product Qualities

Nominal cut range, °F	C_5/350	350/650	650/800	800/975
Density, °API	59.5	28.5	15.4	7.9
Sulfur, wt %	0.72	2.81	4.59	5.21
Nitrogen, wt %	0.007	0.119	0.417	0.674
Conradson carbon, wt %	—	—	2.30	6.37
Nickel, wt ppm	—	—	—	2.30
Vanadium, wt ppm	—	—	—	11.20
H/C weight ratio	0.160	0.137	0.125	0.117
Aniline point, °F	51	85	101	130
Pour point, °F	—	−25	59	88
Bromine no., g/100 g	117	62	41	29
Cetane no.	—	30.2	—	—
Viscosity, cSt:				
At 100°F	0.6	2.2	44.2	1360.6
At 210°F	0.4	0.9	4.2	30.6
Paraffins, olefins, naphthenes, and aromatics, LV %:				
Paraffins	22.3	15.5	—	—
Olefins	59.0	30.0	—	—
Naphthenes	5.5	2.9	—	—
Aromatics	13.2	51.6	—	—
Heater/gasifier low-Btu gas*				
Gas rate, kSCFM		219.5		
Composition, mol %				
CO	20.16			
CO_2	9.51			
H_2	21.47			
N_2	44.54			
CH4	0.80			
H_2S	0.01 (150 vol ppm)			
COS	0.01 (120 vol ppm)			
H_2O	3.50			
Total	100.00			
Gas heating value (sweet dry), Btu/SCF	135			

*Sweet LBG after H_2S removal in the Flexsorb-SE unit.
Note: H/C = hydrocarbon-to-carbon; SCFM = standard cubic feet per minute.

LOW-BTU GAS UTILIZATION

Low-Btu Gas Characteristics

Flexicoking gas (LBG), which is derived from the gasification of coke, has a typical heating value of 120 to 140 Btu/SCF and contains 45 to 55 percent nitrogen. The roughly 20 percent hydrogen content makes it an easy fuel to burn. The LBG properties are well within the range of commercially proven low-Btu fuels. When cleaned up by Flexsorb* amine treating technology and particulate removal, it provides a safe, reliable, and environmentally acceptable fuel source. Table 12.1.11 compares key burning characteristics to other fuels commonly used.

*Trademark of Exxon Corporation.

TABLE 12.1.4 Flexicoking Estimates—Case 1: Estimated Utility Requirements

Steam required, klb/h:	
High pressure (600 lb/in^2 gage, 700°F)^a	3
Low pressure (125 lb/in^2 gage, 475°F)^b	242
High temperature (125 lb/in^2 gage, 950°F)^c	36
Other utilities:	
Boiler feedwater,^{d} gal/min	576
Cooling water,^{e} gal/min	11,633
Electricity, kW	552
Instrument air, SCFM	171
Nitrogen, SCFM	210
Air blower utilities:^{f}	
Compressor brake horsepower	16,181
Steam required (600 lb/in^2 gage, 700°F), klb/h	410
Steam produced (125 lb/in^2 gage, 475°F), klb/h	410
Auxiliary cooling water, gal/min	58
Steam produced, klb/h	
High pressure (600 lb/in^2 gage, saturated)^g	254
Low pressure (125 lb/in^2 gage, saturated)^h	80

^{a}Feed atomization steam.
^{b}Transfer line, heater, and gasifier process steam.
^{c}Reactor stripping and anticoking baffle steam.
^{d}Boiler feedwater includes 10% blowdown.
^{e}Cooling water is used for fractionator overhead cooling and LBG cooling in the heater overhead. Cooling water delta temperature is assumed to be 30°F.
^{f}Air blower utility requirements are broken out separately to facilitate alternative driver studies. An axial compressor with a noncondensing steam turbine driver is assumed.
^{g}Includes steam produced from the heater overhead only. There is no high-pressure steam production in the scrubber pumparound at a fresh feed temperature of 500°F.
^{h}Includes steam produced from the fractionator middle and bottom pumparounds.

Combustion Characteristics

LBG can be utilized in several ways, depending on site-specific conditions and economic factors. LBG is scrubbed to remove H$_2$S and particulates, resulting in a clean burning fuel which can back-out purchased natural gas or refinery gas/liquid fuels. ER&E Flexsorb solvent technology is available to reduce H$_2$S levels in LBG down to less than 10 ppm, providing good potential for environmental credits.

LBG Utilization

The technology for utilizing LBG has advanced considerably since the original units were built. Determining the optimum disposition of LBG requires a study to come up with the correct split between new and existing equipment. Experience has shown that many existing furnaces/heaters within the refinery can be modified in a cost-effective manner. Also, large nearby grass-roots consumers can make a project attractive. The trend is to consider all combustion facilities as potential users of LBG. ER&E can assist in LBG utilization studies.

TABLE 12.1.5 Flexicoking Estimates—Case 1: Investment Estimates*

	Million $
Direct material and labor cost:	
Flexicoking unit	54.9
Air blower	6.8
Primary fractionator	4.9
Subtotal direct costs	66.6
Indirect costs	36.1
Total prime contract cost†	102.7

*Investment for Flexicoking unit only. Does not include Flexsorb-SE facilities. Basis: fourth quarter 1993, U.S. Gulf Coast.

†Total prime contract cost includes all capital facilities inside battery limits, capital spares, vendor shop fabrication of all vessels and piping, direct labor wages, inland freight to site, field labor overheads, contractor's detailed engineering, and loss on surplus materials. Items excluded are warehouse spares, off-sites/utilities, site preparation, sales tax on materials, escalation, catalyst/chemicals, basic engineering costs, Exxon Engineering services, royalties/licensing fees, contractor's engineering/construction fees, vendor shop inspection costs, vendor representatives, start costs, owner's costs, and project contingency for changes.

TABLE 12.1.6 Flexicoking Estimates—Case 2: Feed Rate and Properties

Feed type	Arab Heavy
Nominal cut point, °F	1050+
Feed rate, kB/SD	23.2
Feed properties:	
Density, °API	3.4
Sulfur, wt %	6.0
Nitrogen, wt %	0.3
Conradson carbon, wt %	27.7
Nickel, wt ppm	64
Vanadium, wt ppm	205
LV% below 1000°F	5

Typical Applications for LBG

- Miscellaneous outlets

    Fuel for atmospheric and vacuum pipestill furnaces

    Fuel for H_2 plant (steam reforming)

    Fuel for hot belt system

    Fuel for naphtha reformer

    Steam superheaters

    Waste heat boilers

    Reboilers

TABLE 12.1.7 Flexicoking Estimates—Case 2: Estimated Product Yields

Reactor yields	wt % FF	klb/h	LV% FF	kB/SD
H_2S	1.19	4.22		
COS	0.00	0.01		
CSH	0.05	0.16		
C_2SH	0.04	0.16		
NH_3	0.02	0.09		
CO	0.15	0.53		
CO_2	0.10	0.36		
Total	1.55	5.53		
H_2	0.14	0.50		
Methane	2.58	9.15		
Ethylene	0.60	2.13		
Ethane	2.24	7.94		
Propylene	1.13	4.01		
Propane	1.63	5.78		
Total C_3-	8.32	29.51		
Butadiene	0.05	0.18	0.08	0.02
Butenes	1.39	4.93	2.39	0.55
iC_4	0.19	0.67	0.35	0.08
nC_4	0.99	3.51	1.78	0.42
Total C_4	2.62	9.29	4.60	1.07
C_5/430°F	14.76	52.35	20.36	4.72
430/650°F	10.15	36.00	11.90	2.76
650/975°F	28.37	100.62	29.86	6.93
Total C_5+	53.28	188.97	62.12	14.41
Gross coke*	34.23	121.36		
Total	100.00	354.66	66.72	15.48

*97.3% of gross coke would be gasified, the rest being entrained as fines.

  Fluid catalytic cracker feed preheater
  Steam generation boilers
- Neighboring facilities
  Power plants
  Steel manufacturing
  Others
- Gas turbine generators

  Potential outlet for LBG, although there is no commercial application yet.
  Requires gas compression from 15 to 250 lb/in² gage.
  Overall efficiency of power system for LBG is lower than equivalent high-Btu gas system (28 percent versus 32 percent).
  In combined cycle, plant steam and power needs will dictate cogeneration cycle. A study is required to balance plant fuel, steam, and power needs.

TABLE 12.1.8 Flexicoking Estimates—Case 2: Estimated Product Qualities

Nominal Cut Range, °F	$C_5/430$	430/650	650/975
Density, °API	55.1	25.8	10.5
Sulfur, wt %	1.07	3.84	5.98
Nitrogen, wt %	0.005	0.077	0.290
Conradson carbon, wt %	—	—	4.74
Nickel, wt ppm	—	—	0.87
Vanadium, wt ppm	—	—	2.80
H/C weight ratio	0.159	0.135	0.120
Aniline point, °F	58	88	118
Pour point, °F	—	−9	82
Bromine no., g/100 g	109	58	32
Cetane no.	—	30.8	—
Viscosity, cSt:			
At 100°F	0.6	3.2	365.5
At 210°F	0.4	1.1	12.6
Paraffins, olefins, naphthenes, and aromatics, LV%:			
Paraffins	21.4	14.6	—
Olefins	54.9	26.1	—
Naphthenes	5.1	2.5	—
Aromatics	18.6	56.8	—
Heater/gasifier low-Btu gas:*			
Gas rate, kSCFM	181.0		
Composition, mol %:			
CO	18.64		
CO_2	10.45		
H_2	20.99		
N_2	45.61		
CH_4	0.80		
H_2S	0.00		
COS	0.01		
H_2O	3.50		
Total	100.00		
Gas heating value, Btu/SCF	128.5		

*Sweet LBG after H_2S removal in the Flexsorb-SE unit.

General Guidelines for Utilization

The many choices to be considered are listed below along with some general guidelines.

- New process furnaces/heaters

    Should be given top priority

    Can be designed for high efficiency

    Investment approximately 20 percent over high-Btu-fired furnace.

- New boilers with or without power generation via condensing steam turbine generators

    Provide a large outlet for LBG

    Can be designed for high boiler efficiency

    Investment approximately 10 to 15 percent above conventional boiler

TABLE 12.1.9 Flexicoking Estimates—Case 2: Estimated Utility Requirements

Steam required, klb/h:	
High pressure (600 lb/in² gage, 700°F)[a]	2
Low pressure (125 lb/in² gage, 475°F)[b]	209
High temperature (125 lb/in² gage, 950°F)[c]	30
Other utilities	
Boiler feedwater,[d] gal/min	762
Cooling water,[e] gal/min	8,585
Electricity, kW	493
Instrument air, SCFM	152
Nitrogen, SCFM	188
Air blower utilities[f]	
Compressor brake horsepower	14,785
Steam required (600 lb/in² gage, 700°F), klb/h	409
Steam produced (125 lb/in² gage, 475°F), klb/h	409
Auxiliary cooling water, gal/min	53
Steam produced, klb/h	
High pressure (600 lb/in², saturated)[g]	281
Low pressure (125 lb/in² gage, saturated)[h]	65

[a]Feed atomization steam.
[b]Transfer line, heater, and gasifier process steam.
[c]Reactor stripping and anticoking baffle steam.
[d]Boiler feedwater includes 10% blowdown.
[e]Cooling water is used for fractionator overhead cooling and LBG cooling in the heater overhead. Cooling water delta temperature is assumed to be 30°F.
[f]Air blower utility requirements are broken out separately to facilitate alternative driver studies. An axial compressor with a noncondensing steam turbine driver is assumed.
[g]Includes steam produced from scrubber pumparound and heater overhead.
[h]Includes steam produced from the fractionator middle and bottom pumparounds.

Supplemental fuel, 10 to 20 percent, usually used to facilitate a smooth switch between fuels.

- Revamp existing furnaces/heaters

    Slight efficiency debit expected

    Design study required to determine economic optimum between derating unit and modifications

    Partial use of LBG to be considered

    Revamping costs in the order of $2.5 million for 150 MBtu/h and $3.5 million for 300 MBtu/h

- Revamp existing boilers

    Drop in boiler rating may be required, depending on design.

    Supplemental fuel, 10 to 20 percent, usually used to facilitate a smooth switch between fuels.

    Revamp cost highly sensitive to complexity (i.e., forced draft versus balanced draft, presence of air heaters, split of radiant and convection duties). For boilers, an approximation is to assume costs similar to those of revamping furnace.

TABLE 12.1.10 Flexicoking Estimates—Case 2: Investment Estimates*

	Million dollars
Direct material and labor cost	
Flexicoking unit	49.5
Air blower	6.5
Primary fractionator	4.5
Subtotal direct costs	60.5
Indirect costs	35.0
Total prime contract cost†	95.5

*Investment for Flexicoking unit only. Does not include Flexsorb-SE facilities. Basis: fourth quarter 1993, U.S. Gulf Coast.

†Total prime contract cost includes all capital facilities inside battery limits, capital spares, vendor shop fabrication of all vessels and piping, direct labor wages, inland freight to site, field labor overheads, contractor's detailed engineering, and loss on surplus materials. Items excluded are warehouse spares, off-sites/utilities, site preparation, sales tax on materials, escalation, catalyst/chemicals, basic engineering costs, Exxon Engineering services, royalties/licensing fees, contractor's engineering/construction fees, vendor shop inspection costs, vendor representatives, start costs, owner's costs, and project contingency for changes.

TABLE 12.1.11 Low-Btu Gas Characteristics

Source	Flexicoking	Blast furnace	Typical refinery fuel gas
Composition, mol %:			
CO	18.6	23.3	—
H_2	21.0	3.2	—
$C_1 - C_4$	0.8	0.1	100.0
CO_2	10.4	11.5	—
N_2	45.6	53.7	—
H_2O	3.5	8.2	—
Mol wt	23.7	28.2	18.9
Lower heating value, Btu/SCF	128	83	1050
Pressure, lb/in^2 gage	15	1.5	60
Adiabatic flame temp., °F at 2% excess air	2750	—	3285.0

Example of Flexicoking LBG Utilization Basis:

- A nominal 32-kB/D Flexicoking unit added to an existing refinery
- All LBG utilized within refinery

Places utilizing LBG:

- Grassroots boilers
- Grassroots vacuum pipestill (VPS) furnace

- Grassroots naphtha reforming furnace
- Grassroots steam reforming furnace
- Grassroots steam superheater
- Revamp atmospheric pipestill (APS) furnace
- Revamp boilers
- Revamp light ends reboilers

PURGE COKE UTILIZATION

Purge Coke comes from three locations in the Flexicoking unit: (1) bed purge, (2) dry fines from cyclones, and (3) wet fines from scrubber. It is usually relatively low in sulfur, and its metals concentrate in fines.

Bed coke and dry fines make a good fuel that can be sold and can be considered for on-site fuel. A coke-burning boiler and possibly flue gas cleanup equipment will be needed. Wet fines can be sold as a special fuel or used for metal recovery. Metal reclaiming may be attractive, depending on the metal concentration in coke. Technology is commercially available and is currently applied to the product from one unit.

Disposition of Purge Coke

- *Unit A:* Dry fines currently sold for use as blast furnace coke. Wet fines currently dried to 10 percent moisture and used as auxiliary fuel for a cement kiln.
- *Unit B:* Project under way to use coke products as a boiler fuel.
- *Unit C:* Dry fines sold into metallurgical industry. Wet fines routed to a delayed coker. Bed purge sold to a broker for fuel applications.
- *Unit D:* Blended coke fines sold for metal recovery. Bed purge sold as fuel for cement industry.
- *Unit E:* Bulk of coke products used as fuel in cement industry. Additional sales into other industries when sufficient quantity is available.

FLEXICOKING UNIT SERVICE FACTOR

Flexicoking units can typically achieve 18- to 24-month runs, followed by turnarounds of 45 to 60 days, yielding service factors of 90 percent or better. This is shown in Fig. 12.1.3. Experience with all the units commissioned since 1980 has demonstrated that the mature unit service factor can be achieved from the initial start-up, with no learning curve.

COMMERCIAL FLEXICOKING EXPERIENCE

Exxon has 40 years of commercial coking experience, 20 years of Flexicoking. Table 12.1.12 shows ER&E-designed commercial units.

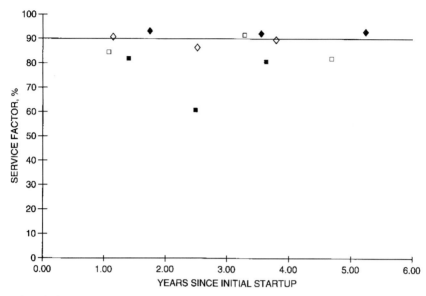

FIGURE 12.1.3 Flexicoking unit service factor.

TABLE 12.1.12 Commercial Flexicoking Units

Company	Locations	Initial design fresh feed rate, B/SD	Initial start-up date	Comments
Licensed unit	Kawasaki, Japan	21,000	Sept. 1976	
Licensed unit	Amuay, Venezuela	52,000	Nov. 1982	Expanded to 65,000 B/SD
Licensed unit	Martinez, California	22,000	March 1983	
Esso Netherlands	Rotterdam, Netherlands	32,000	Aug. 1986	Expanded to 38,000 B/SD
Exxon USA	Baytown, Texas	28,000	Sept. 1986	Expanded to 37,000 B/SD

Figure 12.1.4 shows a Flexicoking unit in the Gulf Coast region of the United States. From left to right are the coke silos, the gasifier vessel, heater vessel, reactor/scrubber, and the fractionator with its associated overhead system. The air blower train is in the lower left foreground. The background includes the heater overhead fines capture and sulfur removal systems.

FLEXICOKING OPTIONS

Flexicoking Reactor Side Process Options

Flexicoking is a versatile process that fits into many refinery/crude upgrading situations. In a fuel refinery, Flexicoking, coupled with hydrotreating, allows the refiner to

FIGURE 12.1.4 Flexicoking unit in the Gulf Coast region of the United States.

produce low-sulfur oil (less than 0.3 wt %) from high-sulfur crudes. Flexicoking can be utilized with catalytic cracking or hydrocracking in a high-conversion refinery to produce high yields of distillates and gasoline. Since Flexicoking is compatible in a fuel oil or high-conversion refinery, it can be utilized by the refiner to stage refinery investment.

Flexicoking offers unique flexibility in processing options to tailor the coker to the refiner's needs:

- *Conventional coking.* Typically converts atmospheric or vacuum residua to 975°F− products. This is shown in Fig. 12.1.5.
- *Once-through coking.* Once-through coking is another operating option in which the bottoms from the reactor scrubber are drawn off as product instead of being recycled to the reactor. The scrubber bottoms have a higher selectivity to coke than the virgin feed. Therefore, by removing this material from the coker, overall coke production can be reduced. This reduction in coke make can lead to a reduced investment for a grassroots unit or to more fresh feed processing capacity in an existing unit limited by coke gasification. The once-through option also provides increased liquid product yield at the expense of coke and gas. Of course, with this option the liquid product has a higher endpoint and may have to have coke particles removed (typically 1 to 2 wt %), depending on its disposition. Once-through coking is illustrated in Fig. 12.1.6.

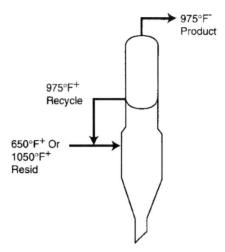

FIGURE 12.1.5 Conventional coking.

- *VPS integrated coking.* The maximum gas oil cut point that can be achieved in a conventional Flexicoking reactor scrubber is about 975°F. This cut point is limited by the maximum temperature that can be maintained in the scrubber without excessive coke deposition. By integrating the coking reactor and scrubber with an upstream vacuum pipestill, the cut point capability can be extended to 1050°F vacuum bottoms that are injected directly into the coking unit reactor. The scrubber bottoms boiling above 975°F, instead of being recycled directly to the reactor, are sent to the vacuum pipestill where the coker gas oil boiling between 975 and 1050°F is recovered. The product boiling above 1050°F is recycled to the reactor with the virgin vacuum bottoms. This effectively increases the coker recycle cut point to

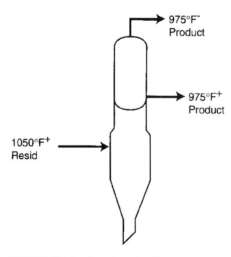

FIGURE 12.1.6 Once-through coking.

1050°F. The integration with the vacuum pipestill processing scheme, illustrated in Fig. 12.1.7, can be utilized in both Fluid Coking and Flexicoking units and can be applied in retrofit as well as grass-roots projects.

- *Combo coking.* Atmospheric resid is fed to the scrubber where gaseous reactor products "lift" vacuum gas oil range material overhead in admixture with coker products. The vacuum pipestill is deleted. This is illustrated in Fig. 12.1.8. Depending on the qualities and quantities of feed sent to the scrubber, the overall liquid yield is somewhat lower than for a VPS/conventional vacuum resid Flexicoking unit combination.

Flexicoking Gasification Options

With *integrated high gasification,* typically 97 percent of gross coke is gasified for most feeds. Typically 90 to 95 percent of gross coke is gasified for high-metal/ash feedstocks. This option results in 97 to 99 percent conversion of coker feed to gaseous and light liquid products.

Partial gasification allows the refiner to vary capacity and yields of coke gas and low-sulfur coke to meet specific operating objectives.

FLUID COKING OPTIONS

Description of Fluid Coking Process

A simplified flow sheet of the Fluid Coking process is shown in Fig. 12.1.9. There are two major fluidized-bed vessels: a reactor and a burner. The heavy hydrocarbon feed is introduced into the scrubber where it exchanges heat with the reactor overhead effluent and condenses the heaviest fraction of the hydrocarbons. The total reactor

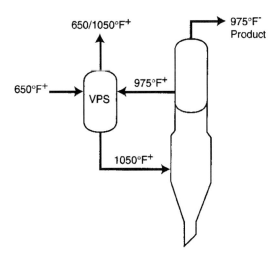

FIGURE 12.1.7 VPS integrated coking.

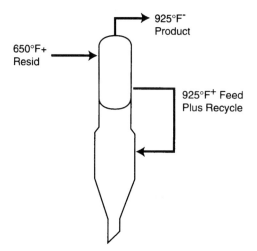

FIGURE 12.1.8 Combo coker.

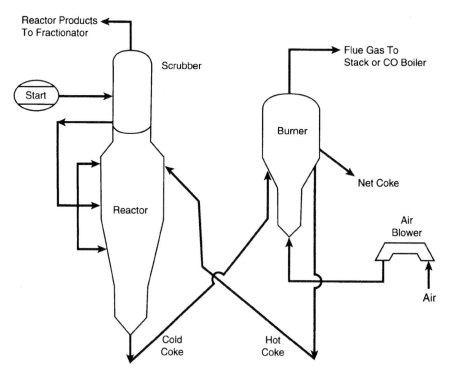

FIGURE 12.1.9 Simplified Fluid Coking flow plan.

feed, including both the fresh feed and the recycle condensed in the scrubber, is injected into a bed of fluidized coke in the reactor, where it is thermally cracked to produce lighter liquids, gas, and coke. The coke is laid down on the fluidized coke particles while the hydrocarbon vapors pass overhead into the scrubber. The reactor overhead is scrubbed for solids removal and the material boiling above 975°F is condensed and recycled to the reactor. The lighter hydrocarbons are sent from the scrubber to conventional fractionation, gas compression, and light ends recovery units.

Heat required to maintain the reactor at coking temperature is supplied by circulating coke between the reactor and burner. A portion of the coke produced in the reactor is burned with air to satisfy the process heat requirements. The excess coke is withdrawn from the burner and sent to storage.

Additional Background on Fluid Coking as a Seasoned Refinery Process

The process was first commercialized in 1954. There are currently nine operating Fluid Coking units processing petroleum residua with a total throughput of about 159 kB/D. Two Fluid Coking units, processing a total of over 200 kB/D of Athabasca oil sands bitumen, have been in operation since 1978.

Product coke has been used as a fuel, either on site or sold to coke merchants. Local environmental regulations may require flue gas desulfurization on boilers/furnaces firing high-sulfur coke. On the other hand, high-sulfur fluid coke has been successfully fired in cement kilns while meeting environmental standards. Current outlets for coke include:

- Aluminum electroplating electrodes if metal content is low (less than 400 ppm vanadium)
- Steel industry if sulfur is low (less than 3 percent sulfur)
- Fuel in steam/power plants, including fluid-bed boilers
- Cement industry fuel if sulfur is high (more than 3 percent sulfur)
- Partial oxidation units

Fluid Coking liquid yields are higher than those of conventional delayed coking. The external energy supply requirements for Fluid Coking are less than those of delayed coking since the heat of reaction is provided by the combustion of coke circulated within the system. Liquid yield and coke make can be varied slightly to fit the product slate. The process is generally insensitive to feed contaminants such as sulfur, nitrogen, metals, and ash.

Representative Fluid Coking Yields

Fluid Coking yields for three feedstocks are shown in Table 12.1.13.

Comparison of Flexicoking and Fluid Coking Yields

Table 12.1.14 presents a comparison of Flexicoking and Fluid Coking yields for a sample feedstock.

TABLE 12.1.13 Representative Fluid Coking Yields*

Feedstock	Middle East 1050°F+		Tia Juana medium 1050°F+		Bachaquero 1050°F+	
Density, °API	5.1		7.9		2.6	
Sulfur, wt %	3.4		3.0		3.7	
Nitrogen, wt %	0.77		0.52		0.81	
Conradson carbon, wt %	21.4		23.3		26.5	
Yields	wt %	LV %	wt %	LV %	wt %	LV %
C_4 – gas	11.8	—	12.1	—	12.6	—
Naphtha (C_5/360°F)	11.0	15.4	10.8	15.1	10.3	14.7
Gas oil (360/975°F)	50.8	55.1	47.9	52.2	44.2	48.3
Reactor coke†	26.4	—	29.2	—	32.9	—
Total	100.0	70.5	100.0	67.3	100.0	63.0

*Reactor yields listed are representative for a given feedstock. Yield pattern may be altered by changing coker reactor operating conditions.
†A portion of the reactor coke must be combusted to provide reactor heat requirements.

TABLE 12.1.14 Comparison of Flexicoking and Fluid Coking Yields

	Arabian 1050°F +	
	Fluid Coking	Flexicoking
Feed Properties:		
Conradson carbon, wt %	←24.4→	
Density, °API	← 4.4→	
Sulfur, wt %	←5.34→	
Feed rate, kB/SD	←25.0→	
Yields, % on fresh feed:		
C_4 –, wt %	←13.1→	
C_5/950°F, LV %	←65.1→	
Gross coke, wt %	←30.7→	
Net coke, wt % (ST/SD)	24.9 (1130)	0.61 (28)
Coke gas, LV % (FOEB)	—	18.80

Note: ST/SD = tons/stream day; FOEB = fuel oil equivalent barrels.

Fluid Coking Service Factor and Mechanical Reliability

A recent extensive study of Fluid Coking unit operating performance covering over 170 years of cumulative experience on 20 operating units provides the following results:

- Some of the units have been on-stream for over 25 years.
- An average mature service factor of about 90.5 percent has been obtained. Service factors of 95 percent are being achieved routinely at some units.
- Up to 2-year run lengths have been planned for and achieved.
- The maximum run length to date is 31 months.

Fluid Coking Commercial Experience

Fluid Coking is a well-established process, with 13 units built through the years since the first commercial coker started up in Billings, Montana, in 1954, as shown in Table 12.1.15. There is over 150,000 B/D of capacity processing conventional crude vacuum bottoms, and two units are processing over 200,000 B/D of Athabasca oil sands bitumen.

Figure 12.1.10 shows a Fluid Coking unit on the west coast of the United States. The burner vessel is in the foreground, with the reactor and scrubber behind. The scrubber overhead line to the fractionator descends on the right side of the reactor/scrubber.

TABLE 12.1.15 Commercial Fluid Coking Units

Company	Locations	Initial design fresh feed rate, B/SD	Initial start-up date	Comments
Exxon USA	Billings, Montana	3,800	Dec. 1954	Currently running 7700 B/SD
Exxon USA	Baltimore, Maryland	10,000	Oct. 1955	Refinery shut down 1957
Licensed unit	Montréal, Quèbec	3,800	Aug. 1956	Shut down 1976
Licensed unit	Detroit, Michigan	4,000	Oct. 1956	Shut down 1970
Licensed unit	Bakersfield, California	4,000	Apr. 1957	Shut down 1984
Licensed unit	Avon, California	42,000	June 1957	
Licensed unit	Delaware City, Delaware	42,000	Aug. 1957	Currently running 45,000 B/SD
Licensed unit	Purvis, Mississippi	4,800	Dec. 1957	Refinery shut down 1994
Licensed unit	Madero, Mexico	10,000	Feb. 1968	
Imperial	Sarnia, Ontario	14,000	Apr. 1968	Currently running 21,000 B/SD
Exxon USA	Benicia, California	16,000	Apr. 1969	Currently running 27,500 B/SD
Licensed unit	Mildred Lake, Alberta	2 × 73,000	July 1978	Current capacity 107,000 B/SD

FIGURE 12.1.10 Fluid Coking unit on the west coast of the United States.

CHAPTER 12.2
FW DELAYED-COKING PROCESS

Howard M. Feintuch
Kenneth M. Negin

Foster Wheeler USA Corporation
Clinton, New Jersey

It has been said that the most formidable processing and economic challenge of the 1980s and 1990s will be the production of clean transportation fuels from heavy, high-sulfur crudes.[1] To accomplish this an extensive use of residue conversion, or so-called bottom-of-the-barrel processing, will be required.

Statistically, in terms of number of units installed and total current operating capacity, it is quite easy to show that delayed coking is the residue-conversion process which is most often used today. In addition, because of its wide commercial acceptance, delayed coking has been referred to as the yardstick against which other, less commercially proven, processes must be measured.[2]

Despite its wide commercial use, only relatively few contractors and refiners are truly knowledgeable in delayed-coking design, so that this process carries with it a "black art" connotation.[3]

The year in which delayed coking was first developed is given in historical listings of petroleum advances as 1928.[4] We know that in early refineries severe thermal cracking of residue would result in the deposit of unwanted coke in the heaters. By evaluation of the art of heater design, methods were found by which it was possible to raise rapidly the temperature of the residue above the coking point without depositing the coke in the heater itself. Provision of an insulated surge drum downstream of the heater so that the coking took place after the heater, but before subsequent processing, resulted in the name "delayed coking."[5]

The next step was to add a second coke drum, which doubled the run length and led to the development of the art of switching coke drums while still maintaining operation.[6] In the early 1930s the drums were limited in size to 10 ft in diameter.[7] Coke drums as large as 28 ft in diameter have now been installed. Figure 12.2.1 shows the general trend in the growth in coke-drum diameter from 1930 to the present.

As of Jan. 1, 1995, there were in operation in the United States 45 delayed cokers with the capacity to process 1,598,000 BPSD of fresh feed.[8] Table 12.2.1 shows coking-plant-capacity statistics for the United States from 1946 to 1995.[9,10] Figures 12.2.2 and 12.2.3 present two delayed cokers, one operating on the United States Gulf Coast and the other in the Netherlands. (Abbreviations are defined in Table 12.2.17 at the end of the chapter.)

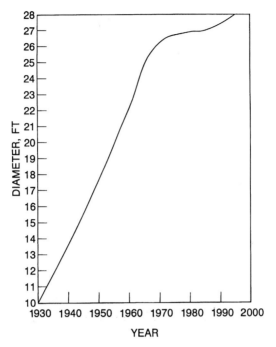

FIGURE 12.2.1 Trend in delayed-coker maximum drum size.

TABLE 12.2.1 United States Coking-Plant Statistics

Year	United States coking capacity, BPSD
1946	102,000
1950	158,000
1960	475,000
1970	835,000
1973	1,008,000
1982	1,118,100
1995	1,598,000

Source: W. L. Nelson, *Oil Gas J.*, **74**, 60 (May 24, 1976); *Oil Gas J.*, **80**, 81, 130 (Mar. 22, 1982); *Oil Gas J.*, **92**, 51, 53 (Dec. 19, 1994).

Since it appears certain that the overall quality of crude oil in terms of API density, sulfur content, and residue content will continue to worsen through the remainder of the twentieth century,[11] the need for delayed coking as an effective residue-conversion process is not likely to diminish.

FIGURE 12.2.2 Delayed coker in the Gulf Coast region of the United States. Center left and right each show four coke drums with drilling platforms for hydraulic decoking. Lower left shows two heaters and four associated stacks. Center right shows two heaters and their associated stacks.

PROCESS DESCRIPTION

Figure 12.2.4 is a process-flow diagram showing the coking, fractionator, coker blowdown, and steam-generation sections of a typical delayed coker. The associated vapor-recovery unit is shown separately in Fig. 12.2.5 and the coke-calcining plant separately in Figs. 12.2.6 and 12.2.7. A brief description of each of these sections is given below.

Coking Section

Reduced-crude or vacuum-residue fresh feed is preheated by exchange against gas oil products before entering the coker-fractionator bottom surge zone. The fresh feed is mixed with recycle condensed in the bottom section of the fractionator and is pumped by the heater charge pump through the coker heater, where the charge is rapidly heated to the desired temperature level for coke formation in the coke drums. Steam is often injected into each of the heater coils to maintain the required minimum velocity and residence time and to suppress the formation of coke in the heater tubes.

The vapor-liquid mixture leaving the furnace enters the coke drum, where the trapped liquid is converted to coke and light-hydrocarbon vapors. The total vapors rise upward through the drum and leave overhead.

A minimum of two drums is required for operation. One drum receives the furnace effluent, which it converts to coke and gas while the other drum is being decoked.

Fractionation Section

The coke-drum overhead vapors flow to the coker fractionator and enter below the shed section. The coke-drum effluent vapors are often "quenched" and "washed" with hot

FIGURE 12.2.3 An advanced needle coker in the Netherlands. Center shows two coke drums with drilling platforms for hydraulic decoking. Coke drums are elevated for gravity-flow, totally enclosed coke handling for environmental reasons. Left shows heater stack.

gas oil pumped back to the trayed wash section above the sheds. These operations clean and cool the effluent-product vapors and condense a recycle stream at the same time. This recycle stream, together with the fresh feed, is pumped from the coker fractionator to the coking furnace. The washed vapors pass to the rectifying section of the tower. A circulating heavy gas oil pumparound stream, withdrawn from the pumparound pan, is used to remove heat from the tower, condensing the major portion of heavy gas oil and cooling the ascending vapors. The hot pumparound stream of heavy gas oil withdrawn from the fractionator can be used to reboil the towers in the vapor-recovery plant, to preheat the charge to the unit, or to generate steam. The heavy gas oil product is partially cooled via exchange with the charge and air-cooled to storage temperature. Light gas oil product is steam-stripped to remove light ends, partially cooled via heat exchange with the charge, and air-cooled to storage temperature.

If a vapor-recovery unit is included in the design, then a sponge-oil system may be required. Lean sponge oil is withdrawn from the fractionator, cooled by heat exchange with the rich sponge oil, and then air-cooled before flowing to the top of the sponge absorber. Rich sponge oil is returned to the top heat-transfer tray above the lean-sponge-oil draw-off tray after preheat by exchange with the lean sponge oil.

The overhead vapors are partially condensed in the fractionator overhead condenser before flowing to the fractionator overhead drum. The vapor is separated from liquid in this vessel. The vapor flows under pressure control to the suction of the gas compressor in the vapor-recovery unit. The top of the fractionator is refluxed with part of the condensed hydrocarbon liquid collected in the overhead drum. The balance of this liquid is sent with the compressed vapors to the vapor-recovery unit. Sour water is withdrawn from the overhead drum and typically pumped to off-site treating facilities.

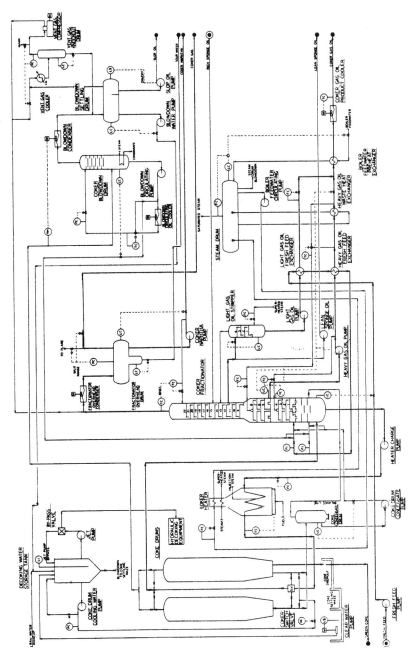

FIGURE 12.2.4 Process-flow diagram for a typical delayed coker.

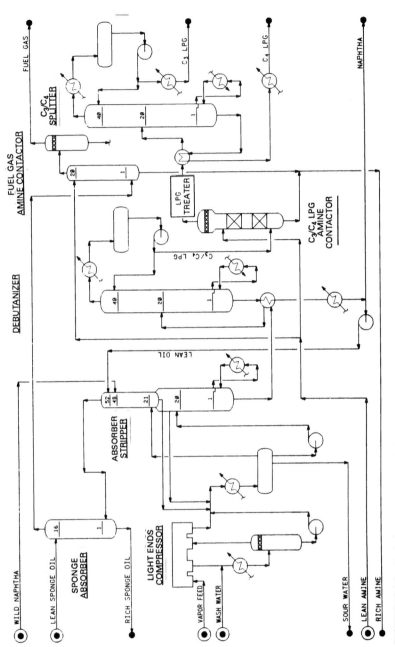

FIGURE 12.2.5 Typical vapor-recovery unit for a delayed coker.

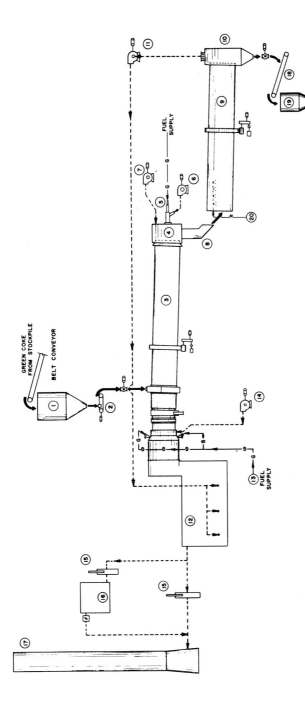

FIGURE 12.2.6 Simplified schematic of a coke-calcining plant; case A: rotary-kiln calciner. 1 = feed bin; 2 = weigh feeder; 3 = rotary kiln with scoop feeder; 4 = kiln firing hood; 5 = firing system; 6 = primary air fan; 7 = secondary air fan; 8 = transfer chute; 9 = rotary cooler; 10 = cooler discharge hood; 11 = cooler exhaust air fan; 12 = incinerator; 13 = incinerator auxiliary burners; 14 = incinerator air fan; 15 = dampers; 16 = waste-heat boiler with fan; 17 = stack; 18 = product conveyer; 19 = product bin; 20 = quench water. (*Courtesy of Kennedy Van Saun Corporation.*)

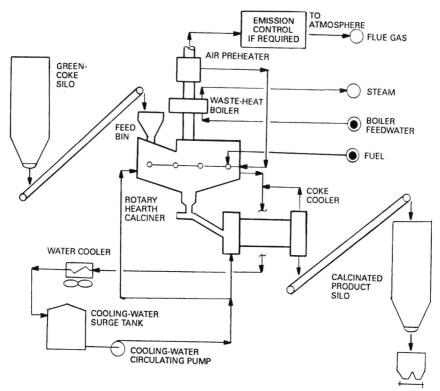

FIGURE 12.2.7 Simplified schematic of a coke-calcining plant; case B: rotary-hearth calciner.

Coker Blowdown

The coke-drum blowdown system serves the purpose of recovering hydrocarbon and steam vapors generated during the quenching and steaming operation. It is designed to minimize air pollution during normal operation. The system[7] includes a coker blowdown drum, blowdown condenser, blowdown settling drum, blowdown circulating oil cooler, vent-gas compressor system, and attendant pumps.

During the coke-drum cooling cycle, steam and wax tailings flow to the coker blowdown drum, where they are condensed by contact with a cooled circulating oil stream. This circulating oil stream also dilutes the wax tailings. The diluted wax tailings are withdrawn from the bottom of the drum and recirculated after cooling in the blowdown circulating-oil cooler. Excess oil is returned to the fractionator. Light gas oil makeup is charged to the coker blowdown drum as required for dilution of the mixture.

Steam and light hydrocarbons from the top of the coker blowdown drum are condensed in the blowdown condenser before flowing to the blowdown settling drum. In the settling drum, oil is separated from condensate. The oil is pumped to refinery slop, while the water is pumped either to off-site treating facilities or to the decoking-water storage tank for reuse.

Light hydrocarbon vapors from the blowdown settling drum are compressed in the vent-gas compressor after being cooled in the vent-gas cooler and separated from the resultant liquid in the vent-gas knockout drum. The recovered vent gas flows to the inlet of the fractionator overhead condenser. Alternatively, it may be sent directly to the fuel-gas-recovery system.

Steam Generation

The heat removed from the fractionator by the heavy gas oil pumparound stream is used to preheat feed and to generate steam. Depending on economics, additional steam may be generated in the convection section of the coker-fired heater. A common steam drum is utilized. Circulation through the steam-generating coil of the heater is provided by the boiler feedwater circulating pump.

Decoking Schedule

The decoking operation consists of the following steps:

1. *Steaming.* The full coke drum is steamed out to remove any residual-oil liquid. This mixture of steam and hydrocarbon is sent first to the fractionator and later to the coker blowdown system, where the hydrocarbons (wax tailings) are recovered.
2. *Cooling.* The coke drum is water-filled, allowing it to cool below 93°C. The steam generated during cooling is condensed in the blowdown system.
3. *Draining.* The cooling water is drained from the drum and recovered for reuse.
4. *Unheading.* The top and bottom heads are removed in preparation for coke removal.
5. *Decoking.* Hydraulic decoking is the most common cutting method. High-pressure water jets are used to cut the coke from the coke drum. The water is separated from the coke fines and reused.
6. *Heading and testing.* After the heads have been replaced, the drum is tightened, purged, and pressure-tested.
7. *Heating up.* Steam and vapors from the hot coke drum are used to heat up the cold coke drum. Condensed water is sent to the blowdown drum. Condensed hydrocarbons are sent to either the coker fractionator or the blowdown drum.
8. *Coking.* The heated coke drum is placed on stream, and the cycle is repeated for the other drum.

Typical coke-drum schedules for two-drum and six-drum delayed cokers are shown in Figs. 12.2.8 and 12.2.9. Although these are both 48-hour coking cycles, composed of 24 hours of coking and 24 hours of decoking, they are often referred to as *24-hour cycles.* Refiners sometimes operate on "short cycles," which have cycle times less than 24 hours. This has an operating advantage in cokers that were designed for a 24-hour cycle. It allows the refiner to increase the unit throughput by filling the coke drums faster. The refiner takes advantage of the inherent design margins in the rest of the unit's equipment to process this increased capacity. If necessary, the rest of the unit may require a revamp to handle the extra capacity, but this can readily be achieved. Refiners have reported short cycles as low as 11 hours in small cokers, but 14 to 16 hours is more typical. By using short cycles for a new design, smaller coke drums would be required with a reduced investment cost.

Vapor-Recovery Unit

The vapor and liquid streams from the fractionator overhead drum are processed further in the vapor-recovery unit. The liquid stream goes directly to the top of the absorber. The vapor stream is compressed and cooled, the resulting vapor and liquid streams are fed to the absorber-stripper, the vapor goes to the bottom of the absorber, and the liquid goes to the top of the stripper.

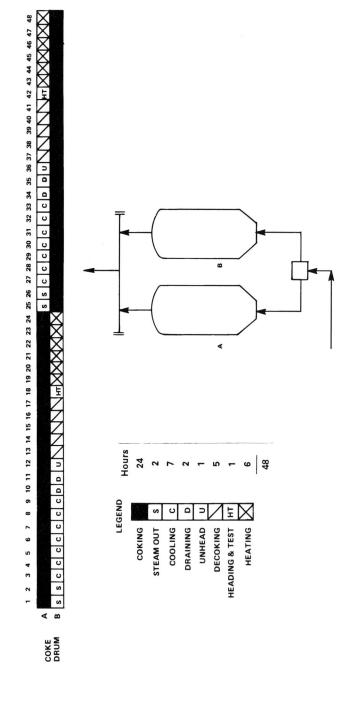

FIGURE 12.2.8 Typical coke-drum cycle for two drums.

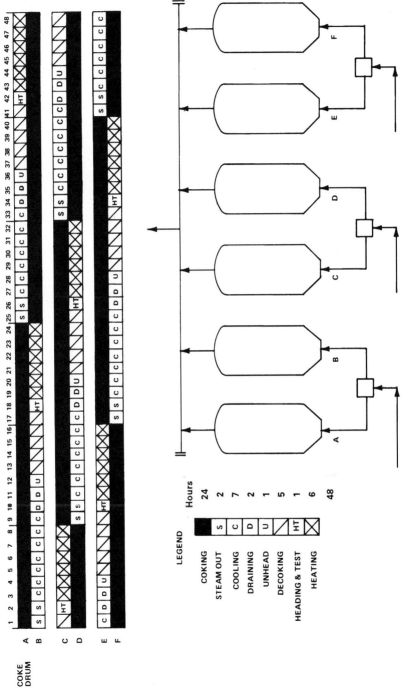

FIGURE 12.2.9 Typical coke-drum cycle for six drums.

The absorber-stripper produces a bottoms stream that contains most of the C_3 and heavier material in the feed. The overhead from the absorber contains the C_2 and lighter portion, plus some unrecovered C_3 and heavier material. This unrecovered C_3 and heavier material is recovered in the sponge absorber and recycled back to the fractionator. The C_2 and lighter portion leaving the top of the sponge absorber passes through an amine absorber, where the hydrogen sulfide is removed, before it goes on to the fuel-gas system. The sponge absorber uses a side cut from the fractionator as an absorbing medium.

The bottoms from the stripper flow to the debutanizer, where the C_3 and C_4 are removed overhead, leaving a stabilized naphtha as a bottoms product. The naphtha can go to product storage or to further processing, as required.

The debutanizer distillate, or C_3-C_4 liquefied petroleum gas (LPG), goes to a "treating" section, where hydrogen sulfide, mercaptans, and other sulfur compounds are removed. This treating section usually has an amine absorber (liquid-liquid contactor), followed by a mercaptan-removal processing facility. From here the stream flows into the C_3-C_4 splitter, where the feed is separated into C_3 and C_4 LPG products.

Coke-Calcining Plant

Two methods for calcining coke are available commercially. They are the rotary-kiln method, as shown in Fig. 12.2.6, and the rotary-hearth method, as shown in Fig. 12.2.7. The rotary-kiln method is the older of the two methods and has been in use for many years. The rotary-hearth method recently has been gaining increased popularity. The two methods are similar in concept but differ in mechanical details. The description given below is specific to the rotary-kiln method.

Coke which has not yet been calcined for removal of excess moisture and volatile matter is referred to as "green" coke. After draining, the coke is charged to a crusher and then to a kiln feed bin or bins. The rate of charge to the kiln is controlled by a continuous-weigh feeder.

In the kiln first the residual moisture and subsequently the volatile matter are removed, as the green coke moves countercurrently to the heat flow. Process heat is supplied to the kiln through a burner which is designed to handle the available fuel. Another source of process heat is combustion of the volatile matter released by the green coke in the kiln.

The calcined coke leaving the kiln is discharged into a rotary cooler, where it is quenched with direct water sprays at the inlet and then cooled further by a stream of ambient air which is pulled through the cooler. The coke is conveyed from the rotary cooler to storage.

A rotary kiln coke calciner is shown in Fig. 12.2.10.

FEEDSTOCKS

Heavy residues such as vacuum residue or occasionally atmospheric residue are the feedstocks which are most commonly used in delayed coking. For special applications in which high-quality needle coke is desired, certain highly aromatic heavy oils or blends of such heavy oils may be used instead. The discussion which follows describes various types of feeds and their characteristics both for routine and for specialized delayed-coking applications.

FIGURE 12.2.10 Coke-calcining plant. The plant is for the Martin Marietta Carbon Co., Wilmington, Calif. The photograph shows a 13½- by 270-ft rotary kiln for a plant processing 300,000 tons/year of calcined coke. (*Photograph courtesy of Martin Marietta Carbon Co. and Kennedy Van Saun Corporation.*)

Regular-Grade Coke Production

The most common type of coke produced by a majority of the delayed cokers in operation today is a regular-grade coke known as *sponge coke*. As we will discuss later in the section "Uses of Petroleum Coke," depending on the impurity levels present, the coke may be suitable for use in the manufacture of electrodes for the aluminum industry or alternatively for use as a fuel.

Petroleum residue from a refinery vacuum tower, less frequently from an atmospheric tower, or sometimes from a mixture of both, is the feed which is typically used in the production of regular-grade coke.

Table 12.2.2 shows the most important feedstock characteristics for several vacuum residues.

Carbon Residue. In determining the quantity of coke that will be produced from any particular feedstock, the most important characteristic to be considered is the carbon residue. The carbon residue may be defined as the carbonaceous residue formed after evaporation and pyrolysis of a petroleum product.[17] Two methods of testing are available. They are the Conradson carbon test (ASTM D 189) and the Ramsbottom coke test (ASTM D 524). For the purposes of our discussion we will be concerned with the Conradson carbon residue (CCR). The higher the CCR, the more coke that will be produced. Since, in most cases, the object of delayed coking is to maximize the produc-

TABLE 12.2.2 Feedstock Characteristics for Various Vacuum Residues

Crude source	African	Southeast Asian	Mexican	Middle East
TBP cut point, °C	482+	482+	538+	538+
Density, °API	12.8	17.1	4.0	8.2
Conradson carbon, wt %	5.2	11.1	22.0	15.6
Sulfur, wt %	0.6	0.5	5.3	3.4
Metals (Ni+V), wt ppm	50	44	910	90

tion of clean liquid products and minimize the production of coke,[7] the higher the CCR the more difficult this is to achieve.

Although CCR values may formerly have ranged from less than 10 wt % to rarely more than 20 wt %, with the trend in recent years toward processing heavier crudes, values of CCR in excess of 20 wt % and sometimes higher than 30 wt % are becoming more common.

Sulfur. Sulfur is an objectionable feed impurity which tends to concentrate in the coke and in the heavy liquid products. In a manner similar to CCR, the trend in recent years, owing to increased processing of less desirable, heavier, higher-sulfur crudes, is for a resultingly higher sulfur content in the feed to a delayed coker. This results in corresponding high levels of sulfur in the coke and in the heavy-liquid products.

Metals. Metals such as nickel and vanadium are objectionable feed impurities which tend to be present in increasing quantities in heavier feeds. The metals present in the feed tend to concentrate almost entirely in the coke.[13] Some heavy feeds contain metals in excess of 1000 wt ppm.

TBP Cut Point. For vacuum residues a typical true boiling point (TBP) cut point is 538°C, but it may be lower or higher depending on the crude. For atmospheric residues a TBP cut point of 343°C is typical. The TBP cut point will define the concentration of CCR, sulfur, and metals in the feed and thereby affect yields and product quality.

Needle Coke Production

Needle coke is a premium coke used in the manufacture of high-quality graphite electrodes for the steel industry. It owes this application to its excellent electrical conductivity, good mechanical strength at high temperatures, low coefficient of thermal expansion, low sulfur content, and low metal content.

In general, vacuum or atmospheric crude residues are not suitable feedstocks for needle coke production. What is needed instead is a heavy feedstock which is highly aromatic and, in addition, is low in sulfur and low in metals. Table 12.2.3 gives a comprehensive list of potential feedstock sources for needle-coke production. Depending upon the specific properties of a particular feedstock in question, it may or may not prove to be suitable for needle coke production.

Table 12.2.4 lists the aromatic content, sulfur content, and CCR for three feedstocks which are known to be suitable for needle coke production. For the feedstocks shown we can see that, in general, the aromatics content is greater than 60 liquid volume percent (LV %), the sulfur content is less than 1 wt %, and the CCR is less than 10 wt %.

Residue Hydrodesulfurization of Feedstocks

To combat the impurity and yield problems which result from using very heavy poor-quality residues as feedstocks, there is a growing trend to employ residue hydrodesulfurization upstream of the delayed-coking unit.[7] When this is done, the metals and CCR, as well as the sulfur level of the feedstock, are reduced. This results in an attractively lower yield of higher-purity coke and a resultingly higher yield of clean liquid products.

To show how beneficial this is, Figs. 12.2.11, 12.2.12, and 12.2.13 illustrate the effects of typical residue hydrodesulfurization on delayed-coker yields.[7] Yields for three alternative operations on Kuwait atmospheric residue were estimated and are presented as follows:

TABLE 12.2.3 Potential Feedstocks for Needle Coke Production

Thermal tars
Vacuum-flashed thermal tars

Decant oils (slurry oils)
Thermally cracked decant oil

Pyrolysis tars
Topped pyrolysis tars
Thermally cracked pyrolysis tar

Lubricating-oil extract
Thermally cracked coker gas oils

Synergistic mixtures
    Decant oil–pyrolysis tar
    Decant oil–pyrolysis tar–vacuum residue
    Decant oil–thermal tar
    Decant oil–thermal tar–vacuum residue
    Thermally cracked vacuum gas oil–coker gas oil
    Thermal tar–pyrolysis tar
    Pyrolysis tar–hydrotreated FCC gas oil

TABLE 12.2.4 Needle Coke Feedstock Characteristics

Item	Slurry oil	Thermal tar no. 1	Thermal tar no. 2
Aromatic content, LV %	61.7	89.8	66.1
Sulfur content, wt %	0.48	0.07	0.56
Conradson carbon residue, wt %	5.7	9.4	8.6

Source: D. H. Stormont, *Oil Gas J.,* **67,** 75 (Mar. 17, 1968).

Figure 12.2.11 shows a scheme in which atmospheric residue is sent to a vacuum flasher and the resulting vacuum residue is fed to a delayed coker. Figure 12.2.12 shows a scheme in which atmospheric residue is sent to a residue hydrodesulfurizer. The resultant 650°F+ residue is then charged to the delayed coker. Figure 12.2.13 is the same scheme as Fig. 12.2.12 except that the 650°F+ desulfurized residue is charged to a vacuum flasher. The desulfurized vacuum residue is then charged to the delayed coker.

A comparison of the overall yields for the three cases is summarized in Table 12.2.5. From this table it is easy to see how residue hydrodesulfurization of the feedstock increases the yield of desirable liquid products.

YIELDS AND PRODUCT PROPERTIES

This section describes the reactions and the types of products which normally are produced by delayed cokers and gives typical yield predictions for these products. Also given is information on product impurities and typical product properties.

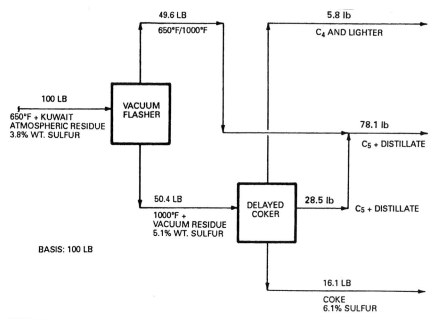

FIGURE 12.2.11 Scheme A: vacuum distillation followed by delayed coking.

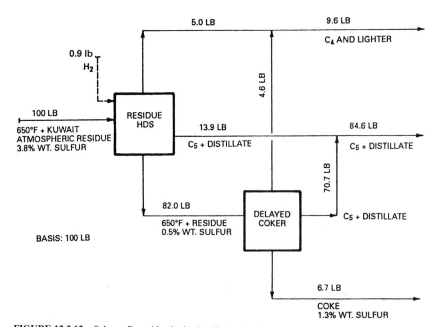

FIGURE 12.2.12 Scheme B: residue hydrodesulfurization followed by delayed coking.

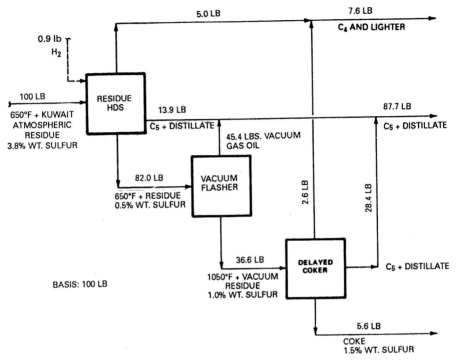

FIGURE 12.2.13 Scheme C: residue hydrodesulfurization followed by vacuum distillation followed by delayed coking.

TABLE 12.2.5 Comparison of Estimated Overall Yields of Delayed Coking plus Residue Hydrodesulfurization

Properties of feedstock	
Crude source	Kuwait
TBP cut point, °C	343+
Density, °API	16.6
Conradson carbon residue wt %	9.0
Sulfur, wt %	3.8

Yields, wt %	Vacuum flasher plus delayed coker	Residue hydrodesulfurizer– plus delayed coker	Residue hydrodesulfurizer– vacuum flasher plus delayed coker
C_4 and lighter	5.8	9.6	7.6
C_5 + distillate	78.1	84.6	87.7
Coke	16.1	6.7	5.6
Total	100.0	100.9	100.9

Reactions

Delayed coking is an endothermic reaction with the furnace supplying the necessary heat to complete the coking reaction in the coke drum. The exact mechanism of delayed coking is so complex that it is not possible to determine all the various chemical reactions that occur, but three distinct steps take place:

1. Partial vaporization and mild cracking of the feed as it passes through the furnace
2. Cracking of the vapor as it passes through the coke drum
3. Successive cracking and polymerization of the heavy liquid trapped in the drum until it is converted to vapor and coke

Products

Four types of products are produced by delayed coking: gas, naphtha, gas oil, and coke. Each of these products is discussed briefly below.

Gas. Gas produced in the coker is fed to a vapor-recovery unit, where LPG and refinery fuel gas are produced. Typically, the LPG, after treatment for H_2S and mercaptan removal, is split into separate C_3 and C_4 products. Coker LPG can also be used as alkylation or polymerization unit feedstock. For this purpose, the coker LPG is often mixed with catalytic cracker LPG.

Naphtha. Light coker naphtha, after stabilization in the vapor-recovery unit, is often mercaptan-sweetened and then used in the gasoline pool. Heavy coker naphtha can be hydrotreated and used either as catalytic-reformer feedstock or directly in the gasoline pool.

Gas Oil. Light coker gas oil can be hydrotreated for color stabilization and used in the refinery distillate blend pool for No. 2 heating oil. The heavy or the total coker gas oil is commonly used as catalytic cracker or hydrocracker feedstock. This use of the coker gas oil can result in a considerable increase of refinery gasoline, jet fuel, or diesel production.

Coke. Depending on the unit feedstock and operating conditions, different types of cokes an be produced. These are discussed in detail in the sections "Feedstocks" and "Uses of Petroleum Coke."

Predicting Yields

Because the correlations used to predict coking yields are, in general, considered to be proprietary information to the companies which have developed these correlations, relatively little information is given in the published literature on how to predict coking yields. Nelson[10] and Gary and Handwerk[5] give some simple correlations which, as shown and discussed below, are adequate to make very rough coking yield estimates. For more precise yield predictions, more sophisticated correlations, developed from nonpublished comprehensive pilot plant data and/or from commercial operating data, must be used instead.

Predicting Yield of Coke. The most important parameter in predicting the yield of coke is the CCR (weight percent) in the feed. Figure 12.2.14, developed by Nelson,[10]

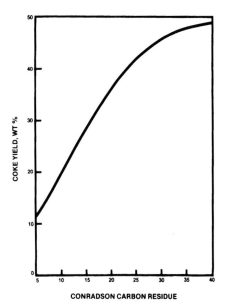

FIGURE 12.2.14 Rough estimation of coke yields from straight-run residue. [*Courtesy of Pennwell Publishing Company, publishers of the* Oil and Gas Journal, *74, 60 (May 24, 1976).*]

shows a very simple correlation which can be used to predict a rough preliminary coke yield based only on CCR.

An even simpler correlation for estimating a rough coke yield is the following equation, as given by Gary and Handwerk,[5]

$$\text{Coke yield (wt \%)} = 1.6 \times \text{CCR} \qquad (12.2.1)$$

Figure 12.2.14 and Eq. (12.2.1) are suitable for predicting preliminary coke yields to within an accuracy of perhaps ±25 percent. More precise predictions can be made only by using more sophisticated proprietary correlations which are a function not only of CCR but of other additional parameters such as the variables described in the section "Operating Variables."

Predicting Yields of Gas and Liquid Products. As in predicting the yield of coke, the CCR is still the most important single parameter for predicting the yields of gas and liquid products. Gary and Handwerk[5] give the following set of equations which can be used to make rough preliminary estimates of the yield of gas and liquid products as a function of CCR:

$$\text{Gas yield (wt \%)} = 7.8 + 0.144 \times \text{CCR} \qquad (12.2.2)$$

$$\text{Naphtha yield (wt \%)} = 11.29 + 0.343 \times \text{CCR} \qquad (12.2.3)$$

$$\text{Gas oil yield (wt \%)} = 100 - \text{coke yield} - \text{gas yield} - \text{naphtha yield} \qquad (12.2.4)$$

Distribution of Impurities among Products

The two impurities in the products from delayed coking which are of greatest concern are sulfur and metals. As a rule of thumb, the weight percent of sulfur in the coke will be somewhat greater than that in the feedstock. The ratio of these two numbers will usually range between 1:1 (or slightly less) and 2:1. The weight percent of sulfur in the other products varies greatly with each particular feedstock, and although some limited information is published,[12] it is nevertheless difficult to make any generalizations as to how the sulfur will be distributed. With regard to metals, the bulk of the metals present in the feedstock generally will concentrate in the coke, with a very small percentage remaining in the heavy gas oil product.[13]

Typical Yields and Product Properties

Table 12.2.6 presents typical delayed-coking yield estimates for the various vacuum-residue feedstocks defined in Table 12.2.2. The feedstocks have been selected to illustrate typical coking yields over a wide range of feedstock characteristics varying from 4.0° API and 22.0 wt % CCR to 12.8° API and 5.2 wt % CCR. The metals content in the feedstocks varies from 44 to 910 wt ppm (Ni + V). All the yields are presented at conditions of constant recycle ratio and pressure. The delayed-coking yields presented in Table 12.2.6 have been estimated by generalized correlations developed by Foster Wheeler on the basis of previous pilot-plant work and commercial operations.

For the typical yields presented in Table 12.2.6 we see that the yield of dry gas varies between 6.2 and 10.5, the yield of naphtha between 17.4 and 21.4, the yield of gas oil between 33.0 and 65.3, and the yield of coke between 10.0 and 35.1. These estimated values are not meant to represent any absolute maximum or minimum for any of the yields given, but rather a typical range over which yields may vary.

TABLE 12.2.6 Typical Yields and Product Properties as Estimated for Various Delayed-Coker Feedstocks at Constant Recycle Ratio and Pressure

	Crude source			
Products	African	Southeast Asian	Mexican	Middle East
Dry gas and C_4, wt %	6.2	7.4	10.5	9.2
C_5- 193°C, naphtha, wt %	18.5	20.4	21.4	17.4
Density, °API	56.1	62.3	54.9	58.3
Sulfur, wt %	0.1	0.2	0.9	0.5
193°C+, gas oil, wt %	65.3	54.5	33.0	48.5
Density, °API	22.4	34.9	20.5	25.3
Sulfur, wt %	0.59	0.42	4.26	2.28
Coke, wt %	10.0	17.7	35.1	24.9
Sulfur, wt %	1.1	0.8	6.4	5.1
Ni + V, wt ppm	500	249	2592	361

TABLE 12.2.7 Estimated Yields and Product Properties for Needle-Coke Production

	Visbroken thermal tar	Decanted oil
Feed:		
°API	2.4	−0.66
Sulfur, wt %	1.0	0.45
Products:		
Dry gas+C_4, wt %	14.4	9.8
C_5 − 193°C, wt %	16.7	8.4
°API	54.9	59.8
Sulfur	0.04	0.01
193°C+, wt %	15.7	41.6
°API	23.3	16.9
Sulfur, wt %	0.7	0.34
Coke, wt %	53.2	40.2
Sulfur, wt %	1.0	0.60

All the yields and properties given in Table 12.2.6 are for vacuum-residue feedstocks which produce regular grade coke. Yields and properties for special feedstocks which produce needle coke are given in Table 12.2.7. The high coke yields shown in Table 12.2.7 are consistent with the philosophy of selecting operating conditions which would favor the production of needle coke for the special feedstocks which are being processed.

OPERATING VARIABLES

Three basic operating variables contribute to the quality and yields of delayed-coking products. They are temperature, pressure, and recycle ratio. Each of these is discussed below, and typical ranges are shown in Table 12.2.8.

Temperature

Temperature is used to control the volatile combustible material (VCM) content of the coke product. The current trend is to produce coke with a VCM ranging between 6.0 and 8.0 wt %. This results in a harder coke and, if structure and impurity levels are acceptable, in a more desirable aluminum-grade coke. At constant pressure and recycle ratio the coke yield decreases as the drum temperature increases. Since delayed coking is an endothermic reaction, the furnace supplies all the necessary heat to promote the coking reaction. If the temperature is too low, the coking reaction does not proceed far enough and pitch or soft-coke formation occurs. When the temperature is

TABLE 12.2.8 Operating Variables: Typical Ranges

Heater outlet temperature, °C	468–524
Top coke-drum pressure, lb/in^2 gage	15–150
Recycle ratio, volume recycle/volume fresh feed	0.05–2

Pressure

At constant temperature and recycle ratio, the effect of increased pressure is to retain more of the heavy hydrocarbons in the coke drum. This increases the coke yield and slightly increases the gas yield while decreasing the pentane and heavier liquid-product yield. The trend in the design of delayed cokers which maximize the yield of clean liquid products is to design for marginally lower operating pressures. This tendency is the result of close scrutiny of conditions which affect refining profit margins. The use of a heavier coker feedstock which produces fuel-grade coke having a market value 15 to 30 percent of that for aluminum-grade coke drives design economics to the absolute minimum coke yield, even though it results in an increased expense for vapor-handling capacity. As a result, units are currently being designed with coke-drum pressures as low as 15 lb/in^2 gage. Table 12.2.9[7] compares the effect on delayed-coker yields of a 15-lb/in^2 gage coke-drum pressure with that of a more traditional 35-lb/in^2 gage pressure for the same feedstock at constant conditions of recycle ratio and temperature.

This tendency to operate at lower pressure applies to most standard operations but does not apply to the special case of needle coke production. For needle coke production, a pressure as high as 150 lb/in^2 gage may be required.[15]

TABLE 12.2.9 Estimated Effect of Pressure on Delayed-Coking Yields

Feedstock	
Crude source	Alaskan North Slope
TBP cut point, °C	566+
Density, °API	7.4
Conradson carbon, wt %	18.1
Sulfur, wt %	2.02

Estimated yields at constant recycle ratio		
	Coke-drum pressure	
Products	15 lb/in^2 gage	35 lb/in^2 gage
Dry gas + C$_4$, wt %	9.1	9.9
C$_5$ − 380°F, naphtha, wt %	12.5	15.0
Density, °API	60.4	57.1
Sulfur, wt %	0.5	0.5
380°F+, gas oil, wt %	51.2	44.9
Density, °API	23.8	26.0
Sulfur, wt %	1.36	1.22
Coke, wt %	27.2	30.2
Sulfur, wt %	2.6	2.6

Source: R. DeBiase and J. D. Elliott, "Recent Trends in Delayed Coking," NPRA Annual Meeting, San Antonio, March 1982.

FW DELAYED-COKING PROCESS

Recycle Ratio

Recycle ratio has the same general effect as pressure on product distribution; i.e., as the recycle ratio is increased, the coke and gas yields increase while the pentane and heavier liquid yield decreases. The recycle ratio is used primarily to control the end-point of the coker gas oil. The same economics which are forcing the operation of cokers to lower operating pressures are also at work on recycle ratios. Units operating at recycle ratios as low as 3 percent have been reported. In general, a refinery operates at as low a recycle ratio as product quality and unit operation will permit.

Other Variables

Although our discussion in this section has been directed solely to operating (or process) variables, it is possible to consider delayed coking as dependent upon three interrelated classes of variables.[16] These are feedstock variables, processing variables,

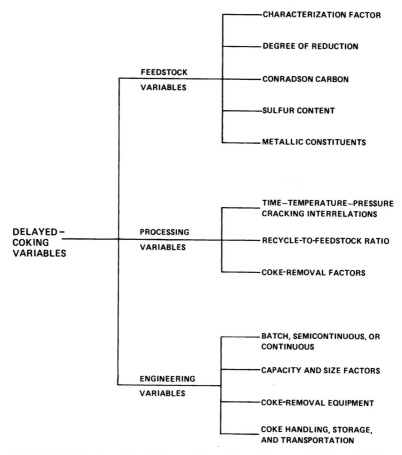

FIGURE 12.2.15 Interrelated delayed-coking variables [*Courtesy of McGraw-Hill Book Company (Virgil B. Guthrie,* Petroleum Products Handbook, *McGraw-Hill, New York, 1960).*]

and engineering variables. Figure 12.2.15 gives an interesting graphical representation of each of these variables.

COKER HEATERS

Careful heater design is critical to successful delayed-coking operation and to the achievement of desirable long run lengths. The major design parameters which influence heater operation are discussed below.

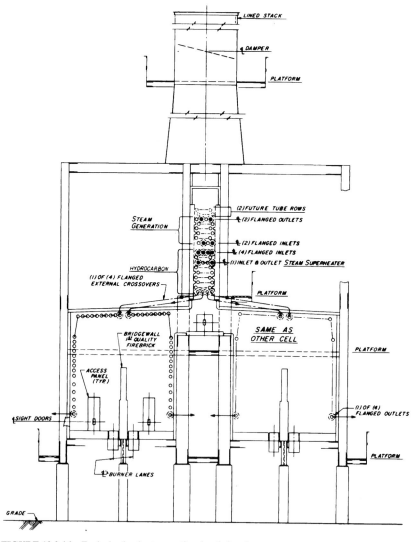

FIGURE 12.2.16 Typical coker heater: sectional end elevation.

Heater Design

Figure 12.2.16 is a simple sketch of a typical delayed-coker heater. Coker heater design has been modified in recent years in response to refiners' needs for longer run lengths between decoking while processing heavier, higher-CCR feedstocks. There is a trend to design for higher cold-oil velocities in the order of 6 ft/s and provide for the multiple injection of steam into the heater coil to adjust coil residence time and velocity.

On a number of recent designs, in addition to more liberal firebox dimensions, there has been a tendency to specify lower allowable average radiant-flux rates in the order of 9000 Btu/(h · ft^2) to provide for longer run lengths, future capacity allowances, and, in general, a more conservative heater design. By way of comparison, traditional allowable average radiant-flux rates range from 10,000 to 12,000 Btu/(h · ft^2).

During the operation of the delayed coker unit, coke slowly deposits on the inside of the heater tubes. This results in higher pressure drop and higher tube metal temperatures. When one or both of these operating variables exceed design levels the heater must be decoked. The duration of time from the time the heater is put on-stream until it is shut down is defined as the *run length*.

Heater run length is affected by feedstock quality, operation conditions and the consistency with which they are maintained, and the frequency and handling of upset operations. Run lengths varying between 9 and 12 months can be expected, with run lengths of 18 months or longer being reported.

On-line spalling is a technique that is sometimes used to extend the heater run length of multipass heaters beyond these values. On-line spalling does not require a heater shutdown with a resultant loss of unit production. In this technique, the passes of the heater are decoked one at a time while the other passes remain in coking service. For the pass being spalled, the hydrocarbon fresh feed is shut off and a spalling medium, either steam or condensate, is immediately introduced to that pass. The rate and temperature of the medium are cycled in a prescribed manner so that the coke is thermally stressed until it breaks off the heater tubes and is swept into the coke drums. The flow of hydrocarbon to the other passes is sometimes increased to compensate for the loss from coking service of the pass being spalled.

The effectiveness of on-line spalling is shown in Fig. 12.2.17, a plot of tube skin temperatures versus time for each pass of a four-pass heater. There are temperatures in a band between 1140 and 1180°F which are the target temperatures for applying the on-line spalling procedure. The figure illustrates the effect of the on-line spalling procedure on skin temperatures when practiced at the end of April and the beginning of August. It shows that the heater run length would have been only 3 months if the on-line spalling procedure had not been used. It has been reported by some refiners that the effectiveness of the procedure deteriorates after every cycle and that ultimately the heater needs to be shut down and decoked; other refiners have reported that the effectiveness is constant and that the heater can be run indefinitely by practicing on-line spalling. In any event, refiners have reported run lengths greater than 2 years.

The alternative to on-line spalling is to shut down the heater and decoke it by either steam-air decoking or by pigging. Steam-air decoking utilizes steam and air to first spall some of the coke from the heater tubes and then to burn off the remaining coke. In pigging, the pig, a semiflexible plug with projections, is inserted into the tubes and circulated in a water stream. The projections scrape the coke off the inside of the tubes.

The effect of energy conservation on modern heater design is discussed in the section "Typical Utility Requirements."

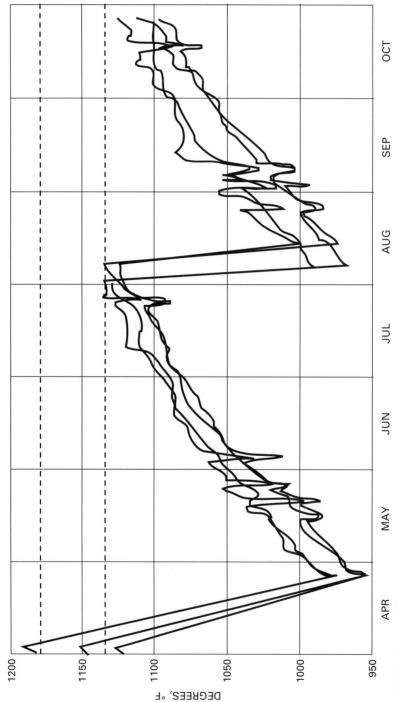

FIGURE 12.2.17 Effectiveness of on-line spalling as measured by tube skin temperatures on a four-pass furnace.

HYDRAULIC DECOKING

Unheading Device

Coke drums have a large, flanged opening approximately 6 ft in diameter at the bottom to facilitate the decoking operation. At most times this opening is closed by a large metal cover containing a nozzle, called a *head*. In order to remove the coke from the drum the head is removed and a chute is attached to the flange. The chute directs the coke away to the coke handling and coke dewatering area. The head is removed after the coke is cooled and the cooling water is drained from the coke drum.

Traditionally, the head was lowered on to an unheading cart. Recently, attention has been paid to facilitating this operation by the development of automatic unheading systems. One such system was developed by Foster Wheeler. The Foster Wheeler bottom coke drum unheading device is designed for remote-controlled hydraulic raising and lowering of the coke drum bottom head and coke chute. It dramatically improves the safety and ease of the unheading operation by making it possible to have all the operators at a safe distance from the coke drum when the bottom cover first separates from the drum as well as when the coke chute is raised to its decoking position.

Unlike other devices, the Foster Wheeler system not only separates the head from the coke drum, but also moves the head out of the way and raises the coke chute up to the drum in preparation for the decoking operation. When the decoking operation is complete, the coke chute is lowered and the bottom cover is raised to prepare for the next coking cycle, all from a remote location.

The Foster Wheeler unheading system operates without a cart. Instead, it uses a skid and cradle assembly to move the head out of the way from under the drum. This is an improvement over previous designs in that it reduces the possibility of cart failure, minimizes floor loadings, and allows for safer operation.

The new unheading system holds the bottom cover in place while the cover bolts are safely removed. The unheading device and vertical cylinders are capable of withstanding the total load of coke and water in the drum that may be applied on the bottom cover.

In this system, the vertical cylinders are attached to the concrete support structure. With this design, the load applied by the vertical cylinders is transferred to the structure directly.

A computer-aided design (CAD) representation of the unheading system is in Fig. 12.2.18. Three views are depicted. The top-left view shows the device in coking position with the head raised to meet the coke drum flange and the inlet piping connected. The top right view shows the device right after unheading with the inlet piping detached and the bottom head moved away from the drum. The bottom view shows the device in the decoking position. The telescopic chute is raised and attached to the bottom flange of the coke drum.

Description of System

In early delayed-coking units, cable decoking was used to remove the coke from the drums. Later a more sophisticated method was developed. This method employed a hydraulically operated mechanical drill to remove the coke from the drums. In the late 1930s hydraulic decoking was introduced and is still in use today.

Today hydraulic decoking utilizing high-impact water jets, which operate at approximately 2000 to 4000 lbf/in^2, is the standard method of removing coke formed in the coke drums. This method has replaced older methods, such as coiled wires and mechanical drills. These older methods were unsuitable for larger drums and were costly in terms of maintenance.

FIGURE 12.2.18 Foster Wheeler's advanced coke drum unheading system.

In hydraulic decoking, the coke is cut as the water jet impacts on the coke. Both boring and cutting tools are used; each tool produces several jets of water from high-pressure nozzles. The coke is removed in essentially two operations:

1. The boring tool, with jet nozzles oriented vertically downward, is used to bore hydraulically a pilot hole, which is typically 2 to 3 ft in diameter, down through the coke from the top. See Fig. 12.2.19 for a typical sketch of a boring tool.
2. The cutting tool, with jet nozzles oriented horizontally, is used to cut hydraulically the coke from the drum after the pilot hole has been drilled. See Fig. 12.2.20 for a sketch of a typical cutting tool.

Effective boring and cutting by the water jets are accomplished by rotating and lowering the respective tools into the drum. The boring and cutting tools are attached to a hollow drill stem which also rotates and supplies high-pressure water to the boring and cutting tools. The drill stem is rotated by an air motor connected to the power swivel. The power swivel is the rotary joint which connects the nonrotating water-supply line to the drill stem.

The high-pressure water is supplied to the cutting assembly by the jet pump and delivered via a piping manifold and rotary drilling hose connected to the power swivel.

The drill stem and power swivel assembly are raised and lowered by an air-motor hoist via a wire rope and a set of sheave blocks. The power swivel is attached to a crosshead which is guided in its vertical travel by a pair of channel-type crosshead guides. A drill rig is used to support the entire assembly. Figure 12.2.21 indicates the relationship of the major components in the hydraulic-decoking system.

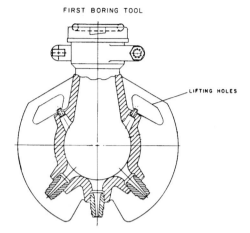

FIGURE 12.2.19 Boring tool for hydraulic decoking. (*Courtesy of Worthington Division, McGraw Edison Co.*)

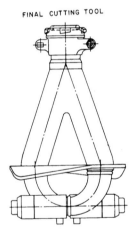

FIGURE 12.2.20 Final cutting tool for hydraulic decoking. (*Courtesy of Worthington Division, McGraw Edison Co.*)

Sequence of Operations

The sequence for cutting the coke out of a drum is shown schematically in Fig. 12.2.22 and is outlined in the following four steps:

1. Hydraulically bore a pilot hole through the coke with high-pressure cutting water.
2. Replace the boring tool with the final cutting tool and widen the original pilot hole.

12.54 VISBREAKING AND COKING

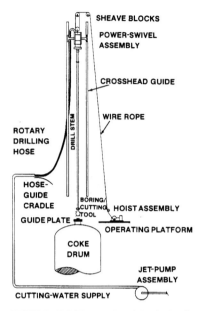

FIGURE 12.2.21 Details of the hydraulic decoking system.

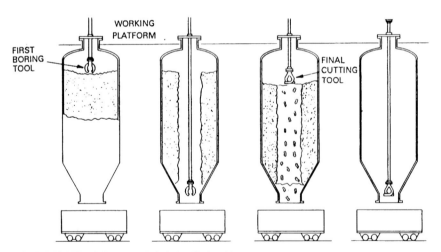

FIGURE 12.2.22 Coke removal via hydraulic decoking.

3. Remove coke from the bottom cone section.
4. Cut coke out of the drum, starting from the top and working downward in layers.

Care must be taken in the procedures employed in carrying out steps 1 and 2 in order to minimize the amount of undesirable coke fines that are produced in the decoking operation.

COKE-HANDLING AND -DEWATERING SYSTEMS

During the decoking operation, large volumes of coke and drilling water drop out of the bottom of the coke drum. The most common coke-dewatering and -handling systems are direct railcar loading, pad loading, pit loading, and dewatering bins.[7] Each of these systems is discussed below.

Direct Railcar Loading

Direct railcar loading is usually restricted to two drum cokers. This system allows coke and water to drop from the coke drum directly into a railcar positioned underneath the coke drum. The water and some coke fines drain from the car and are directed into a sump. Final removal of coke fines from the water before its reuse is generally accomplished by a clarifier. Although direct railcar loading has the lowest investment, the disadvantage of this system is the dependency of the decoking operation on the availability and movement of railcars. Figure 12.2.23[7] depicts a typical schematic of the decoking water system associated with direct railcar loading.

Pad Loading

Pad loading is generally limited to two or four drum cokers. This system allows the coke and water to drop out of the drum, through a chute, directly onto a large concrete pad adjacent to the coke drums. In traditional designs, water and coke fines then flow through a series of ports located at the periphery of the pad. The ports are packed with sized coke as a filtering medium. Most of the coke fines are thus recovered on the pad before the water reaches a settling maze. There, entrained coke fines are allowed to settle out before the clear water is pumped back to the decoking-water surge tank for reuse. The coke is removed from the pad by a front-end loader.

Although pad loading places no constraints on the rate at which the coke can be removed from a drum, within the general limitations of the equipment, the large area required for drainage and short coke-storage time present disadvantages. The front-end-loader operation usually associated with pads can tend to increase the generation of coke fines.

Foster Wheeler has recently developed a potentially patentable innovation in pad design which eliminates the need for packing the maze inlet ports with coke. This design utilizes a sump to trap the fines carried off the pad by the cutting water. The water flows out of the sump into the maze through special removable coke-filled baskets which filter out unsettled fines. Flushing-water circuits are provided for agitating the fines in the sump and backflushing the baskets. A slurry pump is provided to remove the slurried fines trapped in the sump. The fines may be recovered by pumping them into a partially filled railcar or over the coke-storage pile. Figure 12.2.24[17] shows the general layout of this type of pad-loading operation.

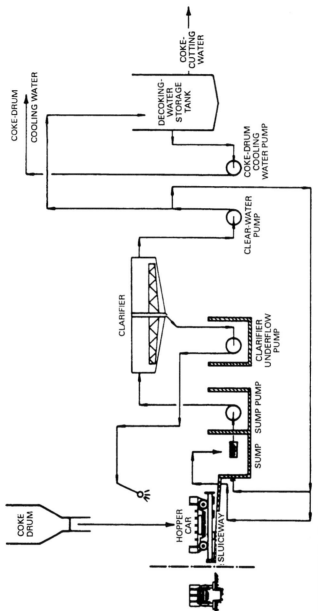

FIGURE 12.2.23 Decoking water system for direct railcar loading.

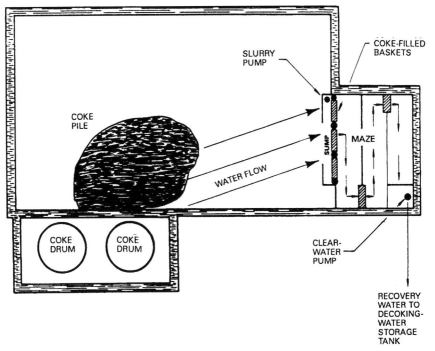

FIGURE 12.2.24 Coke handling: pad loading.

Pit Loading

Pit systems are similar to both direct-railcar-loading and pad-loading operations. In pit systems, however, the coke and water empty into a rectangular concrete pit generally located below grade. The decoking water drains out through ports at one or both ends of the pit, depending on the size of the facility. A "heel" of coke located in front of these ports acts to filter fines from the water. The remaining coke fines settle out in the maze before the clear water is pumped to the decoking-water storage tank.

Pit design inherently provides several days' storage of coke, presenting an advantage over pad loading. An overhead bridge crane with a clamshell bucket is required to remove the coke from the pit. Figure 12.2.25 depicts a general pit-loading operation.

Dewatering Bins

Foster Wheeler has recently developed two different dewatering-bin systems. Both designs evolved from the concept of the traditional slurry-type dewatering-bin system commercially in use for several decades. Dewatering is accomplished through the use of special vessels, known as dewatering bins or drainage silos, for dewatering coke. The two types of dewatering-bin systems are known as *slurry* and *gravity-flow*. In both designs, coke and cutting water pass through a coke crusher. Either system may be totally enclosed to meet exceptional environmental requirements or to prevent coke contamination in areas where sandstorms may present a problem. Conventional unen-

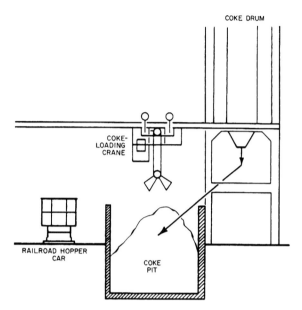

FIGURE 12.2.25 Coke handling: pit operation.

closed slurry-type dewatering-bin systems employing an open sluice have been available for more than 30 years.

Slurry System. The slurry system allows coke and water from the crusher to drop into a sluice, where the mixture is washed into a slurry sump. From this sump, a slurry pump transports the coke and water to the dewatering bin. Here the coke settles, and the water is drained off. Final separation of coke fines from the water is accomplished either by a clarifier or by a special decanter. The dewatered coke is moved from the bin onto a conveyor or directly into railcars or trucks. The slurry system provides a relatively clean operation that allows the coke drums to be located close to the grade. This system, however, requires recirculation of relatively large volumes of water. With appropriate know-how, the slurry system can be adapted as a totally enclosed system which will meet exceptional environmental requirements. Figures 12.2.26 and 12.2.27 provide a schematic of a totally enclosed slurry system and illustrate a typical elevation view of the coke-drum structure.

Gravity System. In the innovative gravity-flow system, coke and water from the crusher drop into a dewatering bin located directly beneath the crusher. The coke-water mixture is allowed to settle, and the water is drained off. Final separation of coke fines from the water is accomplished by special decanters, and the dewatered coke is typically fed from the dewatering bin onto a conveyor. Although this design requires a very tall coke-drum structure, it provides a clean operation that eliminates dependence on the slurry pump and the need for large volumes of recirculated water. Another advantage of the gravity-flow system is that it produces fewer coke fines than the slurry system. Figures 12.2.28 and 12.2.29 provide a schematic and elevation drawing of a totally enclosed gravity-flow dewatering-bin system.

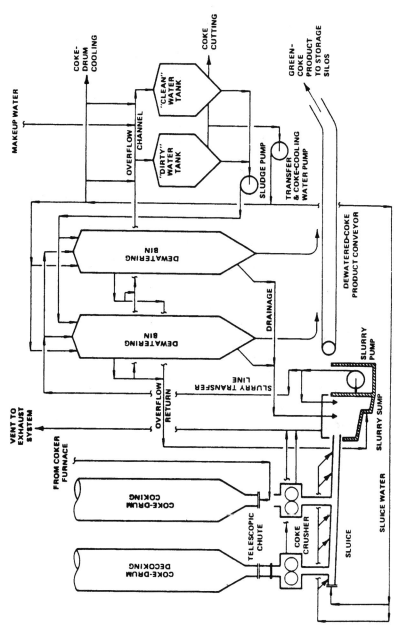

FIGURE 12.2.26 Schematic flow diagram of a totally enclosed slurry dewatering system.

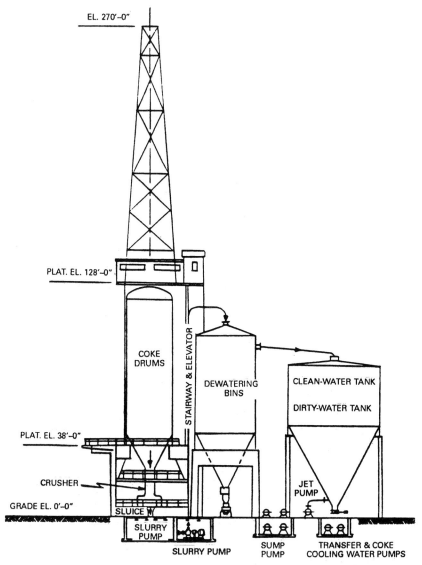

FIGURE 12.2.27 Elevation view of a totally enclosed slurry dewatering system.

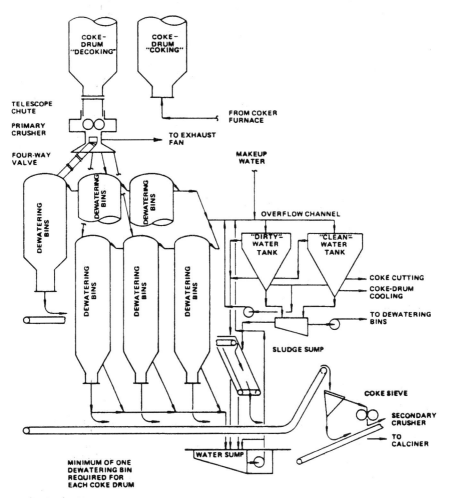

FIGURE 12.2.28 Schematic flow diagram of a gravity-flow dewatering-bin system.

Glossary of Terms

Solids handling requires expertise which traditionally is not part of the refining industry. Table 12.2.10[18] gives a glossary of terms associated with solids handling which should prove useful to those who are normally involved in this area.

USES OF PETROLEUM COKE

Depending on the fundamental type produced and the specific impurity levels present in the final product, petroleum coke is basically used for three types of applications. These applications can be classified as fuel, electrode, and metallurgical. A fourth and

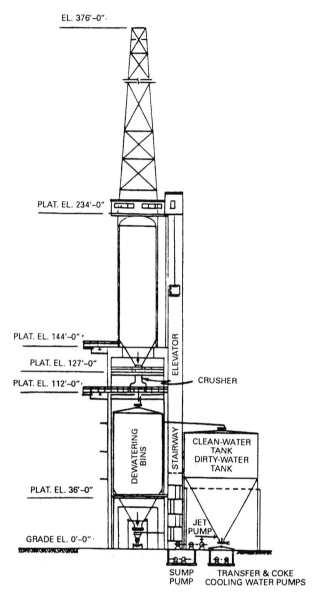

FIGURE 12.2.29 Elevation view of a totally enclosed gravity-flow system.

TABLE 12.2.10 Glossary of Terms

Belt scale	A weighing device on a belt conveyor located under a section of belt carrying material. The scale measures the weight of material on a short length of belt and converts the reading to a unit rate. The scale reading may also give a total weight of material moved during a certain period.
Belt take-up	A mechanical system of pulleys on a belt conveyor used to control excess length of belt and belt tension.
Belt wiper	The edge of an adjustable blade held against a conveyor-belt surface to wipe off any bulk product adhering beyond a normal discharge point.
Boom	A belt conveyor elevated and extended so as to carry bulk product to and from a storage pile or from dock to ship.
Bridge crane	An elevated horizontal beam mounted on a carriage which spans the coke pit across the short dimension. A drive on the beam carriage moves the beam back and forth along the length of the coke pit to give universal positioning for a clamshell-bucket hoist.
Bucket wheel	A circular bulk elevator with multiple scoops. When the wheel turns, the scoops fill at bottom positions and dump at top positions to load bulk product onto reclaim belt conveyors.
Clamshell bucket	A grapple having two vertically hinged jaws used to acquire and lift coke from a pit to a transport or crusher car.
Cleaner plow	An adjustable blade with an edge held against a conveyor-belt surface with its face positioned obliquely to the line of belt travel.
Coke pit	A cavity in the ground adjacent to the delayed-coker vessels used for short-term surge storage of coke. The pit provides a means of draining water from the coke which is used to cut the coke from the coker vessels.
Coker discharge	Bottom opening of a delayed-coker-vessel and the companion chute which diverts coke into the coke pit.
Conveyor belt	An endless strip surface which is supported on rollers and is a means of continuous transporting of bulk material between predetermined points. The surface is flexible and is usually caused to trough by support rollers as a way of limiting spillage.
Crusher	A large machine usually consisting of rolls in juxtaposition. Bulk materials are forced through the space between the rolls so that oversized lumps are broken into smaller size by fracture pressure.
Diverter	A metal surface located across a chute by which the gravity flow of bulk material is caused to go in one direction or another.
Grizzly	A course grid used to separate oversized lumps from acceptable material going to the primary crusher.
Idlers	Multiple small-diameter rollers used to direct and provide an antifriction support for the belt of a belt conveyor.
Metal detector	An electronic device located adjacent to the carrier section of a belt conveyor to sense the presence of magnetic metal debris in the coke being transported.
Metal separator	A magnet system located adjacent to the carrier section of a belt conveyor used to attract and hold magnetic metal debris from the coke on the belt.
Rip detector	A system of electric conductors embedded in the conveyor belt which, if cut or broken as a result of a belt tear, gives an alarm signal or shuts down the affected conveyor.
Stacker	A mobile support frame for belt conveyors which deliver and discharge bulk materials at storage piles.

12.64 VISBREAKING AND COKING

TABLE 12.2.10 Glossary of Terms (*Continued*)

Stringer	A structural member on each side of a belt conveyor to which idlers, covers, and supports are attached.
Tower	A structural support for transfer chutes and related equipment between conveyors.
Tripper	A set of auxiliary pulleys between the head and tail pulleys of a belt conveyor which fold the belt and cause it to discharge at intermediate locations.

Source: Robert C. Howell and Richard C. Kerr, *Hydrocarb. Process.*, **60**(3), 107 (1981).

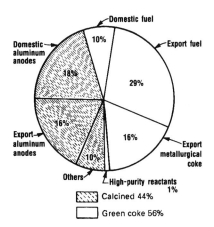

FIGURE 12.2.30 United States petroleum-coke markets in 1980. [*Courtesy of Pennwell Publishing Company, publishers of the* Oil and Gas Journal, **80**, *145 (Oct. 25, 1982), 76 (Nov. 1, 1982), and 198 (Nov. 8, 1982), and the Pace Company, which developed the information contained in the figure.*]

relatively new usage classification, which is gasification, is currently under evaluation by many companies but does not represent a significant application at this time.

Figure 12.2.30 shows how United States coke production was allocated in 1980. Each of the uses is described below.

Types

As described in the section "Feedstocks," sponge coke is the most common type of regular-grade petroleum coke, while needle coke can be made only from special feedstocks. The name *sponge coke* is used because the lumps of coke that are produced are porous and at times resemble spongelike material.[11] Typical sponge-coke specifications both before and after calcination are given in Table 12.2.11.

Figure 12.2.31 shows that, of the total petroleum coke produced in the United States in 1980, 90 percent (14,320,000 short tons) was conventional delayed coke, while 2 percent (318,000 short tons) was needle coke. The other 8 percent (1,275,000 short tons) was fluid coke produced by a totally different process. Figure 12.2.31 also

TABLE 12.2.11 Typical Sponge-Coke Specifications

Item	Green coke	Calcined coke
Moisture	6–14%	0.1%
Volatile matter	8–14%	0.5%
Fixed carbon	86–92%	99.5%
Sulfur	1.0–6.0%	1.0–6.0%
Silicon	0.02%	0.02%
Iron	0.013%	0.02%
Nickel	0.02%	0.03%
Ash	0.25%	0.4%
Vanadium	0.015%	0.03%
Bulk density	45–50 lb/ft^3 (720–800 kg/m^3)	42–45 lb/ft^3 (670–720 kg/m^3)
Real density		2.06 g/cm^3
Grindability (Hardgrove number)		50–100

shows the various ways in which these three types of green coke (uncalcined coke) were used in 1980 for both the green-coke market and the calcined-coke market. It is interesting to note that, unlike liquid petroleum products, of which nearly all are consumed domestically, over 60 percent of the petroleum coke produced by the United States was exported.

So far, we have discussed only the types of coke which are considered to be desirable products. There is, however, another type of coke which is often considered an undesirable product because it may lead to difficulty during the decoking cycle. It is usually produced from very heavy feedstock, especially at low pressure and low recycle ratio. It is known as *shot coke*. Shot coke is spheroid in shape, with sizes that range from as small as buckshot to as large as basketballs.[7] Shot coke may be used as a fuel, but it is less desirable in this usage than is sponge coke.

Use as Fuel

The use of petroleum coke as a fuel generally falls into two major categories, fuel for steam generation and fuel for cement plants. For either of these applications coke is generally blended with bituminous coal or used in combination with oil or gas. In general, coke as a fuel used in combination with bituminous coal has the following advantages over bituminous coal alone:

1. *Grinding.* Coke is easier to grind than bituminous coal, resulting in lower grinding costs and less maintenance.
2. *Heating value.* The heating value of petroleum coke is more than 14,000 Btu/lb, compared with 9000 to 12,500 Btu/lb for coal.
3. *Ash content.* The very low ash content (less than 0.5 wt %) of coke results in lower ash-handling costs.

Steam Generation. Steam generation by coke burning can be accomplished either in specially designed utility boilers or in fluidized-bed boilers.

Utility Boilers. Industry has been firing petroleum coke, typically in combination with other fuels, in large and small boilers for over 50 years. This usage includes in-

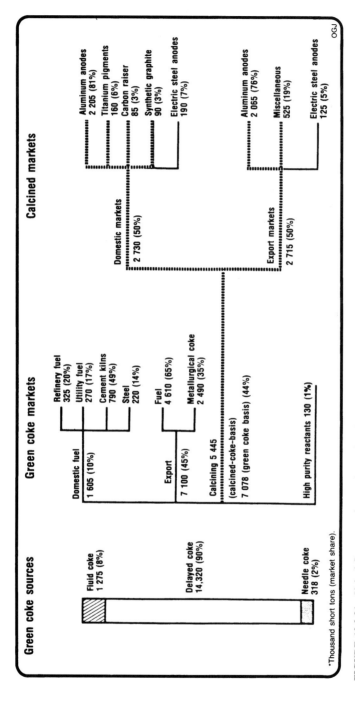

FIGURE 12.2.31 United States petroleum-coke markets in 1980. [*Courtesy of Pennwell Publishing company, publishers of the* Oil and Gas Journal, *80, 145 (Oct. 25, 1982), 76 (Nov. 1, 1982), and 198 (Nov. 8, 1982), and the Pace Company, which developed the information contained in the figure.*]

cludes in-refinery commercial experience producing steam from petroleum coke without supplementary fuel. When used with bituminous coal, coke can be blended within piles, bunkers, burners, or conveyor belts. With the high-sulfur petroleum coke expected from heavier, high-sulfur coker feedstocks, wet flue-gas scrubbing may be required to meet emission requirements for utility boilers.

Fluidized-Bed Boilers. This type of steam generator, which has been developed commercially by Foster Wheeler, allows firing of a wide range of low-cost, low-grade, high-ash, and high-sulfur fuels to produce steam efficiently without any harmful effect to the environment. The fluidized bed is formed by the agitated burning of fuel with limestone. The fluidization medium is combustion air introduced from below the bed. The limestone absorbs the SO_2, and the low combustion temperature inhibits the formation of NO_x. Particulates are removed with a baghouse or, in some cases, by a conventional electrostatic precipitator.

The fluidized-bed boiler lends itself to the combustion of fuels containing vanadium, such as delayed cokes produced from heavy coker feedstocks. Slagging, fouling, and corrosion of steam-generator surfaces are essentially eliminated because the fluidized bed operates at temperatures below the ash-softening point.

Recent evaluations indicate that both for new installations and for revamps fluidized-bed steam generators burning high-sulfur coke, with sulfur capture, can offer a significant advantage both in capital and in operating costs compared with conventional boilers using either high-sulfur coke or oil as a fuel accompanied by wet SO_2-removal systems.[19]

Cement Plants. Coke can be used with coal, natural gas, refinery fuel gas, or oil as a supplementary fuel in fired kilns. Coke by itself does not contain enough volatile material to produce a self-sustaining flame, and as a result it cannot be fired alone in cement kilns. Typical fuel combinations for cement plants are 25 percent oil or gas and 75 percent coke or 70 percent bituminous coal and 30 percent coke.

The sulfur contained in the coke reacts with the cement to form sulfate, which reduces the requirements for calcium sulfate (gypsum) in the cement. Metals (V and Ni) from the coke are not detrimental to the cement. On the basis of a 25 percent oil and 75 percent coke fuel combination, coke consumption for a modern cement plant will be 75 to 115 tons/1000 tons of cement.

Use for Electrodes

Low-sulfur, low-metals sponge coke, after calcination, can be used to manufacture anodes for the aluminum industry. The aluminum industry is the greatest single consumer of coke.[11] Figure 12.2.30 shows the combined domestic and export total as 34 percent in 1980. For every pound of aluminum produced by smelting, nearly ½ lb of calcined coke is consumed. Figure 12.2.32 shows aluminum anodes arranged in a reduction cell for the smelting of aluminum (on the left) and formed in hydraulic presses (on the right). Figure 12.2.33 shows aluminum anodes in storage.

Needle coke is a highly ordered petroleum coke produced from special low-sulfur aromatic feedstocks. The main use of calcined needle coke is in the manufacture of graphite electrodes for electric-arc furnaces in the steel industry. Since these electrodes are subject to extremes in temperature shock, a low coefficient of thermal expansion is very important.[15] Typical properties of needle coke, used for graphite-electrode manufacture, both before and after calcination, are given in Table 12.2.12.

FIGURE 12.2.32 Aluminum electrodes: formation and utilization. (*Courtesy of Noranda Aluminum Inc.*)

FIGURE 12.2.33 Aluminum electrodes in storage. (*Courtesy of Noranda Aluminum Inc.*)

Metallurgical Use

Petroleum coke with a low sulfur content (2.5 wt % or less) can be used in ferrous metallurgy when blended with low-volatility coking coals. Petroleum coke used in foundries or for steel making enhances the properties of coking coals by reducing the total amount of volatiles and increasing the average heating value.

Metal content in the coke does not normally present a problem in the metallurgical industry.

TABLE 12.2.12 Typical Needle-Coke Specifications

Item	Green coke	Calcined coke
Moisture	6–14%	0.1%
Volatile matter	4–7%	0.5%
Sulfur	0.5–1.0%	0.5–1.0%
Silicon	0.02%	0.02%
Iron	0.013%	0.02%
Nickel	0.02%	0.03%
Ash	0.25%	0.4%
Vanadium	0.01%	0.02%
Bulk density	45–50 lb/ft^3 (720–800 kg/m^3)	42–45 lb/ft^3 (670–720 kg/m^3)
Real density		2.11 g/cm^3
Coefficient of thermal expansion (25–130°C), 1/°C		5×10^{-7}

Use for Gasification

The use of delayed coke as gasification feed is currently under investigation by many companies. Gasification to low-Btu gas or syngas can be accomplished through the use of partial-oxidation techniques. Low-Btu gas can be used as a fuel gas in the refinery; syngas can be used for the production of methanol for automotive fuel blending or as a feedstock for other chemical processes. Partial oxidation can also be used to produce the increased hydrogen necessary to refine heavy or higher-sulfur crudes into commercial products. The amount of low-Btu gas that can be handled in existing refineries may be limited by existing equipment. However, in new grass-roots refineries a large part of fuel needs can be satisfied by gasification.

Coke Prices

Past and current prices for various types of petroleum coke are shown in Fig. 12.2.34. The sulfur levels for each of these types of coke are defined in Table 12.2.13. We can see that while coke is generally not produced for its own market value but rather for that of the resulting liquid products, a reasonable market value nevertheless does exist and is predicted[11] to continue. This is the case even though in years to come the quality of most of the available coke will undoubtedly continue to worsen in terms of sulfur level and metal impurities.

INTEGRATION OF DELAYED COKING IN MODERN REFINERIES

One of the basic problems that refiners face is how to select from among the available bottom-of-the-barrel conversion processes the best residual-processing route to meet the needs of their own particular set of refining objectives.[13] Such a decision can be made only after a detailed analysis of the various alternatives. How delayed coking fits in, as one of the available alternatives, is discussed below.

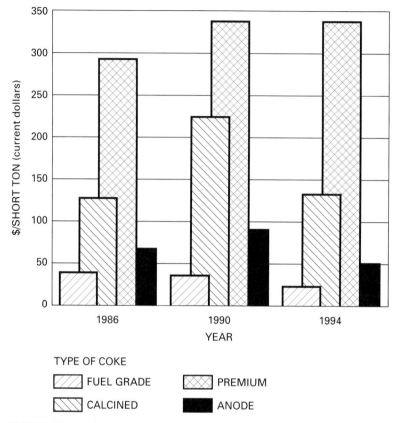

FIGURE 12.2.34 U.S. petroleum coke prices. [Average price per short ton (dry), FOB Gulf Coast.] (*Courtesy of the Pace Company, which developed the data shown.*)

TABLE 12.2.13 Coke Type versus Sulfur Level

Coke type	Sulfur level, wt %
Fuel grade*	4+
Premium grade*	Below 2
Anode grade*	2–3
Calcined coke	2–3

*Green coke.
Source: W. L. Nelson, *Oil Gas J.,* **74,** 60 (May 24, 1976).

Bottom-of-the-Barrel Processing

Although process developments do continue, the questions which refiners face in deciding on a specific residual-conversion-processing route are mainly ones of application rather than ones of development. Therefore, one of the basic problems is to provide a processing route which makes optimal use of the available bottom-of-the-barrel residual-conversion processes.[19] These bottom-of-the-barrel processes can be classified into five groups, as follows:[13]

1. Separation processes

    Vacuum distillation

    Solvent deasphalting

2. Carbon-rejection processes: Thermal processing

    Visbreaking

    Delayed coking

    Fluid coking and flexicoking

    Combination visbreaking and thermal cracking

3. Catalytic conversion: Residue catalytic cracking
4. Hydrogen-addition processes: Residue hydrocracking
5. Combined carbon rejection–hydrogen addition: Thermal–hydrocracking

In these classifications we see that delayed coking is listed as a carbon-rejection thermal process.

Residual-conversion-processing routes should be specially tailored for each refinery depending on considerations including:

- Properties of the crude oils to be processed
- Marketing requirements
- Economics, including operating costs
- Grass roots versus expansion
- Environmental control requirements

Typically, optimization studies using linear programming techniques are utilized during the investigatory phase prior to deciding on a residual-conversion route. It there is one maxim inherent in the analysis of residual-conversion routes it is "There are no generalities."

We mentioned at the beginning of the chapter that delayed coking has been referred to as the yardstick against which other processes must be measured. If we refer to Table 12.2.14,[20] we can get a better appreciation of how delayed coking may be compared with various other available bottom-of-the-barrel processing alternatives simply in terms of investment. Table 12.2.14 is for a specific study on light Arabian crude, which was prepared by the UOP Process Division.[20]

Information similar to that of Table 12.2.14, but this time also showing product yields, is presented in Fig. 12.2.35. From Table 12.2.14 and Fig. 12.2.35 we can conclude that delayed coking has one of the lowest investments of the various alternatives and that the investment increases as the schemes become more complicated in order to yield more liquid products. Although this cost information is somewhat dated, the relative trend should still be correct. Whether or not an additional investment is justified

TABLE 12.2.14 Upgrading-Scheme Cost Summary: Estimated Erected Costs, First Quarter, 1981

Vacuum fractionators	RCD Unibon	BOC unit	Vacuum fractionator	Demex	Visbreaker	Coking	Total, $ million	Total $/bbl*
X							14.5	346
	X		X				99.5	2278
	X		X	X			109.1	2602
	X		X			X	115.7	2760
X		X	X				72.3	1725
X		X	X	X			82.0	1956
X		X	X		X		80.8	1927
X				X			27.8	663
X			X		X		30.6	730
X						X	41.2	983

*Basis: 41,925 bbl/calendar day of reduced crude.

Source: Courtesy of Pennwell Publishing Company, publishers of the *Oil and Gas Journal* [John G. Sikonia, Frank Stolfa, and LeRoi E. Hutchings, *Oil Gas J.,* **79,** 258 (Oct. 19, 1981)] and UOP Process Division, which developed the information contained in the table. Demex, BOC Unibon, and RCD Unibon are registered trademarks and/or service marks of UOP. X = units included in flow scheme.

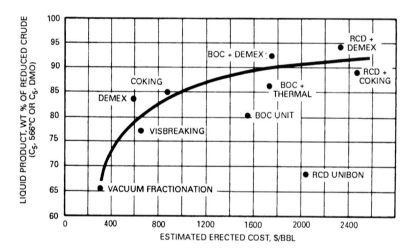

FIGURE 12.2.35 Liquid product versus estimated cost for various residue-processing schemes. Estimated costs are for the first quarter of 1981. Demex, BOC Unibon, and RCD Unibon are registered trademarks and/or service marks of UOP Inc. [*Courtesy of Pennwell Publishing Company, publishers of the* Oil and Gas Journal, **79,** *258 (Oct. 19, 1981), and UOP Process Division, which developed the information contained in the figure.*]

will depend on the specific set of conditions prevailing for the particular refinery in question.

Typical Refinery Schemes Utilizing Delayed Coking

We have shown in a simplified manner, in the section "Feedstocks," how the overall yield from delayed coking may be enhanced if it is employed in conjunction with residue desulfurization. To develop this topic further and to show how delayed coking can be integrated in various possible ways within the structure of a refinery, we refer to the following case study.[19]

Consider a grassroots refinery processing 100,000 BPSD of Alaskan North Slope crude to produce transportation fuels. Additional considerations include:

- Gasoline pool: 40 percent premium unleaded, 40 percent regular unleaded, 20 percent regular leaded
- Refinery fuel oil: 0.5 wt % sulfur maximum
- Consistent approach to the processing of light ends and catalytic re-forming of straight-run naphtha and thermal naphthas for all cases
- Isobutane requirements for alkylation to be met without requiring any outside purchase
- All refinery fuel requirements to be met internally

The base case and two alternative conversion routes for processing a typical North Slope atmospheric residue may be simply described as follows:

Designation	Description
Base case	Vacuum flashing, vacuum gas oil desulfurizing, fluid catalytic cracking
Alternative A	Vacuum flashing, delayed coking, desulfurization of vacuum gas oil and coker gas oil, fluid catalytic cracking
Alternative B	Residue desulfurization, vacuum flashing, delayed coking, fluid catalytic cracking

Simplified block flow diagrams (Figs. 12.2.36, 12.2.37, and 12.2.38) for the base refinery and the two alternative conversion refineries indicate the net product yields from each. The net product yields are given in metric tons per calendar day and, when appropriate, in barrels per calendar day (BPCD). To simplify this presentation, some of the required process units such as LPG and naphtha treaters, amine regenerator, sulfur recovery, and tail-gas-treating unit are not shown on the diagrams. However, the investment costs, operating requirements, and effect on product yields and qualities of these units, as well as the support facilities required, have been taken into consideration throughout.

Base Refinery. The processing route for the base refinery (Fig. 12.2.36) uses a conventional-crude-vacuum-flasher scheme coupled with vacuum gas oil (VGO) desulfurization followed by fluid catalytic cracking (FCC). Straight-run naphtha is catalytically re-formed to improve octane, straight-run middle distillates are desulfurized, and kerosene is hydrotreated to reduce aromatic and naphthalene contents in order to meet Jet A fuel specifications. Olefinic light ends from the FCC are polymerized and also alkylated with internally produced isobutane to produce gasoline blend stocks. In addition to improving FCC product yields, the use of FCC feed desulfurization results

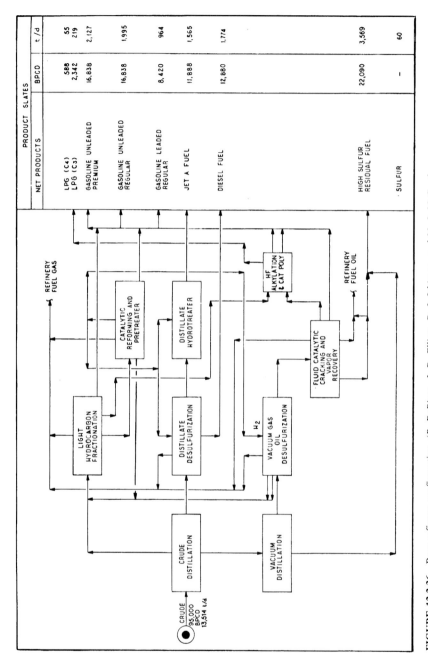

FIGURE 12.2.36 Base refinery configuration. (R. DeBiase, J. D. Elliott, D. I. Ichiman, and M. J. McGrath, "Alternate Conversion Schemes for Residual Feedstocks," AIChE Meeting, Houston, April 1981.)

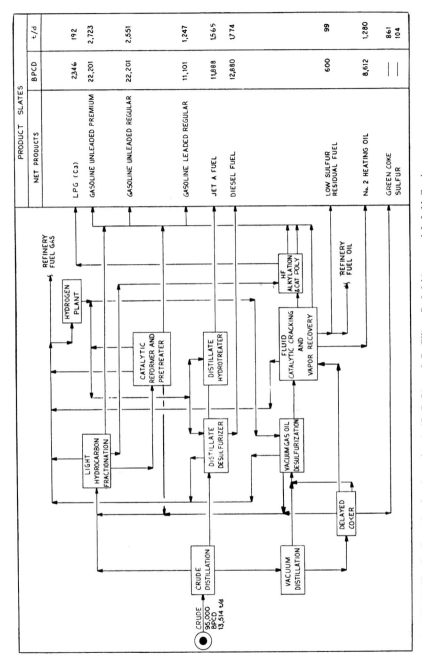

FIGURE 12.2.37 Configuration for alternative A. (*R. DeBiase, J. D. Elliott, D. I. Izhiman, and J. J. McGrath, "Alternate Conversion Schemes for Residual Feedstocks," AIChE Meeting, Houston, April 1981.*)

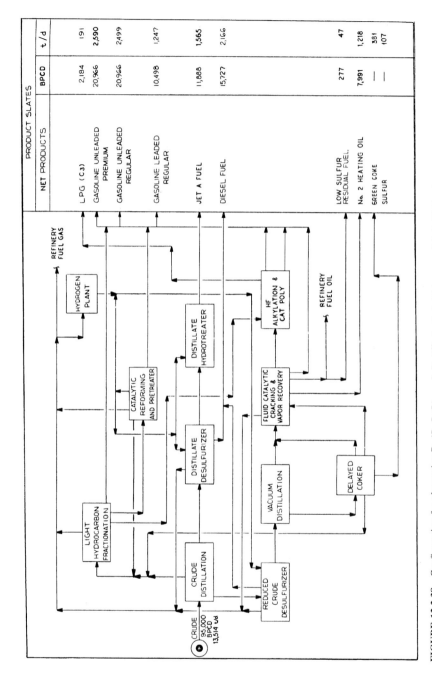

FIGURE 12.2.38 Configuration for alternative B. (R. DeBiase, J. D. Elliott, D. I. Izhiman, and J. J. McGrath, "Alternate Conversion Schemes for Residual Feedstocks," AIChE Meeting, Houston, April 1981.)

in lower sulfur emissions from the FCC regenerator, as well as allowing the FCC decant oil to be used as a low-sulfur refinery fuel oil.

The base refinery does not provide any residual-conversion capability. It should be pointed out that it is not always necessary to provide residual conversion. Depending on crude type, product specifications, and available markets, the refiner may be able to justify no or low residual-conversion rates through the projected sale of residual fuel or asphalts.

Alternative A. The first type of residual-conversion refinery (Fig. 12.2.37) utilizes essentially the same route as the base refinery with a delayed coker converting the vacuum residue to cracked distillates and green coke. Coker gas oil is desulfurized with the VGO before being fed to the FCC unit.

Alternative A is a basic residual-conversion route which is frequently utilized for refinery expansions. The residual-conversion capability of a base refinery is achieved by adding a delayed coker while increasing the capacity of existing downstream units. This route is suitable for processing high-sulfur, high-metal crudes. However, when such is the case, the coke produced usually is not suitable for aluminum anodes and must instead be used for fuel.

The sulfur emissions for alternate A are slightly lower than those of the base case.

Alternative B. Alternative B (Fig. 12.2.38) is presented as a case in which it may be advantageous to produce low-sulfur, green sponge coke for special market requirements. This conversion route utilizes a reduced-crude desulfurizer followed by a vacuum flasher to produce delayed-coker feedstock. The low-sulfur vacuum and coker gas oils are fed to a conventional gas-oil FCC unit. A small amount of net residual fuel oil is produced from FCC cycle and decant oils.

The use of a residue desulfurizer upstream of a delayed coker improves ultimate coke quality by reducing sulfur and metals in the coker feed. The low-sulfur coker gas oil which is produced via this processing scheme will give better FCC yields than untreated coker gas oils from straight-run residues. Because the residue desulfurizer reduces the carbon-residue content of the coker feedstock, the yield of coke from the delayed coker is much less than it would be with untreated reduced crude as the feed. The vacuum unit downstream of the residue desulfurizer further reduces the quantity of low-sulfur coke which is produced.

Comparisons of Base Case and Alternatives. Total refinery investment requirements are indicated for the base case and each of the two alternatives in Table 12.2.15. These investment costs are conceptual-curve-type costs for the fourth quarter of 1994 United States Gulf Coast grassroots construction. They do not include working capital, inventories, start-up expense, and royalties and exclude the cost of land, site preparation, taxes, licenses, permits, and duties.

Incremental operating requirements over those necessary for the base refinery are given for each of the two alternate residual-conversion refineries in Table 12.2.16. In developing utility requirements, it has been assumed that electric power will be pur-

TABLE 12.2.15 Investment Summary, US$1000*

	Base refinery	Alternative A	Alternative B
Process units	421,000	539,000	608,000
Support facilities	251,000	327,000	364,000
Total delivered and erected investment	672,000	866,000	972,000

*Basis: Fourth quarter, 1994, United States Gulf Coast.

TABLE 12.2.16 Incremental Operating Requirements over Base Refinery

	Alternative A	Alternative B
Purchased electric power, kWh/h	2,900	27,500
Catalyst and chemicals, $/day*	1,600	11,300
Operating personnel	18	20

*Does not include product-blending chemicals such as tetraethyl lead.

chased. All fuel requirements have been met internally. This is reflected in the net-product slates given on each of the simplified block flow diagrams.

No conclusions can be reached as to which of three cases presented above is best, since any such conclusion must necessarily depend on the specific set of economic conditions for the refinery under study.

TYPICAL UTILITY REQUIREMENTS

The total utility requirements for any delayed coker may be considered as consisting of two separate parts. One part is the continuous requirement, and the second part is the intermittent requirement. Typical values for each part are given below.

Continuous Utilities

Summarized below are typical continuous utility requirements for a delayed coker. In order to facilitate preliminary evaluations, values are given for the type of unit shown in Fig. 12.2.4 and are listed on the basis of 1000 BPSD of fresh feed or, in the case of raw water, for each short ton per day of green coke produced. Actual utilities will vary from the typical numbers shown below on the basis of individual heat- and material-balance calculations and specific requirements for downstream processing.

Fuel liberated	5,100,000 Btu/(h · 1000 BPSD)
Power consumed	150 kW/1000 BPSD
Steam exported	1700 lb/(h · 1000 BPSD)
Boiler feedwater consumed	2400 lb/(h · 1000 BPSD)
Cooling water, $\Delta t = 14°C$	5–25 gal/(min · 1000 BPSD)*
Raw water consumed	20–35 gal/day per short ton/day coke

*Based on maximum use of air cooling.

Intermittent Utilities

Intermittent utilities are required for the decoking and coke-drum blowdown systems. The utility consumptions given are typical for delayed cokers having coke-drum diameters of 20 ft or greater. The time required for the utilities is typical for a two-drum delayed coker operating under a 24-h coking cycle. Actual utilities will vary with coke production and drum size.

All intermittent utility requirements are consumptions.

Power consumed	h/day	kW
Jet pump	5	2000
Blowdown circulating-oil cooler	8	45
Blowdown condenser	5	207
Coke-drum condensate pump	5	6
Slop-oil pump	6	8
Coke-drum cooling-water pump	6	69
Clear-water pump	4	19
Vent-gas compressor	5	100 average
Overhead bucket crane	7	200
Elevator	...	10
Lights and instruments	...	25
Steam consumed	h/day	lb/h
Coke-drum steam-out to fractionator	1	10,000
Coke-drum steam-out to blowdown drum	1	20,000
Coker blowdown drum	8	750
Blowdown circulating-oil cooler	8	2,000
Cooling-water required	h/day	gal/min
Jet pump	5	25
Plant air consumed	h/day	SCF/min
Hoist	5	600
Rotary motor	5	200

Energy-Conservation Measures

The trend toward increased energy efficiency to reduce utility consumption has affected the design of delayed cokers in much the same way as it has other conventional refinery process units. A review of the methods used in achieving this increased efficiency follows.

Coker-Furnace Air Preheat. The refining industry is moving more and more in the direction of air preheat for process furnaces. Coker furnaces are no exception. Traditionally, delayed cokers have had high furnace inlet temperatures, in excess of 260°C, and have had to rely on the generation of steam to improve fuel efficiency. Preheating of relatively cold boiler-feedwater makeup together with steam generation has not usually proved to be a viable economic alternative to air preheat. Air preheat not only provides fuel efficiencies as high as 92 percent, compared to typical steam-generation efficiencies of 87 percent, but requires less fuel. This is true because it does not achieve its efficiency by increasing the absorbed heat, which would be necessary to generate steam.

Many of the new coker projects include air preheat to improve heater efficiency.

Increased Recovery of Fractionator Heat. Traditionally, the top section of the coker fractionator, above the light gas oil draw-off, was considered to be too cold for the economic recovery of heat. However, this type of conventional design must be evaluated when there is a greater economic incentive for the recovery of low-level heat.

Several recent designs have recovered heat from the fractionator by using a light gas oil pumparound. Heat from this pumparound may be used to generate low-pressure steam, to preheat cold process streams, and to reboil low-temperature vapor-recovery-unit towers.

Conventional recovery of very low level heat may also be accomplished through the use of a hot-water circulating system tied into a central refinery circulating-hot-water belt system. A portion of the heat recovered could be used for tempering cold fresh air to the furnace preheater.

ESTIMATED INVESTMENT COST

For certain processing units, it is possible to develop a rough estimate of investment cost simply from the feed capacity of the unit. For delayed cokers this is not practicable because one must also know the amount of coke that is produced from the particular feedstock in question. For this reason, it is better to cross-correlate the investment cost of delayed cokers with a parameter such as tons per day of product coke as well as barrels per day of feed.

Although a highly accurate investment cost for a delayed coker can be determined only by a detailed definitive estimate, it is often necessary when carrying out economic evaluations to develop a rough, preliminary budget-type estimate. This type of estimate typically has an accuracy of ±30 percent. For a delayed coker a cost in the range $45,000 to $90,000/(short ton-day) of coke produced may be used for preliminary evaluations. This cost excludes the vapor-recovery unit and is based on the following assumptions.

General

United States Gulf Coast location.

Time basis of fourth quarter 1994, with costs reflecting no future escalation.

Coke produced is sponge coke.

Coke handling is via a pit with an overhead crane.

Calcining is not included.

Clear and level site conditions free of above- and below-ground obstructions; 3000 to 4000 lb/ft^2 soil bearing at 4 ft below grade.

Normal engineering design standards and specifications.

Vapor-recovery unit is not included.

Exclusions

Cost of land.

Taxes and owner's insurances.

Licenses, permits, and duties.

Spare parts.

Catalysts and chemicals.

Process royalties and fees (normally none for delayed cokers).

Startup costs.

Interest.

Forward escalation.

Support facilities

ABBREVIATIONS

Abbreviations used in this chapter are listed in Table 12.2.17.

REFERENCES

1. *Oil Gas J.,* **79,** 43 (Jan. 5, 1981).
2. W. J. Rossi, B. S. Deighton, and A. J. MacDonald, *Hydrocarb. Process.,* **56**(5), 105 (1977).
3. S. B. Heck, Jr., "Process Design of a Modern Delayed Coker," American Petroleum Institute Midyear Meeting, New York, May 1972.
4. *Oil Gas J.,* **75,** 340 (August 1977).
5. James H. Gary and Glenn E. Handwerk, *Petroleum Refining: Technology of Petroleum,* Marcel Dekker, New York, 1975.
6. Frank Stolfa, *Hydrocarb. Process.,* **59**(5), 101 (1980).
7. R. DeBiase and J. D. Elliott, "Recent Trends in Delayed Coking," NPRA Annual Meeting, San Antonio, March 1982.
8. *Oil Gas J.,* **92,** 51, 53 (Dec. 19, 1994).
9. *Oil Gas J.,* **80,** 81, 130 (Mar. 22, 1982).
10. W. L. Nelson, *Oil Gas J.,* **74,** 60 (May 24, 1976).
11. Peter Fasullo, John Matson, and Tim Tarrillion, *Oil Gas J.,* **80,** 145 (Oct. 25, 1982), 76 (Nov. 1, 1982), 198 (Nov. 8, 1982).
12. William F. Bland and Robert L. Davidson, *Petroleum Processing Handbook,* McGraw-Hill, New York, 1967.
13. J. A. Bonilla, "Delayed Coking and Solvent Deasphalting: Options for Residue Upgrading," AIChE Meeting, Anaheim, Calif., June 1982.
14. D. H. Stormont, *Oil Gas J.,* **67,** 75 (Mar. 17, 1969).
15. Barend Alberts and Dick P. Zwartbol, *Oil Gas J.,* **77,** 137 (June 4, 1979).
16. Virgil B. Guthrie, *Petroleum Products Handbook,* McGraw-Hill, New York, 1960.

TABLE 12.2.17 Abbreviations

°API	Degrees on American Petroleum Institute scale; °API = (141.5/sp gr) − 131.5	FI	Flow indicator
		HC	Hand controller
		HDS	Hydrodesulfurization
ASTM	American Society for Testing and Materials	HF	Hydrofluoric acid
		LC	Level controller
BOC Unibon	Black-oil conversion Unibon	LI	Level indicator
BPCD	Barrels per calendar day	LPG	Liquefied petroleum gas
BPSD	Barrels per stream day	M	Motor
°C	Degrees Celsius	Ni	Nickel
CCR	Conradson carbon residue (defined in subsection "Regular-Grade Coke Production")	PC	Pressure controller
		RCD Unibon	Reduced-crude desulfurization Unibon
		TBP	True boiling point
CW	Cooling water	TEL	Tetraethyl lead
DMO	Demetallized oil	V	Vanadium
FC	Flow controller	VCM	Volatile combustible material
FCC	Fluid catalytic cracker	WH	Waste heat

17. W. L. Nelson, *Petroleum Refining Engineering,* 4th ed., McGraw-Hill, New York, 1958.
18. Robert C. Howell and Richard C. Kerr, *Hydrocarb. Process.,* **60**(3), 107 (1981).
19. R. DeBiase, J. D. Elliott, D. I. Izhiman, and M. J. McGrath, "Alternate Conversion Schemes for Residual Feedstocks," AIChE Meeting, Houston, April 1981.
20. John G. Sikonia, Frank Stolfa, and LeRoi E. Hutchings, *Oil Gas J.,* **79,** 258 (Oct. 19, 1981).

BIBLIOGRAPHY

DeBiase, R., and J. D. Elliott: *Heat Eng.,* 53 (October–December 1981).
——— *Oil Gas J.,* **80,** 81 (Apr. 19, 1982).
———: *Hydrocarb. Process.,* **61**(5), 9 (1982).
Heat. Eng., 46 (May–June 1967).
Heat. Eng., 12 (January–March 1979).
Hydrocarb. Process., **61**(9), 162 (1982).
Kirk-Othmer Encyclopedia of Chemical Technology, 3d ed., vol. 4, Wiley, New York, 1978.
Nagy, R. L., R. G. Broeker, and R. L. Gamble: "Firing Delayed Coke in a Fluidized Bed Steam Generator," NPRA Annual Meeting, San Francisco, March 1983.
Nelson, W. L.: *Oil Gas J.,* **72,** 70 (Jan. 14, 1974).
——— *Oil Gas J.,* **72,** 118 (Apr. 1, 1974).
——— *Oil Gas J.,* **76,** 71 (Oct. 9, 1978).
Oil Gas J., **78,** 78 (Mar. 24, 1980).
Oil Gas J., **80,** 134 (Dec. 27, 1982).
Rose, K. E.: *Hydrocarb. Process.,* **50**(7), 85 (1971).
Sikonia, John G., Frank Stolfa, and LeRoi E. Hutchings: *Oil Gas J.,* **79,** 141 (Oct. 5, 1981).
Zahnstecher, L. W., R. DeBiase, R. L. Godino, and A. A. Kutler: *Oil Gas J.,* **68,** 92 (Apr. 6, 1970).

CHAPTER 12.3
FW/UOP VISBREAKING

Vincent E. Dominici and Gary M. Sieli
Foster Wheeler USA Corporation
Clinton, New Jersey

INTRODUCTION

Visbreaking is a well-established noncatalytic thermal process that converts atmospheric or vacuum residues to gas, naphtha, distillates, and tar. Visbreaking reduces the quantity of cutter stock required to meet fuel oil specifications while reducing the overall quantity of fuel oil produced.

The conversion of these residues is accomplished by heating the residue material to high temperatures in a furnace. The material is passed through a soaking zone, located either in the heater or in an external drum, under proper temperature and pressure constraints so as to produce the desired products. The heater effluent is then quenched with a quenching medium to stop the reaction.

With refineries today processing heavier crudes and having a greater demand for distillate products, visbreaking offers a low-cost conversion capability to produce incremental gas and distillate products while simultaneously reducing fuel oil viscosity. Visbreaking can be even more attractive if the refiner has idle equipment available that can be modified for this service.

When a visbreaking unit is considered for the upgrading of residual streams, the following objectives are typically identified:

- Viscosity reduction of residual streams which will reduce the quantity of high-quality distillates necessary to produce a fuel oil meeting commercial viscosity specifications.
- Conversion of a portion of the residual feed to distillate products, especially cracking feedstocks. This is achieved by operating a vacuum flasher downstream of a visbreaker to produce a vacuum gas oil cut.
- Reduction of fuel oil production while at the same time reducing pour point and viscosity. This is achieved by utilizing a thermal cracking heater, in addition to a visbreaker heater, which destroys the high wax content of the feedstock.

Specific refining objectives must be defined before a visbreaker is integrated into a refinery, since the overall processing scheme can be varied, affecting the overall economics of the project.

COIL VERSUS SOAKER DESIGN

Two visbreaking processes are commercially available. The first process is the *coil,* or *furnace,* type, which is the type offered through Foster Wheeler and UOP. The coil process achieves conversion by high-temperature cracking within a dedicated soaking coil in the furnace. With conversion primarily achieved as a result of temperature and residence time, coil visbreaking is described as a high-temperature, short-residence-time route. Foster Wheeler has successfully designed many heaters of this type worldwide.

The main advantage of the coil-type design is the two-zone fired heater. This type heater provides for a high degree of flexibility in heat input, resulting in better control of the material being heated. With the coil-type design, decoking of the heater tubes is accomplished more easily by the use of steam-air decoking.

Foster Wheeler's coil-type cracking heater produces a stable fuel oil. A stable visbroken product is particularly important to refiners who do not have many options in blending stocks.

The alternative *soaker* process achieves some conversion within the heater. However, the majority of the conversion occurs in a reaction vessel or soaker which holds the two-phase effluent at an elevated temperature for a predetermined length of time. Soaker visbreaking is described as a low-temperature, high-residence-time route. The soaker process is licensed by Shell. Foster Wheeler has engineered a number of these types of visbreakers as well.

By providing the residence time required to achieve the desired reaction, the soaker drum design allows the heater to operate at a lower outlet temperature. This lower heater outlet temperature results in lower fuel cost. Although there is an apparent fuel savings advantage experienced by the soaker-drum-type design, there are also some disadvantages. The main disadvantage is the decoking operation of the heater and soaker drum. Although decoking requirements of the soaker drum design are not as frequent as those of the coil-type design, the soaker design requires more equipment for coke removal and handling.

The customary practice of removing coke from a drum is to cut it out with high-pressure water. This procedure produces a significant amount of coke-laden water which needs to be handled, filtered, and then recycled for use again. Unlike delayed cokers, visbreakers do not normally include the facilities required to handle coke-laden water. The cost of these facilities can be justified for a coker, where coke cutting occurs every day. However, because of the relatively infrequent decoking operation associated with a visbreaker, this cost cannot be justified.

Product qualities and yields from the coil and soaker drum design are essentially the same at a given severity and are independent of visbreaker configuration.

FEEDSTOCKS

Atmospheric and vacuum residues are normal feedstocks to a visbreaker. These residues will typically achieve a conversion to gas, gasoline, and gas oil in the order of 10 to 50 percent, depending on the severity and feedstock characteristics. This will therefore reduce the requirement for fuel oil cutter stock. The conversion of the residue to distillate and lighter products is commonly used as a measurement of the severity of the visbreaking operation. Percent conversion is determined as the amount of 650°F+ (343°C+) material present in the atmospheric residue feedstock or 900°F+ (482°C+) material present in the vacuum residue feedstock which is visbroken into lighter boiling components.

The extent of conversion is limited by a number of feedstock characteristics, such as asphaltene, sodium, and Conradson carbon content. A feedstock with a high asphaltene content will result in an overall lower conversion than a normal asphaltene feedstock, while maintaining production of a stable fuel oil from the visbreaker bottoms. Also the presence of sodium, as well as higher levels of feed Conradson carbon, can increase the rate of coking in the heater tubes. Minimizing the sodium content to almost a negligible amount and minimizing the Conradson carbon weight percent will result in longer cycle run lengths.

Variations in feedstock quality will impact the level of conversion obtained at a specific severity. Pilot plant analyses of a number of different visbreaker feedstocks have shown that, for a given feedstock, as the severity is increased, the viscosity of the 400°F+ (204°C+) visbroken tar initially decreases and then, at higher severity levels, increases dramatically, indicating the formation of coke precursors.

The point at which this viscosity reversal occurs differs from feed to feed but typically coincides with approximately 120 to 140 standard cubic feet (SCF) of $C_3 -$ gas production per barrel of feed (20.2 to 23.6 normal m^3/m^3). It is believed that this reversal in viscosity defines the point beyond which fuel oil instability will occur. Fuel oil instability is discussed in the next section of this chapter, "Yields and Product Properties."

The data obtained from these pilot tests have been correlated. The viscosity reversal point can be predicted and is used to establish design parameters for a particular

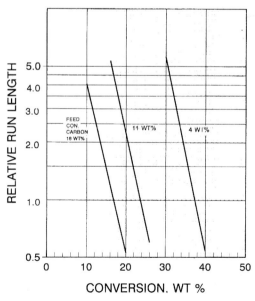

FIGURE 12.3.1. Relative run length versus conversion at various feed qualities.

feedstock to avoid the formation of an unstable fuel oil, while maximizing conversion.

Pilot plant work has also been done relating visbreaker heater run length to conversion and feedstock quality. Figure 12.3.1 graphically represents the decrease in heater run length with increasing feedstock conversion. This graph has been plotted with data for three atmospheric residues with varying feed Conradson carbon. Figure 12.3.1 shows that, for a given percent conversion, as the feed quality diminishes (i.e., as Conradson carbon grows higher), coking of the heater tubes increases, resulting in shorter run lengths.

It has been found that visbreaking susceptibilities bear no firm relationship with API density, which is the usual chargestock property parameter utilized in thermal cracking correlations. However, feedstocks with low n-pentane insolubles and low softening points show good susceptibility to visbreaking, while those having high values for these properties respond poorly. Figure 12.3.2 shows the capability of greater conversion at lower n-pentane insolubles for a 900°F+ (482°C+) vacuum residue.

Residues with low softening points and low n-pentane insolubles contain a greater portion of the heavy distillate, nonasphaltenic oil. It is this heavy oil that cracks into lower boiling and less viscous oils which results in an overall viscosity reduction. The asphaltenes, that fraction which is insoluble in n-pentane, goes through the furnace relatively unaffected at moderate severities. The table below shows the typical normal pentane insolubles content of vacuum residues prepared from base crudes.

Crude-type source of vacuum residue	Range of n-pentane insolubles, wt %
Paraffinic	2–10
Mixed	10–20
Naphthenic	18–28

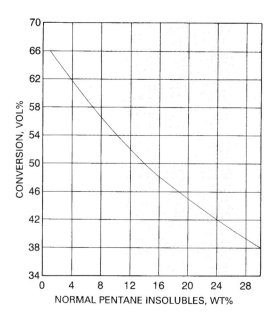

FIGURE 12.3.2. Visbreaking susceptibility (900°F+ charge converted).

YIELDS AND PRODUCT PROPERTIES

Product stability of the visbreaker residue is a main concern in selecting the severity of the visbreaker operating conditions. Severity, or the degree of conversion, if improperly determined, can cause phase separation of the fuel oil even after cutter stock blending. As previously described, increasing visbreaking severity and percent conversion will initially lead to a reduction in the visbroken fuel oil viscosity. However, visbroken fuel oil stability will decrease as the level of severity—and hence conversion—is increased beyond a certain point, dependent on feedstock characteristics.

Until a few years ago, fuel oil stability was measured using the Navy Boiler and Turbine Laboratory (NBTL) heater test. The NBTL test was the accepted test to measure fuel oil stability. However, in the late 1980s there was a general consensus that the NBTL test did not accurately measure fuel oil stability and therefore American Society for Testing and Materials (ASTM) discontinued the test in 1990. Refiners today use the Shell hot filtration test or some variation of it to measure fuel oil stability.

Sulfur in the visbroken fuel oil residue can also be a problem. Typically the sulfur content of the visbreaker residue is approximately 0.5 wt % greater than the sulfur in the feed. Therefore it can be difficult to meet the commercial sulfur specifications of the refinery product residual fuel oil, and blending with low-sulfur cutter stocks may be required.

The development of yields is important in determining the overall economic attractiveness of visbreaking. Foster Wheeler uses its own in-house correlations to determine yield distributions for FW/UOP visbreaking. Our correlations have been based on pilot plant and commercial operating data which allow us to accurately predict the yield distribution for a desired severity while maintaining fuel oil stability. A typical visbreaker yield diagram showing trends of gas and distillate product yields as a function of percent conversion is presented in Fig. 12.3.3.

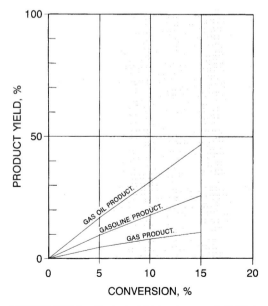

FIGURE 12.3.3. Typical yield trend, gas and distillate products.

TABLE 12.3.1 Typical Yields and Product Properties

Feed properties	Light Arabian atmospheric residue				Light Arabian vacuum residue			
Density, °API	15.9				7.1			
Density, kg/m³	960				1021			
Conradson carbon, wt %	8.5				20.3			
Sulfur, wt %	2.95				4.0			
Viscosity, cSt:								
At 130°F (54°C)	150				30,000			
At 210°F (99°C)	25				900			
Estimated yields	wt %	°API	kg/m³	S, wt %	wt %	°API	kg/m³	S, wt %
H₂S	0.2				0.2			
C₃⁻	2.0				1.5			
C₄'s	0.9				0.7			
C₅–330°F (C₅–166°C)	7.9	57.8	748	0.54	6.0	57.8	748	0.6
330–600°F (166–316°C)	14.5	36.5	842	1.34	15.5*	33.3	859	1.7
600°F+ (316°C+)	74.5	13.5	976	3.48	76.1†	3.5	1048	4.7
	100.0				100.0			

*330–662°F (166–350°C) cut for Light Arabian vacuum residue.
†662°F+ (350°C+) cut for Light Arabian vacuum residue.

Also note in Fig. 12.3.3 that, as the percent conversion increases, the gas, gasoline, and gas oil product yields also increase. However, the conversion can be increased only to a certain point before risking the possible production of an unstable fuel oil. It should also be kept in mind that, at higher percent conversion, some of the gas oil product will further crack and be converted into gas and gasoline products. This will occur particularly when high conversion is achieved at higher heater outlet temperatures.

In Table 12.3.1 we provide typical feed and product properties for light Arabian atmospheric and vacuum residues. These yields are based on a standard severity and single-pass visbreaking while producing a stable visbroken residue. It should be noted that the resultant yield distribution for either a coil or soaker visbreaker is essentially the same for the same conversion.

OPERATING VARIABLES

The main operating variables in visbreaking are temperature, pressure, and residence time. Increasing any one of these three variables will result in an increase in overall severity. To achieve a certain severity, these variables can be interchanged within limits. For a given severity, as measured by conversion, product distribution and quality are virtually unchanged.

An increase in yields of distillate and gaseous hydrocarbons can be achieved by increasing visbreaking severity—for example, by raising the heater outlet temperature. Increasing visbreaker severity will also result in a reduction of cutter stock required to meet fuel oil specifications. However, the higher severities will cause the heavy distillate oils to break down and crack to lighter components. These heavy distillate oils act to solubilize (peptize) the asphaltic constituents. The asphaltic constituents will then tend to separate out of the oil and form coke deposits in the furnace tubes. Visbreaker operation at this level can cause premature unit shutdowns. There is also a tendency to produce unstable fuel oils at these more severe conditions.

PROCESS FLOW SCHEMES

Presented in this section are three visbreaking process schemes, with a diagram and a general description of each:

1. A typical visbreaker unit (Fig. 12.3.4)
2. A typical visbreaker unit with vacuum flasher (Fig. 12.3.5)
3. A typical combination visbreaker and thermal cracker (Fig. 12.3.6)

The first is the most basic scheme, the other two schemes being expanded versions. Figure 12.3.7 is a photograph of a visbreaker designed and built by Foster Wheeler in Spain.

Typical Visbreaker Unit

A typical visbreaker (Fig. 12.3.4) can be employed when viscosity reduction of residual streams is desired so that the need for high-quality distillate cutter stock can be reduced in order to produce a commercial-grade residual fuel oil.

The visbreaker unit is charged with atmospheric or vacuum residue. The unit charge is raised to the proper reaction temperature in the visbreaker heater. The reaction is allowed to continue to the desired degree of conversion in a soaking zone in the heater. Steam is injected into each heater coil to maintain the required minimum velocity and residence time and to suppress the formation of coke in the heater tubes. After leaving the heater soaking zone, the effluent is quenched with a quenching medium to stop the reaction and is sent to the visbreaker fractionator for separation.

The heater effluent enters the fractionator flash zone where the liquid portion flows to the bottom of the tower and is steam-stripped to produce the bottoms product. The vapor portion flows up the tower to the shed and wash section where it is cleaned and cooled with a gas oil wash stream. The washed vapors then continue up the tower. Gas oil stripper feed, as well as pumparound, wash liquid, and the gas oil to quench the charge are all removed on a side drawoff tray. The pumparound can be used to reboil gas plant towers, preheat boiler feedwater, and generate steam. The feed to the gas oil stripper is steam-stripped, and then a portion of it is mixed with visbreaker bottoms to meet viscosity reduction requirements; the remainder is sent to battery limits.

The overhead vapors from the tower are partially condensed and sent to the overhead drum. The vapors flow under pressure control to a gas plant. A portion of the condensed hydrocarbon liquid is used as reflux in the tower and the remainder is sent to a stabilizer. Sour water is withdrawn from the drum and sent to battery limits.

Typical Visbreaker Unit with Vacuum Flasher

The flow scheme for this configuration (Fig. 12.3.5) is similar to the first scheme except that the visbreaker tower bottoms are sent to a vacuum tower where additional distillate products are recovered. This scheme may be desirable since a portion of the residual feed is converted to a cracking feedstock.

In this scheme, the visbreaker bottoms are sent to the vacuum tower flash zone. The liquid portion of the feed falls to the bottom section of the tower, where it is

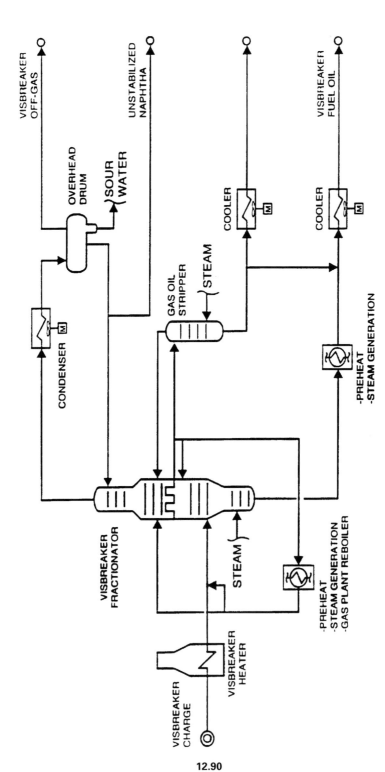

FIGURE 12.3.4. Process schematic for typical visbreaker unit. (*Foster Wheeler and UOP.*)

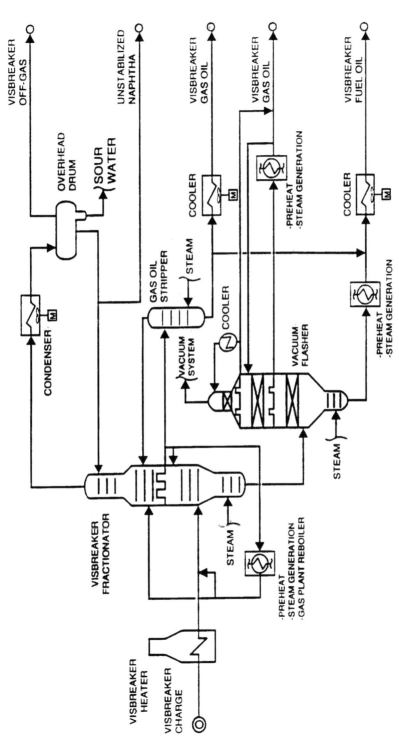

FIGURE 12.3.5. Process schematic for typical visbreaker unit with vacuum flasher. (*Foster Wheeler and UOP.*)

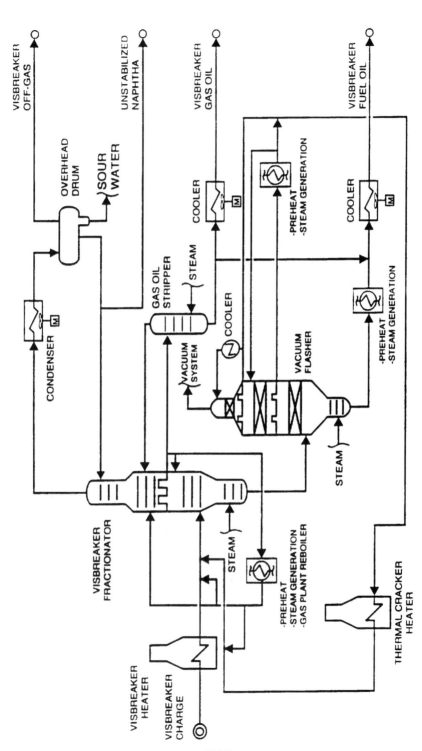

FIGURE 12.3.6. Process schematic for combination visbreaker and thermal cracker. (*Foster Wheeler and UOP.*)

FIGURE 12.3.7. Visbreaker designed and built by Foster Wheeler in Spain.

steam-stripped. The vapor portion rises through the tower wash section and then is partially condensed into distillate products. On this process flow diagram, we have shown two side draws. On the lower drawoff, heavy vacuum gas oil (HVGO) product and pumparound along with wash oil are withdrawn. On the upper one, light vacuum gas oil (LVGO) and reflux are withdrawn. LVGO and HVGO are then combined to form a single vacuum gas oil product which, after visbreaker fuel oil viscosity reduction requirements are met, can be used as a cracker feedstock.

The overhead vapors from the vacuum tower flow to a three-stage vacuum ejector system. Condensed vapor and motivating steam are collected in a condensate accumulator.

Typical Combination Visbreaker and Thermal Cracker

This last scheme is similar to the second except that the vacuum gas oil is routed to a thermal cracker heater instead of to battery limits as a product (Fig. 12.3.6). The vacuum gas oil is cracked and then sent to the visbreaker fractionator along with the visbreaker heater effluent.

A thermal cracking heater is utilized with a visbreaker when maximum light distillate conversion is desired or where extreme pour point reduction is required. Products from this last configuration are a blend of heavy vacuum tar and visbreaker atmospheric gas oil, plus a full range of distillates. Extreme pour point reduction is required for cases in which a high wax content feedstock is processed. The total conversion of the visbreaker vacuum gas oil essentially destroys all of the wax it contains, thus drastically reducing the pour point of the resulting visbreaker fuel oil.

REACTION PRODUCT QUENCHING

In order to maintain a desired degree of conversion, it is necessary to stop the reaction at the heater outlet by quenching. Quenching not only stops the conversion reaction to produce the desired results, but will also prevent production of an unstable bottoms product. For a coil-type visbreaker, quenching of the heater outlet begins from approximately 850 to 910°F (454 to 488°C) depending on the severity. The temperature of the quenched products depends on the overflash requirements and the type of quenching medium used. The overflash requirements are set by the need to maintain a minimum wash liquid rate for keeping the visbreaker fractionator trays wet and preventing excessive coking above the flash zone. Typically, the temperature of the quenched products in the flash zone will vary between approximately 730 and 800°F (388 and 427°C).

Quenching can be accomplished by using different mediums. The most frequently used quenching mediums are gas oil, residue, or a combination of both. These are discussed below. The decision as to which quenching medium is to be used must be made very early in a design. This decision will greatly affect the unit's overall heat and material balance as well as equipment sizing.

Gas oil is the most prevalent medium used for reaction quenching. The gas oil quench works primarily by vaporization and therefore requires a smaller amount of material to stop the conversion reaction than a residue quench. The gas oil quench promotes additional mixing and achieves thermal equilibrium rapidly. The residue quench operates solely by sensible heat transfer rather than the latent heat transfer of the gas oil quench.

The gas oil quench is a clean quench and thus minimizes the degree of unit fouling. It is believed that the use of a residue quench gives way to fouling in the transfer line and fractionator. Also the visbreaker bottoms circuit, from which the residue quench originates, is in itself subject to fouling. The gas oil quench arrangement increases the vapor and liquid loadings in the tower's flash zone, wash section, and pumparound. This will result in a larger tower diameter than if residue quench was used alone.

In order to achieve the same reaction quenching, residue quench flow rates need to be greater than for gas oil quenching. This, as noted above, is because gas oil quenches the reaction by vaporization and residue quenches by sensible heat. In addition, the quenching duty goes up as the percentage of residue quench increases. The actual quenching duty increases because more residue is required in order to achieve the same enthalpy at the flash zone. The use of residue quench means more tower bottoms, product plus recycle, are processed.

Residue quenching provides the potential for additional heat recovery within the unit at a higher temperature level than gas oil quenching. For example, heat recovery from a recycle residue stream may be between 680 and 480°F (360 and 249°C), while heat recovery for a gas oil stream may be from 620 to 480°F (327 to 249°C). With the increase in visbreaker bottoms, additional residue steam stripping is required, which also increases the size of the overhead condenser.

Some visbreaker units designed by Foster Wheeler and UOP employ a combination of both gas oil and residue quenching. It has been found for several visbreaker designs that using a combination quench rather than 100 percent gas oil will shift a significant amount of available heat from steam generation, in the gas oil pumparound, to feed preheat. This is normally preferred as it results in a smaller visbreaker heater and minimizes utility production.

Selection of a combination quench system is preferred for its overall unit flexibility. However, it is more expensive because of duplication of cooling services on the residue and gas oil circuits. It is believed, however, that additional cooling is advantageous since the visbreaking operation can continue by shifting gas oil/residue require-

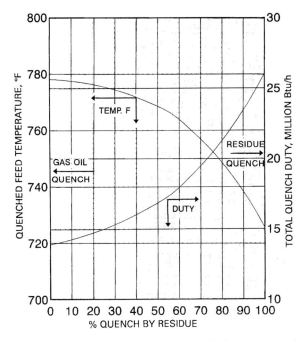

FIGURE 12.3.8. Quench parameters: quench feed temperature and duty versus residue quench.

ments, even if exchangers in the residue circuit become fouled. These exchangers can be bypassed without excessive turndown.

Additionally, the residue and gas oil quench can be used to vary the fractionator flash zone temperature. In visbreaking, many refiners try to keep the flash zone temperature as low as possible in order to minimize the potential for coking. When evaluating the flash zone temperature for a fixed overflash, increasing the percentage of residue will reduce this temperature. The flash zone could vary by as much as 50°F (28°C) between the extremes of total gas oil and residue quench. Figure 12.3.8 shows the basic relationship of the quench feed temperature to the flash zone as a function of the percentage of the reaction quench performed by residue quench. The total reaction quench duty as a function of percent quench by residue for a fixed overflash is also shown.

HEATER DESIGN CONSIDERATIONS

The heater is the heart of the coil-type visbreaker unit. In the design of its visbreakers, Foster Wheeler prefers using a horizontal tube heater for FW/UOP visbreaking to ensure more uniform heating along the tube length. A horizontal-type heater allows the flow pattern for each pass to be as symmetrical as possible. Overheating of one pass can result in thermal degradation of the fluid and eventual coking of that pass. Horizontal-type heaters are also preferred since they have drainable type systems, and liquid pockets cannot develop as in vertical type heaters.

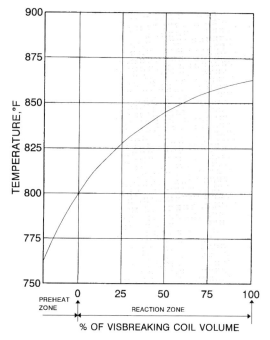

FIGURE 12.3.9. Heater temperature curve.

For the coil-type design, the heater is designed with two independently fired zones. The first is a preheat cell which heats the feed to reaction temperature, approximately 800°F (427°C). The second is a reaction cell which provides the heat input and residence time required for the desired reaction. The visbreaking reaction continues as the fluid leaves the furnace, where it is stopped by quenching. Figure 12.3.9 shows a typical temperature curve for the preheat and reaction zones of a visbreaker heater. This figure shows an 865°F (463°C) heater outlet temperature; however, this temperature can be over 900°F (482°C), depending on the severity of the operation.

In order to achieve the desired residence time in the heater, coil volume in the reaction section is very important. The coil volume will directly affect the cost of the heater. The coil volume specified by Foster Wheeler in designing these heaters is based on previous experience and operating data. During operation, the residence time can be adjusted by controlling the heat input to the reaction cell, the back-pressure on the heater, and the injection steam rate.

Visbreaker heaters typically have a process preheat coil and a steam superheat coil in the convection section. The steam coil is used for superheating steam for residue and gas oil stripping. Steam generation is normally not required in the heater convection section, since the visbreaker produces steam in its bottoms and pumparound circuits.

The heater tube metallurgy is specified as 9% Cr–1% Mo for the main process coil in both the radiant and convection sections. This material is required because of the high heater outlet temperature, regardless of the weight percent sulfur in the feed. Steam superheat coils are usually specified as carbon steel.

Foster Wheeler and UOP typically specify a normal (clean) pressure drop of approximately 300 lb/in^2 gage (20.7 bar gage) and a dirty (fouled) pressure drop of approximately 400 lb/in^2 gage (27.6 bar gage). The elastic design pressure of the heater tubes is based on the shutoff pressure of the charge pump at maximum suction.

Some refiners use a relief valve, located at the heater outlet, to lower the design pressure required for the tubes. Foster Wheeler and UOP do not rely on a relief valve in this service, as the inlet to the valve tends to coke up.

Turndown on a visbreaker heater is typically limited to 60 percent of design capacity. On some projects, clients have installed two heaters, which provide greater unit turndown capability. The additional heater also allows decoking of one heater without shutting down the unit. A two-heater visbreaker may be economically justified for larger units.

The visbreaker heater can be fired on fuel gas, fuel oil, or visbreaker tar. It is economically attractive to fire the visbreaker heater on visbreaker tar since no external fuel source would be needed. However, tar firing requires correctly designed burners to avoid problems of poor combustion. The burners typically require tar at high pressure and low viscosity. Therefore the tar system needs to be maintained at a higher temperature than a normal fuel oil system. Although refiners may prefer to fire the heater with visbreaker tar, many are being forced to burn cleaner fuel gases so as to comply with environmental regulations.

TYPICAL UTILITY REQUIREMENTS

The following represents typical utility consumptions for a coil-type visbreaker:

Power, kW/BPSD [kW/($m^3 \cdot$ h) feed]	0.0358 [0.00938]
Fuel, 10^6 Btu/bbl (kWh/m^3 feed)	0.1195 (220)
Medium-pressure steam, lb/bbl (kg/m^3 feed)	6.4 (18.3)
Cooling water, gal/bbl (m^3/m^3 feed)	71.0 (1.69)

ESTIMATED INVESTMENT COST

Estimated capital costs for a coil-type visbreaker without vacuum flasher or gas plant are $16 million in 10,000 BPSD (66.2 m^3/h) capacity and $31 million in 40,000 BPSD (265 m^3/h) capacity.

These are conceptual estimates with ±30 percent accuracy. They apply to battery-limits process units, based on U.S. Gulf Coast, first quarter 1995, built according to instant execution philosophy, through mechanical completion only. The estimates assume that land is free of aboveground and underground obstructions. Excluded are cost of land, process licensor fees, taxes, royalties, permits, duties, warehouse spare parts, catalysts, forward escalation, support facilities, and all client costs.

BIBLIOGRAPHY

Allan, D., C. Martinez, C. Eng, and W. Barton, *Chemical Engineering Progress,* p. 85, January 1983.

McKetta, J., and W. Cunningham, "Visbreaking Severity Limits," *Petroleum Processing Handbook,* Marcel Dekker, New York, 1992, p. 311.

Rhoe, A., and C. de Blignieres, *Hydrocarbon Processing,* p. 131, January 1979.

PART · 13

OXYGENATES PRODUCTION TECHNOLOGIES

CHAPTER 13.1
HÜLS ETHERS PROCESSES

Scott Davis
UOP
Des Plaines, Illinois

INTRODUCTION

During the 1990s, the oxygenate portion of the gasoline pool has been the fastest-growing gasoline component, and the majority of this growth has been in methyl tertiary butyl ether (MTBE). The major reasons for this growth are generally considered to be environmental concerns and octane upgrades. Countries such as the United States, Korea, and Taiwan have mandated the use of oxygenates in gasoline to promote cleaner-burning fuels. Lead phasedown programs, the introduction of midgrade and higher-octane premium gasolines, and newer, more sophisticated car engines have all contributed to a steadily increasing demand for higher-quality gasoline and thus a continuing need to increase the octane of the refinery gasoline pool.

An important source of MTBE, as well as other ethers, is the refinery. In 1994, installed refinery MTBE capacity of more than 5.57 million metric tons per annum (MTA) [129,000 barrels per stream day (BPSD)] represented about 28 percent of the worldwide MTBE production. Other major sources of MTBE production are from the dehydrogenation of isobutane (see Chap. 5.1), as a by-product from propylene oxide production and from naphtha cracker C_4's.

Although MTBE is the most common ether, it is not the only ether used in gasoline blending today. Tertiary amyl methyl ether (TAME), ethyl tertiary butyl ether (ETBE), and diisopropyl ether (DIPE) are also used as gasoline blending ethers. Table 13.1.1 provides a list of the gasoline blending properties of ethers being used in gasoline pools. In addition to providing a gasoline oxygenate source, these ethers have excellent research and motor blending octanes.

Ethers are generally favored over alcohols in gasoline blending for two reasons: they have a very low solubility in water compared to alcohols, and they have a low blending vapor pressure compared to alcohols.

TABLE 13.1.1 Refinery Oxygenates

Ethers	Blending Octane			Blending RVP		Oxygen, wt %
	RONC	MONC	$(R + M)/2$	kg/cm^2	lb/in^2	
MTBE	118	100	109	0.56–0.70	8–10	18.2
DIPE	112	98	105	0.28–0.35	4–5	15.7
TAME	111	98	105	0.21–0.35	3–5	15.7
ETBE	117	102	110	0.21–0.35	3–5	15.7

Note: RVP = Reid vapor pressure; RONC = research octane number clear; MONC = motor octane number clear; $(R + M)/2$ (RONC + MONC)/2 (sometimes referred to as road octane).

HÜLS ETHERS PROCESS FOR MTBE, ETBE, AND TAME

The Hüls ethers process is marketed under the names *Hüls MTBE process, Hüls ETBE process,* and *Hüls TAME process* and is jointly licensed by Hüls AG in Marl, Germany, and UOP* in Des Plaines, Illinois. This process can be used to produce the ethers for gasoline blending from olefin feedstocks available within a refinery. Depending on the type of hydrocarbon and alcohol feed, the following etherification reactions take place:

$$\underset{\text{Isobutylene}}{CH_2 = C(CH_3)_2} + \underset{\text{Methanol}}{CH_3OH} \rightarrow \underset{\text{MTBE}}{(CH_3)_3\text{-C-O-CH}_3} \quad (13.1.1)$$

$$\underset{\text{Isobutylene}}{CH_2 = C(CH_3)_2} + \underset{\text{Ethanol}}{CH_3CH_2OH} \rightarrow \underset{\text{ETBE}}{(CH_3)_3\text{-C-O-CH}_2CH_3} \quad (13.1.2)$$

$$\underset{\text{Isoamylene}}{CH_3CH = C(CH_3)_2} + \underset{\text{Methanol}}{CH_3OH} \rightarrow \underset{\text{TAME}}{CH_3CH_2(CH_3)_2\text{-C-O-CH}_3} \quad (13.1.3)$$

The reactions proceed in the liquid phase at mild conditions in the presence of a solid acidic catalyst. The catalyst typically is a sulfonic ion-exchange resin. The reaction temperature is kept low and can be adjusted over a fairly broad range. Higher temperatures are possible, but excessive temperatures are not recommended because resin fouling from polymers can occur. Around 130°C (266°F), sulfonic ion-exchange resins become unstable. Operation in the lower temperature range ensures stable operation and long catalyst life.

The reaction of an isoolefin with alcohol is conducted in the presence of a small excess of alcohol relative to that required for the stoichiometric reaction of the isoolefin contained in the hydrocarbon feed. Operation with a small excess of alcohol has a number of advantages and practically no drawbacks because any excess alcohol is recovered and recycled. Some of the advantages are

- The equilibrium is displaced toward the production of ether to favor higher per-pass conversion.
- Production of high-octane ether is maximized, and production of lower-octane oligimers is minimized.
- Process temperature is more efficiently and securely controlled.

---

*Trademark and/or service mark of UOP.

In the absence of a small excess of alcohol, isoolefin dimerization, also exothermic, can take place rapidly. This reaction can result in a sharp temperature rise in the resin bed. Such an increase causes irreversible catalyst fouling, and catalyst destruction can occur if the temperature rise is excessive.

Under proper conditions, the etherification reaction is nearly one hundred percent selective except for minor side reactions resulting from the presence of certain feed impurities. Water contained in the feed results in equivalent amounts of tertiary butyl alcohol (TBA) in the MTBE or ETBE product. Water in isoamylene feed yields tertiary amyl alcohol (TAA) in the TAME product. In small quantities, these alcohol by-products are unimportant. They need not be separated from the ether product because they have high octane values and can be used as gasoline blending agents.

Either a one-stage or two-stage Hüls design can be used for MTBE, TAME, or ETBE production. A simplified flow diagram of the single- and two-stage designs is shown in Figs. 13.1.1 and 13.1.2, respectively. The two-stage unit produces higher conversion levels but costs more compared to the one-stage design.

Because of the lower cost, the Hüls one-stage design is by far the most common inside the refinery. Two-stage units are typically built only when extremely high purity raffinate is required, such as in butene-1 production, or when the raffinate is used in a recycle operation. Typical one-stage olefin conversions are shown in Table 13.1.2.

PROCESS FLOW

The fresh hydrocarbon feed must be treated in a water wash if it comes from a fluid catalytic cracking (FCC) unit. The treatment step is needed to remove basic nitrogen compounds, which are catalyst poisons. This procedure is not necessary if the feed comes from either a steam cracker or from a UOP Oleflex* unit. In the case of TAME production, diolefins must also be removed in a hydrogenation unit (see Chap. 8.2). Figure 13.1.1 is a simplified single-stage process flow diagram. The clean fresh feed is mixed with fresh and recycled alcohol and charged to the reactor section. The reactor can be a tubular reactor, or more typically, two adiabatic reactors with recycle are used. The majority of the reaction occurs in the first reactor. The second reactor completes the reaction of isoolefins to ether. Cooling between reactors is required to maximize the approach to equilibrium in the second reactor.

The product from the reactor section primarily contains ether, excess alcohol (methanol or ethanol), and unreacted C_4 or C_5 hydrocarbons. This stream is sent to a fractionation column, where high-purity MTBE, ETBE, or TAME is recovered from the bottoms. The unreacted hydrocarbon, typically referred to as the *raffinate stream*, and alcohol are taken off the top of the fractionator. Before it leaves the unit, the raffinate is water-washed to remove excess alcohol. The water-alcohol mixture from the water wash is fractionated in the alcohol recovery section. The recovered alcohol is recycled back to the reactor, and the water is recycled back to the water wash.

YIELDS

The yields in Table 13.1.3 are representative of the oxygenate production from FCC olefins using a single-stage Hüls process to separately process the C_4 and C_5 cuts. The TAME is assumed to be pretreated in a diene saturation unit.

---

*Trademark and/or service mark of UOP.

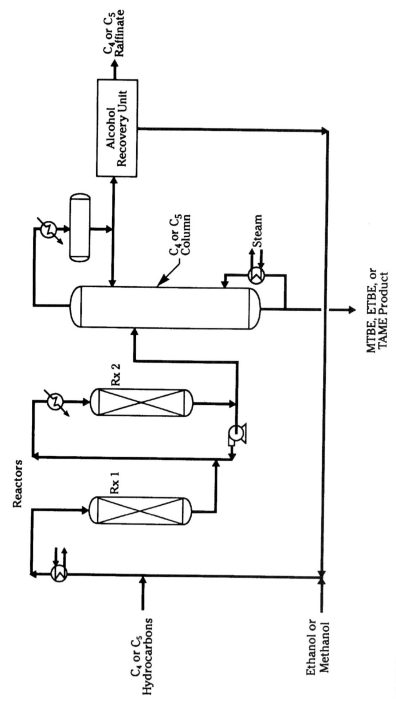

FIGURE 13.1.1 Hüls MTBE, ETBE, and TAME process—single-stage unit.

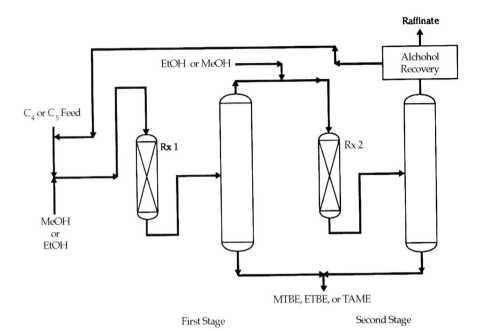

FIGURE 13.1.2 Hüls MTBE, ETBE, and TAME process—two-stage unit.

TABLE 13.1.2 Typical One-Stage Olefin Conversions

Olefin feedstocks	MTBE	TAME	ETBE
Isobutylene	96–97%	—	86–88%
Isoamylene	—	65–70%	—

TABLE 13.1.3 Ethers Production from Single-Stage Units

	Hüls MTBE process, MTA (BPD)	Hüls ETBE process, MTA (BPD)	Hüls TAME process, MTA (BPD)
Total hydrocarbon feed	264,100 (8437)	264,100 (8437)	305,700 (8971)
Reactive components in the feed:			
Hydrocarbons:			
iC_4	39,600 (1247)	39,600 (1247)	—
iC_5	—	—	76,500 (2167)
Alcohols:			
Methanol	22,000 (522)	—	23,700 (562)
Ethanol	—	28,100 (666)	—
Ethers product:			
MTBE	60,000 (1520)	—	—
ETBE	—	62,000 (1573)	—
TAME	—	—	75,000 (1825)

Note: MTA = metric tons per annum; BPD = barrels per day; i = iso.

ECONOMICS AND OPERATING COSTS

The estimated erected cost of a UOP-designed single-stage Hüls MTBE process unit for the production of 60,000 MTA (1520 BPD) of MTBE in 1995 is $8.2 million. This capital estimate is for an inside-battery-limits unit erected on the U.S. Gulf Coast.

The utility requirements for a 60,000-MTA (1520-BPD) MTBE unit and 75,000-MTA (1825-BPD) TAME, respectively, are estimated in Table 13.1.4.

COMMERCIAL EXPERIENCE

Twenty Hüls MTBE units have been brought on-stream. The first unit started up in 1976 in Marl, Germany. Operating plant capacities range up to about 600,000 MTA (15,200 BPSD) of MTBE. The units cover the entire range of feed compositions, product qualities, isobutylene conversions, and end uses for MTBE.

TABLE 13.1.4 Utility Requirements

Utilities	MTBE	TAME
Power, kWh	129	160
Low-pressure steam, MT/h (klb/h)	7.2 (15.8)	12.0 (26.4)
Condensate,* MT/h (klb/h)	7.2 (15.8)	12.0 (26.4)
Cooling water, m^3/h (gal/min)	64 (282)	63 (278)

*Denotes export.
Note: MT/h = metric tons per hour.

CHAPTER 13.2
UOP ETHERMAX PROCESS FOR MTBE, ETBE, AND TAME PRODUCTION

Scott Davis
UOP
Des Plaines, Illinois

PROCESS DESCRIPTION

The Ethermax* process, licensed exclusively by UOP,* can be used to produce methyl tertiary butyl ether (MTBE), tertiary amyl methyl ether (TAME), or ethyl tertiary butyl ether (ETBE). This process combines the Hüls fixed-bed etherification process with advanced RWD† catalytic distillation technology from Koch Engineering Company, Inc. The combined technology overcomes reaction equilibrium limitations inherent in a conventional fixed-bed etherification process.

The Ethermax process reacts tertiary olefins, such as isobutylene and isoamylene, over an acid resin in the presence of alcohol to form an ether. The reaction chemistry and unit operating conditions are essentially the same as those of a conventional ether process, such as the Hüls MTBE process (Chap. 13.1), except that KataMax† packing has been added to increase the overall conversion.

KataMax packing represents a unique and proprietary approach to exposing a solid catalyst to a liquid stream inside a distillation column. The reactive distillation zone of the RWD column uses KataMax packing to overcome reaction equilibrium constraints by continuously fractionating the ether product from unreacted feed components. As the ether product is distilled away, the reacting mixture is no longer at equilibrium. Thus, fractionation in the presence of the catalyst promotes additional conversion of the reactants. Isobutylene conversions of 99 and 97 percent, respectively, for MTBE and ETBE are typical, and isoamylene conversions of up to 94 percent can be

---

*Trademark and/or service mark of UOP.
†Trademark of Koch Engineering Company, Inc.

achieved economically with this process. These design specifications are typical for gasoline blending; however, practically any olefin conversion is achievable by designing a unit to accommodate individual refinery needs. For example, the Ethermax process can be designed to convert 99.9+ percent of the isobutylene when butene-1 production is a design objective.

The flexibility of the Ethermax process provides refiners with many routes to increase oxygenate or octane levels in their gasoline pool. Existing MTBE units can be converted to TAME or ETBE production. Increases in throughput and olefin conversion are possible in an existing ethers unit by revamping it to the Ethermax process. The revamp increases the oxygenate level of the gasoline pool, and the resulting octane improvement gives a refiner the flexibility to optimize gasoline production from other refinery processes.

PROCESS FLOW

The process flow for the Ethermax process is shown in Fig. 13.2.1. The majority of the reaction is carried out in a simple fixed-bed adiabatic reactor. The effluent from this reactor feeds the RWD column, where the ethers are separated from unreacted feed components. The bottoms from the RWD column are the MTBE, ETBE, or TAME product. The unreacted components move up the column and enter the catalytic section of the fractionator for additional conversion. The catalytic section of the RWD column uses KataMax packing to overcome reaction equilibrium constraints by simultaneously reacting the feed component and fractionating the ether product.

The overhead from the RWD column is routed to the alcohol (either methanol or ethanol) recovery section. In this system, water is used to separate the alcohol from

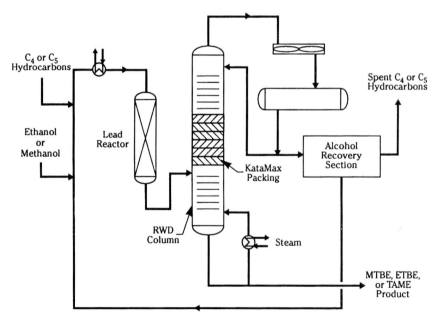

FIGURE 13.2.1 Ethermax process.

the hydrocarbon in a simple countercurrent extraction column, and a distillation column is used to recover the alcohol. The recovered alcohol is recycled to the reactor section. The hydrocarbon raffinate is generally sent downstream to an alkylation unit in the case of C_4's or to gasoline blending in the case of C_5's.

YIELDS

The example in Table 13.2.1 represents oxygenate production from fluid catalytic cracking (FCC) olefins using the Ethermax process to separately process the C_4 and C_5 cuts. The TAME feed is assumed to be pretreated to saturate dienes. All units are in metric tons per year.

OPERATING COST AND ECONOMICS

The estimated utilities for a 60,000-metric ton per annum (MTA) (1520-BPD) Ethermax unit producing MTBE and a 75,000-MTA (1825-BPD) Ethermax unit producing TAME, respectively, are given in Table 13.2.2. The 1995 estimated erected cost for an Ethermax unit to produce 60,000 MTA (1520 BPD) of MTBE is $8 million U.S. based on the inside battery limits of the process unit.

ETHERMAX COMMERCIAL EXPERIENCE

The first Ethermax process unit was commissioned at the Hüls AG, Marl, Germany, facility in March 1992. As of 1995, another four Ethermax process units have been placed on-stream. These five Ethermax units process a wide variety of feedstocks from FCC, stream cracking, and dehydrogenation units. The performance of all five operating units has exceeded representations. An additional eight Ethermax units have

TABLE 13.2.1 Oxygenate Production from Ethermax Process

	MTBE operation, MTA (BPD)	TAME operation, MTA (BPD)
Total hydrocarbon feed	264,100 (8437)	305,700 (8971)
Reactive components in the feed:		
Hydrocarbons:		
iC_4	39,600 (1247)	—
iC_5	—	76,500 (2167)
Alcohols:		
Methanol	22,500 (534)	33,000 (783)
Ethanol	—	—
Ethers product:		
MTBE	61,500 (1558)	—
ETBE	—	—
TAME	—	104,500 (2543)

Note: MTA = metric tons per annum; BPD = barrels per day; i = iso.

TABLE 13.2.2 Utilities for Ethermax Unit Producing MTBE and TAME

Utilities	Ethermax for MTBE	Ethermax for TAME
Electric power, kWh	117	173
Steam, MT/h (klb/h):		
Low pressure	1.1 (2.5)	15.6 (34.3)
Medium pressure	7.5 (16.4)	—
Condensate,* MT/h (klb/h)	8.6 (18.9)	15.6 (34.3)
Cooling water, m^3/h (gal/min)	34 (151)	41 (179)

*Denotes export.
Note: MT/h = metric tons per hour.

been licensed. Together these units represent more than 2.17 MTA (55,000 BPSD) of ethers capacity.

CHAPTER 13.3
UOP OLEFIN ISOMERIZATION

Scott Davis
UOP
Des Plaines, Illinois

INTRODUCTION

The production of ethers, notably methyl tertiary butyl ether (MTBE) and tertiary amyl methyl ether (TAME), is often limited by the isoolefin concentration in a predominately normal olefin stream. Olefin isomerization technology is aimed at increasing the overall yield of ethers from a given feed stream by the skeletal isomerization of normal olefins to reactive isoolefins.

The idea of skeletal isomerization of normal olefins to isoolefins is not new. During the 1960s, UOP scientists, as well as others, worked on catalyst development based on chlorided alumina. However, these previous catalyst systems were not attractive because they showed poor stability and low selectivity. The increasing demand for oxygenates has resulted in a higher demand for isobutylene and isoamylene to be used in ether production. UOP began active research in 1989 for a skeletal olefin isomerization catalyst. The timing of this demand was fortunate because of the emergence of a number of new catalytic materials. The challenge was to develop a catalyst with high selectivity and stability. With its unique expertise in the development of new materials, UOP developed a proprietary catalyst for skeletal isomerization of light normal olefins. The catalyst was further improved, and a successful commercial manufacturing trial run was conducted at the end of 1991.

DESCRIPTION OF THE PENTESOM PROCESS

The UOP* Pentesom* process isomerizes normal C_5 olefins to reactive isoamylene for conversion to TAME. This unit, coupled with an ethers unit, such as the

---

*Trademark and/or service mark of UOP.

Ethermax* process (Chap. 13.2), maximizes the production of TAME derived from fluid catalytic cracking (FCC) unit C_5 olefins. The high-conversion Pentesom unit normally can increase the TAME production from an FCC unit by 1.7 times that of the stand-alone TAME Ethermax unit. The Pentesom-Ethermax flow scheme consumes more than 80 percent of the available C_5 olefins in a typical FCC feed stream. This consumption compares to only about 50 percent C_5 olefin utilization with stand-alone TAME production.

UOP's analysis has shown that, in most cases, operating the Pentesom unit on a once-through basis by adding a second Ethermax unit downstream of the Pentesom unit is preferable to recycling the Pentesom effluent back to a single Ethermax unit. The block flow diagram of this flow scheme is shown in Fig. 13.3.1.

The primary benefit of this flow scheme is a savings in utilities. The FCC unit has a substantial amount of saturate unreactive C_5's contained with the C_5 olefins. When the Ethermax-Pentesom units are operated in a recycle mode, the resulting buildup of normal paraffins consumes both utilities and capacity that are not directed toward TAME production. The normal paraffins must be purged from the recycle loop by a bleed, which also results in a loss of normal pentenes from TAME production. Operation in a once-through flow scheme eliminates these concerns.

Pentesom Process Flow

The Pentesom flow scheme consists of a single reactor containing a high-activity molecular-sieve-based catalyst (Fig. 13.3.2). The Ethermax effluent passes through a fired heater and is combined with a small amount of hydrogen before entering the single, fixed-bed Pentesom reactor. The reactor effluent is cooled and condensed before entering a separator. The separator overhead stream, which is rich in hydrogen, is compressed and recycled to the Pentesom reactor. A small amount of makeup hydrogen is added to the recycle stream. Separator bottoms are routed to a stripper. Stripper bottoms are sent to a second Ethermax unit for additional TAME production.

No feed pretreatment other than that required for the Ethermax unit is required for the Pentesom process. The catalyst operates for 1 year between regenerations. The regeneration is normally conducted in situ using the existing process equipment. No additional regeneration equipment is needed.

---

*Trademark and/or service mark of UOP.

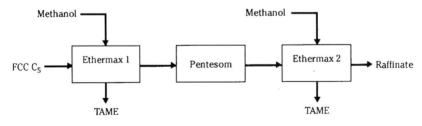

FIGURE 13.3.1 Typical FCCU C_5 processing scheme.

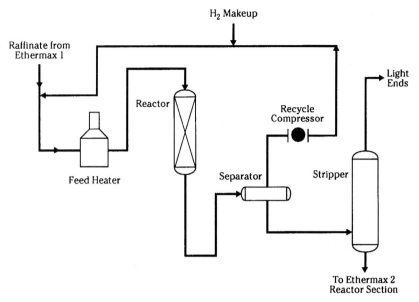

FIGURE 13.3.2 UOP Pentesom process.

DESCRIPTION OF THE BUTESOM PROCESS

The Butesom* process is UOP's C_4 olefin isomerization process. The process isomerizes normal butenes to isobutylenes, which can then be further converted to MTBE. The Butesom unit is operated in a recycle mode with an ethers unit, such as UOP's Ethermax process (Chap. 13.2), for MTBE production.

The Butesom-Ethermax flow scheme is targeted for streams rich in C_4 olefins and low in paraffins. Such a stream is that available from naphtha-based steam crackers. In these situations, the raffinate is typically a relatively low-value product because no other conversion processes are available. The overall olefin utilization for MTBE in these cases typically increases from below 30 to 80 percent with the use of the Butesom-Ethermax flow scheme.

The feedstock from the FCC unit contains a high saturate C_4 concentration, which is normally not a good fit for a Butesom unit. The primary reason that the Butesom unit is more effective on high-olefin feedstocks is that the conversion of normal butene to isobutylene is equilibrium limited to 40 to 50 percent under normal operating conditions. This conversion level results in a recycle operation being required to maximize the normal butene utilization to MTBE. However, the high paraffin concentrations present in FCC feedstocks require a substantial bleed on the recycle loop to purge the paraffins. The recycle bleed drags butenes out of the flow scheme, thereby reducing the availability of butenes for conversion to MTBE.

---

*Trademark and/or service mark of UOP.

Butesom Process Flow

A simplified Butesom flow scheme is shown in Fig. 13.3.3. This simple unit uses a molecular-sieve-based catalyst and swing reactors. The Ethermax effluent passes through a combined feed exchanger and fired heater before entering one of the swing fixed-bed Butesom reactors. The reactors are operated in a swing mode with one reactor on-line and the other in regeneration. The reactor effluent is exchanged with the fresh feed in the combined feed exchanger and cooled before being compressed and condensed in the effluent compressor system. The liquid is then pumped back to the Ethermax unit for conversion to MTBE. The small amount of light ends produced in the Butesom unit are removed in the Ethermax unit.

Regeneration Section

Unlike the UOP C_5 skeletal isomerization system, all C_4 skeletal isomerization catalyst systems have limited stability and require frequent regenerations. During the process cycle, a progressive accumulation of coke on the catalyst occurs. If the process cycle were extended significantly without regeneration, the coke deposited would cause a gradual decrease in catalyst performance. Therefore, the regeneration step is critical to the overall process economics. The Butesom process provides a regeneration system that is simple and low cost. The regeneration consists of a simple carbon burn to remove the coke on the catalyst. Because the burn is conducted in the reactor, less regeneration equipment is required. Consequently, the valving and maintenance problems associated with moving catalyst are eliminated.

The regeneration sequence is as follows:

- Reactor isolation
- Evacuation and N_2 pressure-up
- Carbon burn

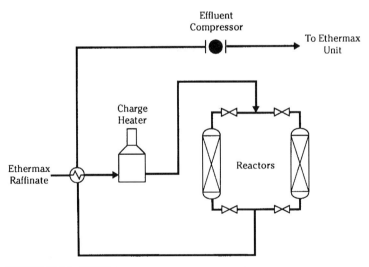

FIGURE 13.3.3 UOP Butesom process.

- Evacuation and N_2 pressure-up
- Reactor on-line

The evacuation and pressure-up steps are accomplished by a common single-stage steam ejector. Purge gas is minimized by designing the system so that only the reactor itself needs to go through the evacuation and purge steps. The carbon burn is the only catalyst regeneration step required to restore the catalyst activity; no promoters or special activators are used on or with the catalyst. The carbon burn is controlled simply by regulating the oxygen content and temperature in the burn zone. Utility air is used as the oxygen source. No costly dryers or special utilities are required.

ECONOMICS

The Pentesom-Ethermax process flow scheme increases the typical FCC TAME production by about 1.7 times compared to a stand-alone Ethermax unit for TAME production. The capital cost for an Ethermax-Pentesom-Ethermax complex built on the U.S. Gulf Coast in 1995 and producing 103,000 metric tons per year (MTA) [2500 barrels per day (BPD)] of TAME from FCC-derived feed is approximately $21 million U.S.

A Butesom-Ethermax complex processing an FCC feedstock can typically produce about 1.7 times the production of a stand-alone Ethermax unit. The estimated 1995 U.S. Gulf Coast erected cost for a complex to produce 86,800 MTA (2200 BPD) of MTBE is $23 million U.S.

COMMERCIAL EXPERIENCE

The Butesom and Pentesom processes are UOP's newest technology to be offered for commercial license. The catalyst systems used in the Pentesom and Butesom processes were under development for almost 5 years. Pilot plant tests included process variable studies as well as contaminant studies. The catalyst was tested under commercial conditions and exposed to multiple regeneration cycles. A commercial manufacturing test run was successfully conducted. The equipment and operating conditions for the Butesom and Pentesom processes were well within the normal refining engineering boundaries. The Butesom and Pentesom designs draw on the expertise gained in these commercial runs as well as on the experience gained in more than 80 years of process commercialization.

CHAPTER 13.4
OXYPRO PROCESS

Scott Davis
UOP
Des Plaines, Illinois

PROCESS DESCRIPTION

The new UOP* Oxypro* process is a unique, low-cost, refinery-based catalytic process for the production of diisopropyl ether (DIPE) from propylene and water. The ether DIPE has high octane, low vapor pressure, and excellent gasoline blending properties.

The Oxypro process is especially well suited for processing propylene derived from the fluid catalytic cracking (FCC) unit within the refinery. After amine and Merox* treating, the FCC-derived propylene is fed directly along with water to the Oxypro process. The propylene and water are converted to DIPE at more than 98 wt % selectivity. The Oxypro product has a purity of more than 98 wt % DIPE and a research octane number clear (RONC) and motor octane number clear (MONC) that are comparable to other ethers, such as MTBE and TAME.

The Oxypro product shows a clear octane advantage over both catalytic polymerization and alkylation of propylene. The DIPE from the Oxypro process generates 112 RONC and 98 MONC compared to only 90 RONC and 89 MONC for C_3 alkylate and 93 RONC and 82 MONC for catalytic polymerization gasoline. The combination of high-octane product and near 100 percent overall conversion gives the Oxypro process superior performance compared to other refinery C_3 alternatives.

PROCESS FLOW SCHEME

A simplified flow scheme of the Oxypro process is shown in Fig. 13.4.1. The amine- and Merox-treated mixed-C_3 steam from the FCC unit enters the unit and is mixed with makeup water and internal recycle streams of propylene, isopropyl alcohol (IPA),

---

*Trademark and/or service mark of UOP.

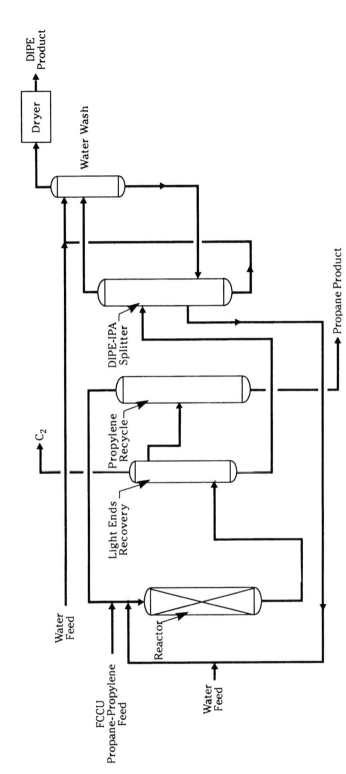

FIGURE 13.4.1 Oxypro flow scheme.

and water. The combined streams are processed downflow in a fixed-bed reactor. The reactor effluent is sent to fractionation, where the light ends and propane are removed. The propane product meets typical liquefied petroleum gas (LPG) specifications of less than 5 wt % propylene. The fractionation product is then sent to product recovery, where water, IPA, and DIPE are separated. The IPA and water are recycled to the reactor to maximize DIPE production. The DIPE product has a purity of more than 98 wt %.

YIELDS

The yields from an Oxypro unit designed to produce 96,000 metric tons per year MTA [2500 barrels per day (BPD)] of DIPE product are shown in Table 13.4.1. The feedstock used to generate these yields is representative of a mixed propane-propylene stream from an FCCU after amine and Merox treating. Specifications typical for an Oxypro DIPE product are shown in Table 13.4.2.

OPERATING COSTS AND ECONOMICS

The estimated inside-battery-limits erected cost of an Oxypro unit built on the U.S. Gulf Coast in 1995 with a capacity to produce 88,500 MTA (2300 BPD) of DIPE is about $13 million. Utility requirements for this unit are given in Table 13.4.3.

TABLE 13.4.1 Oxypro Yield Summary

	BPSD	kg/h
Feed:		
Propylene	2,918	10,050
Propane	1,239	
H_2O	328	4,100
		2,170
Product:		
LPG	1,300	4,310
DIPE	2,500	12,010

Note: BPSD = barrels per stream day.

TABLE 13.4.2 Typical Oxypro Product Specifications

Specific gravity	0.73
DIPE, wt %	98
Water, wt ppm	<100
IPA, wt %	<0.5
C_6+, wt %	<2
Octane:	
RONC	112
MONC	98

TABLE 13.4.3 Operating Utility Requirements

Power, kWh	537
Steam, MT/h (klb/h):	
Low pressure	11.3 (24.9)
High pressure	8.7 (19.1)
Cooling water, m^3/h (gal/min)	336 (1,437)

Note: MT/h = metric tons per hour.

COMMERCIAL EXPERIENCE

The Oxypro process was introduced to the market in 1995. By the end of 1995, one unit had been licensed and was in design and construction. Several other units were under consideration. The Oxypro process equipment and operating conditions are well within normal refining boundaries. The operating conditions are similar to those of hydrotreating, and the reaction chemistry is similar to that of MTBE ethers units. UOP has designed and licensed more than 700 hydrotreaters and more than 30 MTBE units. The Oxypro design draws on the expertise of these designs as well as on experience gained in more than 80 years of UOP process commercialization.

P · A · R · T · 14

HYDROGEN PROCESSING

CHAPTER 14.1
HYDROGEN PROCESSING

Alan G. Bridge
Chevron Research and Technology Company
Richmond, California

INTRODUCTION

General

The potential for applying hydrogenation reactions within the refining industry has been known since the early years of the 20th century. In October 1930 the American Chemical Society conducted a symposium in Cincinnati on the subject of "Industrial High-Pressure Reactions." In one of the papers, Haslam and Russell[37] discussed the five adaptations of hydrogenation which appeared to be of the most immediate importance. Figure 14.1.1 is a copy of the page on which they summarized these five adaptations.

Horne and McAfee[41] in 1958 noted that of these five, the second, third, and fourth were already being practiced and they predicted that the first and fifth would also soon be commercialized. They were right. Such reactions are now commonplace in modern refineries, largely because of the plentiful supply of hydrogen produced either in catalytic reformers or from inexpensive sources of natural gas.

The modern version of distillate hydrocracking was introduced in the United States in the 1960s to convert excess fuel oil into predominantly motor gasoline and some jet fuel. Fluid catalytic cracking (FCC) cycle oils were popular feedstocks at first. The process was then used to upgrade the liquids produced in delayed and fluid coking and solvent deasphalting.

While these residuum conversion schemes were being implemented, a demand for low-sulfur fuel oil (LSFO) developed and fixed-bed residuum hydrotreating was commercialized. Nowadays LSFO is not in such demand, and these hydrotreaters are being used to prepare feed for FCC units. On-stream catalyst replacement technology has been commercially successful in extending the capabilities of fixed-bed residuum hydrotreaters. Several ebullating-bed hydrocrackers are also now operating.

Since the early distillate hydrocrackers were built, the demand for motor gasoline has not grown as quickly as that for middle distillates. The more recent hydrocracking units have been designed, therefore, to make good-quality kerosene and diesel fuel. Besides this trend, hydrocracking is steadily replacing conventional extraction

INDUSTRIAL HIGH-PRESSURE REACTIONS

Presented before a joint session of the Divisions of Industrial and Engineering Chemistry, Gas and Fuel Chemistry, and Petroleum Chemistry at the 80th Meeting of the American Chemical Society, Cincinnati, Ohio, September 8 to 12, 1930

Introduction

THE decision of the AMERICAN CHEMICAL SOCIETY to conduct a symposium on "Industrial High-Pressure Reactions" is timely. To my mind it gives recognition to one of the most important developments in American chemical industry since the war. The coöperation of three such large divisions of the SOCIETY as Industrial and Engineering, Petroleum, and Gas and Fuel is evidence of the wide interest and appeal of this subject.

The tremendous growth of industries based on high-pressure reactions is the most startling development of the decade. Synthetic ammonia is now one of our largest heavy-chemical industries. Synthetic-methanol production is mounting rapidly and oil hydrogenation promises to place the petroleum industry on a better economic basis and bring it into closer contact with more strictly chemical industries. Synthetic phenol and a host of similar miscellaneous developments promise much for the expansion and improvement of existing processes. The industrial chemist of today faces greater opportunity for exploration and accomplishment than ever before.

Not the least of the difficulties met in organizing a symposium such as this is a direct consequence of the novelty of the subject and the present rapid growth and development of processes based on high-pressure technic. The keen industrial competition existing and the comparatively limited number of laboratories from which to solicit contributions combine to enhance the difficulties. The fact that we have been able to secure such an impressive list of papers is a tribute to a developing spirit of coöperation in industrial, government, and university laboratories.

NORMAN W. KRASE, *Chairman*

Hydrogenation of Petroleum[1]

R. T. Haslam[2] and R. P. Russell[3]

STANDARD OIL DEVELOPMENT COMPANY AND HYDRO ENGINEERING & CHEMICAL COMPANY, ELIZABETH, N. J.

THE conditions under which commercial hydrogenation has been practiced since the time of Sabatier have been restricted until the last few years to the use of (1) hydrogen at substantially normal pressure or 2 or 3 atmospheres above normal; (2) hydrogen of a high degree of purity particularly with respect to such catalyst poisons as sulfur, arsenic, and the like; (3) powerful but sensitive catalysts of the type of reduced nickel; and (4) temperatures safely below those at which thermal decomposition of the stock to be hydrogenated takes place. Coal and oil, both always containing sulfur, were not amenable to this type of hydrogenation, and it was therefore restricted to animal and vegetable fats and oils. By eliminating the catalyst and substituting hydrogen pressures one hundred fold greater than had previously been used, a high degree of liquefaction was obtained, but the oils thus produced contained relatively large percentages of oxygenated bodies of the cresolic type, making the oils hard to crack or refine. The able and resourceful research organization of the I. G. Farbenindustrie, through their experimentation, recognized the need of greater hydrogenation intensity than obtainable with hydrogen pressures then commercially permissible and developed a line of sulfur-resistant catalysts

[1] Received September 11, 1930.
[2] Vice president and general manager, Standard Oil Development Co.
[3] General manager, Hydro Engineering & Chemical Co.

This paper deals with some of the recent developments in the hydrogenation of petroleum; and shows the adaptability of the process for converting fuel oil to gasoline and gas oil, increasing the paraffinic nature of kerosenes, burning oils, and lubricants; and discusses the reverse possibility of producing non-paraffinic gasoline. There is pointed out the flexibility of the process, particularly with respect to changes in the characteristics of the product, the handling of a wide variety of charging stock, the elimination of all forms of sulfur, and the conversion of all asphalts to distillate fuels.

There are five adaptations of hydrogenation which appear to be of most immediate importance in oil refining. These are:

(1) The conversion of heavy, high-sulfur, asphaltic crude oils and refinery residues into gasoline and distillates low in sulfur and free from asphalt, without concurrent formation of coke.
(2) The alteration of low-grade lubricating oils, to obtain high yields of lubricating oils of premium quality as to temperature-viscosity relationship, Conradson carbon, flash, and gravity.
(3) The conversion of off-color, inferior-burning oil distillates or light gas oils into high-gravity, low-sulfur, water-white burning oils of excellent burning characteristics, with gasoline being the only other product except for a slight gas formation.
(4) The desulfurization and color- and gum-stabilization of high-sulfur, badly gumming cracked naphthas without marked alteration in distillation range and without major loss in antiknock value. (It is possible to operate so as actually to better the antiknock quality.)
(5) The conversion of paraffinic gas oils into low-sulfur, gum- and color-stable, good antiknock gasolines without the production of coke or heavy product.

which materially speeded up hydrogenation and caused the elimination of all the oxygen from the hydrogenated product. In addition, their long experience in the field of synthetic ammonia enabled them to devise apparatus and methods for better carrying out this type of hydrogenation in a continuous manner.

FIGURE 14.11 Early uses of hydroprocessing.

processes in lube oil base stock manufacture because it can produce much more valuable by-products than the older process. The introduction of Chevron's Isodewaxing process now gives refiners the opportunity to produce unconventional base oils with viscosity indexes greater than 110.

Our understanding of the basic reactions and the subtleties of hydroprocessing scaleup has improved, and hundreds of technical articles have been written on the subject. This chapter summarizes the current understanding. By reviewing the literature references, the reader can explore the subject in more depth. Emphasis will be placed on the hydroprocessing of heavy feedstocks, since this is the field which has seen most of the advances.

For clarity, throughout the chapter we will use the term *hydroprocessing* to describe all the different processes in which hydrocarbons react with hydrogen. Hydrotreating will be used to describe those hydroprocesses dealing predominantly with impurity removal from the hydrocarbon feedstock. Hydrocracking will be used to describe those hydroprocesses which accomplish a significant conversion of the hydrocarbon feedstock into lower-boiling products.

Hydroprocessing Objectives

Hydroprocessing feedstocks—naphthas, atmospheric gas oils, vacuum gas oils (VGOs), and residuum—have widely different boiling character. Within each of these different boiling ranges exist a variety of molecular types. This depends on both the crude oil source and whether or not the material was produced in a cracking reaction or as a straight-run component of the original crude oil. The impurity levels in a variety of crude oils and in their vacuum residua are shown in Table 14.1.1. The vacuum residuum is the lowest-value fraction in the crude oil. Historically it has been blended into heavy fuel oil. The demand for this product, however, has not kept pace with the tremendous increase in demand for transportation fuels. Environment pressures have widened this gap by restricting the use of high-sulfur fuel oil while mandating cleaner light products. The products into which the refiner must convert the bottom of the barrel are summarized in Table 14.1.2.

The introduction of residuum hydroprocessing in the 1960s was a response to an increasing demand for LSFO to replace the high-sulfur heavy fuel. More recently the increased demand for light products has focused attention on converting residuum to higher value products. Products from residuum hydrotreating are often fed to a fluid catalytic cracking unit to produce good-quality motor gasoline. In this case the removal of nitrogen, Conradson carbon, and metal contaminants in the hydrotreater are just as important as sulfur removal. Hydroprocessing feedstocks which boil in the 650 to 1050°F VGO range can be straight-run stocks or stocks produced in cokers, thermal crackers, or visbreakers. Again these are often processed to produce either LSFO or FCC feed. Sometimes they are hydrocracked to produce diesel, kerosene jet fuel, and/or naphtha. The hydrocracked heavy products are also excellent ethylene plant feedstocks or lube oil base stocks because the process removes the less desirable heavy aromatics.

Straight-run or cracked stocks boiling in the atmospheric gas oil range can be hydrotreated to produce good-quality diesel and jet or ethylene plant feedstock. They can be hydrocracked to produce a naphtha which is an excellent feed for a catalytic reformer. Straight-run or cracked naphthas need to be hydrotreated to remove olefins, sulfur, and nitrogen to produce good catalytic reformer feeds. They can also be hydrocracked to give LPG.

TABLE 14.1.1 Inspections of Crude Oils and Vacuum Residua

Source	Arabian Light	Arabian Heavy	Kuwait	Iranian Heavy	Sumatran Light	Venezuelan	Alaskan North Slope	North Sea Ninian	Californian
Crude oil:									
Density, °API	33.3	28.1	31.3	30.8	35.3	33.3	26.3	35.1	20.9
Sulfur, wt %	1.8	2.9	2.5	1.6	0.07	1.2	1.0	0.41	0.94
Nitrogen, wt %	0.16	0.19	0.09	0.18	0.08	0.12	0.22	0.07	0.56
Residuum, 1000°F+ (538°C+):									
Yield, LV %	17.3	28.6	24.8	21.8	24.4	21.2	23.0	17.8	26.1
Density, °API	8.0	4.6	7.4	6.3	20.1	10.9	7.4	13.0	5.4
Sulfur, wt %	3.7	5.6	5.1	3.2	0.18	2.8	2.1	1.3	1.6
Nitrogen, wt %	0.49	0.67	0.38	0.83	0.33	0.56	0.64	0.42	1.33
Asphaltenes, wt %	11.3	20.6	12.0	14.7	7.9	16.0	8.1	6.9	12.0
Nickel+vanadium, ppm	96	220	116	462	41	666	130	28	294
Iron, ppm	—	10	0.9	9	13	5	15	<1	90

TABLE 14.1.2 Hydroprocessing Objectives

Feedstocks	Desired products	Process objectives
Naphthas	Catalytic reformer feed	Removal of S, N, olefins
	LPG	Hydrocracking
Atmospheric gas oils	Diesel	Removal of aromatics and *n*-paraffins
	Jet	Removal of aromatics
	Ethylene feedstock	Removal of aromatics
	Naphtha	Hydrocracking
Vacuum gas oils	LSFO	Removal of S
	FCC feed	Removal of S, N, metals
	Diesel	Removal of S, aromatics
		Hydrocracking
	Kerosene/jet	Removal of S, aromatics
		Hydrocracking
	Naphtha	Hydrocracking
	LPG	Hydrocracking
	Ethylene feedstock	Removal of aromatics
		Hydrocracking
	Lube oil base stock	Removal of aromatics
		Hydrocracking
Residuum	LSFO	Removal of S
	FCC feedstock	Removal of S, N, CCR, and metals
	Coker feedstock	Removal of S, CCR, and metals
	Diesel	Hydrocracking

Note: LPG = liquefied petroleum gas; *n* = normal form; CCR = Conradson carbon residue.

The technical challenge associated with producing these products via hydroprocessing can be illustrated with the use of a chart first proposed by Bruce Stangeland at Chevron Research and since developed by many of his colleagues.[10,12,17,69,76,91] This relates the hydrogen content of a hydrocarbon to its molecular weight. Figure 14.1.2 is a Stangeland chart which shows the regions in which salable products fall. The upper boundary represents the hydrogen content of the paraffinic homologous series. No hydrocarbon exists above this line. The lower line represents aromatic compounds starting with benzene and including the condensed ring compounds, naphthalene, phenanthrene, pyrene, and coronene. These are among the most hydrogen-deficient compounds found in petroleum distillates. All the distillable hydrocarbons used in petroleum refining lie between these two extremes. Though not shown in Fig. 14.1.2, even the hydrocarbons present in the residuum—the nondistillable fraction—can easily be represented on this chart since the molecular weight scale goes up to 10,000, close to the maximum found in petroleum crude oil. Lines showing approximate boiling points have been drawn. These show the well-known fact that aromatic compounds have much lower molecular weights than paraffinic compounds of the same boiling point.

Specification products are shown as regions. Gary and Handwerk[29] described all the specifications of the major petroleum products. It is instructive to discuss the important ones with the aid of the Stangeland chart:

- *Motor gasoline.* This region is quite broad, since high octane numbers can be achieved with either high aromatic levels or high Iso to normal paraffin ratios. Processes like catalytic reforming focus on retaining aromatics, whereas isomeriza-

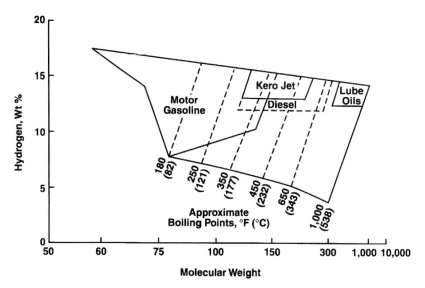

FIGURE 14.1.2 Hydrogen contents of refined products.

tion processes convert normal paraffins into isoparaffins. The initial boiling character of motor gasoline is set by the Reid vapor pressure specification. The back end of the distillation specification is set by engine warm-up and crankcase dilution considerations.

- *Kerosene and aviation jet fuel (kero/jet).* To make these products with acceptably clean burning characteristics, aromatic contents need to be low. The smoke point specification characterizes this quality. The front end of the distillation specification is set by flash point, the back end by freeze point.
- *Diesel fuels.* Here the burning quality is controlled by the cetane number specification which limits the aromatic content. The important cold flow properties are the pour point, cloud point, and cold filter plugging point, and one or more of these set the distillation endpoint specification. As with kero/jet, the front end of the distillation specification is set by flash point.
- *Lube oils.* Aromatic compounds have very low viscosity indexes (VIs) so lube oils must, in general, have limited aromatic levels. Paraffin wax must also be minimized to achieve acceptable pour points, so the more desirable compounds are isoparaffins or molecules containing a combination of naphthenic rings and isoparaffinic side chains. The boiling range of lubricating oils is set by the desired viscosity.
- *Heating and fuel oils.* Hydrogen content is not as important for heating oils or residual fuel oils. However, hydroprocessing is often needed in their production in order to limit sulfur and nickel content. Their boiling range is set by flash point and viscosity considerations.

The Stangeland chart can be used to illustrate the differences between the feedstocks at the refiner's disposal and the required products. Figure 14.1.3 shows the region representing distillable cuts from typical petroleum crude oils. It compares the distillate products from two noncatalytic cracking processes—delayed coking and

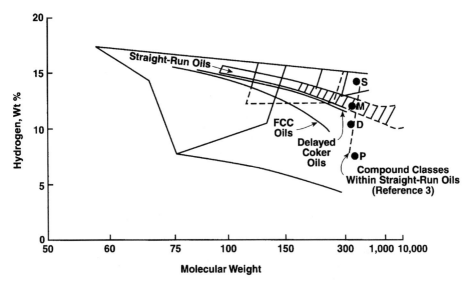

FIGURE 14.1.3 Hydrogen contents of potential feedstocks.

FCC. Neither of these processes consumes hydrogen, but both of them produce high yields of light products. The liquid products are deficient in hydrogen and need further hydroprocessing if they are to become transportation fuels. The light liquid products also contain substantial olefin levels, which can cause unstable products. Straight-run distillates are generally easier to upgrade to finished products.

The Stangeland chart oversimplifies a very complex situation. Modern techniques for characterizing the compounds present in petroleum stocks have helped in our understanding of hydroprocessing reactions, particularly the harmful effects of heavy aromatics. This is illustrated in data measured by M. F. Ali and coworkers[3] on VGOs from Arabian Heavy crude oil. These workers separated the 370 to 535°C distillate fraction into compound-class fractions. They then carried out an elemental analysis on the four major compound classes—saturates, monoaromatics, diaromatics, and polyaromatics. With molecular weights estimated from the other measured physical properties, these four compound classes have been plotted in Fig. 14.1.3. The point labeled P represents the polyaromatics naturally present in this VGO. This compound class represents 22.2 wt % of the total VGO and is more aromatic than either of the cracked stocks referred to previously. The sulfur content of the polyaromatics was reported as 9.83 percent, showing that three-quarters of the sulfur in VGO resides in the polyaromatics. Every polyaromatic molecule contains, on the average, one sulfur atom. The other compound classes in this VGO are shown as points S, M, and D, representing saturates, monocyclics, and dicyclics, respectively.

Boduszynski and Altgelt[8] pointed out average molecular structure determinations for heavy oils give very little indication of the nature of the aromatics in the oil. These aromatics must be upgraded if salable products are to be made from such oils. The amount and character of polyaromatics have a profound effect on the ease of upgrading.

Most modern refineries produce transportation fuels from a blend of components made in different refining processes. Figure 14.1.4 shows the hydrogen content of a variety of diesel boiling range products refined from Arabian crude oils. Components produced by hydrocracking have much higher hydrogen content than those produced

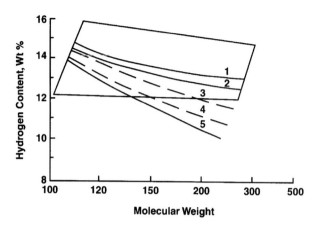

FIGURE 14.1.4 Hydrogen contents of middle distillates.[12] Solid boundaries show 250 to 700°F (121 to 371°C) diesel range. Process: (1) hydrocracker, (2) straight run, (3) delayed coker, (4) fluid coker, and (5) fluid catalytic cracker.

in nonhydrogen processes. Figure 14.1.5 shows a rough correlation exists between jet fuel smoke point and hydrogen content. A 1 percent change in hydrogen content corresponds to a difference of 10 mm in smoke point in the range from 20 to 30 mm. Also consider the difference in hydrogen content between the components shown in Fig. 14.1.4. A 1 percent difference in hydrogen content here represents a difference of 700 standard cubic feet (SCF) of hydrogen consumed per barrel when upgrading to the same product specification. If hydrogen costs between $2 and $3 per thousand SCF, this amounts to a processing cost of $1.50 to $2.25 per barrel of oil processed. The modern refiner must decide where to invest the hydrogen in order to maximize product values. Each refinery is faced with a different situation, depending on processing capabilities and product markets.

The Extent of Commercialization

Of every barrel of crude oil currently refined worldwide, over 45 percent on average receives some hydroprocessing. This percentage varies with geographical area:

North America	63%
Europe	43%
Asia (except People's Republic of China and former Soviet Union)	50%
Southern Hemisphere and Central America	31%
Middle East	26%

Table 14.1.3 shows the crude refining capacity in each region, with the hydroprocessing capacity divided into three categories—hydrocracking, hydrorefining, and hydrotreating. This table divides hydroprocesses according to the amount of hydrogen consumed in them. Typical hydrocrackers consume between 1400 and 2400 standard cubic feet of hydrogen per barrel (SCFB). Naphtha hydrotreaters usually consume less

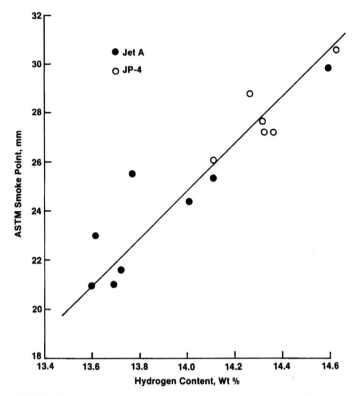

FIGURE 14.1.5 The relationship between kerosene smoke point and hydrogen content.

than 200 SCFB, while hydrorefiners including residuum hydrotreaters typically consume between 500 and 1000 SCFB.

North America uses the most hydroprocessing because residuum conversion is needed to achieve the high ratio of transportation fuels to fuel oil demanded by the market. While delayed coking is the most popular residuum conversion scheme, liquid products produced from it are hydrogen-deficient and, therefore, require further hydroprocessing. In order to produce salable light products, distillate hydrocracking has been widely installed because the proportion of jet fuel and naphtha can be varied to suit seasonal marketing demands. As defined in Table 14.1.3, 8.5 percent of the crude oil refined in North America is upgraded in a hydrocracker.

By contrast, in Japan many residuum and VGO hydrotreating units were installed in the 1960s and 1970s in order to produce LSFO to reduce air pollution from stationary power plants.

Although the Middle East has less overall hydroprocessing than other regions, the Kuwait investment in hydrocracking is substantial; 52 percent of the crude oil refined in Kuwait is processed in a hydrocracker. Middle East hydrocrackers are designed to make middle distillates for export. They consume less hydrogen per barrel of feed than most of the North American distillate hydrocrackers, which are generally operated to produce naphtha and jet fuel.

TABLE 14.1.3 The Geography of Hydroprocessing*

Region	Crude refining capacity, MBPD	Hydrocracking* capacity, MBPD	Hydrorefining† capacity, MBPD	Hydrotreating‡ capacity, MBPD	Hydrogen production capacity		
					Hydrogen plants, 10^6 SCFD	Catalytic reformers, § 10^6 SCFD	Total, 10^6 SCFD
North America	17,227	1462	1975	9124	3612	3245	6857
Asia/Pacific	14,626	508	2289	3497	1278	1555	2833
Western Europe	13,499	619	1881	5136	1941	1382	3323
CIS/Eastern Europe	12,800	57	157	3410	1373	191	1564
Middle East	6044	447	613	1294	571	1131	1702
South/Central America	7172	108	433	1589	441	296	737
Africa	2799	38	138	530	295	65	360

Hydrocracking includes distillate and residuum upgrading and lube oil manufacturing.
†*Hydrorefining* includes residuum and heavy oil desulfurization, FCC feed, cycle oil processing, and mid-distillate processing.
‡*Hydrotreating* includes naphtha processing, atmospheric gas oil processing, and lube oil finishing.
§Assuming all catalytic reformers produce 900 SCF hydrogen/bbl of feed.
Note: BPD = barrels per day; SCFD = standard cubic feet per day; CIS = Commonwealth of Independent States.
Source: *Oil and Gas Journal* Worldwide Refining issue, 1994.

The need for hydrocrackers in the rest of the world has been low. It is instructive to look at the availability of hydrogen in the different regions (see Table 14.1.3). Assuming that catalytic reformers produce an average of 900 SCF of hydrogen per barrel of reformer feed and adding this quantity to the manufactured hydrogen capacity gives a total amount of hydrogen available from these sources that is quite consistent with the hydroprocessing capacities shown in each region. The ratio of manufactured hydrogen to catalytic reformer hydrogen in the different regions varies with end-user needs. Asia and the Middle East have relatively low motor gasoline demands and, therefore, low catalytic reformer capacities. They have to produce more hydrogen in hydrogen plants than they produce in their catalytic reformers. In Europe, Central America, and the Southern Hemisphere there has been no need to invest heavily in hydrogen plants. North America has done so in order to support the residuum conversion processing schemes that have been introduced in the last two decades.

The ever-increasing high price differential between heavy fuel oil and transportation fuels led to the introduction of sophisticated plants which consume large amounts of hydrogen—distillate hydrocrackers, residuum hydrocrackers, and residuum hydrotreaters. In order to support their appetite for hydrogen, large hydrogen manufacturing plants were installed. During periods when crude prices are low and differentials between heavy and light products are small, operation of hydroprocessing units may not be economically justified. But any decision to suspend their operations must be based on the impact of that decision on the whole refinery. For instance, operating a coker may require operating a hydroprocessing unit to obtain marketable products, so that a realistic analysis will recognize the need to run both units if the coker operates.

Figure 14.1.6 shows the growth in capacity of these hydrocracking and residuum processes since Chevron commercialized the first modern distillate hydrocracker in 1958.

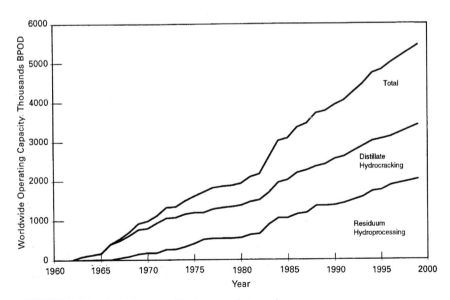

FIGURE 14.1.6 Growth in heavy oil hydroprocessing capacity.

PROCESS FUNDAMENTALS

Chemical Reactions

The impurities which are removed in hydrotreaters are largely concentrated in the aromatic compounds in the feedstocks. Their removal is accomplished therefore by the hydrogenation of these compounds. Simple examples are shown in Fig. 14.1.7. Note that in these examples sulfur is removed without complete saturation of the aromatic ring, whereas nitrogen removal generally involves saturation and destruction of the aromatic ring. Thus, hydrodesulfurization can be accomplished with low hydrogen consumption at low pressures, whereas nitrogen removal needs high hydrogen partial pressures and consumes more hydrogen.

The reactions which occur in hydrocracking are much more complicated. Choudhary and Saraf[18] have written an excellent survey article on early hydrocracking work. The chemistry of hydrocracking is essentially the carbonium ion chemistry of catalytic cracking coupled with the chemistry of hydrogenation. Langlois and Sullivan[49] have reviewed the chemistry of hydrocracking. When the reactants are paraffins, cycloparaffins, and/or alkyl aromatics, the products obtained from both hydrocracking and catalytic cracking are similar, but when the reactant is polycyclic aromatics, wide differences in the product from these two refining processes are obtained. For instance, catalytic cracking of phenanthrene over acidic catalysts produced only coke and small quantities of gas, while hydrocracking of the same gave low-molecular-weight cyclic products.[89] This difference in the product is caused by the hydrogenation component of the catalyst and the excess of hydrogen usually present in hydrocracking. After hydrogenation, these aromatics, which produce coke in catalytic cracking, are converted into readily cracked naphthenes.[95] Di- and polycyclic aromatics joined by only one bond rather than two common carbon atoms are readily cleaved by hydrogen and converted into single-ring aromatics. These enhance the antiknock characteristics of the product gasoline.

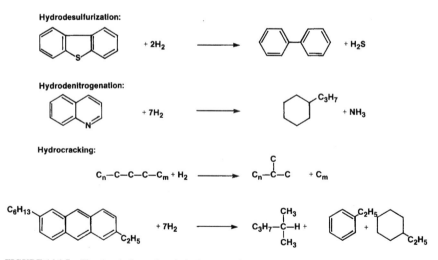

FIGURE 14.1.7 The chemical reactions in hydroprocessing.

Catalysts

Hydrotreating catalysts consist of a hydrogenation component dispersed on a porous, fairly inert, material. Hydrocracking catalysts are dual-functional, containing both hydrogenation and cracking sites. The cracking sites are usually the result of using a porous support of an acidic nature. The best choice of a catalyst for a specific situation requires a particular balance between the cracking and hydrogenation functions.[82] Table 14.1.4 shows catalyst characteristics necessary to accomplish the reactions of industrial importance.

In hydrocracking to LPG and gasoline, strong cracking activity is required. This is achieved by using strongly acidic materials including both amorphous silica-aluminas and crystalline aluminosilicates. The acidity of these materials promotes reactions which lead to high iso-normal ratios in the light paraffin products, low methane and ethane production, and conservation of monocyclic rings. The hydrogenation component reduces the concentration of coke precursors and maintains the effectiveness of the cracking sites. Catalysts can then be operated for long periods (1 to 4 years) at economic processing conditions.

In hydrocracking gas oils to produce jet fuel and middle distillate, catalysts with less acidity and stronger hydrogenation activities are used. This type of catalyst is valuable in producing high-viscosity-index lubricating oils by selectively saturating and converting the heavy aromatics, leaving behind the more valuable compounds. These catalysts are also used for hydrocracking residual fractions such as solvent-deasphalted oils and residua where the high nitrogen content would poison strong cracking activities.

For hydrotreating to remove impurities, catalysts with weak acidity are used, since cracking is usually undesirable. Strong hydrogenation activity is needed, particularly with heavy feedstocks containing high-molecular-weight aromatics. In recent years there has been a great deal of interest in "mild hydrocracking" as refiners try to increase the extent of cracking in their existing hydrotreaters. *Mild hydrocracking* usually refers to hydrocracking operations with less than 40 percent conversion to lighter products but more than the 20 percent that occurs in a simple hydrotreater. This increase in conversion is accomplished by increasing the ratio of the cracking function to the hydrogenation function in the catalyst or catalyst system.

Besides the chemical nature of the catalyst, which dictates the hydrogenation and cracking capabilities, its physical structure is also very important, particularly with heavy feedstocks. With gas oils and residuum feedstocks, the hydrocarbon feedstock is present as a liquid at reacting conditions so that the catalyst pores are filled with liquid. Both the hydrocarbon and the hydrogen reactants must diffuse through this liquid before reaction can take place at the interior surface within the catalyst particle. At high temperatures, reaction rates can be much higher than diffusion rates and concentration gradients can develop within the catalyst particle. This reduces the overall reaction rate and can lead to costly inefficiencies and undesirable side reactions.

The choice of catalyst porosity is, therefore, very important. A high internal surface area gives high local reaction rates; but if, in achieving the high surface area, the catalyst pore size is reduced to a point which hinders reactant diffusion, then the overall performance will suffer.

Certain generalizations can be made about catalyst porosity.[82] For hydrocracking to LPG and gasoline, pore diffusion effects are usually absent. High surface areas (about 300 m²/g) and low to moderate porosity (from 12-Å pore diameter with crystalline acidic components to 50 Å or more with amorphous materials) are used. With reactions involving high-molecular-weight impurities, pore diffusion can exert a large influence. Such processes need catalysts with pore diameters greater than 80 Å.

TABLE 14.1.4 Hydroprocessing Catalyst Characteristics[82]

Desired reaction	Catalyst characteristics			
	Acidity	Hydrogenation activity	Surface area	Porosity
Hydrocracking conversion:				
Naphthas to LPG	Strong	Moderate	High	Low to moderate
Gas oils to gasoline				
HGO to jet and middle distillate	Moderate	Strong	High	Moderate to high
HGO to high VI lubricating oils				
Solvent deasphalted oils and residua to lighter products				
Removal of nonhydrocarbon constituents:				
Sulfur and nitrogen in HGO and LGO	Weak	Strong	Moderate	High
Sulfur and metals in residua				
Aromatics saturation:				
LGO to jet fuel	Weak	Very strong	High	Moderate

Note: HGO = heavy gas oil; LGO = light gas oil.

Reaction Kinetics

The section "Hydroprocessing Objectives," dealt with the difficulty of characterizing hydroprocessing feedstocks.[8] They may contain similar compounds with different boiling points or have similar boiling points for widely different compounds. Knowing the rate of the hydroprocessing reaction is vital in the design of a unit or in deciding how much feedstock can be processed in an existing unit. It determines the size of the reactor required. The rate of reaction is obtained in a pilot plant experiment by measuring the extent of reaction at different residence times and the same temperature. The rate invariably increases with temperature. Designing for high-temperature operation and high reaction rates has to be moderated because undesirable side reactions (including those which deactivate the catalyst) also are faster at high temperature.

Hydrotreating Kinetics. Despite the complexity of hydroprocesses, reaction kinetics can often be expressed in simple terms. Figure 14.1.8 shows the apparent first-order nature of the hydrodenitrification reaction. The data were obtained when a heavy California coker distillate was processed in the pilot plant over a weakly acidic catalyst containing both a Group VI and a Group VIII hydrogenation component. First-order behavior describes the data over a range of product nitrogen covering four orders of magnitude.[82]

Residuum desulfurization and demetalation kinetics are generally not first-order. Chevron Research pilot plant desulfurization and demetalation kinetic data[14] for Arabian Heavy atmospheric residuum are shown in Fig. 14.1.9. The curves drawn through the experimental data are based on a second-order rate expression. Surprisingly, the desulfurization data fit the second-order expression down to product sulfur levels of 0.25 percent. The true mechanism is probably one of the multitude of first-order reactions, of varying rates, with the asphaltene molecules being the least reactive.[7,27] For most design calculations, the second-order expression is a useful simplification.

With high-metal feeds, however, considerable attention needs to be paid to the demetalation kinetics—not only in terms of predicting the product oil metal content, but also in predicting the impact of demetalation on catalyst life. Most of the feed metals react at desulfurization conditions to form metal sulfides. If these reaction products deposit in the interstices of the catalytic bed, then serious bed pressure drop increases can occur. If the reaction occurs inside the catalyst pores, then the sulfide deposit will ultimately deactivate the catalyst. Pilot plant demetalation kinetic data for Arabian Heavy atmospheric residuum are also shown in Fig. 14.1.9. Nickel and vanadium have been added together for this plot although there are subtle differences in their individual behavior. Again, second-order kinetics give the simplest expression capable of describing the data. Just as in the desulfurization reaction, this is probably an oversimplification of the final reaction mechanism. However, it is useful in most aspects of design.

Demetalation of this type is influenced by diffusion of the reactants through the catalyst pores. To explore this phenomenon, experiments with different-sized catalysts were carried out. The results are shown in Fig. 14.1.10. Here desulfurization and demetalation rate constants (second-order) are plotted versus temperature for both $\frac{1}{16}$-in-diameter cylindrical catalyst and the same catalyst crushed to 28 to 60 mesh. The desulfurization data show no significant particle size effect over the temperature range considered. The demetalation data, however, show a substantial pore diffusion limitation at all temperatures above 550°F. Both catalyst activity and activation energy (change of reaction rate with temperature) are higher for the crushed catalyst. Residuum demetalation is a process which usually operates in a diffusion-controlled mode.

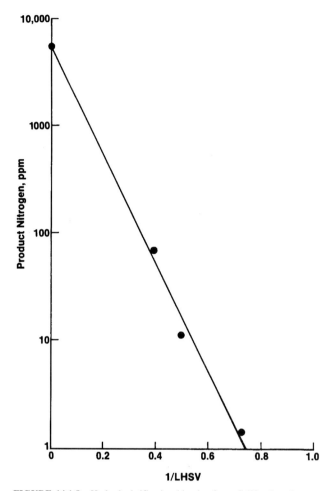

FIGURE 14.1.8 Hydrodenitrification kinetics for a California coker gas oil.[82]

The theory used to describe isothermal reactions in porous catalytic media was developed by Thiele[94] and extended by Wheeler,[102] Weisz,[101] and others (e.g., Refs. 50, 79). It shows that catalyst effectiveness is a function of the ratio of the intrinsic rate of reaction to the rate of reactant diffusion. A Thiele modulus is used to represent this ratio in dimensionless form. Many experimenters[56,57] have compared hydrotreating data with this theory. Chevron Research investigated the effect of catalyst pore size and particle size on the hydrodemetalation of Boscan crude oil.[11] The catalysts used were all of the same composition and all had unimodal microporous pore size distributions. They were each tested at the same pressure level and the same desulfurization severity level. The demetalation data fitted well with pore diffusion theory and predictions outside of the database were possible. Figure 14.1.11 shows the predicted activity versus pore diameter with particle size as a parameter.

This plot, which assumes a catalyst pore volume of 0.5 cm^2/g, shows that the optimum pore diameter for catalyst activity varies with the particle diameter. A small-par-

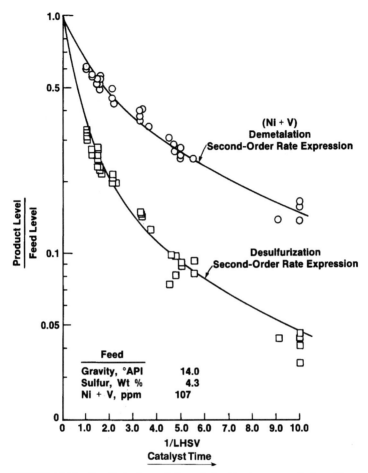

FIGURE 14.1.9 Desulfurization and demetalation kinetics—atmospheric residuum from Arabian Heavy crude.[14]

ticle-size and small-pore-diameter catalyst is the most active. The fact that small-particle-size, high-internal-surface area catalysts are optimum is intuitively obvious for a pore-diffusion-limited reaction. Sometimes, however, one is forced to choose a larger particle size (because of pressure drop considerations in a fixed-bed reactor or fluidizing velocity considerations in a fluidized-bed reactor). In any case, there is an optimum internal surface area and pore diameter for each catalyst size. The amount of hydrogenation component in the initial catalyst is also important.[35] As Spry and Sawyer[86] have pointed out, each crude oil will have a different optimum combination of catalyst size and porosity for maximum activity.

It has been suggested[4] that maximum activity occurs with a combination of narrow pores which create sufficient surface area, and wide pores (above 100 Å) to make this surface accessible. Others[73] have shown that among all catalysts with the same surface area and porosity, the highest activity is attained for catalysts with a uniform pore size.

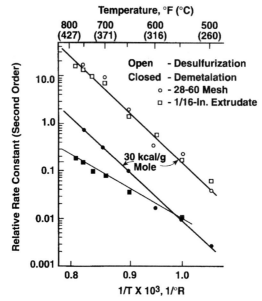

FIGURE 14.1.10 Desulfurization and demetalation kinetics—effects of temperature and particle size on atmospheric residuum from Arabian heavy crude.[14]

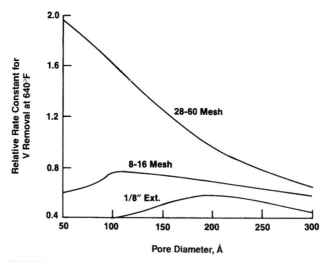

FIGURE 14.1.11 Estimated effect of catalyst size and pore diameter on Boscan demetalation kinetics.[11]

Hydrocracking Kinetics. Considerable work has also been carried out on hydrocracking reaction kinetics. It is generally accepted that the reaction is first-order with respect to the hydrocarbon reactant. It is appropriate to consider two types of hydrocracking catalysts:

- The strongly acidic catalysts are designed to process fairly clean feeds—light naphthas or heavier feedstocks that have already been severely hydrotreated. Zeolites are often used.

- Weakly acidic catalysts have a high ratio of hydrogenation to cracking activity and can hydrocrack raw feedstocks which have not been previously hydrotreated. Amorphous catalysts or catalysts with minor amounts of zeolites are used.

Figure 14.1.12 shows the effect of feed molecular weight on the reaction rates observed with strongly acidic hydrocracking catalysts. These data, which were obtained with an amorphous catalyst, illustrate general trends involving feed character and molecular weight.

First-order reaction rates normalized to a constant temperature and pressure are shown for a variety of pure hydrocarbons. For this display, a line is drawn connecting the points for normal paraffins (n-paraffins). Other points are displayed for isoparaffins (i-paraffins), naphthenes, aromatics, and polycyclics.

The pure compound rate constants were measured with 20 to 28 mesh catalyst particles and reflect intrinsic rates (i.e., rates free from diffusion effects). Estimated pore diffusion thresholds are shown for 1/8-in and 1/16-in catalyst sizes. These curves show the approximate reaction rate constants above which pore diffusion effects may be observed for these two catalyst sizes. These thresholds were calculated using pore diffusion theory for first-order reactions.[79]

The pure compound cracking rates may be compared with typical reaction rates found commercially with wide-boiling petroleum fractions. Commercial naphtha

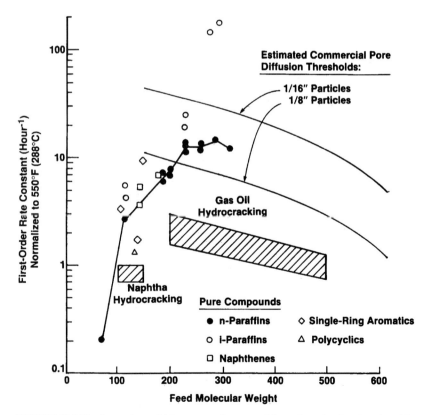

FIGURE 14.1.12 Comparison of laboratory and commercial reaction rates for strongly acidic catalysts.[82]

hydrocracking data are consistent. Gas oil hydrocracking rates are lower and decrease with feedstock molecular weight. This is probably caused by the heavy aromatic molecules inhibiting the acid function of the catalyst. Despite this suppression of reaction rates, careful balancing of hydrogenation and cracking functions produces catalysts which operate efficiently at economical processing conditions. Consistent with the diffusion limit curves, particle size effects have not been observed commercially with these catalysts.

Figure 14.1.13 shows the effect of molecular weight on hydroprocessing rate constants observed with typical catalysts of lower acidity and higher hydrogenation activity. The hydrocracking of residuum is nearly 10 times more difficult than gas oil hydrocracking. This is because of the large asphaltenic molecules present in the residua. The residuum conversion rate constants shown in Figure 14.1.13 represent data for straight-run residua containing a wide range of molecular sizes. Other kinetic experiments have shown[82] that, if the heavy asphaltenic molecules are processed by themselves, much lower reaction rates are observed. Solvent deasphalted oils are correspondingly easier to process than straight-run residua. The reaction rate constants for denitrification of gas oils, and desulfurization and demetalation of residua, are substantially higher than the hydrocracking rate constants. These nonhydrocarbon con-

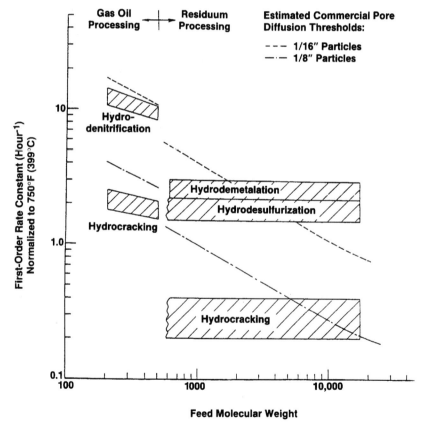

FIGURE 14.1.13 The effect of feed molecular weight on hydroconversion rates—moderately and weakly acidic catalysts.[82]

stituents, therefore, can be removed selectively with minimum hydrocracking of the parent molecule.

Two calculated pore diffusion threshold curves are shown in Fig. 14.1.13. These are the dashed lines which show the rate constants above which pore diffusion controls for both $\frac{1}{16}$-in and $\frac{1}{8}$-in catalyst sizes. For gas oil hydrocracking, the observed reaction rate constants are not high enough to lead to problems; this is supported by commercial hydrocracking experience. The high denitrification rate constants suggest that pore diffusion problems could occur with active catalysts at high temperatures. The estimated diffusion limits for residuum processing with $\frac{1}{16}$-in catalysts confirm that demetalation is influenced markedly and desulfurization to a lesser extent.

Spent Catalyst Analysis. Careful analysis of spent residuum hydrodemetalation catalysts have helped quantify the role of diffusion in the reaction. Examples of the deposition profiles for nickel, vanadium, and iron at both the inlet and outlet of the catalyst bed[93] are shown in Fig. 14.1.14. This catalyst was used to hydrotreat Arabian Heavy residuum. A number of important features are apparent in the spent catalyst results. Iron is found primarily outside the catalyst particle as a thin scale. This is generally

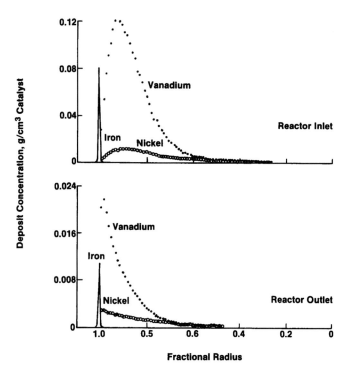

FIGURE 14.1.14 Typical depositional patterns for nickel, vanadium, and iron in residuum hydroprocessing catalyst.[93] [Arabian Heavy atmospheric residuum, reaction temperature = 700°F (371°C), hydrogen partial pressure = 1825 lb/in² abs, $\frac{1}{16}$-in extrudate catalyst.]

the case. Nickel generally seems to penetrate the catalyst to a greater extent than vanadium. These differences in depositional patterns are a result of differences in the reactivities and/or diffusivities of the organometallic molecules. Both nickel and vanadium display a maximum concentration inside the edge of the particle, but the point of maximum concentration approaches the edge of the catalyst near the outlet of the reactor.

The fact that maximum concentrations are found inside the edge of the particle is difficult to explain. It may be due to hydrogen sulfide (H_2S) being a reactant or it may be due to specific reaction intermediates being formed. It complicates data analysis, since pore diffusion theory coupled with a simple reaction mechanism does not predict an internal maximum. Despite this, it is interesting to compare the change in the maximum deposit concentrations from reactor inlet to reactor outlet with the change in concentration of metals in the oil. During the test in which the profiles shown in Fig. 14.1.14 were generated, the average vanadium removal was 58 percent and the average nickel removal was 42 percent. The maximum deposit concentrations of both metals decreased by approximately 80 percent from reactor inlet to outlet, clearly showing that demetalation is not a simple first-order reaction. The change in the maximum deposit considerations is close to what one would predict using second-order kinetics, assuming that the concentrations of metals in the feed and product oil apply to the maxima at the respective ends of the reactor. This result is consistent with the kinetic measurements shown in Fig. 14.1.9.

Hydrogenation-Dehydrogenation Equilibrium

The saturation of aromatic compounds is important in both hydrotreating and hydrocracking. This reaction is reversible and the equilibrium between the forward and reverse reaction can hinder the extent of saturation at normal commercial conditions. Gully and Ballard[36] summarized the early knowledge on aromatics hydrogenation equilibria. The hydrogenation reaction is favored by high hydrogen partial pressures and low operating temperatures. The higher-molecular-weight aromatic compounds need a higher hydrogen partial pressure to achieve the same extent of reaction at the same temperature as the lower-molecular-weight molecules.

In some hydroprocesses, the heaviest naphthenic molecules are dehydrogenated while the lower-boiling ones are undergoing desulfurization. This occurs at end-of-run (EOR) temperatures in low-pressure VGO hydrotreaters. It was also the basis for the Autofining process, developed by British Petroleum, for desulfurization of light oils using no outside source of hydrogen.[72] Hengstebeck[39] has proposed a hydrogenation-dehydrogenation index for correlating experimental data. Yui and Sanford[104] studied the kinetics of aromatics hydrogenation in order to improve the cetane number of an LGO feed so that clean-burning diesel fuel could be produced. They measured the percent aromatics hydrogenation at different temperatures, pressures, and residence times [liquid hourly space velocity (LHSV)]. Data obtained with Arabian Light LGO are shown in Fig. 14.1.15 which is taken from their paper. The results were compared with a kinetic model for aromatics hydrogenation based on a simple first-order reversible reaction. Agreement with the model was excellent. This particular reaction is limited by equilibrium at temperatures above about 360°C when operating pressures of 5 to 10 MPa are used.

For this reason the hydrogen consumption will parallel the extent of saturation. The amount of hydrogen consumed will, therefore, first increase then decrease as operating temperatures are increased. The most important design parameter in such a unit is the hydrogen partial pressure. It should be high enough to allow the achievement of the target cetane number, but not so high as to consume more hydrogen than is absolutely needed.

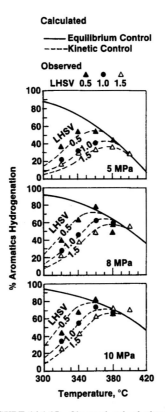

FIGURE 14.1.15 Observed and calculated percent aromatics hydrogenation at various operating conditions (Arabian light gas oil).[104]

Reaction Selectivity

The modern refiner is very interested in controlling selectivity in hydroprocessing. The refiner needs to encourage one chemical reaction while discouraging a number of others. The products must meet certain specifications, without exceeding them and consuming valuable hydrogen that could be used elsewhere in the refinery. Selectivity is influenced by variations in catalyst properties and by variations in operating conditions.

Residuum Processing. Whenever a hydroprocessing unit operates with some of the reactants limited by the rate of diffusion to the active site, there are always opportunities to influence reaction selectivity by modifying the pore size of the catalyst. Hensley and Quick[38] have pointed out that small pore catalysts can be used for selective desulfurization of low-metal feeds with moderate demetalation whereas large pore catalysts can be used to remove metals and asphaltenes with minimum sulfur removal and hydrogen consumption. Other researchers[43] have correlated selectivity with a distribution factor obtained from measuring nickel and vanadium deposition profiles within spent catalyst particles. This factor is similar to the effectiveness factor of pore diffusion theory.

Catalyst operating temperature can influence reaction selectivity also. The activation energy for hydrotreating reactions is much lower than that for the hydrocracking reaction. Raising the temperature in a residuum hydrotreater increases, therefore, the extent of hydrocracking relative to hydrotreating. This, of course, also increases the hydrogen consumption. Figure 14.1.16 illustrates the different operating strategies which have been used when Arabian Light vacuum residuum is hydroprocessed.[13] The region below about 40 percent conversion represents normal residuum hydrotreating which produces LSFO or good-quality FCC feed.

The relative hydrogen consumption is compared with the hydrocracking conversion. These two reactions have different activation energies; i.e., the rates respond to temperature differently. Curves are shown depicting how the hydrogen consumption varies from start to end of a variety of runs carried out at constant but different product sulfur levels. Lines of constant temperature are drawn to show the approximate temperature levels required to achieve at least 6 months' catalyst life at different product sulfur levels (from 1.8 to 0.3 percent). One can see that the hydrocracking conversion increases substantially during the run. The hydrogen consumption increase is less noticeable, showing that the catalyst selectively loses its hydrogenation capability during the run. This operating strategy of maintaining constant product quality tends to minimize hydrogen consumption.

Another operating strategy which is becoming popular in commercial units[78] is to maximize conversion throughout a run cycle. This is represented by a vertical line on Fig. 14.1.16. The start-of-run (SOR) low-temperature condition achieves a very low sulfur product and requires a high hydrogen consumption. As the catalyst fouls, the system moves downward on the chart. The hydrogen consumption drops and the product quality deteriorates. Selectivity differences between SOR and EOR have always been noticeable in residuum hydroprocessing. As refiners strive to maximize hydrocracking in hydrotreating, these differences will be even more striking.

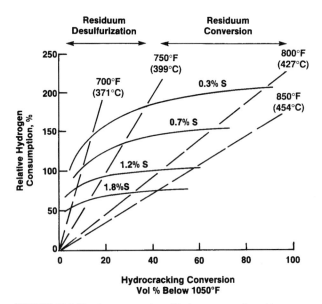

FIGURE 14.1.16 A representation of hydrogen usage in residuum processing.[13] (Arabian Light, 1050°F+ feed, catalyst life greater than six months.)

Distillate Processing. Catalyst properties also influence selectivity in distillate hydrocracking. They can affect the product yield structure and the product qualities. Sullivan and Meyer[90] showed the results of a comparison of seven different catalysts described in Table 14.1.5 (with different relative strengths of acidity to hydrogenation). Since they all can be classified as strongly acidic, they were tested on California heavy gas oils which had already been hydrotreated. The tests were carried out by recycling the heavy product back to the feed so that it was completely converted to jet fuel and lighter products.

The work focused on the octane number of the light product naphtha. A high octane number in the light naphtha is particularly desirable because it is more difficult to upgrade this low-boiling fraction than the higher-boiling naphthas.

Figure 14.1.17 shows the measured jet fuel yields plotted versus the iso to normal ratio measured in the light naphtha product. This ratio correlates well with F-1 clear octane numbers for light naphthas. In general, the catalysts which produce the highest light naphtha octane number produce the lowest jet fuel yields. Catalyst E, based on a crystalline faujasite material, gave somewhat lower jet fuel yields than the amorphous catalysts. The authors were able to influence this selectivity by adding nitrogen and sulfur compounds to the feed. The resulting preferential poisoning of either the acid or the hydrogenation sites showed that the liquid yields and the light naphtha octanes are related to the ratio of the relative strengths of these sites.

TABLE 14.1.5 Experimental Hydrocracking Catalysts[90]

Catalyst identification	Hydrogenation component	Metal content, wt %	Support material
A	Pd	0.5	Activated clay (low acidity)
B	Pd	1.0	Amorphous silica-alumina
C	Pd	0.2	Amorphous silica-alumina
D	Pd	0.5	Activated clay (moderate acidity)
E	Pd	0.5	Faujasite
F	Pd	0.5	Amorphous silica-alumina (activated)
G	Sulfided Ni	10.0	Amorphous silica-alumina

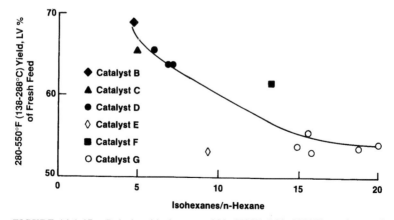

FIGURE 14.1.17 Relationship between 280–550°F (138–288°C) product and isohexanes/*n*-hexane at 580 to 615°F (304–324°C)—hydrocracking of California gas oil.[90]

With mildly acidic hydrocracking catalysts and raw feedstocks, catalyst properties can also influence selectivity for different product yield structures. Stangeland[87,88] has proposed a kinetic model for the prediction of hydrocracker yields. The model represents the large number of different molecules present in the feed as a series of 50°F boiling range cuts. Each of the cuts cracks via a first-order reaction to form a series of lighter cuts. One parameter describes the effect of boiling point on the rate of reaction. Two other parameters determine what products will be generated as each cut cracks. Excellent agreement with experimental data was obtained. Values of the three parameters depend on catalyst type and feed paraffin content. Generally the higher the feed paraffin content, the lower the total liquid yields and the higher the light gas yields.

Hydrocracking catalysts have been developed to produce different ratios of middle distillate to naphtha. Large pore diameters and high alumina-to-silica ratios result in a higher production of middle distillates with less naphtha.[12] An easier way to change the ratio of middle distillate to gasoline in a hydrocracker is to change the operation of the product distillation unit. This will be covered in "Process Capabilities" below.

Catalyst Stability and Life. To quote J. B. Butt:[15]

> The discussion so far has ignored the omnipresent fact of catalyst mortality. Common causes of deactivation are poisoning by strong chemisorption of impurities on the active site, coking or fouling resulting from the formation of hydrogen-deficient carbonaceous residues on the surface in hydrocarbon reactions, and sintering which is the loss of active surface by various processes of agglomeration.

Observations with Residual Feedstocks. In the hydroprocessing of light oils, the catalyst deactivation is usually due to a coking reaction. Coke precursors are the heaviest aromatic compounds in the feed and the coking reaction is favored by high operating temperatures and low pressures. Plants are designed to run at sufficiently high pressure levels so that the coke precursors can be hydrogenated to control the fouling. Long run cycles can then be achieved. With heavy residuum feedstocks, deactivation is thought to be due to poisoning by the feed metals. One of the first commercial observations of this phenomenon was reported by Ozaki, Satomi, and Hisamitsu.[70] In monitoring early operation of the Nippon Mining Company Gulf–designed residuum hydrodesulfurization (HDS) unit, they observed a poisoning wave moving down through the reactor as shown in Fig. 14.1.18. The wave steadily caused the top bed heat release to drop off so that the lower beds had to operate at higher temperatures to compensate.

There have been many studies of this poisoning phenomenon. Chevron[93] analyzed samples of catalyst at both different bed positions and lengths of time on-stream (see Fig. 14.1.20) in a pilot plant run. Figure 14.1.19 shows how the peak concentrations of vanadium varied with time at three different positions.

A simple calculation of monolayer coverage of vanadium sulfide suggests that at the top of the bed the maximum deposit represents 5 to 12 monolayers. If the deposit were V_3S_4 and had the density of the bulk sulfide, such a deposit would be 15 to 40 Å in depth. For a catalyst with a pore diameter in the range of 100 to 200 Å, typical of many residuum hydroprocessing catalysts,[30,67] such a deposit would reduce the diameter of the pores significantly.

The physical obstruction of the pore structure decreases the effective diffusivity for the reactant molecules and, thereby, increases the Thiele modulus for the desired reaction. If the desired reaction were already near the diffusion limit when the catalyst was fresh, it might well be expected to become diffusion-limited when the catalyst is heavily laden with metals. In this case, temperature would have to be raised at an ever-increasing rate to maintain conversion. Such a situation is typical of the later stages of

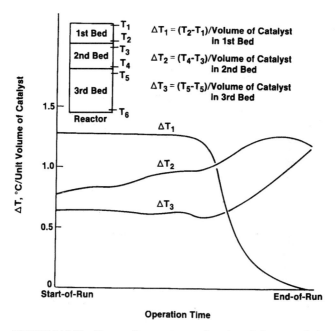

FIGURE 14.1.18 Change of temperature gradient through the reactor beds of a commercial residium hydrotreating plant.[70]

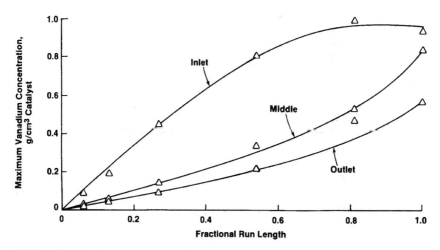

FIGURE 14.1.19 Maximum vanadium deposit concentration as a function of reactor position and time.[93]

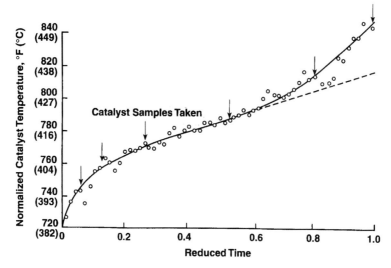

FIGURE 14.1.20 Typical deactivation curve for residuum hydroprocessing catalyst.[93] (Arabian Heavy atmospheric residuum desulfurization, product sulfur = 1 wt %, 1.6-mm. Arabian extrudate catalyst.)

a hydroprocessing run as illustrated in Fig. 14.1.20. The effect of this pore mouth plugging on catalyst activity was measured quantitatively in another experiment. A catalyst bed which had reached a typical EOR condition was divided into six sections, and the second-order desulfurization rate constant was measured independently for each section.

A dramatic activity profile was found (see Fig. 14.1.21). The top one-third of the bed was virtually dead, having little more than one-third the activity of the average bed and less than one-sixth the activity of the bottom of the bed. The bottom one-third of the bed, while significantly deactivated relative to the fresh catalyst, was relatively unaffected by pore plugging and still had sufficient activity to be useful.

This experiment confirms the commercial observation described earlier.[70] Pore plugging occurs as a wave which, after an induction time, moves from the inlet of the reactor toward the outlet.

Factors Affecting Pore Mouth Plugging. The onset of the pore plugging wave and the rapidity with which it moves through the bed are dependent on the details of the catalyst pore structure and the distribution of metals along the length of the catalyst bed. The pore structure directly determines the maximum local deposit buildup which can be tolerated before pore diffusion is adversely affected. The maximum concentration of deposit within a catalyst particle at a given time depends on process and catalyst variables. The more uniform the intraparticle distribution, the lower the maximum concentration will be after a given time, and the later the onset of pore plugging will occur. The rate of advance of the pore plugging wave, on the other hand, is related to the uniformity of the interparticle distribution along the length of the reactor. The more uniform this distribution, the more rapidly the wave will transverse the reactor. This simple principle is illustrated by the following example.

Two catalysts having identical properties, except for their particle size, were used to desulfurize Iranian Heavy atmospheric residuum to an equal extent at identical pro-

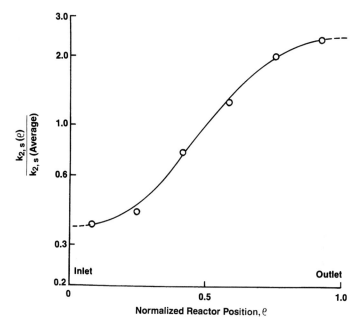

FIGURE 14.1.21 Catalyst activity at end of run as a function of reactor position.[93] (Arabian Light atmospheric residuum desulfurization.)

cessing conditions. Their deactivation curves are compared in Fig. 14.1.22. The onset of pore plugging at the top of the catalyst bed occurred at essentially the same time in these two tests because the porous properties of the catalysts were the same and the processing conditions were the same (with the exception of the subsequent temperature program). However, the speed with which the pore plugging wave moved through the bed is very different. Because a larger fraction of the catalyst volume is accessible to the depositing metals with the small-size catalyst, more metal is accommodated at the top of the bed, and the metal concentration profile down the catalyst bed is steepened. At the decreased concentrations of metal contaminants to which the lower part of the bed is exposed, more time is required for the maximum deposit to reach its limiting value, and the rate of travel of the pore plugging wave is thereby slowed.

Catalyst particle sizes in residuum hydrotreating service have been reduced in some designs in order to maximize life. The pilot plant tests show how important it is to study this deactivation phenomenon over the complete run cycle. Had the tests been terminated after just 20 percent of the time, the relative ranking of the catalysts would have been reversed.

Factors Affecting Initial Catalyst Deactivation. The catalyst deactivation which occurs before the onset of pore mouth plugging is more difficult to characterize, and there is controversy regarding whether it is due primarily to coke or metals deposition.

In the early stages of a hydroprocessing run, a fraction of the catalyst's surface area is converted from its original state to a surface composed of mixed nickel and vanadium sulfides. While these sulfides have catalytic activity for hydrogenolysis, they are considerably less active than the fresh catalysts used in these studies. Under

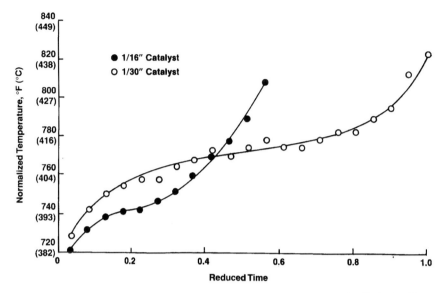

FIGURE 14.1.22 The effect of catalyst particle size on catalyst deactivation.[93] (Iranian Heavy atmospheric residuum, product sulfur = 0.5 wt %.)

these conditions, the catalyst temperature must be raised to hold conversion constant. This form of *partial surface poisoning* may be the major cause of deactivation in the early part of a run. As illustrated in Fig. 14.1.20, the period of initial catalyst deactivation is characterized by a high, but declining, deactivation rate which asymptotically approaches a constant value a quarter of the way through the run. Such behavior is a reasonable consequence of the proposed partial surface poisoning mechanism if multilayers of the contaminant deposit have the same catalytic activity as the initial monolayer.

A high level of coke forms rapidly in an outer annulus of the catalyst. However, it declines slowly as feed metals deposit and is offset by the increase of coke in the interior of the catalyst. Since the initial depositional pattern of coke parallels that of the metals, both probably being due to the presence of high-molecular-weight species, it is difficult to unequivocally assign responsibility for the initial deactivation to either contaminant. However, several arguments favor organometallics as the primary deactivant when the metal content of the feed exceeds about 10 ppm. The length of the initial deactivation period appears to be related to the concentration of organometallics in the feed, but not to the concentration of coke precursors (as measured by Conradson carbon content) in the feed. The period of accelerated coke laydown is short relative to the entire initial deactivation period, and the deposited coke undergoes complex changes throughout this time. On the other hand, the metal deposits build up monotonically, and the time required to achieve monolayer coverage throughout the reactor is comparable in length to the initial deactivation period. There have been a number of both experimental and theoretical studies aimed at developing mathematical models to describe catalyst deactivation in residuum hydroprocessing.[62,64] Commercially the problem is very significant, since large quantities of catalysts are consumed in this service. Increasing catalyst life by 20 percent saves the refiner about 10 to 20 cents per barrel of feedstock processed.

Multicatalyst Systems

As our understanding of the hydrotreating and hydrocracking reactions has improved, the advantages of using multicatalyst systems have been recognized. They are particularly effective in the processing of heavier feeds where impurity removal can be accomplished in a guard bed which protects the major catalyst. They are also used in the hydrotreatment of pyrolysis naphtha where one stage is designed to saturate indenes and diolefins while the second stage saturates olefins and desulfurizes the naphtha.[5]

In residuum hydrotreating, multicatalyst systems are common.[42] In a typical graded system, a metal-selective catalyst protects catalysts in the lower part of the reactor from significant pore mouth plugging. The downstream catalyst (or catalysts) are tailored to be most active for sulfur, Ramsbottom carbon, and molecular weight reduction.

In hydrocracking, the inhibiting effect of heavy aromatics on the cracking reactions also creates a need for a two-catalyst system. The first catalyst is sometimes just a hydrotreating catalyst and sometimes a moderately acidic hydrocracking catalyst. The catalyst saturates aromatics, removes sulfur and nitrogen contaminants, and, if a hydrocracking catalyst is used, causes some cracking to lighter products. The second catalyst is designed to work on clean feeds. Sometimes, if motor gasoline is the major product, a highly acidic catalyst is used. If middle distillate is the required product, then it is better to use a moderately acidic catalyst in the second stage. Examples of product yields achieved with such systems will be given in "Process Capabilities."

Commercial Catalysts

The successful application of hydroprocessing to heavy oil upgrading has been achieved because of emphasis on catalyst development. The *Oil and Gas Journal* October 11, 1993, issue lists the catalysts available to the petroleum industry. Table 14.1.6 summarizes the number of hydroprocessing catalysts on this list and comments on their general characteristics.

The chemical composition and physical properties of catalysts are, in general, proprietary. Various authors[30,67,80] have summarized important aspects of the development of such catalysts. In the case of residuum hydrotreating catalysts, early work centered on the type of alumina used to support the Group VIB and Group VIII metal hydrogenation component.[4] As more pilot plant and commercial data have shown the subtleties of catalyst deactivation, the role that diffusion in the pores plays has become better defined. Smaller particles sizes and unique particle shapes[71] have been developed. Because of the high concentrations of metal poisons in such feedstocks, catalyst consumption is high relative to what can be achieved in distillate processing. Efforts have therefore emphasized low-cost catalysts. Recent work has focused on rejuvenating or regenerating spent residuum hydroprocessing catalysts.[21]

Pore diffusion considerations have also been important in the development of hydrocracking catalysts. The driving force for this approach has been quite different from that in residuum processing. Instead of tailoring catalysts to handle the largest molecules contained in crude oil, the objective has been to find the optimum way to use the small-pore-size crystalline silica-alumina zeolitic materials first reported in the late 1950s. Zeolitic hydrocracking catalysts usually contain noble metal hydrogenation components. They are active in the presence of hydrogen sulfide and, because of the large number of active sites, maintain their activity in the presence of ammonia.[54] They usually make a lighter product than the amorphous catalysts and, because of their small pore size, have difficulty in converting heavy polycyclic aromatics.

TABLE 14.1.6 Commercially Available Hydroprocessing Catalysts

Category	No. of suppliers	No. of catalysts	Comments
Hydrotreating (includes hydrogenation and saturation with no hydrocracking)	13	167	Supports are predominantly Al_2O_3. More than 40% are shaped. NiMo more popular than CoMo. About 7% have noble metals.
Hydrorefining (10% or less hydrocracking)	16	177	About half are for resid feeds. More than 40% are shaped. NiMo more popular than CoMo.
Hydrocracking (50% or more hydrocracking)	14	80	About 35% contain zeolites. Mo more popular than W.
Mild hydrocracking	13	38	NiMo more popular than CoMo.

Note: Shaped catalysts are those with higher surface-to-volume ratios than conventional cylinders.
Source: Oil and Gas Journal Worldwide Catalyst Report, October 11, 1993.

Zeolitic materials have unique selectivity for some reactions because some molecular species are excluded from the pores and therefore cannot react. This is the basis of the first catalytic dewaxing process.[23] Corbett[19] has summarized new zeolite and residuum hydrotreating catalysts.

Based on the hydroprocessing capacity figures shown in Table 14.1.3 and assuming average catalyst lives in barrels of feed processed per pound of catalyst of 200 for hydrotreating, 40 for hydrorefining, and 100 for hydrocracking, the annual catalyst demand in the world (excluding China and the former Soviet Union) exceeds 55,000 short tons. Since residuum hydroprocesses achieve about 10 barrels processed per pound of catalyst, the introduction and growth of residuum processing has increased catalyst demands substantially.

PROCESS DESIGN

Typical Processing Conditions

The conditions under which a hydroprocessing unit operates is a strong function of feedstock. The hydrogen partial pressure must be high enough to accomplish partial saturation of the heavy aromatic molecules. The operating temperature should be sufficiently high to give fast reaction rates but not so high as to promote undesirable side reactions or to exceed the metallurgical limits of the high-pressure vessels. The quantity of catalyst is chosen to give the residence time needed for the reactants to be sufficiently converted at a given operating temperature and pressure.

Typical processing conditions are shown in Table 14.1.7 for a variety of hydroprocesses.[20,99]

Reactor Systems

Light oils are invariably hydroprocessed in reactors containing fixed beds of catalyst. For heavier feedstocks, a wider variety of reactor systems have been developed. They employ smaller catalyst particles in order to take advantage of high reaction rates without diffusion limitations. The more popular reactor systems with particular emphasis on the influence of hydrodynamics on the reaction kinetics are described below.

Fixed-Bed Trickle Reactions. In these reactors both the hydrogen and hydrocarbon streams flow down through one or more catalyst beds. A typical schematic diagram is

TABLE 14.1.7 Typical Hydroprocessing Operating Conditions[20,99]

Process	Hydrogen consumption, SCFB	LHSV	Temperature, °F (°C)	Pressure, lb/in² gage
Naphtha hydrotreating	10–50	2–5	500–650 (260–343)	200–500
Light oil hydrotreating	100–300	2–5	550–750 (288–399)	250–800
Heavy oil hydrotreating	400–1000	1–3	650–800 (343–427)	2000–3000
Residuum hydrotreating	600–1200	0.15–1	650–800 (343–427)	1000–2000
Residuum hydrocracking	1200–1600	0.2–1	750–800 (399–427)	2000–3000
Distillate hydrocracking	1000–2400	0.5–10	500–900 (260–482)	500–3000

Note: LHSV = liquid hourly space velocity.

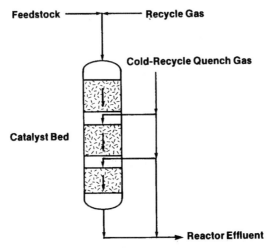

FIGURE 14.1.23 Schematic diagram of trickle-bed reactor.[85]

shown in Fig. 14.1.23. In the Shell hydroprocessing reactors,[97] a liquid collector and distributor tray is located at the top. Each bed is followed by a liquid-collecting quenching tray and a liquid distribution tray. In this way liquid distribution is restored after each bed.

It is important both in commercial units and in pilot plants to see that all the catalyst particles are wetted by the hydrocarbon phase. Also, all the hydrocarbon must be in intimate contact with the hydrogen-rich gas phase in order to keep a uniform concentration of dissolved hydrogen in the hydrocarbon phase. Satterfield[79] has presented a preliminary correlation relating contacting effectiveness with the liquid flow rate per unit cross-sectional area. He recommends a rate of 5 kg/m^2 · s to ensure 100 percent contacting. Hofmann[40] summarizes other work devoted to contacting effectiveness. Van Klinken and Van Dongen[96] suggested an inexpensive way to improve catalyst wetting in pilot plants. They dilute the catalyst bed by filling part of the interstitial volume with small inert particles. In commercial operation, loading catalyst by a dense bed loading technique[84] also helps to ensure uniform contacting and is very popular with distillate feeds. Dense loading is less popular with residuum feedstocks because such feeds are more likely to contain solid contaminants.

Besides the problem of achieving good contacting on a small scale, successful scaleup of trickle beds from pilot plant to commercial scale requires extreme care. Consider, for example, the hydrodynamics of a 1-ft-long pilot plant reactor compared to those of an 80-ft-long commercial reactor. Imagine both operating at a liquid hourly space velocity of 0.5, an inlet gas rate of 5000 SCFB of feed, and a total pressure of 2000 lb/in^2 gage. For these two systems to show the same extent of reaction, the liquid residence times in each reactor must be the same. The residence times of both fluid phases will, to a first approximation, be proportional to the fraction of the particular fluid held up in the reactors. Table 14.1.8 shows typical values for a trickle bed.

The catalyst particles occupy about 0.65 fraction of the reactor volume. The remaining 0.35 of the volume is in the interstices between the particles. The particles are porous so that the space they occupy can be divided into the volume occupied by the catalyst skeleton and the volume of the internal pores. When oil and hydrogen flow uniformly through a reactor, the oil occupies all the volume in the catalyst pores

TABLE 14.1.8 Estimated Phase Holdups in Trickle-Bed Reactors*

	Fraction of reactor space
Catalyst skeleton	0.24
Liquid phase	0.49
Gas phase	0.27

*Assuming extrudate catalyst with ABD = 0.79 g/cm^3 and pore volume = 0.53 cm^3/g, where ABD = apparent bulk density in gm/cm^3 of reactor volume.

plus a fraction of the interstitial volume which, in Table 14.1.8, we have assumed to be 0.2 × 0.35 or 0.07 of the reactor volume. The gas phase flows through the remaining volume. Hydrogen diffuses from this phase into the liquid phase and through it to reach the catalyst internal surface where the reaction takes place.

With these estimates of liquid and gas holdups, the linear velocities of the two fluids and typical reactor pressure drops per foot are as follows:

	Pilot plant	Commercial
Liquid rate, cm/min	0.5	40
Gas rate, cm/min	8	640
Pressure drop, lb/in^2 per foot	0.000002	0.5

Because we are trying to achieve the same reaction rate in reactors differing in length by a factor of 80, the linear velocities differ by this same factor. Since pressure drop ΔP through packed beds is a strong function of linear velocity, the ΔP between the two reactors is extremely large. The pressure drop in the small-scale experiment cannot be measured accurately, since it is dwarfed by reactor end effects. Therefore, small-scale experiments cannot show whether the reactor pressure drop is increasing during a cycle as occasionally happens in commercial units.

As was indicated earlier, it is easier to achieve good contacting at higher linear velocities. Commercial operation is likely to give better results than the pilot plant. If this is not recognized, a refiner may invest more capital in a project than is really needed. Besides the contacting effect, the amount of liquid held up in the interstices of the bed can be a function of linear velocity. Reaction rates will then be a function of the scale of the experiment. In two-phase flow within fixed beds, different flow regimes can exist, depending on the relative rates of gas and liquid, catalyst characteristics, and the scale of the experiments.[40] In some of these regimes pressure drop pulsing can occur, increasing the mean pressure drop substantially.[92] Hydrodynamic studies using liquids like kerosene, desulfurized gas oils, and raw gas oils have also experienced foaming.

These complications have taught process developers to extrapolate the results obtained in small-sized pilot plants with great care. Semicommercial plants are usually required to complete the successful scaleup of novel processes.

Moving-Bed Reactors. Every refinery in the world operates at least one fixed-bed hydroprocessing reactor. They are simple and reliable and quite adequate for handling all the distillate feedstocks which need hydrogen addition.

Since the introduction of residuum hydroprocessing, however, limitations in the application of fixed beds have been recognized. Very high nickel plus vanadium levels in feedstocks require more frequent catalyst changeouts, thereby reducing fixed-bed operating factors. Undissolved particulate matter often is present in residuum feedstocks, and this can increase fixed-bed pressure drops and sometimes reduce plant operating factors.

Since heavy, high-metal crude oils generally cost less than light crudes, refiners need this flexibility for handling tougher residua. Chevron has developed a moving-bed reactor to capture a large fraction of residuum feedstock contaminants so that a downstream residuum hydrotreater can achieve long run cycles. Chevron's on-stream catalyst replacement (OCR) process contains a selective hydrodemetalation catalyst which moves intermittently through a high-pressure vessel. Spent catalyst is withdrawn and fresh catalyst is added weekly. The combination of OCR plus fixed-bed residuum hydrotreating gives the refiner the flexibility to vary crude oil purchases in order to maximize refining margin.

Chevron's OCR process was commercialized in Japan in 1992.[66] A diagram of a Chevron OCR system is shown in Fig. 14.1.24.

Ebullating-Bed Reactors. Another approach toward the problems associated with handling heavy residua in a fixed bed has been the use of ebullating-bed reactors. Two similar residuum hydrocracking processes using ebullating beds have been commercialized. They are the H-Oil process[24] and the LC Fining process.[16] A schematic diagram of an ebullated-bed reactor is shown in Fig. 14.1.25. In such reactors, both the oil and hydrogen flow upward and the catalyst is suspended in the liquid in the form of an expanded bed. The characteristics of such a system is compared with those of a fixed-bed reactor in Table 14.1.9, which is based on the work of Kubo et al.[46]

The hydrodynamics which control the design and operation of ebullating-bed reactors are quite different from those in trickle beds. Since the catalyst particles are suspended by the liquid phase and the gas phase exists as discrete bubbles rising through

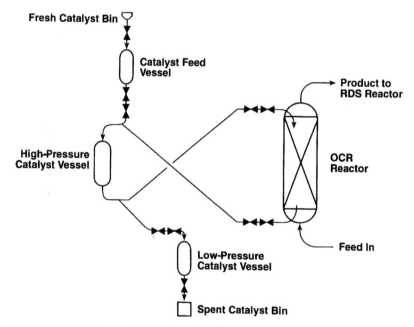

FIGURE 14.1.24 Chevron OCR reactor system.[66]

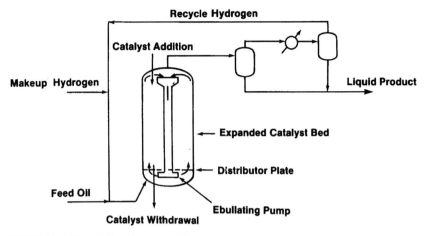

FIGURE 14.1.25 Ebullated-bed reactor.[16]

TABLE 14.1.9 Comparison of Ebullated- and Fixed-Bed Reactors for Hydroprocessing[46]

Features	Ebullated bed	Fixed bed
Continuous catalyst replacement	Yes	Yes
Reactor size	Set by catalyst makeup rate	Set by reaction kinetics
Conversion level per reactor	Limited by back-mixing	Set by reaction kinetics
Catalyst temperature	Constant	Varies with time
Temperature distribution	Uniform	Quench gas needed
Product yields, qualities	Set by catalyst makeup rate	Varies with time
Pressure drop	Constant	Can increase with time
Feedstock treating	None	Desalting, filtering
Auxiliary unit design basis	Normal operation	End-of-run operation
Operability	Complex: fluidization, catalyst addition, catalyst entrainment	Simple

the liquid, particle dynamics predominate. To consider how this affects the scaleup of such units, it is instructive to consider the liquid velocity required to fluidize a catalyst bed.

By using Richardson and Zaki's correlation,[75] fluidizing velocities are estimated as a function of average catalyst particle size from 0.01 to 2 mm in diameter. We have chosen typical residuum hydrocracking reactor conditions and have assumed that the bed is expanded to give a catalyst particle holdup of 0.4. The results are shown in Table 14.1.10.

Besides showing fluidizing velocities, we have estimated the reactor height needed to give the liquid a 1-h residence time (liquid holdup assumed to be 0.5). The results illustrate two points:

1. Unlike the trickle-bed system in which residence times can be closely matched in different-sized reactors (by varying both the reactor height and the fluid velocities at the same time), the ebullating-bed local liquid rates must be maintained in all

TABLE 14.1.10 The Effect of Catalyst Particle Size on Fluidization in Heavy Oil Hydroprocessing*

Catalyst particle diameter, cm	Fluidizing velocity, cm	Reactor height for 1-h residence time, ft
0.001	0.18	0.006
0.01	30.6	1.0
0.1	3640	120
0.2	5470	180

*Based on spherical catalysts, particle density (dry) 1.19 g/cm^3, oil density 0.79, oil viscosity 2.0 cSt; Richardson and Zaki correlation[75]; catalyst particle (wet) holdup 0.4.

reactor sizes if the same size catalyst is to be used in all. From a kinetic standpoint, the same size catalysts should always be used in the scaleup process.

2. For commercial applications the reactor height needed for a reasonable residence time is ridiculously small for small particles and ridiculously large for large particles.

The commercial units operate with ½₂-in extrudate catalysts and have solved the problem of insufficient residence time in a normal-shaped reactor by recycling liquid through the reactor with an ebullating pump located either within the reactor or outside it. This recycling causes complete back-mixing of products with reactants which lowers the reactor efficiency to that of a completely stirred tank.

Back-mixing is also promoted by the flow of the gas phase through the system. In the case of very small catalyst particles, the flow of gas bubbles upward through the expanded bed disturbs the upper surface of the fluidized bed and causes entrainment of catalyst into the product stream; ½₂-in catalyst can give a stable ebullated bed with a well-defined upper level.

The inefficiency of back-mixing is most pronounced when high conversion levels are desired. High conversions can be achieved only if two or more ebullating-bed reactors are used in series. Catalyst makeup strategies from one reactor to the next have been developed to further improve system economics. On the whole, though, fluidized systems are not likely to give any marked improvement in catalyst efficiency over a fixed bed. As Adlington and Thompson[2] have pointed out, other factors, like freedom from pressure drop increases when processing feeds which contain particulate matter or ability to add and withdraw catalyst during operation, are more likely to influence the choice.

Slurry Phase Technology. A variety of heavy feedstock hydroconversion processes based on slurry phase operation have been reported.[22] These include the M-Coke process,[6] the Aurabon process,[1] and the CANMET process.[53] These processes are designed to take advantage of the intrinsic activity of even smaller catalyst particles (0.002 mm) than those used in ebullating beds. Referring to the fluidizing velocities of Table 14.1.10, the liquid flow rates are considerably higher than those required for fluidization. This results in very low catalyst holdups (about 1 percent) in the reactor. Despite this low holdup, the number of particles per unit volume is 10 million times greater than in an ebullating bed.[22]

The main reaction appears to be thermally induced in the liquid between these particles. The catalyst hydrogenates the unstable radicals produced in this thermal reaction.

Scaling up this technology will be difficult. Predicting the holdup of each phase in the reactor and separating the catalyst from the product are two areas requiring particular attention. Regarding the first point, Dautzenberg[22] has suggested that the hydrodynamic studies that have been carried out on two-phase (gas-liquid) systems give some insight into the flow regimes that can be encountered in the more complicated three-phase systems.

Flow Schemes

Hydrocracking. The versatility of the hydrocracking process has been achieved by developing specific families of catalysts and processing schemes which allow these catalysts to function efficiently. Also, optimum refining relationships between hydrocracking and other refining processes like catalytic reforming and fluid catalytic cracking are practiced.

The choice of processing schemes for a given hydrocracking application depends on the quality and quantity of feedstock to be processed and the desired product yield structure and quality. Figure 14.1.26 shows a Chevron two-stage Isocracking process flow arrangement. It converts straight-run heavy gas oils into high yields of diesel fuel, jet fuel, or naphtha.

In the first stage the feed is hydroprocessed to saturate heavy aromatics and remove basic impurities like nitrogen. The second stage hydrocracks this product to extinction recycle. The recycle cut point (RCP) is selected to maximize the yield of a desired product. Distillation facilities may be placed either between the stages or after the second stage. The intermediate distillation option shown in Fig. 14.1.26 reduces the size of the second stage. The tail end option allows more thorough hydrogenation of the light products.

The most costly part of each stage is the equipment in the high-pressure reactor loop. In this section, the feedstock is pumped to high pressure, mixed with recycle hydrogen, and heated in a shell and tube feed/effluent exchanger. The mixture is then passed through a charge furnace and heated to reacting temperature.

The reactor contains many beds of catalyst with quench and redistribution devices between them. The reactor effluent is cooled first in the feed effluent exchanger, then further cooled by exchange with other streams like the product fractionator feed. Finally, an air cooler then brings it to a low enough temperature that the hydrogen flashed off in the high-pressure separator can be recompressed and recycled back to the feed. Makeup hydrogen is added to the loop to maintain system pressure. Water is injected into the effluent before the final air cooler in order to prevent ammonium bisulfide from depositing in the colder section of the cooler. The hydrogen-rich stream from the high-pressure separator is recycled back to the reactor after being scrubbed for H_2S removal if the feedstock contains a high sulfur level.

The hydrocarbon stream from the high-pressure separator is depressured and sent to the fractionation section after passing through a low-pressure separator in which the hydrogen dissolved in oil is flashed overhead and sent to be recovered. The reactor loop in the second stage is similar in principle, although less expensive construction materials can be used because of much lower H_2S levels present in the streams.

The bottoms product from the high-pressure separator is recycled back to feed the second stage. The boiling range of this recycle oil varies, depending on what products are desired from the unit. If a maximum yield of diesel fuel is needed, an RCP of about 700°F is used. If naphtha for aromatics production is in demand, it can be as low as 350°F. In between, jet fuel or catalytic reformer feedstock can be maximized as needed.

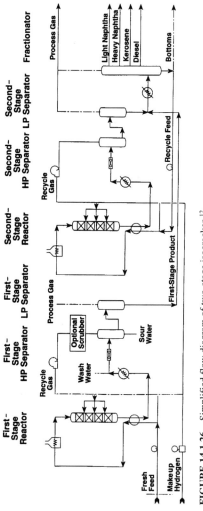

FIGURE 14.1.26 Simplified flow diagram of two-stage isocracker.[12]

There are many variations on flow schemes for two-stage plants. Sometimes the two-phase charge heater is replaced by a single-phase furnace on both hydrogen and hydrocarbon.[16] Sometimes one recycle compressor is used instead of two, in which case the difference in pressure levels in the two stages is small and due only to equipment pressure drops. Speight[85] has summarized the flow schemes of all the major commercial processes.

Single-stage Isocracking plants have also been commercialized. For feed rates of less than 12,000 barrels per operating day (BPOD) and a need for maximum diesel production, a single-stage recycle plant costs less than the corresponding two-stage plant. The flow scheme for such a unit is like the second stage of Fig. 14.1.26. Other refiners operate single-stage once-through Isocrackers which look like the first stage of a two-stage plant. This configuration is the lowest cost option and is attractive if the unconverted (but severely hydrotreated) heavy product is of value. This product can be used as FCC or ethylene plant feedstock, LSFO, or lubricating oil base stock.

Hydrotreating. In those applications where hydrocracking is less important than contaminant removal or where extinction recycle of the heavy product is not practical (as with residuum from heavy crude oils), then a single-stage once-through configuration, like the first stage in Fig. 14.1.26, is used. Residuum feedstocks are always filtered and a different reactor effluent flow scheme is often used. As an example, Fig. 14.1.27 shows the schematic flow for the LC Finer at Salamanca.[16] The hot reactor effluent is flashed in a hot high-pressure separator and only the vapor stream is cooled down to provide the hydrogen-rich recycle stream. This option saves energy by allowing the liquid stream from the reactor to go to product fractionation while it is still very hot. It also eliminates problems which could arise in trying to separate the hydrogen stream from the viscous heavy oil at typical cold high-pressure separator temperatures. The Salamanca unit is equipped with a vacuum column on the product stream. The VGO can therefore be recycled to extinction if product values demand it. Note that Fig. 14.1.27 also shows a plant where the liquid and gaseous reactants are heated

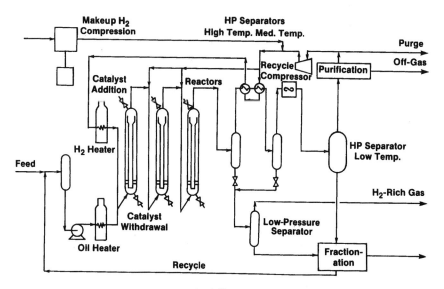

FIGURE 14.1.27 LC fining process flow sketch.[16]

in separate furnaces. Options of the type shown in this plant flow configuration should be explored from the point of view of capital investment, energy usage, and demonstrated reliability in the design phase of a project.

Design Considerations

The decisions made during the design of a hydroprocessing unit are all based on economic analyses—balancing changes in capital investment with corresponding but opposing changes in operating costs. In a few cases there is the more difficult choice of deciding whether to invest more capital in order to add flexibility to plant operation. During the 1970s, the cost of fuel in a refinery changed dramatically so that, for older plants, the design decisions are no longer relevant to the current situation. Ongoing economic studies must therefore be made to keep an existing plant operating in the optimum mode.

Major Design Decisions. The high-pressure items are the most costly in the plant, so most of the important design decisions concern them. The quantity of catalyst needed and the pressure level of the reactor section are based on knowledge of reaction kinetics, catalyst deactivation rates, and makeup hydrogen purity. The design pressure level is also influenced by product quality considerations. As shown in Table 14.1.7, heavier feedstocks generally require higher pressures. The quantity of catalyst is chosen to give reasonably low operating temperatures to avoid undesirable side reactions while staying below the reactor metallurgical limits.

The number of reactors required is based on a variety of considerations:

- The reactor pressure drop must be high enough to promote uniform flow and temperatures but within reason from the point of view of capital investment and operating costs. Once this pressure drop is set, then the total cross-sectional area needed in the reactor section is proportional to the feed rate.
- The refinery may not be able to accept the largest reactors which a fabricator can build. Transporting the reactor to the refinery site may not be possible, in which case field fabrication should be considered. Soil conditions at the refinery may also preclude very heavy reactors.
- The reactor fabricators have limits to the diameter and length of high-pressure vessels. Reactors weighing 1400 metric tons have been built.
- In multireactor plants, particularly with residuum feedstocks, it is common to build more than one reactor train. Sometimes the trains are completely independent plants, since this gives the refiner the most operating flexibility. In other cases the trains share common feed pumps, high-pressure separators, and recycle compressors.

Energy Conservation. Hydroprocessing reactions are exothermic. The design of a process unit must take maximum advantage of this fact. The feed effluent exchanger must recover as much heat as is economically practical in order to minimize the fuel consumption in the charge furnace or furnaces. An accurate estimate of the heat released in the hydroprocessing reaction is essential to achieve this.

Jaffe[44] has proposed a method for predicting heat release by following the chemical bonds formed and consequent hydrogen consumed by important classes of hydrocarbons. There are three important categories: (1) saturates which consume hydrogen with cracking or ring opening and yield 7 to 10 kcal/mol of hydrogen, (2) aromatics saturation which yields 14 to 16 kcal/mol of hydrogen, and (3) olefins saturation which yields 27 to 30 kcal/mol of hydrogen. The total heat release depends on the distribution of hydrogen consumption between saturates, aromatics, and olefins. Modern analytical techniques can be used in small-scale experiments to determine the concentration of bond types in feedstocks and products. Jaffe obtained good agreement between predicted and measured hydrogen consumptions by this method.

There are other important points that must be considered in the overall heat balance of the plant. Quench hydrogen rates between the catalyst beds should be minimized, consistent with safe operation and desired catalyst life. As in any process unit, the size of air coolers should be as small as possible and any steam generation should recognize the refinery's overall steam balance.

Both the capital investment and operating costs of the recycle compressor are dictated by the reactor loop pressure drop. Designing this loop for too low a pressure drop will result in poor heat-transfer coefficients in the heat exchangers and poor flow patterns within the reactor itself. It is important to calculate accurately what the individual equipment pressure drops will be, as well as what flow regimes exist in them. This is particularly important in the reactor where a high fraction of the overall loop pressure drop occurs.

Hofmann[40] has summarized the correlations proposed for predicting pressure drops in trickle-bed reactors: "At low loadings the pressure drop is approximately the same as in single-phase gas flow. At higher gas loadings, the texture of the liquid is modified by gas-phase friction, and the pressure drop rises, together with a decrease of the liquid holdup, in the transition to the pulsing flow region." Figure 14.1.28 shows the different flow regimes that have been characterized in air-water systems. The regions are represented on a plot of liquid Reynolds number versus gas Reynolds number. Also shown on the left-hand side is a contacting efficiency relationship similar to that proposed by Satterfield.[79] Talmor[92] has studied the pulsing region, in which pressure drop oscillations occur. He observed that in a partially pulsing situation, the overall reactor pressure drop increases by as much as a factor of 2. Obviously this region should be avoided whenever possible.

Hydrogen Management. Hydrogen is an important commodity in a modern refinery. There are often a variety of sources, each of different purities. Impurities, like methane

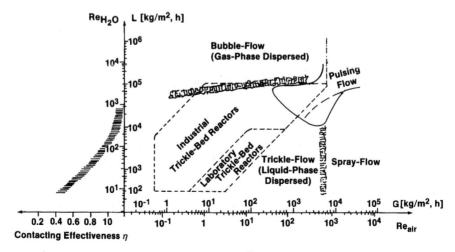

FIGURE 14.1.28 Flow regions in trickle-bed reactors.[40]

in the makeup hydrogen to a hydroprocessing unit, build up in the gas recycle stream until equilibrium develops at the low-pressure separator. The amount of methane dissolved in the liquid effluent equals that coming into the system with the makeup hydrogen plus whatever is produced in the reactor. The effect of this depends on whether a new unit or an existing unit is under consideration. In both cases, the reaction kinetics and catalyst deactivation rate are controlled by the hydrogen partial pressure within the reactor. Impure makeup hydrogen therefore results in new units being built at higher, more costly design pressures. The existing unit's performance is hurt if the makeup hydrogen purity becomes less than design and is helped if its purity can be upgraded. Many approaches have been developed to improve the quality of hydrogen streams so that hydrocrackers and hydrotreaters can operate more profitably. These include pressure-swing absorption[100] and "semipermeable membrane" separators.[9]

Materials. The refining industry has an excellent safety record for operating high-pressure equipment. Special care is placed on the choice of materials of construction, on monitoring the fabrication of critical pieces of equipment, and on using operating procedures which protect the equipment. Most reactors are of the hot-wall variety, and low-alloy steels are required to resist hydrogen attack. Erwin and Kerr[25] have written a comprehensive survey of 25 years of experience with a 2¼ Cr–1 Mo steel in the thick-wall reactor vessels of the petroleum industry. Start-up and shutdown procedures have been developed to account for both temper embrittlement and hydrogen embrittlement in such steels. Nowadays stronger steels using 3 percent chrome and some that are vanadium-modified are available for high-pressure vessels.

When used at temperatures above 500°F, 2¼ Cr–1 Mo has inadequate resistance to H_2S corrosion. Corrosion by H_2S in the presence of hydrogen is far more severe than by H_2S alone.[31] To overcome this, the reactors and hot exchanger shells are made with stainless-steel cladding. In addition, reactor internals and hot exchanger tubes are usually made of stainless steel.

Another part of a hydroprocessing facility which is subject to corrosion is the high-pressure air coolers where H_2S and ammonia exist in the presence of water. Piehl[71] has studied this problem and suggested design and operating guidelines to overcome it.

Emergencies. Design and operating procedures have been developed to minimize temperature runaways within hydroprocessing reactors and take proper action should one start.

Local hot spots are occasionally seen in such reactors. Jaffe[45] has explained them in terms of regions of low flow. He developed a mathematical model which accounts for the temperature rise with rapid reaction of the fluid in the affected region and for temperature drop with the eventual mixing of cooler fluid from the surrounding region. By comparing the model with commercial temperature profiles, he estimated the velocity of the low flow region and its lateral extent. The cause of a low flow region could be the presence of catalyst fines, a physical obstruction, or a failure of the reactor internals.

One of the most serious operating problems for a hydrocracker is an unexpected recycle compressor shutdown. The loss of gas flowing through the reactor results in a sudden increase in the catalyst temperature, since the heat of reaction cannot be transported out of the reactor. Effective procedures have been worked out for handling such a situation. With reference to Fig. 14.1.29, which shows the normal hydrocracker controls, they involve bypassing the feed/effluent exchangers, cutting the charge furnace fires, and possibly stopping the makeup hydrogen flow. Gerdes et al.[33] have described the results of a hydrocracker simulator designed to train operators to handle such emergency situations.

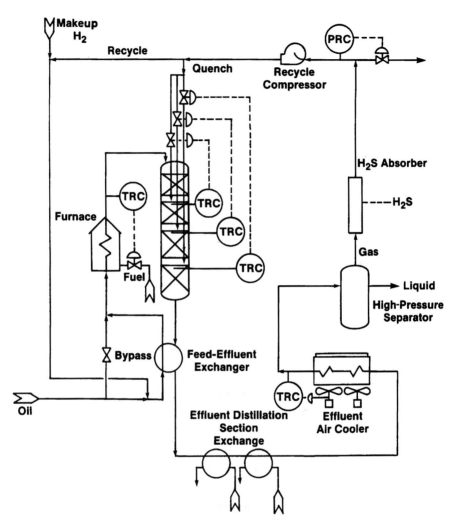

FIGURE 14.1.29 Control of a hydrocracker.[33]

PROCESS CAPABILITIES

Hydrogen Utilization

Except at very low hydrogen consumptions, hydroprocessing results in a volume expansion from feed to product. Figure 14.1.30 shows this for a heavy distillate feedstock in a variety of hydroprocessing situations.[82] The plot shows the percent expansion for the C_5+ product versus the chemical hydrogen consumption. Below a consumption of 500 SCFB, the data points were taken in a hydrotreating mode where selective desulfurization is carried out at relatively lower hydrogen partial pressures

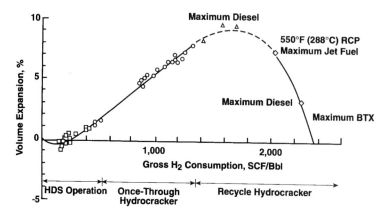

FIGURE 14.1.30 Hydroprocessing Mideast vacuum gas oils.[82]

(see "Typical Processing Conditions," above). A volume shrinkage occurs with the first increment of hydrogen consumed since the initial removal of sulfur is more significant than the production of light products.

Refiners have recently practiced mild hydrocracking in existing VGO hydrotreaters by using catalyst combinations with more acidity. The extra hydrogen consumption is associated with more volume expansion increase, as shown in Fig. 14.1.30. The pressure levels in these units do not allow more hydrogen input because the hydrogenation-dehydrogenation equilibrium limits the desired reaction. At higher pressures, however, substantial hydrocracking can occur so that a once-through hydrocracker on this same feed can consume 1200 SCFB of hydrogen and make 108 volumes of C_5+ product out of 100 volumes of VGO feed. The maximum volume expansion is achieved in a recycle mode operating at a RCP of 700 to 735°F (which maximizes diesel fuel yield). More hydrogen can be consumed in a recycle hydrocracker if the RCP is dropped so that lighter products are made. An RCP of 550°F maximizes the production of jet fuel while consuming about 2000 SCFB of hydrogen. Motor gasoline yield is maximized at 380°F RCP and aromatics (via catalytic reforming) at somewhat lower RCP.

With more aromatic feedstocks, the amount of hydrogen consumed and the amount of expansion will be higher than shown. Scott, Meyer, and Mayer[83] have taken this representation even further by assuming the heavy naphtha produced by hydrocracking will usually be catalytically reformed, so that some hydrogen will be recovered. Also, the light gases produced in the hydrocracker can be used as feedstock to a hydrogen plant. Therefore they are a potential source of hydrogen. If this is accounted for and net hydrogen consumption is the focus instead of the amount used in the hydrocracker, a more complete measure of hydrogen efficiency is given.[83]

Table 14.1.11 shows in tabular form the product yields which correspond to the hydroprocesses covered in Fig. 14.1.30. Again, these are based on experiments with Middle Eastern VGOs. Note that these are just examples of yield structures designed to show what is possible in hydroprocessing. Variations on these yields occur by using different catalysts, process configurations, or operating conditions. No correlation exists which allows *a priori* prediction of product yields. Experimental data are needed to support predictions and extrapolations.

TABLE 14.1.11 Product Yields in Mideast VGO Hydrotreating

Process	VGO Hydrotreating	Once-through	Hydrocracking			
			Recycle	Recycle	Recycle	Recycle
Recycle cut point, °F (°C)			725 (385)	550 (288)	380 (193)	
Product yields:						
C_1–C_3, wt %	0.09	0.63	2.5	3.0	2.64	
C_4's, wt %	0.01	0.83	2.4	9.1	14.07	
Light naphtha, °F (°C)		C_5–180 (C_5–82)	C_5–180	C_5–180	C_5–180	
Yield, LV %		2.0	6.6	15.1	32.7	
Heavy naphtha, °F (°C)	C_5–350 (C_5–177)	180–310 (82–154)	180–250	180–280	180–380	
Yield, LV %	0.2	5.5	7.1	22.5	68.8	
Middle distillate, °F (°C)	350–550	310–655 (154–346)	250–725	280–550		
Yield, LV %	7.1	35.5	96.8	68.8		
Heavy product, °F (°C)	650+ (343+)	655+ (346+)	725+ (385+)	550+ (288+)	380+ (193+)	
Yield, LV %	93.0	61.6	0	0	0	
Chemical hydrogen consumption, SCFB	300	1135	1250	1350	1650	

Note: LV % = liquid volume percent.
Source: Chevron Research pilot plant data.

Product Qualities

Light Oil Hydrotreating. As was shown in "The Extent of Commercialization," above, most hydrotreating capacity worldwide is applied to light oil feedstocks. Attane, Mears, and Young[5] have summarized the capabilities of the Unionfining process on such oils. Table 14.1.12 shows performance on straight-run naphtha and a blend of both straight-run and coker gasoline. These feedstocks need to be desulfurized and denitrified so that a catalytic reformer can upgrade them effectively. Hydrogen consumption is less than 100 SCFB.

Table 14.1.13 shows performance with kerosene fuel, diesel fuel, and cracked heating oil. Kerosene is hydrotreated to remove sulfur and improve color and color stability. The same objectives exist in treating diesel. However, with diesel fuel, producing a clean-burning fuel with less likelihood of emulsification if accidentally contacted by water is also important. Sometimes units are built to treat wide-boiling-range distillates. If so, it may be necessary to overtreat the diesel fraction to obtain satisfactory color stability on the kerosene. Generally this is not a severe penalty because the additional hydrogen requirement is modest and incremental operating costs are small. The cracked heating oil in the example is a blend of FCC heavy gasoline and light cycle oil. The objective is sulfur removal to make the oil acceptable for LSFO use. Low-aromatic diesel fuel is produced in some parts of the world in response to stricter environmental legislation.

Light Oil Hydrocracking. One of the first applications of modern hydrocracking was the conversion of naphtha to liquefied petroleum gas.[81,99] Nowadays the economics for such a conversion are generally poor. However, the process is commercially

TABLE 14.1.12 Unionfining of Naphthas[5]

Feedstock character	Straight run	Coker/straight-run blend
Density, °API	61.1	52.9
ASTM D 86, °F (°C):		
IBP	189 (87)	195 (91)
10	198 (92)	232 (111)
50	245 (119)	300 (149)
90	300 (149)	400 (204)
EP	329 (165)	400 (204)
Sulfur, wt %	0.035	0.14
Nitrogen, ppm		35
Bromine number, g/100 g		10
Hydrogen consumption, SCFB	20	90
Product quality:		
Sulfur, wt %	0.0001	0.0001
Nitrogen, ppm		1
Bromine number, g/100 g		<1
Product yields:		
H_2S, SCFB	1	45
C_1–C_3, SCFB	4	4
C_4, LV %	0.1	0.1
C_5, LV %	0.1	0.2
C_6+, LV %	99.8	99.9

Note: °API = degrees on American Petroleum Institute scale; IBP = initial boiling point; EP = endpoint.

TABLE 14.1.13 Unionfining of Light Oils[5]

Feedstock	Kerosene	Diesel	Cracked heating oil	
Density, °API	40.8	33.8	27.0	
ASTM D 86, °F (°C):				
IBP	384 (196)	451 (232)	328 (165)	
10	397 (202)	511 (267)	377 (192)	
50	425 (218)	580 (304)	438 (226)	
90	487 (252)	637 (336)	535 (280)	
EP	523 (273)	680 (360)	617 (325)	
Sulfur, wt %	0.2	1.85	1.3	
Nitrogen, ppm	3	400	9330	
Smoke point, mm	19			
Bromine number, g/100 g			25.0	
Hydrogen consumption, SCFB	30	175	355	
Liquid Product, °F (°C)	390+ (199+)	450+ (232+)	310–410 (154–210)	410+ (210+)
Yield, LV %	98.8	98.3	38.4	58.9
Sulfur, wt %	0.04	0.15	0.0045	0.044
Nitrogen, ppm	1			
Smoke point, mm	20			
Color, Saybolt	+29			
Color stability, 16 h at 212°F (100°C)	+26			

TABLE 14.1.14 Unionfining for LPG Production[99]

Feedstock—Kuwait naphtha:	
Density, °API	55
Sulfur, wt %	0.097
Nitrogen, ppm	1
ASTM D 86, °F (C°)	
10	250 (121)
50	310 (154)
90	381 (195)
EP	416 (214)
Hydrogen consumption, SCFB	2060
Product yields, LV %	
Propane	77
Butane	62
Total LPG	139

practiced. Typical yields are shown in Table 14.1.14 for extinction recycle Unicracking[99] of a Kuwait naphtha. The total yield of propane and butane is over 135 LV % of feed. Hydrogen consumption to achieve this is 2060 SCFB.

Naphtha can also be hydrocracked to produce isobutane. This can then be used as feedstock to alkylation units for producing high-octane mogas components. Table 14.1.15 illustrates the yields of isobutane obtained by Unicracking[99] a light straight-run gasoline, a Udex raffinate, and a reformer feedstock. In all cases over 45 percent of isobutane were obtained with propane, the next most abundant product.

TABLE 14.1.15 Unicracking for Isobutane Production[99]

Feedstock	Udex raffinate	Light straight-run gasoline	Reformer feedstock
Density, °API	69	68	53
Sulfur, ppm	7	5	3
Nitrogen, ppm	<1	<1	<1
ASTM D 86, °F (°C):			
10	209 (98)	216 (102)	244 (118)
50	220 (104)	232 (111)	289 (142)
90	252 (122)	263 (129)	351 (178)
EP	288 (142)	294 (146)	382 (194)
Isobutane yield, LV %	49–54	50+	45–50

Heavy Distillate Hydrotreating. In the 1960s, as more high-sulfur Middle Eastern crudes were processed in the world's refineries, hydrodesulfurization facilities were added to produce the LSFO that local and national governments were demanding. This was particularly the case in Japan. In the United States, some companies decided to invest immediately in direct atmospheric residuum hydrotreating plants. Others took advantage of the fact that LSFO sulfur specifications were going to drop according to a fixed timetable. These refiners built in stages—VGO hydrotreaters at first, the product from which was blended with the virgin vacuum residuum to meet the specifications for the immediate future. Later the vacuum residuum was upgraded so that the long-term specifications could be achieved. The VGO hydrotreaters were designed to remove at least 90 percent of the feed sulfur with minimum hydrogen consumption. Most were designed at 600 to 1000 lb/in^2 hydrogen partial pressure.[28]

Today's refiner is faced with the need to convert the heavier components of the crude barrel into lighter, more valuable products.[32] An inexpensive and immediate step to achieve residuum conversion is to convert the operation of VGO hydrotreaters to VGO mild hydrocrackers. Table 14.1.16 consists of commercial data showing three different types of operation achieved in the same Chevron VGO hydrotreater.[28] Conventional desulfurization is compared with severe desulfurization (achieved by accepting a reduction of catalyst life). Both are then compared with a mild hydrocracking operation using a blend of desulfurization and hydrocracking catalysts. The LGO product meets Japanese diesel specifications for sulfur, cetane index, pour point, and distillation. The yield of diesel produced can be varied by operating at different conversions. Figure 14.1.31 shows this variation. Note that synthetic conversion is total conversion corrected for the light straight-run distillate material in the feed. Incremental conversion by mild hydrocracking preferentially produces diesel fuel.

Heavy Distillate Hydrocracking. A single-stage once-through hydrocracker can achieve higher conversions to lighter products than the lower pressure units which were originally designed for desulfurization.[34] Table 14.1.17 shows Isocracking process product yields[77] at three different conversion levels—40, 55, and 70 percent below 640°F. In this table the yields are expressed by two different sets of product cut points to show the flexibility for varying the relative motor gasoline to diesel fuel yields. With this Arabian Light VGO feedstock, the diesel yield peaks at the lower conversions. At the 70 percent level, the diesel yield has dropped because some of it has been converted to naphtha. At all conversion levels, the ratio of gasoline to diesel can be varied by about a factor of 2 by adjusting product cut points.

The product inspections are shown for the gasoline and diesel cuts in Table 14.1.18. The 180–390°F cut naphtha produced at 70 percent conversion is a high-quality reformer feed because of its high naphthene and aromatics content. The 640°F+

TABLE 14.1.16 Chevron VGO Hydrotreating and Mild Hydrocracking[28]

Operation	Conventional desulfurization	Severe desulfurization	Mild Isocracking
% HDS	90.0	99.8	99.6
Yields, LV %:			
Naphtha	0.2	1.5	3.5
Light Isomate	17.2	30.8	37.1
Heavy Isomate	84.0	70.0	62.5
Feed:			
Density, °API	22.6	22.6	23.0
Sulfur, wt %	2.67	2.67	2.57
Nitrogen, ppm	720	720	617
Ni+V, ppm	0.2	0.2	—
Distillation, ASTM, °F (°C)	579–993 (303–534)	579–993 (303–534)	552–1031 (289–555)
Light isomate			
Density, °API	30.9	37.8	34.0
Sulfur, wt %	0.07	0.002	0.005
Nitrogen, ppm	90	20	20
Pour point, °F	18	14	18
Cetane index	51.5	53.0	53.5
Distillation, ASTM, °F (°C)	433–648	298–658	311–683
Heavy isomate			
Density, °API	27.1	29.2	30.7
Sulfur, wt %	0.26	0.009	0.013
Nitrogen, ppm	400	60	47
Viscosity, cSt at 122°F (50°C)	26.2	19.8	17.2
Distillation, ASTM, °F (°C)	689–990	691–977	613–1026

Note: HDS = hydrodesulfurization.

bottoms stream is a unique product. It not only is an ultralow-sulfur blend stock for fuel oil but a prime feed component to the FCC unit, a superb source of lube oil base stock, and an attractive ethylene plant feed.

Table 14.1.19 shows the heavy product quality at about 50 percent conversion. Inspections are presented for the 700°F+ portion of this bottoms stream, before and after dewaxing. Lube oil base stocks with viscosity indexes of 110 to 120 can be obtained when running in this conversion range. In 1985, Chevron U.S.A. started up a lube oil complex based on the Chevron Isocracking process.[26] In connection with this lube oil hydrocracking complex, Chevron has upgraded its dewaxing facilities by installing its novel catalytic dewaxing process, Isodewaxing.[103] This process isomerizes the wax to make higher yields of high-viscosity-index base stocks. It allows the combination of hydrocracking and Isodewaxing to produce unconventional lubes with viscosity indexes greater than 130.[55]

For ethylene plant feedstock, the product at 70 percent conversion is more attractive. The high paraffin and naphthenic content of this product is consistent with a Bureau of Mines Correlation Index (BMCI) of about 10. This type of feed should produce a high yield of ethylene and a low yield of pyrolysis fuel oil.

In cases where the bottoms product has low value, and during periods of low crude run when hydrocracker feed rate is less than design, the unit can be run in a partial or total recycle mode. This shifts the product slate more toward diesel than gasoline. Table 14.1.20 compares the yields obtained in a once-through operation using gasoline mode cut points to a total recycle mode operating to maximize diesel fuel.

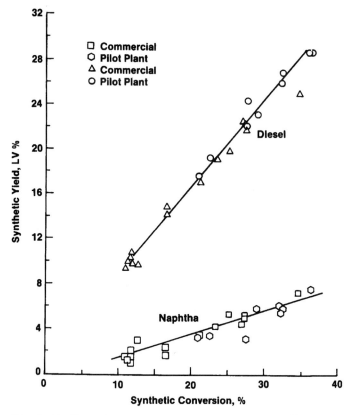

FIGURE 14.1.31 Distillate product yields—VGO mild isocracking.[28]

TABLE 14.1.17 Single-Stage Isocracking Yields* for Gasoline and Diesel Production[77]

	Conversion level, LV %		
	40	55	70
Chemical H_2 consumption, SCFB	1050	1200	1350
Yields (gasoline mode), LV %:			
C_5–180°F (C_5–82°C)	5.0	9.0	16.0
180–390°F (82–199°C)	13.0	24.0	36.5
390–640°F (199–338°C)	26.0	27.0	23.5
640°F+ (338°C+)	60.0	45.0	30.0
Gasoline/diesel ratio	0.7	1.2	2.2
Yields (diesel mode), LV %:			
C_5–180°F (C_5–82°C)	5.0	9.0	16.0
180–310°F (82–154°C)	7.0	15.0	23.0
310–655°F (154–346°C)	34.0	38.0	35.5
655°F+ (346°C+)	58.0	43.0	31.5
Gasoline/diesel ratio	0.4	0.6	1.1

*Arabian Light VGO feed (670–1020°F, 354–549°C); converted below 640°F.

TABLE 14.1.18 Single-Stage Isocracking Light Product Qualities*[77]

	Product Cut		
Inspection	C_5–180°F (C_5–82°C)	180–390°F (82–199°C)	390–640°F (199–338°C)
Density, °API	81	51.5	38
Aniline point, °F		114	143
Paraffins, vol %		37	36
Naphthenes, vol %		48	37
Aromatics, vol %		15	27
Pour point, °F			−20
Octane, F-1 clear	80.5	61	

*Arabian Light VGO feed (670–1020°F, 354–549°C); 70% conversion below 640°F.

TABLE 14.1.19 Single-Stage Isocracking Heavy Product Qualities*[77]

Conversion level, LV %	50	50	70
Product boiling range, °F+ (°C+)	700 (371)	700 (371)	640 (338)
Dewaxed	No	Yes	No
Density, °API	34.8		38.7
Aniline point, °F (°C)			229 (109)
Pour point, °F (°C)	95 (35)	5 (−15)	80 (27)
Paraffins, vol %	37.3	30.0	62.5
Naphthenes, vol %	52.1	58.4	30.8
Aromatics, vol %	10.6	11.3	6.7
Sulfur, ppm			15
Viscosity, cSt:			
At 100°F (38°C)		30.6	
At 210°F (99°C)		5.32	3.3
Viscosity index		118	
ASTM D 1160, °F (°C):			
IBP		700 (371)	675 (357)
10		755 (402)	700 (371)
50		805 (430)	733 (390)
90		933 (501)	820 (438)
EP		965 (519)	960 (516)

*Arabian Light VGO feed (670–1020°F, 354–549°C); 70% conversion below 640°F.

The hydrocrackers operating in North America upgrade feed blends containing a variety of cracked feedstocks (for example, FCC cycle oils[81] and both delayed[10] and fluid coker gas oils.[48] These feeds give more aromatic products than a corresponding straight-run feed would give. Jet fuel smoke point suffers slightly, but motor gasoline quality is better. Operation with high-endpoint cracked stocks has been noticed to cause heavy polycyclic aromatics to precipitate out in the air coolers in recycle operation.[48]

Outside North America, the predominant hydrocracker feedstock is still the straight-run VGO and the most popular operating mode is to maximize middle distillate production. Catalysts are now available which can produce over 95 LV % of good-quality middle distillates from recycle hydrocrackers.[12,54,68] Typical yields and product qualities for such an Isocracking operation are given in Table 4.1.21 The feed-

TABLE 14.1.20 Isocracking Yield Comparison—Once-Through Versus Recycle*[77]

Type of Operation	Once-through	Recycle
Conversion below 640°F, (338°C), LV %	70	
Per pass conversion, LV %		50
Recycle cut point, °F (°C)		650
Yields:		
Gasoline, LV %		
C_5–390°F (C_5–199°C)	52.5	
C_5–310°F (C_5–154°C)		43.0
Diesel fuel, LV %		
390–640°F (199–338°C)	23.5	
310–650°F (154–343°C)		64.0
Bottoms product, LV %	30.0	
Gasoline/diesel ratio	2.2	0.7

*Arabian Light VGO feed (670–1020°F, 354–549°C).

TABLE 14.1.21 Two-Stage Isocracking Middle Distillate Yields and Qualities*[12]

Product cut, °F (°C)	C_5–180 (C_5–82)	180–250 (82–121)	250–525 (121–274)	525–725 (247–385)
Yields, LV %	6.6	7.1	47.9	48.9
Density, °API	80.5	60.8	45.4	39.4
Octane, F-1 clear	76	68		
Paraffin/naphthenes/aromatics, vol %		46/50/4	25/64/11	
Smoke point, mm		25		
Flash point, °F (°C)		100 (38)		
Freeze point, °F (°C)		−75 (−59)		
Pour point, °F (°C)			−10 (−23)	
Cloud point, °F (°C)			0 (−18)	
Aniline point, °F °C)			180 (82)	
Diesel index			69	
Cetane number			61	
Viscosity at 122°F, (50°C), cSt			5.3	

*Feedstock 700–1000°F (371–538°C) Arabian VGO.

stock was 700 to 1000°F straight-run Arabian VGO. The operation produces a yield of 47.9 percent kerosene and 48.9 percent diesel, both meeting normal specifications. Achieving selectivity for middle distillate production depends on a large number of factors including the process configuration, the choice of catalyst, the feedstock to be used, and a number of other design parameters.[12] Amorphous catalysts give higher middle distillate yields than catalysts which contain zeolite components. Highly paraffinic feedstocks give lower yields than less paraffinic ones.

Besides these characteristics of the reactor section, various aspects of the product fractionation section can affect the yield and quality of the middle distillate products. Figure 14.1.32 shows what can be achieved by varying the cut point between jet and diesel while keeping the jet initial cut point at flash point specification and the RCP at 725°F. Varying this intermediate cut point from 450 to 550°F gives a range of jet to diesel production ratios from 0.5 to 1.2.

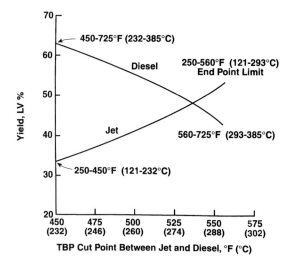

FIGURE 14.1.32 Estimated Chevron Isocracking yields, ICR 120—effect of changing cut point on jet and diesel production (constant total jet plus diesel).[12]

The paraffinicity of the feedstock is even more critical if aromatics are a desired product. Table 4.1.22 shows the typical yields of naphthenes plus aromatics (N + A) in Unicracker naphthas as a function of fresh feed quality.[52] Ring content of the feed is roughly indexed by its UOP characterization factor (UOPK). For low UOPK factors (high feed ring contents), the yields of benzene, toluene, and xylene (BTX) precursors become very high, reaching 69 to 77 vol % of feed for C_6-C_8/C_6-C_9 N + A from a 9.96 UOPK steam-cracked gas oil. The most paraffinic feedstock shown (from Libyan crude oil) gives yields of only 22 to 27 vol %.

Residuum Hydrotreating. As shown in Fig. 14.1.16, the product from a residuum hydrotreater becomes lighter from SOR to EOR if the product sulfur level is held constant. This product yield structure variation is shown in Table 14.1.23 for a Kuwait vacuum residuum feedstock.[17] Saito,[78] with help from Chiyoda Corporation, has described experience with operating the Okinawa Oil Company Gulf–designed residuum hydrotreater at a consistently high hydrocracking conversion. Product yields are plotted versus 1000°F conversion in Fig. 14.1.33. Unstable fuel oil production was observed at high conversions. A sludge formed which limited the maximum conversion that could be achieved with a particular feedstock because it deposited in flash drums, the fractionator, the product rundown heat exchangers, and product lines. The product qualities which were achieved at 43 percent conversion are shown in Table 14.1.24.

Residuum hydrocracking has been practiced commercially in ebullating-bed reactors in which product qualities can be kept constant by continually replacing a small fraction of the catalyst inventory. Typical product yields and qualities are shown in Table 14.1.25 for both long and short residuum feedstocks.[16] Also shown are the product yields achieved in the first operating fixed-bed unit designed specifically for residuum conversion—Natref's BOC Isomax unit[98] designed by UOP.

TABLE 14.1.22 Ring Content of Unicracker Naphtha[52]

Fresh feed:					
UOPK	9.69	10.22	11.40	11.79	11.89	12.45
50% point, °F (°C)	477 (247)	484 (251)	665 (352)	814 (435)	784 (418)	716 (380)
Unicracker stages	1	1	1	2	1	2
Noble metal catalyst	No	Yes	No	No	Yes	Yes
Product naphtha boiling range, °F (°C)	150–348 (66–176)	150–334 (66–168)	150–373 (66–190)	150–360 (66–182)	150–352 (66–178)	150–340 (66–171)
N+A, yield in C_6–C_8	68.8	61.2	39.7	29.4	30.8	21.7
N+A, yield in C_6–C_9	76.6	71.1	52.7	40.9	40.0	27.4
N+A, yield in total naphtha	83.7	74.4	60.5	49.9	44.8	28.5

Note: N+A yields are shown as LV % of fresh feed.

TABLE 14.1.23 Effect of Operating Temperature on Chevron VRDS Hydrotreating Yields[17]

	VRDS start-of-run 1050°F+ (566°C+)	VRDS end-of-run 1050°F+ (566°C+)
Feed	Kuwait	Kuwait
Product		
C_1–C_4, wt %	0.6	3.8
C_5–350°F, LV % (C_5–177°C)	1.3	5.1
350–650°F, LV % (177–343°C)	2.8	20.8
650–1050°F, LV % (343–566°C)	12.8	32.9
1050°F+, LV % (566°C+)	86.1	44.7
H_2 consumption, SCFB	1180	1320

Note: VRDS = vacuum residuum desulfurization.

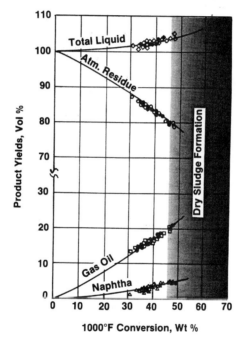

FIGURE 14.1.33 Conversion versus product yields—residuum hydrotreating.[78]

Catalyst Consumption

As feedstocks have become heavier, catalyst consumption per barrel of feedstock has increased. With residuum hydroprocessing, the cost of the catalyst is a significant operating expense. During the scoping studies and the design of a new unit, then, it is important that catalyst life be predicted accurately so that the size of the reactor and its pressure level are well-chosen. Improved catalysts and improved catalyst combinations are continually being developed by the licensors of this technology. They have tested these catalysts in long pilot plant runs and have correlated the data with their

TABLE 14.1.24 Typical Product Yields and Properties—Residuum Hydrotreating

Feed	
Nominal cut point, °F (°C)	800 (427)
Specific gravity, 60/60°F (15.6/15.6°C)	1.0044
Viscosity, cSt at 122°F (50°C)	11,500
Sulfur, wt %	4.18
Nitrogen, wt %	0.40
Metals, wt ppm:	
Vanadium	136
Nickel	40
CCR, wt %	16.2
nC_7 insoluble, wt %	7.0
Product	
Yields, vol %	
Naphtha	3.4
Gas oil	18.4
Atm. residue	82.2
Total liquid	104.0
Properties:	
Naphtha	
Specific gravity, 60/60°F (15.6/15.6°C)	0.7200
Sulfur, wt ppm	40
Nitrogen, wt ppm	20
Gas oil:	
Specific gravity, 60/60°F (15.6/15.6°C)	0.8510
Viscosity, cSt at 122°F (50°C)	2.4
Sulfur, wt %	0.05
Nitrogen, wt %	0.05
Atm. residue:	
Specific gravity, 60/60°F (15.6/15.6°C)	0.9490
Viscosity, cSt at 122°F (50°C)	360
Sulfur, wt %	0.56
Nitrogen, wt %	0.27
CCR, wt %	8.02
1000°F+ (538°C+) conversion, wt %	43
Chemical H_2 consumption, SCFB	1,130

Source: Okinawa Oil Company, Gulf RHDS Unit.

other pilot plant and commercial experience. Nelson[59] has provided general information on this subject, and Fig. 14.1.34 shows his estimated catalyst consumptions for a number of residuum feedstocks at different fuel oil sulfur levels. As one would expect, the nickel and vanadium level in the feedstocks is an important parameter.

Nelson[58] has also commented on another important phenomenon which can limit catalyst life—catalyst bed plugging. This can occur in any fixed-bed reactor within a refinery, since feedstocks are often fed from tanks in which solids can accumulate. Also, shutdowns and start-ups for maintenance can disturb corrosion scale and introduce it into the hydroprocessing feed system. With residuum feeds, however, catalyst bed plugging is much more likely. To counter it, crude oil is commonly double-desalt-

TABLE 14.1.25 Typical Product Yields in Residuum Hydrocracking

	Process		
	LC-Fining[16]		BOC Isomax[98]
	Long residuum	Short residuum	Short residuum
Feedstock			
Density, °API	15.7	10.7	10.0
Sulfur, wt %	2.7	3.2	4.15
1050°F+ (566°C+), LV %	45	66.8	80.0
Hydrogen consumption, SCFB	985	1310	860
Product yields and qualities			
C_3, SCFB	350	590	
C_1–C_4, wt %			2.72
C_4–650°F (343°C) LV %	47.7	36.3	
C_5–650°F (343°C)			8.3
650–1050°F (343–566°C), LV %	37.1	36.1	35.1
Density, °API	22.4		
Sulfur, wt %	0.6	1.0	1.4
1050°F+ (566°C+), LV %	20.0	32.3	54.4
Density, °API	7.0	7.0	
Sulfur, wt %	2.3	2.25	1.77

ed, and the hydroprocessing feedstock is filtered in a sophisticated unit having backflush capabilities.

With some feedstocks, particularly the naphthenic ones from California crude oils, even these precautions are insufficient. Chevron U.S.A.'s Richmond, California, deasphalted oil (DAO) hydrotreater, which has processed a California deasphalted oil containing 33 ppm nickel plus vanadium plus iron since 1966, has seen regular plugging problems because of the reactive soluble iron in the feed.[14] The effect on unit operating factor has not been serious, however, because of improvements in catalyst grading and because the plant was designed with two trains which can be operated independently. Refiners and specialist contractors have developed techniques for unloading reactors quickly and safely. Sometimes when bed plugging has occurred, only the upper bed or beds are skimmed[63] so that the lower catalyst, which is still active, can still be used.

It is instructive to analyze the spent catalyst taken from a residuum hydrotreating unit. Table 14.1.26 shows carbon, nickel, vanadium, and iron levels found[74] after a run in which catalyst consumption corresponded to 14.9 barrels of feed per pound of catalyst. This unit was designed with one guard bed and four main reactors in each train. The axial metal profiles which were measured on spent catalysts showed the iron depositing largely in the guard bed and the nickel and vanadium deposition being spread out as one would expect if second-order kinetics were governing. The coke axial profile is the opposite of the metals, the later reactors having more than the earlier ones. These commercial results are consistent with pilot plant studies (Figs. 14.1.19 to 14.1.21).

When feedstocks that do not contain metal contaminants are hydroprocessed, coke deposition is the normal deactivation culprit. In these cases, the combination of design operating pressure and temperature is chosen to keep the rate of surface coking low enough to achieve an acceptable catalyst life. Refiners usually want 12 months minimum, but some prefer to design for even longer lives. The operating hydrogen partial pressure must be kept at its maximum in order to achieve the design life.

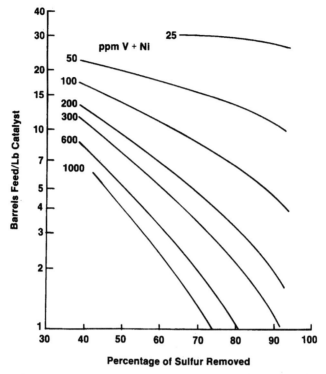

FIGURE 14.1.34 Pattern of catalyst consumption in desulfurizing residua.[59]

TABLE 14.1.26 Deposition of Coke and Metals—Residuum Hydrotreating

	Sample source				
	Guard chamber	Reactor number			
		1	2	3	5
Composition on basis of fresh catalyst, wt %:					
Carbon	8.6	13.4	13.4	16.9	17.4
Nickel	6.1	2.8	4.5	1.5	1.2
Vanadium	28.0	10.5	20.0	5.5	4.4
Iron	4.0	0.7	0.6	0.3	0.7
Total metals, wt %	38.1	25.1	14.0	7.3	6.3
Total metals plus coke, wt %	46.7	38.5	27.4	24.2	23.7
Relative HDS activity (fresh RF-11 = 100)	5	6	32	38	26

Source: Maruzen Oil Company,[74] Unicracking/HDS unit.

With these clean feeds, catalyst regeneration is commonly practiced so that ultimate catalyst lives of 5 years or more have been achieved. Table 14.1.27 shows the experience of the Kuwait National Petroleum Company (KNPC) in achieving long cycle lives in their Chevron Isocracking process unit.[10] KNPC regenerates catalyst in situ. Refiners have the option of ex situ regeneration as well.[47] The choice is based on economic considerations dealing with many factors including the time involved and whether a spare charge of catalyst is on hand.

Hydrogen Consumption

As Nelson has pointed out,[60] the single largest item in the cost of desulfurizing residua is the cost of hydrogen. Only in handling high-metal feeds does catalyst cost approach that of hydrogen. He has correlated the data available up to 1977 as a function of feedstock quality,[61] operating hydrogen partial pressure, product quality, and catalyst age. Nelson's basic correlation is shown as Fig. 14.1.35.

Utilities

As hydroprocessing has been applied more to heavy feedstocks and particularly to the conversion of such feedstocks, the utility balances have become more complicated. The reactor section of the plant and the product fractionation section must be considered as a whole, since the amount of steam generated in the latter is often more than enough to supply the needs of the former. Table 14.1.28 shows typical utility requirements for a two-stage Chevron Isocracker like the one depicted in Fig. 14.1.26.

Most of the cooling water shown is used to condense exhaust steam from the recycle compressor drivers. This could be eliminated in favor of air cooling at somewhat greater capital expense and power requirement. The utility balance above assumes the reactor charge pumps are driven by electric motors with power recovery turbines.

The hydrocracker produces a significant quantity of 200 lb/in^2 gage steam by recovering heat from the reactor effluent streams and the reactor charge furnace convection sections. The larger distillation furnace was assumed to have air preheat.

TABLE 14.1.27 Isocracking Catalyst Regeneration Experience

Catalyst	Cycle no.*	Length of cycle, months	Isocracker stage
ICR 300†	1	14	Single
	2	12	Single
	3	8	Single
ICR 106	1	30	Single
	2	14	Single
ICR 106	1	23	Single‡
	2	25	Second
	3	26	Second
ICR 106	1	60+	First

*In situ regeneration between cycles.
†Early version of ICR 106.
‡The catalyst was regenerated before the change to two-stage operation, although it was still relatively active.
Source: Kuwait National Petroleum Company Isocracker unit.[10]

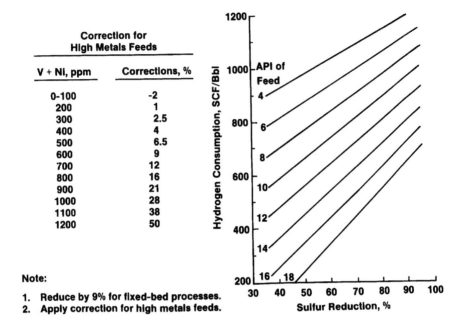

FIGURE 14.1.35 Chemical hydrogen consumption in desulfurization of residua.[61]

TABLE 14.1.28 Typical Utility Requirements

Two-stage Isocracking (see Fig. 14.1.26).

	Reactor section	Distillation section
Feed rate, BPOD	39,000	
Fuel fired, BPOD-EFO*	310	640
Steam at 200 lb/in² gage, lb/h:		
Generated	112,000	3,000
Consumed	61,000	35,000
Net	(51,000)	32,000
Cooling water [15°F (8.3°C) ΔT], gal/min	8000	4000
BFW, gal/min	234	6
Condensate, gal/min (returned)	120	—
Power, kW	10,800	1500

*Barrel of equivalent fuel oil (EFO) is equal to 6,000,000 Btu.

Summary

At the 1930 American Chemical Society symposium on "Industrial High-Pressure Reactions," the chairman of the symposium, Norman W. Krase, stated in his opening address that "oil hydrogenation promises to place the petroleum industry on a better economic basis." This promise has been fulfilled. New technological advances are continually being developed and commercialized. The modern version of distillate hydrocracking was introduced in the United States in the 1960s to convert excess fuel oil into motor gasoline and some jet fuel by using hydrogen produced with the natural gas which was in plentiful supply. FCC cycle oils were popular feedstocks at first. The process was then used to upgrade the liquids produced in delayed and fluid coking and solvent deasphalting.

While these residuum conversion schemes were being implemented, a demand for low-sulfur fuel oil developed and fixed-bed residuum hydrotreating was commercialized. Since then, LSFO demand has dropped, and these hydrotreaters are being used as residuum mild hydrocrackers or as FCC feed pretreaters. The Chevron on-stream catalyst replacement technology has been commercialized to extend the capabilities of fixed-bed residuum hydrotreaters. Four ebullating-bed hydrocrackers are also now operating.

The early distillate hydrocrackers were built in North America, where the demand for motor gasoline is high. The more recent units have been designed for overseas refiners who need to make good-quality kerosene and diesel fuel. Besides this trend, hydrocracking is steadily replacing conventional extraction processes in lube oil base stock manufacture. It results in much more valuable by-products than the older process. The introduction of Chevron's Isodewaxing process now gives refiners the opportunity to produce unconventional base oils with viscosity indexes greater than 110. Hydroprocesses, in general, are clean processes in that feedstock sulfur and nitrogen end up as H_2S and ammonia in a water stream that can easily be purified without harming the environment.

This interest in hydroprocessing has caused a tremendous increase in catalyst consumption in the petroleum industry. The character of the catalysts has changed. With heavier feedstocks and more severe conditions, diffusion limits have been reached and smaller-size catalysts are now used. Also, catalysts of unique shapes have been introduced. Novel reactor systems which use even smaller catalysts are also being considered.

Decisions to construct new hydroprocessing facilities involve some risk in view of uncertainty in future crude oil prices and price differentials between light and heavy products. Refiners, therefore, need to optimize the operation of existing process units. They also want to take advantage of the poorer-quality, lower-priced crudes. Applying the knowledge and understanding of hydroprocessing that has been developed in the past few decades will help them achieve these goals through the efficient use of hydrogen.

ACKNOWLEDGMENTS

I wish to acknowledge two companies for their help with this chapter. I thank my colleagues at Chevron Research and Technology Company for reviewing and helping to improve it. The other company, Gulf Research and Development Corporation (GR&DC), no longer exists since it was merged with Chevron Corporation in 1985. In researching the literature while writing this chapter, I was struck by the enormous contribution that GR&DC scientists have made in this field. It is impossible for me to list all the individual contributors but let me at least acknowledge the debt that the refining industry owes to the Gulf research efforts as a whole.

REFERENCES

1. F. H. Adams, J. G. Gatsis, and J. G. Sikonia, *Proc. First Int. Conf. Future of Heavy Crude Oils and Tar Sands,* McGraw Hill (1981).
2. D. Adlington and E. Thompson, Third European Symposium on Chemical Reaction Engineering, Amsterdam (1964).
3. M. F. Ali, M. V. Hasan, A. M. Bukhari, and M. Saleem, *Hydrocarbon Processing,* 83–86 (February 1985).
4. W. F. Arey, N. E. Blackwell, and A. D. Reichle, Seventh World Petroleum Congress, Mexico City (1967).
5. E. C. Attane, D. E. Mears, and B. J. Young, API Meeting, San Francisco (1971).
6. R. Bearden and C. L. Alderidge, *Energy Progr. 1 (1–4):* 44 (1981).
7. H. Beuther and B. K. Schmid, Sixth World Petroleum Conference, Frankfurt, III-20 (June 1963).
8. M. M. Boduszynski and K. H. Altgelt, *Composition and Analysis of Heavy Petroleum Fractions,* Marcel Dekker, New York (1994).
9. W. A. Bollinger, S. P. Long, and T. R. Metzger, *Chem. Eng. Progr.,* 51–57 (May 1984).
10. A. G. Bridge, G. D. Gould, and J. F. Berkman, *Oil and Gas Journal,* 85–98 (January 19, 1981).
11. A. G. Bridge and D. C. Green, ACS Meeting, Washington (September 1979).
12. A. G. Bridge, J. Jaffe, B. E. Powell, and R. F. Sullivan, API Meeting, Los Angeles (May 1983).
13. A. G. Bridge, E. M. Reed, and J. W. Scott, API Meeting, Chicago (May 1975).
14. A. G. Bridge, E. M. Reed, P. W. Tamm, and D. R. Cash, AIChE Meeting, New Orleans (March 1973).
15. J. B. Butt, *AIChE Journal, 22:* 1–26 (1976).
16. J. A. Celestinos, R. G. Zermeno, R. P. Driesen, and E. D. Wysocki, *Oil and Gas Journal:* 127–134 (December 1, 1975).
17. H. C. Chen and A. G. Bridge, Modern Engineering and Technology Seminar, Taiwan (July 3–21, 1978).
18. N. Choudhary and D. N. Saraf, *Ind. Eng. Chem., Prod. Res. Dev., 14:* 2 (1975).
19. R. A. Corbett, *Oil and Gas Journal,* 73–119 (October 14, 1985).
20. H. G. Cornell and F. J. Heinzelmann, *Hydrocarbon Processing,* 85–90 (August 1980).
21. F. M. Dautzenberg, S. E. George, C. Owerkerk, and S. T. Sie, Advances in Catalytic Chemistry II, Salt Lake City (May 1982).
22. F. M. Dautzenberg and J. C. DeDeken, *Catal. Rev.-Sci Eng., 26* (3,4): 421–444 (1984).
23. S. P. Donnelly, and J. R. Green, Japan Petroleum Institute Meeting, Tokyo (October 1980).
24. R. M. Eccles, A. M. Gray, and W. B. Livingston, *Oil and Gas Journal, 80:* 121 (1982).
25. W. E. Erwin and J. G. Kerr, bulletin 275, Welding Research Council, New York (February 1982).
26. T. R. Farrell and J. A. Zakarian, *Oil and Gas Journal* (May 19, 1986).
27. R. A. Flinn, H. Beuther, and B. K. Schmid, *Petrol. Refin, 40(4):* 139–143 (April 1961).
28. M. J. Fuchs, B. E. Powell, R. S. Tolberg, and Y. Saito, API Meeting, New Orleans (May 1984).
29. J. H. Gary and G. E. Handwerk, *Petroleum Refining,* Marcel Dekker, New York (1984).
30. B. C. Gates, J. R. Katzer, and G. C. A. Shuit, *Chemistry of Catalytic Process,* McGraw Hill, New York (1979).
31. H. G. Geerlings and D. H. Van Nieuwenhuizen, Royal Institution of Engineers in the Netherlands (May 19, 1972).

32. V. A. Gembicki, R. E. Andermann, D. G. Tajbl, *Oil and Gas Journal,* 116–127 (February 21, 1983).
33. K. F. Gerdes, B. E. Stangeland, G. T. S. Chen, and R. J. Gumermann, AIChE Meeting, Chicago (November 1976).
34. G. D. Gould, D. A. Bea, J. F. Mayer, and D. R. Sageman, NPRA Meeting, San Francisco (March 1977).
35. D. C. Green and D. H. Broderick., *Chem. Eng. Prog., 33* (December 1981).
36. A. J. Gully and W. P. Ballard, *J. Adv. Petr. Chem. and Ref. III* (1961).
37. R. T. Haslam and R. P. Russell, *Ind. Eng. Chem., 22* (1930).
38. A. L. Hensley and L. M. Quick, AIChE Meeting, Philadelphia (June 1980).
39. R. J. Hengstebeck, *Hydrocarbon Processing,* 100–102 (August 1970).
40. H. P. Hofmann, *Catal. Rev.-Sci. Eng., 17(1):* 71–117 (1978).
41. W. A. Horne and J. McAfee, *J. Adv. Petr. Chem & Ref., III: 192–277* (1960).
42. R. L. Howell, C. Hung, K. R. Gibson, and H. C. Chen, *Oil and Gas Journal,* 121–128 (July 29, 1985).
43. C. Hung, R. L. Howell, and D. R. Johnson, *Chem. Eng. Progr.,* 57–61 (March 1986).
44. S. B. Jaffe, *Ind. Eng. Chem. Proc. Des. Develop, 13:* 1 (1974).
45. S. B. Jaffe, *Ind. Eng. Chem. Proc. Des. Develop, 15:* 3 (1976).
46. J. Kubo, Y. Oguchi, and H. Nomura, *Bull. JPI, 17:* 100–107 (1975).
47. R. C. Kunzelman, R. E. Ellingham, and R. M. Van Rappard, NPRA Meeting, New Orleans (1980).
48. J. L. Lafferty, D. A. Young, W. W. Burleigh, and C. H. Peters, API Meeting, Chicago (1977).
49. G. E. Langlois and R. F. Sullivan, *Adv. Chem. Ser., 97:* 38 (1970).
50. D. Luss and N. R. Amundson, *AIChE J., 13(4):* 759 (1967).
51. D. C. Matarresse, *Oil and Gas Journal, 81(24):* 111–114 (1983).
52. V. T. Mavity, J. W. Ward, and K. E. Whitehead, API Meeting, Toronto (May 1978).
53. M. A. Menzies, T. F. Scott, and J. A. Denis, *Proc. Intersoc. Energy Conv. Eng. Conf., 15(1):* 26 (1980).
54. J. W. Miller, M. Skripek, T. L. Carlson, and D. Ackelson, AIChE Meeting, New Orleans (November 8, 1981).
55. S. J. Miller, *Microporous Materials, 2:* 439–449 (1994).
56. A. A. Montagna, Y. T. Shah, and J. A. Paraskos, *Ind. Eng. Chem. Process Des. Devel. 16, (1):* 152 (1977).
57. A. A. Montagna and Y. T. Shah, *Ind. Eng. Chem. Process Des. Devel., 14(4):* 479 (1975).
58. W. L. Nelson, *Oil and Gas Journal,* 80–81 (November 18, 1974).
59. W. L. Nelson, *Oil and Gas Journal,* 72–74 (November 15, 1976).
60. W. L. Nelson, *Oil and Gas Journal,* 66–67 (February 21, 1977).
61. W. L. Nelson, *Oil and Gas Journal,* 126–128 (February 28, 1977).
62. E. J. Newson, *Ind. Eng. Chem. Proc. Des. Dev., 14:* 27 (1975).
63. L. D. Nicholson, *Oil and Gas Journal,* 59–63 (March 28, 1977).
64. E. J. Nitta, T. Takatsuka, S. Kodama, and T. Yokoyama, AIChE Meeting, Houston (April 1979).
65. W. J. Novak, Japan Petroleum Institute Meeting, Tokyo (October 1980).
66. S. J. Nutting and P. E. Davis, The Institute of Petroleum, London (April 1995).
67. T. Ohtsuka, *Catal. Rev.-Sci. Eng., 16(2):* 291–325 (1977).
68. R. K. Olson and M. E. Reno, NPRA Meeting, San Francisco (March 1983).

69. D. J. O'Rear, H. A. Frumkin, and R. F. Sullivan, API Meeting, New York (1982).
70. H. Ozaki, Y. Satomi, and T. Hisamitsu, 9th World Petroleum Congress, Tokyo (1975).
71. R. L. Piehl, *Materials Performance,* 15–20 (January 1976).
72. F. W. B. Porter and E. C. Housam, *Inst. Chem. Engrs. Yorkshire Meeting,* 9–18 (June 1963).
73. K. Rajagopalan and D. Luss, *Ind. Eng. Chem. Proc. Des. Devel., 18(3):* 459 (1979).
74. R. L. Richardson, F. C. Riddick, and M. Ishikawa, *Oil and Gas Journal,* 80–94 (May 28, 1979).
75. J. F. Richardson and W. N. Zaki, *Trans. I.Ch.E (London), 32:* 35 (1954).
76. J. W. Rosenthal, S. Beret, and D. C. Green, API Meeting, Los Angeles (May 10, 1983).
77. W. J. Rossi, J. F. Mayer, and B. E. Powell, *Hydrocarbon Processing* (May 1978).
78. K. Saito, S. Shinuzym, Y. Fukui, and H. Hashimoto, AIChE Meeting, San Francisco (November 1984).
79. C. N. Satterfield, *Mass Transfer in Heterogenous Catalysis,* MIT Press, Cambridge, Mass. (1970).
80. S. C. Schuman and H. Shalit, *Catal. Rev., 4(2):* 245–318 (1970).
81. J. W. Scott and N. J. Paterson, Seventh World Petroleum Congress, IV-7, Mexico City (June 1967).
82. J. W. Scott and A. G. Bridge, *Adv. Chem. Ser., 103,* ACS (1971).
83. J. W. Scott, J. F. Mayer, and J. A. Meyer, Modern Engineering and Technology Seminar, Taiwan (July 1980).
84. A. I. Snow and M. P. Grosboll, *Oil and Gas Journal,* 61–65 (May 23, 1977).
85. J. G. Speight, *The Desulfurization of Heavy Oils and Residua,* Marcel Dekker, New York (1981).
86. J. C. Spry and W. H. Sawyer, AIChE Meeting, Los Angeles (1975).
87. B. E. Stangeland and J. R. Kittrell, *Ind. Eng. Chem. Process Des. Develop., 11:* 15–20 (1972).
88. B. E. Stangeland, *Ind. Eng. Chem. Process Des. Develop., 13:* 71–75 (1974).
89. R. F. Sullivan, C. J. Egan, G. E. Langlois, and R. P. Sieg, *J. Am. Chem. Soc., 83:* 1156 (1961).
90. R. F. Sullivan and J. A. Meyer, ACS Meeting, Philadelphia (April 1975).
91. R. F. Sullivan, *Advances in Catal. Chem. I, Meeting,* F.V. Hanson, Chairman, Univ. Utah, Snowbird, Utah (October 1979).
92. E. Talmor, *AIChE Journal, 23:* 868–878 (1977).
93. P. W. Tamm, H. F. Harnsberger, and A. G. Bridge, *Ind. Eng. Chem. Proc. Des. Dev., 20:* 262 (1981).
94. E. W. Thiele, *Ind. Eng. Chem., 31:* 916 (1939).
95. C. L. Thomas and E. J. McNelis, *Proc. 7th World Petroleum Congress 1B,* 161 (1967).
96. J. Van Klinken and R. H. Van Dongen, *Chem. Eng. Sci., 35:* 59–66 (1980).
97. D. Van Zooner, and C. T. Douwes, *J. Inst. Petrol, 49(480):* 383–391 (1963).
98. L. Walliser, *Oil and Gas Journal, 78:* 12 (March 24, 1980).
99. J. W. Ward, *Hydrocarbon Processing,* 101–106 (September 1975).
100. A. M. Watson, *Hydrocarbon Processing,* 91–95 (March 1983).
101. P. B. Weisz, *Chem. Eng. Prog. Symp. Ser. No. 25:* 105–158 (1954).
102. A. Wheeler, *Catalysis,* vol. II, Paul H. Emmett (ed.), Reinhold Publishing Corp. (1955), pp. 105–158.
103. M. W. Wilson, K. L. Eiden, T. A. Mueller, S. D. Case, and G. W. Kraft, NPRA Meeting, Houston (November 1994).
104. S. M. Yui and E. C. Sanford, API Meeting, Kansas City (May 1985).

GLOSSARY

alicyclic hydrocarbons See **naphthenes**.

aliphatic Pertaining to a straight (carbon)-chain hydrocarbon.

alkanes See **paraffins**.

alkenes See **olefins**.

barrel Volumetric measure of refinery feedstocks and products equal to 42 U.S. gallons.

cycloalkanes See **naphthenes**.

motor octane number (MON) Measure of uniformity of burning (resistance to knocking) of gasoline under laboratory conditions which simulate highway driving conditions.

naphtha Petroleum distillate with a boiling range of approximately 20 to 200°C; also an aromatic solvent obtained from coal tar.

naphthenes (alicyclic hydrocarbons or cycloalkanes) Alkylcyclohexanes and alkylcyclopentanes found in crude petroleum.

olefins (alkenes) Unsaturated hydrocarbons with one double bond, having the molecular formula C_nH_{2n}. They may be thought of as derivatives of ethylene.

Oligomer Low-molecular-weight polymer made up of two to four monomer units.

paraffins (alkanes) Saturated aliphatic hydrocarbons with the molecular formula C_nH_{2n+2}.

raffinate Portion of a treated stream that is not removed.

reforming Conversion of naphtha into more volatile products of higher octane via simultaneous combination of polymerization, cracking, dehydrogenation, and isomerization.

research octane number (RON) Measure of uniformity of burning (resistance to knocking) of gasoline under laboratory conditions which simulate city driving conditions.

resid (residuum) Undistilled portion of a crude oil, usually the atmospheric or vacuum tower bottom streams.

saturated hydrocarbons   See **paraffins**.

ABBREVIATIONS AND ACRONYMS

AR	Atmospheric residue	DSD	Desorbent surge drum
AC	Adsorbent chamber	E	Extractor
ASO	Acid-soluble oils	EC	Extraction column
ASTM	American Society for Testing and Materials	FC	Flow controller
		FCC(U)	Fluid catalytic cracking (unit)
B/D or BPD	Barrels per day	FF	Fresh feed
		FI	Flow indicator
BOC	Black-oil conversion	GO	Gas oil
BPSD or B/SD	Barrels per stream day	GPM or gal/min	Gallons per minute
BT	Benzene and toluene	HC	Hand controller
BTX	Benzene, toluene, and xylene	HDM	Hydrodemetallization
CCR	Conradson carbon residue	HDS	Hydrodesulfurization
CFD	Cold flash drum	HFD	Hot flash drum
CHPS	Cold, high-pressure separator	HHPS	Hot, high-pressure separator
COD	Chemical oxygen demand	HS	Hot separator
CS	Cold separator or carbon steel	HVGO	Heavy vacuum gas oil
CT	Clay tower	IBP	Initial boiling point
CW	Cooling water	ID	Inside diameter
Deg API (or °API)	Degree on arbitrary scale for density of liquid petroleum products	KO	Knockout (pot)
		LAB	Linear alkylbenzene
		LC	Level controller
DAO	Deasphalted oil	LCO	Light cycle oil
DCC	Deep catalytic cracking	LGO	Light gas oil
DMB	Dimethylbutane	LHSV	Liquid hourly space velocity
DMO	Demetallized oil	LI	Level indicator

A.1

A.2 ABBREVIATIONS AND ACRONYMS

LIAHL	Liquid-indicator alarm: high or low	RVP	Reid vapor pressure (of gasoline)
LNG	Liquified natural gas	SCF	Standard cubic feet
LPG	Liquefied petroleum gas	SCFD	Standard cubic feet per day
LR	Level recorder	SCFM	Standard cubic feet per minute
LSFO	Low sulfur fuel oil	SD	Solvent drum
LSR	Light straight run (naphtha)	SCR	Selective catalytic reduction
LV	Liquid volume	SDA	Solvent deasphalting
MCP	Methylcyclopentane	SRU	Sulfur recovery unit
MCR	Microcarbon residue	SSU	Seconds Saybolt Universal (viscosity)
MON	Motor octane number		
MONC	Motor octane number clear	TAA	Tertiary amyl alcohol
MP	Methylpentane	TAEE	Tertiary amyl ethyl ether
MSCF	Million standard cubic feet	TAME	Tertiary amyl methyl ether
MTBE	Methyl tertiary butyl ether	TBA	Tertiary butyl alcohol
NPSH	Net positive suction head	TBP	True boiling point
PC	Pressure controller	TC	Temperature controller
PDU	Process development unit	TDS	Total dissolved solids
POX	Partial oxidation	TGA	Thermogravimetric analyzer
RC	Raffinate column	THDA	Thermal hydrodealkylation
RCD	Reduced-crude desulfurization	TPSD	Tons per stream day
		VCM	Volatile combustible material
RCR	Ramsbottom carbon residue	VGO	Vacuum gas oil
RDS	Residuum desulfurization	VPS	Vacuum pipestill
RFCC	Residuum fluid catalytic cracker	VR	Vacuum residue
		VRDS	Vacuum residuum desulfurization
RFG	Reformulated gasoline		
RON	Research octane number	WGS	Wet gas scrubbing unit
RONC	Research octane number clear	WH	Waste heat
RV	Rotary valve	WHSV	Weight hourly space velocity

INDEX

Alkad technology, **1**.50–**1**.52
Alkymax, post fractionation option, **9**.3–**9**.4
Alpha-methylstyrene, **1**.67
ARCO LAB process, **1**.54
Aromatics-cetane relationship, **8**.57
Aromatics complexes, **2**.3–**2**.11
   case study, **2**.9–**2**.10
   commercial experience, **2**.10–**2**.11
   configurations, **2**.4–**2**.6
   feedstock considerations, **2**.8–**2**.9
   introduction, **2**.3–**2**.4
   process flow, **2**.6–**2**.8
Atomax feed injection system, **3**.33–**3**.34

BenSat process:
   compared to Penex-Plus, **9**.19
   fractionation for benzene production, **9**.3
   naphtha splitter combination, **9**.3–**9**.4
   postfractionation option, **9**.3–**9**.4
   (*See also* UOP BenSat process)
Benzene:
   as feed to cumene production, **1**.17–**1**.18
   from alkylbenzenes, **2**.23–**2**.26
   recovery from reformate, **2**.13–**2**.22
   world consumption, **2**.3
Benzene, toluene, and xylene production, **2**.3–**2**.11
Biodegradable soaps, **1**.53
BP-UOP Cyclar process, **2**.27–**2**.35
   case study, **2**.33
   commercial experience, **2**.33–**2**.34
   description of the process flow, **2**.29–**2**.31
   equipment considerations, **2**.32–**2**.33
   feedstock considerations, **2**.31
   introduction, **2**.27
   process chemistry, **2**.28–**2**.29
   process performance, **2**.32

BTX (*see* Benzene, toluene, and xylene production)
Butamer process (*see* UOP Butamer process)
Butane, in alkylation process, **1**.7
Butylene, in alkylation, **1**.4, **1**.6

Carom process, **2**.10, **2**.21–**2**.22
   postfractionation option, **9**.3–**9**.4
Cat Feed Hydrotreating (Go-fining), **3**.4
Cat-Poly process, **1**.21
Catalyst Chemicals Ind. Co. Ltd. (CCIC), hydrotreating catalysts, **8**.39
Catalytic reforming, comparison with Once-Through Zeolitic Isomerization, **9**.33
Caustic-Free Merox process, **11**.40–**11**.41
CBA sub-dew-point reactor, **11**.10
CCR Continuous Catalyst Regeneration technology, **5**.12
CCR Platforming process:
   compared to Pacol unit, **5**.16
   description, **4**.18–**4**.26
   design used in BP-UOP Cyclar process, **2**.27, **2**.29
   design used in Olefiex process, **2**.33
   coordinated with Parex unit, **2**.45, **2**.46, **2**.48
   in aromatics complex, **2**.14
   in conjunction with Parex-Isomar loop, **2**.41
   production of aromatics from naphtha at high severity, **2**.5–**2**.10
   (*See also* UOP Platforming process)
Cetane number versus hydrocarbon type, **8**.56
Chevron Isocracking process, **7**.21–**7**.40, **14**.53
   catalyst regeneration, **14**.63
   importance of hydrogen, **7**.22–**7**.24
   investment and operating expenses, **7**.36–**7**.37
   isocracking catalysts, **7**.24–**7**.28
   isocracking chemistry, **7**.21–**7**.22

INDEX

Chevron Isocracking process (*Cont.*):
  isocracking configurations, **7**.24
  product yields and qualities, **7**.28–**7**.36
  summary, **7**.37–**7**.38
  utilities requirements, **14**.63–**14**.64
Chevron Isodewaxing process, **7**.22, **7**.33–**7**.35
Chevron on-stream catalyst replacement (OCR) process, **10**.3–**10**.13, **14**.38–**14**.39
  applications, **10**.9–**10**.11
  commercial operation, **10**.7–**10**.9
  development history, **10**.3–**10**.4
  economic benefits of OCR, **10**.11–**10**.13
  introduction, **10**.3
  process description, **10**.4–**10**.8
Chevron RDS/VRDS Hydrotreating process, **8**.3–**8**.26
  catalysts, **8**.14–**8**.15
  commercial application, **8**.18–**8**.22
  feed processing capability, **8**.17–**8**.18
  future, **8**.21
  history, **8**.4–**8**.6
  introduction, **8**.3–**8**.4
  metals removal from feed, **10**.3–**10**.13
  process chemistry, **8**.9–**8**.14
  process description, **8**.6–**8**.8
  VRDS Hydrotreating, **8**.15–**8**.16
Chevron's Isodewaxing process, **14**.5
Chevron's vertical RDS (VRDS) process, **3**.87
Chevron VGO hydrotreater, **14**.52
Claus process, **11**.9–**11**.12
Coal gasification, (*see* KRW fluidized-bed gasification process)
Cobalt molybdenum catalyst composition, **8**.34
Continuous stirred reactors, in alkylation, **1**.5–**1**.11
Costs:
  amine regeneration unit, **11**.7
  aromatic complexes, **2**.9–**2**.10
  BP-UOP Cyclar process, **2**.33–**2**.34
  Chevron Isocracking, **7**.21–**7**.40
  Chevron RDS/VRDS Process, **8**.21, **8**.23–**8**.25
  Chevron's On-Stream Catalyst Replacement technology, **10**.11, **10**.13
  Dow-Kellogg Cumene process, **1**.19
  Exxon Flexicoking process, **12**.10, **12**.14
  Exxon Flexicracking IIIR Fluid Catalytic Cracking technology, **3**.9–**3**.10, **3**.17, **3**.22, **3**.23, **3**.24
  Exxon Sulfuric Acid Alkylation technology, **1**.13

Costs (*Cont.*):
  FW Delayed-Coking process, **12**.80–**12**.82
  FW hydrogen production, **6**.48–**6**.52
  FW Solvent Deasphalting, **10**.24, **10**.26, **10**.42, **10**.43
  Hüls selective hydrogenation process, **8**.28
  Kerosene Isosiv process for production of normal paraffins, **10**.72–**10**.73
  linear alkybenzene (LAB) manufacture, **1**.61–**1**.62
  M.W. Kellogg Company Fluid Catalytic Crackling process, **3**.53
  sour water strippers, **11**.9
  UOP BenSat process, **9**.6
  UOP Catalytic Condensation process, **1**.25–**1**.26, **1**.28–**1**.29
  UOP Catalytic Dewaxing process, **8**.53
  UOP Demex process, **10**.60
  UOP Isomar process, **2**.43
  UOP Isosiv process, **10**.65–**10**.66
  UOP Merox process, **11**.45
  UOP Molex process for production of normal paraffins, **10**.76–**10**.77
  UOP Oleflex process, **5**.7–**5**.10
  UOP Once-Through Zeolitic Isomerization process, **9**.35, **9**.36
  UOP Pacol Dehydrogenation process, **5**.18–**5**.19
  UOP Parex process, **2**.52
  UOP Penex process, **9**.24, **9**.27
  UOP Platforming process, **4**.23–**4**.26
  UOP Q-MAX process, **1**.68–**1**.69
  UOP Sulfolane process, **2**.21
  UOP Tatoray process, **2**.62
  UOP Thermal Hydrodealkylation process, **2**.25–**2**.26
  UOP TIP process, **9**.39
  UOP Unicracking, **7**.48–**7**.49
  UOP Unionfining RCD process, **8**.48
  UOP Unionfining technology, **8**.37
  wet gas scrubbing for FCCU, **11**.27
Cracking, during alkylation, **1**.4
Cresex, separated streams, **10**.45
Crude oil distillate qualities, **7**.23, **7**.24
Crude oils and vacuum residua, inspections of, **14**.6
Cumene, as aviation gasoline, **1**.15
Cumene production (*see* Dow-Kellogg Cumene process)
Cyclar process (*see* BP-UOP Cyclar process)
Cymex, separated streams, **10**.45

INDEX

Deep Catalytic Cracking, **3.**101–**3.**112
DeFine process (*see* UOP DeFine process)
Dehydrogenation, UOP Oleflex process, **5.**3–**5.**10
Deisobutanizer (DIB) column, **5.**6–**5.**7
Delayed Coking (*see* FW delayed-coking process)
Demex process:
   with RCD unionfining, **8.**46
   (*See also* UOP Demex process)
Detal process (*see* UOP Detal process)
Detergent manufacture, **1.**53
Disproportionation, during alkylation, **1,**4
Dow-Kellogg Cumene process, **1.**15–**1.**20
   economics, **1.**19
   features, **1.**17
   history, **1.**15–**1.**17
   introduction, **1.**15
   plants, **1.**16
   process description, **1.**17–**1.**18
   product specifications, **1.**19
   wastes and emissions, **1.**20
   yields and balance, **1.**19

Economics (*see* Costs)
Elf Aquitaine, molten sulfur degas process, **11.**9
Environmental control, UOP HF Alkylation technology, **1.**41–**1.**52
ER&E trickle valves, **3.**7
ETBE (*see* Ethyl tertiary butyl ether)
Ethyl tertiary butyl ether, from ethanol and butanes, **5.**6
Exxon Diesel Oil Deep Desulfurization (DODD), **8.**63–**8.**69
   background, **8.**63
   database, **8.**66–**8.**69
   hydrofining characteristics, **8.**63–**8.**65
   summary, **8.**68
   technology, **8.**65
Exxon Flexicoking including Fluid Coking, **12.**3–**12.**24
   commercial flexicoking experience, **12.**15–**12.**17
   Flexicoking options, **12.**16–**12.**19
   Flexicoking unit service factor, **12.**15–**12.**16
   introduction, **12.**3–**12.**4
   investment and operating expenses, **7.**36–**7.**38
   low-Btu gas utilization, **12.**8–**12.**15
   process description, **12.**4–**12.**5
   purge coke utilization, **12.**15

Exxon Flexicoking including Fluid Coking (*Cont.*):
   specific process estimates, **12.**5–**12.**14
   summary, **7.**37–**7.**39
   yields and product dispositions, **12.**5–**12.**6
Exxon Flexicracking IIIR Fluid Catalytic Cracking technology, **3.**3–**3.**28
   economics, **3.**22–**3.**24
   ER&E-designed commercial FCC units, **3.**22–**3.**28
   evolution of technology, **3.**3–**3.**5
   major process features, **3.**7–**3.**17
   process description, **3.**5–**3.**7
   reliability, **3.**17–**3.**19
   resid considerations, **3.**19–**3.**21
   selection rationale, **3.**3
   summary, **3.**23, **3.**28
   upgrading, **3.**21–**3.**22
Exxon Sulfuric Acid Alkylation technology, **1.**3–**1.**14
   advantages of the ER&E reactor, **1.**11–**1.**12
   alkylation is a key processing unit, **1.**4
   balancing process variables for efficiency, **1.**7–**1.**8
   chemistry, **1.**4–**1.**5
   chemistry overview, **1.**4–**1.**5
   commercial experience, **1.**13–**1.**14
   comparisons, **1.**11–**1.**12
   economics, **1.**13
   introduction, **1.**3
   process description, **1.**5–**1.**7
   process variables, **1.**7–**1.**8
   reaction staging results, **1.**10–**1.**11
   reactor cooling, **1.**8–**1.**10
      via autorefrigeration, **1.**8–**1.**9
   reactor improvements, **1.**11–**1.**12
   reactor staging, **1.**10–**1.**11
   units built, **1.**14
Exxon Wet Gas Scrubbing technology, **11.**15–**11.**27
   aboveground PTU: the latest generation, **11.**20–**11.**22
   advantages, **11.**25–**11.**27
   clean gas emission, **11.**19
   EPA testing, **11.**22–**11.**24
   flue gas and scrubber liquid, **11.**17–**11.**18
   introduction, **11.**15–**11.**16
   meeting environmental goals, **11.**22
   operation, **11.**16–**11.**17
   particulate and SO_2 removal, **11.**18–**11.**19
   PTU designs, **11.**19–**11.**20
   purge liquid receives treatment, **11.**19

Exxon Wet Gas Scrubbing technology (*Cont.*):
  separation of the scrubber liquid from the clean flue gas, **11.**19
  summary, **11.**27
  WGS background, **11.**24–**11.**25

FCC unit (*see* Fluid catalytic cracking unit)
Fixed-bed residuum hydroprocessing, growth of, **8.**6–**8.**7
Flexicoking (*see* Exxon Flexicoking including Fluid Coking)
Flexicracking (*see* Exxon Flexicracking IIIR Fluid Catalytic Cracking technology)
Flexsorb amine treating technology, **12.**8
Fluid catalytic cracking unit, wet gas scrubbing for, **11.**15–**11.**27
Fluid Coking (*see* Exxon Flexicoking including Fluid Coking)
Fluidized-Bed Gasification process (*see* KRW Fluidized-Bed Gasification process)
Foster Wheeler's Terrace Wall reformer, **6.**36–**6.**39
FW delayed-coking process, **12.**25–**12.**82
  coke handling and dewatering, **12.**55–**12.**60
  coke uses, **12.**61–**12.**69
  costs, **12.**80–**12.**82
  feedstocks, **12.**27–**12.**36
  heaters, **12.**48–**12.**55
  integration in refineries, **12.**69–**12.**77
  operating variables, **12.**45–**12.**48
  process description, **12.**36–**12.**39
  utilities, **12.**78–**12.**79
  yields and product properties, **12.**39–**12.**45
FW hydrogen production, **6.**21–**6.**52
  economics, **6.**48–**6.**51
  heat recovery, **6.**46–**6.**48
  hydrogen production, **6.**22–**6.**41
  integration into the modern refinery, **6.**41–**6.**46
  introduction, **6.**21
  uses of hydrogen, **6.**21–**6.**22
  utility requirements, **6.**51–**6.**52
FW Solvent Deasphalting, **10.**15–**10.**44
  asphalt properties and uses, **10.**37–**10.**38
  DAO yields and properties, **10.**28–**10.**33
  extraction system, **10.**20–**10.**23
  feedstocks, **10.**19–**10.**20
  integration in modern refineries, **10.**38–**10.**42
  investment cost, **10.**42–**10.**43
  operating variables, **10.**34–**10.**37
  process description, **10.**16–**10.**19

FW Solvent Deasphalting (*Cont.*)
  solvent-recovery systems, **10.**23–**10.**27
  utility requirements, **10.**41–**10.**42

Gas Research Institute (GRI), **6.**3, **6.**4
Gasification (*see* KRW Fluidized-Bed Gasification process)
Gasoline blending components, **1.**3
Gasoline components, properties of common, **9.**29–**9.**30, **10.**62
GDR Parex process, **1.**54

HF Detergent Alkylate process, **1.**56, **1.**57, **5.**13, **5.**14
Hüls Selective Hydrogenation process, **8.**27–**8.**28, **5.**5, **5.**6, **5.**10
  commercial experience, **8.**28
  investment and operating requirements, **8.**28
  process description, **8.**27
  process flow, **8.**28
Hydrofining, diesel oil, **8.**63–**8.**69
Hydrogen consumption in refineries, **6.**22
Hydrogen fluoride, alkylation catalysis by, **1.**31–**1.**51
Hydrogen processing, **14.**3–**14.**67
  capabilities, **14.**47–**14.**65
  design, **14.**35–**14.**47
  fundamentals, **14.**14–**14.**35
  introduction, **14.**3–**14.**13
Hydrogen Production:
  by pressure-swing adsorption unit, **5.**8
  (*See also* FW hydrogen production)
Hydrogen transfer, during alkylation, **1.**4
Hydrogen usage in residuum processing, **14.**26
Hydroprocessing, by geographical area, **14.**10–**14.**12
Hydroprocessing catalysts, commercially available, **14.**34

Institut Francais Du Petrole (*see* Stone & Webster–Institut Francais Du Petrole RFCC process)
Institute of Petroleum Processing (RIPP) (*see* Deep Catalytic cracking)
Instituto Mexicano del Petroleo (IMP) (*see* UOP Demex process)
Isobutane, in alkylation, **1.**4, **1.**5, **1.**6, **1.**10, **1.**11, **1.**12
Isomar process (*see* UOP Isomar process)

INDEX

Isomar unit, **2.**46, **2.**47
   aromatics complex integration, **2.**14, **2.**55
   Parex unit linkage, **2.**46, **2.**47
   xylenes isomerization and the conversion of ethylbenzene, **2.**5, 28
Isopropyl benzene (*see* Cumene)
Isosiv process (*see* Kerosene Isosiv process; UOP Isosiv process)

Kellogg Orthoflow FCC Converter, **3.**31–**3.**33, **3.**36
Kerosene Isosiv process, **10.**67–**10.**73
   detailed process description, **10.**69–**10.**71
   economics, **10.**72–**10.**73
   general process description, **10.**68
   process perspective, **10.**68–**10.**69
   waste and emissions, **10.**71–**10.**72
Kerosene smoke point and hydrogen content, relationship between, **14.**11
KRW fluidized-bed gasification process, **6.**3–**6.**19
   application, **6.**17–**6.**18
   commercial-scale design, **6.**16–**6.**17
   conclusions, **6.**18–**6.**19
   history, **6.**3–**6.**4
   introduction, **6.**3
   KRW single-stage gasification process, **6.**4–**6.**8
   process development unit, **6.**8–**6.**12
   test results obtained in the PDU, **6.**12–**6.**16

LAB (*see* Linear alkybenzene)
Linear alkylbenzene:
   consumption, **1.**64, **1.**65
   demand, **1.**63
Linear alkylbenzene (LAB) manufacture, **1.**55–**1.**66, **5.**14, **5.**15, **5.**17, **5.**18
   common experience, **1.**55–**1.**60
   conclusions, **1.**65–**1.**66
   economics, **1.**61–**1.**62
   environmental safety, **1.**63–**1.**65
   introduction, **1.**53
   markets, **1.**62–**1.**63
   process experience, **1.**55–**1.**60
   product quality, **1.**60–**1.**61
   routes for production, **1.**54–**1.**55
   technology background, **1.**54–**1.**55
   world production, **1.**55
Linear alkylbenzene sulfonate, **1.**53
   use of, **5.**11

Linear internal olefins:
   production cost, **5.**19
   recovery of in Pacol process, **5.**18
   separation in Olex process, **5.**18
   yield in Pacol process, **5.**14
LIO (*see* Linear internal olefins)
Liquefied petroleum gas:
   conversion to BTX, **2.**27–**2.**35
   converted to gasoline, **1.**4
   feed to alkylation process, **1.**21, **1.**24
LPG (*see* Liquefied petroleum gas)

M.W. Kellogg Company Fluid Catalytic Cracking process, **3.**29–**3.**54
   catalyst and chemical consumption, **3.**51–**3.**53
   feedstocks, **3.**29–**3.**30
   introduction, **3.**29
   investment and utilities costs, **3.**53
   process control, **3.**49–**3.**51
   process description, **3.**31–**3.**44
   process variables, **3.**44–**3.**49
   products, **3.**30–**3.**31
M.W. Kellogg Company refinery sulfur management, **11.**3–**11.**14
   amine, **11.**4.–**11.**7
   introduction, **11.**3–**11.**4
   sour water stripping, **11.**7–**11.**9
   sulfur recovery, **11.**9–**11.**12
   tail gas cleanup, **11.**12–**11.**14
MCRC sub-dew-point reactor, **11.**10
MD distillation trays, **2.**51, **2.**52
Merox process (*see* UOP Merox process)
Methanation, **6.**33–**6.**34
Methyl tertiary butyl ether:
   driver for isobutane demand, **9.**12
   from butanes and methanol, **5.**6–**5.**7
Molex process (*see* UOP Molex process)
Molten sulfur production, **11.**9–**11.**12
Moving Bed technology [*see* Chevron on-stream catalyst replacement (OCR) process]
MTBE (*see* Methyl tertiary butyl ether)
MX Sorbex unit, separated streams, **10.**45

Naphtha, feed to UOP Platforming process, **4.**3, **4.**7, **4.**8
Naphtha reforming, **1.**67
Naphthalene, from alkylnaphthalenes, **2.**23–**2.**26
Natural gas composition, **6.**44

O-T Zeolitic Isomerization process (*see* UOP TIP and Once-Through Zeolitic Isomerization processes)
OCR (*see* Chevron on-stream catalyst replacement process)
Olefin feeds, to alkylation, **1**.31
Oleflex process (*see* UOP Oleflex process)
Olex process (*see* UOP Olex process)
On-Stream Catalyst Replacement (OCR), added to RDS/VRDS hydroheaters, **8**.4
Oxo Tecnologies, **5**.14–**5**.15

Pacol Dehydrogenation process (*see* UOP Pacol Dehydrogenation process)
Parex process (*see* UOP Parex process)
Partial oxidation, **6**.29–**6**.31
Penex, combined with naphtha splitter, **9**.3–**9**.4
Penex-Plus, post fractionation option, **9**.3–**9**.4
Penex process (*see* UOP Penex process)
PETRESA Involvement in Detal process, **1**.54
Petroleum coke gasification, **6**.15–**6**.18
Phenol, production from cumene, **1**.15
Platforming (*see* UOP Platforming process)
Polymerization, as side reaction of alkylation, **1**.4, **1**.7
Pressure-swing adsorption (PSA) unit, in UOP Oleflex process, **5**.8
Propane:
  conversion to propylene, **5**.5–**5**.6
  in alkylation processing, **1**.5
Propylene:
  UOP Oleflex process, **5**.3–**5**.10
  as feed to cumene production, **1**.17–**1**.18

Q-MAX process (*see* UOP Q-MAX process)

R-130 CCR Platforming Catalyst Series, **4**.14–**4**.15
RECOVERY PLUS system, **4**.20
Refrigeration in alkylation, **1**.3, **1**.8–**1**.10

Sarex process, separated streams, **10**.45
SCOT/BSR-MDEA (or clone) TGCU, **11**.13
SDA (*see* FW Solvent Deasphalting)
Shell CDC process, **1**.54
Shell Claus Offgas Treating/Beavon Sulfur Reduction-MDEA (SCOT/BSR-MDEA), **11**.12–**11**.13
Shell Higher Olefins (SHOP) process, **1**.54
Shell hydroprocessing reactors, **14**.36
Shell Hysomer process (*see* UOP TIP and Once-Through Zeolitic Isomerization processes)
Shell Molten Sulfur Degas process, **11**.9
Shift conversion, **6**.28–**6**.29, **6**.33
Sinopec International (*see* Deep Catalytic Cracking)
Soap manufacture, **1**.53
Solvent Deasphalting (*see* FW Solvent Deasphalting)
Sorbex process (*see* UOP Sorbex family of technologies)
Sorbex units, licensed worldwide, **2**.52
SR Platforming process, **4**.3, **4**.4, **4**.18–**4**.19, **4**.22–**4**.26
Steam Reforming, **6**.24–**6**.27, **6**.35
Stone & Webster–Institut Francais Du Petrole RFCC process, **3**.79–**3**.100
  catalyst, **3**.89–**3**.90
  FCC revamp to RFCC (second-stage regenerator addition), **3**.99–**3**.100
  feedstocks, **3**.88
  history, **3**.79–**3**.81
  mechanical design features, **3**.98–**3**.99
  operating conditions, **3**.88–**3**.89
  process description, **3**.81–**3**.87
  S&W-IFP technology features, **3**.93–**3**.97
  two-stage regeneration, **3**.90–**3**.93
Strangeland Diagram, **7**.23, **14**.7–**14**.10
Sulfolane process (*see* UOP Sulfolane process)
Sulfreen sub-dew-point reactor, **11**.10
Sulfur dioxide, formation in acid alkylation, **1**.5, **1**.8
Sulfur management (*see* M.W. Kellogg Company refinery sulfur management)
Sulfuric acid in alkylation, **1**.3–**1**.12
Superclaus oxidation catalyst, **11**.10

TAEE (*see* Tertiary amyl ethyl ether)
TAME (*see* Tertiary amyl methyl ether)
Tatoray process (*see* UOP Tatoray process)
Tertiary amyl ethyl ether, from ethanol and isopentane, **5**.6
Tertiary amyl methyl ether, from methanol and isopentane, **5**.6
Tetra process, **2**.10, **2**.21–**2**.22
Texaco-UOP Alkad process, **1**.48–**1**.51
Thermal Hydrodealkylation (THDA) process [*see* UOP Thermal Hydrodealkylation (THDA) process]

INDEX

TIP-Plus, **9.**37
  postfractionation option, **9.**3–**9.**4
Toluene:
  conversion to benzene, **2.**23–**2.**26
  recovery from reformate, **2.**13–**2.**22
Total Petroleum Inc. (*see* Stone & Webster–Institut Francais Du Petrole RFCC process)

Udex process, **2.**10, **2.**21–**2.**22
Unicracking process (*see* UOP Unicracking process for hydrocracking)
Unicracking/DW process (*see* UOP Catalytic Dewaxing process)
Unionfining RCD process (*see* UOP Unionfining RCD process)
Unionfining technology (*see* UOP Unionfining technology)
Unionfining-Unisar integration process flow, **8.**60–**8.**61
Unisar process (*see* UOP Unisar process)
Unocal:
  UOP merger, **8.**39
  UOP Unisar process for saturation of aromatics, **8.**55–**8.**62
UOP-BenSat process, **9.**3–**9.**6
  catalyst and chemistry, **9.**5–**9.**6
  commercial experience, **9.**6
  feedstock requirements, **9.**6
  process discussion, **9.**4–**9.**5
  process flow, **9.**5
  (*See also* BenSat process)
UOP-BP Cyclar process, **2.**8, **2.**9, **2.**10
UOP Butamer process, **9.**7–**9.**13
  chemistry, **9.**8–**9.**9
  commercial experience, **9.**12–**9.**13
  contaminants, **9.**10
  introduction, **9.**7–**9.**8
  isomerization reactors, **9.**10–**9.**11
  process description, **9.**8
  process flow scheme, **9.**11–**9.**12
  process variables, **9.**9–**9.**10
UOP Catalyst Cooler design, **3.**20
UOP Catalytic Condensation process for cumene production, **1.**67
UOP Catalytic Condensation process for transportation fuels, **1.**21–**1.**29
  catalytic condensation, **1.**21
  commercial experience, **1.**29
  distillate fuels production, **1.**26–**1.**28
  economics, **1.**25–**1.**26, **1.**28–**1.**29

UOP Catalytic Condensation process for transportation fuels (*Cont.*):
  history of, **1.**21–**1.**22
  introduction, **1.**21
  jet fuel production, **1.**26–**1.**29
  process chemistry, **1.**22–**1.**23
  process description, **1.**23–**1.**26
  process thermodynamics, **1.**23
  product properties, **1.**24
  production of distillate-type fuels, **1.**26–**1.**29
  thermodynamics, **1.**23
  yields, **1.**24–**1.**25, **1.**27–**1.**28
UOP Catalytic Dewaxing process, **8.**49–**8.**53
  catalyst, **8.**50–**8.**51
  commercial experience, **8.**53
  introduction, **8.**49
  investment and operating expenses, **8.**52–**8.**53
  process chemistry, **8.**50
  process flow, **8.**51–**8.**52
  yield patterns, **8.**52
UOP DeFine process, **1.**56, **1.**59, **1.**60, **5.**13–**5.**14, **5.**18–**5.**19
UOP Demex process, **10.**53–**10.**60
  DMO processing, **10.**58–**10.**60
  introduction, **10.**53
  process description, **10.**53–**10.**55
  process economics, **10.**60
  process status, **10.**60
  process variables, **10.**56–**10.**58
  product yields and quality, **10.**55–**10.**56
  (*See also* Demex process)
UOP Detal process, **1.**55, **1.**56, **1.**58, **1.**60, **1.**62, **5.**13–**5.**14, **5.**18–**5.**19
UOP fluid catalytic cracking process, **3.**55–**3.**78
  catalyst history, **3.**64
  development history, **3.**56–**3.**60
  FCC unit, **3.**70–**3.**72
  feedstock variability, **3.**72–**3.**76
  introduction, **3.**55
  market situation, **3.**78
  process chemistry, **3.**60–**3.**63
  process costs, **3.**76–**3.**77
  process description, **3.**64–**3.**70
  thermodynamics of catalytic cracking, **3.**63
UOP HF Alkylation technology, **1.**31–**1.**52
  chemistry, **1.**32–**1.**34
  commercial information, **1.**41–**1.**42
  design, **1.**39–**1.**41
  economics, **1.**41–**1.**42
  environmental considerations, **1.**41–**1.**48
  introduction, **1.**31–**1.**32

UOP HF Alkylation technology (*Cont.*):
  mitigating HF releases, **1.48–1.51**
  process description, **1.34–1.39**
  process development, **1.50–1.52**
UOP HF Detergent Alkylate process, **1.54**
UOP High Flux Tubing, **2.51**
UOP Isomar process, **2.37–2.44**
  case study, **2.43**
  chemistry, **2.37–2.40**
  commercial experience, **2.43–2.44**
  equipment, **2.42–2.43**
  feedstock, **2.41**
  introduction, **2.37**
  Parex unit coordination, **2.45, 2.46**
  performance, **2.41–2.42**
  process flow, **2.40–2.41**
  (*See also* Isomar unit)
UOP Isosiv process, **1.54, 10.61–10.66**
  detailed process description, **10.64–10.65**
  economics, **10.65–10.66**
  general process description, **10.63–10.64**
  introduction, **10.61–10.63**
  process perspective, **10.64**
  product and by-product specifications, **10.65**
  waste and emissions, **10.65**
UOP Merox process, **11.31–11.41**
  introduction, **11.31**
  Minalk technology, **11.33–11.34**
  process chemistry, **11.37–11.39**
  process description, **11.32–11.37**
  process economics, **11.39–11.41**
  process status and outlook, **11.39–11.41**
  product specifications, **11.39–11.40**
UOP Molex process, **10.75–10.81**
  combined with UOP Molex process, **9.17, 9.18, 9.25**
  commercial experience, **10.77**
  comparisons, **1.54**
  dehydrogenation, **5.3–5.10**
  discussion, **10.75–10.76**
  economics, **10.76–10.77**
  separated streams, **10.45**
  yield structure, **10.76**
UOP Naphtha IsoSiv process, combined with O-T Zeolitic Isomerization process, **9.36**
UOP Oleflex process, **5.3–5.10, 5.12**
  dehydrogenation complexes, **5.5–5.7**
  introduction, **5.3**
  process description, **5.3–5.5**
  propylene production economics, **5.7–5.10**

UOP Olex process, **5.14, 10.79–10.81**
  commercial experience, **10.81**
  discussion, **10.79–10.81**
  economics, **10.81**
  LIO Separation unit, **5.18**
  separated streams, **10.45**
UOP Pacol Dehydrogenation process, **1.54, 1.56, 1.58, 1.60, 5.11–5.19**
  catalysts, **5.12**
  commercial experience, **5.18**
  economics, **5.18–5.19**
  history, **5.11–5.12**
  introduction, **5.11–5.12**
  process description, **5.12–5.15**
  process improvements, **5.15–5.17**
  yield structure, **5.17–5.18**
UOP Parex process, **1.54, 2.45–2.53**
  aromatics complex integration, **2.14, 2.55, 2.61**
  case study, **2.52**
  commercial experience, **2.52**
  crystallization comparison, **2.46–2.47**
  equipment, **2.51–2.52**
  feedstock, **2.48**
  introduction, **2.45**
  Isomar unit combination, **2.40**
  para-xylene recovery by continuous adsorptive separation, **2.5–2.11**
  performance, **2.48**
  process flow, **2.49–2.51**
  raffinate as feed to Isomar unit, **2.37, 2.40**
  separated streams, **10.45**
UOP Penex process, **9.15–9.27**
  applications, **9.19–9.20**
  commercial experience, **9.24–9.27**
  feedstock requirements, **9.23**
  introduction, **9.15**
  process discussion, **9.16**
  process flow, **9.16–9.19**
  thermodynamics, catalysts, and chemistry, **9.20–9.23**
  upgrade on LSR feedstock, **9.29**
UOP Platforming process, **1.31, 4.3–4.26**
  aromatics complex integration, **2.55**
  case studies, **4.22–4.24**
  catalysts, **4.13–4.15, 4.17, 4.18, 9.7**
  chemistry, **4.7–4.15**
  commercial experience, **4.24–4.26**
  continuous process, **4.18–4.22**
  economics, **4.23–4.26**
  heats of reaction, **4.13**
  naphtha splitter combination, **9.3–9.4**

UOP Platforming process (*Cont.*):
   Oleflex comparison, **5.**4
   process evolution, **4.**3–**4.**7
   reformate splitter combination, **9.**3–**9.**4
   variables, **4.**15–**4.**18
   yields and properties, **4.**21, **4.**23
UOP Q-MAX process, **1.**67–**1.**69
   catalysts for, **1.**67
   commercial experience, **1.**69
   economics, **1.**68–**1.**69
   process, **1.**67–**1.**68
   propylene feed, **1.**67
   quality of, **1.**67
   transalkylation, **1.**68
   yield, **1.**68
UOP Sorbex family of technologies, **1.**54, **10.**45–**10.**51
   adsorptive separation principles, **10.**46–**10.**47
   commercial experience, **10.**51
   fixed-bed adsorption comparison, **10.**50–**10.**51
   introduction, **10.**45–**10.**46
   Molex process relation, **10.**75
   Parex process comparison, **2.**49
   process flow, **10.**48–**10.**50
   Sorbex concept, **10.**47–**10.**48
   (*See also* Sorbex units)
UOP Sulfolane process, **2.**13–**2.**22
   aromatics complex integration, **2.**14, **2.**55
   benzene and toluene extraction, **2.**5, **2.**7, **2.**10
   case study, **2.**21
   commercial experience, **2.**21–**2.**22
   concept, **2.**15–**2.**17
   equipment, **2.**20
   feedstock, **2.**19
   introduction, **2.**13–**2.**14
   licensed units, **2.**10
   Parex-Isomar loop coordination, **2.**41
   Parex unit coordination, **2.**46
   performance, **2.**20
   process flow, **2.**17–**2.**19
   solvent selection, **2.**15
UOP Tatoray process, **2.**55–**2.**62
   aromatics complex integration, **2.**14
   case study, **2.**62
   chemistry, **2.**56–**2.**57
   commercial experience, **2.**62
   equipment, **2.**61–**2.**62
   feedstock, **2.**59–**2.**60
   introduction, **2.**55–**2.**56

UOP Tatoray process (*Cont.*):
   Parex unit integration, **2.**45, **2.**46
   performance, **2.**60–**2.**61
   process flow, **2.**57–**2.**59
   toluene and heavy aromatics conversion to xylenes and benzene, **2.**5–**2.**8
UOP Thermal Hydrodealkylation (THDA) process, **2.**23–**2.**26
   economics, **2.**25–**2.**26
   introduction, **2.**23
   process description, **2.**24–**2.**25
UOP TIP and Once-Through Zeolitic Isomerization processes, **9.**29–**9.**39
   introduction, **9.**29–**9.**31
   O-T Zeolitic Isomerization process, **9.**31–**9.**35
   TIP process, **9.**35–**9.**39
UOP Unicracking process for hydrocracking, **7.**41–**7.**49
   applications, **7.**42
   introduction, **7.**41
   investment and operating expenses, **7.**48–**7.**49
   process description, **7.**42
   yield patterns, **7.**48
UOP Unionfining RCD process, **8.**39–**8.**48
   catalyst, **8.**40–**8.**42
   chemistry, **8.**42–**8.**44
   commercial installations, **8.**48
   introduction, **8.**39
   market drivers for RCD unionfining, **8.**39–**8.**40
   operating data, **8.**47–**8.**48
   process description, **8.**42–**8.**47
UOP Unionfining technology, **8.**29–**8.**38
   applications, **8.**35–**8.**37
   catalyst, **8.**34–**8.**35
   chemistry, **8.**29–**8.**34
   introduction, **8.**29
   investment, **8.**37
   process flow, **8.**35
   Unionfining Unit, **8.**59–**8.**60
   UOP hydroprocessing experience, **8.**37
UOP Unisar process, **8.**55–**8.**62
   applications, **8.**60–**8.**62
   diesel fuels, **8.**56–**8.**57
   introduction, **8.**55
   process description, **8.**57–**8.**61

Vacuum residua properties, **10.**55
VSS separation system, **3.**58

Wastes and emissions:
   in the Dow-Kellogg Cumene process,
      1.20
   UOP HF Alkylation technology,
      1.41–**1.**52
Westinghouse Electric Corporation,
   6.3–**6.**4

Wet gas scrubbing (*see* Exxon Wet Gas Scrubbing technology)

Xylenes:
   recovery from reformate, **2.**19
   world consumption, **2.**4

ABOUT THE EDITOR

Robert A. Meyers is president of Ramtech Limited, a consulting firm specializing in technology development and information services for the chemical process industries. Formerly, he was manager of New Process Development for TRW. Dr. Meyers holds 12 U.S. patents in the field of chemical technology and is the author of numerous scientific papers and books, including McGraw-Hill's *Handbook of Synfuels Technology* and *Handbook of Chemicals Production Processes*. He received his Ph.D. in chemistry from UCLA and was a postdoctoral fellow and faculty member at the California Institute of Technology. His biography appears in *Who's Who in the World*. He resides in Tarzana, California.